# Mobile Robots Navigation

Edited by
Alejandra Barrera

**Mobile Robots Navigation**

Edited by Alejandra Barrera

**Republished by InTech**

Janeza Trdine 9, 51000 Rijeka, Croatia



**Technical Editor** Goran Bajac

Additional hard copies can be obtained from orders@intechopen.com

Mobile Robots Navigation,
Edited by Alejandra Barrera
p. cm.
ISBN 978-953-307-076-6

For EU product safety concerns:
IN TECH d.o.o., Prolaz Marije Krucifikse Kozulić 3, 51000 Rijeka, Croatia,
info@intechopen.com or visit our website at intechopen.com.

# Meet the editor

Alejandra Barrera is a Full Professor at the Computer Engineering Department in Mexico's Autonomous Institute of Technology (ITAM), where she directs the Biorobotics Laboratory. She obtained the Computer Engineering degree and the Master degree in Information Technology and Management from ITAM; and her PhD in Biorobotics from Mexico's Autonomous National University (UNAM). Since 2008. she has been a member of Mexico's National Research System (SNI). Her research interests include biorobotics, bioinformatics, and data visualization. She is the author of several research papers, and the editor of the book untitled "Mobile Robots Navigation" published by InTech.

# Preface

Mobile robots navigation includes different interrelated activities: (i) perception, as obtaining and interpreting sensory information; (ii) exploration, as the strategy that guides the robot to select the next direction to go; (iii) mapping, involving the construction of a spatial representation by using the sensory information perceived; (iv) localization, as the strategy to estimate the robot position within the spatial map; (v) path planning, as the strategy to find a path towards a goal location being optimal or not; and (vi) path execution, where motor actions are determined and adapted to environmental changes.

The book addresses those activities by integrating results from the research work of several authors all over the world. Research cases are documented in 32 chapters organized within 7 categories next described.

**Sensory perception**

The accurate perception of sensory information by the robot is critical to support the correct construction of spatial representations to be exploited with navigational purposes. Different types of sensor devices are introduced in this part of the book together with interpretation methods of the acquired sensory information. Specifically, Chapter 1 presents the design of a sensor combining omni-directional and stereoscopic vision to facilitate the 3D reconstruction of the environment.

Chapter 2 describes the prototype of an optical azimuth angular sensor based on infrared linear polarization to compute the robot's position while navigating within an indoor arena.

Chapter 3 depicts the design of a stereoscopic vision module for a wheeled robot, where left and right images from the same scene are captured, and one of two appearance-based pixel descriptors for surface ground extraction are employed, luminance or Hue, depending on the environment particular characteristics. This vision module also detects obstacle edges and provides the reconstruction of the scene based on the stereo image analysis.

Chapter 4 presents a sensor setup for a 3D scanner to promote a fast 3D perception of those regions in the robot's vicinity that are relevant for collision avoidance. The acquired 3D data is projected into the XY-plane in which the robot is moving and used to construct and update egocentric 2.5D maps storing either the coordinates of closest obstacles or environmental structures.

Closing this first part of the book, Chapter 5 depicts a sensor fusion technique where perceived data are optimized and fully used to build navigation rules.

**Robot localization**

In order to perform successful navigation through any given environment, robots need to localize themselves within the corresponding spatial representation. A proper localization allows the robot to exploit the map to plan a trajectory to navigate towards a goal destination. In the second part of the book, four chapters address the problem of robot localization from visual perception. In particular, Chapter 6 describes a localization algorithm using information from a monocular camera and relying on separate estimations of rotation and translation to provide an uncertainty feedback for both motion components while the robot navigates in outdoor environments.

Chapter 7 proposes a self-localization method using a single visual image, where the relationship between artificial or natural landmarks and known global reference points is identified by a parallel projection model.

Chapter 8 presents computer simulations of robot heading and position estimation by using a single vision sensor system to complement the encoders' function during robot motion.

By means of experiments with a robotic wheelchair, Chapter 9 demonstrates the localization ability within a topological map built by using only an omni-directional camera, where environmental locations are recognized by identifying natural landmarks in the scene.

**Path planning**

Several chapters focus on discussing path planning algorithms within static and dynamic environments, and two of them deal with multiple robots. In this way, Chapter 10 presents a path planning algorithm based on the use of a neural network to build up a collision penalty function. Results from simulations show proper obstacle avoidance in both static and dynamic arenas.

Chapter 11 proposes a path planning algorithm avoiding obstacles by classifying them according to their size to decide the next robot navigation action. The algorithm starts by considering the shortest path, which is then expanded on either side spreading out by considering the obstacles type and proximity.

In the context of indoor semi-structured environments full of corridors connecting offices and laboratories, Chapter 12 compares several approaches developed for door identification based on handle recognition, where doors are defined as goals for the robot during the path planning process. The chapter describes a two-step multi-classifier that combines region detection and feature extraction to increase the computational efficiency of the object recognition procedure.

In the context of planetary exploration vehicles, Chapter 13 describes a path planning and navigation system based on the recognition of occluded areas in a local map. Experimental results show the performance of a vehicle navigating through an irregular rocky terrain by perceiving its environment, determining the next sensing position that maximizes the non-occluded region within each local map, and executing the local path generated.

Chapter 14 presents a robotic architecture based on the integration of diverse computation and communication processes to support the path planning and navigation of service robots.

Applied to the flock traffic navigation context, Chapter 15 introduces an algorithm capable of planning paths for multiple agents on partially known and changing environments.

Chapter 16 studies the problem of path planning and navigation of robot formations in static environments, where a formation is defined, composed and repaired according to a proposed mereological method.

**Obstacle avoidance**

One of the basic capabilities that mobile robots need to exhibit in navigating within any given environment is obstacle detection and avoidance. This part of the book is dedicated to review diverse mechanisms to deal with obstacles, being static and/or dynamic, implemented on robots with different purposes, from service robots in domestic or office-like environments to car-like vehicles in outdoors arenas. Specifically, Chapter 17 proposes an approach to reactive obstacle avoidance for service robots by using the concept of artificial protection field, which is understood as a dynamic geometrical neighborhood of the robot and a set of situation assessment rules that determine if the robot needs to evade an object not present in its map when its path was planned.

Chapter 18 describes a hierarchical action-control method for omni-directional mobile robots to achieve a smooth obstacle avoidance ensuring safety in the presence of moving obstacles including humans.

Chapter 19 presents a contour-following controller to allow a wheeled robot to follow discontinuous walls contours. This controller is integrated by a standard wall-following controller and two complementary controllers to avoid collisions and find lost contours.

Chapter 20 introduces a fuzzy decision-making method to control the motion of car-like vehicles in dynamic environments showing their ability to park in spatial configurations with different placement of static obstacles, to run with the presence of dynamic obstacles, and to achieve a final target from a given arbitrary initial position.

Chapter 21 presents a qualitative vision-based method to follow a path avoiding obstacles.

**Analysis of navigational behavior**

A correct evaluation of the navigational behavior of a mobile robotic system is required prior its use solving real tasks in real-life scenarios. This part of the book stresses the importance of employing qualitative and quantitative measures to analyze the robot performance. From diverse perspectives, five chapters provide analysis metrics and/or results from comparative analysis of existing methods to assess different behavioral aspects, from positioning underwater vehicles to transmitting video signals from tele-operated robots.

From an information theory perspective, Chapter 22 studies the robot learning performance in terms of the diversity of information available during training. Authors employ motivational measures and entropy-based environmental measures to analyze the outcome of several robotic navigation experiments.

Chapter 23 focuses on the study of positioning as a navigation problem where GPS reception is limited or non-existent in the case of autonomous underwater vehicles that are forced to use deadreckoning in between GPS sightings in order to navigate accurately. Authors provide an analysis of different position estimators aiming at allowing vehicle designers to improve performance and efficiency, as well as reduce vehicle instrumentation costs.

Chapter 24 provides results from analyzing several performance metrics to contrast mobile robots navigation algorithms including safety, dimension and smoothness of the planned trajectory.

Chapter 25 analyses the performance of different codecs in transmitting video signals from a teleoperated mobile robot. Results are shown from robot tests in an indoor scenario.

With an aim at supporting educational and research activities, in Chapter 26, authors provide a virtual environment to develop mobile robot systems including tools to simulate kinematics, dynamics and control conditions, and monitor in real time the robot performance during navigation tasks.

**Inspiration from nature**

Research cycles involving living organisms' studies, computational modeling, and robotic experimentation, have inspired for many years the understanding of the underlying physiology and psychology of biological systems while also inspiring new robotic architectures and applications. This part of the book describes two different studies that have taken inspiration from nature to design and implement robotic systems exhibiting navigational capabilities, from visual perception and map building to place recognition and goal-directed behavior. Firstly, Chapter 27 presents a computational system-level model of rat spatial cognition relating rat learning and memory processes by interaction of different brain structures to endow a mobile robot with skills associated to global and relative positioning in space, integration of the traveled path, use of kinesthetic and visual cues during orientation, generation of topological-metric spatial representation of the unknown environment, management of rewards, learning and unlearning of goal locations, navigation towards the goal from any given departure location, and on-line adaptation of the cognitive map to changes in the physical configuration of the environment. From a biological perspective, this work aims at providing to neurobiologists/neuroethologists a technological platform to test with robots biological experiments whose results can predict rodents' spatial behavior.

Secondly, Chapter 28 proposes an approach inspired after developmental psychology and some findings in neuroscience that allows a robot to use motor representations for learning a complex task through imitation. This framework relies on development, understood as the process where the robot acquires sophisticated capabilities over time as a sequence of simpler learning steps. At the first level, the robot learns about sensory-motor coordination. Then, motor actions are identified based on lower level, raw signals. Finally, these motor actions are stored in a topological map and retrieved during navigation.

**Navigation applications**

The book concludes by introducing different contexts and real scenarios where mobile robots have been employed to solve diverse navigational tasks.

Presenting successful results along four testing years, Chapter 29 provides a mechatronic description of an autonomous robot for agricultural tasks in greenhouses emphasizing the use of specialized sensors during the development of control strategies of plants spraying and robot navigation.

A sociological application is introduced in Chapter 30, consisting on providing electric-powered wheelchairs able to predict and avoid risky situations and navigate safely through congested areas and confined spaces in the public transportation environment. Authors propose a high-level architecture that facilitates terrain surveillance and intelligence gathering through laser sensors implanted in the wheelchair in order to anticipate accidents by identifying obstacles and unusual patters of movement.

Chapter 31 describes the communication, sensory, and artificial intelligence systems implemented on the CAESAR (Contractible Arms Elevating Search And Rescue) robot, which supplies rescuers with critical information about the environment, such as gas detection, before they enter and risk their lives in unstable conditions.

Finally, another monitoring system is depicted by Chapter 32. A mobile robot being controlled by this system is able to perform a measuring task of physical variables, such as high temperatures being potentially hazardous for humans, while navigating within a known environment by following a predefined path.

The successful research cases included in this book demonstrate the progress of devices, systems, models and architectures in supporting the navigational behavior of mobile robots while performing tasks within several contexts. With no doubt, the overview of the state of the art provided by the book may be a good starting point to acquire knowledge of intelligent mobile robotics.

Alejandra Barrera
*Mexico's Autonomous Technological Institute (ITAM)*
*Mexico*

# Contents

1

# A 3D Omnidirectional Sensor For Mobile Robot Applications

Rémi Boutteau, Xavier Savatier, Jean-Yves Ertaud and Bélahcène Mazari
*Institut de Recherche en Systèmes Electroniques Embarqués (IRSEEM)*
*France*

## 1. Introduction

In most of the missions a mobile robot has to achieve - intervention in hostile environments, preparation of military intervention, mapping, etc - two main tasks have to be completed: navigation and 3D environment perception. Therefore, vision based solutions have been widely used in autonomous robotics because they provide a large amount of information useful for detection, tracking, pattern recognition and scene understanding. Nevertheless, the main limitations of this kind of system are the limited field of view and the loss of the depth perception.

A 360 degree field of view offers many advantages for navigation such as easiest motion estimation using specific properties on optical flow (Mouaddib, 2005) and more robust feature extraction and tracking. The interest for omnidirectional vision has therefore been growing up significantly over the past few years and several methods are being explored to obtain a panoramic image: rotating cameras (Benosman & Devars, 1998), muti-camera systems and catadioptric sensors (Baker & Nayar, 1999). Catadioptric sensors, i.e. the combination of a camera and a mirror with revolution shape, are nevertheless the only system that can provide a panoramic image instantaneously without moving parts, and are thus well-adapted for mobile robot applications.

The depth perception can be retrieved using a set of images taken from at least two different viewpoints either by moving the camera or by using several cameras at different positions.

The use of the camera motion to recover the geometrical structure of the scene and the camera's positions is known as Structure From Motion (SFM). Excellent results have been obtained during the last years with SFM approaches (Pollefeys et al., 2004; Nister, 2001), but with off-line algorithms that need to process all the images simultaneous. SFM is consequently not well-adapted to the exploration of an unknown environment because the robot needs to build the map and to localize itself in this map during its world exploration.

The in-line approach, known as SLAM (Simultaneous Localization and Mapping), is one of the most active research areas in robotics since it can provide a real autonomy to a mobile robot. Some interesting results have been obtained in the last few years but principally to build 2D maps of indoor environments using laser range-finders. A survey of these algorithms can be found in the tutorials of Durrant-Whyte and Bailey (Durrant-Whyte & Bailey, 2006; Bailey & Durrant-Whyte, 2006).

Vision-based SLAM algorithms are generally dedicated to monocular systems which are cheaper, less bulky, and easier to implement than stereoscopic ones. Stereoscopic systems have, however, the advantage to work in dynamic environments since they can grab simultaneously two images. Calibration of the stereoscopic sensor enables, moreover, to recover the Euclidean structure of the scene which is not always possible with only one camera.
In this chapter, we propose the design of an omnidirectional stereoscopic system dedicated to mobile robot applications, and a complete scheme for localization and 3D reconstruction. This chapter is organized as follows. Section 2 describes our 3D omnidirectional sensor. Section 3 is dedicated to the modelling and the calibration of the sensor. Our main contribution, a Simultaneous Localization and Mapping algorithm for an omnidirectional stereoscopic sensor, is then presented in section 4. The results of the experimental evaluation of each step, from calibration to SLAM, are then exposed in section 5. Finally, conclusions and future works are presented section 6.

## 2. System overview

### 2.1 Sensor description

Among all possible configurations of central catadioptric sensors described by Nayar and Baker (Baker & Nayar, 1999), the combination of a hyperbolic mirror and a camera is preferable for the sake of compactness since a parabolic mirror needs a bulky telecentric lens.
Although it is possible to reconstruct the environment with only one camera, a stereoscopic sensor can produce a 3D reconstruction instantaneously (without displacement) and will give better results in dynamic scenes. For these reasons, we developed a stereoscopic system dedicated to mobile robot applications using two catadioptric sensors as shown in Figure 1.

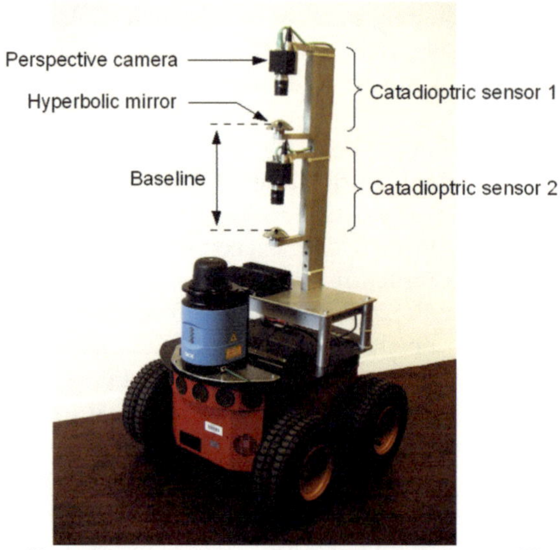


Fig. 1. View of our catadioptric stereovision sensor mounted on a Pioneer robot. Baseline is around 20cm for indoor environments and can be extended for outdoor environments. The overall height of the sensor is 40cm.

### 2.2 Imposing the Single-Viewpoint (SVP) Constraint

The formation of images with catadioptric sensors is based on the Single-Viewpoint theory (Baker & Nayer, 1999). The respect of the SVP constraint permits the generation of geometrically correct perspective images. In the case of a hyperbolic mirror, the optical center of the camera has to coincide with the second focus **F'** of the hyperbola located at a distance of *2e* from the mirror focus as illustrated in Figure 2. The eccentricity *e* is a parameter of the mirror given by the manufacturer.

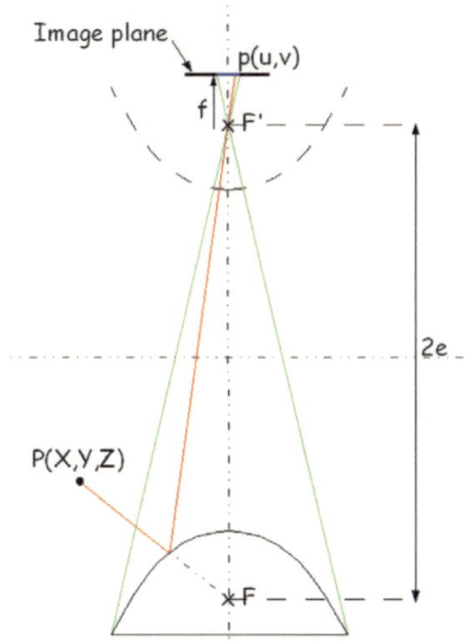


Fig. 2. Image formation with a hyperbolic mirror. The camera center has to be located at *2e* from the mirror focus to respect the SVP constraint.

A key step in designing a catadioptric sensor is to respect this constraint as much as possible. To achieve this, we first calibrate our camera with a standard calibration tool to determine the central point and the focal length. Knowing the parameters of both the mirror and the camera, the image of the mirror on the image plane can be easily predicted if the SVP constraint is respected as illustrated in Figure 2. The expected mirror boundaries are superposed on the image and the mirror has then to be moved manually to fit this estimation as shown in Figure 3.

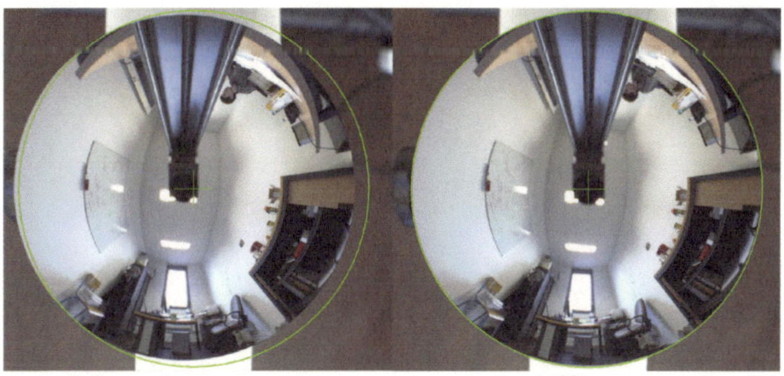

Fig. 3. Adjustment of the mirror position to respect the SVP constraint. The mirror border has to fit the estimation (green circle).

## 3. Modelling of the sensor

The modelling of the sensor is a necessary step to obtain metric information about the scene from the camera. It establishes the relationship between the 3D points of the scene and their projections into the image (pixel coordinates). Although there are many calibration methods, they can be classified into two main categories: parametric and non-parametric. The first family consists in finding an appropriate model for the projection of a 3D point onto the image plane. Non-parametric methods associate one projection ray to each pixel (Ramalingram et al., 2005) and provide a "black box model" of the sensor. They are well adapted for general purposes but they make more difficult the minimization algorithms commonly used in computer vision (gradient descent, Gauss-Newton, Levenberg-Marquardt, *etc*).

### 3.1 Projection model

Using a parametric method requires the choice of a model, which is very important because it has an effect on the complexity and the precision of the calibration process. Several models are available for catadioptric sensors: complete model, polynomial approximation of the projection function and generic model.

The complete model relies on the mirror equation, the camera parameters and the rigid transformation between them to calculate the projection function (Gonzalez-Barbosa & Lacroix, 2005). The large number of parameters to be estimated leads to an error function which is difficult to minimize because of numerous local minima (Mei & Rives, 2007). The polynomial approximation of the projection function was introduced by Scaramuzza (Scaramuzza et al., 2006), who proposed a calibration toolbox for his model. The generic model, also known as the unified model, was introduced by Geyer (Geyer & Daniilidis, 2000) and Barreto (Barreto, 2006), who proved its validity for all central catadioptric systems. This model was then modified by Mei (Mei & Rives, 2007), who generalized the projection matrix and also took into account the distortions. We chose to work with the unified model introduced by Mei because any catadioptric system can be used and the number of parameters to be estimated is quite reasonable.

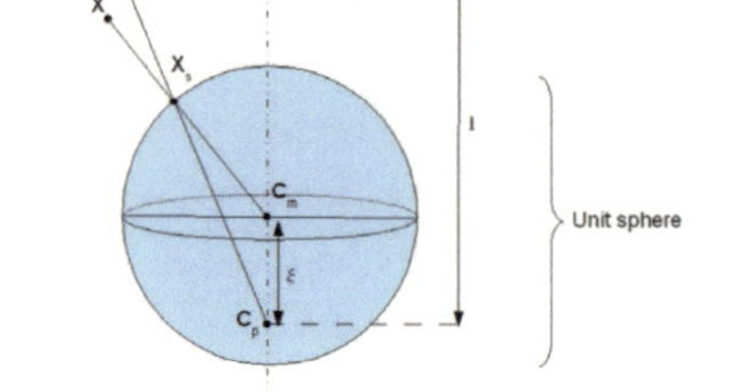


Fig. 4. Unified projection model.

As shown in Figure 4, the projection $\mathbf{p} = [u \quad v]^{\mathrm{T}}$ of a 3D point $\mathbf{X}$ with coordinates $[X_w \quad Y_w \quad Z_w]^{\mathrm{T}}$ in the world frame can be computed using the following steps:

- The coordinates of the point $\mathbf{X}$ are first expressed in the sensor frame by the rigid transformation $W$ which depends on the seven parameters of the vector $\mathbf{V_1} = [q_w \quad q_x \quad q_y \quad q_z \quad t_x \quad t_y \quad t_z]^{\mathrm{T}}$. The first four parameters are the rotation $\mathbf{R}$ parameterised by a quaternion and the three others are those of the translation $\mathbf{T}$. The coordinates of $\mathbf{X}$ in the mirror frame are thus given by:

$$\begin{bmatrix} X \\ Y \\ Z \end{bmatrix} = \mathbf{R} \begin{bmatrix} X_w \\ Y_w \\ Z_w \end{bmatrix} + \mathbf{T} \quad (1)$$

- The point $\mathbf{X} = [X \quad Y \quad Z]^{\mathrm{T}}$ in the mirror frame is projected from the center onto the unit sphere giving $\mathbf{X_S} = [X_S \quad Y_S \quad Z_S]^{\mathrm{T}}$. This point is then projected onto the normalized plane from a point located at a distance $\xi$ from the center of the sphere. These transformations are combined in the function $H$ which depends on only one parameter: $\mathbf{V_2} = [\xi]$. The projection onto the normalized plane, written $\mathbf{m} = [x \quad y]^{\mathrm{T}}$ is consequently obtained by:

$$\begin{bmatrix} x \\ y \end{bmatrix} = \begin{bmatrix} \dfrac{X_S}{Z_S + \xi} \\ \dfrac{Y_S}{Z_S + \xi} \end{bmatrix}, \text{with} \begin{bmatrix} X_S \\ Y_S \\ Z_S \end{bmatrix} = \begin{bmatrix} \dfrac{X}{\sqrt{X^2 + Y^2 + Z^2}} \\ \dfrac{Y}{\sqrt{X^2 + Y^2 + Z^2}} \\ \dfrac{Z}{\sqrt{X^2 + Y^2 + Z^2}} \end{bmatrix} \quad (2)$$

- Distortions are then added to the point $\mathbf{m}$ using the distortion function $D$, which depends on 5 coefficients: 3 for radial distortions and 2 for tangential distortions. Let $\mathbf{V_3} = [k_1 \quad k_2 \quad k_3 \quad k_4 \quad k_5]^{\mathrm{T}}$ be the parameter vector containing these coefficients, $\rho = \sqrt{x^2 + y^2}$, and $\mathbf{m_d} = [x_d \quad y_d]^{\mathrm{T}}$ the resulting point. Its coordinates are obtained by the following equation:

$$\mathbf{m_d} = \begin{bmatrix} x(1 + k_1\rho^2 + k_2\rho^4 + k_5\rho^6) + 2k_3xy + k_4(\rho^2 + 2x^2) \\ y(1 + k_1\rho^2 + k_2\rho^4 + k_5\rho^6) + 2k_4xy + k_3(\rho^2 + 2y^2) \end{bmatrix} \quad (3)$$

- Final projection is a perspective projection involving the projection matrix $\mathbf{K}$. This matrix contains 5 parameters: the generalized focal lengths $\gamma_u$ and $\gamma_v$, the coordinates of the principal point $u_0$ and $v_0$, and the skew $\alpha$. Let $K$ be this projection function, and $\mathbf{V_4} = \begin{bmatrix}\alpha & \gamma_u & \gamma_v & u_0 & v_0\end{bmatrix}^{\mathrm{T}}$ be the parameter vector. The projection $\mathbf{p} = \begin{bmatrix}u & v\end{bmatrix}^{\mathrm{T}}$ of the 3D point $\mathbf{X}$ is given by equation (4).

$$\mathbf{p} = \mathbf{K} \cdot \mathbf{m_d} = \begin{bmatrix}\gamma_u & \gamma_u\alpha & u_0 \\ 0 & \gamma_v & v_0 \\ 0 & 0 & 1\end{bmatrix}\mathbf{m_d} \tag{4}$$

Let $\mathbf{V}$ be the global parameter vector, i.e. $\mathbf{V} = \begin{bmatrix}\mathbf{V_1}^{\mathrm{T}} & \mathbf{V_2}^{\mathrm{T}} & \mathbf{V_3}^{\mathrm{T}} & \mathbf{V_4}^{\mathrm{T}}\end{bmatrix}^{\mathrm{T}}$. The global projection function of a 3D point $\mathbf{X}$, written $P(\mathbf{V},\mathbf{X})$, is obtained by chain composition of the different functions presented before:

$$P(\mathbf{V},\mathbf{X}) = K \circ D \circ H \circ W(\mathbf{V},\mathbf{X}) \tag{5}$$

These steps allow the computation of the projection onto the image plane of a 3D point knowing its coordinates in the 3D space. In a 3D reconstruction framework, it is necessary to do the inverse operation, i.e. to compute the direction of the luminous ray corresponding to a pixel. This step consists in computing the coordinates of the point $\mathbf{X_S}$ belonging to the sphere given the coordinates $\begin{bmatrix}x & y\end{bmatrix}^{\mathrm{T}}$ of a 2D point on the normalized plane. This step of retro projection, also known as lifting, is achieved using formula (6).

$$\mathbf{X_S} = \begin{bmatrix} \dfrac{\xi + \sqrt{1 + (1-\xi^2)(x^2+y^2)}}{x^2+y^2+1}\,x \\ \dfrac{\xi + \sqrt{1 + (1-\xi^2)(x^2+y^2)}}{x^2+y^2+1}\,y \\ \dfrac{\xi + \sqrt{1 + (1-\xi^2)(x^2+y^2)}}{x^2+y^2+1} - \xi \end{bmatrix} \tag{6}$$

### 3.2 Calibration

Calibration consists in the estimation of the parameters of the model which will be used for 3D reconstruction algorithms. Calibration is commonly achieved by observing a planar pattern at different positions. With the tool we have designed, the pattern can be freely moved (the motion does not need to be known) and the user only needs to select the four corners of the pattern. Our calibration process is similar to that of Mei (Mei & Rives, 2007). It consists of a minimization over all the model parameters of an error function between the estimated projection of the pattern corners and the measured projection using the Levenberg-Marquardt algorithm (Levenberg, 1944; Marquardt, 1963).

If $n$ is the number of 3D points $\mathbf{X_i}$, $\mathbf{x_i}$ their projections in the images, we are looking for the parameter vector $\mathbf{V}$ which minimizes the cost function $E(\mathbf{V})$:

$$E(\mathbf{V}) = \frac{1}{2}\sum_{i=1}^{n}\left[P(\mathbf{V},\mathbf{X_i}) - \mathbf{x_i}\right]^2 \qquad (7)$$

Our calibration tool was developed in C++ using the computer vision library OpenCV and can be freely downloaded from our website (Boutteau, 2009). It does not require any commercial software and a particular attention has been given to the optimization of the computation time since a calibration with 10 images does not exceed 2 minutes.

### 3.3 Relative pose estimation

The estimation of the intrinsic parameters presented in the previous section allows to establish the relationship between 3D points and theirs projections for each sensor of the stereoscopic system. To obtain metric information from the scene, for example by triangulation, the relative pose of the two sensors has to be known.

This step is generally performed by a pixel matching between both images followed by the estimation of the essential matrix. This matrix, originally introduced by Longuet-Higgins (Longuet-Higgins, 1981), has the property to contain information on the epipolar geometry of the sensor. It is then possible to decompose this matrix into a rotation matrix and a translation vector, but the last one can only be determined up to a scale factor (Bunschoten & Kröse, 2003). The geometrical structure of the scene can consequently be recovered only up to this scale factor.

Although in some applications, especially for 3D visualization, the scale factor is not needed, it is required for preparation of intervention or for navigation. To accomplish these tasks, the size of the objects and their distance from the robot must be determined. The 3D reconstruction has therefore to be Euclidean.

Thus, we suggest in this section a method to estimate the relative pose of the two sensors, with a particular attention to the estimation of the scale factor. The estimation of the relative pose of two vision sensors requires a partial knowledge of the environment to determine the scale factor. For this reason, we propose a method based on the use of a calibration pattern whose dimensions are known and which must be visible simultaneously by both sensors.

Let $(\mathbf{C_1}, \mathbf{x_1}, \mathbf{y_1}, \mathbf{z_1})$ and $(\mathbf{C_2}, \mathbf{x_2}, \mathbf{y_2}, \mathbf{z_2})$ be the frames associated with the two sensors of the stereoscopic system, and $\mathbf{M}$ be a 3D point, as shown in Figure 5.

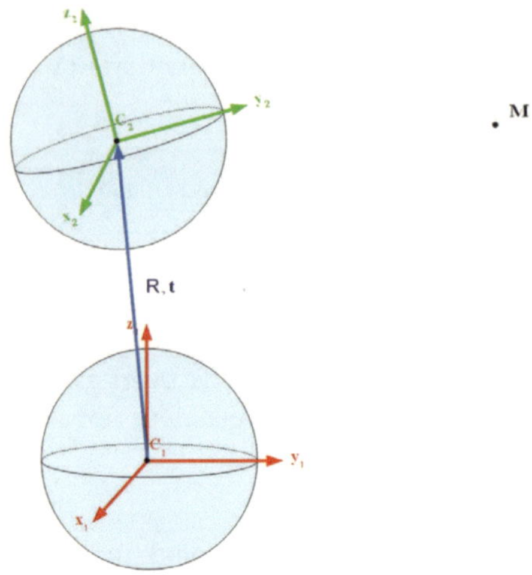


Fig. 5. Relative pose estimation principle.

The point **M** with coordinates $\begin{bmatrix} x_2 & y_2 & z_2 & 1 \end{bmatrix}^{\mathrm{T}}$ in the frame associated with the second sensor has for coordinates in the first sensor frame:

$$\begin{bmatrix} x_1 \\ y_1 \\ z_1 \\ 1 \end{bmatrix} = \begin{bmatrix} r_{11} & r_{12} & r_{13} & t_x \\ r_{21} & r_{22} & r_{23} & t_y \\ r_{31} & r_{32} & r_{33} & t_z \\ 0 & 0 & 0 & 1 \end{bmatrix} \cdot \begin{bmatrix} x_2 \\ y_2 \\ z_2 \\ 1 \end{bmatrix} \tag{8}$$

where $\mathbf{t} = \begin{bmatrix} t_x & t_y & t_z \end{bmatrix}^{\mathrm{T}}$ and $\mathbf{R} = \begin{bmatrix} r_{11} & r_{12} & r_{13} \\ r_{21} & r_{22} & r_{23} \\ r_{31} & r_{32} & r_{33} \end{bmatrix}$ correspond to the pose of the second sensor with respect to the first one. With $n$ control points, equation (8) yields to the following system:

$$\begin{bmatrix} x_1^1 & y_1^1 & z_1^1 & 0 & 0 & 0 & 0 & 0 & 0 & 1 & 0 & 0 \\ 0 & 0 & 0 & x_1^1 & y_1^1 & z_1^1 & 0 & 0 & 0 & 0 & 1 & 0 \\ 0 & 0 & 0 & 0 & 0 & 0 & x_1^1 & y_1^1 & z_1^1 & 0 & 0 & 1 \\ \vdots & \vdots & \vdots & \vdots & \vdots & \vdots & \vdots & \vdots & \vdots & \vdots & \vdots & \vdots \\ x_1^n & y_1^n & z_1^n & 0 & 0 & 0 & 0 & 0 & 0 & 1 & 0 & 0 \\ 0 & 0 & 0 & x_1^n & y_1^n & z_1^n & 0 & 0 & 0 & 0 & 1 & 0 \\ 0 & 0 & 0 & 0 & 0 & 0 & x_1^n & y_1^n & z_1^n & 0 & 0 & 1 \end{bmatrix} \cdot \begin{bmatrix} r_{11} \\ r_{12} \\ r_{13} \\ r_{21} \\ r_{22} \\ r_{23} \\ r_{31} \\ r_{32} \\ r_{33} \\ t_x \\ t_y \\ t_z \end{bmatrix} = \begin{bmatrix} x_2^1 \\ y_2^1 \\ z_2^1 \\ \vdots \\ x_2^n \\ y_2^n \\ z_2^n \end{bmatrix} \tag{9}$$

The principle of the relative pose estimation consists therefore in computing the rigid transformation between the two sensors, knowing the coordinates of a set of points in the two frames. The coordinates of the points are however not known directly since the location of the pattern is not known. The pose of the pattern has to be determined in the two frames using the Levenberg-Marquardt algorithm to minimize the error between estimated projection and projections extracted from the images. The resolution of equation (9) then gives the relative pose $(\mathbf{R}, \mathbf{t})$ of the two sensors including the scale factor.

## 4. Simultaneous Localization and Mapping

3D reconstruction of a scene using the motion of a sensor addresses two problems: localization and mapping. Localization consists in estimating the trajectory of a robot in a known map (Thrun et al., 2001). The aim of mapping is the creation of a map of the environment using measurements from the sensors embedded on a robot knowing its trajectory (Thrun, 2003). When neither the trajectory of the robot, nor the map of the environment are known, localization and mapping problems have to be considered simultaneously: it is the issue of SLAM (Simultaneous Localization And Mapping).
The first approach to solve the SLAM problem is to assume the motion known (by odometry or by the command law) even if this one is corrupted by noise. The position of visual landmarks can consequently be predicted. Sensors embedded on the mobile robot, for example laser range-finders, provide measurements of its environment. These observations are then used to update the model containing the coordinates of the visual landmarks and the positions of the robot. These steps (prediction/observation/update) are implemented using the Kalman filter or one of its derivatives.
The second approach is to optimize the geometrical structure of the scene and the positions of the sensor using the bundle adjustment method. A synthesis on bundle adjustment algorithms was published by Triggs (Triggs et al., 1999). Bundle adjustment provides more accurate results than Kalman filters (Mouragnon et al., 2009) but needs more computing time. In most of the applications, this algorithm is used off-line to obtain a very accurate model, but it is also possible to apply it iteratively. Although bundle adjustment is commonly used with conventional cameras, there are very few works on its adaptation to omnidirectional sensors. The main works in this field are those of Lhuillier (Lhuillier, 2005) and Mouragnon (Mouragnon et al., 2009) who suggest to find the structure of the scene and the motion of a catadioptric sensor by a local bundle adjustment followed by a global one to obtain more accurate results. Their works highlight the difficulty of estimating the scale factor, although it is theoretically possible with a non-central sensor.
The contribution of this section deals with the application of a bundle adjustment algorithm to an omnidirectional stereoscopic sensor previously calibrated to solve the ambiguity on the scale factor. Bundle adjustment relies on the non-linear minimization of a criterion, so a first estimation of the parameters as to be found to ensure the convergence. Before the presentation of the bundle adjustment algorithm, we thus expose our initialization step.

### 4.1 Initialization

Estimating the camera motion, also called ego-motion, requires to relate the images of the sequence grabbed during the motion. Relating images consists in localizing in the images the projections of a same 3D point of the scene. This step is decisive because the precision of the motion relies on the quality of this matching. The visual landmarks are detected in a majority of works by the Harris corner detector since it has good performances on luminosity change. This detector is however very sensitive to scale change and it can thus fail with large motion. To avoid this, Lowe (Lowe, 2004) has presented a very interesting approach for the detection and the matching of regions of interest: the Scale Invariant Feature Transform (SIFT). The SIFT principle is to detect features from images which are invariant to scale change, rotation, and small point of view change. A descriptor, which corresponds to the orientation histogram, is then associated to the features and the matching can be achieved by comparison of their Euclidean distances.

Once the images are related, the epipolar geometry can be estimated between the two times $k-1$ and $k$. The epipolar geometry is interesting since it gives information on the relative pose of two vision sensors. Several works are dedicated to the estimation of the epipolar geometry for catadioptric sensors. Some of them (Pajdla et al., 2001; Gonzalez-Barbosa & Lacroix, 2005; Mariottini & Prattichizzo, 2005) give analytical solution to the estimation of the epipolar curves. Their methods need nevertheless to introduce the mirror equations and the proposed solutions are thus specific to the kind of sensor used.

Other works (Bunschoten & Kröse, 2003; Negishi, 2004) rely on the use of panoramic images, i.e. unwrapped images, and consider the epipolar curves as the intersection of the epipolar plane with a cylinder representing the image plane. This approach is interesting because the idea is to generalize the notion of epipolar geometry to panoramic sensors.

We suggest to generalize the epipolar geometry for all central sensors using the model of the equivalence sphere. With this model, the coplanarity constraint initially defined for perspective cameras (Longuet-Higgins, 1981) can be transposed to all central sensors. As shown in Figure 6, if the points $\mathbf{X_{S1}}$ and $\mathbf{X_{S2}}$ correspond to the projection of the same 3D point $\mathbf{X}$ onto the two spheres, then $\mathbf{C_1}$, $\mathbf{C_2}$, $\mathbf{X_{S1}}$, $\mathbf{X_{S2}}$ and $\mathbf{X}$ lie in the same plane.

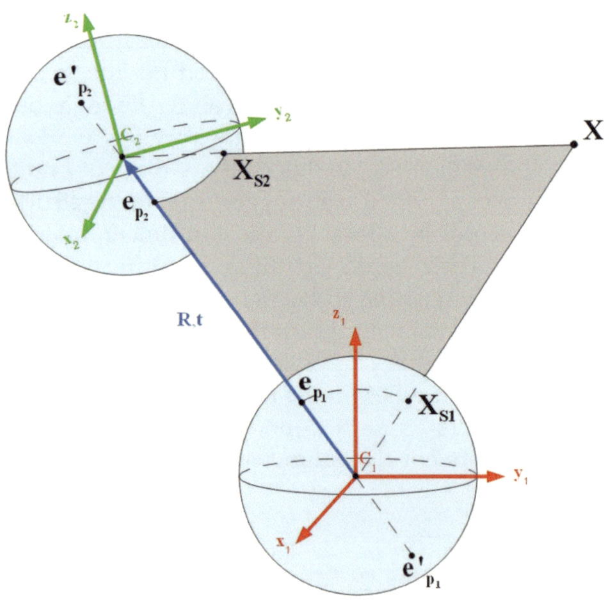


Fig. 6. Epipolar geometry for central sensors.

Let $\mathbf{t}$ be the translation vector of the sensor and $\mathbf{R}$ be the rotation matrix, the coplanarity constraint can be expressed in the coordinate system $(\mathbf{C_2}, \mathbf{x_2}, \mathbf{y_2}, \mathbf{z_2})$ as:

$$\mathbf{X}_{\mathbf{S2}}^{\mathrm{T}}\mathbf{R}(\mathbf{t} \times \mathbf{X}_{\mathbf{S1}}) = 0 \tag{10}$$

The coplanarity constraint (10) can be rewritten in the matrix form as:

$$\mathbf{X}_{\mathbf{S2}}^{\mathrm{T}}\mathbf{E}\mathbf{X}_{\mathbf{S1}} = 0 \tag{11}$$

where $\mathbf{E} = \mathbf{RS}$ is the essential matrix first introduced by Longuet-Higgins (Longuet-Higgins, 1981) and $\mathbf{S}$ is an antisymmetric matrix characterizing the translation:

$$\mathbf{S} = \begin{bmatrix} 0 & -t_z & t_y \\ t_z & 0 & -t_x \\ -t_y & t_x & 0 \end{bmatrix} \tag{12}$$

The essential matrix $\mathbf{E}$ can be estimated from a set of matched points using the eight-point algorithm (Longuet-Higgins, 1981). Given two lifted points $\mathbf{X}_{\mathbf{S1}} = \begin{bmatrix} x_1 & y_1 & z_1 \end{bmatrix}^{\mathrm{T}}$ and $\mathbf{X}_{\mathbf{S2}} = \begin{bmatrix} x_2 & y_2 & z_2 \end{bmatrix}^{\mathrm{T}}$ corresponding to the same 3D point $\mathbf{X}$, (11) becomes for each pair of matched points:

$$x_2x_1e_{11} + x_2y_1e_{12} + x_2z_1e_{31} + \cdots + z_2z_1e_{33} = 0 \tag{13}$$

where $\mathbf{E} = \begin{bmatrix} e_{11} & e_{12} & e_{13} \\ e_{21} & e_{22} & e_{23} \\ e_{31} & e_{32} & e_{33} \end{bmatrix}$.

Introducing the vector $e = \begin{bmatrix} e_{11} & e_{12} & e_{13} & e_{21} & e_{22} & e_{23} & e_{31} & e_{32} & e_{33} \end{bmatrix}^{\mathrm{T}}$ and using $n$ pairs of matched points, the set of equations (13) can be expressed in the matrix form as:

$$\mathbf{Ae} = 0 \tag{14}$$

With more than eight points, a least squares solution can be found by singular value decomposition (SVD) of $\mathbf{A}$. Because of the least squares estimation, the estimated matrix may not respect the two constraints of an essential matrix: two of its singular values have to be equal, and the third has to be zero (Hartley & Zisserman, 2004). A constraint enforcement step is therefore necessary and is achieved by the method described by Hartley (Hartley & Zisserman, 2004). Given a 3x3 matrix $\mathbf{E} = \mathbf{UDV}^{\mathrm{T}}$, where $\mathbf{D} = diag(a,b,c)$ with $a \geq b \geq c$. The closest essential matrix to $\mathbf{E}$ in Frobenius norm is given by:

$$\hat{\mathbf{E}} = \mathbf{U}\hat{\mathbf{D}}\mathbf{V}^{\mathrm{T}} \tag{15}$$

where $\hat{\mathbf{D}} = diag((a+b)/2, (a+b)/2, 0)$.

The estimation of the essential matrix may fail when there is one outlier among the set of points used in equation (14). The solution is therefore found by using the RANSAC algorithm. The main difference between perspective cameras and omnidirectional sensors comes from the computation of the error to determine if a point belongs to the consensus set. In the perspective case, the error is defined as the distance $d$ between the point $\mathbf{X_2}$ and the epipolar line $\mathbf{l}$ corresponding to the point $\mathbf{X_1}$ as shown in Figure 7. In the omnidirectional case, the computation of the error is more difficult since the epipolar line becomes an epipolar curve (see Figure 8). We therefore suggest to work on the equivalence sphere using the coplanarity constraint defined by equation (11) to compute this error. Given a point $\mathbf{X_{S1}}$ on the first sphere, the normal of the epipolar plane in the second sensor frame is given by $\mathbf{N_2} = \mathbf{EX_{S1}}$. The vectors $\mathbf{X_{S2}}$ and $\mathbf{N_2}$ are orthogonal when the coplanarity constraint is respected. The error $e$ can consequently be computed as the angular error given by:

$$e = \mathbf{X_{S2}^T N_2} \tag{16}$$

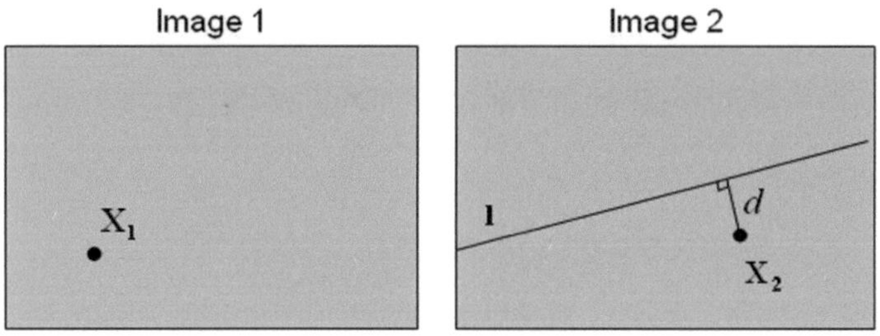


Fig. 7. Computation of the error in the perspective case.

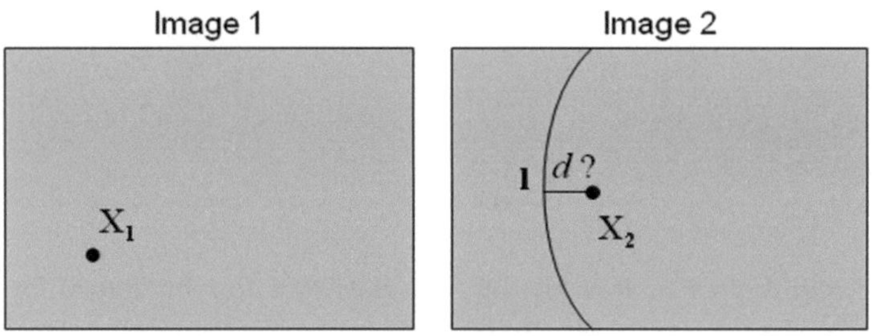


Fig. 8. Computation of the error in the omnidirectional case.

The essential matrix does not give directly the displacement of the sensor since it is a combination of the rotation matrix $\mathbf{R}$ and the antisymmetric matrix $\mathbf{S}$ (see equation 12). The essential matrix has therefore to be decomposed to retrieve the rotation $\mathbf{R}$ and the translation $\mathbf{t}$. After the constraint enforcement step, the essential matrix can be written as $\mathbf{E} = \mathbf{U}\text{diag}(1,1,0)\mathbf{V}^T$. Let $\mathbf{u_i}$ be the i[th] column of $\mathbf{U}$ and $\mathbf{v_i}$ the i[th] column of $\mathbf{V}$, there are four possible solutions for the rotation $\mathbf{R}$ and two solutions for the translation vector $\mathbf{t}$ (Wang & Tsui, 2000):

$$\begin{cases} \mathbf{R_1} = (-\mathbf{u_2}, \mathbf{u_1}, \mathbf{u_3})\mathbf{V}^T \\ \mathbf{R_2} = (-\mathbf{u_2}, \mathbf{u_1}, -\mathbf{u_3})\mathbf{V}^T \\ \mathbf{R_3} = (\mathbf{u_2}, -\mathbf{u_1}, \mathbf{u_3})\mathbf{V}^T \\ \mathbf{R_4} = (\mathbf{u_2}, -\mathbf{u_1}, -\mathbf{u_3})\mathbf{V}^T \\ \mathbf{t_1} = \mathbf{v_3} \\ \mathbf{t_2} = -\mathbf{v_3} \end{cases} \quad (17)$$

Two solutions for the rotation matrix can be easily eliminated because they do not respect the property of a rotation matrix $\det(\mathbf{R}) = 1$. It remains consequently four possible combinations of rotation and translation. In the perspective case, the right solution is trivial since it can be obtained after the triangulation of a point by checking the reconstructed point is in front of both cameras. This notion of "in front of the cameras" does not exist anymore in the omnidirectional case since points from the entire 3D space can have a projection onto the image plane. The right solution could be recovered using other sensors, for example odometers, but it is preferable to use only visual data to be totally independent from the mobile robot. For each possible combination, the right solution is found by triangulating the points, and then reprojecting them onto the image plane. The computation of the reprojection error allows the determination of the correct solution since the others give totally aberrant projections. The right solution is the one which gives the minimal reprojection error since the others are totally aberrant as shown in Figure 9.

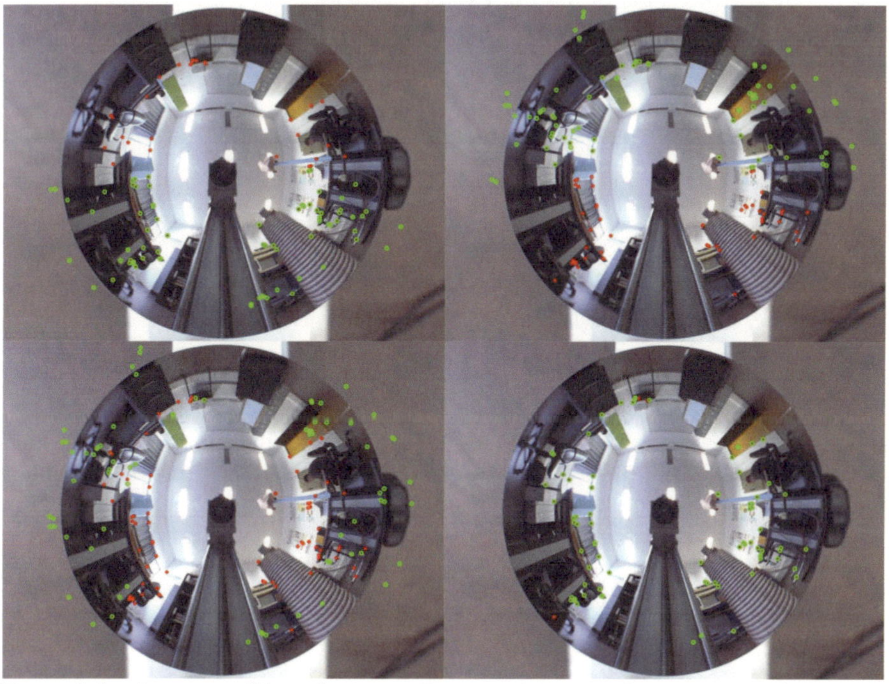

Fig. 9. Test of the four possible solutions. Green points are the reprojected points and red points are the real 2D projections.

The estimation of the essential matrix, followed by its decomposition, enables to retrieve the motion of the system, but only up to a scale factor. In a large majority of works, this ambiguity is removed by the use of odometry data (Bunschoten & Kröse, 2003). This approach is interesting since it gives directly the result, but it is preferable to use only visual data to avoid any communication with the mobile robot. The stereoscopic structure of our system can be used to retrieve the scale factor since the baseline is known. The 3D coordinates $\mathbf{X_{k-1}}$ and $\mathbf{X_k}$ of a set of points can actually be computed at times $k-1$ and $k$ and the scale factor can be retrieved by the computation of the norm of $\mathbf{t}$, given by:

$$\mathbf{t} = \mathbf{X_k} - \mathbf{R}\mathbf{X_{k-1}} \tag{18}$$

### 4.2 Bundle Adjustment

The algorithms presented in the previous section provide initial values for the motion of the sensor and for the 3D coordinates of points. These values are however not sufficiently accurate to be used directly. An excessive error will provide, because of its accumulation, a wrong 3D model. An extra optimisation step, the bundle adjustment, is therefore necessary. This method is well known for perspective cameras (Triggs et al., 2000) but only few works are dedicated to omnidirectional sensors. In this section, we propose a bundle adjustment algorithm which takes into account the specificity of our sensor: its omnidirectional field of view and its stereoscopic structure.

Bundle adjustment consists in minimizing the cost function corresponding to the error between estimated projections of 3D points and their measured projections. Let $m$ be the number of positions of the stereoscopic system and $n$ be the number of 3D points, the cost function can be written as:

$$F(\mathbf{V},\mathbf{X}) = \frac{1}{2}\sum_{j=1}^{m}\sum_{i=1}^{n}\left[P(\mathbf{V}_j,\mathbf{X}_i) - \mathbf{x}_{ij}\right]^2 \tag{19}$$

where $\mathbf{V}_j$ is the parameter vector of the $j^{th}$ camera, $\mathbf{X}_i$ is the coordinates of the $i^{th}$ 3D point in the world frame, $\mathbf{x}_{ij}$ is the projection of the $i^{th}$ 3D point in the image of the $j^{th}$ camera and $P(\mathbf{V}_j,\mathbf{X}_i)$ is the predicted projection of point $i$ in image $j$.

Using a stereoscopic sensor allows to recover the scale factor, which is impossible with a monocular sensor. The projection function defined by equation (5) has therefore to be modified to take into account the stereoscopic structure of the sensor. Thus, the projection function provides now four elements: the coordinates $u_{low}$ and $v_{low}$ of the projection of $\mathbf{X}$ onto the image plane of the first sensor, and the coordinates $u_{high}$ and $v_{high}$ of its projection onto the image plane of the second sensor. The system is no longer considered as two separate sensors but as a global device. The relative pose of the two sensors $(\mathbf{R},\mathbf{t})$ is consequently added to the model parameters as shown in Figure 10.

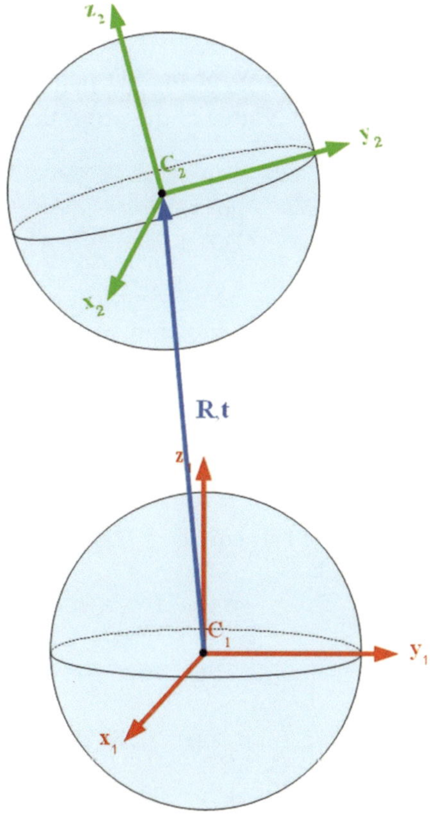


Fig. 10. Rigid transformation between the two sensors of the stereoscopic system.

Let $\mathbf{V_5} = \begin{bmatrix} q_{w12} & q_{x12} & q_{y12} & q_{z12} & t_{x12} & t_{y12} & t_{z12} \end{bmatrix}^{\mathrm{T}}$ be the new parameter vector, the rotation being parameterised by a quaternion. An extra rigid transformation, written $C(\mathbf{V_5})$, is added to the initial projection function to express coordinates either in the low sensor frame or in the high sensor frame:

$$P(\mathbf{V},\mathbf{X}) = K \circ D \circ H \circ C \circ W(\mathbf{V},\mathbf{X}) \tag{20}$$

where $\mathbf{V} = \begin{bmatrix} \mathbf{V}_1^{\mathrm{T}} & \mathbf{V}_2^{\mathrm{T}} & \mathbf{V}_3^{\mathrm{T}} & \mathbf{V}_4^{\mathrm{T}} & \mathbf{V}_5^{\mathrm{T}} \end{bmatrix}^{\mathrm{T}}$

The minimization of the reprojection error defined by equation (19) is carried out by the Levenberg-Marquardt algorithm. The key step of this algorithm lies in the resolution of the augmented normal equation:

$$\left(\mathbf{J}^{\mathrm{T}}\mathbf{J} + \lambda\mathbf{I}\right)\mathbf{\Delta} = -\mathbf{J}^{\mathrm{T}}\mathbf{e} \tag{21}$$

where $\lambda$ is a real number varying from iteration to iteration and $\mathbf{I}$ is the identity matrix.

The resolution of equation (21) requires the computation of the Jacobian $\mathbf{J}$ of the projection function. The sensor is calibrated so the parameters that have to be estimated are the poses $\mathbf{V}_1^j$ of the cameras and the coordinates $\mathbf{X}_i$ of the 3D points. Thus, the partial derivatives which have to be estimated are the derivative of the projection function whith respect to the pose of the system:

$$\frac{\partial \mathbf{P}}{\partial \mathbf{V}_1^j} \tag{22}$$

as well as the derivative of the projection function with respect to the coordinates of the points:

$$\frac{\partial \mathbf{P}}{\partial \mathbf{X}_i} \tag{23}$$

The form of the Jacobian matrix used in equation (21) is illustrated in Figure 11 for three poses of the sensors and four points.

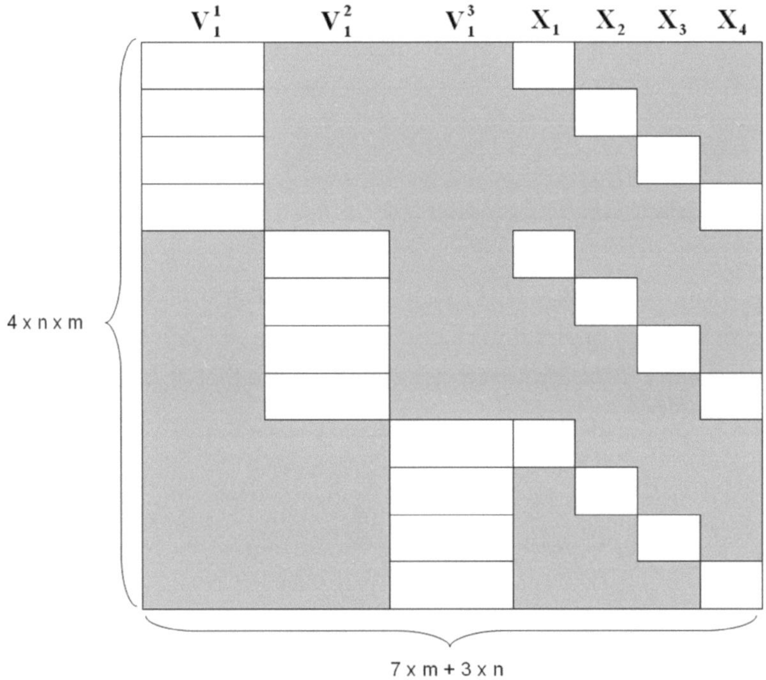


Fig. 11. Form of the Jacobian matrix for 3 poses and 4 points.

The iterative resolution of the augmented normal equation as described in (Levenberg, 1944; Marquardt, 1963) minimizes the reprojection error and provides a good estimate of the poses of the system and of the coordinates of the 3D points.

## 5. Experimental results

The stereoscopic omnidirectional sensor was mounted on a Pioneer robot with a Sick LD-PDS laser range-finder used to provide the ground truth. Each step, from calibration to 3D reconstruction and motion estimation, was evaluated on real images and without prior knowledge to evaluate the system in real conditions. Some results obtained with synthetic images are also presented to validate results that need a perfect knowledge of the 3D points coordinates in the sensor frame.

### 5.1 Calibration of the sensor

The calibration step was evaluated by computing the Root of Mean Squares (RMS) distances between estimated and theoretical projections of a set of 3D points. As it is very difficult to know precisely the coordinates of a 3D point in the sensor frame, we used synthetic images. The sensor was simulated in POV-Ray, a ray-tracing software, to generate omnidirectional images containing a calibration pattern as shown in Figure 12 and these images were used to calibrate the sensor.

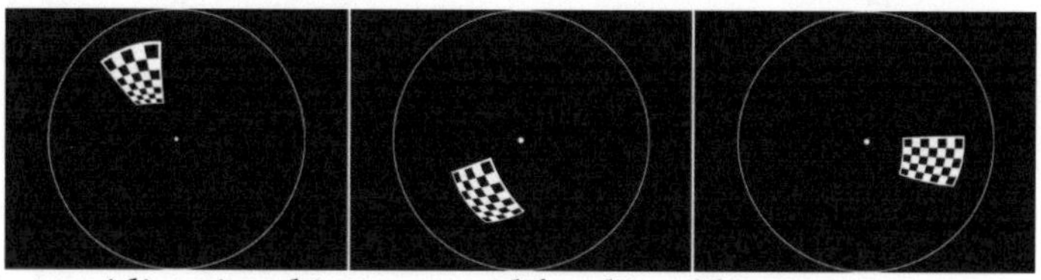

Fig. 12. Synthetic omnidirectional images used for the calibration

As the projection model and the 3D coordinates of reference points are perfectly known in this case, their theoretical projection can easily be computed and compared to the projection obtained by the estimated model. Table 1 shows the mean error and the standard deviation obtained on a set of 150 points.

| | |
|---|---|
| Mean error (pixels) | 0.24 |
| Standard deviation (pixels) | 0.11 |

Table 1. Calibration results on synthetic images

The calibration was then evaluated on real images. Two sets of omnidirectional images containing a calibration pattern were taken. We used the first set to calibrate the sensor as described in section. The second set was used to compute the error between estimated projections of the grids points computed using estimated model and their measured projections extracted from the images. Table 2 summarizes the results obtained on this set of real images.

| | |
|---|---|
| Mean error (pixels) | 0.44 |
| Standard deviation (pixels) | 0.26 |

Table 2. Calibration results on real images

The error on real images is greater than the error on synthetic images, but both are very low since they are less than half a pixel. These results highlight the good estimation of the model parameters even with noisy data since we used a real system which is not perfect. This can be explained by the introduction of the distortions into the model which allows some misalignments between the mirror and the camera.

### 5.2 Relative pose estimation

The estimation of the relative pose was evaluated on synthetic images for a vertical setup of the two sensors, i.e. the same setup as the real system. Several pairs of images containing a calibration were taken as shown in Figure 13 and the relative pose was evaluated with the method presented in section 3.3. The distance between the two sensors was defined with 10cm and they are supposed perfectly aligned. The result of the relative pose estimation is presented in Table 3.

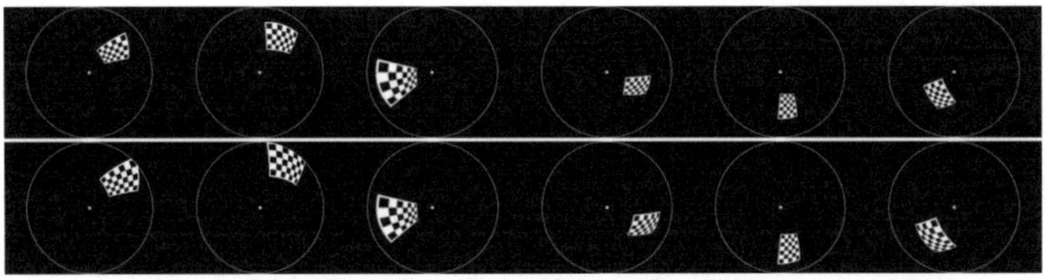

Fig. 13. Pairs of images used for the estimation of the relative pose

| | |
|---|---|
| Rotation | $\mathbf{R} = \begin{bmatrix} 1.0000 & -0.0001 & 0.0006 \\ 0.0001 & 1.0000 & 0.0005 \\ -0.0006 & -0.0005 & 1.0000 \end{bmatrix}$ |
| Translation | $\mathbf{T} = \begin{bmatrix} -0.413 \\ 0.142 \\ 99.948 \end{bmatrix}$ |

Table 3. Results of the estimation of the relative pose on synthetic images

The rotation matrix is almost an identity matrix since the sensors are perfectly aligned. The norm of the translation vector is 99.95mm instead of 100mm, so an error of 0.05%. These results are particularly good especially since they involve the calibration results needed for the estimation of the pose of the calibration pattern.

The estimation of the relative pose was then evaluated on real images. The sensor was mounted on a graduated rail and was moved by steps of 10cm. At each position, an omnidirectional image was grabbed with the aim of computing the displacement of the sensor according to the first position using five calibration patterns placed in the room as shown in Figure 14. Table 4 summarizes the results.

| Displacement (mm) | 100 | 200 | 300 | 400 | 500 | 600 |
|---|---|---|---|---|---|---|
| Estimation (mm) | 100.13 | 200.77 | 296.48 | 393.66 | 494.70 | 594.60 |

Table 4. Calibration results on real images

The average error is greater than the one obtained with synthetic images but it is less than 0.9%. This increase can be explained by the noise in real images but also by the accumulation of the errors because the estimation of the relative pose involves the calibration values.
Once the relative pose is estimated, the essential matrix can be computed and used to check the epipolar geometry properties. In Figure 14, for each pixel on the left image (red crosses), the corresponding epipolar curve (green curves) is drawn on the right image and vice versa.

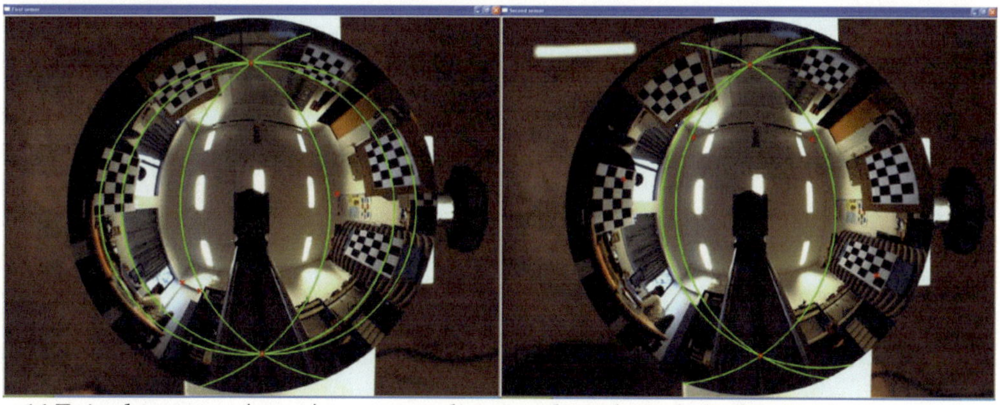

Fig. 14.Epipolar curves (green) corresponding to selected pixels (red crosses)

### 5.3 Triangulation

The combination of the calibration of the two sensors and of their relative pose estimation was evaluated by triangulation. The coordinates of the 3D points of five grids disposed on three walls of a room were estimated. Images used for this step are presented in Figure 15.

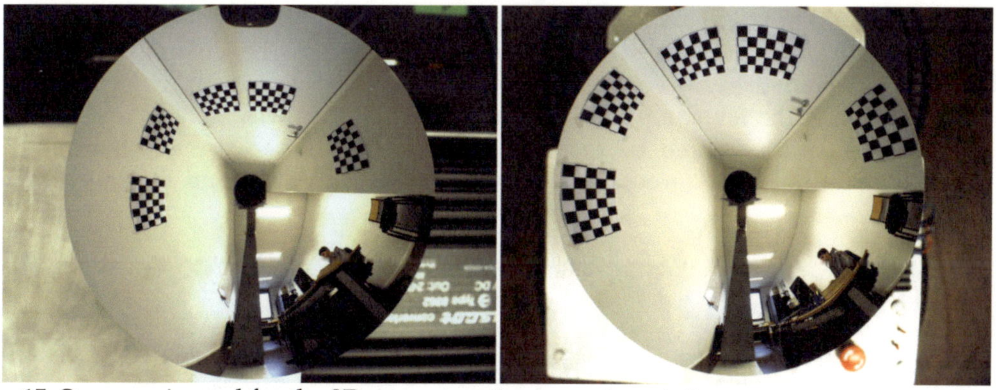

Fig. 15. Stereo pair used for the 3D reconstruction experimentation.

For each grid, the four corners were selected manually on the two images and an automatic corners detector was applied to extract all grid points. The 3D position of these points was evaluated by triangulation and is displayed in Figure 16. The position of the points is compared to the ground truth obtained by the laser range-finder. A linear regression was performed on raw laser data (in blue) to obtain the position of the walls (red lines).

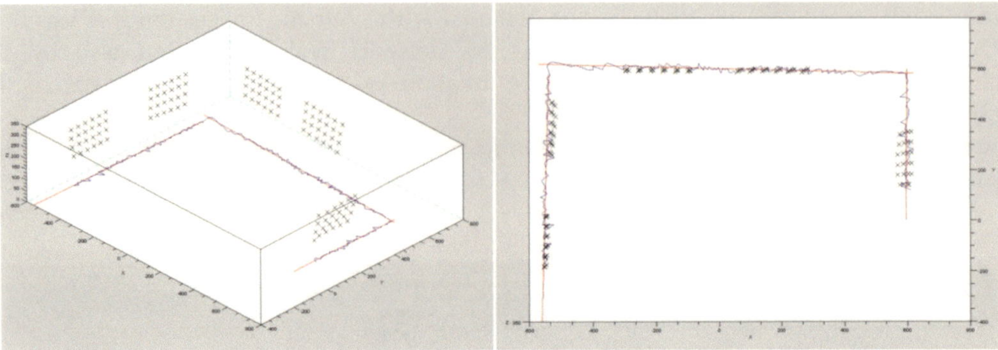

Fig. 16. Position of the five grids (crosses) compared to ground truth (blue and red lines).

For each grid, the mean error and the standard deviation of the position of its points were computed and summarized in Table 5.

| Grid number | 1 | 2 | 3 | 4 | 5 |
|---|---|---|---|---|---|
| Mean error (mm) | 5.843 | 15.529 | 13.182 | 3.794 | 12.872 |
| Standard deviation (mm) | 2.616 | 4.500 | 2.420 | 2.504 | 7.485 |

Table 5. Triangulation results.

The 3D error is very low, around 1% at 1 meter. The good results obtained in triangulation imply a reliable estimation of the sensor parameters and of the relative pose. Due to the resolution of catadioptric sensors, this error will nevertheless increase with the distance and will be around 10% at 10 meters.

### 5.4 Simultaneous Localization And Mapping

Our SLAM algorithm was evaluated on a real video sequence. The mobile robot was moved along the front of a building on a distance of 30 meters. The environment used is particularly rich since there are both manmade and natural items as shown in Figure 17.
At each position of the robot, a pair of omnidirectional images was grabbed and a map of the environment was acquired by the laser range-finder. These maps are then used to compute the real trajectory (ground truth) of the robot since this kind of sensor is very accurate.

Fig. 17. Some of the pictures of the video sequence used for the Simultaneous Localization and Mapping.

The knowledge of the poses of the robot allows merging the local maps to obtain the global map of the environment, as shown in Figure 18. Thus, a comparison between the estimations provided by the laser range-finder, by the SLAM algorithm and by odometry can be achieved by observation of the global map, as shown in Figure 18.

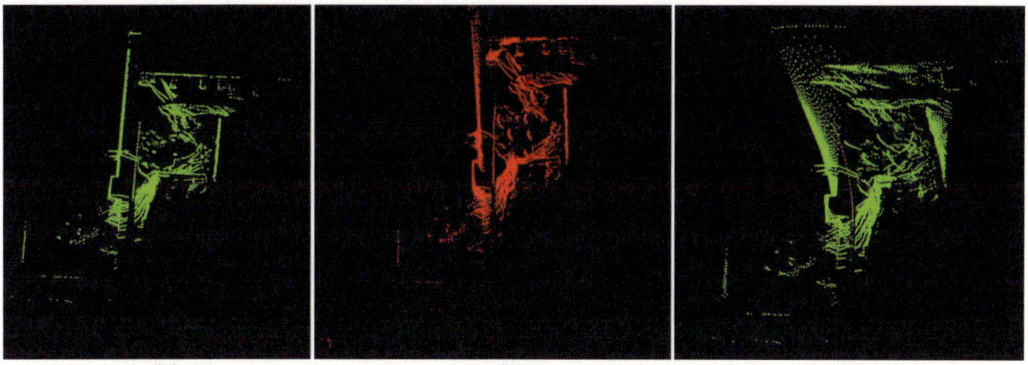

(a) (b) (c)

Fig. 18. Global map of the environment using: (a) laser data, (b) SLAM algorithm, (c) odometry data

As shown in Figure 18, the global map of the environment obtained by the laser range-finder is the best result since local maps are perfectly coincident. Our SLAM algorithm provides a slight deviation but the global map is far better than the map that can be obtained by odometry. The mean error on this sequence is around 12%.

## 6. Conclusion and future work

This paper has been devoted to the design of an innovative vision sensor dedicated to mobile robot application. Combining omnidirectional and stereoscopic vision offers many advantages for 3D reconstruction and navigation that are the two main tasks a robot has to achieve. In this article we have highlighted that the best stereoscopic configuration is the vertical one as it simplifies the pixel matching between images.

A special care has been put on the sensor calibration to make it flexible since it only requires the use of a planar calibration pattern. Experimental results highlight a high accuracy, which foreshadows good results for the following algorithms.

The fundamental issue of Simultaneous Localization And Mapping (SLAM) was then addressed. Our solution to this problem relies on a bundle adjustment between two displacements of the robot. The initialization of the motion and the coordinates of 3D points is a prerequisite since bundle adjustment is based on a non-linear minimization. This step is a tricky problem to which we answered by the generalization of the epipolar geometry for central sensors using the unified model. Our experimental results on SLAM are promising but the error is higher than the one expected further to the calibration results.

Our future work will focus on the improvement of our SLAM method by adding a global bundle adjustment to avoid error accumulation. A lot of challenges in SLAM are moreover always open, for instance SLAM based only on vision systems, SLAM taking into account six degrees of freedom, or SLAM for large-scale mapping.

## 7. References


Bailey, T. & Durrant-Whyte, H (2006). Simultaneous Localization and Mapping (SLAM): Part II. *Robotics and Automation Magazine,* Vol. 13, No. 3, (September 2006) 108-117.

Baker, S. & Nayar, S.K. (1999). A Theory of Single-Viewpoint Catadioptric Image Formation. *International Journal of Computer Vision (IJCV),* Vol. 35, No. 2, (November 1999) 175-196, ISSN 0920-5691

Barreto, J.P. (2006). A Unifying Geometric Representation for Central Projection Systems. *Computer Vision and Image Understanding,* Vol. 103, No. 3, (September 2006) 208-217, ISSN 1077-3142

Benosman, R. & Devars, J. (1998). Panoramic Stereovision Sensor, *Proceedings of the International Conference on Pattern Recognition (ICPR),* pp. 767-769, Brisbane, Australia, August 1998, IEEE Computer Society, Los Alamitos

Boutteau, R. (2009). 3D Reconstruction for Autonomous Navigation, *http://omni3d.esigelec.fr/doku.php/thesis/r3d/start*

Bunschoten, R. & Kröse, B. (2003). Robust Scene Reconstruction from an Omnidirectional Vision System. *IEEE Transactions on Robotics and Automation,* Vol. 19, No. 2, (April 2003) 351-357, ISSN 1042-296X

Durrant-Whyte, H & Bailey, T. (2006). Simultaneous Localization and mapping: Part I. *Robotics and Automation Magazine,* Vol. 13, No. 2, (June 2006) 99-110, ISSN 1070-9932

Geyer, C. & Daniilidis, K. (2000). A Unifying Theory for Central Panoramic Systems and Practical Implications, *Proceedings of the European Conference on Computer Vision*, pp. 445-461, ISBN 3-540-67685-6, Dublin, Ireland, June 2000, Springer, Berlin

Gonzalez-Barbosa, J.J. & Lacroix, S. (2005). Fast Dense Panoramic Stereovision, *Proceedings of the International Conference on Robotics and Automation (ICRA),* pp. 1210-1215, ISBN 0-7803-8914-X, Barcelona, Spain, April 2005, IEEE

Harley, R. & Zisserman, A. (2004). *Multiple View Geometry in Computer Vision,* Cambridge University Press, ISBN 0521540518

Levenberg, K. (1944). A method for the solution of certain problems in least squares. *Quarterly of Applied Mathematics,* Vol. 2, 164-168

Lhuillier, M. (2005). Automatic Structure and Motion using a Catadioptric Camera, *Proceedings of the 6th Workshop on Omnidirectional Vision, Camera Networks and Non-classical* Cameras, pp. 1-8, ISBN, Beijing, China, October 2005

Longuet-Higgins, H.C. (1981). A computer algorithm for reconstructing a scene from two projections. *Nature,* Vol. 293, (September 1981) 133-135

Lowe, D.G. (2004). A Distinctive image features from scale-invariant keypoints. *International Journal of Computer Vision (IJCV),* Vol. 60, No. 2, (November 2004) 91-110, ISSN 0920-5691

Mariottini, G.L. & Prattichizzo, D. (2005). The Epipolar Geometry Toolbox : multiple view geometry and visual servoing for MATLAB. *International IEEE Robotics and Automation Magazine,* Vol. 12, No. 4, (December 2005) 26-39, ISSN 1070-9932

Marquardt, D.W. (1963). An Algorithm for Least-Squares Estimation of Nonlinear Parameters. *Journal of the Society for Industrial and Applied Mathematics,* Vol. 11, No. 2, 431-441

Mei, C. & Rives, P. (2007). Single View Point Omnidirectional Camera Calibration from Planar Grids, *Proceedings of the International Conference on Robotics and Automation (ICRA),* pp. 3945-3950, ISBN 1-4244-0601-3, Roma, Italy, April 2007, IEEE

Mouaddib, E.M. (2005). Introduction à la vision panoramique catadioptrique. *Traitement du Signal,* Vol. 22, No. 5, (September 2005) 409-417, ISSN 0765-0019

Mouragnon, E.; Lhuillier, M.; Dhome, M.; Dekeyser, F.; Sayd, P. (2009). Generic and Real-Time Structure from Motion using Local Bundle Adjustment. *Image and Vision Computing Journal,* Vol. 27, No. 8, (July 2009) 1178-1193, ISSN 0262-8856

Negishi, Y.; Miura, J.; Shirai, Y. (2004). Calibration of Omnidirectional Stereo for Mobile Robots, *Proceedings of the IEEE/RSJ International Conference on Intelligent Robots and Systems (IROS),* pp. 2600-2605, ISBN 0-7803-8463-6, Sendai, Japan, September 2004

Nister, D. (2001). Automatic Dense Reconstruction from Uncalibrated Video Sequence. *PhD Thesis, Royal Institute of Technology KTH, Stockholm, Sweden*

Pajdla, T.; Svoboda, T.; Hlavac, V. (2001). Epipolar Geometry of Central Panoramic Catadioptrics Cameras, In: *Panoramic Vision: Sensors, Theory and Applications,* (Gries, D. & Schneider, F.B.), (73-102), Springer, ISBN 978-0-387-95111-9, Berlin, Germany

Pollefeys, M; Gool, L.; Vergauwen, M; Verbiest, F.; Cornelis, K. ; Tops, J. (2004). Visual Modeling with a Hand-Held Camera. *International Journal of Computer Vision (IJCV),* Vol. 59, No. 3, (September-October 2004) 207-232, ISSN 0920-5691

Ramalingram, R.; Sturm, P.; Lodha, S.K. (2005). Towards Complete Generic Camera Calibration, *Proceedings of the International Conference on Computer Vision and Pattern Recognition (CVPR),* pp. 767-769, ISBN 0-7695-2372-2, San Diego, USA, June 2005, IEEE Computer Society, Los Alamitos

Scaramuzza, D.; Martinelli, A.; Siegwart, R. (2006). A Flexible Technique for Accurate Omnidirectional Camera Calibration and Structure From Motion, *Proceedings of the International Conference on Computer Vision Systems (ICVS),* pp. 45-52, ISBN 0-7695-2506-7, New York, USA, January 2006, IEEE Computer Society, Los Alamitos

Thrun, S.; Fox, D.; Burgard, W.; Dellaert, F. (2001). Robust Monte Carlo Localization for Mobile Robots. *Artificial Intelligence,* Vol. 128, No. 1-2, (May 2001) 99-141

Thrun, S. (2003). Robotic mapping: A survey, In: *Exploring Artificial Intelligence in the New Millenium,* (Kakemeyer, G. & Nebel, B.), (1-29), Morgan Kaufmann, ISBN 1-55860-811-7, San Francisco, USA

Triggs, B.; McLauchlan, P.; Hartley, R.; Fitzgibbon, A. (1999). Bundle Adjustment - A modern Synthesis. *Vision Algorithms: Theory and Practice,* Vol. 1883, (September 1999) 298-372, ISSN 0302-9743

Wang, W. & Tsui, H.T. (2000). A SVD decomposition of essential matrix with eight solutions for the relative positions of two perspective cameras, *Proceedings of the International Conference on Pattern Recognition (ICPR),* pp. 362-365, ISBN, Barcelona, Spain, September 2000, IEEE Computer Society, Los Alamitos

2

# Optical Azimuth Sensor for Indoor Mobile Robot Navigation

Keita Atsuumi and Manabu Sano
*Graduate School of Information Sciences, Hiroshima City University*
*Japan*

## 1. Introduction

A new type of azimuth angular sensor for an indoor mobile robot navigation is developed. This is a kind of optical sensors using infrared linear polarization. Since an angle is measured without contact, it is applicable to the navigation of an autonomous mobile robot. In the indoor environment, the navigation system by using GPS cannot be used. In dead reckoning, the accumulation of measurement error occurs a serious problem. If we use this sensor, we can measure a position only by installing one landmark all over the working space. We can acquire simultaneously not only the distance from a landmark but an angle of direction. Like a gyroscope or a geomagnetism sensor, the large drift depending on environment or time does not occur by this sensor. We make a prototype of the sensor based on this technique and conduct the measurement experiment. The accuracy of azimuth angle error is about 4% and the radial error is 93(mm) at 3(m) distance from the landmark.
An autonomous mobile robot needs to acquire self-position information at the time of moving in the working space. On actuators, such as a robot's mechanism and an arm with moving, there are many constraints in flexibility. Therefore, the position of the robot in the coordinate system fixed to working space and an azimuth angle are very important for moving in the working space. When working space is the outdoors, a self-position can be known with high precision using the positioning system (GPS) by an artificial satellite. It is also comparatively easy to measure geomagnetism using a flux-gate type or a semiconductor type magnetometric sensor, and to acquire an angle of direction. However, it is difficult to use such sensors in the indoor environment, for example, in an office, a factory and an inside of a tunnel and so on. Since an electric wave decreases the performance of the guidance control in the indoor autonomous mobile robot by existing the roof and the surface of a wall, the self-position measurement using GPS is difficult in those environment. The iron desk and the iron bookshelf which exist in the narrow space like office deteriorate the accuracy of a geomagnetism sensor. And though an angle of direction can be acquired using an angular velocity sensor (gyroscope), the error of measurement is accumulated with time. Although the method of marking beforehand on the floor and acquiring a self-position is technically easy, a lot of marks must be prepared manually, it is troublesome and inflexible. The specified of coordinates point in the working space is called a landmark. If working space with a two-dimensional spread is assumed, the point in coordinates cannot be

determined a specified point only using the single landmark. The measuring method of the robot position using a infrared landmark was reported (N.Ushimi , M.Yamamoto et. al., 1999-2000). It is interesting that their method is simple in the structure and the position is derived from angles. Even though they use a landmark, it is disadvantage that they cannot determine the absolute azimuth. This research proposes a special idea to the landmark and shows how to acquire a position and an azimuth angle of direction simultaneously using a single landmark. The sensor system of the special infrared landmark is using linear polarization. The performance of the sensor is shown by the experiment.

## 2. Basic Measureing Principle by Linear Polarizer

### 2.1 Linear Polarizer

Generally, the random polarizing component is containing in the natural ray. Linear polarizer is a kind of optical filters that can pass through only the oscillating component of an e-vector along a single polarizing plane. There are two manufacturing methods: (a) Spattering method of the fine material fibers to a unique direction on the glass basis. In this case, it is superior to antienvironmental properties, the characteristics of wave lengths and the uniformity of the obtained polarizer ray. However it is difficult to handle and manufacture practically. (b) Stretching method of the colored high-polymer with the Iodine and unifying the orientation of the molecule and binding protecting by matrices. Even if this method has a disadvantage to minor changes of temperature and humidity, it is easy to cut and bind in the manufacturing process because it is thin, light and more flexible. In this study, we are focusing to the latter merit in particular, we realize a new angular sensor by using the layered polarizer and manufacturing to cone-like threedimensional shape.

### 2.2 Polarizing Angle Measurement

We use the modulation technique that the angular displacement can be measured as the angle of polarizing plane. We arrange the light source and the polarizer and the photodetector in Fig. 1. The polarizer(a) is fixed and the polarizer(b) is rotated around the axis. The light from the source is passed through both polarizers and arrived at the photo detector.

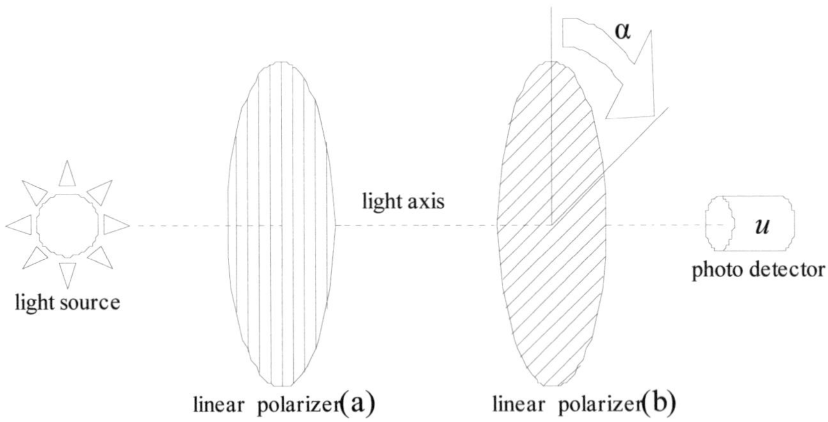


Fig. 1. Basic measuring principle by linear polarizer

When we assume that the angle between the polarizer(a) and the polarizer(b) is α and the intensity of the light is u, we can get the following expression,

$$u = A \cos^2 \alpha. \tag{1}$$

Here the A is an attenuation ratio which is positive constant less than 1.0 given by the material of polarizer or optical wavelength. The change of the polarizer(b) depends on the relative angle around the light axis between two polarizers. Since the polarizer(a) is located to the closer position with the measuring object and the polarizer(b) is near to the sensor, the measuring resolution of the angle is not influenced principally by the arrangement with a distance of two polarizers. Many measuring techniques on the angle of polarized plane were reported and these techniques can be divided to two categories. One is the measuring method using some anglefixed polarizers and the photo detector coupled to the former polarizers. The other is the one using the rotated polarizer and the photo detector. We refered those works for measuring the angle of polarized plane; we apply the latter measuring technique to our system. We can measure the angle of polarizing plane on coming by the relative intensity of the arrived light to the photo detector. In our system, the polarizer(b) is rotated by using the small DC motor. If we calculate the phase from the time-varing angle change of the polarizer(b) and the time-varing change of the intensity of arrived light, we can estimate the angle on the polarizing plane. Thus, if we get the observing intensity of the light on the photo detector is u*, the estimated value of the angle on the polarizing plane α* is given by (2a-c).

$$x^* = \int_{t=0}^{360} u^*(t)\cos(2t)\, dt \tag{2a}$$

$$y^* = \int_{t=0}^{360} u^*(t)\sin(2t)\, dt \tag{2b}$$

$$\alpha^* = \tan^{-1} \frac{y^*}{x^*} \tag{2c}$$

## 3. Cone Shape Linear Polarizer

### 3.1 Extension of Measurable Angle Range

A certain kind of insect can return to a nest by assisting polarization in the sky. It is familiar that outdoors, weak polarization occurs in the sky by air correlate with sunrays. Such as an insect's behavior is applied to robot's action (D.Lambrinos et.al., 1997). At first, we have to understand the basic property of linear polarizing light. While the polarizer rotates one turn, same polarizing states appear twice. The range of an azimuth angle is 360(deg). On the contrary, the range of polarizing angle is 180(deg). This means that azimuth of 0(deg) and that of 180(deg) are undistinguished. Polarizing planes cannot discriminate the front surface and the rear surface. Since a linear polarizer is flat sheet in the original shape, there are some problems when we use it for the angle sensor. For solving this problem, we invent a new form of the linear polarizer for the angular sensor. We cut out the semi-circular sheet from the flat sheet of the linear polarizer. We make the cone-shaped linear polarizer from the semi-circular flat sheet by attaching each straight edge mutually around the center of the straight line as Fig. 2. We assume that the apex of the cone-wise plane is the point O, the center axis is Z, the angle along the cone-wise plane from the line OA to an arbitrary position C is $\theta$, the rotating angle of the cone-wise plane around the Z-axis is $\varphi$, the angle between the line of vision (ray axis) and the ground is $\gamma$, and the inclined angle of the mother line is $\gamma_0$. We can obtain next formulas from the above geometrical relation,

$$\cos \varphi = \cos (2\theta) \tag{3a}$$

$$\sin \varphi = \cos ( \gamma - \gamma_0 ) \sin (2\theta) \tag{3b}$$

If we assume the relation $\gamma = \gamma_0$ , we can get

$$\varphi = 2\theta . \tag{4}$$

In the above relation, the angle $\theta$ means the angle of the polarizing plane based on the line OA. Hence we can extend the polarizing angle $\theta$ in the range of 180(deg) to the rotating angle of the cone-wise plane $\varphi$ in the range of 360(deg). Since both angles are one-to-one correspondence, we can determine the azimuth angle uniquely. When the sharpness of the cone-wise plane is the angle $\gamma_0$, the extending shape of the cone-wise plane is determined uniquely as shown in Fig. 2. and we can get the next relation

$$\gamma_0 = 30(\text{deg}) . \tag{5}$$

Then we can show that the rotating angle of the cone-wise plane around the Z-axis can be modulated as the angle of the polarizing plane around the ray axis. That is, if we set a light source inside the cone-wise polarizer and observe from outside the cone-wise plane, the angle of the observed polarizing plane is proportional to the rotating angle of the cone-wise plane.

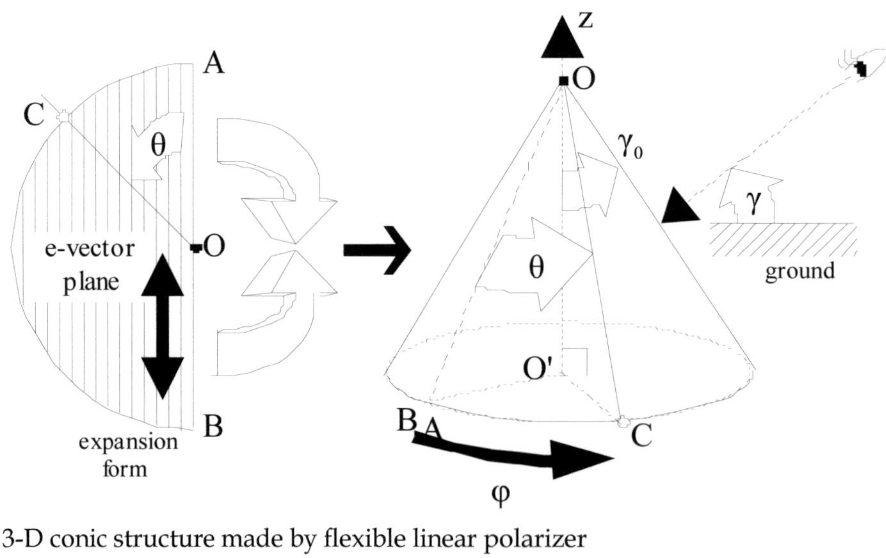


Fig. 2. 3-D conic structure made by flexible linear polarizer

### 3.2 Compensation of Discontinuity

There is an adhered part that is parallel to the mother line in the manufactured cone-wise polarizer structurally. If we use a single cone-wise polarizer, sensing information is discontinuous and influenced polarizing properties at the junction. So we use two cone-wise polarizers as shown in Fig. 3. and assign them at distorted angles each other for compensating discontinuity.

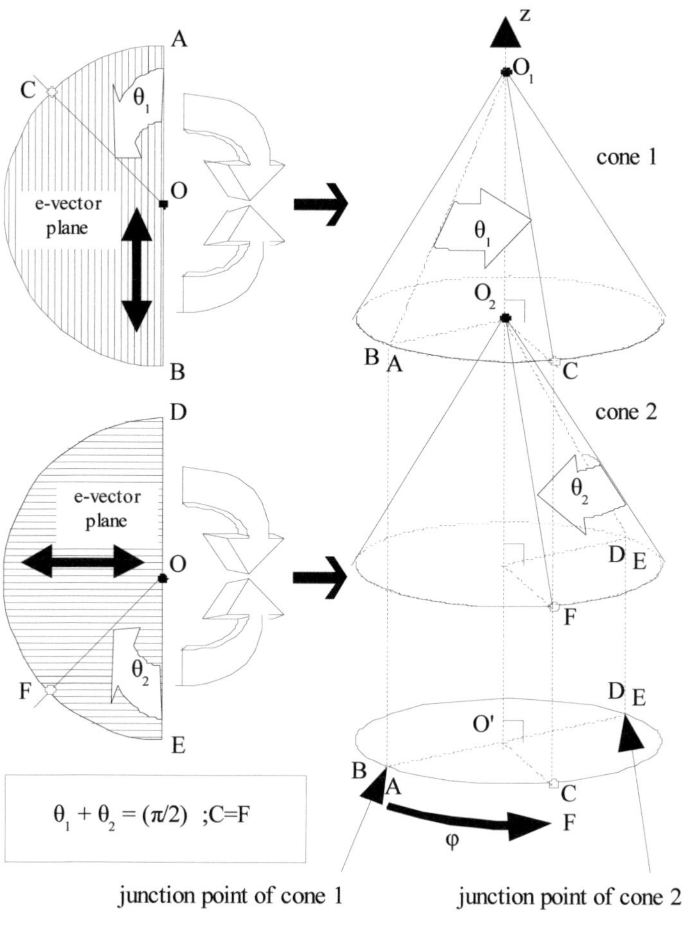


Fig. 3. Elimination of constructional discontinuity

## 4. Measuring Principle of Distance

### 4.1 Distance Determination by Elevation Angle

At indoor environment, we image the working space with the constant distance from the floor to the ceiling. We set an infrared landmark to a ceiling and we assume that the landmark is located at the arbitrary ceiling in the working space. The light coming from a landmark is correctly caught on the directive principal axis by the sensor attached to the robot. If it assumes that the height H from floor to the ceiling is constant, the horizontal distance from the transmitter to the receiver can be calculated from the elevation of the receiver as shown in Fig. 4. If the elevation angle β is measured, the horizontal distance L between a transmitter and a receiver will be obtained by

$$L = \frac{H}{\tan \beta} \,. \tag{6}$$

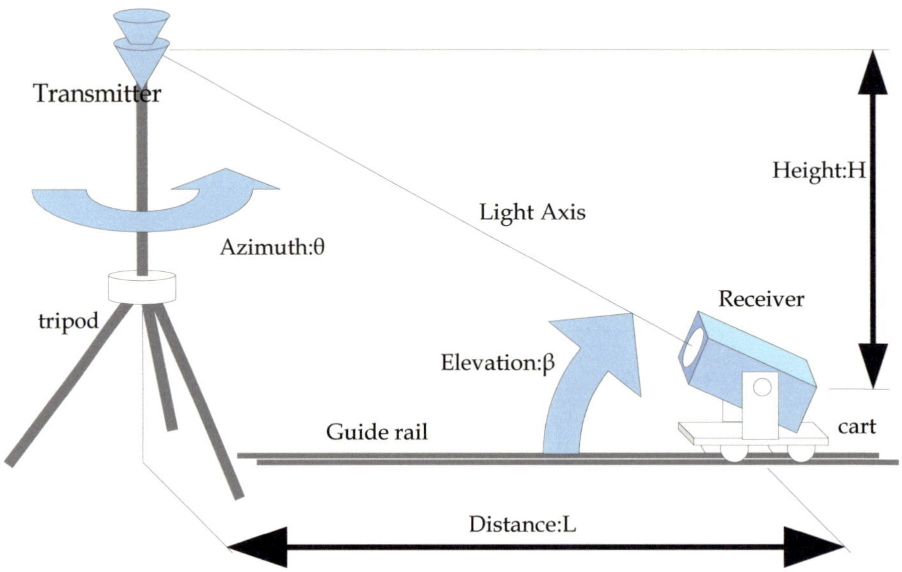


Fig. 4. Experimental setup

#### 4.2 Elevation Angle Measurement using Accelerometer

An elevation angle is measured from the plane of the floor. With the ideal level, a floor is not always horzontal but there is an inclination and unsmoothness. When the cart of the mobile robot with the sensor may incline, the measured result of the elevation angle might be included some errors. So we measure the elevation angle on the basis of gravity acceleration. We measure gravity acceleration using a semiconductor-type acceleration sensor and acquire an elevation angle from the ratio of gravity acceleration which acts on each axis. If the robot is stationary, downward gravity acceleration will act on a sensor. An acceleration sensor has specific axes which shows higher sensitivity. In this research, we call them sensitivity axes.

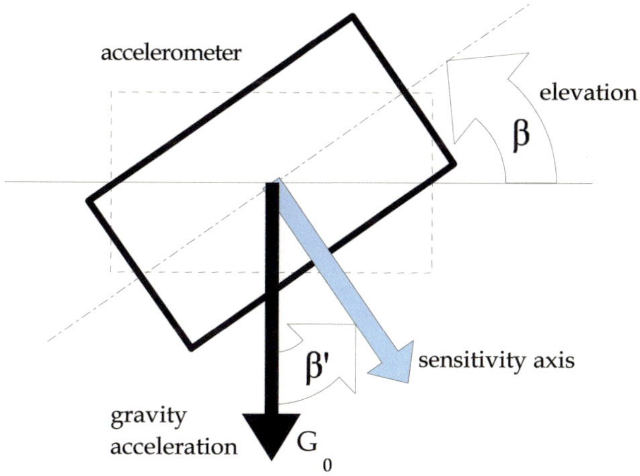


Fig. 5. Accelerometer and sensitivity axis

We set a sensitivity axis perpendicular to downward direction with β' as the preparation of measurements as shown in Fig. 5. An output voltage from gravity acceleration $V_{out}$ which acts along a single sensitivity axis is expressed in the following

$$V_{out} = K_s G_0 \cos \beta' + V_{offset}. \tag{7}$$

Here $K_s$ is the sensor gain , $G_0$ is constant gravity accerelation and $V_{offset}$ is offset voltage of the sensor which adjust to zero in advance. Differentiating (7) about β', we get

$$V_{diff} = - K_s G_0 \sin \beta' . \tag{8}$$

We know, the closer a sensitivity axis approaches vertically from horizontal axis, the worse the sensitivity of an acceleration sensor becomes.

### 4.3 Improvement of the measureing precision

When the elevation angle β in eq.(6) is include the measuring error Δβ , we get

$$L + \Delta L = \frac{H}{\tan(\beta + \Delta\beta)} = H\left(\frac{1 - \tan\beta \tan\Delta\beta}{\tan\beta + \tan\Delta\beta}\right) \cdot \tag{9}$$

By Eqs.(6) and (9), the distance error ΔL is shown as follows

$$\Delta L = H\left(\frac{1 - \tan\beta \tan\Delta\beta}{\tan\beta + \tan\Delta\beta} - \frac{1}{\tan\beta}\right) \cdot \tag{10}$$

In the height H is constant value(H=2.0(m)), the relation between distance error and elevation angle error is shown as Fig. 6, in the case of $d = 0.1, 1, 5$ and 10 (m).

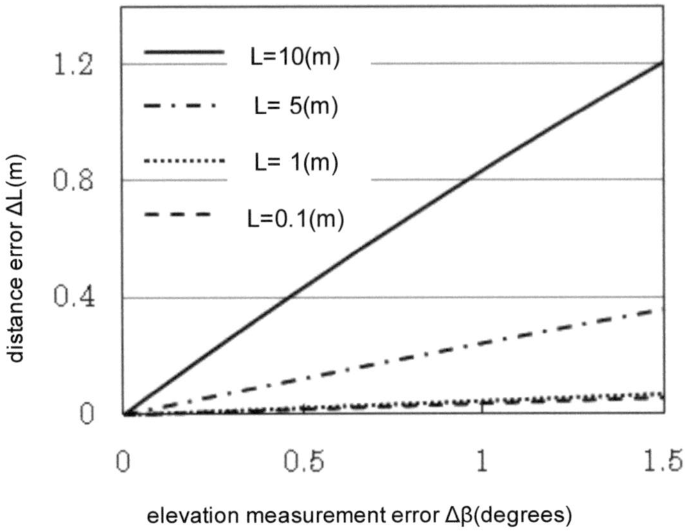


Fig. 6. Distance error vs elevation measurement error

Gravity acceralation on the stationally body is always constantly downward 1.0(G). If we assume components of the acceleration $X_G$(G) and $Y_G$(G), with the inclination angle β(deg) of mutually orthogonal principle axes of accelerations, **β*** is satisfied with the following

$$\beta_x^* = \cos^{-1}\left(\frac{V_{out\ x}}{K_{sx} G_0}\right). \tag{11a}$$

$$\beta_y^* = \cos^{-1}\left(\frac{V_{out\ y}}{K_{sx} G_0}\right). \tag{11b}$$

When we measure the acceleration among which axis, we get the elevation on that axis.
In the elevation angle measurement of using single axis, the measureing presition is remarkably decreased by the non-linearity of the trigonometric function at the specified angles.
The sensitivity of the gravity acceleration affects on that of the elevation angle at the proximity of the angle which the principle axis is parallel to the vartical axis. We compensate the elevation angle measurement by using multi axes in the following two approaches so that we consider the angle range of confirming more presize measurement.

(a) Right angle switching method; for excluding the angle range of principle axes with the remarkably worth precision, we use the single axes of more suitable axis. Such the angle $\beta^*$(deg) is including the range of $0 \le \beta < 45$, $135 \le \beta < 225$ and $315 \le \beta < 360$, we use the angle on the X-axis and otherwise we use the angle on the Y-axis. i.e.

$$\beta_{(a)}^* = \begin{cases} \beta_x^* & ; 0 \le \beta < 45 \,,\, 135 \le \beta < 225 \,,\, 315 \le \beta < 360 \\ \beta_y^* & ; \quad 45 \le \beta < 135 \,,\, 225 \le \beta < 315 \end{cases} . \tag{12}$$

(b) Weighting method; the way of measureing angle without switching axis that we put the more weight as more principle axis is closer to horizontal direction and vice versa. If we make the voltage $V_x$(V),$V_y$(V) and the angle of $\beta_x^*$(deg),$\beta_y^*$(deg)of each axis, we can get the weighting average as follows,

$$\beta_{(b)}^* = \beta_x^* \left( \frac{V_y}{V_x + V_y} \right) + \beta_y^* \left( \frac{V_x}{V_x + V_y} \right) . \tag{13}$$

We use the electric capacity type 3-axes semiconductor acceleration sensor (Kionix KXM52-1050). Sensitivity axes of this sensor cross orthogonal mutually. We measured the elevation angle using two of three sensitivity axes $V_{out\,x}$ and $V_{out\,y}$.
An X-axis positive direction is defined as 0 (deg), Y-axis positive direction is 90 (deg) and Z-axis is perpendicular to the X-Y plane. That is used as rotation axis in this experiment. Next, we adjust offset and gain of an accelerometer that X-axis output voltage $V_{out\,x}$ to 0(V) when the $X_0$ axis is 0(deg), Y-axis output voltage $V_{out\,y}$ to 0(V) when the X-axis is 90(deg). We regard the angle set by the angle-setup apparatus as the true value in X-axis direction. Then we adjust each 5(deg) in the range of 0 ~ 355(deg) and we compare the error between the angle calculated from accelerometer output (angle measurement) and that of the evaluated value. In addition, we compare and evaluate on two ways that uses two axes.

| method | $\beta^*_x$ only | $\beta^*_y$ only | Right angle switching | Weighting |
|---|---|---|---|---|
| maximum error (deg) | 1.479 | 2.860 | 0.321 | 0.294 |
| variance | 6.37e-2 | 1.88e-1 | 3.26e-2 | 5.49e-2 |
| S.D. | 2.52e-1 | 4.33e-1 | 1.81e-1 | 2.34e-1 |

Table 1. Relationship of measurement error

The table1 summarize the angle error of the masured angle and the true one by calculating measured data on one axis or two axes. In all items, the two-axis measureing accuracy is better than that of the data by single axis. Additionally in two-axis measurement by using a weighting method, the maximum error seems to be improved. By using the weighting method, the maximum error in the best case is suppressed under the 1/10 value when compared with the single axis. The maximum elevation error obtained by weighting method is 0.294(deg). If we calculate the distance error from this result, eventhough the distance L from the experimental apparatus in Fig. 4 to the target point is 10(m), the error of the distance is about 0.3(m). Therefore, this measuring apparatus and technique can confirm the high precision on the distance measurement. We employed the weighting method for distance measurement.

## 5. Experimental Setup

### 5.1 Introduction

The proposed sensor consists of two parts. One is the transmitter which emits polarized-modulating infrared rays. Another is the receiver which demodulates infrared rays from the transmitter. Thus we can acquire the heading of the receiver's azimuth. By using the transmitter as the landmark, we measure a self-position and absolute azimuth of the receiver. A schematic view of an experimental setup is shown in Fig. 4. The transmitter is attached to a tripod at the height of 1700(mm) from the floor. The vertex direction of conic polarizer corresponds to downward. Since the transmitter can rotate arround a perpendicular axis, it can set arbitrary azimuth angle. The receiver is installed to the cart on the straight rail for setting arbitarary horizontal distance. Setting the height of the receiver to 200(mm) with the cart, we can get the height of H=1500(mm) between the transmitter and the receiver.

### 5.2 Transmitter

The transmitter plays the most important role in this sensor. It consists of an infrared LED array, a pulse generating circuit, and a conic linear polarizer. If the LED is driven by a narrow duty pulse with a subcarrier frequency of 1(MHz), momentary optical power can be increased and we make the signal easy to distinguish from the disturbance light which comes out of lighting apparatus. The infrared rays emitted from LED have strong directivity, and the light intensity which comes out of single LED is weak. In order to keep the strong light intensity, seven LEDs have been arranged over 1 turn. Since the polarizing plane is discontinuous at the jointed line on the cone, we want to shut off the light at this point. We employed a special idea and made a sophisticated device that we used to combine two modules with mutually different direction of jointed line. The block diagram of the transmitter is shown in Fig. 7. Actual size of conic linear polarizer is about 50(mm) diameter.

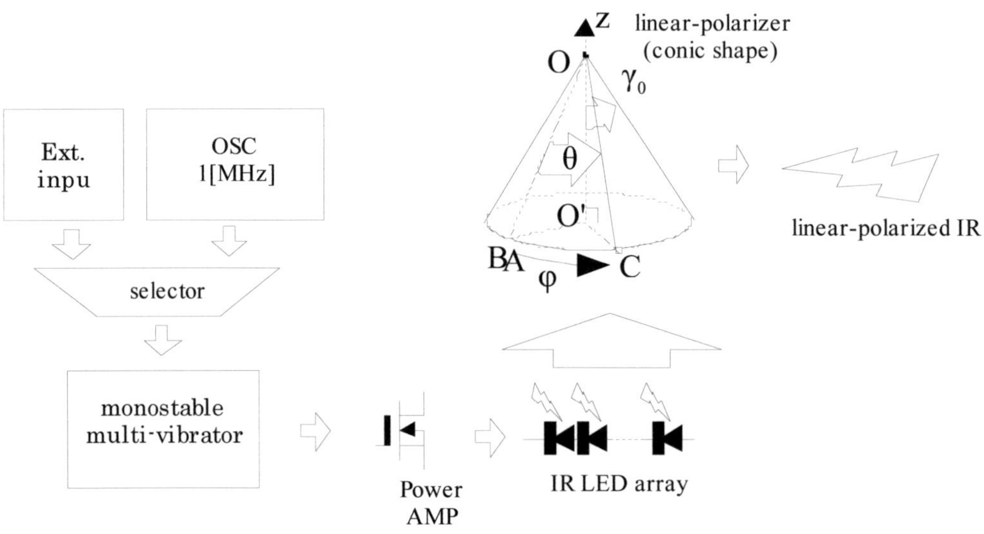


Fig. 7. Block diagram of the transmitter

### 5.3 Receiver

A receiver is constituted by a convex lens, a rotating light polarizer, a photo detector, a preamplifier, and an AM receiving circuit. The mechanical structure of a receiver is shown in Fig. 8. The light coming from the transmitter is first condensed with a convex lens about 35(mm) diameter. Next, the light is passed through the polarizer attached to the small DC motor, which made an amplitude modulation (AM) signal, and a photo detector. The motor has a rotation-synchronized pulse generator. The light which entered into the photo detector is changed into an electric signal, and is inputted into the AM receiving circuit through a preamplifier. The AM receiving circuit is equivalent to the AM middle wave radio in of a superheterodyne system. Thus, the light signal is convert to the phase between the synchronized pulse and the sine wave depend on the angle of polarizing plane. The signal frequency is twice of a motor speed as an AF band approximately 400(Hz). A received signal is taken into PC from A/D conversion through after a low pass filter. Based on the pulse outputted from a motor, the phase of a received sine wave-like signal is proportional to the angle of the linear polarization of the light from the transmitter. Therefore, we will be obtained that the angle around the perpendicular axis of a transmitter by calculate the phase difference.

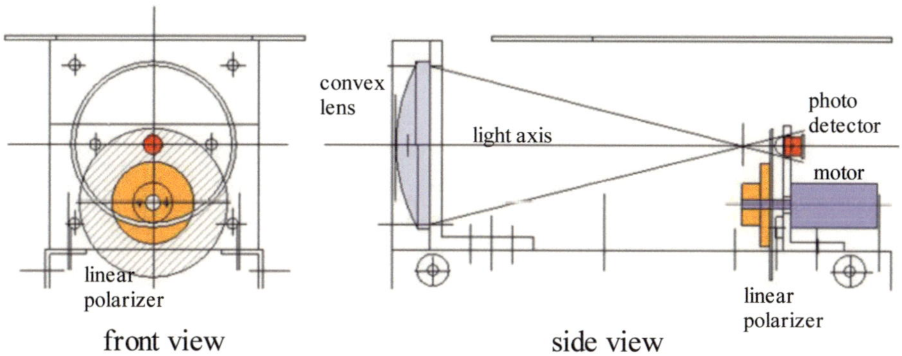


Fig. 8. Mechanical structure of the receiver (fragment)

## 6. Experimental Result and Discussion

### 6.1 Azimuth Measurement

A transmitter is rotated every 10 degrees and azimuth angles at specified ones among 1 turn are measured. The distance from a transmitter to a receiver is influenced by the measuring error of angles. When we change the distance L as the Fig. 4 from 1000(mm) to 3000(mm) at each 500(mm), the measured results of the angle is shown in Fig. 9. Also Fig. 10 shows the measurement angle error. The alignment relation is obtained within 4% at all over the distance range. When the linear polarized light is transmitted inside of the free space where a refractivity does not change, a polarizing plane is maintained. Therefore, angle measurement is not influenced even if distance changes theoretically. However, if the distance from a transmitter to a receiver increases, as a result of a signal declination, in a long distance, the S/N ratio may deteriorate and angle measurement may be affected. The receiver for an experiment rotates the polarizer using the motor, and can obtain the angle of polarization from the phase. If the rotation speed of a motor changes, since the generated delay in LPF will change relatively, the measurement accuracy of a phase deteriorates.

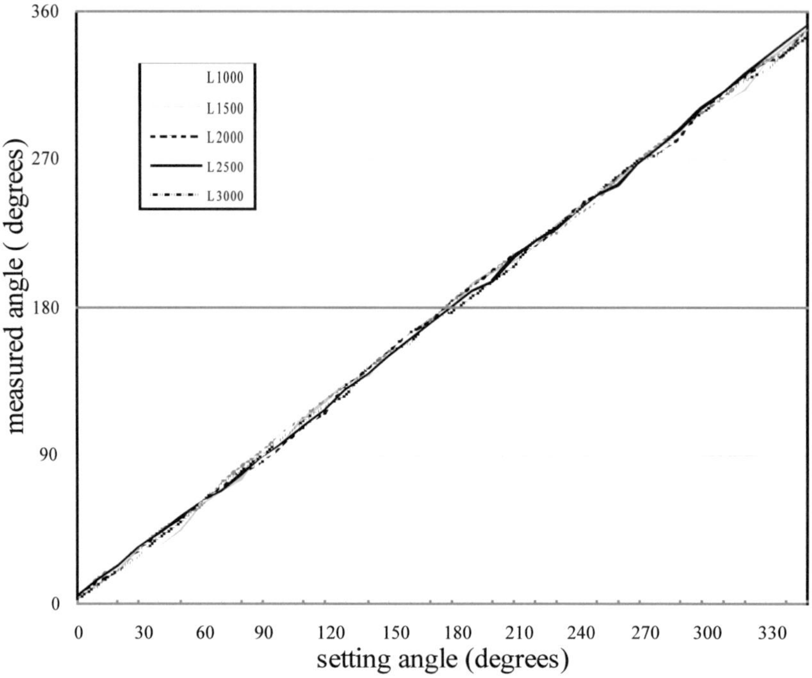


Fig. 9. Measured angle vs set azimuth angle (L:mm)

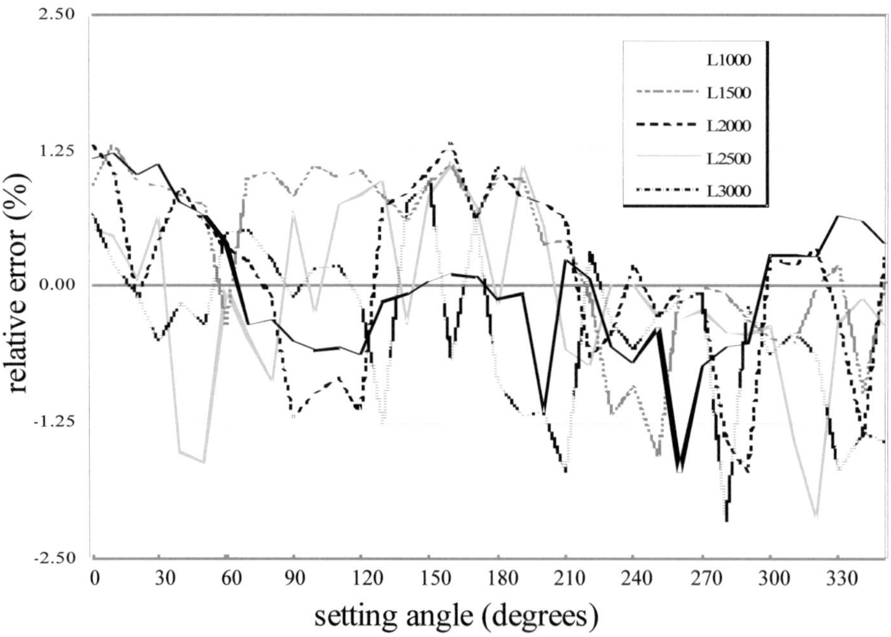


Fig. 10. Measurement error of azimuth angle (L:mm)

A polarizing plate is not moved mechanically but the detecting method of an angle from the strength of the signal from two or more sensors is also discussed (see references). While, in order to receive the light of a transmitter from several meters away, we have to set light axis precisely. It is difficult to configure two or more sensors with same properties exactly. If we employ the two or more divided type monolithic photodiode, it may solve to the problem to some extent. However, we have to attach the polarizing plate adjusted to the angle in front of each element. Our system should be considered as only one optical sensor in total. If the speed of a motor can be stabilized more accurately we expect the measurement accuracy of the direction angle to increase.

### 6.2 Localization Measurement

Fig. 11 depicts the distance measurement result. Relation between setting distance and measured one is linear. The latter shows less than the former. In this experiment, the absolute maximum error is 93(mm) at set distanse of 3000(mm). Finally, we get Fig. 12 which is whole result of the experiment. This r-θ plot illustrates that estimated position of a mobile robot using our sensor system. Of course, the center is a landmark.

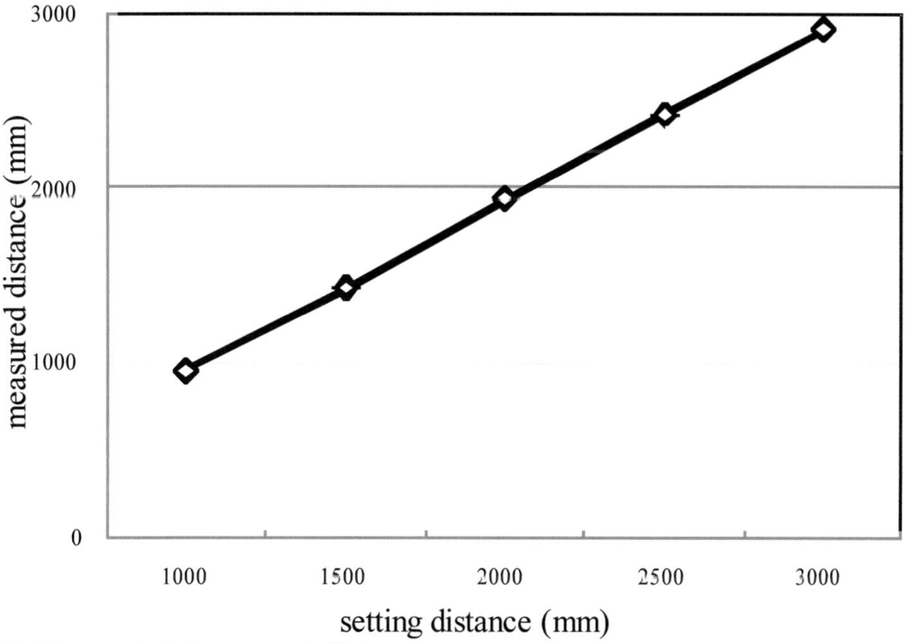


Fig. 11. Measured distance vs set distance

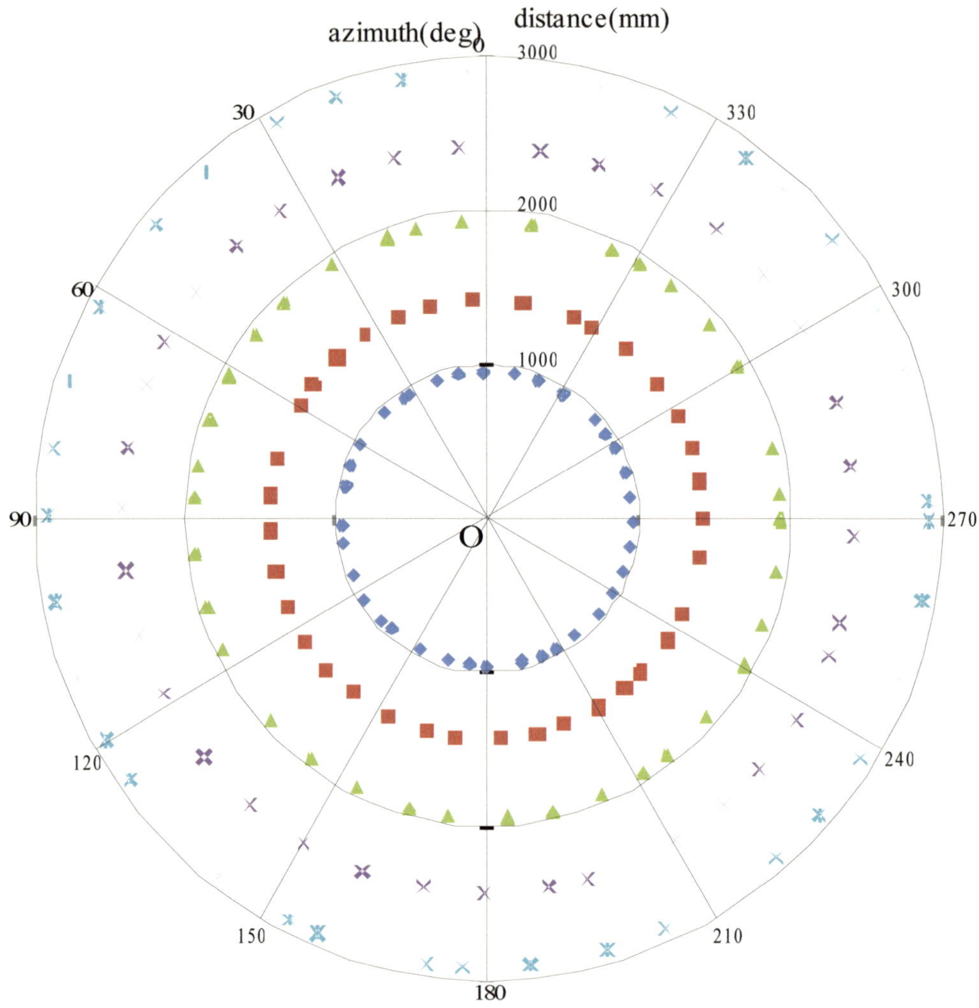


Fig. 12. Localization result of our sensor system.

## 7. Conclusion

We can acquire simultaneously both the azimuth angle and the distance from the target position by using a single landmark. This sensing system is constructed by combination of the optical sensor using infrared linear polarizer which developed by authors and the commercial semiconductor type of acceleration sensor. By using a semiconductor acceleration sensor, experiments on the elevation angle were measured based on the direction of gravity. It is very useful to acquire the information on the position and the angle in the indoor environment.

## 8. References

D.Lambrinos, M.Maris, H.Kobayashi, T.Labhart, R.Pfeifer and R.Wehner (1997), "*An Autonomous Agent Navigating with a Polarized Light Compass*", Adaptive behavior, Vol.6-No.1 ,pp.131-161

K.Atsuumi, M.Hashimoto and M.Sano(2008). "*Optical Azimuth Sensor for Indoor Mobile Robot Navigation*", The 2008 International Conference on Computer Engineering & Systems(ICCES'08), ISBN:978-1-4244-2116-9, Cairo, Egypt.

M.Yamamoto, N.Ushimi and A.Mohri(1999-Mar), "*Navigation Algorithm for Mobile Robots using Information of Target Direction*", Trans.JSME,Vol.65-No.631,pp.1013-1020. (in Japanese)

N.Ushimi, M.Yamamoto and A.Mohri(2000-Mar), "*Development of a Two Degree-of-Freedom Target Direction Sensor System for Localization of Mobile Robots*", Trans.JSME, Vol.66-No.643, pp.877-884. (in Japanese)

Japanese patent No.2001-221660(2001) (in Japanese)

Japanese patent No.H08-340475(1996) (in Japanese)

3

# Vision Based Obstacle Detection Module for a Wheeled Mobile Robot

Oscar Montiel, Alfredo González and Roberto Sepúlveda
*Centro de Investigación y Desarrollo de Tecnología Digital del Instituto Politécnico Nacional. México*

## 1. Introduction

Navigation in mobile robotic ambit is a methodology that allows guiding a mobile robot (MR) to accomplish a mission through an environment with obstacles in a good and safe way, and it is one of the most challenging competence required of the MR. The success of this task requires a good coordination of the four main blocks involved in navigation: perception, localization, cognition, and motion control. The perception block allows the MR to acquire knowledge about its environment using sensors. The localization block must determine the position of the MR in the environment. Using the cognition block the robot will select a strategy for achieving its goals. The motion control block contains the kinematic controller, its objective is to follow a trajectory described by its position (Siegwart & Nourbakhsh, 2004). The MR should possess an architecture able to coordinate the on board navigation elements in order to achieve correctly the different objectives specified in the mission with efficiency that can be carried out either in indoor or outdoor environments.

In general, global planning methods complemented with local methods are used for indoor missions since the environments are known or partially known; whereas, for outdoor missions local planning methods are more suitable, becoming global planning methods, a complement because of the scant information of the environment.

In previous work, we developed a path planning method for wheeled MR navigation using a novel proposal of ant colony optimization named SACOdm (Simple Ant Colony Optimization with distance (d) optimization, and memory (m) capability), considering obstacle avoidance into a dynamic environment (Porta et al., 2009). In order to evaluate the algorithm we used virtual obstacle generation, being indispensable for real world application to have a way of sensing the environment.

There are several kind of sensors, broadly speaking they can be classified as passive and active sensors. Passive sensors measure the environmental energy that the sensor receives, in this classification some examples are microphones, tactile sensors, and vision based sensors. Active sensors, emit energy into the environment with the purpose of measuring the environmental reaction. It is common that an MR have several passive and/or active sensors; in our MR, for example, the gear motors use optical quadrature encoders, it uses a high precision GPS for localization, and two video cameras to implement a stereoscopic vision system for object recognition and localization of obstacles for map building and map reconstruction.

This work presents a proposal to achieve stereoscopic vision for MR application, and its development and implementation into VLSI technology to obtain high performance computation to improve local and global planning, obtaining a faster navigation by means of reducing idle times due to slow computations. Navigation using ant colony environment is based on map building and map reconfiguration; in this model, every ant is a virtual MR. The MR system, composed by the MR and the global planner in the main computer, see Fig 1, has the task to construct the map based on a representation of the environment scene, avoiding the use of Landmarks to make the system more versatile. The MR stereo vision transforms the visual information of two 2D images of the same scene environment into deep measure data. Hence, the MR sends this data via RF to the global planner in the main computer; this data is a 3D representation of the MR scene environment and its local position and orientation. By this way, the optimal path in the environment is constantly updated by the global planner.
The MR stereo vision has the advantage, with respect to other navigation techniques, that depth can be inferred with no prior knowledge of the observed scene, in particular the scene may contain unknown moving objects and not only motionless background elements.
For the environment map construction and reconfiguration, the MR makes an inference of the three dimensional structure of a scene from its two dimensional 2D projections. The 3D description of the scene is obtained from different viewpoints. With this 3D description we are able to recreate the environment map for use in robot navigation.
In general, in any stereoscopic vision system after the initial camera calibration, correspondence is found among a set of points in the multiple images by using a feature based approach. Disparity computation for the matched points is then performed. Establishing correspondences between point locations in images acquired from multiple views (matching) is one of the key tasks in the scene reconstruction based on stereoscopic image analysis. This feature based approach involves detecting the feature points and tracking their positions in multiple views of the scene. Aggarwal presented a review of the problem in which they discussed the developments in establishing stereoscopic correspondence for the extraction of the 3D structure (Aggarwal et al., 2000). A few well-known algorithms representing widely different approaches were presented, the focus of the review was stereoscopic matching.
For map construction or reconfiguration of the MR obstacles environment there is not necessary to reconstruct an exact scene of the environment. There are other works in the same line, in (Calisi et al., 2007) is presented an approach that integrates appearance models and stereoscopic vision for decision people tracking in domestic environments. In (Abellatif, 2008) the author used a vision system for obstacle detection and avoidance, it was proposed a method to integrate the behavior decisions by using potential field theory (Khatib, 1985) with fuzzy logic variables. It was used Hue, Saturation, and Intensity (HSI) color since it is perceptually uniform. In (Cao, 2001) was presented an omnidirectional vision camera system that produces spherical field of view of an environment, the continuation of this work was presented in (Cao et al., 2008) where the authors explained several important issues to consider for using fisheye lens in omnidirectional vision, some of them are the lens camera calibration, rectification of the lens distortion, the use of a particle filter for tracking, as well as the algorithms and the hardware configuration that they implemented.
Recently, the company "Mobile Robots" announced a heavy duty high speed stereoscopic vision system for robots called "MobileRanger StereoVision System", that is able to provide processed images at a maximal rate of 60 fps (frames per second) with a resolution of $752 \times 480$ pixels.

The proposed method has some advantages over existing methods, for example it does not need camera calibration for depth (distance) estimation measurements; an improved efficiency in the stereoscopic correspondence for block matching; adaptive candidate matching window concept is introduced for stereoscopic correspondence for block matching resulting in improved efficiency by reducing calculation time, also improves matching accuracy assuring corresponding process only in the areas where there are vertical or corners arranged pixels corresponding to the obstacles selected features. The calculation process is reduced in average 40% corresponding to the surface ground image content which is previously extracted from every image. The areas between edges in the obstacles itself are also excluded from matching process. This is an additional increase in the method efficiency by reducing calculation for matching process. This feature provides the optimal choice of the best component of the video signal giving improvements in precision of architecture based on FPGA implementation of a vision module for obstacle detection, for map building and dynamic map reconfiguration as an extension research of the ant colony environment model described in a previous work (Porta et al., 2009).
This work is organized as follows: In section 2 the general system architecture is explained. Section 3 is dedicated to give a description of the process to extract the surface ground and obstacle edge detection using luminance components, as well as the process when we include the Hue to obtain the ground surface, moreover, in this section we comment some advantages obtained with the implementation of the vision module into an FPGA. In Section 4 some important concepts about stereoscopic vision are given. In Section 5 is explained how the modification of the road map is achieved. Finally, in Section 6 are the conclusions.

## 2. General System Overview

Figure 1 shows the two main components of the system architecture, the computer, and the MR:

1. The computer contains the global planner based on the SACOdm algorithm, and the communication system.
2. The MR is a three wheels system with frontal differential tracking, it has six main subsystems:
    (a) The stereoscopic vision includes parallel arrange dedicated purpose video decoders controlled via IIC by the FPGA.
    (b) The Spartan-3 FPGA controller board that contains embedded the Microblaze microcontroller, as well as the motors and tracking controllers that were coded in VHDL hardware description language software.
    (c) The power module consists of a high capacity group of rechargeable batteries (not shown in the figure), two H-bridges motor drivers, and the two Pittman DC geared-motor model GM9236S025-R1.
    (d) The communication system based on the XbeePro RF, integrated WiFi communication module.
    (e) A high accuracy GPS module with 1 cm of resolution, 0.05% of accuracy, such as the VBOX 3i from Racelogic (VBOX, 2009), or similar.
    (f) An electromagnetic custom made compass IIC bus compatible, based on the LIS3LV02DL integrated circuit from STMicroelectronics.

The communication between the MR and the computer is achieved using the XBeePro RF Modules that meets the IEEE 802.15.4 standards, the modules operates within the ISM (Industrial Scientific and Medical) 2.4 GHz frequency band. The range of application for indoor/urban range is 100 meters (m), and for outdoor applications with RF line of sight the range is about 1500 m. The serial data rate is in between 1200 bits per second (bps) to 250 kilo bits per second (kbps) (XBee XBee OEM RF Modules, 2007). With no hardware modification it is possible to change the RF module to the XBee-Pro XSC to improve the communication range from 370 m for indoor/urban applications, and 9.6 Km for outdoor line sight applications.

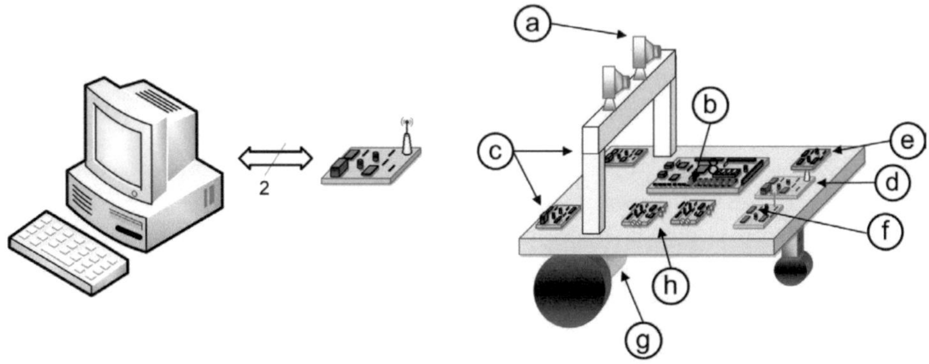


(1) (2)

Fig. 1. The global planner is in the computer communicated through RF with the MR, this is shown in 1). In 2) is the MR with its main components: a) the cameras, b) FPGA system board, c) H bridge motor drivers, d) RF communication system based on the Zigbee technology, e) Magnetic Compass, f) GPS module, g) Gear Pittman DC-motors, h) NTSC Composite video to RGB converter cards.

In Fig. 2 a more detailed description of the stereoscopic vision system is given, each video camera is connected to a conversion board from NTSC composite video to RGB 24 bits video signals, which in turn are controlled by the FPGA based controller board using IIC communication. The video cards send the video information to the controller board where it is processed.

Fig. 3 shows the Microblaze processor, it is a 32 bit soft core processor with Harvard architecture embedded into a Xilinx FPGA. The Microblaze allows to customize its architecture for a specific application. It can manage 4 GB of memory. The 32 bits Local Memory Bus (LMB) connects the processor's core to the RAM Memory Blocks (BRAM) for data (DLMB) and instruction (ILMB) handling. The Microblaze uses the Processor Local Bus (PLB) also called On-Chip Peripheral Bus (OPB) to connect different slave peripherals (SPLB) with the CPU, for data and instruction exchange it uses the DPLB and IPLB, respectively. In the figure are connected also to the Microblaze core: The peripherals PWM, RS232, IIC, Timer, etc. These last modules were designed for specific application and glued to the Microblaze architecture. An important feature of this processor is that also contains the Microprocessor Debug Module (MDM) that gives the possibility to achieve real time debugging using the JTAG interface. The stereoscopic vision module was programmed using ANSI C/C++ language.

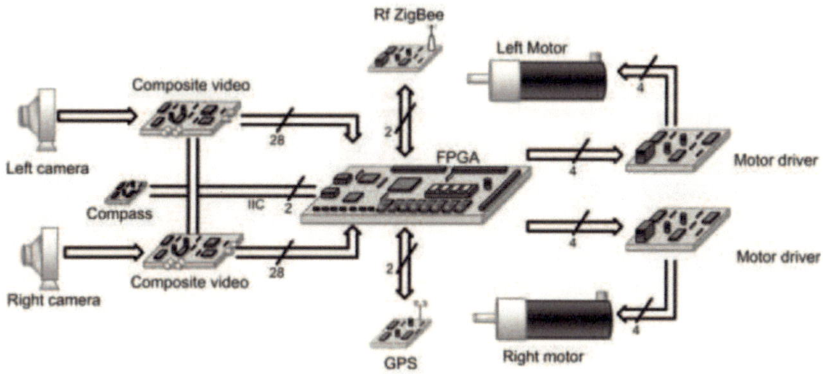


Fig. 2. Detailed overview of subsystems of the Stereoscopic vision stage on board of the MR.

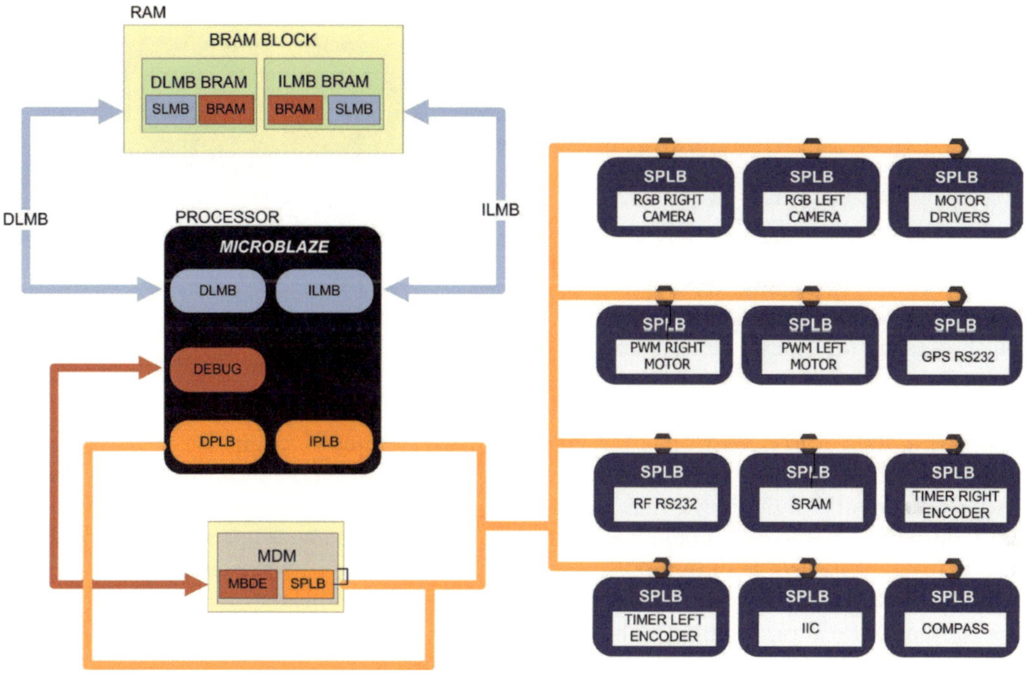


Fig. 3. Microblaze processor embedded into Xilinx FPGA and system peripherals.

## 3. Description of the Detection Module with Stereoscopic Vision

The navigation task is achieve using the relative depth representation of the obstacles based on stereoscopic vision and the epipolar geometry. The map represents the status at the time of drawing the map, not necessarily consistent with the actual status of the environment at the time of using the map. Mapping is the problem of integrating the information gathered in this case by the MR sensors into a complex model and depicting with a given representation. Stereo images obtained from the environment are supplied to the MR, by applying disparity algorithm on stereo image pairs, depth map for the current view is obtained. A cognitive map of the environment is updated gradually with the depth information extracted while the MR

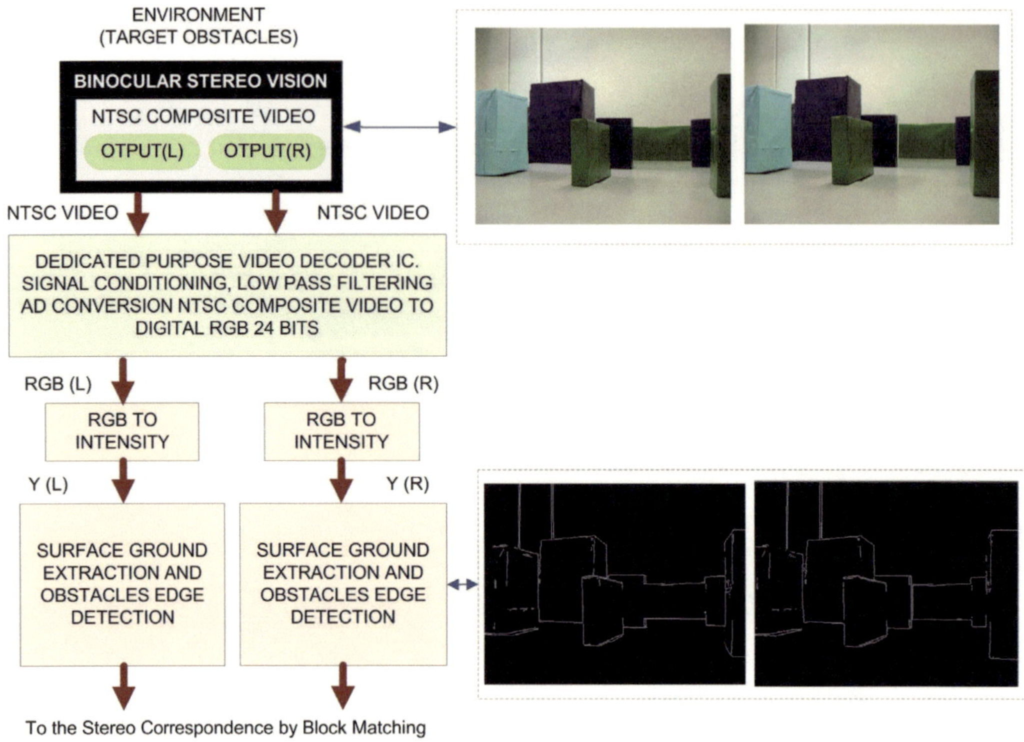


Fig. 4. Process in the detection module for surface ground extraction, and obstacles edge detection using luminance component.

is exploring the environment. The MR explores its environment using the current views, if an obstacle in its path is observed, the information of the target obstacles in the path will be send to the global planner in the main computer. After each movement of the MR in the environment, stereo images are obtained and processed in order to extract depth information. For this purpose, obstacle's feature points, which are obstacle edges, are extracted from the images. Corresponding pairs are found by matching the edge points, i.e., pixel's features which have similar vertical orientation. After performing the stereo epipolar geometry calculation, depth for the current view is extracted. By knowing the camera parameters, location, and orientation, the map can be updated with the current depth information.

### 3.1 Surface Ground and Obstacles Detection Using Luminance and Hue

The vision based obstacle detection module classifies each individual image pixel as belonging either to an obstacle or the ground. Appearance base method is used for surface ground classification and extraction from the MR vision module captured images, see Fig. 4. Any pixel that differs in appearance from the ground is classified as an obstacle. After surface ground extraction, remaining image content are only obstacles. A combination of pixel appearance and feature base method is used for individual obstacle detection and edge extraction. Obstacles edges are more suitable for stereo correspondence block matching in order to determine the disparity between left and right images. For ground surface extraction purpose, two assumptions were established that are reasonable for a variety of indoor and

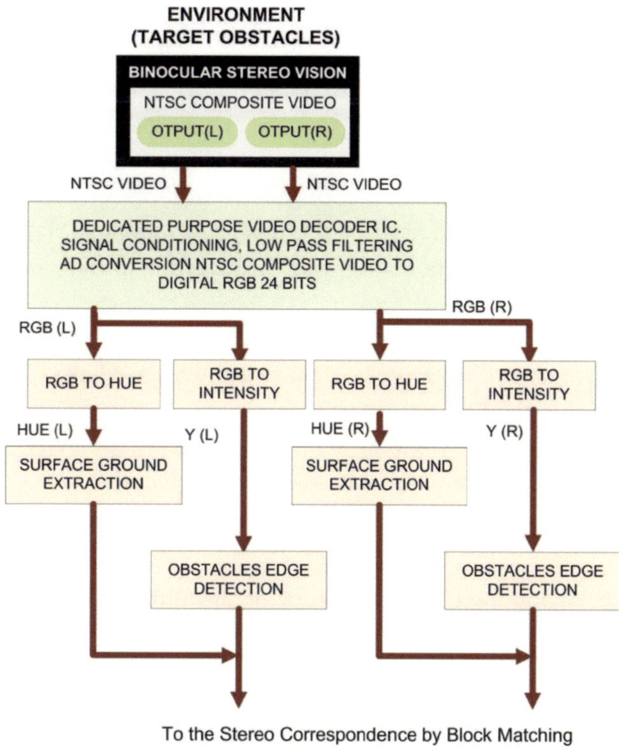


Fig. 5. Process in the detection module for surface ground extraction using Hue, and obstacles edge detection using luminance components.

outdoor environments:

1. The ground is relatively flat.
2. Obstacles differ in color appearance from the ground. This difference is reasonable and can be subjectively measured as Just Noticeably Difference (JND), which is reasonable for a real environment.

Above assumptions allow us to distinguish obstacles from the ground and to estimate the distances between detected obstacles from the vision based system. The classification of a pixel as representing an obstacle or the surface ground can be based on local visual attributes: Intensity, Hue, edges, and corners. Selected attributes must provide information so that the system performs reliably in a variety of environments. Selected attributes should also require low computation time so that real time system performance can be achieved. The less computational cost has the attribute, the obstacle detection update rate is greater, and consequently the MR travel faster and safer.

For appearance classification we used Hue as a primary attribute for ground surface detection and extraction, see Fig. 5. Hue provides more stable information than color or luminance based on pixel gray level. Color saturation and luminance perceived from an object is affected by changes in incident and reflected lightness. Also compared to texture, Hue is more local attribute and faster to calculate. In general, Hue is one of the main properties of a color, defined as the degree of perceived stimulus described as Red, Green, and Blue. When a pixel is classified as an obstacle, its distance from the MR stereo vision cameras system is estimated. The considerations for the surface ground extraction and obstacle edge detection for correspondences block matching are:

1. Color image from each video camera is converted from NTSC composite video to RGB 24 bits color space.
2. A typical ground area in front of the MR is used as a reference. The Hue attributes from the pixels inside this area are histogrammed in order to determine its Hue attribute statistics.
3. Surface ground is extracted from the scene captured by the MR stereo vision by means of a comparison against the reference of point 2 above, and based on Hue attribute. Hue limits are based in JND units.
4. Remaining content in images are only obstacles. Edges are extracted from individual obstacles based on feature and appearance pixel's attributes.
5. Correspondence for block matching is established in pixels from the obstacle vertical edges.
6. Disparity map is obtained from the sum of absolute differences (SAD) correlation method.

### 3.2 Vision System Module FPGA Implementation

When a robot has to react immediately to real-world events detected by a vision system, high speed processing is required. Vision is part of the MR control loop during navigation. Sensors and processing system should ideally respond within one robot control cycle in order to not limit their MR dynamic. An MR vision system equipped, requires high computational power and data throughput which computation time often exceed their abilities to properly react. In the ant colony environment model, every ant is a virtual MR full equipped, trying to find the optimal route, eventually, weather there exist, it will be obtained. Of course, the ACO based planner will give the best route found, and the real ant, the MR, which is equipped on board with the vision system, will update the global map in the planner. There are many tasks to do at the same time, however, a good feature of using FPGAs is that they allow concurrently implementation of the different tasks, this is a desirable quality for processing high speed vision. High parallelism is comprised with high use of the FPGA resources; so a balance between parallelization of task, and serial execution of some of them will depend on the specific necessities.

The vision system consists of stereoscopic vision module implemented in VHDL and C codes operating in a Xilinx based FPGA, hence a balanced used of resources were used. Video information is processed in a stereo vision system and video interface. The NTSC composite video signals from each camera after properly low pass filtering and level conditioning, are converted to RGB 24 bits color space by a state of the art video interface system HDTV capable. The rest of the video stage was programmed in C for the Microblaze system embedded into the FPGA. Other tasks, such as the motion control block are parallel implementation to the video system.

## 4. Design of the Stereoscopic Vision Module

The two stereo cameras parallel aligned, capture images of the same obstacle from different positions. The 2D images on the plane of projection represent the object from camera view. These two images contain the encrypted depth distance information. This depth distance information can be used for a 3D representation in the ant colony environment in order to build a map.

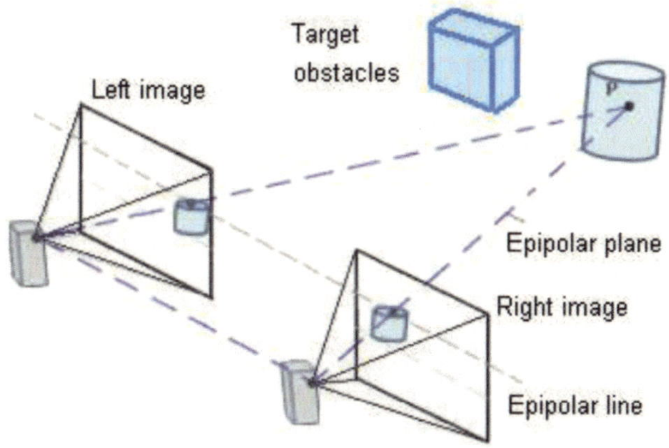


Fig. 6. Projection of one point into left and right images from parallel arrange stereo cameras.

### 4.1 Stereoscopic Vision

The MR using its side by side left and right cameras see the scene environment from different positions in a similar way as human eyes, see Fig. 6. The FPGA based processing system finds corresponding points in the two images and compares them in a correspondence matching process. Images are compared by shifting a small pixels block "window". The result is a comparison of the two images together over top of each other to find the pixels of the obstacle that best match. The shifted amount between the same pixel in the two images is called disparity, which is related to the obstacle depth distance. The higher disparity means that the obstacle containing that pixel is closer to the cameras. The less disparity means the object is far from the cameras, if the object is very far away, the disparity is zero, that means the object on the left image is the same pixel location on the right image.
Figure 7 shows the geometrical basis for stereoscopic vision by using two identical cameras, which are fixed on the same plane and turned in the same direction, parallax sight. The position of the cameras is different in the $X$ axis. The image planes are presented in front of the cameras to model the projection easier. Consider the point $P$ on the object, whose perspective projections on the image planes are located at $P_L$ and $P_R$ from left and right cameras respectively. These perspective projections are constructed by drawing straight lines from the point to the center lens of the left and right cameras. The intersection of the line and image plane is the projection point. The left camera's projection point $P_L$ is shift from the center, while the right camera's projection point $P_R$ is at the center. This shift of the corresponding point on left and right camera can be computed to get the depth information of the obstacle.

### 4.2 Depth Measure from Stereo Image

In order to calculate the depth measure of the obstacles in the scene, the first step is to determine the points of interest for correspondence matching between the two images. This corresponding points are selected based on the obstacle edge feature. Then calculate the depth distance based on the shifting "disparity". The disparity is calculated based on the amount of pixel's shifting in a particular corresponding point. There are stereo image constraints to be assume for solving the correspondence problem:

1. Uniqueness. Each point has at most one match in the other image.
2. Similarity. Each intensity color area matches a similar intensity color area in the other image.
3. Ordering. The order of points in two images is usually the same.
4. Continuity. Disparity changes vary slowly across a surface, except at depth edges.
5. Epipolar constraint. Given a point in the image, the matching point in the other image must lie along a single line.

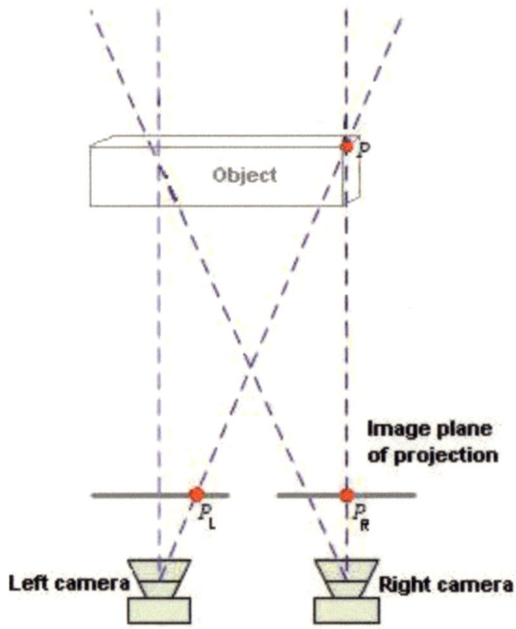


Fig. 7. Points $P_L$ and $P_R$ are the perspective projections of $P$ in left and right views.

## 5. Modifying Road Maps

The modification of the Road Maps is achieved using the information of disparity in pixels, where the distance of the MR from the obstacle is estimated using disparity measures, the less disparity measure means that the obstacle is far from the visual system of the MR as can be seen in Fig. 8. Moreover, the MR uses a high accuracy GPS and a digital compass. For every capture scene, the MR sends the location, orientation $(x, y, \theta)$ and the corresponding disparity map with all the necessary $(x, y, d)$ coordinates and corresponding disparities, which in reality are a 3D representation of the 2D obstacles images captured from the stereoscopic visual system. After pixel's scaling and coordinates translation, the global planner is able to update the environment, its representation includes the visual shape and geographical coordinates. Once the global planner in the main computer has been modified using the new information about new obstacles and current position of the MR, the global planner performs calculations using ACO to obtain an updated optimized path, which is sent to the MR to achieve the navigation. The MR has the ability to send new information every $100ms$ via RF from every scene captured; however, times in the global planner are bigger since it is based on a natural optimization method, and it depends on the actual position of MR with respect to the goal. Hence, most of times a new path can be obtained every 3 seconds.

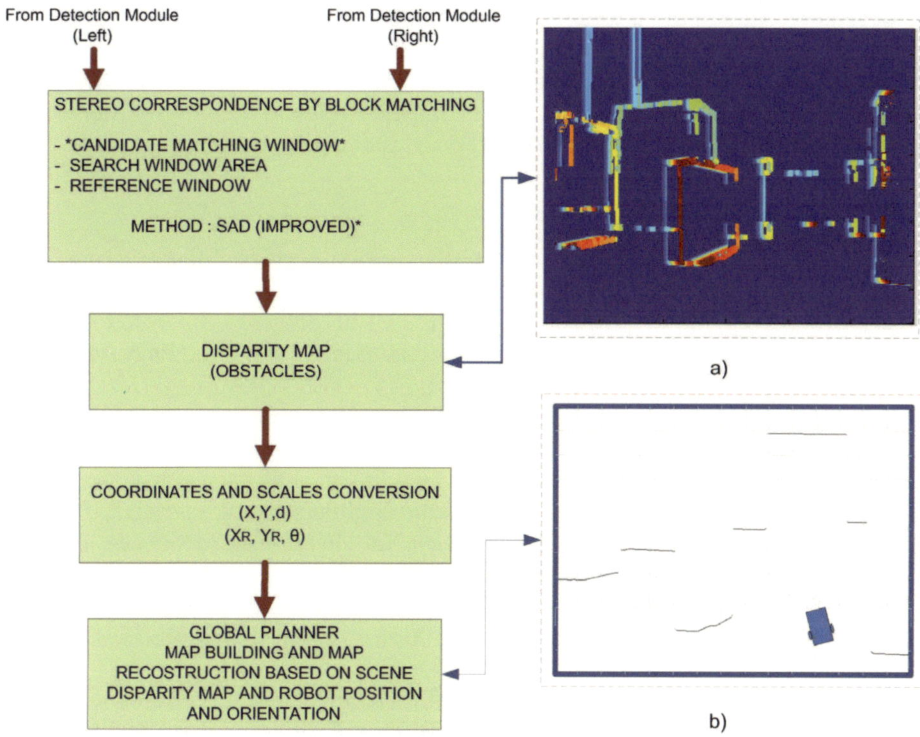


Fig. 8. Process for map building and map reconfiguration.

## 6. Conclusion

In this work was shown the design of an stereoscopic vision module for a wheeled mobile robot, suitable to be implemented into an FPGA. The main purpose of the onboard system of the MR is to provide the necessary elements for perception, obstacles detection, map building and map reconfiguration in a tough environment where there are no landmarks or references. The stereoscopic vision system captures left and right images from the same MR scene, the system is capable of using both appearance based pixel descriptors for surface ground extraction, luminance or Hue depending of the environment particular characteristics. In an environment with constant lightness, minimum reflections and proper setting in the edge detector threshold level, luminance can be suitable because surface ground and obstacles edge detection can be performed at the same time. For environment with variable light conditions or uncertain, Hue is the primary attribute for pixel appearance descriptor in the surface ground extraction process due to its invariance to changes in luminance and color saturation. After surface ground extraction and obstacles edge detection, stereoscopic corresponding by block matching is performed, the correspondence is found among a set of points in the left and right images by using a feature based approach. Disparity computation for the matched points is then performed. Establishing correspondences between point locations in images acquired from multiple views (matching) is one of the key tasks in the reconstruction based on stereo image analysis. This feature based approach, involves detecting the feature points and tracking their positions in multiple views of the environment. Stereoscopic camera calibration is not required due to the improvements in matching process. Disparity maps which are the depth measure of the obstacles position in the environment are obtained after the stereo correspondence process. The MR sends this data, including its position and orientation via RF to the global planner located in the main computer outside the environment. With this information the global planner is able to constantly update the environment map.

## 7. References


Abellatif M. (2008). Behavior Fusion for Visually-Guided Service Robots, Xiong Zhihui *In:* In-Teh, *Computer Vision*, Croatia, pp. 1-12.

Aggarwal J. K., Zhao H., Mandal C., Bemuri B. C. (2000). 3D Shape Reconstruction from Multiple Views, *in* Alan C. Bovik, Editor, *Handbook of Image and Video Processing*, Academic Press, pp. 243-257.

Calisi D., Iocci L., Leone G. R. (2007). Person Following through Appearence Models and Stereo Vision using a Mobile Robot, *Proc. of International Workshop on Robot Vision*, pp. 46-56.

Cao Z. L. (2001). Ommi-vision based Autonomous Mobile Robotic Platform, *Proceedings of SPIE Intelligent Robots and Computer Vision XX: Algorithms, Techniques, and Active Vision*, Vol. 4572, Newton USA, pp. 51-60.

Cao Z., Meng X., & Liu S. (2008). Dynamic Omnidirectional Vision Localization Using a Beacon Tracker Based on Particle Filter, *In:* Xiong Zhihui, *Computer Vision*, Ed. In-Teh, Croatia, pp. 13-28.

Khatib O. (1985). Real-Time Obstacle Avoidance for Manipulators and Mobile Robots, *Procedings of IEEE International conference on Robotics and Automation*, pp. 500-505.

Porta García M. A., Montiel O., Castillo O., Sepúlveda R., Melin P. (2009). Path planning for autonomous mobile robot navigation with ant colony optimization and fuzzy cost function evaluation, *Applied Soft Computing*, Vol. 9 (No. 3): 1102-1110.

Siegwart R., & Nourbakhsh I. R. (2004). *Introduction to Autonomous Mobile Robots, A Bradford Book*, The MIT Press, Cambridge Massachusetts, London, England.

Tsai R. Y. (1986). An efficient and accurate camera calibration technique for 3D machine vision, *IEEE Conference on Computer Vision and Pattern recognition*. pp. 364-374.

VBOX product (2009). Web page available at: http://www.racelogic.co.uk/?show=VBOX

Woods A., Docherty T., Koch R. (1993). Image distortions in stereoscopic video systems, *Proceedings of the SPIE*, San Jose, Ca. USA., Vol. 1925.

Xbee XBee-Pro OEM RF Modules (2007) , Product Manual v1.xAx - 802.15.4 Protocol, MaxStream, Inc.

4

# Fast 3D Perception for Collision Avoidance and SLAM in Domestic Environments

Dirk Holz, David Droeschel and Sven Behnke
*Department of Computer Science VI, University of Bonn*
*Germany*

Stefan May
*Institut National de Recherche en Informatique et en Automatique, Sophia-Antipolis*
*France*

Hartmut Surmann
*Fraunhofer Institute Intelligent Analysis and Information Systems, Sankt Augustin*
*Germany*

## 1. Introduction

Autonomous service robots that assist in housekeeping, serve as butlers, guide visitors through exhibitions in museums and trade fairs, or provide care to elderly and disabled people could substantially ease everyday life for many people and present an enormous economic potential (Haegele et al., 2001; Pollack et al., 2002; Siegwart et al., 2003). Moreover, regarding the aging society in most industrialized countries the application of service robots in (elderly) health care might not only be helpful but necessary in the future. However, these service robots face the challenging task of operating in real-world indoor and domestic environments. Domestic environments tend to be cluttered, dynamic and populated by humans and domestic animals. In order to adequately react to sudden dynamic changes and avoid collisions, these robots need to be able to constantly acquire and process, in real-time, information about their environment. Furthermore, in order to act in a goal-directed manner, plan actions and navigate effectively, autonomous mobile robots need an internal representation or *map* of their environment. Nature and complexity of these representations highly depend on the robot's task and workspace.

When operating in preliminary unknown environments, e.g., when it is unfeasible (or simply uncomfortable) to manually model the environment beforehand, the robot needs to construct an internal environment model on its own. Moreover, in dynamic environments the robot further needs to be able to continuously acquire and integrate new sensory information to update the internal environment model in regions where changes have taken place. As integrating new information into the model (*mapping*) requires knowledge about the robot's pose (position and orientation in the environment) and determining the robot's pose requires a map of the environment, these two problems need to be considered jointly and the problem

of constructing or updating an internal environment model is commonly referred to as Simultaneous Localization and Mapping (SLAM). In fact, SLAM is regarded as one of the major prerequisites for truly autonomous robots (Wang, 2004).
Both, collision avoidance and SLAM are well understood in the two-dimensional case, e.g., when acquiring geometric information about surrounding environmental structures with 2D laser range finders. However, as will be shown in the following section, this information is not sufficient in order to adequately navigate in cluttered and dynamic, domestic environments. Presented in this work are methods and means for a fast 3D perception of the robot's surrounding environment as well as for extracting and processing relevant information in the context of robust collision avoidance and SLAM.
Section 3 will provide an overview on methods and means for acquiring three-dimensional information using laser range finders as well as related work. Extracting relevant information from the continuous 3D data stream for the purpose of collision avoidance and SLAM is described in Section 4. Efficient algorithms for collision avoidance based on this information as well as SLAM for constructing two-dimensional and three-dimensional environment models are described in Sections 5 and 6. Extensions for using recent Time-of-Flight cameras are presented in Section 7. Section 8 will contain some concluding remarks and an outlook on future work.

## 2. 2D-Perception in Domestic Environments

2D laser range-finders became the de facto standard sensor to tackle the problems of SLAM and collision avoidance. These sensors measure, with high frequency and accuracy, the distances to environmental structures surrounding the robot. They emit a laser range beam and measure the time until the emitted beam is received after being reflected on the surface of an object in the robot's vicinity. By means of a rotating mirror these beams are emitted over a two-dimensional plane differing, depending on the used sensor, in apex angle and angular resolution. Typically, 2D laser range finders are mounted horizontally in order to measure distances to surrounding objects in a plane being parallel to the floor. A laser scan $S$ is a set of 2-tuples $(d, \theta)$ where $d$ is a distance measurement and $\theta$ the angle under which the measurement has been taken, i.e.,

$$S = \{(d_i, \theta_i) \mid i \in [1, N_s]\}. \tag{1}$$

Each tuple $(d_i, \theta_i)$ in $S$ forms the polar coordinates of a point $\mathbf{p}_i$ measured on the surface of an object in the surrounding environment, i.e.,

$$\forall i \in [1, N_s] : \quad \mathbf{p}_i = \begin{pmatrix} d_i \cos\theta_i \\ d_i \sin\theta_i \\ 0 \end{pmatrix} = \begin{pmatrix} x_i. \\ y_i \\ z_i \end{pmatrix} \tag{2}$$

using a right-handed coordinate frame. That is, the robot is looking along the $x$-axis with the $y$-axis extending to the left. The $z$-axis points upwards and represents the height of objects. Depending on the measurement principle, $S$ can be a totally ordered set with respect to, respectively, $i$ and $\theta_i$ in either clockwise or anti-clockwise direction, i.e., for $i < j : \theta_i < \theta_j$ or $\theta_i > \theta_j$. A 2D laser range scan is exemplarily depicted in Figure 1(a).
The inherent drawback of 2D laser range finders, in the context of simultaneous localization and mapping (SLAM) as well as collision avoidance, is that objects not intersecting the scanner's measurement plane are not perceived. Consider for example the couch table in Figure

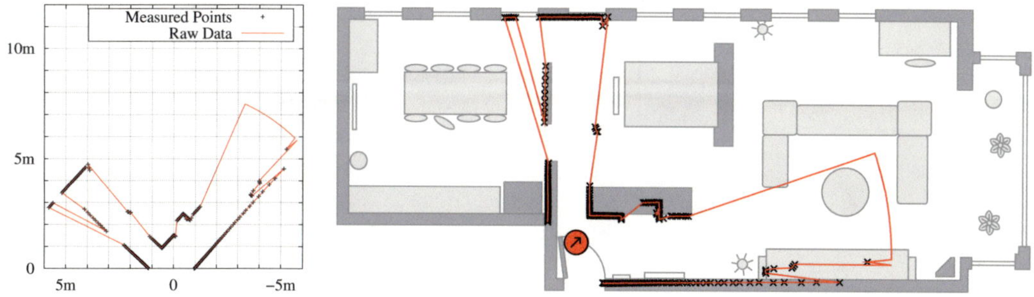


(a) Acquired 2D scan (b) Scan superimposed on a sketch of the environment

Fig. 1. 2D laser range scan in an example scenario. Depicted is a range scan (a) from a data set recorded by Zivkovic et al. (2007) together with an approximate floor plan of the scenario (b). Note, that the couch table has not been sensed at all since it did not intersect the scanner's measurement plane.

1(b). The 2D laser range scan taken in this example scenario does adequately model surrounding environmental structures whereas not a single measurement has been taken on the surface of the couch table. That is, the couch table is not at all perceived. Hence, it cannot be modeled in the robot's internal environment representation. As there is no obstacle in the corresponding model region, the robot might plan a path that directly leads through the couch table. Furthermore, a collision with the couch table cannot be avoided as it does not intersect the scanner's measurement plane. Even when standing directly in front of the table not a single measurement would be reflected. In this example, the scanner is mounted too high so that even the legs of the table do not intersect the measurement plane. However, even when intersecting the scanner's measurement plane, especially table and chair legs are not always perceivable. Depending on material and shape of table legs, e.g., round metal rods, only a portion of emitted laser beams are reflected in a way so that they are received by the scanner. The same holds true for objects whose surface is less reflective. That is, primary reasons for not perceiving an object with a 2D laser ranger finder are:

1. **The object does not intersect the 2D scan plane**. That is, objects below or above the two-dimensional measurement plane cannot be perceived.
2. **The surface of the object is less reflective** or reflects incoming range beams in directions other than the emitting range scanner. Black surfaces, for example, absorb a larger portion of the incoming light. On the other hand, metallic objects with round shape, like for instance table or chair legs, might cause diffuse reflections or completely diffract the emitted beam.

Tables and chairs are not the only objects in domestic environments that are hard to perceive with 2D laser range finders. Objects like for instance table tops, open drawers, small objects lying on the ground or stairs might not be appropriately perceivable by the robot and modeled in its internal environment representation (see Figure 2). When mounting the scanner in another height to perceive a specific class of obstacles, other types of obstacles are still not perceivable. Even the usage of several 2D range scanners in different heights does not appropriately solve this problem. For adequately handling all kinds of obstacles in a cluttered and dynamic environment 3D information becomes crucial.

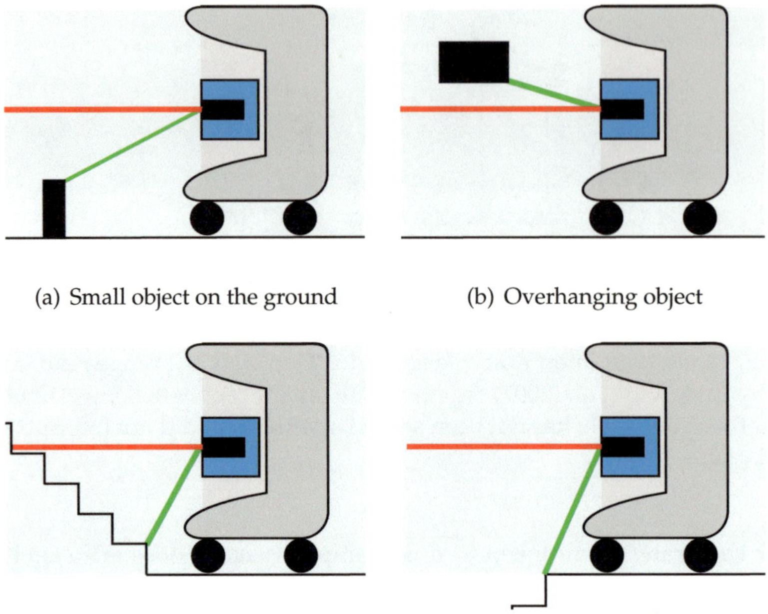

(a) Small object on the ground (b) Overhanging object

(c) Ascending Stairs (d) Descending Stairs

Fig. 2. Different types of obstacles. Shown are four different examples of obstacles in a robot's workspace. They have in common that the robot is not able to reliably avoid them by means of simple 2D perception. The measurement plane of a standard 2D laser range finder is depicted in red only intersecting the ascending stair. By means of 3D perception, a single distance measurement (green line) would allow the robot for detecting the obstacles.

## 3. Using Laser Range Finders for 3D Perception

Important criteria for the choice of a particular sensor in the design of mobile robots are size, weight, power consumption and prize. Several approaches for acquiring 3D information have been proposed that differ, amongst others, in the type, number and setup of sensors. Here, we will focus on approaches that are based on range sensors.

### 3.1 Related Work

Thrun et al. (2000) use two 2D laser range finders. One scanner is mounted horizontally and used for localizing the robot with three degrees-of-freedom. The other scanner is mounted vertically. The data of the latter is used to compute a volumetric model of the scene based on the poses determined by using the horizontal scanner. However, whereas this approach allows for constructing three-dimensional environment models, it only provides 3D information about objects that are currently passed by the robot and intersect the measurement plane of the vertically mounted scanner. That is, objects in front of the robot are only, if at all, perceivable by the horizontally mounted 2D scanner.

Zhao & Shibasaki (2001) follow a similar approach but use multiple vertically mounted scanners in order to reduce the size of these occlusions. In fact, the usage of multiple 2D range scanners became a commonly found sensor setup in the context of the DARPA Grand and Urban Challenges. Thrun et al. (2006) mounted five 2D laser range scanners on the top of Stanley,

the robot that won the DARPA Grand Challenge. The scanners were mounted in driving direction (just like horizontally mounted 2D scanners), but under different pitch/tilt angles to perceive the surface of the ground in different distances. Similar setups can be found in many other participating teams of the Grand and the Urban Challenge. However, such a battery of 2D range scanners cannot be mounted on smaller household service robots.
Another possibility for acquiring three-dimensional information is the application of commercially available 3D laser scanners as used, e.g., in land surveying. Sequeira et al. (1998) use a RIEGL[1] laser scanner on an autonomous robot to construct 3D models of indoor environments. Allen et al. (2001) use a Leica CYRAX[2] laser scanner on a car to construct three-dimensional models or urban environments. A Zoller+Fröhlich[3] scanner is used by Huber et al. (2000) for reconstructing volumetric models of indoor environments. Urmson et al. (2008) use a Velodyne[4] laser scanner in addition to a battery of 2D laser range finders for detecting and avoiding obstacles in the immediate vicinity of an autonomous car in the DARPA Urban Challenge. Commercially available 3D laser scanners directly provide highly accurate three-dimensional point clouds. However, compared to 2D laser range finders, they are quite expensive and unwieldy for the application on mobile household service robots.
Jet another possibility to acquire three-dimensional data is to mount a single 2D laser range finder on a mechanical actuator to gain an additional degree of freedom. That is, in addition to the rotating mirror for scanning two-dimensional planes, the actuator rotates the complete scanner. Taking multiple 2D scans at different rotation angles allows for constructing locally consistent 3D point clouds. Different setups have been proposed that differ primarily in field-of-view (FOV) and spatial measurement density. The highest point density lies around the rotation axis. Amongst others, Surmann et al. (2003) and Hähnel et al. (2002) started using a horizontally mounted scanner where a rotation angle of 0° corresponds to acquiring a regular 2D laser range scan, i.e., with the measurement plane being parallel to the floor. Standard servo motors or pan-tilt units are used to rotate, respectively, the scanner and the measurement plane upwards and downwards in a nodding-like fashion. Wulf & Wagner (2003) evaluate this and other sensor setups and refer it to as a *pitching scanner* due to the rotation about the $y$-axis in a right-handed coordinate frame. For acquiring a locally consistent 3D laser scan, Surmann et al. stop the robot and rotate the scanner over the complete vertical aperture angle of 120°. Multiple 3D scans are then matched and registered into a global coordinate frame to construct a three-dimensional point model (Surmann et al., 2003). Strand & Dillmann (2008) let the scanner continuously rotate about the $x$-axis (*rolling scanner*) and use the acquired data for exploring and mapping indoor environments.

### 3.2 Continuously Rotating Laser Range Finders

The aforementioned approaches have in common that the robot is stopped in order to acquire a locally consistent 3D point cloud by rotating the scanner. The resulting behavior of the robot is thus composed of *stop-scan-move* cycles. Wulf et al. (2006) started using a continuously yawing scanner for acquiring 3D data while moving. They segment the data stream into individual point clouds and use the acquired information to localize the robot based on ceiling structures that are normally not occluded by people or objects. Cole & Newman (2006) use a pitching scanner that is continuously rotated over the complete vertical aperture angle

[1] RIEGL Laser Measurement Systems: `http://www.riegl.com`
[2] Leica Geosystems: `http://www.leica-geosystems.com`
[3] Zoller+Fröhlich: `http://www.zf-laser.com`
[4] Velodyne: `http://www.velodyne.com/lidar`

in a nodding-like fashion. Their robot is, however, not autonomously controlled and the acquired 3D data is solely used to construct three-dimensional environment models. However, augmenting a commercial 2D laser range finder with an additional degree of freedom seems to be the most appropriate approach for acquiring three-dimensional information on smaller mobile robots.

For safe navigation in dynamic and cluttered environments it is crucial to detect, as fast as possible, all obstacles with which the robot could eventually collide. Hence, the area in the robot's movement direction and in the height spanned by the robot's three-dimensional bounding box is of special interest. Since the continuously yawing scanner of Wulf et al. (2006) scans only vertical planes by means of a single 2D range scanner, objects in the robot's movement direction get out of sight when the scanner is sensing environmental structures behind the robot. For the DARPA Urban Challenge we have extended this setup by using two antipodally mounted 2D laser range scanners (Maurelli et al., 2009; Rojo et al., 2007). The two scanners are, furthermore, not mounted vertically but can be adjusted in the rotation angle. The resulting scanner, the Fraunhofer IAIS 3DLS-K is shown in Figure 3.a and consists of two SICK LMS 291 2D laser range finders mounted on a rotatable carrier. This carrier is continuously rotated around the vertical axis. Depending on the current orientation, the 2D laser range scans of the scanners are transformed into a sensor-centric coordinate frame. The transformed scans are aggregated to form a local 3D point cloud (see Figure 3.b). However, even when not waiting for a complete point cloud, but processing every single 2D scan as it arrives, it remains the drawback of having larger areas in the robot's vicinity not in sight until the other scanner arrives at the corresponding rotation angle. Furthermore, larger portions of the acquired information is obtained in regions that do not pose a threat to the robot, e.g., ceiling structures. This is, in fact, the reason why the two scanners are not mounted vertically as for the normal acquisition of 3D laser scans, but almost diagonally. This decreases the size of the unseen region when processing the acquired information scan-wise.

However, for further reducing the sensed regions to those that are relevant for collision avoidance, a pitching 3D laser scanner seems to be more appropriate (Holz et al., 2008; Wulf & Wagner, 2003).

(a) (b) (c)

Fig. 3. Continuously rotating 3D laser scanner (a) and example data (b) taken in front of the Robotics Pavilion at Fraunhofer IAIS. A photo of the scene is shown in (c). The color of the 3D points in (b) corresponds to the received remission values.

### 3.3 A Continuously Pitching Laser Scanner for Collision Avoidance and SLAM

In a *pitching scanner* setup the 2D laser range finder is mounted horizontally and rotated around the $y$-axis. A pitching scanner, the IAIS 3DLS, is shown in Figure 4. For the RoboCup@Home world championship in Atlanta 2007, it was mounted on a three-wheeled VolksBot RT3 platform allowing to acquire 3D scans of the arena. With the additional rotation axis, driven by a standard servo motor, the scanner has a vertical aperture angle of up to $\Theta_{\text{pitch}} = 120°$ with a maximum angular resolution of $\Delta\theta_{\text{pitch}} = 0.25°$. Taking a 3D scan by rotating the scanner over the complete vertical range and using a horizontal angular resolution of $\Delta\theta_{\text{yaw}} = 0.25°$ results in 3D point clouds containing 346 080 points. However, as a relatively low angular resolution is sufficient for robust collision avoidance and has benefits in terms of speed concerns while still providing a sufficient detail for mapping purposes, an angular resolution of $\Delta\theta_{\text{yaw}} = 1°$ is preferable. For the used SICK LMS 2xx range scanners, a single 2D laser scan of 181 distance measurements is read in approximately 13.32 ms ($\approx$ 75 Hz) in this operating mode. Reliable navigation in domestic environments requires for a fast and continuous 3D perception of surrounding environmental structures and obstacles. Therefore, the scanner is continuously pitched around its horizontal axis in a nodding-like fashion allowing the robot to perceive surrounding environmental structures in 3D while moving through its workspace. Since a rotation over the complete aperture angle $\Theta_{\text{pitch}}$ might yield a couple of single 2D laser scans primarily containing information being not relevant for collision avoidance, e.g., only floor points if the scanner is directed downwards or ceiling structures if the scanner is directed upwards, we defined an *area of interest (AOI)* (Holz et al., 2008). This area restricts the range of used rotation angles $\theta_{\text{pitch}}$ to the interval $[\theta_{\text{pitch, min}}, \theta_{\text{pitch, max}}]$ so that it contains primarily relevant information.

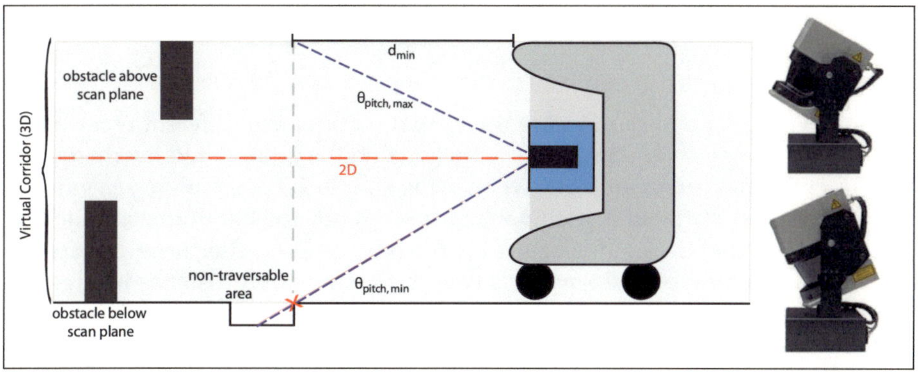


Fig. 4. Continuously pitching scanner and virtual corridor. By continuously rotating the scanner in a nodding-like fashion over the area of interest (AOI), all obstacles in the virtual corridor are perceived.

To be able to react to suddenly appearing obstacles in front of the robot even a restricted but fixed area might be too large to timely perceive especially small objects intersecting only a single measurement plane during one rotation. Therefore, the boundaries of the AOI ($\theta_{\text{pitch, min}}$ and $\theta_{\text{pitch, max}}$) as well as the scanner's pitch rate ($\Delta\theta_{\text{pitch}}$/13.32 ms corresponding to the number of consecutive 2D laser scans taken during one pitch movement), can be adjusted, e.g., to depend on the robot's current velocity or special characteristics of the environment. Increasing the upper bound $\theta_{\text{pitch, max}}$ when moving slow allows to perceive more information about the environmental boundaries such as walls due to the height of the measured points.

For moving faster it can be decreased so that only the volume corresponding to the robot's height is sensed. We refer to this minimal area as the *virtual corridor*. It extends the idea of *virtual roadways* (Lingemann et al., 2005a) to the third dimension, i.e., with respect to the robot's boundaries (in 3D) and thus possible areas of collision. The concept of the virtual corridor is depicted in Figure 4. Here lower bound $\theta_{\text{pitch, min}}$ and upper bound $\theta_{\text{pitch, max}}$ of the AOI correspond to the size of the virtual corridor and thus to the robot's boundaries (marked with the slightly colored background). The length $d_{\text{min}}$ corresponds to the distance from which on the full virtual corridor can be perceived during one pitch movement. It has to be chosen appropriately, e.g., for a dense or narrow environment it has to be rather small whereas $d_{\text{min}} = 1\,\text{m}$ is absolutely sufficient when driving fast along an uncluttered corridor. The minimum size of the AOI for driving fast covers exactly the virtual corridor while the maximum size corresponds to a complete 3D scan over the full 120° of $\Theta_{\text{pitch}}$. This allows to construct *complete* 3D models of the environment containing all perceivable information as in (Surmann et al., 2003). For the pitching scanner, a scan point is represented by the tuple $(d_i, \theta_{\text{yaw},i}, \theta_{\text{pitch}})$ with $d_i$ being the $i$-th distance measurement in the latest 2D laser scan while $\theta_{\text{yaw},i}$ and $\theta_{\text{pitch}}$ are the $i$-th measurement angle and the current pitch angle of the laser scanner respectively. The coordinates of the 3D points corresponding to acquired distance measurements result from rotating the 2D measurement plane by the current pitch angle $\theta_{\text{pitch}}$. Here, the scanner's position on the robot (w.r.t. the robot's center of rotation) is taken into account with the translational part $\mathbf{t}_s = (\mathbf{t}_s^x, \mathbf{t}_s^y, \mathbf{t}_s^z)^T$. Note that the scanner's orientation with respect to the robot's movement direction, has to be taken into account using additional rotations (not necessary here).

$$\mathbf{p}_i = \begin{pmatrix} x \\ y \\ z \end{pmatrix} = \underbrace{\begin{pmatrix} \cos\theta_{\text{pitch}} & 0 & \sin\theta_{\text{pitch}} \\ 0 & 1 & 0 \\ -\sin\theta_{\text{pitch}} & 0 & \cos\theta_{\text{pitch}} \end{pmatrix}}_{\mathbf{R}_y(\theta_{\text{pitch}})} \underbrace{\begin{pmatrix} d_i\cos\theta_{\text{yaw},i} \\ d_i\sin\theta_{\text{yaw},i} \\ 0 \end{pmatrix}}_{\mathbf{R}_z(\theta_{\text{yaw},i})} + \mathbf{t}_s \tag{3}$$

By continuously sensing and monitoring the virtual corridor the different types of obstacles (see Figure 2) can be perceived. Small obstacles lying on the ground and overhanging objects do not intersect the measurement plane when using simple 2D perception, i.e., when holding the scanner in a fixed horizontal position (red line). With the continuously pitching scanner they are perceived and can thus be avoided. The ascending stair is perceivable with 2D perception but depending on the scanner's height, the measured distance is larger than the distance to the first step. With 3D perception, the first step is perceived just like a small object lying on the ground. For descending stairs, however, a little trick needs to be applied that is presented in the following section.

## 4. Extracting Relevant Information from 3D Data

Real-time applicability does not only necessitate a fast acquisition of information but also to efficiently process the acquired information. Due to the larger amount of data and the higher dimensionality of information, directly processing raw 3D data is not feasible in many applications. Especially in the context of navigation, existing state-of-the-art approaches that show real-time applicability normally perform on less complex and less information bearing 2D laser data (see e.g. Lingemann et al., 2005a). In order to combine these well-studied and well-performing algorithms with the rich continuously gathered 3D data it is suggestive to break down the three-dimensionality of the data into a slim two-dimensional representation that still holds all necessary 3D information but is nevertheless efficient enough to apply these efficient algorithms. For this purpose we are using *virtual maps* that have been initially presented in (Holz et al., 2008) and that are based on the idea of virtual 2D scans (Wulf et al., 2004). These egocentric maps, in which, respectively, the robot and the sensor form the origin of the coordinate frame, store only relevant information that has been extracted from 3D data. Two types of virtual 2D maps are distinguished: *2D obstacle maps* and *2D structure maps*. Both are generated from consecutive single 2D laser range scans acquired during the continuous pitching movement of the laser scanner presented above or extracted, for example, from one depth image of a Time-of-Flight camera (see Section 7). Note that the following descriptions will focus on continuously pitching lasers scanners. The actual implementation, however, is the same for all kinds of 3D sensors.

### 4.1 Structure of Virtual Maps

To be able to apply the same algorithms for collision avoidance, mapping and localization purposes to both standard 2D laser scans and virtual 2D maps constructed by means of 3D perception, the representation of the virtual maps is chosen to extend the representation of standard laser scans. That is, they are organized as a vector of distance measurements $d_i$ ordered by the discretized measurement angle ($\theta_{\text{yaw},i}$). This extended representation has, compared to a 2D laser scanner, a variable aperture angle $\Theta \in [0°, \ldots, 360°]$ and a variable angular resolution $\Delta\theta_{\text{yaw}}$. It is implemented as a vector of $N = \Theta/\Delta\theta_{\text{yaw}}$ points indexed by the accordingly discretized angle in which the measured point is lying from the robot's perspective. Furthermore, each point is represented by means of Cartesian and polar coordinates to avoid algorithm-dependent transformations. An example of a virtual map modeling nearest obstacles in a cluttered environment is visualized in Figure 5.

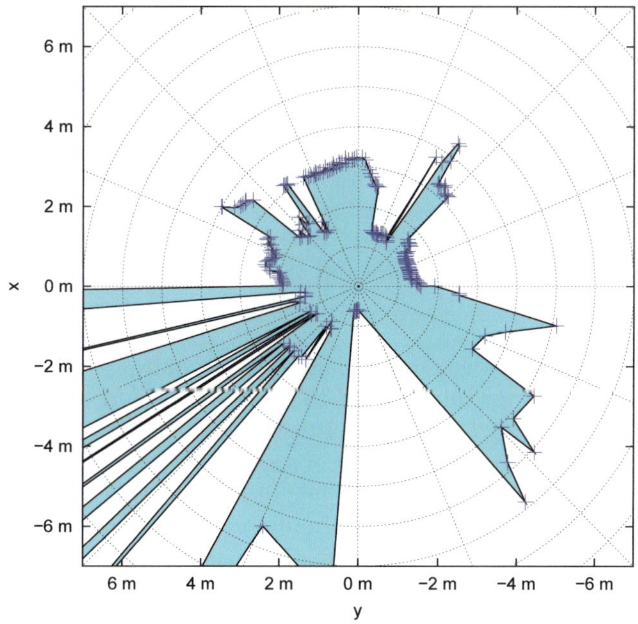


Fig. 5. Visualization of a 360° obstacle map containing information about nearest obstacles aggregated while moving through a cluttered indoor environment.

The virtual maps are ego-centric, i.e., the robot's center of rotation forms the origin of the map's coordinate frame and the coordinates of points stored in the maps are referenced to the base coordinate frame $\{B\}$ of the robot. In addition to being an efficient representation of relevant 2.5D information from 3D data, the maps can also be used to fuse information from several sensors, e.g., different 2D laser range finders or a combination of 2D and 3D laser scanners. In fact, the concept of virtual maps has recently been adopted by Stiene & Hertzberg (2009) to fuse information of different 2D range scans. Furthermore, as the representation (and the actually used implementation) does not differ from that of a standard 2D laser range scan, basically all algorithms working on 2D laser data can be applied to the aggregated or fused information in a virtual map. The remainder of this section will describe the two types of virtual maps as well as the procedures for updating them.

### 4.2 2D Obstacle Maps

In the case of the obstacle maps, the point corresponding to the *minimum* distance in each scan direction ($\theta_{\text{yaw},i}$), projected into the $xy$-plane in which the robot is moving, is extracted and inserted into the virtual map. These measurements correspond to the closest objects or obstacles in that particular direction regardless of the actual pitch angle $\theta_{\text{pitch}}$ of the scanner. Of course, only those points whose height above ground would intersect, respectively, the robot's bounds and the virtual corridor are inserted into the map. This explicitly includes obstacles like small objects lying on the ground or overhanging objects like open drawers as shown in Figure 4. Objects not intersecting the virtual corridor pose no threat to the robot and can thus be ignored. In the update procedure of *2D structure maps* they are, of course, used since a lot of information especially on higher environmental structures would be neglected otherwise.

In order to represent non-traversable areas and especially areas that correspond to holes in the ground, like for instance descending stairs in the examples of Figure 2, artificial obstacles are inserted into the map. Such non-traversable areas are characterized by those distance measurements that correspond to points in the real environment that are located below floor level. Note that the robot is assumed to be only able to traverse almost flat floor areas what is, after all, a feasible assumption for domestic indoor environments. Once a measurement below floor level occurs in an acquired laser scan its intersection with the floor plane is computed and an artificial measurement is added to the obstacle map at exactly this point. Thereby, the robot is able to perceive descending stairs and stop before the first step is reached. Such an intersection point and artificial distance measurement representing the stair as an obstacle is depicted with a red cross in Figure 4.

An important issue when constructing virtual obstacle maps is to not add points to the map that correspond to traversable surface as the robot would otherwise avoid to move through that particular region. Under the assumption of a perfect flat floor and minimum measurement inaccuracies, the most straightforward way for determining which points belong to the floor plane, is to apply a simple height thresholding. That is, all points whose height lies approximately at $z = 0$ are filtered out and ignored in the update procedure, i.e., $floor(\mathbf{p}_i) = \mathbf{p}_i^z \in [-\epsilon_z, \epsilon_z]$ where $\epsilon_z$ represents a tolerance according to the sensor's accuracy. Amongst others, Yuan et al. (2009) follow this approach. A value of $\epsilon_z = 2.5$ cm, for example, is an appropriate tolerance for the aforementioned sensor setup leading to a robust removal of floor points but would also neglect measurements on the surface of small objects lying on the ground if their height does not exceed 2.5 cm.

A more reasonable, still simple and efficient, way for determining floor points is to evaluate the neighborhood of individual measurements. Nüchter et al. (2005), for example, apply a segmentation algorithm that exploits the order of points in a range scan, the order of range scans in the 3D point cloud and the 3D sensor setup to classify points regarding their correspondence to floor, wall, object or ceiling structures. It originates from the work in (Wulf et al., 2004) and will be used in the remainder of this section.
By the aforementioned means, i.e., filtering out minimum distances in each direction while ignoring points belonging to traversable surfaces, the robot obtains an egocentric map containing all obstacles and non-traversable areas close to the robot. An obstacle map that exemplarily shows how small objects are perceived is depicted in Figure 6.a.

(a) Obstacle map (b) Example scenario (c) Structure map

Fig. 6. Demonstration of the virtual 2D map types in an example scenario. The *obstacle map* (a) is generated by extracting minimum distances (projected into 2D) in the continuously acquired 3D data. Extracting maximum distances results in the *structure map* (c).

### 4.3 2D Structure Maps

In the case of virtual structure maps, the *maximum* distance in each scan direction ($\theta_{\text{yaw},i}$), projected into the $xy$-plane, is extracted and inserted into the map. Extracting maximum distances automatically filters out all objects that do not extend over the full height of the AOI since the scanner will eventually look above or beneath these objects. The robot thereby replaces a previously measured smaller distance value with the newly obtained larger distance reading in that direction. The resulting map will thus only contain those points that most probably correspond to the environmental bounds while all points that belong to smaller or overhanging obstacles are filtered out as are those that belong to dynamic obstacles. Such a structure map is exemplarily depicted in Figure 6.c. Whereas the obstacle map shows the red and blue cans representing the obstacle type of small objects lying on the ground, the structure map only contains the environmental boundaries. When updating structure maps points belonging to the floor or to traversable surfaces do not need to be ignored. They are inherently replaced by points being measured on environmental structures farther away from the robot. Instead, maximum range readings need to be sorted out as they would otherwise replace shorter but valid measurements on environmental structures. However, compared to determining floor points, an according procedure is straightforward.
While the obstacle maps are very valuable when it comes to local collision avoidance, the structure maps are, for instance, very suitable for robotic self-localization, i.e., for tasks that need large scale information about an environment. When sensing un-occluded parts of walls

and ceilings, using structure maps in a localization algorithm such as Monte-Carlo Localization will have similar effects as localization based on ceiling structures (Wulf et al., 2006). The obstacle maps will surely not be appropriate for such purposes as they would miss a lot of information about environmental structures.

### 4.4 Update Procedures of Obstacle and Structure Maps

The steps to keep both representations egocentric and to update them, according to newly acquired range scans and the definitions above, are:

1.) **Transformation** of the map to keep it sensor-centric (e.g., according to odometry or a known pose).

2.) **Removal** of obsolete points to handle dynamics and inaccurate pose shift estimates.

3.) **Replacement** of already saved points using more relevant points from the current laser scan.

If the robot stands still and no pose shift has been estimated respectively, steps 1.) and 2.) are skipped. The same holds true if the virtual maps are used as efficient representations of single 3D sensor readings, e.g., as obtained from 3D cameras (see Section 7). In its initial state, the map is filled with *dummy points* that are chosen in a way that they are immediately replaced during in the first update, i.e., points corresponding to the maximum measurable distance for obstacle maps and distances of 0 m for structure maps. As soon as a valid measurement has been taken in the direction of the dummy point, it replaces the dummy. As these measurements correspond to either 0 m or maximum range readings, they are naturally ignored in algorithms processing 2D laser range scans (and the virtual maps respectively).

#### 4.4.1 Transformation of the map to keep it egocentric

According to the robot's movement the pose shift between the current and the last map update (e.g. current and last reception of a laser scan) consists of a rotation $\mathbf{R}_{\Delta\theta}$ around the $z$-axis by an angle $\Delta\theta$ and a translation $(\Delta x, \Delta y)^T$. The egocentric maps need to be transformed according to:

$$\begin{pmatrix} x_{i,t+1} \\ y_{i,t+1} \end{pmatrix} = \begin{pmatrix} \cos\Delta\theta & -\sin\Delta\theta \\ \sin\Delta\theta & \cos\Delta\theta \end{pmatrix} \begin{pmatrix} x_{i,t} \\ y_{i,t} \end{pmatrix} + \begin{pmatrix} \Delta x \\ \Delta y \end{pmatrix} \quad (4)$$

where $t$ and $(t+1)$ represent discrete points in time.

As Eq. (4) transforms the map based on Cartesian coordinates, the values of the polar coordinates have to be adjusted accordingly. Note, that the polar coordinates are not only stored and kept up to date in order to avoid unnecessary transformations in individual algorithms but also to avoid repetitive re-calculations in the update procedure itself. That is, even if not a single algorithm operates on the polar coordinates of the points stored in the virtual maps, this step cannot be skipped without a degradation of performance.

Due to the discretization of the $N = \Theta/\Delta\theta_{\text{yaw}}$ valid angles two points can potentially fall into the same vector index. In this specific case the point being more relevant with respect to the map type has priority. That is, in an obstacle map, for example, a point with a smaller distance replaces a point in the same angular interval that is farther away from the robot. Furthermore, vector indices being unassigned after the transformation are filled with dummy points.

#### 4.4.2 Removal of obsolete Points to handle Dynamics

If the maps are not intended to only represent the gathered 3D information of one rotation over the AOI or a single range image as acquired by a 3D camera, but to be used endlessly, i.e., updated with every sensor reading, it is suggestive to remove points after a certain while. Therefore, the number of transformations applied during step 1.) is stored for every single point. To deal with dynamic obstacles, a saved point is removed and replaced by a dummy point after its count of transformations exceeds a certain threshold (e.g., 500 transformations, approx. 5 s in the case of the continuously pitching IAIS 3DLS). This is because an obstacle passing by or crossing the robot's path leaves a trace of non-existent points in the obstacle map. This is not a drawback of the approach but a simple accounting for the uncertainty in the obstacle's movement. Points being removed by this means that correspond to static obstacles will immediately be measured again if the obstacle is still in the virtual corridor and thus still originates a possible source for a collision. The effect of applying this step can be seen in the region behind the robot in Figure 5. Here the point density is smaller than in the angular region in front of the robot, as some object points have already been removed and not sensed again since the removal.

Furthermore, if pure odometry is used to estimate the robot's pose shift for the transformation in step 1.) instead of pose tracking and scan matching to accurately determine the pose shift, errors and inconsistencies in both types of maps may arise from imprecise odometry. These are, in the same way, removed from the map. As a side note, it is to remark that even the rotated single 2D laser scans during the nodding-like movement of the sensor can be used for fast pose tracking algorithms like the one presented in (Lingemann et al., 2005b) if floor points are filtered out and the number of residual points is still sufficient for matching a newly acquired scan against the last one. Here, we apply the scan matching algorithm that is going to be described in Section 6.

#### 4.4.3 Replacement of already saved Points

The final update procedure highly depends on the map type as described above. In a nutshell, a point $\mathbf{p}_i$ stored in an obstacle map is replaced by a point $\mathbf{s}_i$ in the current laser scan $S$ if the angle of acquisition $\mathbf{s}_i^\theta$ falls into the discretized angular interval of $\mathbf{p}_i^\theta$ and the measured distance $\mathbf{s}_i^d$ is less than or equal to $\mathbf{p}_i^d$. In the same way a point $\mathbf{p}_i$ stored in a structure map is overwritten with $\mathbf{s}_i$ if $\mathbf{s}_i^\theta = \mathbf{p}_i^\theta$ and $\mathbf{s}_i^d \geq \mathbf{p}_i^d$. The height $\mathbf{s}_i^z$ of an acquired point in a perceived environmental structure is used as an additional information in both types of maps resulting in a 2.5D representation. That is, the virtual maps store for each discrete angle the polar coordinates $(d_i, \theta_{\text{yaw},i})$ as well as the Cartesian coordinates $(\mathbf{p}_i^x, \mathbf{p}_i^y, \mathbf{p}_i^z)$. This will be extended in future work to not only store the particular height of the most recent points but to store minimum and maximum height of all points measured within a range of approximately 10 cm around that most recent point or in a way comparable to *Multi-Level Surface Maps* (Triebel et al., 2006). By this simple extension a complete egocentric 3D model of the surrounding environmental structures can be reconstructed on the basis of the virtual 2D maps. In the case of obstacle maps the height information $\mathbf{p}_i^z$ of an acquired point $\mathbf{p}_i$ is also used to neglect those points that do not lie within the virtual corridor and are hence not relevant for representing nearby obstacles. This has the effect, that the robot can, for example, underpass a table as long as the table top is higher than the highest point on the robot and the passage between the table legs is not too narrow.

## 5. Simple Reactive Collision Avoidance using Virtual Obstacle Maps

In addition to the goal-directed motion control of a mobile robot, e.g., to reach a certain position, reactive collision avoidance is important in dynamic and human-populated environments. That is, the motion of the robot needs to be adapted in the presence of obstacles suddenly appearing in the robot's vicinity. The virtual obstacle maps can be used with any collision avoidance procedure known from processing standard 2D range scans. As a simple example, we use a set of three simple reactive behaviors controlling and adapting the robot's translational and rotational velocities based on the robot's current movement direction and speed as well as surrounding objects modeled in the obstacle map. The obstacle map is constructed and updated during navigation. These behaviors and the corresponding algorithms have been initially introduced in (Lingemann et al., 2005a).
The first behavior slows down the robot if obstacles appear, respectively, in the virtual corridor and in front of the robot in the virtual obstacle map. If the distance to the nearest obstacle in the virtual corridor falls below a certain threshold, the robot is completely stopped. Another behavior turns the robot on the spot once it has been completely stopped. This avoids that the robot gets caught in dead ends or corners. Alternatively, the robot can be moved backwards so that it is positioned in free space again. Then an alternative path can be planned to reach the position that is to be approached.
The third behavior is more complex compared to the aforementioned ones. It slightly adapts the rotational velocity of the robot so that it prefers moving along free space. Consider for example, the robot has planned a path along a longer corridor. As this path results from searching for the shortest path between two positions, it can run directly along one of the walls. When exactly following such a path, this third behavior causes that the robot is not directly moving along the wall, but instead along the center of the corridor. The concept of the behavior is to steer the robot towards a *freespace* orientation.

### 5.1 Determining the Freespace Orientation

The origin of determining the freespace orientation lies in the early work of Surmann & Peters (2001) for fuzzy-based control of autonomous mobile robots. A comparable behavior was obtained by applying a fuzzy controller with fuzzy rules like the following:

```
IF COMMAND is straight-ahead
  AND IF FRONT-SENSOR is very-near
     AND FRONT-LEFT-SENSOR is very-near
     AND FRONT-RIGHT-SENSOR is near
THEN SPEED is positive-small, ANGLE is negative-small
```

An adaption to 2D laser range scans has been presented in (Lingemann et al., 2005a). Here the following fuzzy rule is applied to every single distance measurement:

```
IF (angle_i is in driving direction) AND (distance_i is large)
THEN drive in this direction.
```

The actual driving direction of the robot, the wanted freespace orientation $\alpha_{\text{free}}$, further adapted to meet our requirements, results as follows and is based on the above rule.

$$\alpha_{\text{free}} = \operatorname{atan2}\left(\sum_{i=1}^{N} \sin \mathbf{s}_i^{\theta} \cdot f_{\theta}(\mathbf{s}_i^{\theta}) \cdot f_d(\mathbf{s}_i^{d}), \sum_{i=1}^{N} \cos \mathbf{s}_i^{\theta} \cdot f_{\theta}(\mathbf{s}_i^{\theta}) \cdot f_d(\mathbf{s}_i^{d})\right) \tag{5}$$

The functions $f_\theta(\mathbf{s}_i^\theta)$ and $f_d(\mathbf{s}_i^d)$ relate the $i$-th range reading in the form of the polar coordinates $(\mathbf{s}_i^\theta, \mathbf{s}_i^d)_{i=1\ldots N}$ as obtained from a 2D laser scanner or an obstacle map to the fuzzy sets *"angle is in driving direction"* and *"distance is large"*. $N$ is the number of points in the map and the laser range scan respectively.

$$f_\theta(\theta) = \cos\left(\frac{\theta}{1.2}\right) \quad (6)$$

$$f_d(d) = \frac{1}{1+\exp\left(-\left(\frac{d-dto_{\max}}{dto_{\min}}\right)\right)} \quad (7)$$

Here $dto_{\min}$ and $dto_{\max}$ determining the slope and the inflection point of the exponential, correspond to the thresholds of the behavior that slows down the robot. That is, if an object appears in front of the robot within a range $dto_{\max}$, the behavior starts slowing down the robot according to the distance to that object. If the distance to the object falls below $dto_{\min}$, the robot is completely stopped or moved backwards. Plots of the weighting functions $f_\theta(\theta)$ and $f_d(d)$ as well as the resulting application of the fuzzy AND by means of a multiplication are shown in Figure 7.

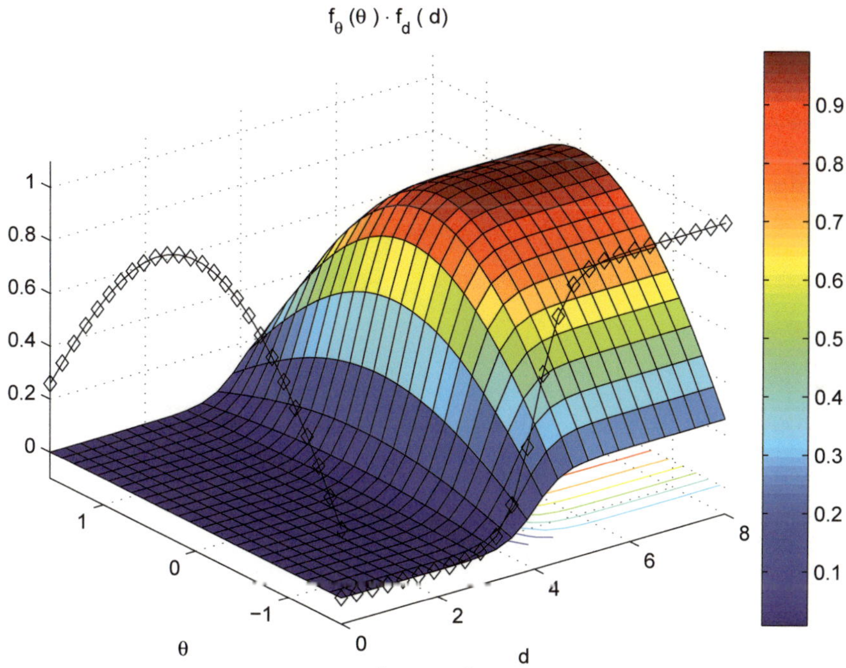


Fig. 7. Composed weighting function $f_\theta(\mathbf{s}_i^\theta) \cdot f_d(\mathbf{s}_i^d)$ to determine the freespace orientation in a laser scan or obstacle map $S$.

The behavior simply adapts the rotational velocity, e.g., as set by a motion controller for following a planned path, so that the robot slightly moves towards free space thereby swerving to avoid collisions. The influence of the behavior can be adapted and is kept rather small so that the robot can enter narrow passages and follow paths that lead away from the maximally free space in the robot's workspace. Referring to the resulting weighting function $f_\theta(\theta) \cdot f_d(d)$, the robot prefers moving straight and not adjusting its translational velocity.

### 5.2 Typical Results

A typical result of applying the three behaviors during navigation is shown in Figure 8. The robot was put into an example scenario bounded by movable walls with an exit in the opposite corner. The experiment was repeated two times. In the first run the scanner was held in a horizontal position (2D perception) comparable to a standard 2D laser range finder. In the second run, the scanner was continuously rotated over the aforementioned area of interest in a nodding-like fashion (3D perception). In both experiments, a constant translational velocity of $0.3\,\mathrm{m\,s^{-1}}$ was set with no rotational velocity. That is, the robot was commanded to move forward. The application of the behaviors successfully moved the robot through the exit and outside the scenario. With 2D perception, small test objects lying on the ground were not perceived. The robot crushed into these objects and pushed them through the exit. With 3D perception, the objects have been perceived and the robot successfully avoided them. Although it is not explicitly covered in this example, it is to note that the robot robustly perceives sudden dynamic changes in the environment due to the fast pitch rate of the scanner while driving. It therefore detects and avoids dynamic obstacles like, for instance, suddenly opened drawers or people passing by.

What can also be seen in the figure is that simply commanding the robot to move forward while enabling the three collision avoidance behaviors leads to an emergent behavior of wandering around. This strategy can, amongst other applications, be used to perform a random exploration.

(a) 2D Perception (b) Example scenario (c) 3D Perception

Fig. 8. Behavior-based collision avoidance in an example scenario by means of 2D perception (2D laser range finder in a fixed horizontal position) and continuous 3D perception together with the concept of obstacles maps. The small test objects are successfully avoided by the robot when using 3D perception. A video showing the robot perform in a similar experiment is available under `http://www.b-it-bots.de/media`.

## 6. Simultaneous Localization and Mapping

In the previous sections we showed that continuous 3D environment sensing together with the concepts of an area of interest and the virtual corridor enables an autonomous mobile robot to perceive and react to various obstacles being typical for domestic environments, like for instance open drawers, small obstacles lying on the floor or descending stairs. In this section we will show how the same continuous 3D data flow of the pitching laser scanner can be used for 2D and 3D *Simultaneous Localization and Mapping (SLAM)*, i.e., constructing two-dimensional and three-dimensional environment representations and localizing the robot with three and six degrees-of-freedom (DOF) respectively. Central questions in this context are:

1. **Which information** can be used for 3DOF- and 6DOF SLAM respectively?
2. **Which SLAM algorithms** can be used to process this information?

### 6.1 Extracting Information and Forming Local Point Clouds for SLAM

The pitching laser scanner delivers a continuous stream of 3D data acquired during the robot's movement through the environment. The first question addresses the segmentation of this data stream and the extraction of locally consistent information about environmental structures usable in a SLAM approach. In the two-dimensional case (3DOF-SLAM) this is quite straightforward. By storing maximum distances in each measurement direction the virtual structure maps store aggregated information about boundaries of the environment, like for instance walls, as well as larger static obstacles. A these maps are already represented in the form of a 2D laser range scan, the contained information can be used with any known SLAM algorithm working on range scans. Constructing individual structure maps for every full rotation over the area of interest (AOI) results in locally consistent virtual range scans. Due to the fact that the rotational velocity of the scanner as well as the size of the AOI are adjusted according to the robot's current velocities, these virtual range scans have a sufficient overlap, e.g., for applying scan matching algorithms. However, the range scans are only locally consistent under the assumption that the pose shift estimates, used for keeping the structure maps egocentric during construction, are not too inaccurate. Solely transforming the maps based on inaccurate or erroneous odometric pose shift estimates can lead to inconsistent information, distorted environmental structures when turning fast for example. To account for that, we interrupt the construction of a structure map when the pose shift estimates suggest that the robot is turning fast and start building a new structure map after that turn. A similar procedure is followed by Cole & Newman (2006). However, matching consecutive scans, as described later in this section, can correct false or inaccurate pose shift estimates and the aforementioned interruption is only applied when solely using odometry information.
Another straightforward way of extracting information for 3DOF-SLAM is to take those 2D range scan that have a measurement plane being parallel to the ground, i.e., scans taken at $\theta_{\text{pitch}} = 0$ for a flat floor. Such a scan can be taken as is and is comparable to a range scan acquired by a fixed horizontally mounted 2D laser range finder. Both the horizontal laser scans as well as the constructed structure maps can be used with any SLAM algorithm processing range scans, e.g., Rao-Blackwellized Particle Filters (Grisetti et al., 2007).
For being able to construct three-dimensional environment models and to localize the robot with six degrees-of-freedom (6DOF SLAM) we need to extract locally consistent 3D point clouds out of the continuous data stream delivered by the pitching scanner while moving. That is, we want to construct individual 3D point clouds that are each referenced to a distinct vehicle pose, the base pose $\mathcal{P}_b$, just like 3D range scans acquired while standing (as in Surmann et al., 2003). We therefore strip the robot's movement in space during one complete pitch movement by transforming successively the thereby gathered 2D scans according to the estimated relative pose shift between the current robot pose and $\mathcal{P}_b$. The scans are then combined to form one 3D point cloud that, depending on the currently used pitch rate and size of the AOI, consists of 15 to 500 single 2D laser scans. A point cloud being generated by this means is shown in Figure 9.
The figure also shows the results of different processing steps that are performed for every 2D laser range scan acquired during the nodding-like movement of the scanner: 1.) reducing

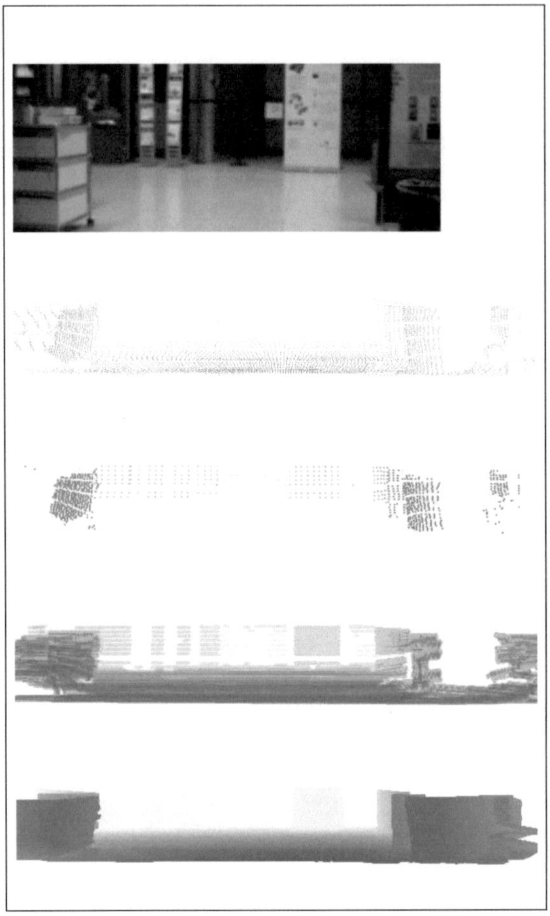

Fig. 9. Example of a generated point cloud. The figure shows (from top to bottom): a photo of the captured scene, the range measurements acquired during movement, the reduced set of points (without floor points), detected lines and a rendered depth image.

the point density in the scan by replacing clusters of points, whose distance to each other falls below some threshold, by their centroid, 2.) detecting line segments in the residual points, 3.) merging line segments to planes and 4.) classifying points whether or not they belong to the floor. All these algorithm are described, in detail, in (Surmann et al., 2003).

### 6.2 A Brief Overview on SLAM Algorithms

Performing SLAM to build maps and localizing in preliminary built maps are major preconditions for the autonomous operation of mobile robots in changing or preliminary unknown environments. Approaches addressing mapping and localization differ, amongst others, in formulating the problem, the means to cope with the addressed problem and in representing the environment. A majority of SLAM algorithms is probabilistic and based on formulations using *Extended Kalman Filters* (EKFs, Leonard & Feder, 1999), *Unscented Kalman Filters* (UKFs, Chekhlov et al., 2006), Sparse Extended Information Filters (SEIFs, Thrun et al., 2004) or *Rao-Blackwellized Particle Filters* (RBPFs, Grisetti et al., 2007). Graph-based SLAM algorithms

(Grisetti et al., 2008; Olson et al., 2006) address SLAM in terms of nonlinear optimization and are often used as generic back-ends to distribute accumulated errors after having detected loops in the robot's trajectory or to better align graphs of robot poses.
Another way of formulating SLAM is that by interpreting it as a problem of *multi-view range image registration*, i.e., integrating and aggregating range measurements taken at different positions and orientations in the environment into a common coordinate frame forming the environment model. This problem can be formulated as follows: Given two sets of geometrical features, a data set $D$ or *scene* and a data set $M$ or *model*, find a transformation $\mathbf{T}$ that minimizes the alignment error between the two sets and correctly maps $D$ onto $M$. The goal is to transform $D$ in a way that it "fits best" with $M$. Normally, range images are taken of non-deformable, static objects with the same sensor but from diverse viewpoints, i.e., from different positions and under different orientations. In this case, $\mathbf{T}$ is a rigid transformation that includes only translation and rotation.

### 6.3 Incremental Registration using the ICP Algorithm and Sparse Point Maps

A widely used solution to the registration problem is the Iterative Closest Point (ICP) algorithm by Besl & McKay (1992), which determines $\mathbf{T}$ in an iterative way. In each iteration step, the ICP algorithm determines pairs of corresponding points from $D$ and $M$ using a nearest-neighbor search. These correspondences are used to quantify and minimize the alignment error $E$:

$$E(\mathbf{R},\mathbf{t}) = \sum_{i=1}^{|M|}\sum_{j=1}^{|D|} w_{i,j} \|\mathbf{m}_i - \left(\mathbf{R}\mathbf{d}_j + \mathbf{t}\right)\|^2, \quad w_{i,j} = \begin{cases} 1, & \mathbf{m}_i \text{ corresponds to } \mathbf{d}_j \\ 0, & \text{otherwise.} \end{cases} \tag{8}$$

$$\mathbf{T} = \begin{pmatrix} \mathbf{R}_{ICP} & \mathbf{t}_{ICP} \\ 0\,0\,0 & 1 \end{pmatrix} \quad \text{with } (\mathbf{R}_{ICP}, \mathbf{t}_{ICP}) = \arg\min_{\mathbf{R},\mathbf{t}} E(\mathbf{R},\mathbf{t}) \tag{9}$$

Finding the nearest neighbors and determining the correspondences is the computationally most expensive step in the ICP algorithm ($O(|D|\,|M|)$ for a brute-force implementation), since for every point $\mathbf{d}_j \in D$ the closest point $\mathbf{m}_i \in M$ needs to be determined. Here, we use an approximate *kd*-tree search (Arya et al., 1998), which reduces the complexity of the algorithm to $O(|D| \log |M|)$.
To estimate the rigid transformation $\mathbf{T}$, consisting of a rotation $\mathbf{R}$ and a translation $\mathbf{t}$, that minimizes Eq. (8) there are closed form solutions in both the two- and three-dimensional case (Lorusso et al., 1995). In order to cope with only partially overlapping sets, we reject correspondence pairs for which the point-to-point distance exceeds a certain threshold. This threshold exponentially decays during the registration process. While initially permitting larger distances between corresponding points guarantees fast convergence of $E(\mathbf{R},\mathbf{t})$, smaller distances in later iteration steps allow fine-tuning the registration result. Furthermore, we reject pairs that contain the same model point and only keep the pair with the closest point-to-point distance (Zinßer et al., 2003).
For registering multiple range scans and constructing a consistent map that models environmental surfaces, an incremental registration procedure is used. The first laser scan $D_0$ is used as the initial environment model $M_0$. Thus, the local coordinate frame of $D_0$ forms the coordinate frame for the overall map. All subsequent scans $D_i, i > 0$ are matched against $M_{i-1}$. The resulting transformation $\mathbf{T}_i$ is used to correct the position of all points contained in $D_i$, yielding the transformed point set $\check{D}_i = \{\check{\mathbf{d}}_{i,j} \mid \check{\mathbf{d}}_{i,j} = \mathbf{R}\mathbf{d}_{i,j} + \mathbf{t}\}$. As an initial estimate $\hat{\mathbf{T}}_i$ for $\mathbf{T}_i$ in this incremental registration we use the transformation from the last registration, i.e.,

$\hat{\mathbf{T}}_i = \mathbf{T}_{i-1}$. This speeds up the convergence in the ICP algorithm and drastically reduces the probability of converging to a local minimum possibly resulting in an incorrect registration result. If odometry information is available, the estimate $\hat{\mathbf{T}}_i$ is further corrected taking into account the estimated pose shift between the acquisition of $D_{i-1}$ and $D_i$. Furthermore, we only register a new range scan $D_i$ if the robot traversed more than a certain distance, for example 50 cm, or turned more than a certain angle, for example 25°. Such a practice is quite common in a variety of recent SLAM algorithms.

To account for possibly new information in $D_i$, the transformed points are than added to $M_{i-1}$. That is, after matching range image $D_i$, the model set $M_{i-1}$ computed so far is updated in step $i$ to:

$$M_i = M_{i-1} \cup \{\check{\mathbf{d}}_{i,j} \mid \check{\mathbf{d}}_{i,j} \in \check{D}_i\}. \tag{10}$$

Thus, a model $M_N$, constructed by incrementally registering $N$ range images, contains all points measured in the environment, i.e.

$$M_N = \bigcup_{i=[0,N]} \{\check{\mathbf{d}}_{i,j} \mid \check{\mathbf{d}}_{i,j} \in \check{D}_i\}. \tag{11}$$

The main problem of this incremental registration approach is its scalability with respect to the size of the environment and the number of range images taken. To fully cover a large environment, a lot of range images might be needed. When registering and adding all acquired range images, the model set $M$ can get quite large, e.g. several million points for 3D scans taken in a large outdoor environment (Nüchter et al., 2007). However, when acquiring range images in parts of the environment which are already mapped, lots of points would be added to $M$ without providing new information about the environment. This is exploited by the following improvement to our SLAM approach, which makes the point clouds sparse.

The key idea of *sparse point maps* is to avoid duplicate storage of points, and thereby minimize the amount of memory used by the map, by conducting an additional correspondence search. Correspondence is, thereby, defined just like in the ICP algorithm. That is, a point $\check{\mathbf{d}}_{i,j} \in \check{D}_i$ is not added to $M_{i-1}$, if the point-to-point distance to its closest point $\mathbf{m}_{i-1,k} \in M_{i-1}$ is smaller than a minimum allowable distance $\epsilon_{\text{D}}$.

$$M_i = M_{i-1} \cup \{\check{\mathbf{d}}_{i,j} \mid \check{\mathbf{d}}_{i,j} \in D_i, \nexists \mathbf{m}_{i-1,k} \in M_{i-1} : \|\check{\mathbf{d}}_{i,j} - \mathbf{m}_{i-1,k}\| < \epsilon_{\text{D}}\} \tag{12}$$

The threshold $\epsilon_{\text{D}}$ spans regions in the model in which the number of points is limited to 1, thereby providing an upper bound on the point density in a sparse point map $M$. Choosing a value of $\epsilon_{\text{D}}$ according to the accuracy of the range sensor used will exactly neglect duplicate storage of one and the same point assuming correct alignment of range images. Choosing, however, a larger value allows to reduce the number of points stored in the map. Although some details of the environment might not get modeled, a map constructed in this manner still provides a coarse-grained model of the environment. In the actual implementation, the additional correspondence search is carried out on the *kd*-tree built for the ICP algorithm but using $\epsilon_{\text{D}}$ as the distance threshold in the pair rejection step. However, here the rejected pairs are used to determine the points in $\check{D}_i$ that need to be added to $M_{i-1}$.

### 6.4 3DOF-SLAM and 6DOF-SLAM in an Example Scenario

The following figures show typical results from applying the above described algorithms for extracting 2D and 3D laser range scans out of the continuous 3D data stream and the incremental registration using the ICP algorithm. In this example the robot was manually driven through a laboratory environment. The earlier described methods and means for avoiding collisions were applied in order to change the manually set velocities in order to automatically swerve around obstacles. Out of the stream of rotated 2D range scans delivered by the continuously pitching 3D scanner both virtual structure maps as well as locally consistent 3D point clouds have been extracted. These point sets were then incrementally registered using the ICP algorithm and the aforementioned extensions. The two-dimensional sparse point map obtained from matching the structure maps is shown in Figure 10(a) as well as the estimated trajectory of the robot. Incrementally registering the generated 3D point clouds results in the 3D point model shown in Figure 10(b) in a top-down perspective with points classified as corresponding to floor not being visualized. Views on the complete 3D model including floor points are shown in Figure 11.

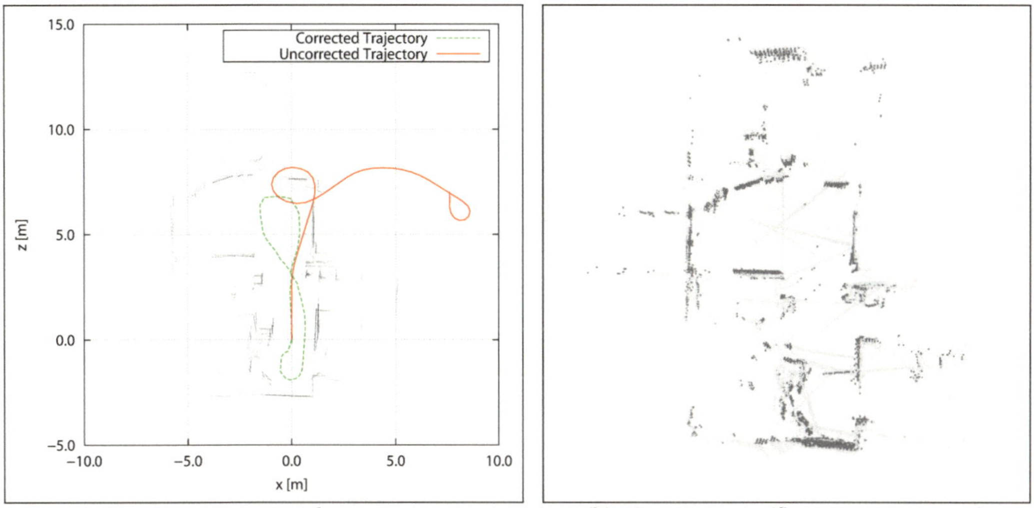


(a) 2D point map and trajectory (b) 3D point map (floor points removed)

Fig. 10. Shown here are a two-dimensional sparse point map (a) as well as a three-dimensional sparse point map (b) of an example scenario. Measurements corresponding to flat floor have been removed.

## 7. Extensions for using recent Time-of-Flight Cameras

An inherent drawback of the nodding-like movement of the continuously pitching laser scanner is the high mechanical load of the actuator especially when accelerating the scanner at the lower and upper bound of the AOI, i.e., when changing the rotation direction. Longer operations can cause an increase in gear backlash. Estimating the pitch angle solely based on the current position of the servo motor can result in inaccuracies, i.e., the estimated angle may deviate from the actual rotation angle of the pitching scanner. This effect can be seen in Figure 12(a). In this example a 3D scanner, that had already been used for several years without changing motor or gear, was rotated from $-30°$ to $15°$ and back to $-30°$. A vertical slice of the generated point cloud is extracted that contains measurements on the planar floor as well as on a vertical plane. Due to the inaccuracies in the estimation of the pitch angle, the resulting geometric structure is distorted. When explicitly measuring the pitch angle, like for

(a)

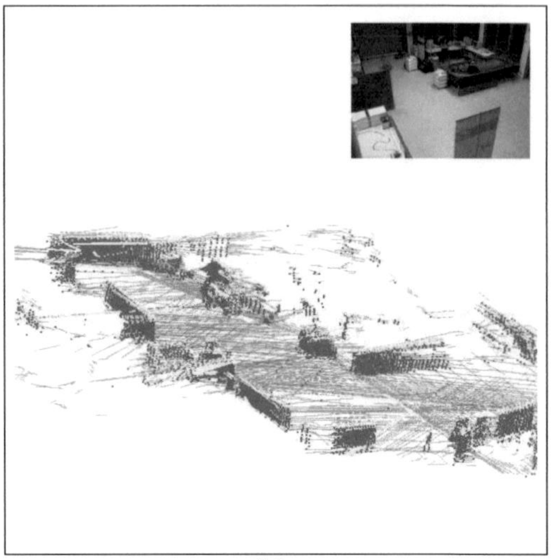

(b)

Fig. 11. Different views on the complete 3D model. It can be seen that the map does only model environmental structures in a height of the robot. That is, only those regions are modeled with which the robot can interact.

instance using gyroscopes or an inertial measurement unit, allows for correctly reconstructing the geometric structure as shown in Figure 12(b).

In contrast to custom built 3D scanners as the continuously pitching scanner used here, Time-of-Flight (ToF) cameras directly deliver 3D point clouds without a mechanical actuator. These solid-state sensors emit an amplitude-modulated signal using an array of near-infrared light emitting diodes (LEDs). The on-chip calculation of depth information is based on the phase shift between the emitted and the received signal. ToF cameras provide depth images at high frame rates while preserving a compact size, a feature that has to be balanced with measurement accuracy and precision. Besides the depth measurements, they also provide the amplitude of the reflected signal, which correlates to the reflectivity of the measured object.

Depending on external interfering factors (sunlight for example) and scene configurations, i.e., distances, surface orientations, and reflectivity, distance measurements from different perspectives of the same scene can entail large fluctuations. Furthermore, these sensors have, compared to 3D laser scanners, a restricted field of view, e.g., only 43° (H) × 34° (V) for a SwissRanger SR4000 from Mesa Imaging[5].

One of the first applications in robotics considering ToF cameras as an alternative to laser scanning, was presented by Weingarten et al. (2004). They evaluated a SwissRanger SR-2 camera in terms of basic obstacle avoidance and local path planning capabilities. In order to cope with the poor data quality, Weingarten et al. determined the parameters for the perspective projection into the image plane in a photogrammetrical calibration as well as scaling and offset values for correcting systematic errors in individual measurements. In order to detect obstacles they apply a simple thresholding to remove floor points. As a result they presented simple examples where an autonomous mobile robot using the SwissRanger camera was able

[5] Mesa Imaging SwissRanger cameras: `http://www.mesa-imaging.ch`

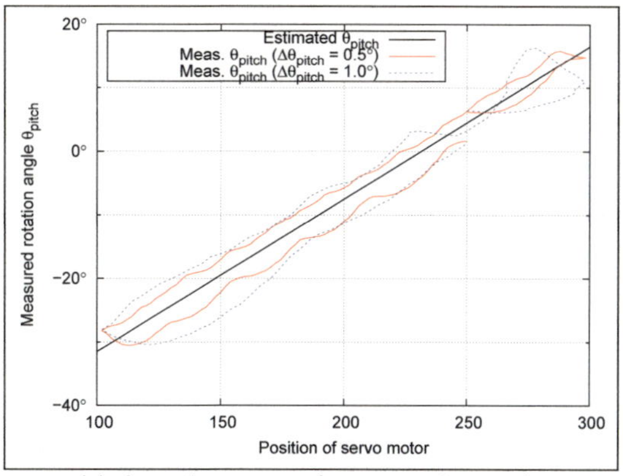


(a) $\theta_{\text{pitch}}$ in one up-down rotation cycle

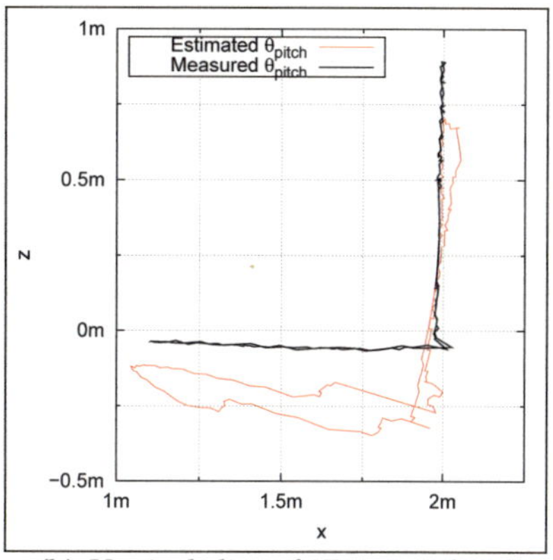


(b) Vertical slice of 3D point cloud

Fig. 12. Estimated and measured pitch angles. Depending on the gear backlash, pitch angles estimated based on the position of the servo motor may deviate from the actual pitch angle (a). Measuring the pitch angle, e.g., using gyroscopes still yields accurate and correct geometric information (b).

to stop in front of a table that would have not been perceivable using a 2D laser range scanner. However, they also mentioned that some objects and environmental structures have not been perceivable with the SwissRanger due to the smaller field of view (Weingarten et al., 2004). Sheh et al. (2006) also perform a per-pixel calibration similar to Weingarten et al. (2004) and determine a lookup table storing an offset and a multiplying factor for every distance measurement. In addition, Sheh et al. explicitly handle defect pixels. In order to handle the smaller field of view in the context of 3D mapping, they stop the robot, rotate the camera and acquire multiple range images under different rotation angles. The individual range images are then merged into a single point cloud and registered into a global point map. The actual registration is thereby assisted by a human operator. Swadzba et al. (2007) address the poor data quality of the sensor by using a sequence of filtering operations and register successive range images using a combination of feature tracking and registration with the ICP algorithm. Recently, Yuan et al. (2009) adopted the ideas of virtual scans (Wulf et al., 2004) and virtual obstacle maps (Holz et al., 2008) to fuse the sensory information of a ToF camera with that of a 2D laser range scanner. They apply the same pre-processing steps as Swadzba et al. to filter out and correct inaccurate or erroneous measurements. However, due to that fact the camera is fixed on the robot, they suffer from the same problem with the narrow field of view as Weingarten et al., namely that not all obstacles in the robot's vicinity can be perceived. Furthermore, just like Weingarten et al., they only perform a simple thresholding to classify floor points and detect obstacles. However, this procedure can cause different problems as already mentioned in Section 4.

The narrow field of view and the poor data quality necessitate several extensions when using a ToF camera in the context of SLAM and collision avoidance.

1. Due to inaccuracies and erroneous measurements, the acquired 3D point clouds need to be filtered and corrected in order to obtain accurate and correct geometric information.
2. Furthermore, the poor data quality necessitates a more robust classification of measurements for reliably detecting obstacles.

3. The narrow field of view needs to be explicitly taken into account in the registration with the ICP algorithm in order to avoid false correspondence when single range images contain less geometric features. When all measurements in the point cloud have been acquired on a single planar surface for example, the registration problem is underconstrained.

4. Again caused by the narrow field of view, the robot does not have all objects in sight when the camera is mounted at a fixed position. In order to adequately perceive obstacles in the robot's vicinity and its movement direction respectively, the camera needs to be rotated.

### 7.1 Filtering and Correction of Depth Measurements

Measurements from Time-of-Flight cameras are subject to different error sources (Lange, 2000). They can be divided into systematic and random errors. Systematic errors can be corrected by calibration (Fuchs & Hirzinger, 2008), whereas random errors can be coped with by means of filtering. A common way in filtering random errors is to neglect measurements based on their amplitude. Thresholding the amplitude neglects measurements from poorly reflecting objects, objects being farther away from the robot and objects in the peripheral area of the measurement volume.

Another source of errors in distance measurements are so called *jump-edges*. They occur at transitions from one shape to another. The shapes seem to be connected due to spurious measurements in between. These spurious measurements result from multiple-ways reflections which also cause that hollows and corners appear rounded off (Lange, 2000; May et al., 2009). Jump edges can be filtered by means of local neighborhood relations (May et al., 2009). From a set of 3D points $P = \{\mathbf{p}_i \in \mathbb{R}^3 \mid i = 1, \cdots, N_p\}$, jump edges $J$ can be selected by comparing the opposing angles $\theta_{i,n}$ of the triangle spanned by the focal point $\mathbf{f} = \mathbf{0}$, point $\mathbf{p}_i$ and its eight neighbors $P_n = \{\mathbf{p}_{i,n} \mid i = 1, \ldots, N_p : n = 1, \cdots, 8\}$ with a threshold $\epsilon_\theta$:

$$\theta_i = \max \arcsin \left( \frac{|| \mathbf{p}_{i,n} ||}{|| \mathbf{p}_{i,n} - \mathbf{p}_i ||} \sin \varphi \right), \tag{13}$$

$$J = \{\mathbf{p}_i \mid \theta_i > \epsilon_\theta\}, \tag{14}$$

where $\varphi$ is the angle between two neighboring distance measurements. Since this filter is sensitive to noise, we apply a median filter to the depth image beforehand.

### 7.2 Detecting Obstacles and Information Fusion in Obstacle Maps

The filtered depth measurements are transformed to the robot's Cartesian coordinate frame by the extrinsic camera parameters, taking into account the sensor's height and orientation. A typical example of a resulting point cloud taken in an indoor environment is shown in Figure 13(a). The colors of the points correspond to the distance of a point from the sensor, brighter colors relate to shorter distances. This point cloud can be used to build a so-called *height image* as shown in Figure 13(b). The gray scale value of every pixel in the height value corresponds to the height, i.e., the $z$-coordinate of the respective point in the point cloud. A point $\mathbf{p}_{i,j}$ is classified as belonging to an obstacle if

$$(W_{\max} - W_{\min}) > \epsilon_H \tag{15}$$

where $W_{\max}$ and $W_{\min}$ are the maximum and minimum height values from a local window $W$, spanned by the Moore neighborhood around $\mathbf{p}_{i,j}$. The threshold $\epsilon_H$ thereby corresponds to the minimum tolerable height of an obstacle. It needs to be chosen appropriately since it should not be smaller than the sensor's measurement accuracy. Due to evaluating a point's local neighborhood, floor points are inherently not considered as obstacles. The result of this filter is shown in Figure 13.

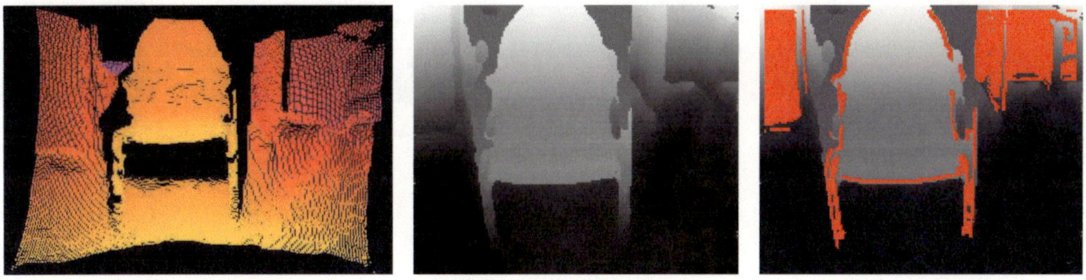

(a) Raw point cloud (b) Rendered height image (c) Obstacles

Fig. 13. Detecting obstacles in range images. Based on the acquired range image (a), a height image (b) is constructed. Filtering the height image results in the set of obstacle points (c).

The resulting obstacle points are used to extract a two-dimensional *virtual scan* similar to an obstacle map by 1.) projecting the 3D data into the $xy$-plane and 2.) extracting relevant information. The number of range readings in the virtual scan as well as its apex angle and resolution correspond to the acquired 3D data. For the SR4000, the number of range readings is 176, which is the number of columns in the image array. The apex angle and the angular resolution are 43° and 0.23°, which corresponds to the camera's horizontal apex angle and resolution. For every column of the ToF camera's distance image, the obstacle point with the shortest Euclidean distance to the robot is chosen similar to the update procedure of obstacle maps in Section 4. This distance constitutes the range reading in the scan. If no obstacle point is detected in a column, the scan point is marked invalid, by setting it to the maximum measurable distance of the sensor.

Figure 14(a) shows an example scene of an indoor environment. The point cloud which results from the ToF camera's depth image is shown in Figure 13. The result of the filtering and obstacle detection step is depicted in Figure 14(b). Points with a low amplitude are removed from the cloud. Points classified as belonging to obstacles and the extracted virtual scan are shown in Figure 14(c). Obstacle points are marked white and the obstacle points that contribute to the *virtual scan* are marked red. The remaining points are marked green.

The resulting *virtual scan* is fused with a 2D laser range scan yielding a common *obstacle map* modeling the closest objects in both sensors. The obstacle map for the aforementioned example scenario is visualized in Figure 15. Measurements from the laser range scan are illustrated by the blue points. The red points illustrate the *virtual scan*. The chair shows only a few points in the 2D laser range scan since only the legs of the chair are in the scan plane, whereas the *virtual scan* outlines the contour of the chair. By fusing the information of both sensors, the robot possesses correct information about traversable free space (light blue region) in its immediate vicinity. The obstacle maps can be used for collision avoidance just like the obstacle maps obtained from extracting relevant information from the continuous data stream of the pitching 3D laser scanner.

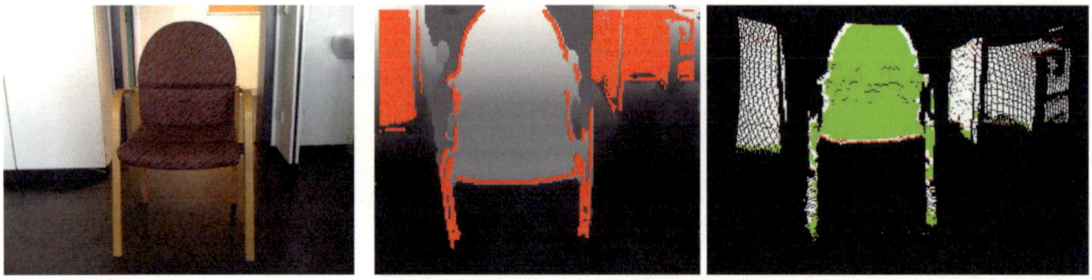

(a) Photo of the scene (b) Obstacles in height image (c) Point cloud and extracted scan

Fig. 14. (a) An example scene of an indoor environment. (b) The height image with detected obstacle points (red) . (c) The result of the filtering and obstacle detection step. Points with a low amplitude are removed from the point cloud. Obstacle points are marked white and the obstacle points that contribute to the virtual scan are marked red. The remaining points are marked green.

### 7.3 Frustum Culling as an Extension to the ICP Algorithm

The SLAM approach based on the ICP algorithm and described in Section 6 is quite sensitive to false correspondences, especially when the point sets only partially overlap. Points in a correspondence pair that do not model the same point in the physical environment can negatively affect the registration result possibly leading to an incorrect local minimum. Due to the narrow field of view of the SwissRanger camera, the overlap of the latest range image and the so far built model has to be handled more explicit than solely using the pair rejection extensions mentioned earlier. Here we apply a technique called *frustum culling* which known from 3D computer graphics. The *frustum* defines the volume that has been in the range of vision while acquiring the model point set $M$. Points in the data set $D$ that do not lie within the model frustum, are neglected in the correspondence search as they cannot form a valid correspondence pair (May et al., 2009). Luck et al. (2000) use frustum culling in a preprocessing step to sort out points by means of an initial pose esimate. The registration is then conducted on the residual points. In contrast to that, we do not rely on an initial pose estimate and apply frustum culling in every iteration step of ICP algorithm. This takes into account that points in the data set are moved inside or outside the frustum during registration. Applying frustum culling in every iteration step effectively reduces the number of false correspondences in the registration and the total number of misregistrations caused by false correspondences. A detailed description of this approach as well as the aforementioned calibration and filtering procedures can be found in (May et al., 2009).

The procedure of neglecting points from the data set $D$ that do not lie within the model frustum is visualized in Figure 16(a). Points of the model set $M$ are shown in green. Those points of $D$ that lie within the frustum and that are considered in, respectively, the nearest neighbor search and the registration of $D$ are shown in red. Points from $D$ that are neglected because of not lying in the model frustum are shown in blue. A point map constructed by incrementally registering range images using the aforementioned extensions and the frustum culling is visualized in Figures 16(b) and 16(c).

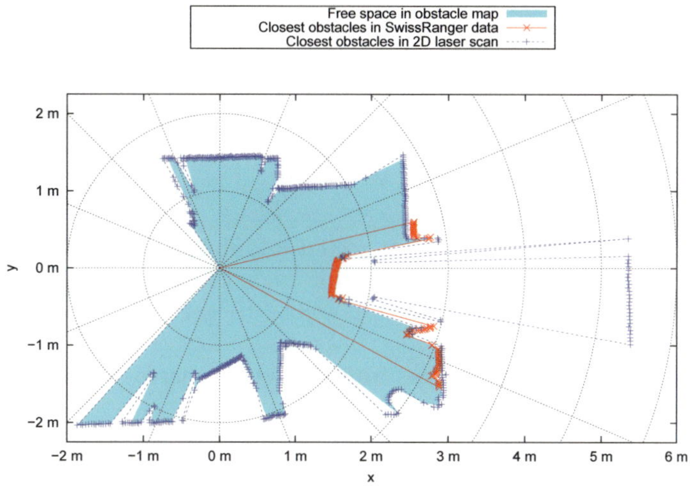


Fig. 15. The resulting *virtual scan* of the scene is compared with the base laser scan. The base laser scan is illustrated by the dashed blue line. The red line illustrates the virtual laser scan. The chair shows only a few points in the base laser scan since only the legs of the chair are in the scan plane, whereas the *virtual scan* outlines the contour of the chair.

(a) (b) (c)

Fig. 16. Visualization of the frustum culling in the ICP algorithm (a) and two views on a constructed 3D model (b+c).

### 7.4 Gaze Control

As already mentioned, the narrow field of view is a major issue in the context of collision avoidance as not all obstacles in the robot's vicinity can adequately be perceived. Here, we propose to mount the SwissRanger camera on some mechanical actuator, e.g., a pan-tilt unit, and to rotate the camera in a way that the robot perceives all obstacles in its movement direction. Rotating the camera is especially relevant when the robot is able to move sidewards or to turn fast.

In order to rotate the camera and control the viewing direction or *gaze*, we have mounted the SwissRanger on the head pan-tilt of an antropomorphic mobile service robot (see Figure 17(a)). As described in Section 7.2, the range measurements of the SwissRanger camera are transformed into the robot's coordinate frame and fused with the sensory information of a 2D laser scanner. In order to perceive all obstacles in the robot's vicinity and its movement direction, we determine an appropriate camera orientation for looking into and perceiving all types of obstacles in the robot's movement direction which is given by the translational velocities $(\mathbf{v}^x\ \mathbf{v}^y)^T$ and the rotational velocity $\omega$. We compute the camera's look at point $\mathbf{g} = (\mathbf{g}^x\ \mathbf{g}^y\ \mathbf{g}^z)^T$ using a simple controller.

$$\begin{pmatrix} \mathbf{g}^x \\ \mathbf{g}^y \\ \mathbf{g}^z \end{pmatrix} = \alpha \begin{pmatrix} \cos \beta\omega & -\sin \beta\omega & 0 \\ \sin \beta\omega & \cos \beta\omega & 0 \\ 0 & 0 & 1 \end{pmatrix} \left( \frac{d_{\min}}{\|\mathbf{v}\|} + \gamma \right) \begin{pmatrix} \mathbf{v}^x \\ \mathbf{v}^y \\ 0 \end{pmatrix} \quad (16)$$

with $d_{\min}$ being the minimum distance (projected into the $xy$-plane) that can be perceived. It is typically limited by the minimum pitch/tilt angle of the actuator. The scaling factors $\alpha$ and $\beta$ as well as the offset $\gamma$ can be used to adjust the gaze vector according to a specific robot platform. We do not scale the function but use an offset of $\gamma \geq 1$ thereby preferring the perception of obstacles being farther away from the robot. That is, we assume that there is no single obstacle directly in front (touching) the robot.

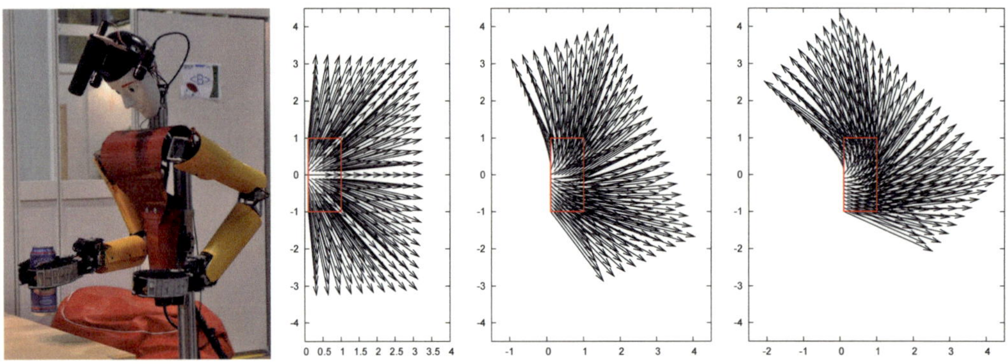

(a) (b) $\omega = 0$ (c) $\omega = 0.125\pi s^{-1}$ (d) $\omega = 0.25\pi s^{-1}$

Fig. 17. The anthropomorphic service robot Dynamaid (a) and look at points $\mathbf{g}$ for different configurations of $\mathbf{v}^x$, $\mathbf{v}^y$ and $\omega$.

The result of applying this simple controller yields a behavior of smoothly adjusting the orientation of the robot's head towards the current movement direction. Figure 17 shows the camera on top of the head of the antropomorphic robot as well as possible outcomes of the gaze controller described above. Although this behavior appears quite reasonable for a human spectator and allows for perceiving obstacles that would have not been perceivable with a fixed camera, it should be noted that elaborating a more sophisticated strategy for controlling the viewing direction is a matter of future work.

## 8. Conclusion and Future Work

With a focus on navigation in dynamic and cluttered domestic environments, we have presented a sensor setup for a 3D scanner that is especially appropriate for a fast 3D perception of those regions in the robot's vicinity that are relevant for collision avoidance. The 3D scanner is continuously pitched in a nodding-like fashion where the range of rotation angles (area of interest) is adapted in size just as the angular velocity with which the scanner is rotated. The acquired 3D data is projected into the $xy$-plane in which the robot is moving and used to construct and update egocentric 2.5D maps storing either the coordinates of closest obstacles or environmental structures. The representation of these maps is defined in a way that

they can be used with any algorithm for SLAM or collision avoidance that is operating on 2D laser range scans. The obstacle maps have been used in a set of simple behaviors for reactive collision avoidance. By means of the continuously pitching laser scanner and the concept of the obstacle maps, an autonomous mobile robot is able to successfully avoid different obstacles that can typically be found in domestic environments, like for instance table tops, open drawers or stairs. The structure maps have been used in an ICP-based SLAM approach to incrementally build two-dimensional point maps modeling the environmental structures of the robot's workspace and to localize the robot with three degrees of freedom. Furthermore, we segmented the continuously acquired 3D data stream into locally consistent 3D point clouds and used this information in the same SLAM approach to build three-dimensional point maps and to localize the robot with six degrees of freedom.
Current research and future work will focus on the application of recent Time-of-Flight cameras. Several extensions have been described that allow the application of these cameras in mapping tasks as well as for reactive collision avoidance by explicitly handling their narrow field-of-view as well as inaccuracies and erroneous measurements.

## Acknowledgements

This work has been partially funded by the German Research Foundation (DFG) under contract number BE 2556/2-2 and the European Commission under contract numbers FP6-004381-MACS and FP7-247870-NIFTI. Furthermore, we would like to thank our current and former colleagues Christopher Lörken, Andreas Nüchter, Kai Lingemann, Joachim Hertzberg, Kai Pervölz, Erich Rome, Jörg Stückler, Michael Schreiber, Matthias Nieuwenhuisen and Stefach Fuchs.

## 9. References

Allen, P., Stamos, I., Gueorguiev, A., Gold, E. & Blaer, P. (2001). AVENUE: Automated Site Modeling in Urban Environments, *Proceedings of the International Conference on 3D Digital Imaging and Modeling*, Quebec, Canada, pp. 357–364.

Arya, S., Mount, D. M., Netanyahu, N. S., Silverman, R. & Wu, A. Y. (1998). An optimal algorithm for approximate nearest neighbor searching fixed dimensions, *Journal of the ACM (JACM)* **45**(6): 891–923.

Besl, P. J. & McKay, N. D. (1992). A Method for Registration of 3-D Shapes, *IEEE Transactions on Pattern Analysis and Machine Intelligence* **14**(2): 239–256.

Chekhlov, D., Pupilli, M., Mayol-Cuevas, W. & Calway, A. (2006). Real-Time and Robust Monocular SLAM Using Predictive Multi-resolution Descriptors, *Proceedings of the 2nd International Symposium on Visual Computing*, Lake Tahoe, Nevada, USA, pp. 276–285.

Cole, D. & Newman, P. (2006). Using Laser Range Data for 3D SLAM in Outdoor Environments, *Proceedings of the IEEE International Conference on Robotics and Automation (ICRA)*, Orlando, Florida, USA, pp. 1556–1563.

Fuchs, S. & Hirzinger, G. (2008). Extrinsic and Depth Calibration of ToF-Cameras, *Proceedings of the IEEE International Conference on Computer Vision and Pattern Recognition (CVPR)*, Anchorage, Alaska.

Grisetti, G., Rizzini, D. L., Stachniss, C., Olson, E. & Burgard, W. (2008). Online Constraint Network Optimization for Efficient Maximum Likelihood Map Learning, *Proceedings of the IEEE International Conference on Robotics and Automation (ICRA)*, Pasadena, California, USA, pp. 1880–1885.

Grisetti, G., Stachniss, C. & Burgard, W. (2007). Improved Techniques for Grid Mapping with Rao-Blackwellized Particle Filters, *IEEE Transactions on Robotics* **23**(1): 34–46.

Haegele, M., Neugebauer, J. & Schraft, R. (2001). From Robots to Robot Assistants, *Proceedings of the 32nd International Symposium on Robotics (ISR)*, Seoul, South Korea, pp. 404–409.

Hähnel, D., Schulz, D. & Burgard, W. (2002). Map building with mobile robots in populated environments, *Proceedings of the IEEE/RSJ International Conference on Intelligent Robots and System (IROS)*, Lausanne, Switzerland, pp. 496–501.

Holz, D., Lörken, C. & Surmann, H. (2008). Continuous 3D Sensing for Navigation and SLAM in Cluttered and Dynamic Environments, *Proceedings of the International Conference on Information Fusion (FUSION)*, Cologne, Germany, pp. 1469–1475.

Huber, D., Carmichael, O. & Hebert, M. (2000). 3-D Map Reconstruction from Range Data, *Proceedings of the IEEE International Conference on Robotics and Automation (ICRA)*, San Francisco, California, USA, pp. 891–897.

Lange, R. (2000). *3D time-of-flight distance measurement with custom solid-state image sensors in CMOS/CCD-technology*, Phd thesis, University Siegen.

Leonard, J. & Feder, H. (1999). A computationally efficient method for large-scale concurrent mapping and localization, *in* D. K. J. Hollerbach (ed.), *International Symposium on Robotics Research*, Snowbird, Utah, USA.

Lingemann, K., Nüchter, A., Hertzberg, J. & Surmann, H. (2005a). About the Control of High Speed Mobile Indoor Robots, *Proceedings of the Second European Conference on Mobile Robots (ECMR)*, Ancona, Italy, pp. 218–223.

Lingemann, K., Nüchter, A., Hertzberg, J. & Surmann, H. (2005b). High-Speed Laser Localization for Mobile Robots, *Robotics and Autonomous Systems* **51**(4): 275–296.

Lorusso, A., Eggert, D. W. & Fisher, R. B. (1995). A Comparison of Four Algorithms for estimating 3-D Rigid Transformations, *Proceedings of the British conference on Machine vision BMVC*, pp. 237–246.

Luck, J., Little, C. & Hoff, W. (2000). Registration of Range Data Using a Hybrid Simulated Annealing and Iterative Clostest Point Algorithm, *Proceedings of the IEEE International Conference on Robotics and Automation (ICRA)*, San Francisco, California, USA, pp. 3739–3744.

Maurelli, F., Droeschel, D., Wisspeintner, T., May, S. & Surmann, H. (2009). A 3d laser scanner system for autonomous vehicle navigation, *Proceedings of the 14th International Conference on Advanced Robotics (ICAR)*, Munich, Germany.

May, S., Droeschel, D., Holz, D., Fuchs, S. & Nüchter, A. (2009). Robust 3D-Mapping with Time-of-Flight Cameras, *Proceedings of the IEEE/RSJ International Conference on Intelligent Robots and Systems (IROS)*, St. Louis, Missouri, USA. To appear.

Nüchter, A., Lingemann, K., Hertzberg, J. & Surmann, H. (2007). 6D SLAM – 3D Mapping Outdoor Environments, *Journal of Field Robotics (JFR), Special Issue on Quantitative Performance Evaluation of Robotic and Intelligent Systems* **24**(8–9): 699–722.

Nüchter, A., Lingemann, K., Hertzberg, J., Wulf, O., Wagner, B. & Surmann, H. (2005). 3D Mapping with Semantic Knowledge, *Proceedings of the RoboCup International Symposium 2005: Robot Soccer World Cup IX*, Osaka, Japan, pp. 335–346.

Olson, E., Leonard, J. & Teller, S. (2006). Fast Iterative Alignment of Pose Graphs with Poor Estimates, *Proceedings of the IEEE International Conference on Robotics and Automation (ICRA)*, Orlando, Florida, USA, pp. 2262–2269.

Pollack, M. E., Engberg, S., Matthews, J. T., Thrun, S., Brown, L., Colbry, D., Orosz, C., Peintner, B., Ramakrishnan, S., Dunbar-Jacob, J., McCarthy, C., M. Montemerlo, J. P. & Roy, N. (2002). Pearl: A Mobile Robotic Assistant for the Elderly, *Proceedings of the AAAI Workshop "Automation as Caregiver: the Role of Intelligent Technology in Elder Care"*, Edmonton, Canada, pp. 85–92.

Rojo, J., Rojas, R., Gunnarsson, K., Simon, M., Wiesel, F., Ruff, F., Wolter, L., Zilly, F., Santrač, N., Ganjineh, T., Sarkohi, A., Ulbrich, F., Latotzky, D., Jankovic, B., Hohl, G., Wisspeintner, T., May, S., Pervölz, K., Nowak, W., Maurelli, F. & Dröschel, D. (2007). A 3d laser scanner system for autonomous vehicle navigation, *Team Descriptions Papers of the DARTPA Urban Challenge 2007*. Available online: `http://www.darpa.mil/GRANDCHALLENGE/TechPapers/Team_Berlin.pdf`.

Sequeira, V., Ng, K. C., Wolfart, E., Goncalves, J. G. & Hogg, D. C. (1998). Automated 3D reconstruction of interiors with multiple scan views, *Society of Photo-Optical Instrumentation Engineers (SPIE) Conference Series*, Vol. 3641, pp. 106–117.

Sheh, R., Kadous, M. W. & Sammut, C. (2006). On building 3D maps using a Range Camera: Applications to Rescue Robotics, *Techreport*, ARC Centre of Excellence for Autonomous Systems, School of Computer Science and Engineering, University of New South Wales, Sydney, Australia.

Siegwart, R., Arras, K. O., Bouabdallah, S., Burnier, D., Froidevaux, G., Greppin, X., Jensen, B., Lorotte, A., Mayor, L., Meisser, M., Philippsen, R., Piguet, R., Ramel, G., Terrien, G. & Tomatis, N. (2003). Robox at Expo.02: A large-scale installation of personal robots, *Robotics and Autonomous Systems* **42**(3-4): 203–222.

Stiene, S. & Hertzberg, J. (2009). Virtual Range Scan for Avoiding 3D Obstacles Using 2D Tools, *Proceedings of the International Conference on Advanced Robotics (ICAR)*, Munich, Germany.

Strand, M. & Dillmann, R. (2008). Using an attributed 2D-grid for next-best-view planning on 3D environment data for an autonomous robot, *Proceedings of the IEEE International Conference on Information and Automation (ICIA)*, Zhangjiajie, Hunan, China, pp. 314–319.

Surmann, H., Nüchter, A. & Hertzberg, J. (2003). An autonomous mobile robot with a 3D laser range finder for 3D exploration and digitalization of indoor environments, *Journal Robotics and Autonomous Systems (JRAS)* **45**(3-4): 181–198.

Surmann, H. & Peters, L. (2001). *Fuzzy Logic Techniques for Autonomous Vehicle Navigation*, Physica-Verlag, chapter MORIA - A Robot with Fuzzy Controlled Behaviour, pp. 343–365.

Swadzba, A., Liu, B., Penne, J., Jesorsky, O. & Kompe, R. (2007). A Comprehensive System for 3D Modeling from Range Images Acquired from a 3D ToF Sensor, *Proceedings of the International Conference on Computer Vision Systems (ICVS)*, Bielefeld, Germany.

Thrun, S., Burgard, W. & Fox, D. (2000). A real-time algorithm for mobile robot mapping with applications to multi-robot and 3D mapping, *Proceedings of the IEEE International Conference on Robotics and Automation (ICRA)*, San Francisco, California, USA, pp. 321–328.

Thrun, S., Liu, Y., Koller, D., Ng, A., Ghahramani, Z. & Durrant-Whyte, H. (2004). Simultaneous Localization and Mapping with Sparse Extended Information Filters, *International Journal of Robotics Research* **23**(7-8): 693–716.

Thrun, S., Montemerlo, M., Dahlkamp, H., Stavens, D., Aron, A., Diebel, J., Fong, P., Gale, J., Halpenny, M., Hoffmann, G., Lau, K., Oakley, C., Palatucci, M., Pratt, V., Stang, P.,

Strohband, S., Dupont, C., Jendrossek, L.-E., Koelen, C., Markey, C., Rummel, C., van Niekerk, J., Jensen, E., Alessandrini, P., Bradski, G., Davies, B., Ettinger, S., Kaehler, A., Nefian, A. & Mahoney, P. (2006). Stanley: The Robot that Won the DARPA Grand Challenge, *Journal of Field Robotics* **23**(9): 661–692.

Triebel, R., Pfaff, P. & Burgard, W. (2006). Multi-Level Surface Maps for Outdoor Terrain Mapping and Loop Closing, *Proceedings of the IEEE/RSJ International Conference on Intelligent Robots and Systems (IROS)*, Beijing, China, pp. 2276–2282.

Urmson, C., Anhalt, J., Bae, H., Bagnell, J. A., Baker, C., Bittner, R. E., Brown, T., Clark, M. N., Darms, M., Demitrish, D., Dolan, J., Duggins, D., Ferguson, D., Galatali, T., Geyer, C. M., Gittleman, M., Harbaugh, S., Hebert, M., Howard, T., Kolski, S., Likhachev, M., Litkouhi, B., Kelly, A., McNaughton, M., Miller, N., Nickolaou, J., Peterson, K., Pilnick, B., Rajkumar, R., Rybski, P., Sadekar, V., Salesky, B., Seo, Y.-W., Singh, S., Snider, J. M., Struble, J. C., Stentz, A., Taylor, M., Whittaker, W. L., Wolkowicki, Z., Zhang, W., & Ziglar, J. (2008). Autonomous driving in urban environments: Boss and the urban challenge, *Journal of Field Robotics Special Issue on the 2007 DARPA Urban Challenge, Part I* **25**(1): 425–466.

Wang, C.-C. (2004). Simultaneous localization, mapping and moving object tracking, *PhD Thesis CMU-RI-TR-04-23*, Robotics Institute, Carnegie Mellon University, Pittsburgh, Pennsylvania, USA.

Weingarten, J., Gruener, G. & Siegwart, R. (2004). A state-of-the-art 3D sensor for robot navigation, *Proceedings of the IEEE/RSJ International Conference on Intelligent Robots and Systems (IROS)*, Sendai, Japan, pp. 2155–2160.

Wulf, O., Arras, K. O., Christensen, H. I. & Wagner, B. (2004). 2D Mapping of Cluttered Indoor Environments by Means of 3D Perception, *Proceedings of the IEEE/RAS International Conference on Robotics and Automation (ICRA)*, New Orleans, Louisiana, USA, pp. 4204– 4209.

Wulf, O., Lecking, D. & Wagner, B. (2006). Robust self-localization in industrial environments based on 3d ceiling structures, *Proceedings of the IEEE/RSJ International Conference on Intelligent Robots and Systems (IROS)*, Beijing, China, pp. 1530–1534.

Wulf, O. & Wagner, B. (2003). Fast 3D-Scanning Methods for Laser Measurement Systems, *Proceedings of the International Conference on Control Systems and Computer Science (CSCS14)*, Bucharest, Romania.

Yuan, F., Swadzba, A., Philippsen, R., Engin, O., Hanheide, M. & Wachsmuth, S. (2009). Laser-based Navigation Enhanced with 3D Time-of-Flight Data, *Proceedings of the IEEE International Conference on Robotics and Automation (ICRA)*, Kobe, Japan, pp. 2844–2850.

Zhao, H. & Shibasaki, R. (2001). Reconstructing textured cad model of urban environment using vehicle-borne laser range scanners and line cameras, *Proceedings of the Second International Workshop on Computer Vision Systems (ICVS)*, London, UK, pp. 284–297.

Zinßer, T., Schmidt, J. & Niemann, H. (2003). A Refined ICP Algorithm for Robust 3-D Correspondence Estimation, *Proceedings of the International Conference on Image Processing (ICIP)*, Barcelona, Spain, pp. 695–698.

Zivkovic, Z., Booij, O. & Kröse, B. (2007). From images to rooms, *Robotics and Autonomous Systems* **55**(5): 411–418.

5

# Sensors Fusion Technique for Mobile Robot Navigation using Fuzzy Logic Control System

S.Parasuraman[1], Bijan Shirinzadeh[2] and V.Ganapathy[1]
[1]*Monash University Malaysia*
[2]*Monash University Australia*

## 1. Introduction

The mobile robot navigation with complex environment needs more input space to match the environmental data into robot outputs in order to perform a realistic task. At the same time, the number of rules at the rule base needs to be optimized to reduce the computing time and to provide the possibilities for real time operation. In this paper, the optimization of fuzzy rules using a Modified Fuzzy Associative Memory (MFAM) is designed and implemented. MFAM provides good flexibility to use multiple input space and reduction of rule base for robot navigation. This paper presents the MFAM model to generate the rule base for robot navigation. The behavior rules obtained from MFAM model are tested using simulation and real world experiments, and the results are discussed in the paper and compared with the existing methods.

In most of the researches in the area of fuzzy logic based mobile robot systems, use only few inputs to simplify the navigation process in order to test the hypothesis. In order to perform realistic tasks with a mobile robot in complex surroundings, the number of inputs should not be so limited. This raises some important questions to be explored. They are: How can the fuzzy control system be scaling up without throwing out useful data or decreasing the input space by non-fuzzy means? How can a reactive fuzzy system respond to a complex environment? The schemes used by Sugeno's car [M. Sugeno, Murofushi, T. Mori, T.Tatematasu, J. Tanaka (1989)] the extraction of specific state variables from sensor data before the fuzzy control stage. Thus, the sensor data have been matched to a world model. The resulting effect is to reduce the size of the input space to the rule set. The disadvantages in the robot described by Pin, et al. [F. G. Pin H. Watanabe, J.R. Symon, and R.S. Pattay (1992)] features 15 ultrasonic rangefinders, without matching data to a world model. However, the sonar data is bunched together into three inputs, each being the minimum of a neighborhood of five, before being sent to the controller. Another disadvantage of their system is the need for special computer hardware for real-time operation. Papers cited in [(M. Balzarotti and G. Ulivi(1996), A. Ollero, A. Garcia-Cerezo et al (1997), J. Pereira and J.B Bowles (1994), E. Tunstel, et al (1994) and C. Voudouris et.al,1995)] are mainly focused on some kind of environmental features like walls, road edges, white lines on the floor, avoid obstacles etc. In the research article cited in [P. Althauas et.al, 2001], sonar sensors are used to map the structured indoor environment to navigate the robot with simple behaviors

called obstacle avoiding and corridor following. The input space is restricted only for these behaviors and not to the entire mapping of the environments.
In the past, several works relating to FAM have been done for control applications. Kosko's FAM [B. Kosko (1992)] is one of the earliest attempts to integrate the fuzzy sets, and neural network is used to learn the mapping of the inputs to output. Kosko's FAM is restricted to limited rule based applications. The FAM models have been successfully applied to problems like backing up a truck-and-trailer [S.G Kong and B. Kosko, (1992)], target tracking [Hirohide Ushida (1999)] and voice-cell control in ATM networks [T. D. Ndousse (1994)], where distinctive features like modularity, robustness, and adaptability have been demonstrated. Despite of aforementioned feature, FAM does not provide acceptable solution when there are a large number of fuzzy inputs. In the proposed approach, an improved FAM is established and the number of rule bases is optimized without throwing away any useful inputs.

Based on the MFAM, the most influential navigation rules of the fuzzy system inputs are arranged in the first level. The outputs of the first level together with the next most important variables are arranged in the second level and so on. In each level, the fuzzy system output is modified according to the degree of importance of the corresponding input. Finally, the control rule is activated, when the measurement of the obstacle distance exactly matches the rule condition part ('if' part of the rule). The final decision made is based on the corresponding action part ('then' part of the rule) of the matched rule and not based on the aggregated output of the entire rules. Thus, the computational complexity to estimate the control output is minimized in each control cycle using the proposed MFAM. The rules obtained from FAM model are experimented using Active Media Pioneer Robot. The input and output data of the robot control system are observed and investigated and the results are provided in this paper.

## 2. Theoretical work

### 2.1 MFAM model

The fuzzy relation formed by the collection of rules is represented as the fuzzy set and denoted as R. In a fuzzy control system, the control rules in the collection R are first matched with the available data (context), then a matched rule is fired, thereby providing a control action. Usually the context would be the measured outputs of the process, and these are crisp quantities, and the control action that drives the process is also a crisp quantity. However, for general considerations, the context data are denoted by a fuzzy set D and the control action is denoted by a fuzzy set C. The compositional rule of inference states that [S. Parasuraman, V.Ganapathy, Bijan Shirinzadeh (2005)]:

$$C = D \circ R = \bigvee_{x \varepsilon X} \wedge \left\{ \mu_D(x),\ \mu_R(x_1, x_2, \ldots x_n) \right\} \tag{1}$$

Based on the equation (1), the fuzzy relation formed by the collection of rules is represented as the fuzzy set and denoted as R as given below.

$$R = \left\{ R_1, R_2\, R_3 \ldots\ldots R_i, \ldots\ldots\ldots R_k \right\} \tag{2}$$

where $R_k$ is the total number of rule inference and Ri is the ith rule of the FIS and is expressed in equation (2) as

$$\text{IF } X_1 \text{ is } A_{1i} \text{ and } X_2 \text{ is } A_{2i} \text{ and.......and } X_n \text{ is } A_{ni} \text{ Then Z is } C_i \tag{3}$$

where $X_1, X_2, \ldots X_n$ are the input variables (sensor data), $A_{1i}, A_{2i} \ldots A_{mn}$ are the input fuzzy linguistic variables, $C_i$ is the output linguistic variable of the model, Z is the output, n is the dimension of the input vector and m is the fuzzy set. The following fuzzy relation is used to implement the ith rule Ri.

$$R_i(X_1, X_2, \ldots, X_n, Z) = \left[A_{1i}(X_1) \wedge A_{2i}(X_2) \wedge \ldots \wedge A_{mn}(X_n)\right] \rightarrow C_i(Z) \tag{4}$$

By applying compositional rule of inference as defined in the equation (1), an n dimensional fuzzy input vector $\overline{X}$, with $(\overline{X}_{10}, \overline{X}_{20}, \ldots \overline{X}_{i0}, \ldots \overline{X}_{n0})$, is generated. During the process of compositional rule of inference the context inputs need to compose the input vector $\overline{X}$ with the fuzzy relation Ri to produce the output $C_i'$ and is given by

$$C_i' = \left(\overline{X}_{10}, \overline{X}_{20}, \ldots\ldots\ldots \overline{X_{i0,}} \ldots\ldots \overline{X}_{mn}\right) \circ R_i \tag{5}$$

where $\overline{X}_{i0}$ is the fuzzified crisp value of $X_{i0}$ and $C_i'$ is the defuzzified output of the ith rule and defined as follows.

$$C_1'(Z) = \left[A_{1i}(X_{10}) \wedge A_{2i}(X_{20}) \wedge \ldots\ldots \wedge A_{mi}(X_{m0})\right] \rightarrow C_i(Z) \tag{6}$$

The overall system output is obtained by using min and max aggregation operators based on the equation (6) as follows:

$$C = \bigcup_{i=1}^{k} C_1' = \bigcup_{i=1}^{k} \left(\left[A_{1i}(X_{10}) \wedge A_{2i}(X_{20}) \wedge \ldots\ldots \wedge A_{mi}(X_{m0})\right] \rightarrow C_i(Z)\right) \tag{7}$$

where k is the number of rules in the system. In order to relate the mth fuzzy set of the ith fuzzy rule, the fuzzy implication model by Mamdani's min operator is used to interpret the logical rules. [C. T. Lin and C.S.G.Lee (1991), J. S. R Jang (1993), H. Ushida, T.Yamaguchi, and T. Takagi (1995), F. G. Pin and Y.Watanabe (1993)] Combining the equations (4) and (7), it can be rewritten as follows:

$$\mu_{R_i}(X_{1i,} X_{2i}, \ldots\ldots X_{mi}) = \min_{i=1}^{k}[\mu_{Amn}(X_{mn}) \tag{8}$$

The final output membership function is

$$\mu_{C'}(Z) \text{ is } y_c(z) = \max_{i=1}^{k}[\min[\mu_{Amn}(x_{mn}), \mu_{R_i}(X_{1i,} X_{2i}, \ldots\ldots X_{mi}]] \tag{9}$$

In equation (9), $\mu_{R_i}(X_{1i}, X_{2i}, \ldots\ldots X_{mi})$ is the membership function of ith rule and the value of $i = 1$ to k, and k is the total number of rules. The total number of rules in the rule base depends on the input variables $(X_{1i}, X_{2i}, \ldots\ldots X_{mi}, \ldots X_{mn})$. Here the input variables are the linguistic fuzzy sets of sensor values.

### 2.2 Modified Fuzzy Associative Memory

If the number of fuzzy inputs and linguistic variables of each fuzzy set increases, the number of fuzzy rules grow exponentially. As an example, a model with Active Media

Pioneer Robot uses eight input and two output fuzzy sets. If each input fuzzy set is represented by three fuzzy linguistic variables and each output fuzzy set is represented by seven fuzzy linguistic variables, then a single layer of inference will lead to determining $3^8$ = 6,561 rules that would be difficult to evaluate, time consuming and making real time operation difficult. In order to reduce the number of rules, without reducing any of the fuzzy inputs, the new methodology is proposed, which uses compositional rule of inference as described by equations (1) and (9). The following section describes the proposed methodology MFAM to integrate the multiple sensors. The FIS is generated based on the MFAM rules. MFAM is a process of encoding and mapping the input fuzzy sets to output fuzzy sets. The relationships between the fuzzy sets and rules are shown as a Modified Fuzzy Associative Memory. A fuzzy-logic rule as per the equation (10) is given below.

$$R1: (Ai, Bi) \rightarrow Ci \tag{10}$$

This is called a "fuzzy association." A Fuzzy Associative Memory (FAM) is formed by partitioning the universe of discourse of each condition variable (i.e., Ai and Bi in the above example) according to the level of fuzzy resolution chosen for these antecedents, thereby generating a grid of FAM elements [Bo Hyeun Wang, George Vachtsevanos, 1990)]. The entry at each grid element of the FAM corresponds to the fuzzy action (Ci in the above example). The equation (10) is good enough to get the FAM rules if only two input fuzzy sets and one output fuzzy set are present in the process. If the number of input fuzzy sets and linguistic variables of each fuzzy set is more than two then the proposed Modified Fuzzy Associative Memory (MFAM) is suitable to optimize the number of rules and integrate the entire input variables to the process effectively. The proposed MFAM is defined based on the following FAM rule reduction theorem.

### 2.3 MFAM rule reduction theorem

MFAM rule reduction theorem is defined as follows: Let a fuzzy implicative rule for a two input system is defined of the form, "if $X_{1i}$ and $X_{2i}$ then $Z_i$", then the fuzzy implicative rules for more than two input system is defined using compositional rule of inference and is given below.

$$(X_{1i+1}, X_{2i+1}) \circ (X_{1i} \text{ and } X_{2i} \rightarrow Z_i) = Z_i'$$

Based on the law of compositional rule of inference and associativity [19] the output Zi' is expressed as follows:

$$Z_{i'} = [(X_{1i+1}) \circ (X_{1i} \rightarrow Z_i)] \wedge [(X_{1i+1}) \circ (X_{1i} \text{ and } X_{2i} \rightarrow Z_i)] \wedge [(X_{2i+1}) \circ (X_{2i} \rightarrow Z_i)] \wedge [(X_{2i+1}) \circ (X_{1i} \text{ and } X_{2i} \rightarrow Z_i)] \tag{11}$$

where $X_{1i}$, $X_{2i}$, $X_{1i+1}$, $X_{2i+1}$ are the fuzzy inputs ( i = 0, 1,2, …,n) with fuzzy sets m (m=1,2,3). The final control output is obtained by combining equations (11) and (9) and is as given below.

$$\mu_C(z) = \max_{\substack{i=1 \\ x \in U}}^{k} [[(X_{1i+1}) \circ (X_{1i} \rightarrow Z_i)] \wedge [(X_{1i+1}) \circ (X_{1i} \text{ and } X_{2i} \rightarrow Z_i)] \wedge [(X_{2i+1}) \circ (X_{2i} \rightarrow Z_i)] \wedge [(X_{2i+1}) \circ (X_{1i} \text{ and } X_{2i} \rightarrow Z_i)]] \tag{12}$$

Applying min and max compositional inferences, the equation (12) is written as

$$\mu_C(z) = \max_{\substack{i=1 \\ x \in U}}^{k} [\min [(X_{1i+1}) \circ (X_{1i} \rightarrow Z_i)], [(X_{1i+1}) \circ (X_{1i} \text{ and } X_{2i} \rightarrow Z_i)], [(X_{2i+1}) \circ (X_{2i} \rightarrow Z_i)], [(X_{2i+1}) \circ (X_{1i} \text{ and } X_{2i} \rightarrow Z_i)]] \quad (13)$$

The MFAM matrix is established based on the above output membership function as stated in the equation (13).

### 2.4 Establishment of FIS using MFAM Rule

The Establishment of MFAM matrix tables involves the following assumptions:

- The variables ($X_{11}$, $X_{21}$, $X_{31}$), ($X_{12}$, $X_{22}$, $X_{32}$)...($X_{16}$, $X_{26}$, $X_{36}$) are the crisp values of the distance measurement obtained from the sensors (S1 to S6) of the mobile robot.
- Similarly, the variables ($Z_1$, $Z_2$, $Z_3$ and $Z_4$) are the crisp values of turn angles obtained after defuzzification, which are used by the mobile robot.
- The Mamdani-Style inference [C. W. de Silva, (1995)] system is used to perform the fuzzification, rule evaluation and defuzzification process.
- During the rule evaluation, the composition inference is performed to obtain the behavior rules in such a way that the rule combinations will have a tendency to select the direction of the robot that is closest to the front direction, so that the robot does not make unnecessary rotations.
- Rule combinations are obtained based on the human expertise and experience
- The rule evaluation is performed in two steps. (a) when there are two fuzzy variables; the equation (10) is perfectly applied and (b) when there are more than two inputs, then the compositional rule inference equation (13) is applied.

During the design of behaviour rules, the MFAM uses the fuzzy sets and their corresponding linguistic variables as indices to a lookup table. The MFAM rule structure has been proposed using the following steps:

- The most influential navigation rules of the fuzzy system inputs are from the front sensors, which are located in the front direction of the robot. These rules are arranged in the first level based on the Equation (10) and referred to as Preferred Front sector rules (PF).
- The outputs of the first level together with the next most important variables are arranged in the second level and the navigation rules are formulated. The composition of the rules are based on the MFAM theorem and the rules are referred to as Preferred Right (PR) and Preferred Left (PL) and
- In each level, the fuzzy system output is modified according to the degree of importance of the corresponding inputs. Finally, as per the compositional rule of inference as defined by the Equation (13), MFAM has applied the steps (a) and (b), and the resultant rules are tabulated as shown in the Table 1.

The FIS uses the MFAM rule matrix, which involves the processes of fuzzification, rule evaluation and defuzzification procedures.

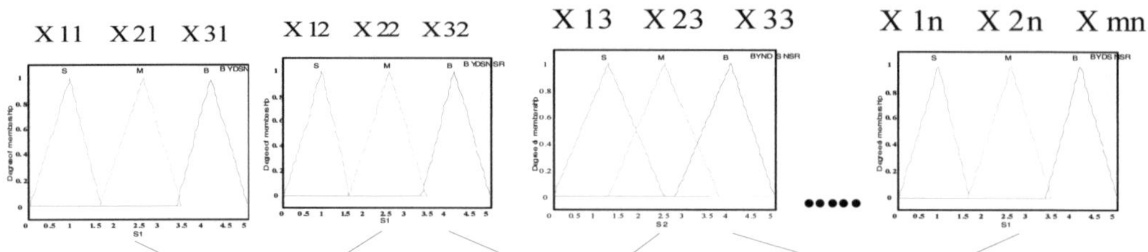


Level: 1

| n,m | X 12 | X 22 | X 32 |
|---|---|---|---|
| X 11 | Z 4 | Z 4 | Z 4 |
| X 12 | Z 4 | Z 4 | Z 4 |
| X 13 | Z 4 | Z 4 | Z 4 |

| n,m | X 13 | X 23 | X 33 |
|---|---|---|---|
| X 12 | Z 4 | Z 4 | Z |
| X 22 | Z 4 | Z | Z |
| X 32 | Z | Z | Z |

| n,m | X 14 | X 24 | X 34 |
|---|---|---|---|
| X 13 | Z 2 | Z 5 | Z 4 |
| X 23 | Z 2 | Z | Z 4 |
| X 33 | Z1 | Z1 | Z |

Level: 2

| n,m | X 15 | | X 25 | X 35 |
|---|---|---|---|---|
| X 11 X 12 | Z 1 | | Z 1 | Z |
| X11 X22 | Z 1 | | Z | Z |
| | | | | |
| X11 X32 | Z | | Z | Z |

| n,m | X 15 | | X 25 | X 35 |
|---|---|---|---|---|
| X 11 X 12 | Z 1 | | Z 1 | Z |
| | | | | |
| X11 X22 | Z 1 | | Z | Z |
| | | | | |
| X11 X32 | Z | | Z | Z |

Level: 3

| n,m | | X 16 | X 26 | | X 36 |
|---|---|---|---|---|---|
| | Z2 | | | Z4 | |
| X 15 | | Z 1 | Z 1 | | Z 1 |
| X 25 | | Z 1 | Z 1 | | Z 1 |
| | .. | | | | |
| X 36 | .. | Z 1 | Z 1 | | Z 1 |

Table 1. MFAM Matrix Table

## 3. Experimental works

The proposed MFAM methodology for mobile robot navigation is validated using the simulation study. The simulation environment consists of simulated obstacles such as walls, blocks, lines, circles, and partitions that are drawn in an unstructured manner as shown in Figure 1. In the simulation environment, sensor locations of the robot are defined circumferentially and the sensors S3, S4, S5, S6, S7, S8, S9, and S10 are set as the front direction sensors. The input and output parameter settings and their corresponding linguistic terms are defined based on the real world data of the Active Media Pioneer Robot considered for this simulation study. These parameters are then encoded using the proposed MFAM based FIS

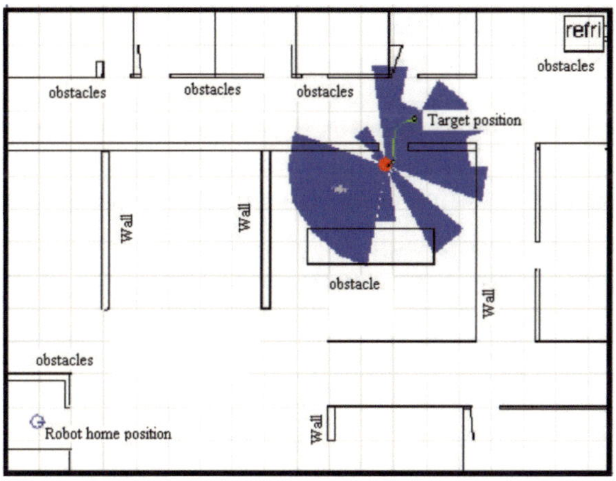


Fig. 1. Simulation environment for robot navigation.

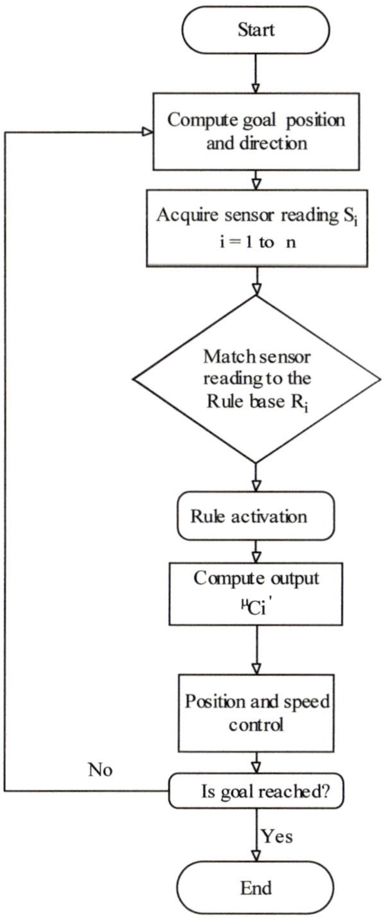


Fig. 2. Implementation of MFAM for obstacle avoidance behavior while the robot reaching the goal position.

A procedure is established for robot navigation and it is shown using the flowchart given in Figure 2. The obstacle positions in the environment are obtained using sensors S1 to S10. The final defuzzified output is fed to the controller to navigate the robot towards the target position. Figures 3 (a to h) show the various situations with multiple obstacles that a robot encounters during navigation. Based on these observations, the effectiveness of the use of multiple inputs and their roles during behaviour rule design are discussed below.

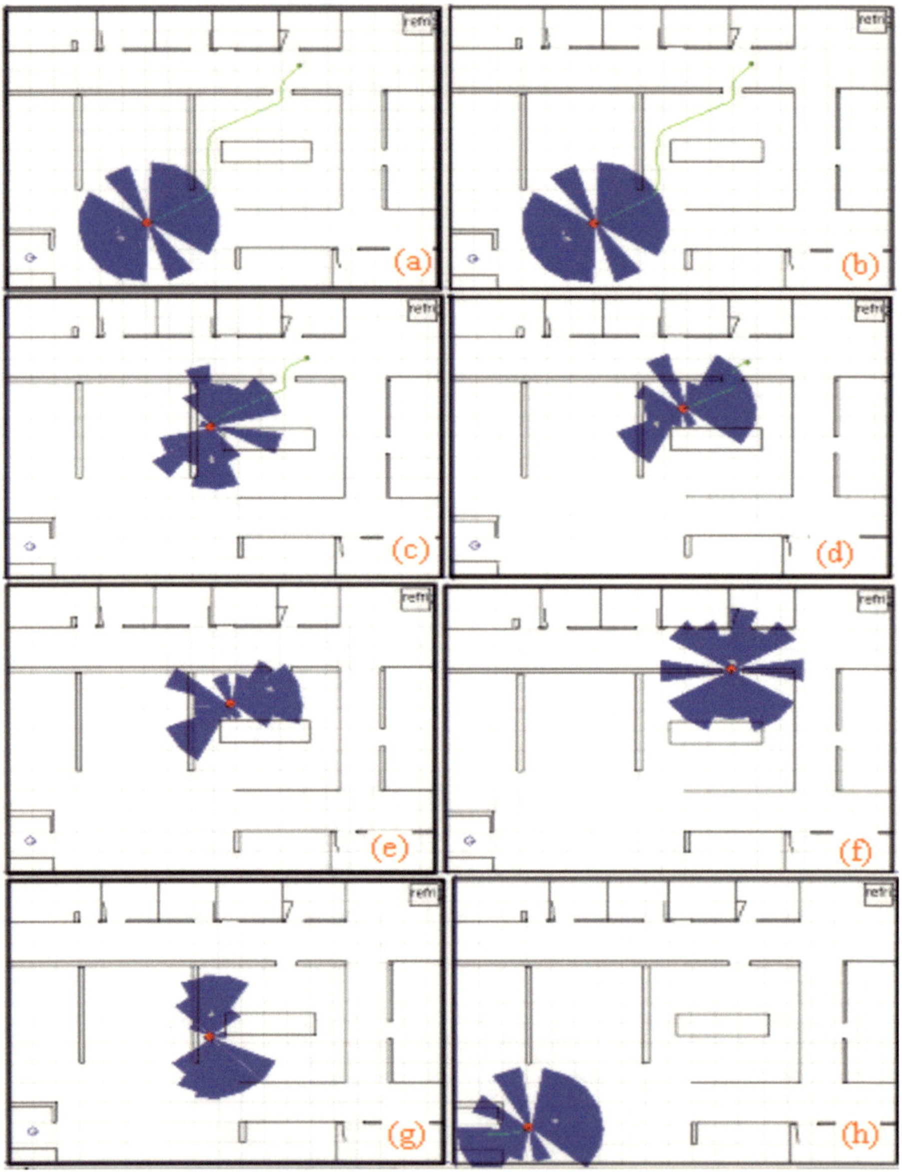


Fig. 3. (a)-(h) Simulation environment with obstacle avoidance behavior while reaching to the goal position.

The results of the simulation consist of inputs (sensor readings from sensor 0 to 14) and outputs (turn angle and speed), which are further analyzed to measure the effectiveness of MFAM. The example of the turn rule activation at a particular situation is illustrated in Figure 4. The surface view illustrates the turn rule activation based on the sensors S8 and S9.

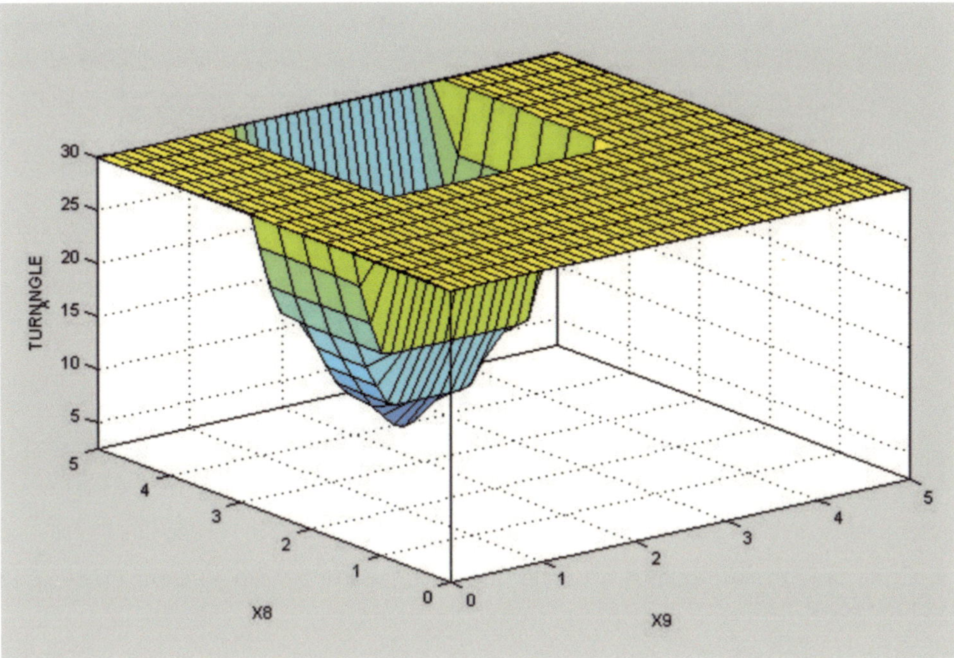


Fig. 4. Example of the rule activation, based on the sensor values received by sensors S8 (X8) and S9 (X9).

### 3.1 Results and Discussions of the Simulation Studies

The MFAM rules are applied and simulations are carried out for various starting and end positions with different target locations. The simulation results are recorded in all cases, when the robot encounters obstacles while moving from the initial position to the goal position. Based on these observations, the effectiveness of the use of multiple inputs and their roles during behaviour rule design are discussed below.

Figure 5 shows the sensors' responses of the environment in terms of the robot controller cycle time plotted against obstacle distances measured by sensors S3 to S10 and the steering angle in degrees. This plot illustrates various obstacle distances from the robot, which are perceived by the sensors S3 to S10 while the robot is in navigation. These multiple input data are then fuzzified using the compositional rule of inference as described in the proposed MFAM. In order to describe the role of each sensor, some critical environment in the plot shown in Figure 5 is enlarged and obstacle distances perceived by various sensors are shown in Figure 6 between the cycle time 200 x 20 ms to 300 x 20 ms. This plot clearly shows the multiple obstacle distances as perceived by sensors S4, S5, S6, S7 S9 and S8 and the robot steering angle during 240 x 20 ms to 295 x 20 ms time period. This implies that the environment consists of obstacles, which are detected in the respective sensor directions. In these situations the robot is required to turn and avoid obstacles. The corresponding turn angle is computed using the proposed MFAM based FIS and the computed value is then applied to the control system.

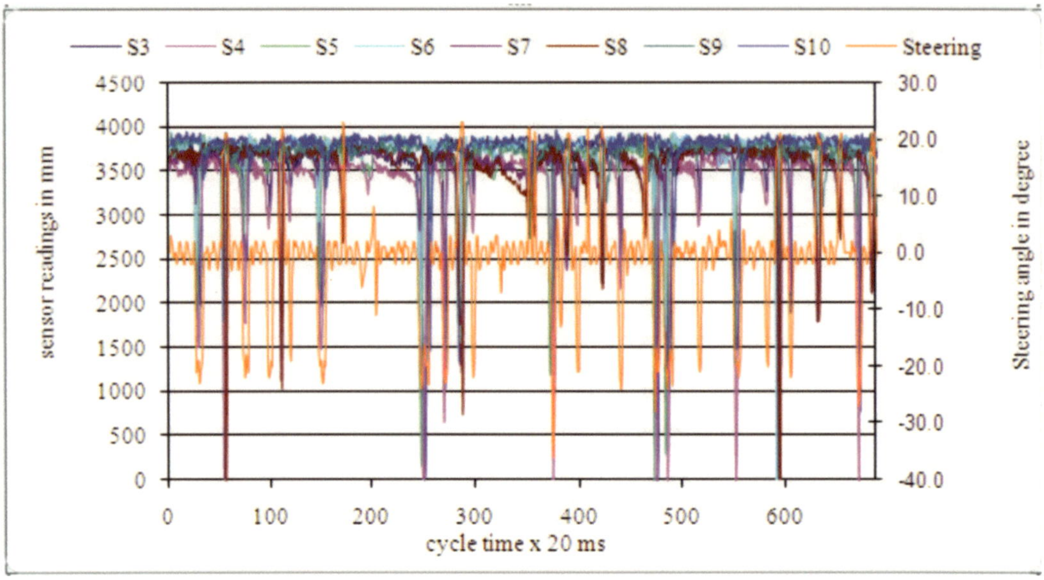


Fig. 5. Cycle time drawn against obstacle distances (sensor readings) from robot and turn angle of the robot.

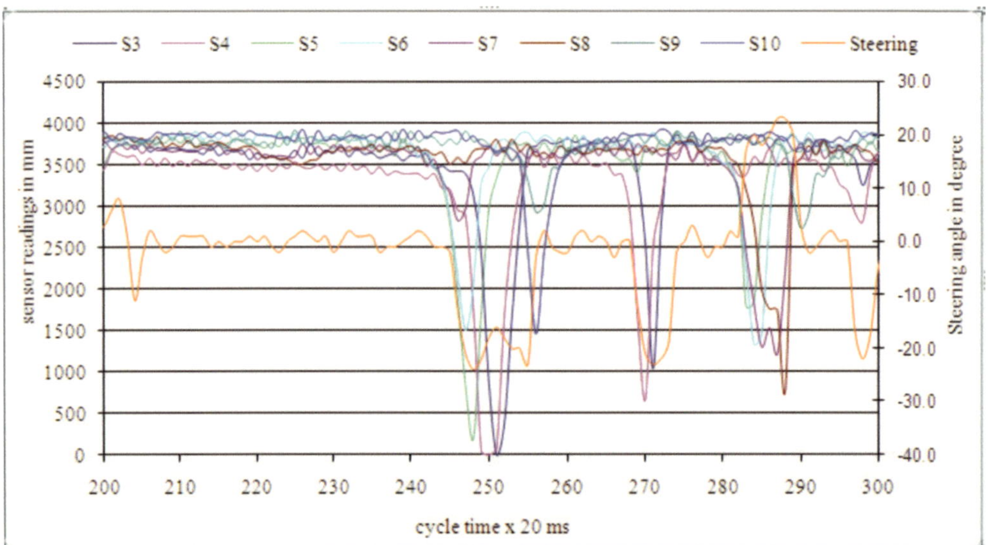


Fig. 6. Enlarged region of Figure 5 between the cycle time 200 x 20 to 300 x 20 ms plotted against obstacle distances from robot and the steering angle of the robot

<table>
<tr><td rowspan="2">Sensor</td><td colspan="2">Obstacle details obtained from sensors</td><td colspan="3">Robot output</td></tr>
<tr><td>Obstacle distances in mm</td><td>Direction of obstacle location from robot in degrees</td><td>Right wheel</td><td>Left wheel</td><td>Steering angle</td></tr>
<tr><td>S3</td><td>28</td><td>67</td><td rowspan="5">-18 deg</td><td rowspan="5">7 deg</td><td rowspan="5">-25 deg</td></tr>
<tr><td>S4</td><td>40</td><td>45</td></tr>
<tr><td>S5</td><td>108</td><td>22</td></tr>
<tr><td>S6</td><td>1500</td><td>0</td></tr>
<tr><td>S7</td><td>2800</td><td>-45</td></tr>
</table>

Table 2. Results provide the obstacle distances as perceived by multiple sensors and the corresponding turn angle between the cycle time 240 x 20 ms and 260x 20 ms.

Table 2 provides the number of sensors involved to perceive the obstacle environment in a particular situation and the corresponding steering angle of the robot. The tabulated data are obtained from the plot shown in Figure 4.12 between the cycle times 240 x 20 ms to 260 x 20 ms for the selected critical situations. In these situations the multiple sensor fusion using the proposed MFAM and the corresponding rule activation is illustrated as follows:
"IF X13 and X14 and X15 and X26 then Z4".

where X13 is Sensor 3 and fuzzy distance measure is small (1) fuzzy set,

X14 is Sensor 4 and the fuzzy distance measure is small (1) fuzzy set,

X15 is Sensor 5 and the fuzzy distance measure is small (1) fuzzy set,

X26 is Sensor 6 and the fuzzy distance measure is medium (2) fuzzy set and

Z4 is Robot output and the fuzzy measure is Small Positive (MP)

The illustrated situation is shown in Figure 3 (e). From the simulation studies and the observed results, it has been found that the behaviour rules obtained from linguistic variables of multiple sensors can be effectively integrated using the proposed MFAM and used for robot navigation.

## 4. Experimental Studies and Investigations of MFAM using Active Media Pioneer Robot

The behavior rules obtained using the proposed MFAM methodology is demonstrated and investigated using Active Media Pioneer Robot. In the previous section, simulation study uses the front sensor data S3 to S10 to guide the robot for navigation. Similarly, in real world experiment, the same sensor configurations are used and these sensors are sensors S0, S1, S2, S3, S4, S5, S6, and S7. The experimental environment consists of real obstacles such as walls, doors, and obstacles that are situated in an unstructured manner as shown in Figures 8. The experimental studies are performed using various combinations among sensors (S1, S2, S3, S4, S5 and S6) without changing the environment and with the same starting and goal positions. In order to illustrate the importance and significance of multiple sensor fusion using the proposed MFAM for mobile robot navigation, experiments are carried out with the selected combination of sensors as S3-S4, S2-S3-S4-S5, and S1-S2-S3-S4-S5-S6. These combinations are chosen randomly from the robot as shown in Figure 7.

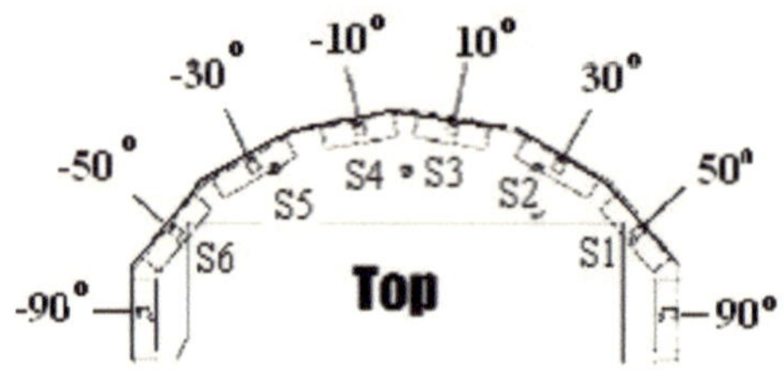


Fig. 7. Location of sensors in Active Media Pioneer Robot.

In each of the experiments, the obstacle distances perceived by each sensor and the direction of robot movements (output activation values) are obtained. These results are investigated using the graph showing the control cycle time plotted against various sensors perception data and the robot steering angles.

### 4.1 Experimental Studies using Sensors S3 and S4

The most influential navigation rules of the robot navigation system comes from the front sensors S3 and S4, which are located in the front direction of the robot. In this experiment, only front sensors S3 and S4 are taken into account to perceive the environment and build the navigation rules. These navigation rules are applied to the robot control system and the experiments are performed. The experimental environments are shown in Figures 8 (a) - (d) while the robot is in navigation.

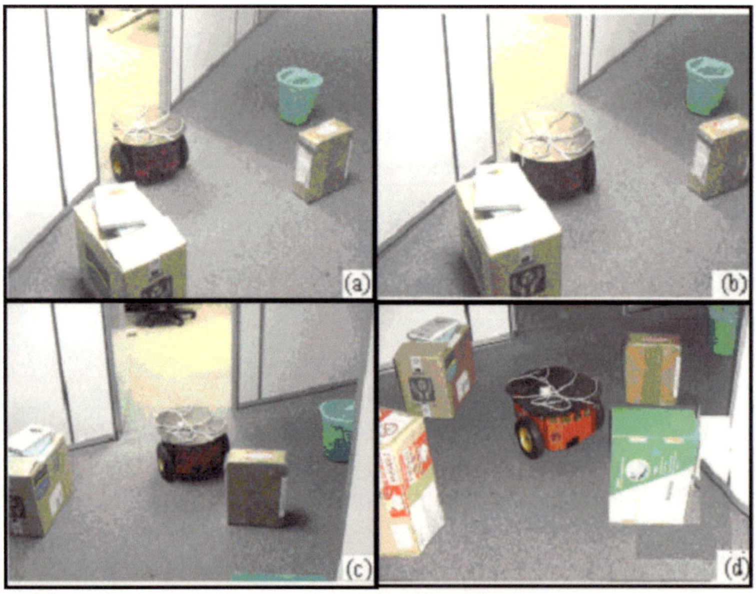


Fig. 8. Real world experiments using only sensor S3 and S4.

Figure 8 (a) is the initial position of the robot. As the robot is starting to navigate (Figure 8 (b)), sensor S3 detects obstacle, and based on this encountered obstacle, robot deviates and navigates towards the target position. Again, the robot is detecting another obstacle from the direction facing the sensor S4 as shown in the Figure 8 (c). After deviating from the encountered obstacle in the sensor direction S4, the robot is encountering a large obstacle, which is perceived by sensors other than S3 and S4. Since in this study, sensors S3 and S4 are only considered into account to build the navigation rules, further navigation of the robot to avoid the encountered obstacles is not viable. The experimental results of the above situations are obtained, and the performance of the proposed methodology is investigated and the results are discussed. The experiments are repeated several times with the same environment and with the same starting and goal positions.

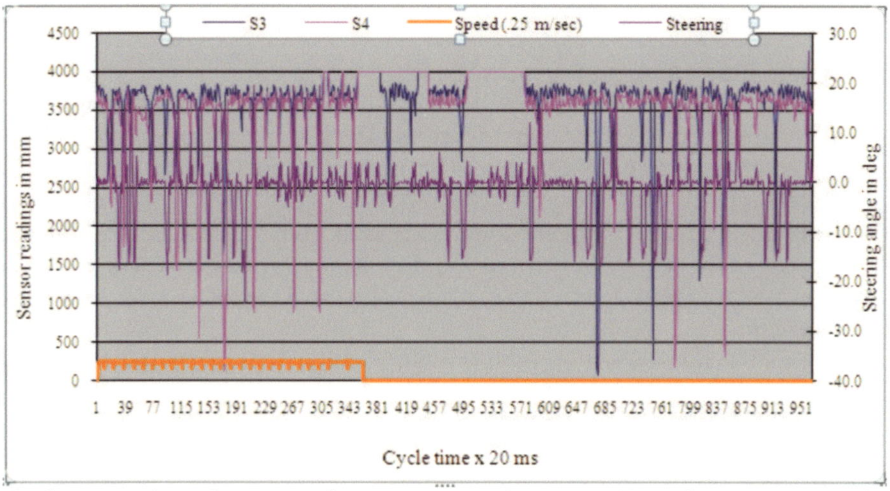


Fig. 9. Cycle time plotted against the responses of sensors S3 and S4 in terms of obstacle distances from robot, steering angle, and speed of the robot.

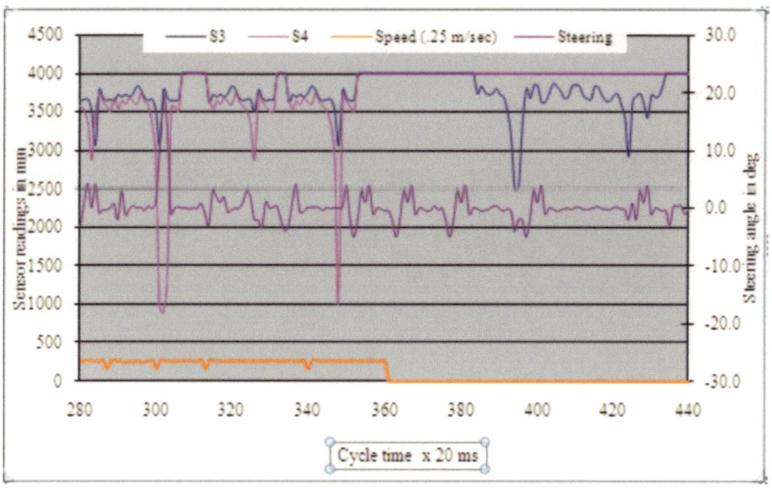


Fig. 10. Enlarged portion of Figure 9 between cycle times 280 X 20 to 440 x 20 ms plotted against obstacle distances from robot, steering angle and speed of the robot.

The plot shown in Figure 10 is the cycle time drawn against the obstacle distances measured by the sensors, speed of the robot and robot steering angles. For ease of visualization and explanations, a section of the results plotted in the Figure 9 are shown enlarged in Figure 10. It is found from the plot that the robot navigates autonomously with a constant speed of 250 mm/sec and it avoids obstacles, and navigates autonomously up to 1.785 meter of the span. The mobile robot stops after traveling 1.785 meter due to the obstacles in front of robot facing other sensors as shown in Figure 8 (d). During this period, (between 358 x 20 ms to 448 x 20ms) there are no obstacles in the proximity of the front of sensors S3 and S4. Whereas the other sensors on the right hand side of the robot as shown in Figure 8 (d) detect obstacles. From the observations of the navigation results, it is found that the insufficient knowledge of the environment causes the robot to stop. Similar experiments have been carried out for the following combinations of selected sensors.

### 4.2 Experimental Studies using Sensors S2, S3, S4, and S5.

In this experimental study, front sensors S2, S3, S4, and S5 are taken into account to acquire the environment details and build the navigation rules. The experiment is repeated several times with the same environment and with the same starting and goal positions. The environment shown in Figures 11 (a) - (d) illustrates the robot navigation while the robot encounters obstacles. In Figure 11 (d), the robot encounters obstacle from the direction of sensors S1 and S2 (right hand side of the robot), and as a result, the robot stops. This is due to the sensors S1's input, which is not considered into account to build the navigation rule. In the following section, the experimental results are discussed and illustrated to show the autonomous performance of mobile robot while the mobile robot uses the front sensors S2, S3, S4, and S5.

Fig. 11. Real world experiments using sensors S2, S3, S4 and S5

The Figure 12 is the cycle time plotted against the obstacle distances measured by sensors, speed of the robot, and robot steering angle. Since the number of sensors involved in the Figure 12 (S2, S3, S4 and S5) is more, it is difficult to visualize the graph. Hence, a particular portion of the graph (which shows the critical situation during navigation) is enlarged and shown in Figure 13. It is found from the plot that the robot navigates autonomously with the constant speed of 250 mm/sec and avoids encountered obstacles upto 3.005 meters of the span. The mobile robot stops after traversing 3.005 m, due to the obstacles in front of robot facing the other sensor (sensor S1) as shown in Figure 11 (d). During the non-acting time of the robot controller, (between 601 x 20 ms and 648 x 20ms) there are no obstacles near the sensors S2, S3, S4, and S5. Whereas the other sensor on the right hand side of the robot (S1) as shown in Figure 11 (d) detects obstacles. From the observations of the navigation results and investigations, it is found that there are insufficient sensor perceptions again and because of this reason, the robot stops.

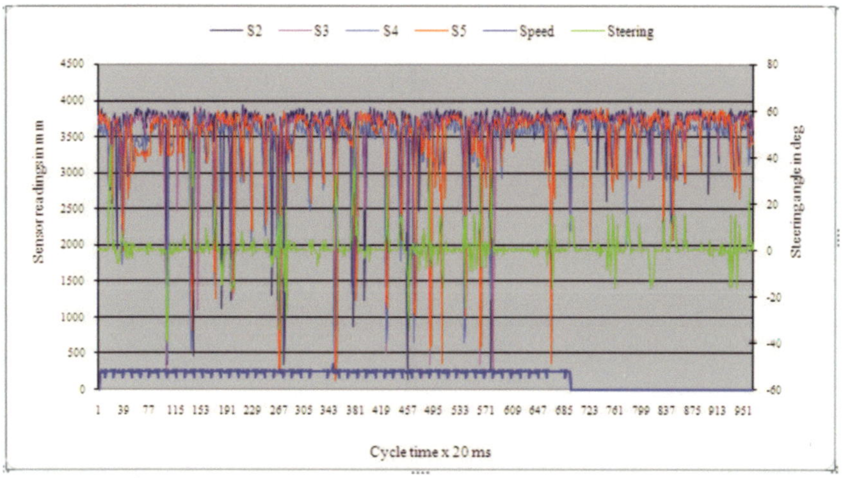


Fig. 12. Cycle time drawn against the responses of sensors S2, S3, S4, and S5 in terms of obstacle distances from robot, steering angle, and speed.

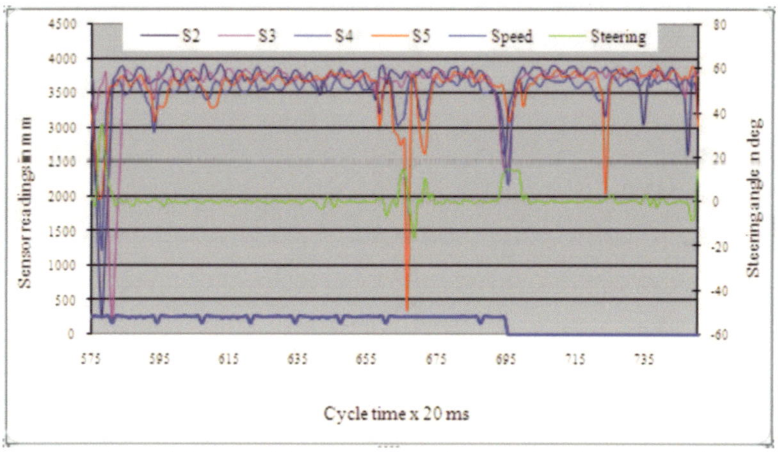


Fig. 13. Enlarged region of Figure 12 between cycle time 550 X 20 and 700 x 20 ms drawn against obstacle distances from robot, steering angle, and speed of the robot.

### 4.3 Experimental Studies using Sensors S1, S2, S3, S4, S5 and S6

In this experiment, front sensors S1, S2, S3, S4, S5, and S6 are taken into account to obtain the environment details and establish the navigation rules. The experiment is repeated several times with the same environment and with the same starting and goal positions. Figures 14 (a) - (e) illustrate the robot navigation through obstacles. In this experiment, the robot encounters many obstacles and in all the situations, the robot navigates successfully without any collisions and reaches the goal position. The experimental results are analysed and the autonomous performance of mobile robot is evaluated while the mobile robot uses the front sensors S1, S2, S3, S4, S5, S6, and the results are given in the following section.

Fig. 14. Experimental studies in Active Media Pioneer Robot using sensors S1, S2, S3, S4, S5 and S6

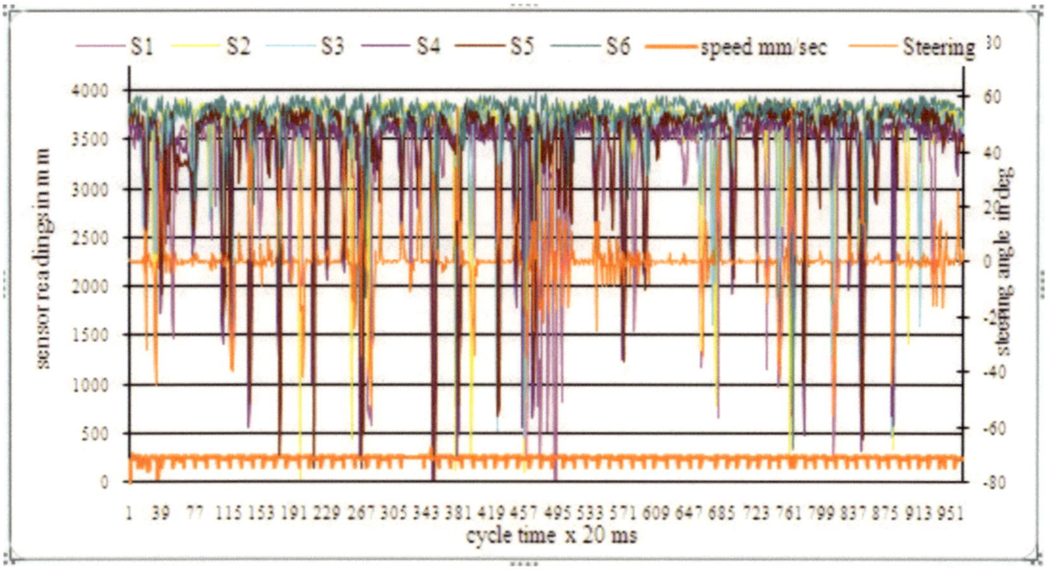


Fig. 15. Cycle time drawn against the responses of sensors S1, S2, S3, S4, S5, and S6 from robot, steering angle, and speed.

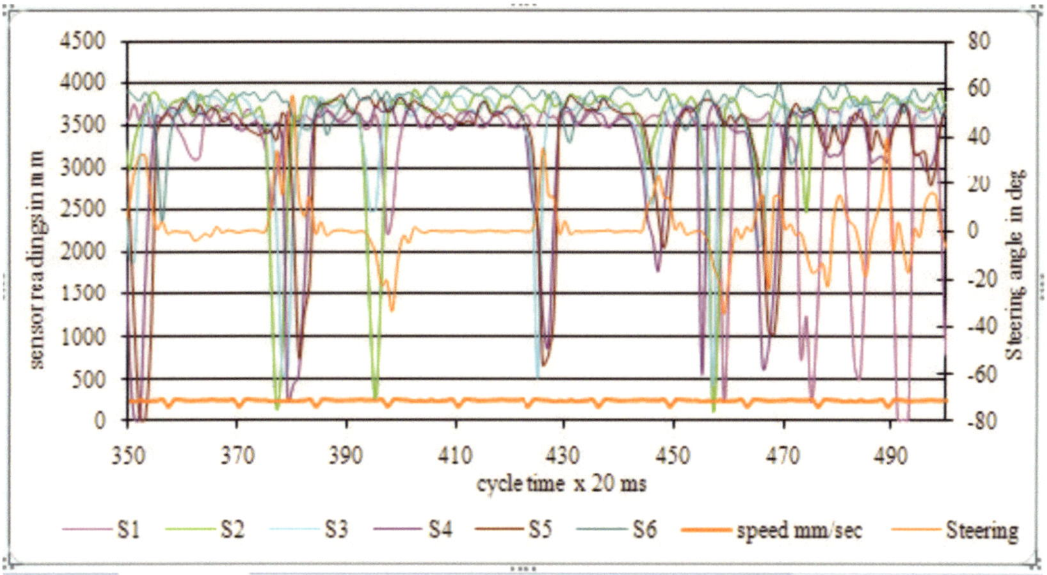


Fig. 16. Enlarged region of Figure 15 between cycle time 350 x 20 and 500 x 20 ms plotted against obstacle distances from robot, steering angle and speed of the robot.

The graph shown in Figure 15 is the cycle time plotted against the obstacle distances measured by sensors, speed of the robot and robot steering angle. Since the number of sensors used for navigation is greater than the previous experiments, the sensor results shown in Figure 15 cannot be seen clearly. Therefore, the selected critical portion of the graph is enlarged and shown in Figure 16. It is observed from the enlarged graph that the robot navigates autonomously with a constant speed of 250 mm/sec and avoids the encountered obstacles up to the target position (4.83 meters from the starting position). This plot shows the continuous responses of sensors S1 to S6 while navigating from start to the goal position. For example, when the robot traverses a distance of 1.9 meters (380 x 20 ms), it encountered many obstacles, which are in the close vicinity of the robot (350 to 550 mm from robot reference point) in the directions of 50˚, 30˚, and 10˚and -10˚. This implies that the environment consists of obstacles, which are identified in the respective sensor directions. In these situations the robot is required to turn and avoid those obstacles. The required defuzzified output (turn angle) is obtained from MFAM based FIS and is provided to the robot control system. The turn angle deviations are obtained from control system outputs in terms of X-Pos (right wheel's encoder data) and Y-Pos (Left wheel encoder data).

| Sensors | Output results of robot navigation in various selected situations as illustrated in the Figure 4.22 | | | | | |
|---|---|---|---|---|---|---|
| | Time 370x20 to 390x20 ms | | Time 420 x 20 to 430 x 20 ms | | Time 450 x 20 to 470 x 20 ms | |
| | Obstacle distances | Turn angle | Obstacle distances | Turn angle | Obstacle distances | Turn angle |
| S1 | 2528 | | 3532 | | 459 | |
| S2 | 124 | | 521 | | 108 | |
| S3 | 554 | | ---- | | 457 | |
| S4 | 384 | 56 deg | 897 | 36 deg | 564 | |
| S5 | 789 | | 678 | | 3834 | -35 deg |
| S6 | 3767 | | 3908 | | 3783 | |

Table 3. Output results of multiple sensors and the corresponding turn angle in various situations while using all the sensors for the robot navigation (reference Figure 16)

Table 3 provides the result of the selected situations during robot navigation using the proposed MFAM rules. The Table 3 shows the number of sensors involved to perceive the obstacle environment of the selected situations and the corresponding steering angle of the robot. The tabulated data are obtained from the experimental investigation using the sensors S1, S2, S3, S4, S5 and S6. Example of navigation rule activation as per the MFAM matrix for the selected situation is as follows:

"IF X13 and X14 and X15 and X26 then Z4".

Where, X13 is Sensor 3 and fuzzy distance measure is small (1) fuzzy set,

X14 is Sensor 4 and the fuzzy distance measure is small (1) fuzzy set,

X15 is Sensor 5 and the fuzzy distance measure is small (1) fuzzy set,

X26 is Sensor 6 and the fuzzy distance measure is medium (2) fuzzy set and

Z4 is Robot output and the fuzzy measure is Small Positive (MP)

It is observed that the robot speed from starting to goal position is constant and the robot does not collide with any of the obstacles. From the present experimental studies, it is found that more sensor data are required to ensure full robot autonomy that is achieved while using all the sensors in the sensor space, which are considered for navigation during mobile robot navigation.

## 5. Results and Discussions

The experimental investigations as studied above are summarised and shown in the chart of Figure 17. The experimental studies are performed using the navigation rules, which include the various combinations among the available sensor inputs (S1, S2, S3, S4, S5 and S6). In all the experiments, the environment is kept unchanged with the same starting and goal positions.

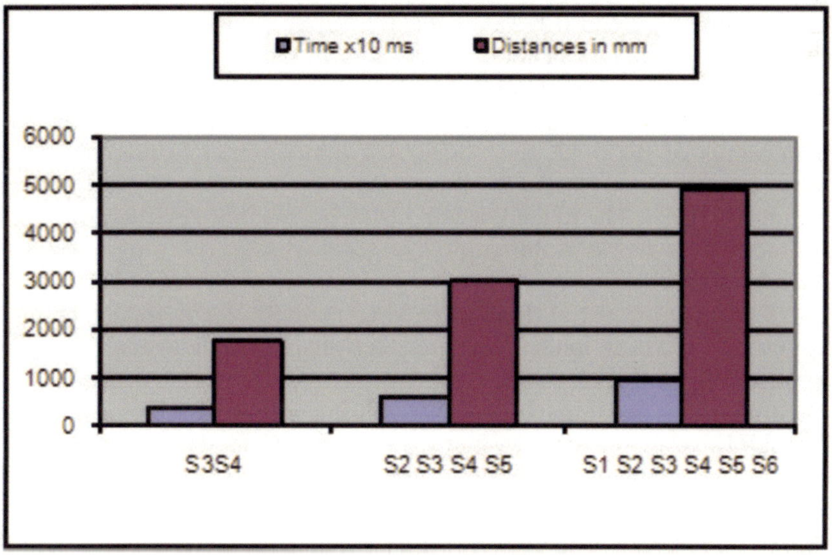


Fig. 17. Effectiveness of sensors and their fusion during robot navigation.

The effectiveness of MFAM is demonstrated by evaluating the autonomous performance of the mobile robot using the experimental results as illustrated in the chart shown in Figure 8. The chart shows the various combinations of sensors considered into account for robot navigation plotted against the distance travelled and the time of robot traversal. As seen from the Figure 8, it is found that that when all the sensors S0 to S6 are used for robot navigation, the complete autonomous performance is achieved, whereas the use of less data with less number of sensors for robot navigation results in an incomplete autonomous performance during navigation. Since the proposed MFAM rules effectively uses all sensor data to find the navigation path, the performance of navigation is improved considerably.

## 6. Conclusion

In this paper the optimization of fuzzy rules using modified Fuzzy Associative memory (FAM) is designed and implemented. The behavior rules obtained from FAM is tested in a simulation environment and validated by conducting real world experiments on a popular robot. The experimental results clearly indicate the mapping of multiple inputs to outputs with optimum path in every control cycle of the robot navigation. This approach involves the natural way of dealing with the environments using simple linguistic logic rules without using any mathematical model. The knowledge base of each behavior rule is easy to comprehend, because it captures the behavior rules in a linguistic form. The strength of the proposed methodology is the mapping of the inputs to the output through compositional association of multiple input variables, thus reducing the number of rules without elimination of any of the sensor's input. The robot navigation used in this research is purely perception based and perceived data are optimized and fully used for building navigation rules. Utilization of the proposed FAM methodology for other applications requires minimum modifications during setting of input and output linguistic variables.

## 7. References

M. Sugeno T. Murofushi, T. Mori, T. Tatematasu, J. Tanaka, "Fuzzy Algorithmic Control of a Model Car by Oral Instructions," Fuzzy Sets and Systems, No. 32, pg. 207-219, 1989.

F. G. Pin H. Watanabe, J.R. Symon, and R.S. Pattay "Using Custom-Designed VLSI Fuzzy Inferencing Chips for the Autonomous Navigation of a Mobile Robot" in Proceedings of IROS 92, the IEEE/RSJ International Conference on Intelligent Robots and Systems, North Carolina, 1992.

M. Balzarotti and G. Ulivi. The fuzzy horizon obstacle avoidance method for mobile robots. In Procs. of the world Automation Congress (WAC), vol 3, pg 51-57, Montpellier, FR, TSI press, 1996.

A. Ollero, A. Garcia-Cerezo et al. Fuzzy tracking methods for mobile robots chapter 17, Application of fuzzy logics: Towards high machine intelligence quotient systems, prentice hall, New Jersey, 1997.

J. Pereira and J.B Bowles. A comparison of PID and fuzzy control of a model car. In Procs. of the IEEE Int. Con. on Fuzzy Systems, pg 849-854, Orlando, FL, 1994.

E. Tunstel, et al, Autonomous navigation using an adaptive hierarchy of multiple fuzzy behaviors. In Procs. of the IEEE Int. Sym. on Computational Intelligence in Robotics and Automation, Monterery, CA, 1997.

C. Voudouris et.al, "Hierarchical behavioral control for autonomous vehicles", In Procs. of the 2nd IFAC Conf. on Intelligent Autonomous Vehicles, pg 267-272, Helsinki, FI, 1995.

P. Althauas et.al, Using the Dynamical System Approach to Navigate in Realistic Real-world Environments In Procs. of IEEE/RSJ International conference on Robotics and Intelligence systems, pg 1023-1029, 2001.

B. Kosko, neural networks and fuzzy systems. Prentice-Hall, Englewood Cliffs, 1992

S.G Kong and B. Kosko, "Adaptive fuzzy systems for baking up a truck-and-trailer", IEEE Transaction on Neural networks, vol 3,no:2, pg 211-223, 1992.

Hirohide Ushida, Toru Yamaguchi, and Tomohiro Takagi "Fuzzy-Associative-Memory-Based Knowledge Construction with an Application a Human-Machine Interface" IEEE Transactions on Industrial Electronics, vol 46, No.4, August 1999.

T. D. Ndousse, "Fuzzy neural control of voice cell in ATM networks," IEEE Journal on Select. Areas Communication, vol 12, no.9, pg.1488-1494, Dec.1994.

C. T. Lin and C.S.G.Lee, "Neural network based fuzzy logic control and decision systems," IEEE Transaction on Computations, vol 40, pg.120-1336, Dec.1991

J. S. R Jang, "ANFIS: Adaptive-network based fuzzy inference system," IEEE Transactions on Systems, Man, Cybernetics, vol 23, pg.665-684, May/June, 1993.

H. Ushida, T.Yamaguchi, and T. Takagi, "Human -motion recognition via a fuzzy associative memory system, "Syst.Comput.Jpn., vol26, no.9, pg.90-104,1995.

F. G. Pin and Y.Watanabe. "Steps towards sensor based vehicle navigation in outdoor environments using fuzzy logic behaviorist approach. Journal of Intelligent and Fuzzy Systems, pg :95-107, 1993.

Bo Hyeun Wang, George Vachtsevanos, "Fuzzy Associative Memories: Identification and Control of Complex Systems" Proceeding of 5th IEEE symposium on intelligent control, , pg 910-915 vol2, USA., 1990.

S. Parasuraman, V.Ganapathy, Bijan Shirinzadeh, "Behavior based Mobile Robot Navigation by AI Techniques: Behavior selection and resolving behavior conflicts using Alpha level Fuzzy Inference System" Proceeding of the ICGST International conference on Automation, Robotics and Autonomous System, ARAS2005, P1130533120, Egypt., 2005.

C. W. de Silva 'Intelligent control: Fuzzy logic Applications, CRC press, New York, US.1995.

# $Z_\infty$ - Monocular Localization Algorithm with Uncertainty Analysis for Outdoor Applications

Elmar Mair and Darius Burschka
*Technische Universität München (TUM)*
*Germany*

## 1. Motivation

Localization is an essential task for a meaningful application of a robotic system. The knowledge about its 6DoF *pose* represented as a 3D position and the orientation in space allows a system to execute tasks defined in the 3D workspace and to prevent collisions using the knowledge about obstacles from a model of the environment. A system needs to be capable of estimating its absolute location in the world and to track the changes in the position during its motion. We speak in this context of global and relative localization capabilities. The localization task performs a registration of the sensor perception to a reference. The two above localization alternatives differ in the way how this reference is defined. The global localization requires an a-priori knowledge about the environment stored usually as a geometric model or as an image database that needs to be registered to the current sensory perception, see (Thrun, 2002) for a comprehensive survey.

Since early work in the 1970's, such as SRI's Shakey (Nilsson, 1984) and Moravec's Cart (Moravec, 1983), there have been great strides in the development of vision-based navigation methods for mobile robots operating both indoors and outdoors. Much of the efficiency and robustness of the recent systems can be attributed to the use of special purpose architectures and algorithms that are tailored to exploit domain specific image cues. For example, road

Fig. 1. The $Z_\infty$ algorithm estimates the rotation (red landmarks) and the translation up to scale (blue landmarks) based on a monocular camera in an efficient way. Therefore, it fits well for resource-limited systems like e.g. flying mobile robots.

followers rely on finding the road boundary and lane markers (Dickmanns & Graefe, 1988; Hebert et al., 1995) or landmarks (Fennema et al., 1990; Kuhnert, 1990; Lazanas & Latombe, 1992; Levitt et al., 1987) whereas mobile robots navigating in hallways have exploited uniform texture of the floor (Horswill, 1992), floor/wall features (Kim & Navatia, 1995; Kriegman et al., 1989), and overhead lights (Fukuda et al., 1995). Although these domain specializations lead to impressive performance, they do so by imposing particular sensor cues and representations on low-level navigation. As a result, a system that works in one domain may require substantial redesign before it can be used in another.

Another interesting localization approach for mobile navigation on predefined paths are the *Vision-Based Control* approaches. In many cases it is not necessary to use sophisticated map-based systems to control the paths of the robot—instead a simple teaching phase may be sufficient to specify the robot's nominal pathways. Consider for example mobile robots performing delivery in office environments, serving as AGV's in an industrial setting, acting as a tour guide, or operating as a security guard or military sentry. In these situations, the robot repeatedly follows the same nominal path, except for minor perturbations due to control/sensing errors, slippage, or transient/dynamic obstacles. Such a system has to be *walked* in a teaching step through the environment. During this teaching phase the robot learns the path based on sensor perception and then later repeats this path using the same sensors together with previously stored information. Teaching (showing) a robot its nominal pathways has been considered by others including (Matsumoto et al., 1996; 1999; Ohno et al., 1996; Tang & Yuta, 2001). One approach is to use a stored two or three-dimensional representation (map) of the environment together with sensing. A learned path can be replayed by first constructing a map during training and then continuously localizing the robot with respect to the map during playback. However, it is not clear that building a metrically accurate map is in fact necessary for navigation tasks which only involve following the same path continuously. Another approach would be to use no prior information, but rather to generate the control signals directly from only currently sensed data. In this case no path specification at all is possible. An approach based on an Image Jacobian was presented in (Burschka & Hager, 2001).

On the other hand, relative localization approaches track merely the incremental changes in the pose of the robot and do not necessarily require any a-priori knowledge. In relative localization approaches, the initial or sensor perception from the previous time step is used as a reference.

There are several methods, how the *pose* of a camera system can be estimated. We can distinguish here: multi-camera and monocular approaches. Since a multi-camera system has a calibrated reference position of the mounted cameras, stereo reconstruction algorithms (Brown et al., 2003; Hirschmüller, 2008) can be used to calculate the three-dimensional information from the camera information and the resulting 3D data can be matched to an a-priori model of the environment. Consequently, there has been considerable effort on the problem of mobile robot localization and mapping. This problem is known as simultaneous localization and mapping (SLAM) and there is a vast amount of literature on this topic (see e.g., (Thrun, 2002) for a comprehensive survey). SLAM has been especially succesful in indoor structured environments (Gonzalez-Banos & Latombe, 2002; Konolige, 2004; Tards et al., 2002).

Monocular navigation needs to solve an additional problem of the dimensionality reduction in the perceived data due to the camera projection. A 6DoF pose needs to be estimated from 2D images. There exist solutions to pose estimation for 3 point correspondences for most traditional camera models, such as for example orthographic, weak perspective (Alter, 1994), affine, projective (Faugeras, 1993; Hartley & Zisserman, 2000) and calibrated perspective (Har-

alick et al., 1994). These approaches constrain the possible poses of the camera to up to four pairs of solutions in the case of a calibrated perspective camera. At most one solution from each pair is valid according to the orientation constraints and the other solution is the reflection of the camera center across the plane of the three points.
In the work from Nister (Nister, 2004), an approach sampling for the correct solution along the rays of projection solving an octic polynomial to find the actual camera pose is presented that is limited to exactly 3 points neglecting any possible additional information. A solution provided by Davison consists of building a probabilistic 3D map with a sparse set of good landmarks to track (Davison, 2003). Klein was able to achieve even more accurate results as the EKF based approach of Davison by efficiently separating the tracking and the mapping routines (Klein & Murray, 2008).
In this chapter we address the problem of the robust relative localization with monocular video cameras. We propose a localization algorithm that does not require any a-priori knowledge about the environment and that is capable not only of estimation of the 6 motion parameters together but also the uncertainty values describing the accuracy of the estimates. The output of the system is an abstraction of a monocular camera to a relative motion sensing unit that outputs the motion parameters together with the accuracy estimates for the current reading. Only this additional accuracy information allows a meaningful fusion of the sensor output in SLAM and other filtering approaches that are based on Kalman Filters.

## 2. $Z_\infty$: monocular motion estimation

One problem in estimating an arbitrary motion in $3D$ from a real sensor with noise and outliers is to quantify the error and to suppress outliers that deteriorate the result. Most known approaches try to find all six degrees of freedom simultaneously. The error can occur in any dimension and, therefore, it is difficult in such methods to weight or isolate bad measurements from the data set. The erroneous data can be detected and rejected more effectively if the error is estimated separately along all parameters instead of a global value. Thus, a separation of the rotation estimation from the translation simplifies the computation and the suppression of error prone data immensely. This is one major advantage of our $Z_\infty$ algorithm presented in this chapter. We use the usually undesirable quantization effects of the camera projection to separate translation-invariant from translation-dependent landmarks in the image. In fact, we determine the rotational component of the present motion without ever considering the translational one. Fast closed form solutions for the rotation and translation from optical flow exist if the influences are separable. These allow an efficient and globally optimal motion estimation without any a-priori information.
In this section, we describe how we detect translation-invariant landmarks for rotation estimation and how the whole algorithm is applied to an image sequence.

### 2.1 Principle of $Z_\infty$

Talking about the projective transformation and the pixel discretization of digital cameras, these characteristics are usually considered to be the barriers for image based motion estimation. However, splitting the motion in a rotational and translational part is based on just these effects - $Z_\infty$ uses the projective transformation and pixel discretization to segment the tracked features into a translation-invariant and a translation-dependent component.
We see from the basic camera projection equation

$$\begin{pmatrix} x \\ y \end{pmatrix} = \frac{f}{Z} \begin{pmatrix} X \\ Y \end{pmatrix} \tag{1}$$

with $f$ representing the focal length, that the $X$ and $Y$, being $3D$ world coordinates of the imaged point, are both divided by the landmark's distance $Z$.
According to the projective transformation and the motion equation, which applies a rotation $R$ and a translation $T$ to a point $P_i$,

$$P_i' = R^T P_i - T \tag{2}$$

the camera motion affects the landmarks mapped onto the image plane as follows.
Each point in the image of a calibrated camera can be interpreted as a ray from the optical center to the 3D landmark intersecting the camera image at a point.

$$P_i = \lambda_i \vec{n}_i \tag{3}$$

The length $\lambda_i$ of the vector $\vec{n}_i$ to the landmark $i$ is unknown in the camera projection.
Applying a rotation to the camera appears to rotate the displayed landmarks in an opposite direction by the angle of rotation . The rotation is not affected by the distance to a landmark. This is not the case for translations. A translational motion of the camera results in a different motion for each landmark depending on its distance to the camera.

$$\lambda_i' \vec{n}_i' = R\,(\lambda_i \vec{n}_i + \vec{T}) \tag{4}$$

Due to the pixel discretization, the translation cannot be measured anymore if a landmark exceeds a certain distance to the camera. This distance

$$z_\infty = \frac{f}{s_m} T_m \tag{5}$$

depends on the translational component parallel to the camera plane, the focal length $f$ and the pixel size $s_m$. $T_m$ is the translation between frames of the video. Depending on the translational motion between two images $z_\infty$ can be quite close to the camera, e.g., if we move a standard camera with a focal length of 8.6 $mm$, a pixel size of 9.6 $\mu m$ and a framerate of 30 $Hz$ with a translational component parallel to the image plane of about $2\,km/h$ this results in a translation invariant distance threshold of 16.6 $m$. Fig. 2 illustrates these characteristics of the camera projections.
The typical observed motion $T_m$ is smaller than the value assumed above for a motion observed by a camera looking parallel to the motion vector $T$. This would correspond to a camera looking to the side during a forward or backward motion. An important observation is that $T_m$ is not only scaled by the distance to the observed point but that the measurable translation vector depends also on the angle $\Psi$, included by the motion vector and the perpendicular to the projection ray, and the angle $\varphi$, included by the projection ray and the optical axis (Fig. 3). We see directly from Equation 6 that a radial motion as it is the case for points along the optical center but also any other the line of projection lets a point become a part of the $\{P_{k\infty}\}$ set of points, from which the translation cannot be calculated (Fig. 3).

$$\begin{gathered} \Delta T_{s_m} = \cos\psi_m \Delta T \\ \Delta T_m = \frac{\Delta T_{s_m}}{\cos\varphi_m} = \frac{\cos\psi_m}{\cos\varphi_m}\,\Delta T \\ \Delta T_x = \frac{\cos\psi_x}{\cos\varphi_x}\,\Delta T \quad \wedge \quad \Delta T_y = \frac{\cos\psi_y}{\cos\varphi_y}\,\Delta T \end{gathered} \tag{6}$$

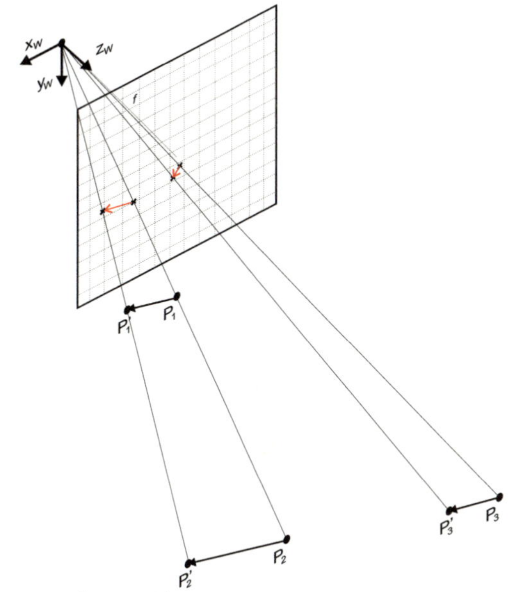


Fig. 2. Optical flow vector 1 ( $P_1$, $P_1'$ ) is shorter than vector 2 but it is closer to the optical center. Therefore, both vectors result in the same projection. Vector 3 has the same length as vector 1, but it is further from the image plane - therefore, the projection is much shorter. It becomes so small, that the projection of $P_3$ and $P_3'$ lie within the same pixel. If such a projection results only from a translation then it would be translation-invariant, because the translation has no measurable influence on it.

Therefore, image features can be split into a set which is translation-invariant and a set which is translation-dependent. Tests on outdoor pictures have shown that the contrast of the sky to the ground at the horizon generates many good features to track. Indeed, an average percentage of 60 % of the selected features lies in outdoor images at the horizon.
But how can we identify which feature corresponds to which set? We have no information about the rotation, the translation or the distance of the landmarks visible in the image. The solution is provided by the algorithm described in the next section.

### 2.2 RANSAC revisited

The **Ran**dom **S**ampling **C**onsensus algorithm (RANSAC) is an iterative framework to find parameters for a given model from an data-set including outliers (Fischler & Bolles, 1981). In each iteration, an as small as possible data subset is randomly chosen to calculate the model parameters. This model is than applied to the residual data set and the elements are split into fitting elements (inliers) and non-fitting elements (outliers). This step is repeated several times and the best model, according to the number of inliers and their residual error, is chosen. Preemptive RANSAC is a variant which allows the algorithm to leave the loop in case that a certain quality criterion applies also before the maximum number of iterations is reached.
In our case, the estimated model is a rotation matrix for which the entries have to be calculated. Therefore, we take three corresponding points from the two subsequent images and we estimate the rotation matrix based on them, as it is explained in Section 2.3. The estimated rotation is applied to the residual data set. In case that the initial three correspondences were

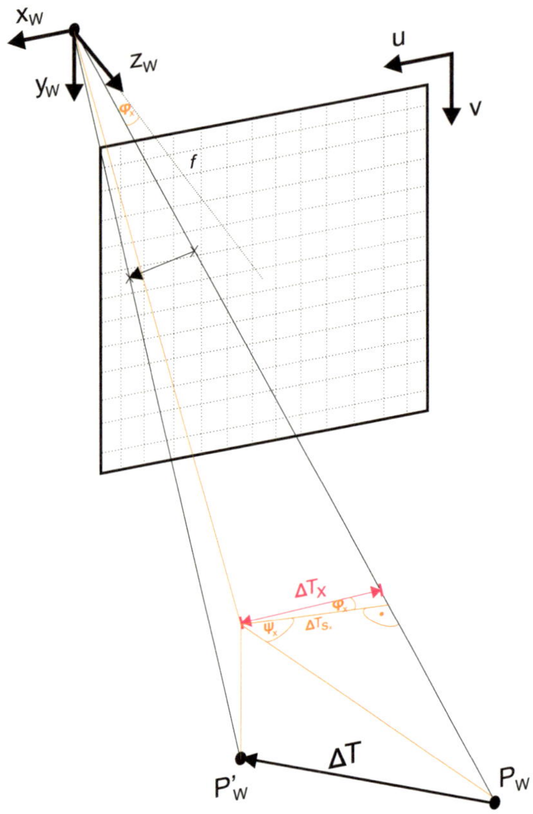


Fig. 3. This drawing visualizes how the measurable translational component $\Delta T_x$ can be calculated. In general the optical flow vectors become longer the more parallel they become to the image plane and the closer they get to the image border. The orange auxiliary line illustrates the projections onto the x-z plane.

from points in the world at a distance further than $z_\infty$ and so represent only the rotational part, the true rotation is calculated. Otherwise, the translational component of the OF-vectors results in a wrong rotation matrix. Applying the resulting rotation on the remaining data set shows if the initial points where truly translational invariant.

In case that we find other feature sets resulting from the same rotation we can assume that the first three vectors where translation-independent and we found a first estimate for a possible rotation. Another indication for a correct rotation estimate is that the back-rotated and therefore purely translational vectors intersect all in one point in the image, the epipole. A degenerated intersection point is also the infinity, where all resulting vectors are parallel in the image. This is a result of a motion parallel to the image. If there are no further OF-pairs agreeing with the calculated rotation, we can expect, that we had at least one translation-dependent element. In case of success, a more accurate and general rotation is estimated on all inlier found by the first estimate. The average error of the back-projection of the vector-endpoints according to the result and the number of found inlier gives an accuracy measure and permits a comparison of the results of each iteration.

The vectors identified as rotation-inliers overlap after the compensation of rotation in both

images and are, therefore, translation-invariant. The rotation-outliers represent translation-dependent optical flow vectors and real mismatches. Again, we use RANSAC to suppress outliers for a robust translation estimation as described in Section 2.4.
The probability to find an initial set of three translation-invariant optical flow elements depends on the percentage of such vectors in the data set. As mentioned earlier, tests have shown, that an average of 60 % of the features are far enough to be translation-invariant. The lower 5 % quantile was about 30 %. These results in approximately 37 iterations necessary to assure that in 95 % of the cases RANSAC can determine the rotation and divide the data set. Usually several rotation inliers can also be tracked in the next image. Such features are then preferred for rotation estimation, because it can be assumed that these landmarks are still further than $z_\infty$. Thus, the number of RANSAC iterations for rotation estimation is reduced to one - in practice a few iterations are done to improve the robustness.

### 2.3 Rotation estimation

The rotation of a point cloud can be estimated in closed form based on the direction vectors to the different landmarks. Therefore, three different registration methods exist: the rotation can be calculated using quaternions (Walker et al., 1991), by singular value decomposition (SVD) (Arun et al., 1987) or Eigenvalue decomposition (EVD) (Horn, 1987). We use the so called Arun's Algorithm, based on SVD (Arun et al., 1987). It is analogue to the more familiar approach presented by Horn, which uses an EVD.
The corresponding points used for rotation estimation belong all to translation-invariant points. Therefore, Equation 4 can be abbreviated to

$$P_i' = \lambda_i' \vec{n}_i' = R * P_i = R\,\lambda_i\,\vec{n}_i = \lambda_i\,R\,\vec{n}_i, \quad \textbf{with} \quad \lambda_i' \approx \lambda_i \Rightarrow \vec{n}_i' = R\,\vec{n}_i \tag{7}$$

Thus, we must solve following least squares problem

$$min \sum \|RP_i - P_i'\|^2 \tag{8}$$

This is achieved by the Arun's Algorithm which works as follows. The input for the algorithm are the two corresponding point clouds $\{P_i\}$ and $\{P_i'\}$, which are only rotated and whose rotation we want to estimate. First the origin of the coordinate frame has to be moved to the centroid of the point cloud:

$$\bar{P} = \frac{1}{n}\sum_{i=1}^{n} P_i, \quad \bar{P}' = \frac{1}{n}\sum_{i=1}^{n} P_i' \tag{9}$$

$$P_i^* = P_i - \bar{P}, \quad P_i'^* = P_i' - \bar{P}' \tag{10}$$

Now, the non scaled sample cross-covariance matrix for these point clouds is calculated to

$$\tilde{M} = \sum_{i=1}^{n} P_i'^*\, {P_i^*}^T \tag{11}$$

Therefore, $\frac{1}{n}\tilde{M}$ is the sample cross-covariance matrix between $\{P_i\}$ and $\{P_i'\}$. It can be shown that the rotation matrix which minimizes Equation 8 also fulfills

$$\tilde{R} = argmax_{\tilde{R}}\; tr(R\tilde{M}) \tag{12}$$

Is $(\tilde{U}, \tilde{\Sigma}, \tilde{V})$ the SVD of $\tilde{M}$

$$\tilde{M} = (\tilde{U}, \tilde{\Sigma}, \tilde{V}) \tag{13}$$

then $\tilde{R}$ can be calculated to

$$\tilde{R} = \tilde{V}\,\tilde{U}^T \tag{14}$$

$\tilde{R}$ is orthonormal, symmetric and positive definite. However, it can happen that all features lie in a plane. In that case not the rotation matrix, put a mirroring matrix is calculated. Such a result can be recognized by the determinant of $R$, if $det(\tilde{R}) = -1$ instead of $+1$. The rotation matrix can then be calculated to

$$\tilde{R}' = \tilde{V}'\tilde{U}^T, \quad \text{with } \tilde{V}' = \begin{pmatrix} v_1 \\ v_2 \\ -v_3 \end{pmatrix} \tag{15}$$

$v_1$, $v_2$ and $v_3$ are the rows in $V$ corresponding to the singular values $\lambda_1$, $\lambda_2$ and $\lambda_3$, whereas $\lambda_1 > \lambda_2 > \lambda_3 = 0$.
The uncertainty estimate of the rotation estimation consists in the average reprojection error and the percentage of rotation inliers in the data set.
Actually, the Arun's Algorithm is thought to estimate both, rotation and translation, but to estimate latter the origin of both point clouds must be the same, which is not the case for a moving camera.

### 2.4 Translation estimation

Compared to the rotation estimation, the calculation of the translation is rather trivial. All the points in $\{P_i'\}$ are rotated back by $R^T$ and consist afterwards only of the translational part of the motion.

$$\widehat{P'} = R^T P' = R^T (R(P + \vec{T})) = P + \vec{T} \tag{16}$$

$\vec{T} \neq! \, 0$ because only translation-dependent features are used.
We know from epipolar constraints that these back-rotated optical flow vectors $\widehat{P'} - P$ meet all in the epipole in theory. Due to noise, approximation and discretization issues this will not be the case in real data. Therefore, we calculate the point cloud of all intersections. The centroid of this point cloud is supposed to be the epipole. However, there are also several short vectors from very distant points which contain only a small observable translational component. These vectors are inaccurate for direction indication. Further, there are several almost parallel vectors which are also ill-conditioned to calculate their intersection. It seems reasonable to weight the intersection points by a quality criterion resulting from the angle between the rays which form the intersection and their length.
As uncertainty value for the translation the euclidean distance between the calculated epipole and the intersection of the optical flow vectors is used, as well as the weights used to calculate the intersections' centroid.

## 3. Experiments

In the following section some simulation and experimental results are presented. First, some explanations to the experimental setup and the visualization of the $Z_\infty$ output are given. Then, a few simulation results are shown and finally two experiments with real data are demonstrated. Amongst others, the method is compared to an other visual localization approach and to the estimates from an inertial measurement unit (IMU).

### 3.1 Explanation of the Visualization Encoding

In the following, the color encoding of the visualization is briefly explained. The landmarks are marked green if they could be tracked from the last image or red if they were lost and replaced by new good features to track (see Fig. 4(a)). The optical flow is represented as green vectors, where the original position of the feature is marked by a circle and the tracked location by a cross. The results of the rotation estimation are colored red. The endpoints of the optical flow vectors are rotated back according to the computed rotation – they are marked as crosses too. Thereby, translation-invariant features are represented by red vectors, while the outlier of the rotation computation are magenta (see Fig. 4(b)). The translation estimation is illustrated blue. Similar to the rotational result, the vectors, which were used for epipole estimation, are dark blue, while the outlier and therefore wrong tracked features are light blue. The black star (circle and cross) represents the computed epipole (see Fig. 4(c)). A result of the obstacle avoidance is shown in Fig. 4(d) and consists in two parts: the main image shows the optical flow vectors which are used for obstacle detection and the small sub-image shows the computed obstacle map. The vectors are mapped regarding their angle to the calculated epipole and a red line illustrates the suggested swerve direction (swerve angle) (please refer also to Section 3.6).

(a) Tracking output.

(b) Rotation estimation.

(c) Translation estimation.

(d) Obstacle detection.

Fig. 4. Visualization example for the $Z_\infty$ output.

### 3.2 Simulation results

The $Z_\infty$ algorithm has also been tested extensively with artificial, controllable data. Matlab has been used as simulation environment. Six parameter sets have been chosen to show the algorithms reliability and insensibility to white noise and outlier (see Table 1). The conditions of this test are as follows:

| | white noise ($\sigma^2$) | outlier (%) | pixel-discretization |
|---|---|---|---|
| parameter set 1 | 0 | 0 | no |
| parameter set 2 | 0.05 | 0 | no |
| parameter set 3 | 0.2 | 0 | no |
| parameter set 4 | 0 | 5 | no |
| parameter set 5 | 0 | 0 | yes |
| parameter set 6 | 0.05 | 2 | yes |

Table 1. Characteristics of the parameter sets used for simulation.

A standard camera with 8.6 $mm$ focal length, 8.6 $\mu m$ pixel size, and a resolution of $768 \times 576$ pixel is simulated. The parameters are transformed to a unit focal lenght camera model first. The simulated translation can randomly vary between $-0.05$ to $+0.05\,m$ in X- and Y-direction and twice as much along the optical axis. Further, a rotation of up to 0.5° per axis is allowed. At maximum 100 points are simulated in the field of view and they are located within a distance of 1 $km$ and 100 $m$ altitude. The camera is simulated to be in 50 $m$ altitude. The translation computation is executed every $10^{th}$ frame. Fig. 5 and 6 show an example of such a rotation and translation estimation on artificial data.
The simulated white noise is added after projection onto the camera plane and the outlier can have a size of up to 10 pixels – a local tracker would not find any further correspondences neither. Each parameter set is tested on a simulated image sequence of 2000 images. The most important parameters during the simulation are:

- maximum number of iterations for rotation estimation: 40
- maximum number of iterations for translation estimation: 20
- maximum number of steps per translation estimation: 5

| | average rotational error (°) | average transl. dir. error (°) | failed rotation calculations | failed transl. calculations |
|---|---|---|---|---|
| parameter set 1 | 0.0109 | 1.71 | 0 | 0 |
| parameter set 2 | 0.0150 | 2.59 | 0 | 1 |
| parameter set 3 | 0.0295 | 4.74 | 1 | 1 |
| parameter set 4 | 0.0098 | 3.82 | 2 | 10 |
| parameter set 5 | 0.0188 | 2.82 | 0 | 2 |
| parameter set 6 | 0.0213 | 6.10 | 1 | 1 |

Table 2. Average rotation and direction of translation error and the number of failed rotation and translation computations for the simulation of the six parameter sets of Table 1.

The results of the simulation are shown in Table 2. While the rotation estimation is rather robust against noise and outliers, the translation computation suffers from these characteristics. Too many outliers even prevent a successful translation estimation.

Also other ill-conditioning circumstances can be considered. In the following, we depict a few insights which could be verified based on the results of the simulations:

- The estimation of the rotation about the optical axis is ill-conditioned. This results in an approximately twice as large error compared to the other two axes.
- While the rotation estimation does not depend at all on the translation, an error in the computation of the rotation will also be reflected in the translational result. Large errors at the rotation estimation even prevent the translation estimation – this is also a reason for the large number of failed translation computations in Table 2.
- If the translation is quite small it is more probable to misguide the rotation estimation because the translational component is within the error tolerance of the rotation computation. This fact is also noticeable in the experiment of Section 3.4.
- If the camera motion is along the optical axis, the estimation of the direction of translation is conditioned best (see also Fig. 6). If the camera translation becomes perpendicular to the direction of view, small errors in the rotation estimation, noise or a few outliers may yield a large error. An exact parallel motion would imply that the translational optical flow vectors meet at infinity, which is computationally ill-conditioned. This fact should be highly recommended when choosing a camera setup for the $Z_\infty$ algorithm. It works best if the cameras are mounted in the direction of motion, if such a preference exists.

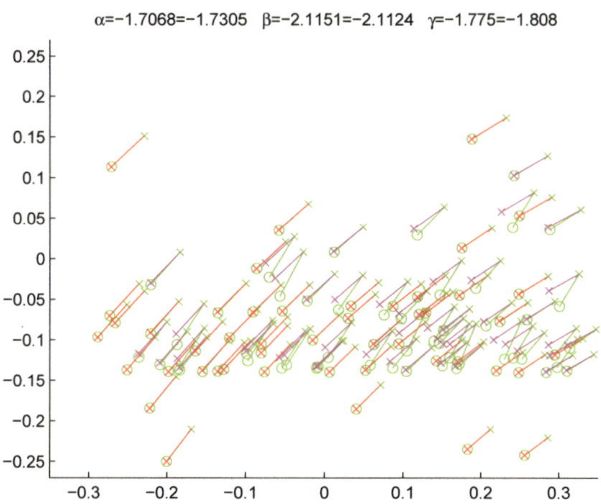


Fig. 5. Simulation of the rotation estimation in Matlab.

### 3.3 Experimental setup

To test our algorithm, first feature correspondences have to be found. Therefore, the Kanade-Lucas-Tomasi (KLT) tracker has been used (Lucas & Kanade, 1981). Good features are selected according to (Shi & Tomasi, 1994). The tracker is a local tracker with subpixel accuracy. It has been chosen due to its performance and robustness. Fig. 4(a) shows an example output of the tracker with 500 features. However, also global tracker, like SURF have been successfully tested (see Section 3.7).

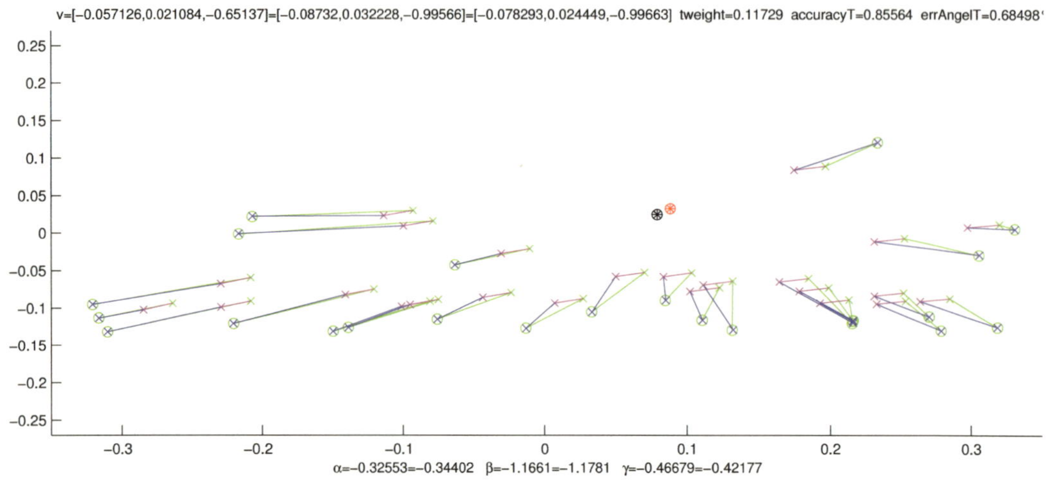


Fig. 6. Simulation of the translation estimation in Matlab.

Fig. 7 shows the processing times of a C++-implementation of the algorithm on a 1.6 GHz Intel Core Duo processor T2300. The same parameters as in the simulations (Section 3.2) were used. The code is not optimized for performance and the processing is only single-threaded. The time was measured with the "gprof" profiling tool.

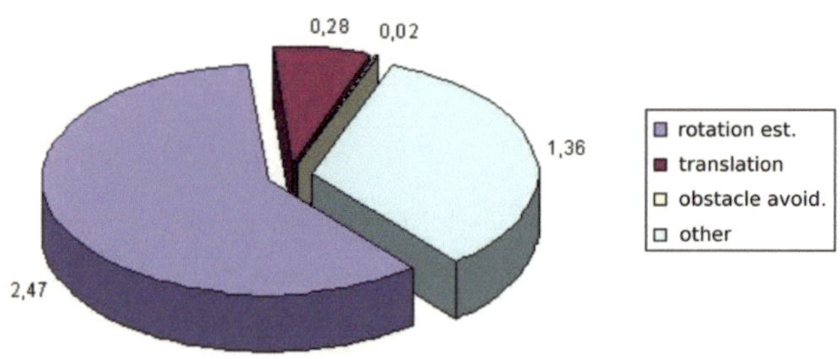


Fig. 7. The computation time for each module of the $Z_\infty$ algorithm in milliseconds. In total an average time of 4.13 ms on a 4 years old notebook is required.

### 3.4 Comparison of $Z_\infty$ to stereo-supported visual localization (SSVL)

In this experiment the $Z_\infty$ algorithm is compared to an algorithm, which has proven to be accurate enough even for short-range 3D-modeling (Strobl et al., 2009). The method has been designed for such short-distance applications, but it has been shown to be also accurate and reliable for mobile robots (Meier et al., 2009). Comparing this indoor and outdoor localization algorithm, we want to highlight the advantages and disadvantages, the bottlenecks and the limits of both approaches.

In the next section the modules of this SSVL approach are briefly described to improve the understanding of the reader.

### 3.4.1 Stereo-supported visual navigation

This visual localization method is indeed also based on monocular image processing. However, usually it is supported by a second camera. In a nutshell: A patch-based KLT tracker is used to select good features and track them from image to image. Those features have to be initialized for the subsequent pose estimation. Therefore, three different initialization possibilities exist: structure from motion, structure from reference or structure from stereo. In order to get the proper scale and to be independent from any modifications of the environment, a fast subpixel-accurate stereo initialization has been invented and is typically used. The pose is calculated by the so called robustified visual GPS (RVGPS) algorithm (Burschka & Hager, 2003). Every time the features get lost or are moving outside of the field of view, new features are initialized using the stereo images. However, the old feature sets are not dropped, but maintained by an intelligent feature set management. As soon as the camera comes back to a location it has already been, the features are refound and reused for tracking. Thus, the accumulated bias gets reduced again and the error is kept as small as possible. This method does not provide any probabilistic filtering like Kalman or particle filters. Nevertheless, any of these methods can be used to smooth the localization output and reject outlier. For further details to this algorithm refer to (Mair et al., 2009).

### 3.4.2 Comparison results: $Z_\infty$ - SSVL

For this experiment two Marlin F046C cameras from Allied Vision have been mounted on a stereo rig with $9\,cm$ baseline on the top of a Pioneer3-DX from MobileRobots Inc. (Fig. 8). The robot has been programmed to follow a circular path with $3\,m$ diameter. During the run 1920 stereo image pairs were acquired. The cameras have been slightly tilted to the ground to provide closer areas in the images, while keeping the horizon and the structures at infinity well visible. Although, the odometry is rather imprecise, the Pioneer has been able to close the circle with only a few centimeters of error.

Fig. 8. The Pioneer3-DX carrying the two Marlin cameras.

Fig. 9 shows the trajectory computed by the stereo supported localization. It is apparent that there is no bundle adjustment applied: wrong initialized features lead to a wrong pose estimation, but are usually immediately removed by the M-estimator in RVGPS. However, if there are too many wrong features, it can last some images until they get removed from the current

feature set. Such a behavior is visible in the lower half of the circle, where the computed trajectory leaves the circular path and jumps back on it again after a while. Nevertheless, such outlier could also be suppressed by a probabilistic filter, like e.g. a Kalman filter.
To provide an accurate stereo initialization with this baseline, all landmarks have to lie within a range of approximately 10 $m$. This image sequence is therefore ill conditioned for RVGPS-based localization due to several facts: only features in the lower half of the image can be used for pose estimation, due to the large rotation the feature sets are leaving the field of view quickly and the ground on which the robot is operating has only little structure which eases miss-matches and the loss of features during tracking (see Fig. 11).
The large rotational component of the motion compared to the amount of translation prevents the $Z_\infty$ algorithm to determine the direction of translation accurately. The small translation yields in a $z_\infty$-range of 10 cm per image. Therefore, the translation can only be evaluated after several images, when it gets accumulated to a measurable length. Further, the small translational component in the optical flow vectors yields the features to be tested as translational invariant and therefore to be misleadingly used for rotation estimation and to be rejected from further translation estimation. A large number of images for translation accumulation increases also the probability of the features to get lost by the tracker – this is also favored by the poor structured ground.

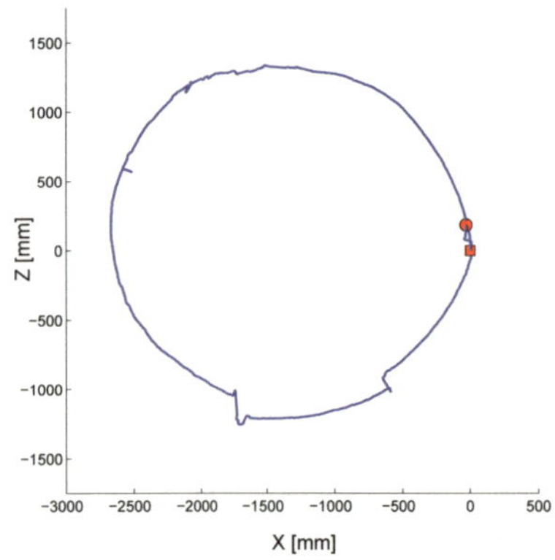


Fig. 9. The computed trajectory by the SSVL algorithm. The red square is the starting point and the red circle the endpoint of the path.

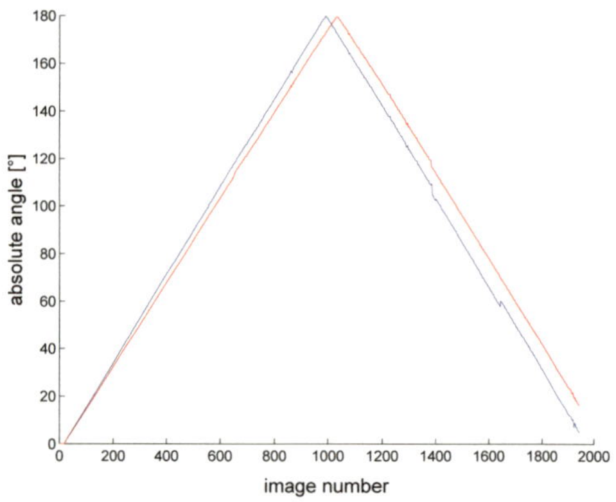


Fig. 10. These figure compares the absolute angle estimated by the $Z_\infty$ (red) and the SSVL (blue) algorithm.

Fig. 10 illustrates the computed angle of the $Z_\infty$ and SSVL algorithm and Table 3 shows some key-values regarding the result. The SSVL method proves its accuracy with only 3.8° absolute error, even though the ill-conditioned images. On the other hand, $Z_\infty$ accumulates an absolute error of 16° during the sequence. This seems to be a large error before taking into account two facts: The difference between both rotation estimations is not zero-mean as expected. This can be easily explained reminding again the problem to detect translation-variant landmarks at such a small translational motion. Close landmarks are also used for rotation estimation, as shown in Figure 12, and because the translation is always done in the same direction on a circular path, the estimation error corresponds to a measurement bias.
The other advantage of the SSVL approach is its bias reduction strategy. As explained in

Section 3.4.1, the pose estimation is based on initialized feature sets. The distance of each feature is estimated by stereo triangulation and is not updated anymore. The poses are always calculated respective to the image, where the landmarks were initialized. Thus, the only time, where a bias can be accumulated is when SSVL switches to a new feature set. The memory consumption grows continuously if the camera does not operate in a restricted environment. That prevents the SSVL algorithm to be used on platforms operating in large workspaces like outdoor mobile robots. In this experiment 116 sets were necessary. On the other hand, the $Z_\infty$ method does not need to provide a feature management. It estimates the pose from image to image, which makes it adequate for outdoor applications on resource-limited platforms without environment restrictions. As a drawback, it suffers from an error accumulation like all non-map-based algorithms without a global reference. Considering the relative error in the sense of error per accumulation step, we achieve a much smaller error for the $Z_\infty$ algorithm than the SSVL approach. For the $Z_\infty$ method with 1920 steps, an average error of $0.008°$ arises. Despite of the error accumulated by the mistakenly used translation-variant features, this is almost 4 times less than with SSVL, where 116 steps yield an error of $0.03°$ each. In Fig. 13 the computed angles for each frame are plotted. Only a few outliers can be identified, whereas $Z_\infty$ has less outlier than SSVL. This is not so problematic for SSVL due to its global pose estimation approach as described above. Fig. 14 shows again the differences between the estimated rotations of both methods for each image. The mean difference corresponds to the explained bias of $-0.02°$ with a standard deviation of $0.14°$.

Fig. 11. A screenshot of the SSVL algorithm at work. Only close landmarks are used.

Fig. 12. The small translations in that example yield also close features to be used for rotation estimation.

| | abs. X-rot. error (°) | abs. Y-rot. error (°) | abs. Z-rot. error (°) | no. of error accumulations | aver. Y-rot. error per acc. (°) | average rotation (°) |
|---|---|---|---|---|---|---|
| SSVL | 3.8 | 2.87 | -0.01 | 116 | 0.03 | 0.21 |
| $Z_\infty$ | 16.0 | -1.14 | 1.37 | 1920 | 0.008 | 0.19 |

Table 3. Keyvalues of the comparison between $Z_\infty$ and SSVL for the circle experiment. The absolute rotation error for all three axes, the number of error accumulations of each method, the average rotational error about the Y-axis for each error accumulation and the average rotation per frame are listed.

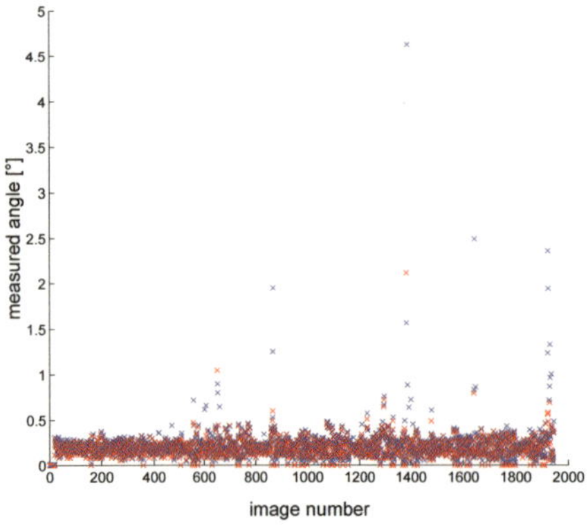


Fig. 13. This figures shows the angles computed for each frame - red the $Z_\infty$ and blue the SSVL results. The crosses mark the angle about the Y-axis. Only these marks can be identified in the plot.

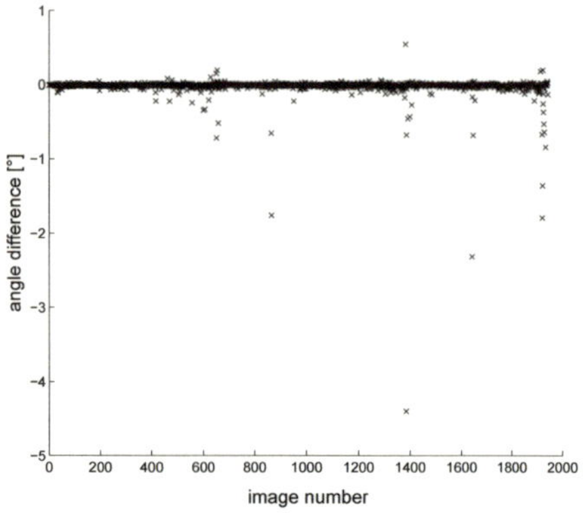


Fig. 14. This plot shows the difference between the $Z_\infty$ and SSVL rotation estimates along the Y-axis. Only a few outlier can be identified, which are detected by the SSVL's M-estimator. Hence, they do not have any effect on the final result.

### 3.5 Comparison of $Z_\infty$ to an inertial measurement unit

For this experiment the inertial measurement unit (IMU) "MTi" from Xsens has been mounted on a Guppy F046C camera from Allied Vision (see Fig. 15). The IMU provides the absolute rotation, based on the earth's magnetic field and three gyroscopes, and the acceleration of the device. The camera images and the IMU data were acquired with 25 $Hz$. Before the data can be evaluated, a camera to IMU calibration is necessary.

Fig. 15. The Xsens IMU is mounted on a Guppy camera. The data sets were acquired by holding the camera-IMU system out of a car.

#### 3.5.1 Camera to IMU calibration

Again, we benefit from the $Z_\infty$ algorithm – this time to calibrate the camera to the IMU. To speed up the calibration procedure we use a global tracker, namely SURF. The camera-IMU setup was rotated along all the axes while logging the sensor's data. From that data four images were chosen for the final calibration (see Fig. 16). The relative rotations $R_{imurel}$ and

$R_{camrel}$ were computed and the rotation between the two coordinate frames ${}^{imu}R_{cam}$ can be determined using the Euler axis-angle representation. The axis of the rotation between the frames is the cross-product of the rotation axes of the two relative rotations, while the angle corresponds to their dot-product.
An other way to calibrate the two coordinate frames would be to solve following equation

$$ {}^{imu}R_{cam}\, R_{camrel}\, {}^{imu}R_{cam}^{T} = R_{imurel} \tag{17}$$

This problem is known as $AX = XB$ problem and is not trivial – therefore, the upper approach has been used.

Fig. 16. These four images were used for camera to IMU calibration.

### 3.5.2 Comparison results: $Z_\infty$ - IMU

An IMU provides accurate information about the rotational speed. Combined with a magnetic field sensor, which measures the earth's magnetic field, it becomes a powerful angle sensor. The Xsens IMU comes up with an internal Kalman filter which fuses these two sensors. However, the accelerometer is a bad translation sensor, due to several facts. First, the measured acceleration has to be integrated twice in order to get the translational motion and second, this integration has to start when the device is static in order to get the proper velocity. An other problem are the forces which are also measured by an accelerometer like the earth's gravitation and the centrifugal force in curves.
Fig. 17 visualizes two data sets[1] acquired while holding the camera-IMU system out of a car. The left image shows a sine trajectory on a slightly left bent street. The right image describes the path through a turnaround. Again, this street is slightly left bent. Both runs prove the $Z_\infty$ based direction of translation estimation to outperform the IMU based acceleration integration. The main problem are the centrifugal forces, which, ones integrated, they are not removed from the velocity vector anymore and cause the estimated pose to drift away.
One of the main disadvantages of the $Z_\infty$ translation estimation is the lack of scale. Tests, where we tried to keep at least the scale constant over a run, failed because of the numerical conditions of the problem. Therefore, in this experiment the same scale resulting from the IMU measurement is used for weighting the $Z_\infty$ based translation vector.
Fig. 18 shows the angles measured along the axis perpendicular to the street, when driving the sinuous line illustrated in the left image of Fig. 17. Even though the IMU has a Kalman filter integrated, the $Z_\infty$ algorithm provides a much smooth measurement with less outlier. However, both, $Z_\infty$ and IMU have the problem to miss very small rotations, because the pixel displacement is to small to be tracked or the rotation lies within the noise of the gyroscopes respectively. This sensitivity problem can be solved at least for the $Z_\infty$ approach by simply keeping the same image as reference frame instead of changing it with each iteration. The algorithm can then swap respective to the magnitude of the measured rotation.

[1] The pictures at the right hand-side are "Google maps" screen-shots: http://maps.google.com

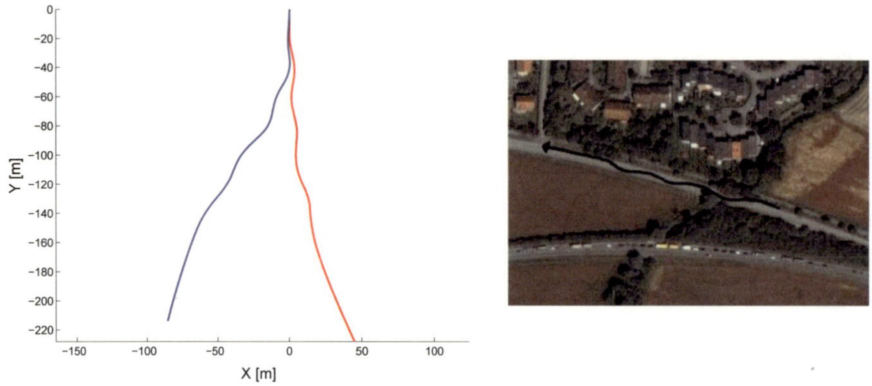


(a) Driving a sinuous line.

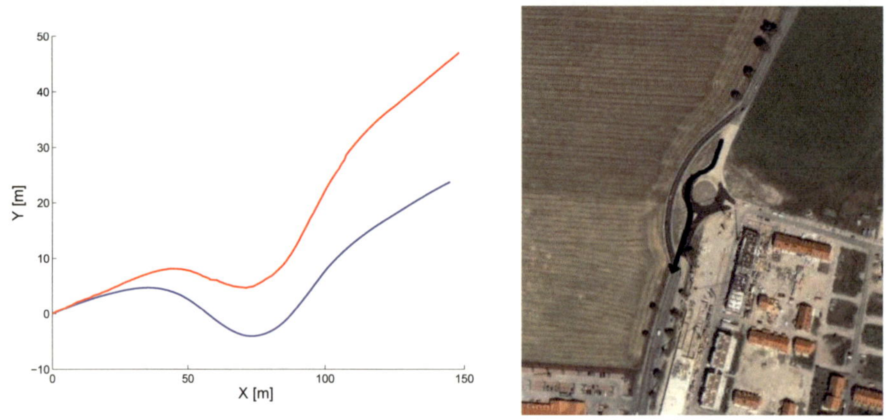


(b) Passing a turnaround.

Fig. 17. Two trajectories measured by the $Z_\infty$ algorithm (red) and the IMU (blue). In the pictures at the right-hand side the driven trajectories are sketched in black.

### 3.6 Obstacle Avoidance

Potential obstacles can be detected by evaluating the optical flow. Long vectors correspond to close or/and fast objects. The closer or the faster an object is moving respective to the camera, the larger the optical flow vectors become. For obstacle avoidance this relation is advantageous in a twofold manner. A certain length is identified, depending on the camera and the application, which defines a "dangerous" object. In a static environment, where only the camera is moving, such objects should be detected earlier if the motion of the camera is fast. This gives enough time to consider the obstacle for trajectory planning. In a dynamic environment, fast objects are much more dangerous than slow ones and, therefore, these should be detected earlier.

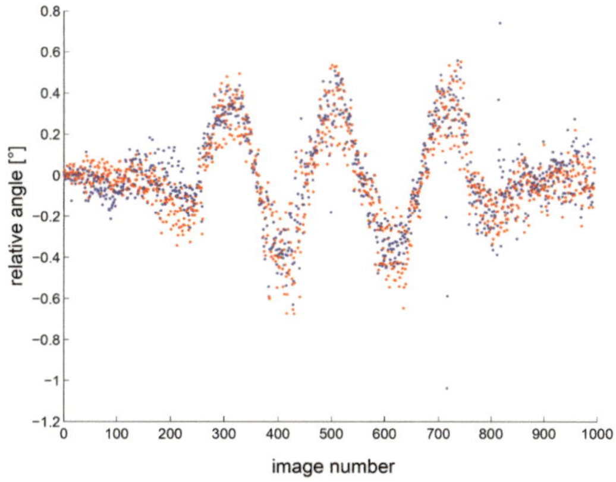


Fig. 18. This image shows the relative angle measured during the same run as in the left image of Fig. 17. The $Z_\infty$ rotation estimation is red, while the IMU data is blue.

(a) Input images for the obstacle detection: the black star is the epipole and the green vectors are the optical flow.

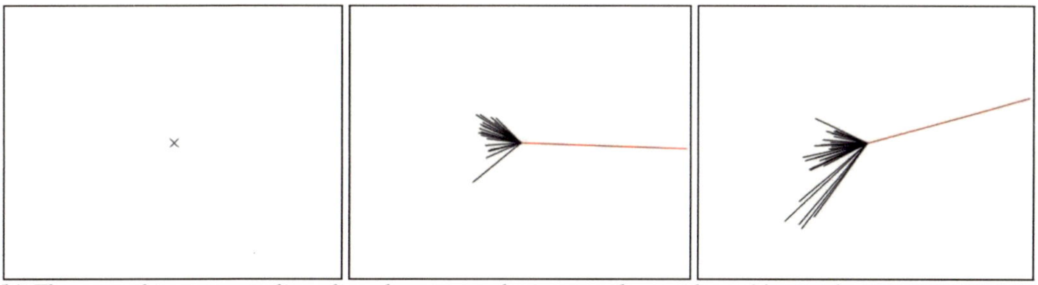

(b) These are the corresponding obstacle maps to the images above - the red line is the suggested swerve direction for obstacle avoidance.

Fig. 19. The $Z_\infty$ algorithm provides also an obstacle detection and their avoidance according to the built obstacle map.

The question remains how to determine where the objects are located. Without a global reference it is only possible to describe them relative to the camera. Therefore, the calculated epipole is used and objects are detected respective to the direction of translation. An obstacle map shows the characteristics of obstacles: their angle respective to the epipole, their level of danger as length and a suggested swerve direction for obstacle avoidance. Fig. 19 shows the output of the obstacle avoidance module on an image sequence acquired by the Pioneer equipped with a Marlin F046C camera as in the experiment of Section 3.4. The robot was programmed to move straight forward for the time of acquisition.

### 3.7 SURF based $Z_\infty$

The $Z_\infty$ algorithm has also been successfully tested with the global tracker SURF. Like every global tracker, SURF also provides much more outlier and the accuracy is in general worse than with local trackers. We used SURF when processing images acquired by an Octocopter, taking 10 megapixel pictures of the church in Seefeld, Germany. Fig. 20 shows the rotation and translation estimation result for one such image pair. The images where made in quite large distances. Thus, SURF features have been extracted and the rotation and also translation was computed for each feature set. Despite the high percentage of outlier and the poor accuracy, the $Z_\infty$ algorithm could successfully process the image sequence.

Fig. 20. Motion estimation based on SURF features.

## 4. Conclusion

We have presented a localization algorithm which is based on a monocular camera. It relies on the principle of separate estimation of rotation and translation, which significantly simplifies the computational problem. Instead of solving an octal polynomial equation, like it is the case with the 8-point algorithm, the SVD of a $3 \times 3$ matrix and the calculation of the intersection point cloud and its centroid are sufficient for motion estimation. Further, this separation allows an own uncertainty feedback for both motion components. No aprori knowledge is required for the algorithms nor any information needs to be carried between the image pairs. The algorithm requires very little processing time and is very memory-efficient. Nevertheless, we achieve comparable results to other localization methods and we have shown its accuracy in simulations and real world experiments. The drawback of the algorithm is its restriction to outdoor applications. However, especially for such applications the processing time and memory consumption is crucial and the proposed approach seems to fit the requirements for outdoor localization on mobile resource-limited platforms very well. Although, the rotation and translation is calculated analytically, the algorithm to separate these components, RANSAC, is an iterative method. Nevertheless, as explained in Section 2.3, from the second motion estimation on, the number of iterations necessary to split the data into translation-dependent and translation-invariant is reduced to a few. Therefore, $Z_\infty$ is a motion estimation

algorithm based on a monocular camera for outdoor mobile robots applications. Unlike other state of the art approaches, this image based navigation can run also on embedded, resource-limited systems with small memory and little processing power, keeping a high level of accuracy and robustness.
In the future we plan to adapt the algorithm to work also indoor by down-sampling the images and provide a hierarchical, iterative rotation estimation. An other point of research is the fusion of IMU and camera data, because experiments have shown that a camera and an IMU are two sensors which complement each other very well.

## Acknowledgment

We thank the Institute of Robotics and Mechatronics of the German Aerospace Center (DLR) for making available a Pioneer 3-DX and a Xsens IMU for our experiments. Further, we want to acknowledge the permission to process their image sequence of the church of Seefeld, Germany.
This work has been supported by the Technische Universität München (TUM) and the German Aerospace Center (DLR).

## 5. References

Alter, T. (1994). 3D Pose from 3 Points Using Weak Perspective., *IEEE PAMI, Vol. 16, No. 8* pp. 802–808.

Arun, K. S., Huang, T. S. & Blostein, S. D. (1987). Least-squares fitting of two 3-d point sets, *IEEE Trans. Pattern Anal. Mach. Intell.* **9**(5): 698–700.

Brown, M. Z., Burschka, D. & Hager, G. D. (2003). Advances in Computational Stereo, *IEEE Transactions on Pattern Analysis and Machine Intelligence* **25**(8): 993–1008.

Burschka, D. & Hager, G. (2001). Dynamic composition of tracking primitives for interactive vision-guided navigation, *Proc. of SPIE 2001, Mobile Robots XVI*, pp. 114–125.

Burschka, D. & Hager, G. D. (2003). V-GPS - image-based control for 3d guidance systems, *Proceedings of the IEEE/RSJ International Conference on Intelligent Robots and Systems.*

Davison, A. J. (2003). Real-time simultaneous localisation and mapping with a single camera, *International Conference on Computer Vision*, Vol. 2.

Dickmanns, E. D. & Graefe, V. (1988). Dynamic monocular machine vision, *Machine Vision and Applications* **v**: 223–240.

Faugeras, O. (1993). *Three-Dimensional Computer Vision*, The MIT Press.

Fennema, C., Hanson, A., Riseman, E., Beveridge, J. & Kumar, R. (1990). Model-directed mobile robot navigation, *IEEE Trans. on Robotics and Automation* **20**(6): 1352–69.

Fischler, M. A. & Bolles, R. C. (1981). Random sample consensus: a paradigm for model fitting with applications to image analysis and automated cartography, *Commun. ACM* **24**(6): 381–395.

Fukuda, T., Ito, S., Arai, F. & Yokoyama, Y. (1995). Navigation system based on ceiling landmark recogntion for autonomous mobile robot, *IEEE Int. Workshop on Intelligent Robots and Systems*, pp. 150–155.

Gonzalez-Banos, H. H. & Latombe, J. C. (2002). Navigation Strategies for Exploring Indoor Environments, *International Journal of Robotics Research, 21(10-11)* pp. 829–848.

Haralick, R., Lee, C., Ottenberg, K. & Nölle, M. (1994). Review and Analysis of Solutions of the Three Point Perspective Pose Estimation Problem, *International Journal on Computer Vision, 13(3)* pp. 331–356.

Hartley, R. & Zisserman, A. (2000). *Multiple View Geometry in Computer Vision*, Cambridge University Press.

Hebert, M., Pomerleau, D., Stentz, A. & Thorpe, C. (1995). Computer vision for navigation; the cmu ugv project, *Proceedings of the Workshop on Vision for Robots*, IEEE Computer Society Press, pp. 87–96.

Hirschmüller, H. (2008). Stereo processing by semiglobal matching and mutual information, *IEEE Transactions on Pattern Analysis and Machine Intelligence* **30**(2): 328–341.

Horn, B. K. P. (1987). Closed-formsolution of absolute orientation using unit quaternion, *Journal of the Optical Society of America A, Vol. 4, page 629-642* .

Horswill, I. (1992). Polly: A vision-based artificial agent, *AAAI*, pp. 824–829.

Kim, D. & Navatia, R. (1995). Symbolic navigation with a generic map, *Proceedings of the Workshop on Vision for Robots*, IEEE Computer Society Press, pp. 136–145.

Klein, G. & Murray, D. (2008). Improving the agility of keyframe-based SLAM, *10th European Conference on Computer Vision (ECCV'08)*, pp. 802–815.

Konolige, K. (2004). Large-scale Map-making, *In Proceedings of the National Conference on AI (AAAI)* .

Kriegman, D. J., Triendl, E. & Binford, T. O. (1989). Stereo vision and navigation in buildings for mobile robots, *IEEE Trans. on Robotics and Automation* **5**(6): 792–803.

Kuhnert, K.-D. (1990). Fusing dynamic vision and landmark navigation for autonomous driving, *IEEE Int. Workshop on Intelligent Robots and Systems*, pp. 113–119.

Lazanas, A. & Latombe, J.-C. (1992). Landmark-based robot navigation, *Proc. Am. Assoc. Art. Intell.*

Levitt, T. S., Lawton, D. T., Chelberg, D. M. & Nelson, P. C. (1987). Qualitative landmark-based path planning and following, *Proceedings of AAAI-87*, Morgan Kaufman, Los Altos, pp. 689–694.

Lucas, B. D. & Kanade, T. (1981). An iterative image registration technique with an application to stereo vision, *Proceedings of the International Joint Conference on Artificial Intelligence.*

Mair, E., Strobl, K. H., Suppa, M. & Burschka, D. (2009). Efficient camera-based pose estimation for real-time applications, *Proceedings of the IEEE/RSJ International Conference on Intelligent Robots and Systems.*

Matsumoto, Inaba, M. & Inoue, H. (1996). Visual navigation using view-sequenced route representation, *IEEE Conf. on Robotics and Automation*, pp. 83–88.

Matsumoto, Y., Ikeda, K., Inaba, M. & Inoue, H. (1999). Visual Navigation using Omniderctional View Sequence, *IEEE Int. Workshop on Intelligent Robots and Systems*, Kyongju, Korea, pp. 317–322.

Meier, W., Mair, E., Burschka, D. & Steinbach, E. (2009). Visual homing and surprise detection in cognitive mobile robots using image-based environment representations, *Proceedings of the IEEE International Conference on Robotics and Automation.*

Moravec, H. P. (1983). The Stanford cart and the CMU rover, *Proceedings of the IEEE* **71**(7).

Nilsson, N. J. (1984). Shakey the robot, *Technical Report 323*, SRI Int.

Nister, D. (2004). A Minimal Solution to the Generalised 3-Point Pose Problem, *CVPR* .

Ohno, T., Ohya, A. & Yuta, S. (1996). Autonomous Navigation for Mobile Robots Referring Pre-recorded Image Sequences, *IEEE Int. Workshop on Intelligent Robots and Systems*, Osaka, Japan, pp. 672–679.

Shi, J. & Tomasi, C. (1994). Good features to track, *Proceedings of the IEEE Conference on Computer Vision and Pattern Recognition.*

Strobl, K. H., Mair, E., Bodenmüller, T., Kielhöfer, S., Sepp, W., Suppa, M., Burschka, D. & Hirzinger, G. (2009). The self-referenced DLR 3D-modeler, *Proceedings of the IEEE/RSJ International Conference on Intelligent Robots and Systems*.

Tang, L. & Yuta, S. (2001). Vision Based Navigation for Mobile Robots in Indoor Environments by Teaching and Playing-back Scheme, *ICRA*, pp. 3072–3077.

Tards, J. D., Neira, J., Newman, P. M. & Leonard, J. J. (2002). Robust Mapping and Localization in Indoor Environments using Sonar Data, *International Journal of Robotics Research, 21(4)* pp. 311–330.

Thrun, S. (2002). Robotic mapping: A survey, *In G. Lakemeyer and B. Nebel, editors, Exploring Artificial Intelligence in the New Millenium. Morgan Kaufmann* .

Walker, M., Shao, L. & Voltz, R. (1991). Estimating 3-D location parameters using dual number quaternions, *CVGIP: Image Understanding* **54**(3): 358–367.

# Parallel Projection Based Self Localization Method for Mobile Navigation Applications

Shung Han Cho, Yuntai Kyong, Yunyoung Nam and Sangjin Hong
*Stony Brook University*
*USA*
We-Duke Cho
*Ajou University*
*KOREA*

## 1. Introduction

In navigation application of mobile robot, position is estimated based on odometry and the measurement obtained by peripheral devices such as ultra-sonar, range finder, and visual sensor. Odometry is usually unreliable due to the internal factor of encoder device and the external factor of slippage. These kinds of errors accumulated in navigation continuously affect the estimation accuracy (Borenstein & Feng, 1996b; Martinelli et al., 2007). Range finder and sonar are often used in navigation (Borenstein et al., 1996a; Borenstein et al., 1997; Groβmann & Poli, 2001; Demirli & Molhim, 2004). However such measurement is not reliable in highly-dynamic environments where radar or sonar beams can be frequently blocked or confused by moving objects such as people. They are not applicable to localization in large area either because of their limited range. Also, passive sensor requires active landmarks such as beacon, which requires modification of environment and is not practical especially in an outdoor environment. Moreover, interference among several mobile sensors causes inability to properly localize their locations.
Visual-information-based self-localization has advantages over above methods in two-folds. First, it does not require active landmarks (i.e., reference objects) such as beacon and also natural objects can serve as landmarks. The other advantage is that it is more effective and reliable in dynamic environment as the sensible range is not limited by the line-of-sight. These advantages have fostered much effort in research community of navigation application (Jang et al., 2002; Hayet et al., 2007; Se et al., 2002).
Reference objects can be either artificial or natural objects (Betke & Gurvits, 1997; Briggs et al., 2000; Thrun, 1998; Dao et al., 2004). We assume that reference objects can be uniquely identified by a mobile robot. The global coordinates of landmarks or reference objects are known to mobile robot. There have been extensive researches about robot localization with known reference objects. The conventional self localization algorithms use the bearings of the landmarks relative to each other. This is called "visual angle" formed by the rays from a mobile robot to each reference point (Sutherland & Thompson, 1993). The location of a mobile robot is estimated by finding the intersection point of the circles passing the

reference point. In the ideal case, the mobile robot is localized with three landmarks. However, although the matched landmarks from the image are found in the known map, the visual angle is usually distorted by the nonlinear property of a camera lens. Thus, the solution is estimated by minimizing the error of all possible landmarks pairs and the estimation error is minimized in proportion to the number of landmarks pairs (Betke & Gurvits, 1997; Cohen & Koss, 1993). Moreover, multiple solutions exist when all landmarks form a circle. Another approach uses perspective projection model to identify the relationship between the view point and the landmark. Although this method is simple in calculation, the performance is also worsened by the projection method without calibrating the camera nonlinearity (Dao et al., 2004; Liu & Zhou, 2007).

In this chapter, we propose a self localization method using a single visual image with the simple iteration technique. We assume that reference objects can be reliably extracted and identified. The proposed method identifies the relationship between the landmarks on the image and the known global reference points by the parallel projection model. The parallel projection model calibrates the non-linearity of optical lens distortion without computational complexity. The coordinates and the orientations are estimated with minimum relation equations by the simple iteration method. Our method can be used in large area with artificial or natural references as long as their coordinates are known and they can be reliably identified. The possible error source of the self localization method is explained and analyzed in terms of the performance of the self localization.

The rest of this chapter is organized as follows. Section 2 discusses background of self localization and problem description. In Section 3, a parallel projection model and its basic concept are discussed. Section 4 proposes a self localization algorithm for determining the coordinates and the orientation from external reference points. In Section 5, we present an experiment and the analysis of simulation results with extensive analysis of the error effect of the measurement on the performance of the algorithm. Section 6 concludes the chapter.

## 2. Background and Problem Description

### 2.1 Related Work on Self-Localization

In general, self localization with visual information is related to photometry. (Yuan, 1989) presents a general method for determining the three-dimensional position and orientation of an object relative to a camera based on a two-dimensional image of known feature points located on the object. (Horaud et al., 1989) analytically deals with the perspective $n$-point (P$n$P) problems with four correspondences of scene objects. Our approach does not analytically solve matrix transformation, but calculates the orientation and the location using a computationally efficient iterative algorithm.

A simple method for localization which allows a robot to determine its absolute position with a view of single landmark in one image is presented in (Dao et al., 2004). The landmark is chosen as the intersection of natural lines easily found in indoor environment such as edge of doors or walls. Another localization algorithm which is based on comparing the images taken in advance and taken during navigation is discussed in (Betke & Gurvits, 1997). In this scheme, the shape and the coordinate of images are stored in memory efficient format for quick retrieval and comparison. A similar method is presented where planar landmarks such as posters are used in visual localization of a mobile robot in indoor environment (Ayala et al., 2000). This algorithm has a restriction on the shape of landmark

and is not suitable in an open area. Similar to these approaches, our assumption requires a map of global coordinates of reference objects that can either be natural or artificial objects.
Scale Invariant Feature Transform (SIFT) developed for image feature generation in object recognition application is used for robot localization in (Se et al., 2001). The invariant characteristics of SIFT are captured by three images and stereo-matched to elect landmark that is later used to compute 3-D world coordinates relative to the robot. This algorithm does not require modification of the environment or map of reference objects, but needs more than two cameras with expensive computation to compare the features in the image.
In this chapter, we use a parallel projection model to simplify the computational complexity in determining the location and the orientation of a mobile robot. In the parallel projection model, a *virtual viewable plane* is defined to formulate the relationship between a real object and an actual camera. By using this model, the minimum of relation equations between the detected landmarks on the image and known corresponding reference points are obtained. This enables decrease of the computational complexity in finding the solution from all possible landmark pairs with the minimum mean square errors.
The parallel projection model is similar to the pinhole camera model in perspective projection model since the camera angle is relatively narrow and the size of the image obtained by the sensor is much smaller than real object area (Swaminathan et al., 2003). However, in the perspective projection model, the calibration process usually uses a flat plate with a regular pattern or the known several reference points (Lenz & Tsai, 1988; Heikkila & Silven, 1997; Lv et al., 2006). In addition, the calibration information should be updated for projection accuracy whenever the camera status is changed. This is not appropriate for highly dynamic application such as the robot navigation. On the other hand, the parallel projection model uses zoom factor instead of focal length and scale factor. This model simplifies the calibration process of the camera non-linear property by using the pre-obtained calibration table. Therefore, it is easily applied to the reference projection in the robot navigation. As a result, the relation equations are less affected by the camera model. This minimizes the self localization error due to the projection model. Fig.1 illustrates the difference of our parallel projection model and the perspective projection model (Lepetit & Fua, 2006) where $d_1$ and $d_2$ denote the distance of the camera and *Object Plane*. *Object Plane* contains the object and is parallel to *Actual Camera Plane*.

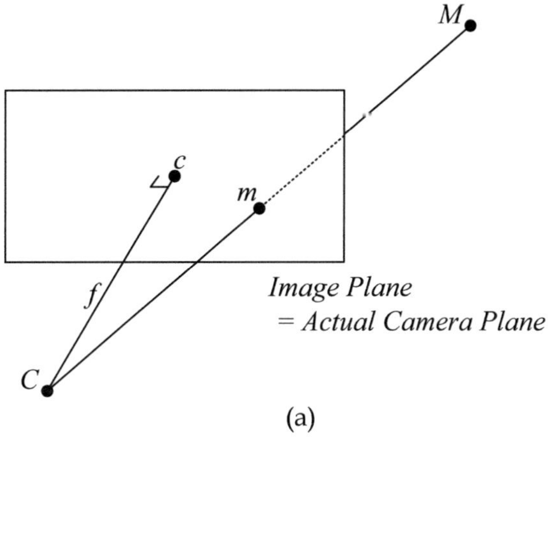


(a)

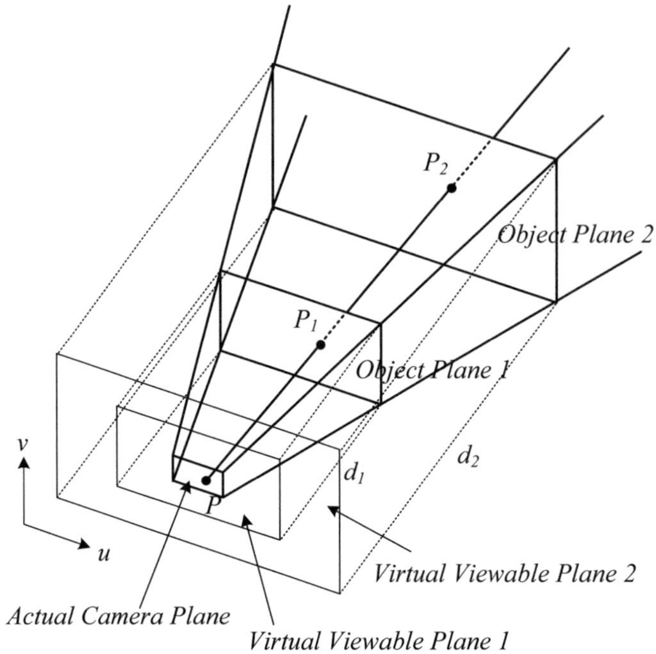


(b)

Fig. 1. Difference of perspective projection model and parallel projection model. (a) Perspective projection model. (b) Parallel projection model.

### 2.2 Problem Overview

In this chapter, we define the term "mobile sensor" to describe a mobile robot with visual sensor. We assume that mobile sensor can identify the reference objects and their coordinates are known (i.e., stored in a database). We limit our discussion to the self localization problem, and the method of how to identify such objects is not considered. The mobile sensor navigates by itself and visual image is obtained periodically. Based on the captured image data, the self localization comprises of determining both the orientation and the location of the mobile sensor. We use the global coordinate system, the origin of which is arbitrarily chosen but used to represent the coordinates of the reference points and the location of the mobile sensor. The objective is to utilize the projected reference points to determine the location of the mobile sensor. In this chapter, we focus on two aspects of the proposed method. The first aspect is to maintain the accuracy of the self localization and the second aspect is to maintain computational efficiency.

Since our method uses captured image data through the application of a typical digital imaging device, several sources of error are possible. Since the proposed approach relies on the point that is chosen from an area of pixels which is the projected image of the reference object, there can be inherent errors from image processing that selects one point from the area of an object image. This error can vary depending on many factors such as distance from mobile sensor to reference objects, distance between reference objects, etc. In addition, the non-linearity of the lens of imaging device causes shifting of projected point when the distance to reference points is not known. This shifting also affects the fidelity of self localization if compensation is not done.

Since the mobile sensor changes its location and orientation continuously, the reference points may be changed accordingly. The self location method should be computationally efficient by effectively utilizing available reference points. As we will show later in this chapter, the selection of reference points affects the self localization errors. When more than three reference points are inside the viewable range of the mobile sensor at the same time, the mobile sensor has freedom to choose the reference objects in such a way that can minimize such errors. Therefore, multiple reference objects should be strategically distributed to harness self localization of individual mobile sensor. A computationally efficient iterative algorithm using the relationship between the locations of reference points is proposed.

## 3. Characterization of Viewable Images

### 3.1 Basic Concept of Parallel Projection Model

In this subsection, we introduce the *parallel projection model*. In order to simplify the process of projected image on the camera device, we define three planes: the *object plane*, the *virtual viewable plane*, and the *actual camera plane* as shown in Fig.2. An object $P$, which is in the viewable area of a mobile sensor, is considered to be on the *object plane*. As opposed to the traditional model of a camera, in a parallel projection model, the object $P$ is projected in parallel onto *virtual viewable plane* and the projected point is denoted as $P_p$. The *virtual viewable plane* is parallel to the *object plane* with distance $d_p$. $L_c$ denotes the length of the *object plane*, which is the length of viewable area at distance $d_p$. $L_s$ denotes the length of the *actual camera plane* on which the measurement of projected image is done. The position of the projected object on a *virtual viewable plane* and an *actual camera plane* is denoted as $u_{pp}$ and $u_p$, respectively.

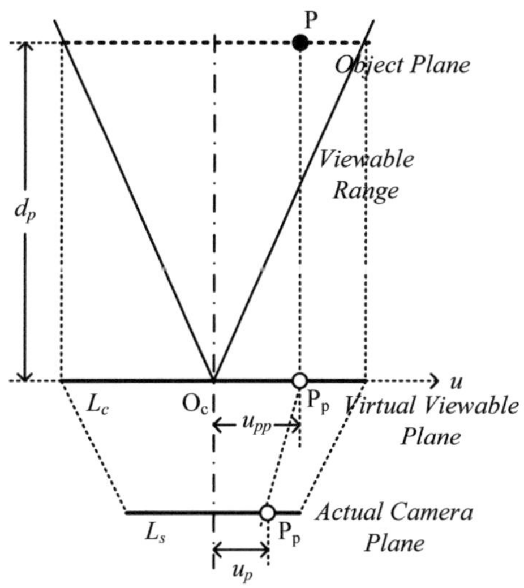


Fig. 2. Parallel projection model and relationship with actual image projected on visual sensor (Camera).

In the parallel projection model, a real object located on an *object plane* is projected to an *actual camera plane* through a *virtual viewable plane*. Hence, as formulated in (1), $u_{pp}$ is expressed as $L_c$, $L_s$ and $u_p$ using the proportionality of the length of a *virtual viewable plane* and an *actual camera plane*.

$$u_{pp} = u_p\left(\frac{L_c}{L_s}\right) \tag{1}$$

The position of the real object can be obtained from $u_{pp}$ and $d_p$, once the ratio of $L_c$ and $d_p$ is known. This ratio is defined to be a zoom factor, $z$, which is the property of the image device.

$$z = \frac{d_p}{L_c} \tag{2}$$

**3.2 Relationship of Reference Points on Different Planes**

Given the parameters of visual sensor, $z$ and $L_s$, we can derive the relationship between a projected reference point on the *virtual viewable plane* and one on the *actual camera plane* with their distance to the origin of each plane. The origin of each plane is defined to be the cross point between a plane and its perpendicular line, the view axis, which also crosses the location of the mobile sensor. Specifically, the origin of the *actual camera plane* is the axis of panning. In Fig. 3, the origin on the actual camera plane is denoted as $O_c$ and the origins of the virtual viewable planes are denoted as $O_{v1}$ and $O_{v2}$, respectively. Even though the planes are rotated as visual sensor is panned, the relationship based on the distance on each plane remains unchanged. When $p_1$ and $p_2$ denote the distance to the view axis on each virtual plane, and $i_1$ and $i_2$ denote the corresponding measurement on the *actual image plane*, using (1) and (2), we can derive

$$\frac{p_1}{i_1} = \frac{L_{p1}}{L_s} = \frac{D_1}{z_1 L_s}, \qquad \frac{p_2}{i_2} = \frac{L_{p2}}{L_s} = \frac{D_2}{z_2 L_s}, \tag{3}$$

where $z_1$ and $z_2$ are the zoom factors of the mobile sensor corresponding to distance, $D_1$ and $D_2$, from the *actual camera plane* to the *object plane* for each reference point. $L_s$ is the size of the image on which $i_1$ and $i_2$ are measured.

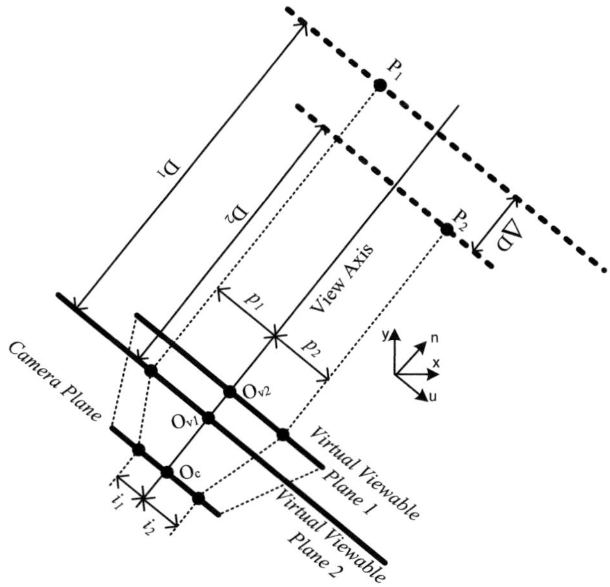


Fig. 3. Self localization with two reference points.

In practice, the location of a projected point on the image device is obtained from the image processing of the target objects such as edge detection and/or feature extraction. Thus, the projected point on the image device usually contains some uncertainty. In later sections, how this uncertainty affects self localization algorithm is discussed in detail.

## 4. Self Localization Algorithm

### 4.1 Self Localization with Known Orientation

In this subsection, we introduce self localization when two reference points and the orientation of visual sensor are known. We define $\theta$ as the angle formed between the camera plane and global $x$ axis. We define a unit vector $\hat{u}$ to have the direction of the camera plane and $\hat{n}$ to be the unit vector along the view axis, the direction to which the visual sensor is facing. Therefore, $\theta$ is the angle between the $x$ axis and $\hat{u}$. Using the following equations, we can obtain values $p_1$, $p_2$, $D_1$ and $D_2$ in (3). In the previous section, we described $i_1$ and $i_2$ as the distance to the view axis on the camera plane, but, from now on, they are considered to be able to have negative values when the projected reference point is in the left side of the view axis. It does not change the distance relationship described in the previous section by allowing $p_1$ and $p_2$ to have negative values as well when they are also in the left side of the view axis.

For $p_1$ and $p_2$,

$$p_1 = \overrightarrow{CP_1} \cdot \hat{u},$$
$$p_2 = \overrightarrow{CP_2} \cdot \hat{u}. \tag{4}$$

For $D_1$ and $D_2$,

$$D_1 = \overrightarrow{CP_1} \cdot \hat{n},$$
$$D_2 = \overrightarrow{CP_2} \cdot \hat{n}. \tag{5}$$

Here, $\overrightarrow{CP_1}$ and $\overrightarrow{CP_2}$ denote the vectors from the location of the mobile sensor, $O_c$, to each reference point. Above the two sets of equations are simply the decomposition of $\overrightarrow{CP_1}$ and $\overrightarrow{CP_2}$ to the $\hat{u}$ and $\hat{n}$ components.

Then, from (3), we have

$$\frac{p_1}{i_1} = \frac{\overrightarrow{CP_1} \cdot \hat{u}}{i_1} = \frac{\overrightarrow{CP_1} \cdot \hat{n}}{zL_s},$$
$$\frac{p_2}{i_2} = \frac{\overrightarrow{CP_2} \cdot \hat{u}}{i_2} = \frac{\overrightarrow{CP_2} \cdot \hat{n}}{zL_s}. \tag{6}$$

or

$$\overrightarrow{CP_1} \cdot (\hat{u} - \frac{i_1}{zL_s} \hat{n}) = 0,$$
$$\overrightarrow{CP_2} \cdot (\hat{u} - \frac{i_2}{zL_s} \hat{n}) = 0. \tag{7}$$

In (6), $\overrightarrow{CP_1}$ and $\overrightarrow{CP_2}$ can be expressed with their $x$ and $y$ components in global $x$-$y$ coordinate system as

$$\overrightarrow{CP_1} = (P_{1x} - x_c)\hat{x} + (P_{1y} - y_c)\hat{y},$$
$$\overrightarrow{CP_2} = (P_{2x} - x_c)\hat{x} + (P_{2y} - y_c)\hat{y}, \tag{8}$$

where $P_{1x}$ and $P_{2x}$ are the $x$ components of $P_1$ and $P_2$, respectively, and $P_{1y}$ and $P_{2y}$ are the $y$ components of $P_1$ and $P_2$, respectively.

The $x$ and $y$ components of the dot products are expressed by

$$
\begin{aligned}
(\hat{u} - \frac{i_1}{zL_s}\hat{n})\cdot\hat{x} &= \cos\theta + \frac{i_1}{zL_s}\sin\theta, \\
(\hat{u} - \frac{i_1}{zL_s}\hat{n})\cdot\hat{x} &= \sin\theta - \frac{i_1}{zL_s}\cos\theta, \\
(\hat{u} - \frac{i_2}{zL_s}\hat{n})\cdot\hat{x} &= \cos\theta + \frac{i_2}{zL_s}\sin\theta, \\
(\hat{u} - \frac{i_2}{zL_s}\hat{n})\cdot\hat{x} &= \sin\theta - \frac{i_2}{zL_s}\cos\theta.
\end{aligned}
\tag{9}
$$

Then, (7) are equivalent to

$$
\begin{aligned}
(P_{1x} - x_c)(\cos\theta + \frac{i_1}{z_1 L_s}\sin\theta) + (P_{1y} - y_c)(\sin\theta - \frac{i_1}{z_1 L_s}\cos\theta) = 0, \\
(P_{2x} - x_c)(\cos\theta + \frac{i_2}{z_2 L_s}\sin\theta) + (P_{2y} - y_c)(\sin\theta - \frac{i_2}{z_2 L_s}\cos\theta) = 0.
\end{aligned}
$$

Let us introduce intermediate variables to simplify the final equations for $x_c$ and $x_y$. They are

$$
\begin{aligned}
\alpha_1 &= \cos\theta + \frac{i_1 \sin\theta}{zL_s}, \\
\beta_1 &= \sin\theta - \frac{i_1 \cos\theta}{zL_s}, \\
\alpha_2 &= \cos\theta + \frac{i_2 \sin\theta}{zL_s}, \\
\beta_2 &= \sin\theta \quad \frac{i_2 \cos\theta}{zL_s}.
\end{aligned}
$$

Thus, we can obtain the coordinate of mobile sensor expressed as

$$
\begin{aligned}
x_c &= -\frac{\alpha_1\beta_2 P_{1x} + \beta_1\beta_2 P_{1y} - \beta_1\alpha_2 P_{2x} - \beta_1\beta_2 P_{2y}}{\beta_1\alpha_2 - \alpha_1\beta_2}, \\
y_c &= -\frac{\alpha_1\alpha_2 P_{1x} - \alpha_2\beta_1 P_{1y} + \alpha_1\alpha_2 P_{2x} + \alpha_1\beta_2 P_{2y}}{\beta_1\alpha_2 - \alpha_1\beta_2}.
\end{aligned}
\tag{10}
$$

Since the reference object is projected onto the camera plane, the coordinates of reference objects correspond to the centroid of the reference objects. Then, we can obtain the coordinate of mobile sensor using (10). However, even though the coordinates of reference points are accurately known in advance, the measurement $i_1$ and $i_2$ on the image may not be corresponding to the true reference points. Possible sources of the uncertainties may arise from the pixel resolution of the image planes as well as incorrect determination of the centroid of the detected reference shape. This uncertainty is evident even with perfect lens view characteristics. We will introduce the effect of non-linearity of camera lens in the later section.

### 4.2 Orientation Determination

Thus far, we have considered determining the position of the mobile sensor when its orientation is given. However, it is necessary to determine the orientation of the mobile sensor as well as its position. Determining both position and orientation concurrently requires a third reference point. From the parallel projection model, using (3), we can obtain the angle of the line that crosses the center of the camera plane and the reference point, where the angle is formed between the line and the camera plane. With two reference points, we have two lines with known angle respect to the camera plane, and we know each reference point is on one of them. Since there are infinite numbers of ways to position a line segment having two reference points as vertexes sitting on those lines, we cannot determine the position and the orientation of the mobile sensor with two reference points. With one more reference point, the problem becomes to position three vertexes of a triangle with known length onto three lines with a known angle. There is only one way to position the triangle in such way if we limit the orientation of the mobile sensor to $180^\circ$ range. From the above, we can conclude that three reference points are enough for determining both the orientation and the location of the mobile sensor when the general directions of the reference points are assumed in the following discussion.
We can find a solution by solving three simultaneous solutions using (10), but its non-linearity requires large computational complexity to be implemented on resource limited devices, such as mobile robot. Instead, we developed an effective iteration algorithm which involves solving only two simultaneous equations and the solution is given in (10). In our iteration approach, we determine the orientation of the mobile sensor. Once we found the orientation, we obtain the location of the mobile sensor using (10).
For a given orientation angle, $\theta$, using (10), we can obtain two sets of coordinates, $(x_{c1}, y_{c1})$ and $(x_{c2}, y_{c2})$ using two pairs of reference points out of three. When three reference points, $P_1$, $P_2$ and $P_3$ are chosen for self-localization, using $P_1$ and $P_2$, we have

$$x_{c1} = -\frac{\alpha_1\beta_2 P_{1x} + \beta_1\beta_2 P_{1y} - \beta_1\alpha_2 P_{2x} - \beta_1\beta_2 P_{2y}}{\beta_1\alpha_2 - \alpha_1\beta_2},$$
$$y_{c1} = -\frac{\alpha_1\beta_2 P_{1x} + \beta_1\beta_2 P_{1y} - \beta_1\alpha_2 P_{2x} - \beta_1\beta_2 P_{2y}}{\beta_1\alpha_2 - \alpha_1\beta_2}, \quad (11)$$

and by using another pair, $P_2$ and $P_3$, we have

$$x_{c2} = -\frac{\alpha_2\beta_3P_{2x} + \beta_2\beta_3P_{2y} - \beta_2\alpha_3P_{3x} - \beta_2\beta_3P_{3y}}{\beta_2\alpha_3 - \alpha_2\beta_3},$$
$$y_{c2} = -\frac{\alpha_2\beta_3P_{2x} + \beta_2\beta_3P_{2y} - \beta_2\alpha_3P_{3x} - \beta_2\beta_3P_{3y}}{\beta_2\alpha_3 - \alpha_2\beta_3}. \quad (12)$$

In order to develop an effective iterative strategy, we investigate the behavior of the differences of the two coordinates, $d_{cx} = x_{c1} - x_{c2}$ and $d_{cy} = y_{c1} - y_{c2}$ while varying the angle of orientation. We define *error_distance* as

$$error\_distance(\theta) = \sqrt{(x1 - x2)^2 + (y1 - y2)^2} \quad (13)$$

where $\theta$ is the angle of the mobile sensor's orientation. Fig. 4 shows the behavior of this function, as $\theta$ varies from 0° to 180°. The figure shows a case when the true orientation of the mobile sensor is 80°, and, at this angle, *error_distance*($\theta$) becomes zero. We call this angle *solution point*. Around this point, the function is found to be symmetric and periodical with 180°. Based on the characteristics of the simulated distance error function, there is always one global minimum between 0° to 180°. Thus, there will be two possible orientations between 0° to 360°. The good initial point is approximated by the direction which a robot moves toward.

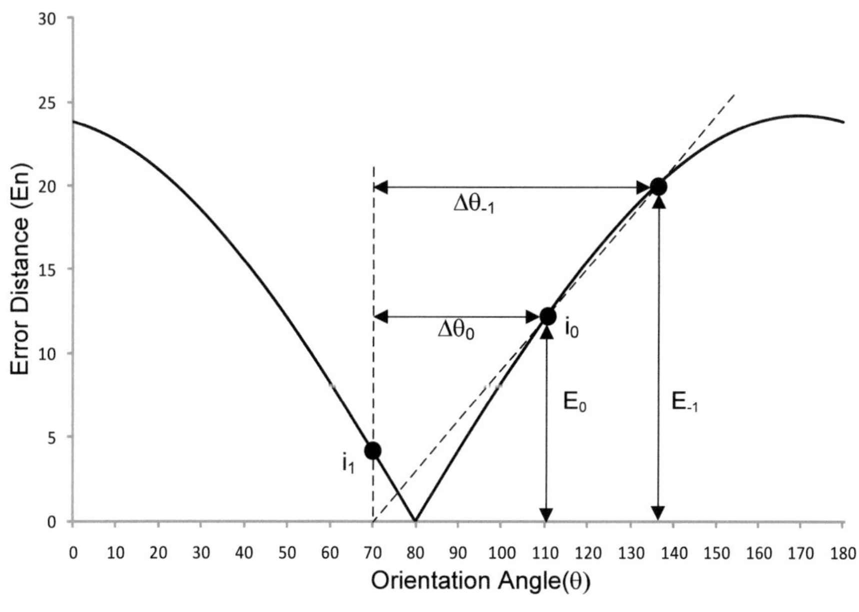


Fig. 4. Distance error function as a function of orientation error. The slope estimations of initial iteration points are shown.

If we start iteration inside 45° range from *solution point*, and if we follow down the slope, it is guaranteed to find the solution. In order to find such an initial iteration point, $i_0$, inside the range, we arbitrarily choose two angles separated with 90°. Since one of them will be inside 45° range from the *solution point* and the other one will be outside, we simply choose the angle that gives smaller *error_distance*($\theta$) as our initial iteration angle, $\theta_0$.
Once we choose an initial point, we have the initial iteration point, $i_0$, determined by $\theta$ and *error_distance*($\theta$). In order to estimate the slope at that point, another *error_distance* function is evaluated using $\theta_{-1}$ which is chosen to be very close to $\theta_0$ such as 0°. We call this estimated slope as $slope_0$ and the relationship of the initial iteration variables are

$$\begin{aligned} \Delta\theta_0 &= \theta_0 - \theta_{-1}, \\ \Delta E_0 &= E_0 - E_{-1}, \\ slope_0 &= \frac{\Delta E_0}{\Delta\theta_0}, \end{aligned} \tag{14}$$

where $E_n = error\_distance(\theta_n)$.
Depending on the sign of the estimated slope, we choose the direction of the iteration, *dir*. If $slope_0 > 0$, we set $dir_0 = -1$, and, swap $\theta_0$ with $\theta_{-1}$, and, $E_0$ with $E_{-1}$. Otherwise, $dir_0 = 1$.
First, by assuming the slope at our initial point being close to be linear, we choose the next angle where the slope line crosses *x*-axis. Since the slope increases as approaching to the *solution point*, the next iteration step will overshoot albeit very close to *the solution point*. As shown in Fig. 4, the *error_distance* function evaluated at $\theta_1$ is the other side of the *solution point*.
From the second step, instead of using the slope, we choose the angle of next iteration step based on the two previous angle and *error_distance* evaluated with them. In this case, since the two triangles shown in Fig. 5 are approximately proportional near to the solution point, the angle for the next step is evaluated by the following equations. From

$$\frac{\Delta\theta_n - \Delta\theta_{n+1}}{E_{n-1}} = \frac{\Delta\theta_{n+1}}{E_n}, \tag{15}$$

the next iteration angle is calculated as

$$\Delta\theta_{n+1} = dir \times \frac{E_n}{E_n + E_{n-1}} \times \Delta\theta_n. \tag{16}$$

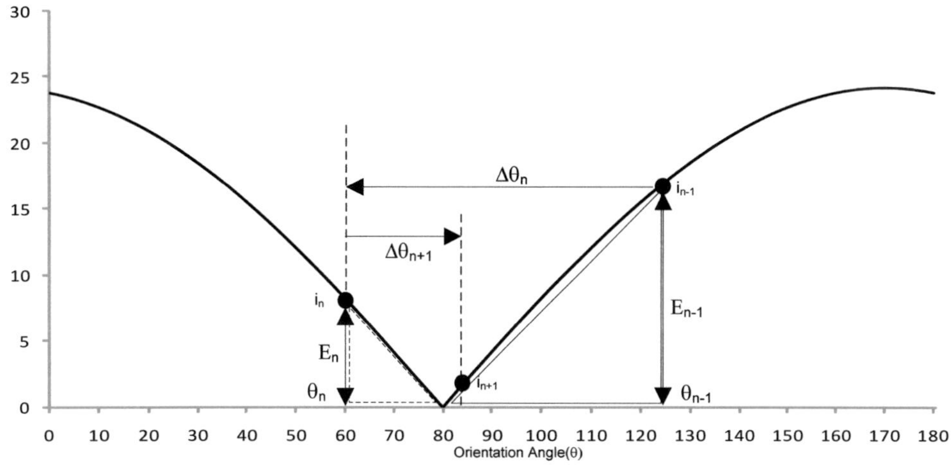


Fig. 5. Convergence steps of the iteration algorithm.

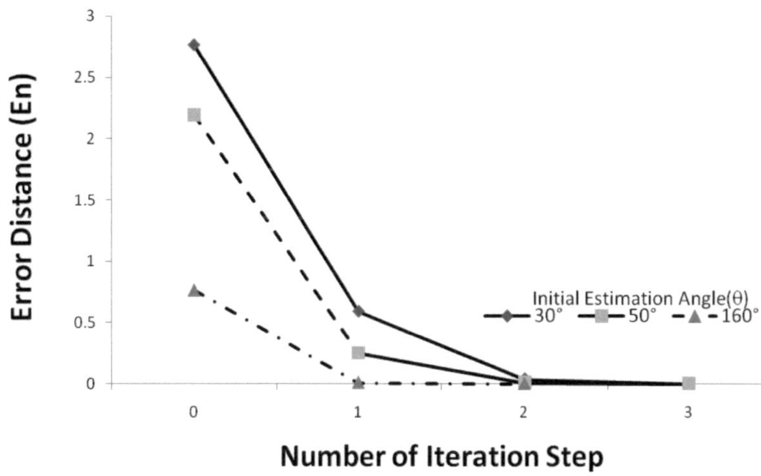


Fig. 6. Convergence of the iteration algorithm as a function of the number of iterations.

The iteration continues until, the change of estimated orientation, $\Delta\theta_n$, becomes smaller than the threshold value, $\varepsilon$. Otherwise, we change the direction, $dir = dir \times (-1)$ and continue. Fig. 6 shows that the algorithm is converging very rapidly. The figure shows the iteration algorithm is applied when three initial estimation angles are used, 10°, 40° and 80°. The value of *error_disance*($\theta$) is plotted at each iteration step. Note that the iteration starts with two initial iterations (i.e., as shown in the figure, the iteration starts at —1 index).
Fig. 7 illustrates the importance of the orientation error on the localization. The plot shows the displacement error for several orientation errors. Throughout this chapter, the displacement error is defined as

$$\sqrt{(x_{c,true} - x_{c,est})^2 + (y_{c,true} - x_{y,est})^2}, \tag{18}$$

where $(x_{c,\text{true}}, y_{c,\text{true}})$ is the true coordinate and $(x_{c,\text{est}}, y_{c,\text{est}})$ is the estimated coordinate. The results are plotted as a function of $\Delta P$ and $D_{\max}$ where $\Delta P$ represents the separation (in parallel to the projected plane) between the reference points and the $D_{\max}$ represents the largest distance (perpendicular to the projected plane) of the reference points. The angle determination is very critical since the displacement is computed after the orientation is determined. Thus, if the orientation computation is not accurate, the localization may not successfully estimate the coordinates.

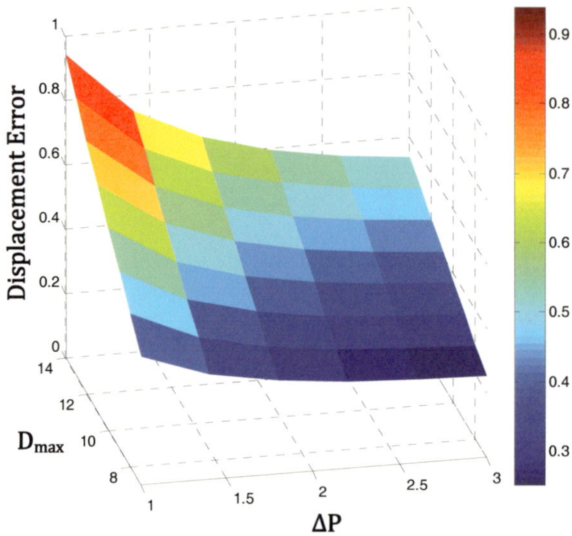


(a)

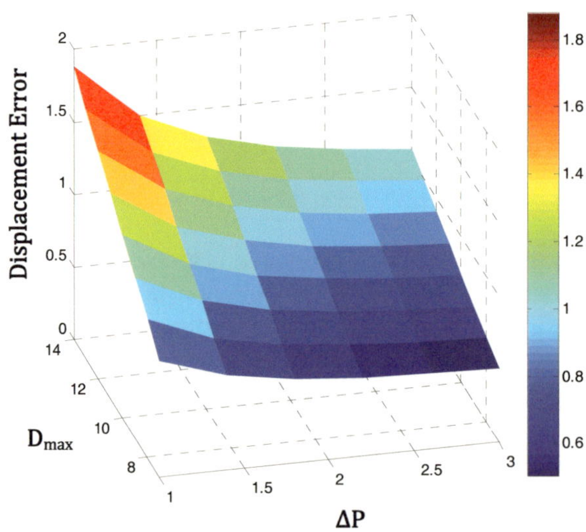


(b)

Fig. 7. Displacement error as a function of the orientation error $\Delta\theta$. (a) Orientation error, $\Delta\theta = 2$. (b) Orientation error, $\Delta\theta = 4$.

### 4.3 Lens Distortion

The non-linear distortion of non-ideal lens affects the scale in the parallel projection model. Since the distances between the mobile sensor and the references are not known, we compute the coordinate of the mobile sensor using the value of $z$ corresponding to the value when the distance is large (i.e., the value of $z$ converges to a specific value). Once initial value of the coordinate is obtained, we use specific values of $z$ (i.e., the one for the first reference and the other for the second reference) to compensate for the non-linearity to obtain the final coordinate of the mobile sensor. Note that the zoom factor depends on the distance from the imaging device, as well as the distance from the axis of the lens. The calibration table for the zoom factor is constructed according to the distance from the imaging device and the distance from the axis of the lens at the interval of 0.3m (Park et al., 2008).

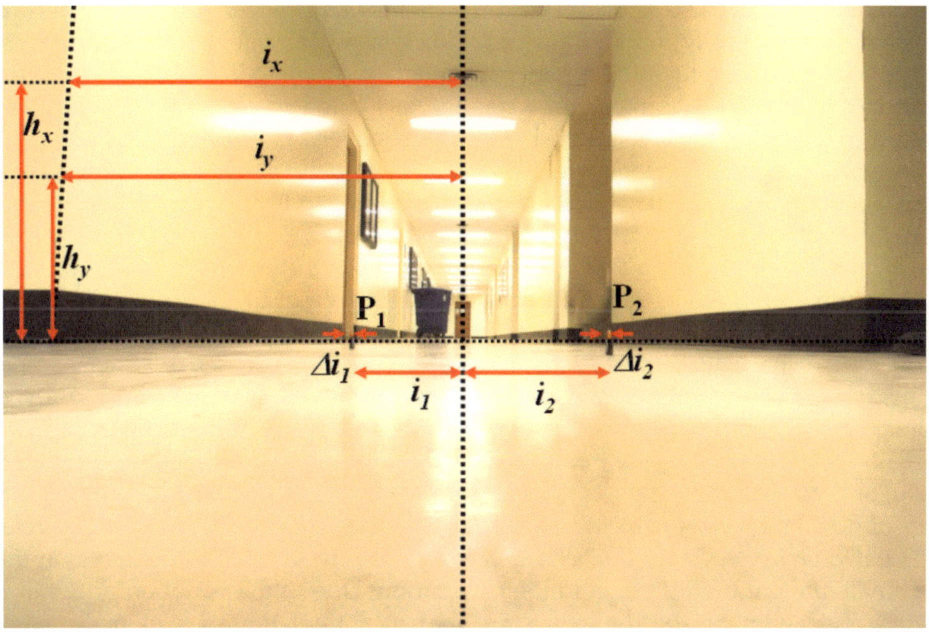


Fig. 8. Nonlinear effect of the lens on estimation error.

Fig. 8 illustrates the nonlinear effect on the coordinate determination. In the figure, the orientation is chosen to be 0 (i.e., view axis is perpendicular to the $x$ axis). The camera is located at (4.8m, 3.0m) with respect to the global coordinate (0, 0) and the first reference $P_1$ is located at (4.5m, 4.5m) and the second reference $P_2$ is located at (5.1m, 5.1m). The lens is set at 17mm zoom range. The value of $L_s$ is 18.4cm (i.e., the image size in $x$ direction). The projected position of the first reference $i_1$ is at 3.85cm and the projected position of the second reference $i_2$ is at 2.45cm from the center of the image. These positions are from the center of the reference objects to the center of the image. The reference objects both have finite widths of 0.16cm and 0.12cm corresponding to $\Delta i_1$ = 0.0087 and $\Delta i_2$ = 0.0065, respectively. In this chapter, $\Delta i$ is defined as the uncertainty range or the measurement error with the respect to the overall image size (i.e., 18.4cm in the example). Since the centroid of the reference points are determined from the image, potential measurement errors will be within $\Delta i$. The actual zoom factor corresponding to the first reference $z_1$ is 0.8238 and the zoom factor corresponding to the second reference $z_2$ is 0.8119. In the initial estimation, the zoom factor corresponding to the infinite distance of 0.8 is used. The estimated coordinate without compensation is (4.8017m, 3.03815m) which is 3.87cm off from the true position of the mobile sensor.

### 4.4 Effects of Reference Measurement Uncertainty

The measurement error directly affects the accuracy of the localization including the orientation. Since the reference object is usually projected as an area on the image, for applying the parallel projection model, one point should be determined from the area. In the parallel projection model, we only take the horizontal component of the determined point. If we designate the coordinate of the reference object as its centroid, we can choose its position on the camera plane, $i$, as the center of the projected area. However, if the shape of reference object is not symmetrically round, there is always certain amount of error in determining $i$. This type of error is usually influenced by the image processing in finding the boundary of the projected references. Quantization error due to limited resolution of the visual sensor may affect the accuracy but it is not the biggest source of error.

Another source of measurement error is that the reference object is not projected in the center of the horizontal line (i.e., illustrated as dotted line in Fig. 8). This is because the lens distorts the projected objects. In the figure, the edge of a wall has one coordinate value. However, multiple reference values can be obtained for the edge of a wall. For example, both $i_x$ measured at $h_x$ and $i_y$ measured at $h_y$ should represent the same coordinate, but the projected values are different. The difference between $i_x$ and $i_y$ contributes as $\Delta i$ in the localization. However, (Park et al., 2008) shows that the projection error is compensated by using the parallel projection model if the height of a landmark is known. Since the height of a landmark can be known, the localization performance is not worsened in this case.

The orientation errors due to the incorrect determination of $i_1$ and $i_2$ are illustrated in Fig. 9. Figs. 9(a) and 9 (b) show the results for two different values of the measurement errors. The average of orientation error is measured using all the possible combinations of the reference points located on 5m by 5m grid. Each reference point is located at 50cm interval in the grid with a small amount of additional random variation. Due to the variation of $i_1$ and $i_2$, the estimated orientation can be different from the true orientation of a mobile sensor. The figures show that overall range of error and standard deviation are larger when $\Delta i$ = 0.03 than when $\Delta i$ = 0.05. Also, when the mobile sensor is closer to the reference points, the orientation error is very critical since the actual coordinates are obtained by computing the orientation first.

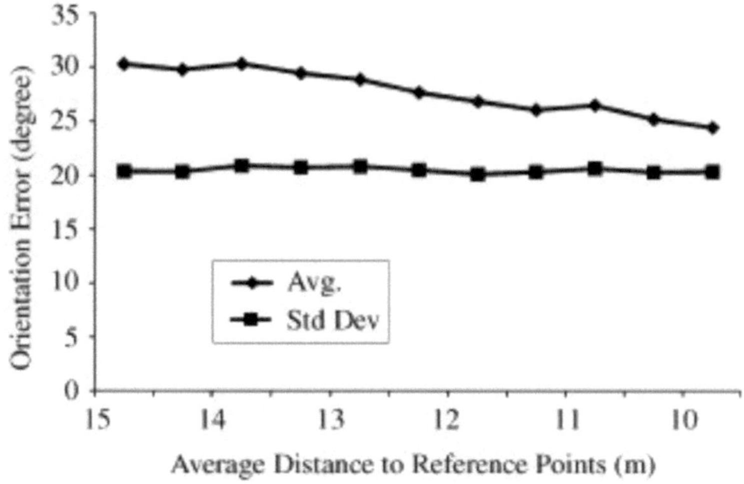


(a)

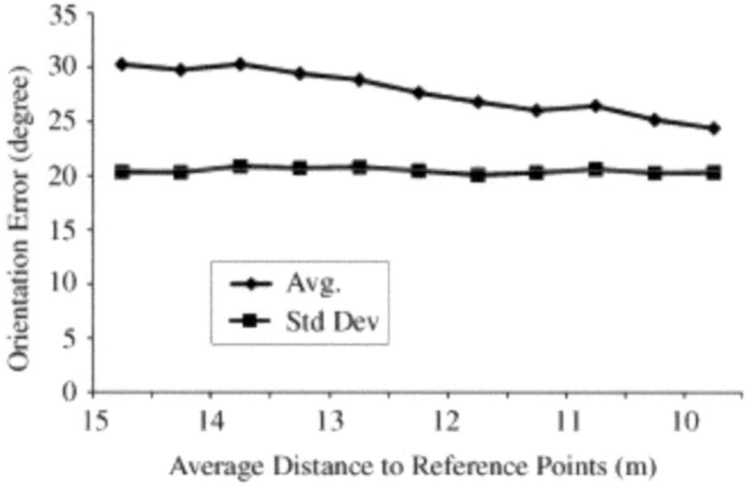


(b)

Fig. 9. Orientation error, $\Delta\theta$, as a function of the projected measurement error $\Delta i$. (a) Orientation error, $\Delta i = 0.03$. (b) Orientation error, $\Delta i = 0.005$.

The similar results are obtained for the displacement errors in Fig. 10. Similarly, Figs. 10(a) and 10(b) show the results for two different values of the measurement errors. The displacement error is plotted as a function of two variables, $D_{max}$, the largest distance from the mobile sensor to the reference points, and $\Delta P$, the distance between two most separated reference points measured from the *view axis* of the mobile sensor. As the figure shows, the overall error range increases as $\Delta i$ increases.

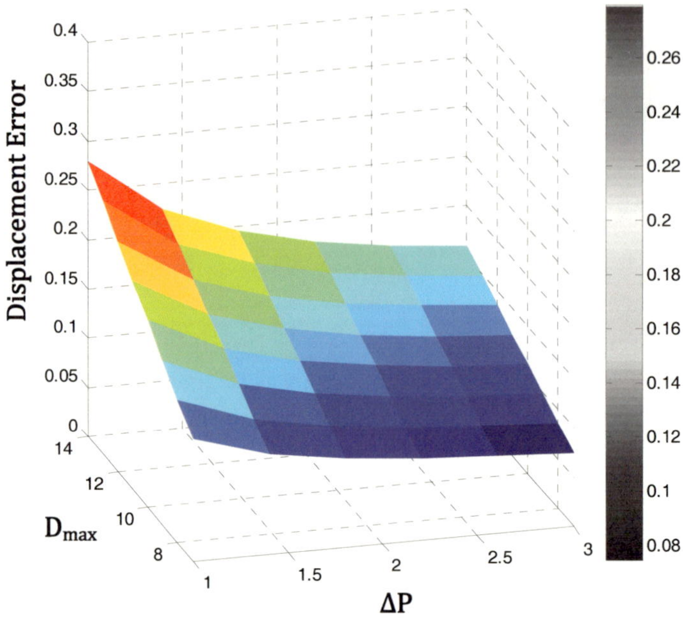


(a)

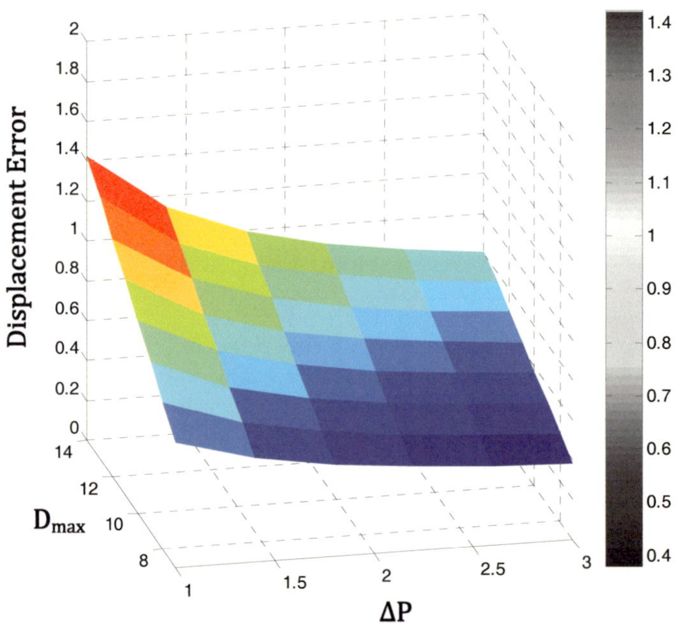


(b)

Fig. 10. Displacement error as a function of projected measurement error $\Delta i$. (a) Measurement error, $\Delta i = 0.01$. (b) Measurement error, $\Delta i = 0.05$.

Both results show that the algorithm is more prone to error when distance from the mobile sensor to the reference points is larger, and when the references are closer to one another. From Fig. 9 and Fig. 10, we know that estimation error is smaller when the distance between the reference objects along the camera plane is larger. Since our iteration algorithm uses two pairs of reference objects out of three pairs that can be made from three reference objects, given three reference points, $R_1$, $R_2$ and $R_3$, we can choose two pairs that give maximum distance on the camera plane to minimize error. This selection criterion can be applied also when there are more than three reference objects viewable and three of them need to be selected for self localization.

## 5. Analysis and Simulation

### 5.1 Experimental Setup

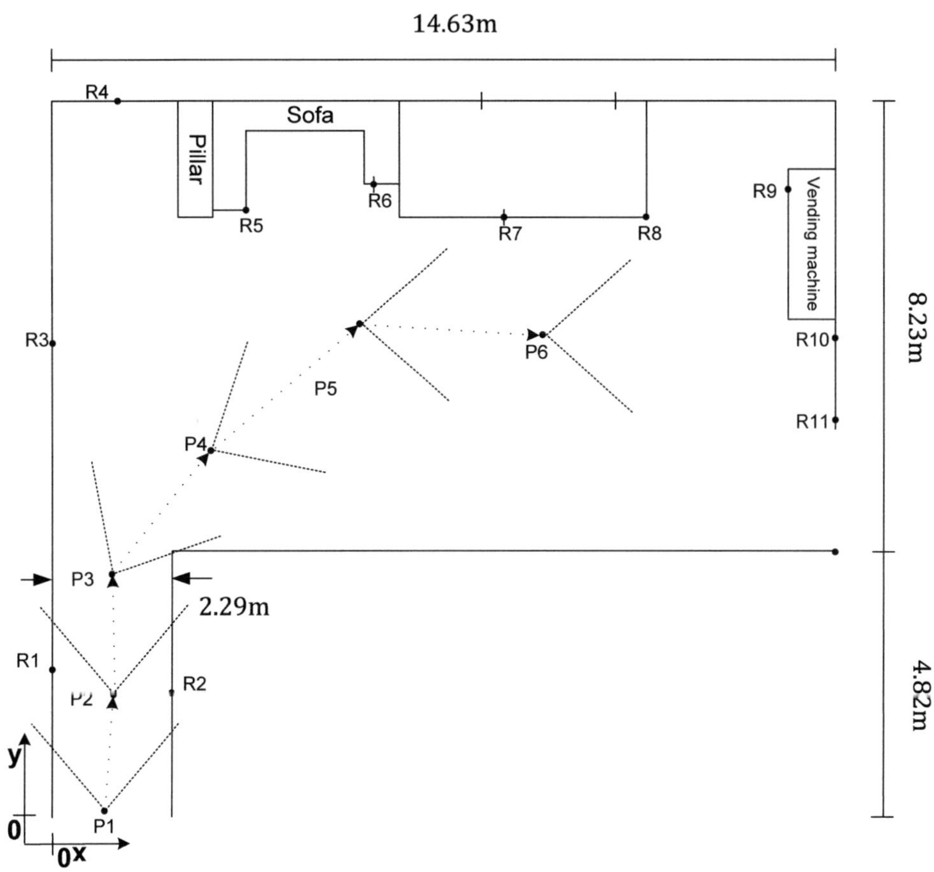


Fig. 11. Experimental setup used in the self localization illustration. 10 reference points are used by the mobile sensor located at 6 distinct coordinates.

Fig. 11 shows the experimental setup for the verification of the proposed method. A mobile sensor is placed at several positions indicated by $P_1$, $P_2$, … , $P_6$ with $P_1$ as the starting position. Several reference points, $R_1$, $R_2$, ... , $R_{11}$, are designated and measured beforehand. At each position, the mobile sensor estimates the position and orientation before moving on to the next position. Navigation application utilizing our method can employ two strategies as to when to run the algorithm. In the first case, the mobile sensor can move until three reference points are in its sight before running the algorithm. In the other case, the mobile sensor can move to a designated position directed by a navigation algorithm, and search for three reference points by rotating before running the localization algorithm. Three reference points are extracted using edge detection and color matching. We evaluate two cases. The first case assumes the orientation is known and the displacement errors are obtained. In the second case, the orientation is computed first, and the displacement errors are obtained from the orientation.

### 5.2 Localization Performance with Known Orientation

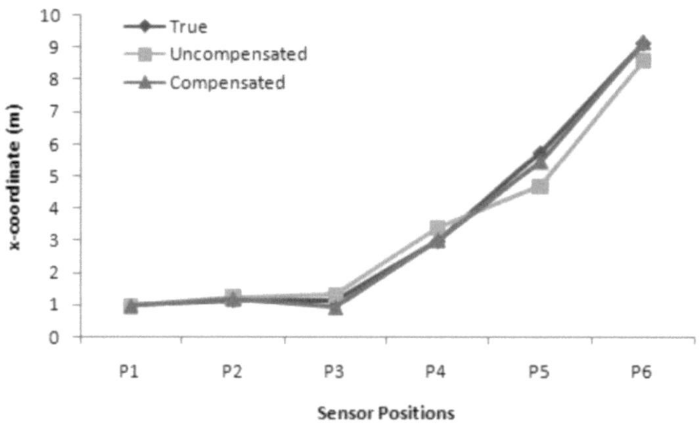


(a)

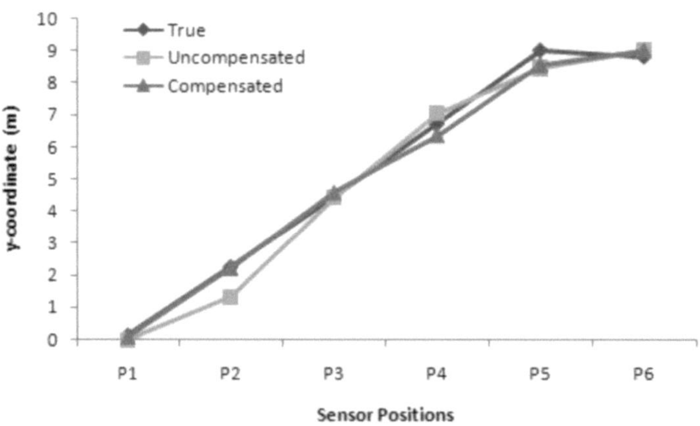


(b)

Fig. 12. Displacement error at each mobile sensor location with known orientation. Both compensated and uncompensated coordinates are compared to those of true coordinates. (a) $x$- direction. (b) $y$-direction.

We first assume that mobile visual sensor knows its orientation. Figs. 12(a) and 12(b) show the true position of the mobile sensor, in $x$-direction and $y$-direction separately, and their estimated trajectory obtained from using the algorithm. The deviation from the true position is shown as the distance error from the true position to the estimated position of the mobile sensor. Two estimated positions are plotted to show the effect of Zoom factor compensation. For the uncompensated estimation, the average value of the zoom factors is used. While displacement error in $x$-direction as shown in Fig. 12(a) is negligible, the displacement error in $y$-direction as shown in Fig. 12(b) illustrates that the uncompensated estimation deviates from the true position as much as 0.5m. It indicates the zooming factor is very sensitive to the distance from the visual sensor to the reference points. It is because the zoom factor has non-linear property only along the $y$-direction or the distance from the mobile sensor to the reference objects. However, when the zoom factors are compensated for within the algorithm, the distance error in $y$-direction disappears.

### 5.3 Localization Performance with Unknown Orientation

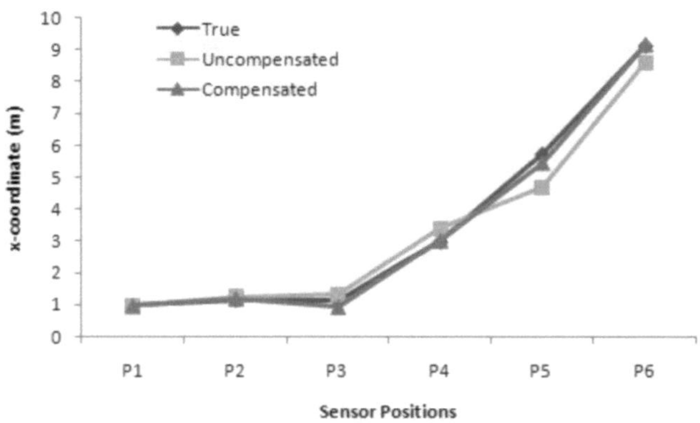


(a)

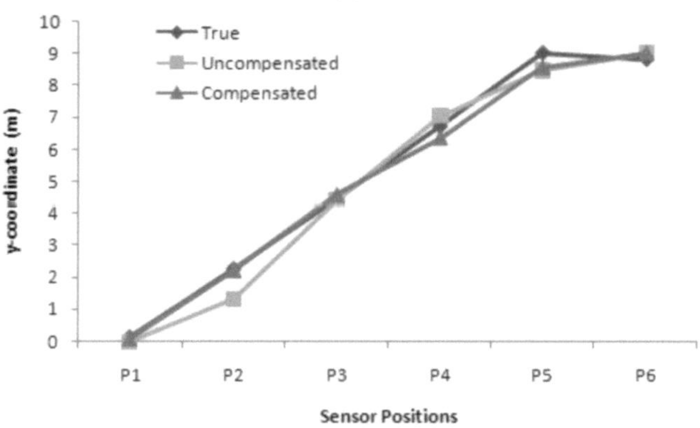


(b)

Fig. 13. Displacement error at each mobile sensor location with unknown orientation. Both compensated and uncompensated coordinates are compared to those of true coordinates. (a) $x$-direction. (b) $y$-direction.

In practice, the mobile sensor estimates the orientation as well as the coordinates. The proposed localization algorithm can determine the orientation of the sensor from the reference points. Since the coordinate of the mobile sensor is determined from the estimated orientation, the minimization of the error in estimating the orientation is very critical.
Fig.13(a) and Fig.13(b) illustrate the displacement error in $x$-direction and $y$-direction, respectively. As before, the displacement error in $x$-direction is negligible even without the compensation. Similar result is also shown in $y$-direction. The difference between these two figures and the previous figures obtained for the known orientation is that the displacement error is computed after the orientation is determined.

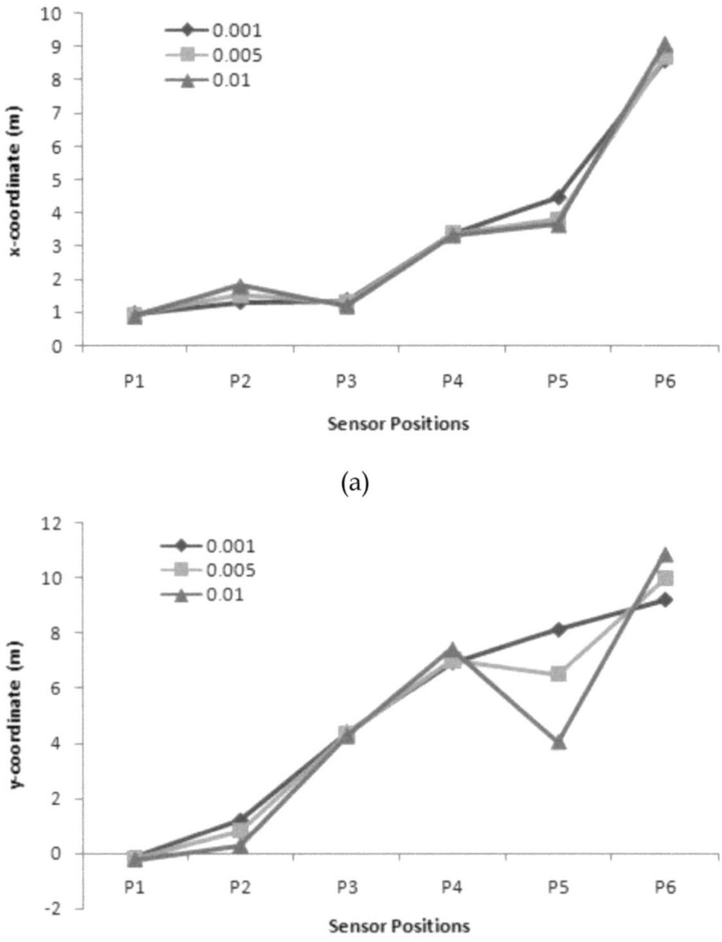


(b)

Fig. 14. Displacement error for different measurement errors for unknown orientation. The displacement error is obtained after the orientation estimation with the measurement errors. (a) $x$- direction. (b) $y$-direction.

Fig. 14 shows the effect of $\Delta i$ on the error of coordinates of the mobile sensor. For this simulation, we used maximal separation criterion for reference objects selection. As shown in the figure, when the mobile sensor is farther from the reference objects, the coordinate error is more sensitive to $\Delta i$. In our experimental setup, the mobile sensor is the closest to its selected reference points at the position $P_4$. In the figure, at $P_4$, the coordinate error is the least sensitive to $\Delta i$. When the mobile sensor at $P_5$, where its distance to the reference objects is the farthest, the coordinate error is very sensitive to $\Delta i$. Especially, the $y$-direction error at $P_5$, in Fig. 14(b) shows the large sensitivity to $\Delta i$ of the coordinate error in $y$-direction. It is because the captured image does not contain any depth information, the variation to $i$ can be mapped to large range of location of the mobile sensor in $y$-direction. 0.01 as $\Delta i$ value, is unrealistic considering the real dimension of the image.

## 6. Conclusion

In this chapter, we present a novel self localization method using parallel projection model for mobile sensor in navigation applications. The algorithm estimates the coordinate and the orientation of mobile sensor using projected references on single visual image. The camera lens non-linearity is compensated for using lens specific calibration table. The method utilizes a simple iterative algorithm, which is very accurate with low computational demand. We identify various sources of measurement error that affect the estimation accuracy. We demonstrate with an example that the algorithm can be utilized in robot navigation as well as positioning application where accurate self localization is necessary.

## 7. References


Ayala, V.; Hayet, J. B.; Lerasle, F.; *et al.* (2000). Visual localization of a mobile robot in indoor environments using planar landmarks, *Proceedings of IEEE/RSJ International Conference on Intelligent Robots and Systems*, pp.275-280, 0-7803-6348-5, Takamatsu, Japan, Oct. 31-Nov. 5, 2000.

Betke, M. & Gurvits, L. (1997). Mobile robot localization using landmarks, *Transactions on Robotics and Automation*, Vol. 13, No. 2, 251-263, 1042-296X.

Borenstein, J.; Everett, B. & Feng, L. (1996a). *Navigating Mobile Robots: Systems and Techniques*, AK Peters, 1-5688-1058-X, MA.

Borenstein, J. & Feng, L. (1996b). Measurement and correction of systematic odometry errors in mobile robots, *IEEE Transactions on Robotics and Automation*, Vol. 12, No. 6, 869-880, 1042-296X.

Borenstein, J.; Everett, H.; Feng, L *et al.* (1997). Mobile robot positioning: Sensors and techniques, *Journal of Robotic Systems*, Vol. 14, No.4, 231-249.

Briggs, A.; Scharstein, D.; Braziunas, D. *et al.* (2000). Mobile robot navigation using self-similar landmarks, *Proceedings of IEEE International Conference on Robotics and Automation*, pp.1428-1434, 0-7803-5886-4, San Francisco, USA, April 24-28, 2000.

Cohen, C. & Koss, F. V. (1993). A comprehensive study of three object triangulation, *Proceedings of SPIE Conference on Mobile Robots*, pp.95-106, Boston, USA, May 4, 1993.

Dao, N. X.; You, B. J.; Oh, S. R. *et al.* (2004). Simple visual self-localization for indoor mobile robots using single video camera, *Proceedings of IEEE/RSJ International Conference on Intelligent Robots and Systems*, pp.3767-3772, 0-7803-8463-6, Sendai, Japan, Sept. 28-Oct. 2, 2004.

Demirli, K. & Molhim, M. (2004). Fuzzy dynamic localization for mobile robots, *Fuzzy Sets and Systems*, Vol. 144, No. 2, 251-283.

Groβmann, A. & Poli, R. (2001). Robust mobile robot localisation from sparse and noisy proximity readings using Hough transform and probability grids abstract, *Robotics and Autonomous Systems*, Vol. 37, No. 1, 1-18.

Hayet, J. B.; Lerasle, F. & Devy, M. (2007). A visual landmark framework for mobile robot navigation, *Image and Vision Computing*, Vol. 25, No, 8, 1341-1351.

Heikkila, J. & Silven, O. (1997). A four-step camera calibration procedure with implicit image correlation, *Proceedings of IEEE Computer Society Conference on Computer Vision and Pattern Recognition*, pp.1106-1112, San Juan, Puerto Rico, June 17-19, 1997.

Horaud, R.; Conio, B.; Leboulleux, O.; *et al.* (1989). An analytic solution for the perspective 4-point problem. *Computer Vision, Graphics, and Image Processing*, Vol. 47, No. 1, 33-44, 0734-189X.

Jang, G.; Kim, S.; Lee, W. *et al.* (2002). Color landmark-based self-localization for indoor mobile robots, *Proceedings of IEEE International Conference on Robotics and Automation*, pp. 1037-1042, 0-7803-7272-7, Washington, USA, May 11-15, 2002.

Lenz, R. & Tsai, R. (1988). Techniques for calibration of the scale factor and image center for high accuracy 3-D machine vision metrology, *IEEE Transactions on Pattern Analysis and Machine Intelligence*, Vol. 10, No. 5, 713-720, 0162-8828.

Lepetit, V. & Fua, P. (2006). Monocular model-based 3D tracking of rigid objects: A survey. *Foundations and Trends in Computer Graphics and Vision*, Vol. 1, No. 1, 1-89.

Liu, W. & Zhou, Y. (2007). Robot self-localization based on a single image of identified landmarks, *Proceedings of IEEE International Symposium on Computational Intelligence in Robotics and Automation*, pp.248-253, 1-4244-0790-7, Jacksonville, USA, June 20-23, 2007.

Lv, F.; Zhao, T. & Nevatia, R. (2006). Camera calibration from video of a walking human, *IEEE Transactions on Pattern Analysis and Machine Intelligence*, Vol. 28, No. 9, 1513-1518, 1051-4651.

Martinelli, A.; Tomatis, N. & Siegwart, R. (2007). Simultaneous localization and odometry self calibration for mobile robot, *Autonomous Robots*, Vol. 22, No. 1, 75-85, 0929-5593.

Park, K.; Lee, J.; Stanaćević, M. *et al.* (2008). Iterative object localization algorithm using visual images with a reference coordinate, *EURASIP Journal on Image and Video Processing*, Article ID 256896.

Se, S.; Owe D. & Little, J. (2001). Vision-based Mobile robot localization and mapping using scale-invariant features, *Proceedings of IEEE International Conference on Robotics and Automation*, pp.2051-2058, 0-7803-6576-3, Seoul, Korea, May 21-26, 2001.

Se, S.; Lowe, D. G. & Little, J. (2002). Global localization using distinctive visual features, *Proceedings of International Conference on Intelligent Robots and Systems*, pp. 226-231, Switzerland, Sept. 30-Oct. 4, 2002.

Sutherland, K. T. & Thompson, W. B. (1993). Inexact navigation, *Proceedings of IEEE International Conference on Robotics and Automation*, pp.1-7, 0-8186-3450-2, Atlanta, USA, May 2-6, 1993.

Swaminathan, R.; Grossberg, M. D. & Nayar, S. K. (2003). A perspective on distortions, *Proceedings of IEEE Conference on Computer Vision and Pattern Recognition*, pp.594-602, 0-7695-1900-8, Madison, Wisconsin, USA, June 18-20, 2003.

Thrun, S. (1998). Finding landmarks for mobile robot navigation, *Proceedings of IEEE International Conference on Robotics and Automation*, pp.958-963, 0-7803-4300-X, Leuven, Belgium, May 16-20, 1998.

Yuan, J. (1989). A general photogrammetric method for determining object position and orientation, *IEEE Transactions on Robotics and Automation*, Vol. 5, No. 2, 129-142, 1042-296X.

# 8

# Vision Based SLAM for Mobile Robot Navigation Using Distributed Filters

Young Jae Lee and Sankyung Sung
*Konkuk University*
*Seoul, Korea*

## 1. Introduction

Simultaneous Localization and Mapping (SLAM) has been developed for mobile robot applications to obtain navigation information. SLAM is a strategy for building up maps in unknown environments and figuring out its position without any prior map information. To implement SLAM, various kinds of sensors such as laser range finders, ultrasonic sensors, vision sensors, etc. (Sheng Fu et al., 2007) can be considered. Each configuration often depends upon the required performance and cost condition.

In this chapter, we suggest a vision-based SLAM method to improve the navigation performance of mobile robot. It consists of a set of vision sensors and encoders which are equipped on mobile robot's wheels (Leonard & Durrant-Whyte, 1991). The heading and position of a mobile robot can be independently estimated with only 2 encoders. However the result often yields large inaccuracies due to errors and the noise of sensors (i.e. a vision sensor and encoders).

On the other hand, as the vision sensor is incorporated, a feature point tracking algorithm must be used (it's not mentioned in this chapter, because it's out of focus). In the integrated system, the relative position between feature points and the mobile robot is continuously estimated using information (i.e. feature points), so the position and heading errors are compensated using the filter estimation results (Hugh Durrant-Whyte & Tim Bailey, 2006).

The Extended Kalman Filter (EKF) has been widely used for SLAM (Jing Wu & Hong Zhang, 2007). Additionally, a particle filter is occasionally used with landmark or geometric constraints to enhance performance (Sukhan Lee & Seongsoo Lee, 2006), because the measurement model of the vision sensor is not linear. In comparison with previous works, this chapter takes a distributed particle filter approach without landmark information and geometric constraints. The distributed particle filter is known to have powerful performance under nonlinear, multi-modal models and features points in varying conditions (Ristic et al., 2004).

In addition to a vision sensor, range sensors are generally used to compute the 3-D position of feature points, because it is difficult to estimate the position in case of a single vision sensor system. This chapter uses a delayed initialization method instead of range sensors to compute the feature point position using single vision sensor system (Philip J. Schneider & David H. Eberly, 2003).

Computer simulations are conducted to demonstrate the performance of the suggested algorithm. The navigation performance is enhanced compared to the encoder only system during simulation. It depends upon the number of particles, feature points and the angle of vision sensor. Furthermore, using landmarks, which are from a previous known position, also improve the performance.

## 2. SLAM

Many vision sensor applications have been developed in recent years due to low costs, low power consumption and the development of high speed processors. Vision sensors can provide continuous image data. If there are some points that are not changed in the whole images during the image acquisition time, they can be continuously distinguishable from other objects (background or moving object) for some periods of time. Then the distinguishable points in the image data are referred to as feature points. Through tracking feature points, the amount of vision sensor's moving can be estimated.
There are many kinds of image types (gray scale, RGB, etc). In this chapter, 8 bits gray scale image is used, which expresses an image with 256 step intensity (Rafael C & Gonzalez, 2004). On image plan, intensity of pixel is changed by location (x, y) and acquisition time (t). Thus, the image (I) can be denoted by the following expression (Carlo Tomasi & Takeo Kanade, 1991).

$$I(x, y, t) \quad (1)$$

Let's assume that there are displacements ($d = (\xi, \eta)$) and the time difference ($\tau$) is small on continuous 2 images, then the relation can be expressed as below.

$$I(x, y, t+\tau) = I(x-\xi, \quad y-\eta, \quad t) \quad (2)$$

It is observed that displacements are proportional to the movement of the vision sensor. If the displacements of a feature point are estimated, the movement of vision sensor can be obtained. Usually, feature point can be selected by using Harris Corner Detection (Konstantinos G. Derpanis, 2004), SUZAN (S. M. Smith & J. M. Brady, 1997) or Fast Corner Detection (Edward Rosten and Tom Drummond, 2005), which can be further used for the continuous tracking of feature point positions on the whole image.

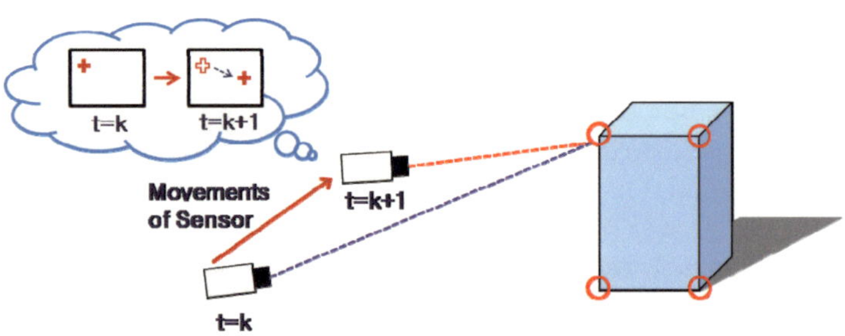


Fig. 1. Concpt of vision based SLAM (vSLAM)

Conventional SLAM algorithms using the vision sensor are adapted for the system integration. Using the vision sensor, feature points are selected and tracked on continuous frames. The vision sensor also provides bearing, elevation or range information from the feature point, which is deduced from the feature points tracking data. Other sensor output goes to SLAM block with vision sensor data. Navigation information (position, velocity, attitude, etc.) and errors in sensors are estimated by integrating information from vision and other sensor. Assuming feature points are fixed and not movable in the local coordinate frame, navigation errors come mainly from sensor outputs. Thus, by compensating estimated errors from sensor output, navigation data can be precisely calculated.

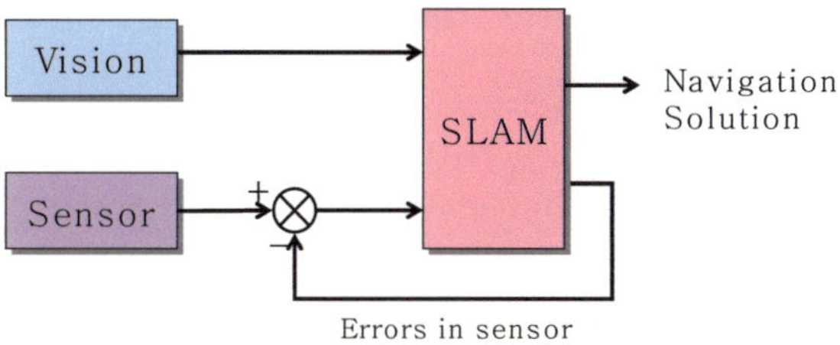


Fig. 2. Basic algorithm of a vision based SLAM

## 3. System Modeling

Figure 3 shows a simple mobile robot platform having only planar dynamics. It has a vision sensor and 2 encoders. A vision sensor acquires image continuously, then feature points are selected and tracked. The encoders are equipped on wheels and provide wheel rotation data. Given the information about wheel radius, the distance between the 2 wheels and pulses per rotation of encoder, range and heading information can be numerically computed. There are, however, various kinds of errors; wheel radius error, wheel distance error, slips error, conversion factor error, etc. The overall effect from the above mentioned sources resulted in accumulated errors and degraded navigation performance, which necessitated error compensation using aided sensors and filters.

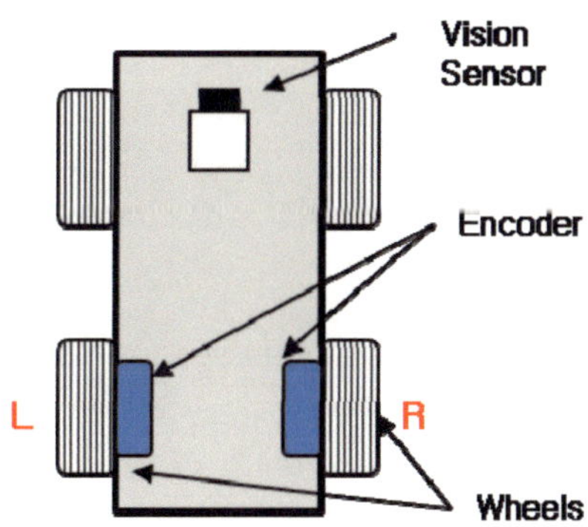


Fig. 3. Mobile robot equippd with encoders and vision sensor

### 3.1 Dynamic Model

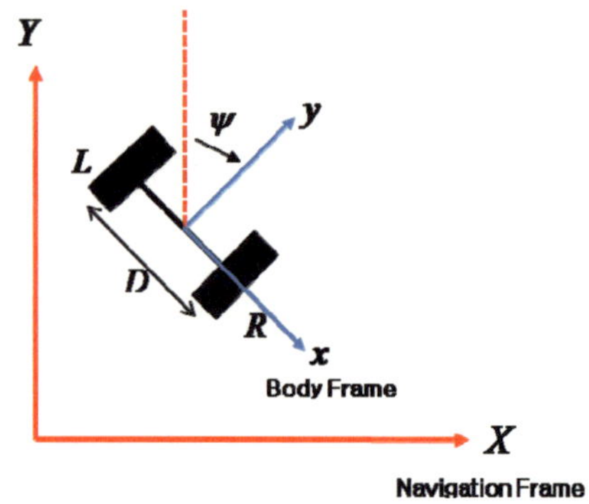


Fig. 4. Coordinate system of mobile robot

In Figure 4, X-Y denotes navigation frame and x-y denotes body frame on Figure 4. Position and heading of mobile robot can be expressed using the following equations.

$$\begin{aligned} X_{k+1} &= X_k + inc \times \sin\psi_k \\ Y_{k+1} &= Y_k + inc \times \cos\psi_k \\ \psi_{k+1} &= \psi_k + \frac{L-R}{D} \\ inc &= \frac{R+L}{2} \end{aligned} \tag{3}$$

L and R are movements of the left and right wheel, respectively. X, Y denotes position in each axis and ψ is heading. D is the distance between two wheels. $S_R, S_L$ are scale factor errors of the encoder, so it is a kind of bias. We consider $S_R, S_L, \delta D$ as a random constant.

$$\begin{aligned} S_{R_{K+1}} &= S_{R_K} \\ S_{L_{K+1}} &= S_{L_K} \\ \delta D_{K+1} &= \delta D_K \end{aligned} \tag{4}$$

To estimate navigation error, the system model was changed into the following form using perturbation.

$$\begin{aligned} \hat{X}_{k+1} &= \hat{X}_k + \sin\hat{\psi}_k \times \left( \frac{R(1+S_{R_k}) + L(1+S_{L_k})}{2} \right) \\ \hat{Y}_{k+1} &= \hat{Y}_k + \cos\hat{\psi}_k \times \left( \frac{R(1+S_{R_k}) + L(1+S_{L_k})}{2} \right) \\ \hat{\psi}_{k+1} &= \hat{\psi}_k + \left( \frac{L(1+S_{L_k}) - R(1+S_{R_k})}{D+\delta D_k} \right) \\ \hat{X}_k &= X_k + \delta X_k,\ \hat{Y}_k = Y_k + \delta Y_k,\ \hat{\psi}_k = \psi_k + \delta\psi_k \end{aligned} \tag{5}$$

Applying a perturbed model and small angle assumption, (3) can be changed into (4). State vector consists of position error (X, Y), heading error, scale factor error (R, L) and wheel distance error.
The following state space equation is an error model of the system.

$$\mathbf{X} = [\delta Y \ \delta X \ \delta\psi \ S_R \ S_L \ \delta D]^T$$

$$\mathbf{X_{k+1}} = \mathbf{\Phi_k X_k} + \mathbf{w_k}$$

$$\mathbf{\Phi_k} = \begin{bmatrix} 1 & 0 & -\frac{R+L}{2}\sin\psi_k & \frac{R}{2}\cos\psi_k & \frac{L}{2}\cos\psi_k & 0 \\ 0 & 1 & \frac{R+L}{2}\cos\psi_k & \frac{R}{2}\sin\psi_k & \frac{L}{2}\sin\psi_k & 0 \\ 0 & 0 & 1 & -\frac{R}{D} & \frac{L}{D} & \frac{R-L}{D^2} \\ 0 & 0 & 0 & 1 & 0 & 0 \\ 0 & 0 & 0 & 0 & 1 & 0 \\ 0 & 0 & 0 & 0 & 0 & 1 \end{bmatrix} \quad (6)$$

### 3.2 Measurement Model

A vision sensor provides sequential images. Feature points are selected and tracked through these images. The feature point position is not changed on the navigation frame. So a measured feature point position is used for reference, when the mobile robot position is estimated.

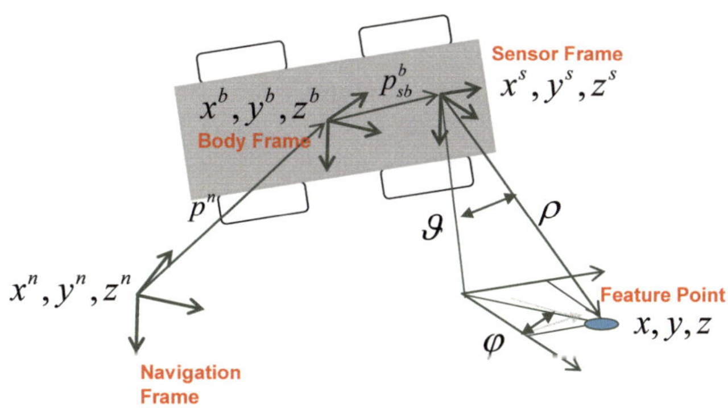


Fig. 5. Coordinate of vision based SLAM

Figure 5 shows the relation between the feature point and frames. The reference frame is a navigation frame; however, the feature point is measured on the sensor frame, so a coordinate transform is needed.
Equation (7) denotes a feature point position on the sensor frame. Using equation (6) and (7), the feature point position can be predicted.

$$x_{FP}^{s} = C_{b}^{s}\left(C_{n}^{b}\left(x_{FP}^{n} - p_{sb}^{b}\right) - p^{n}\right) \tag{7}$$

$C_b^s$ : Coordinate transform (body to sensor frame)

$C_n^b$ : Coordinate transform (navigation to body frame)

$p_{sb}^b$ : Lever arm (between sensor and body frame)

$p^n$ : Lever arm (between body and navigation frame)

Equation (8) is the measurement model. Measurement vector contains bearing and elevation angle, which are readily calculated from the feature point position information.

$$z = \begin{bmatrix} \tan^{-1}\left(\dfrac{y}{x}\right) \\ \tan^{-1}\left(\dfrac{z}{\sqrt{x^2+y^2}}\right) \end{bmatrix} \tag{8}$$

$$x_{fi}^{s} = \begin{bmatrix} x \\ y \\ z \end{bmatrix},\ z = \begin{bmatrix} \phi \\ \theta \end{bmatrix} \quad \begin{pmatrix} \phi : \text{Feature Point Bearing} \\ \theta : \text{Feature Point Elevation} \end{pmatrix}$$

## 4. Integration Filter

### 4.1 Particle Filter

In the filter system implementation, the measurement model of a vision sensor is observed to be nonlinear, which requires for a nonlinear filter approach. Extended Kalman Filter, Unscented Kalman Filter (UKF) and Gaussian Sum Filter are common nonlinear filter (Ristic et al., 2004). In this chapter, a particle filter (PF) is used for state estimation, which is a technique for implementing a recursive Bayesian filter by Monte Carlo simulation. The key point of PF is to represent the posterior density function by a set of random samples with its associated weight and to compute estimates based on these samples and weights.

It can be shown that, at the time instant k, the particles $\{x_{k|k-1}(i) : i = 1, 2 \ldots, N\}$ and $\{x_k(i) : i = 1, 2 \ldots, N\}$ can be recursively obtained by the following algorithm (Ristic et al., 2004):

1. Assume that there is a set of random samples (i.e. particles) $\{x_{k-1}(i): i = 1,2,\ldots,N\}$ from the pdf $p(x_{k-1} \mid Y_{k-1})$.
2. Prediction: Sample $N$ values $\{w_{k-1}(i): i = 1,2,\ldots,N\}$ from the pdf of system noise $w_{k-1}$. Use these samples to generate new swarm of points $\{x_{k|k-1}(i): i = 1,2,\ldots,N\}$ which approximates the predicted pdf $p(x_k \mid Y_{k-1})$ where
$$x_{k|k-1}(i) = f_{k-1}(x_{k-1}(i), w_{k-1}(i))$$
3. Update: Assign each $x_{k|k-1}(i)$ a weight $w_k(i)$ for $i = 1,2,\ldots,N$, after measurement $y_k$ is received. The weights are given by
$$w_k(i) = \frac{p(y_k \mid x_{k|k-1}(i))}{\sum_{j=1}^{N} p(y_k \mid x_{k|k-1}(j))}$$
This defines a discrete distribution over $\{x_{k|k-1}(i): i = 1,2,\ldots,N\}$, which assigns probability mass $w_k(i)$ to the element $x_{k|k-1}(i)$ and results in the posterior pdf $p(x_k \mid Y_k)$ being represented in terms of weighted samples (i.e. particles).
4. Resample: Resample independently $N$ times from the above discrete distribution. The resulting particles $\{x_k(i): i = 1,2,\ldots,N\}$ which satisfies
$$\Pr\{x_k(i) = x_{k|k-1}(j)\} = w_k(j)$$
form an appropriate sample (with equal weight to each element) from the posterior pdf $p(x_k \mid Y_k)$.
5. The prediction, update and resample step form a single iteration and is recursively applied each time $k$.

### 4.2 Integration Filter

The following figure shows the structure of a vision based SLAM. An indirect filter structure is used to reduce computation time and power

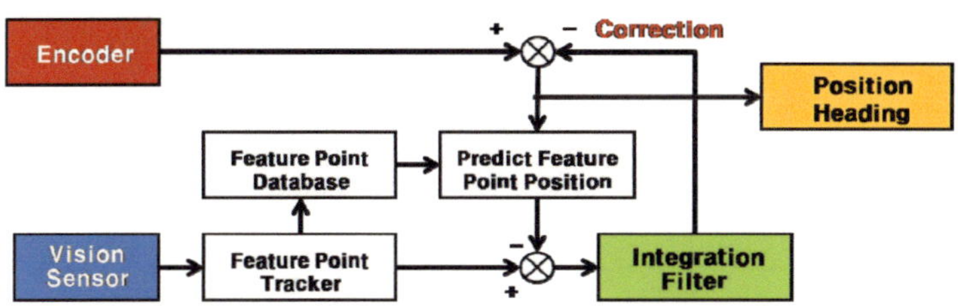


Fig. 6. Structure of vision based SLAM

In the encoder block, the position is locally calculated using large sensor errors. In the vision sensor block, the image data is acquired and feature points are selected. Then, the feature points are tracked via feature point tracker block, where the tracking information is stored in the database and sent to integration filter block simultaneously. Using encoder measurements and feature point database, the next feature point position can be predicted on the image frame. In the integration filter block, the measured feature point position is compared with the predicted position, and navigation errors are estimated. Estimated errors go to the encoder block, and finally compensate the navigation data.
Figure 7 shows the structure of the distributed filter which has several local filters. The distributed filter has good performance with respect to computation time and power, fault detection, isolation, variation in measurements number, etc. The number of local filters is varying in proportion to the number of feature points, which depends on visible environment. The estimated errors from each local filter are combined in the master filter for data integration. Finally, the estimated errors are fed into an encoder to compensate for the position data. In the chapter, error estimate and covariance of master filter is not fed into the local filters for simplicity.

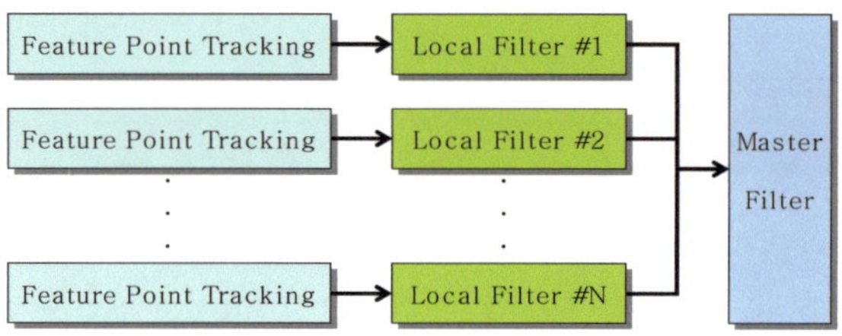


Fig. 7. Structure of distributed filter

## 5. Simulation Results

Computer simulation is done to demonstrate the performance of the suggested algorithm. For simulation, 10m x 10m square track and 10 feature points are used. The time step for each epoch is set to 0.01sec with total epoch number of 4790. A detailed description is shown in Table 1.

| Filter | Particle filter (number of particle: 200) |
|---|---|
| Wheel radius | 50 ± 1 mm |
| Encoder error | ± 6.3 mm/rotation |
| Velocity | 2 m/sec |
| Track | 10m x 10m |
| Feature points | 200 in 40m x 40m |
| Simulation frequency | Positioning: 100 Hz<br>Error Compensation: 10 Hz |

Table 1. Simulation paraeters

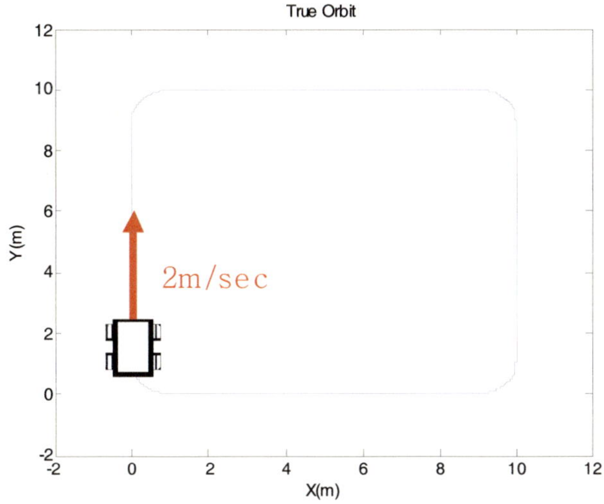


Fig. 8. Simulation track

The navigation performance of a vision based SLAM integrated system is compared together with that of a standalone encoder system. First, Figure 9 is the simulation result showing the navigation performance of an encoder only system. The red line denotes the true trajectory and the blue line denotes the estimated trajectory. As time increases, cumulative errors are increased, which results in an unbounded deviation from the true track. Error characteristics illustrates that the dominant error source of the encoder is scale factor and the distance between two wheels.

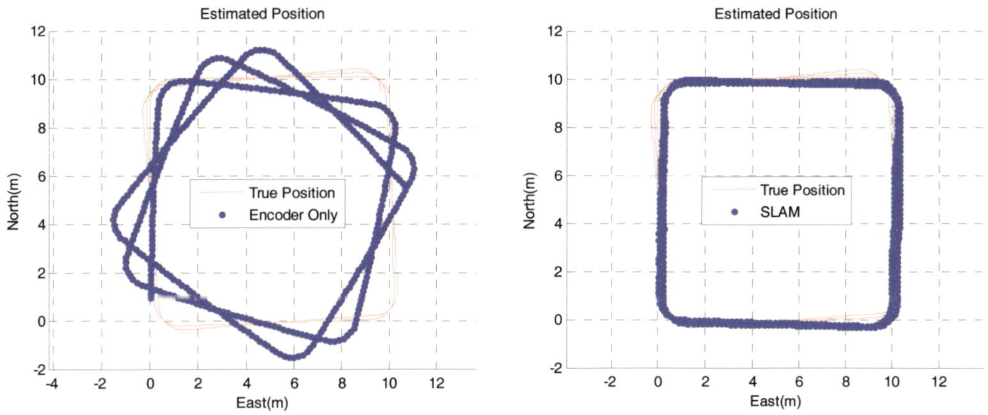


Fig. 9. Estimated position of encoder only navigation (left) and vision based SLAM (right)

In the vision based SLAM case, the navigation performance is greatly enhanced. The estimated position has a bounded error from the true trajectory, which does not diverge as time increases. Throughout the trajectory, properly deployed feature points compensate error accumulation from the encoder and the initial heading bias produced a slightly deviated square trajectory.

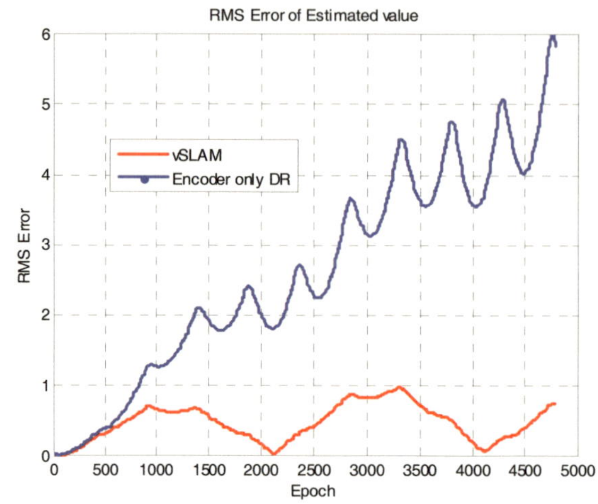


Fig. 10. Position error comparison between encoder only navigation and vision based SLAM

Figure 10 shows the position error of the two methods. In case of the encoder, the error diverges with sinusoid-like fluctuation. The fluctuation comes from the encoder output scale factor error at every turn. The estimation error of the SLAM aided integration system is error bound by about 1m.

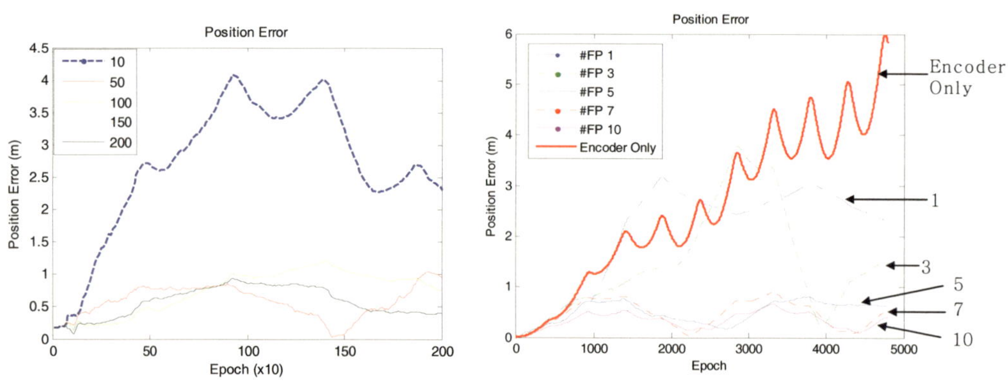


Fig. 11. Position error through changing the number of particles (left) and feature points (right)

In implementing the navigation computer, the computing time highly depends on the particle number and the feature point number, which constrains the particle size in each particle filter. Figure 11 shows position error when the numbers of particles on filter changes. The number of feature points is fixed as 10. When there are particles of more than 50, the error is below 1.5m during operation times. The number of feature points is changed from 1 to 10, and the number of particles is fixed at 100. In the figure, even though #FP 1 and 3 cases have large error boundaries, the position error does not diverge and shows better estimation performance than the encoder only system. The maximum position error is observed to be smaller than 1m when there are more than 5 feature points. Table 2 summarizes the data at the epoch 1000, 2000, 3000 and 4000 in Figure 11.

| *Epoch* / *#FP* | *1000* | *2000* | *3000* | *4000* | *Mean* | *Maximum Error* |
|---|---|---|---|---|---|---|
| *Encoder Only* | *1.26* | *1.98* | *3.14* | *3.56* | *2.58* | *5.99* |
| *1* | ***0.81*** | ***3.00*** | ***2.49*** | ***2.89*** | ***2.12*** | ***3.17*** |
| *3* | ***0.82*** | ***1.41*** | ***3.50*** | ***0.31*** | ***1.49*** | ***3.79*** |
| *5* | ***0.71*** | ***0.38*** | ***0.40*** | ***0.68*** | ***0.50*** | ***0.81*** |
| *7* | ***0.77*** | ***0.27*** | ***0.73*** | ***0.33*** | ***0.48*** | ***0.90*** |
| *10* | ***0.47*** | ***0.28*** | ***0.48*** | ***0.30*** | ***0.34*** | ***0.67*** |

Table 2. Position error through changing the number of feature points

In #FP 1 and 3 cases, the position error is larger than the other cases at early stage, thus these large errors affect all over the stage, so estimated trajectory become inaccurate. In other cases, the error is decreased and has boundary. Compared with the mean error of the encoder only case at 4000 epoch, the #FP 7 case has 82% smaller error and in the #FP 10 case has an 87 % smaller error than the encoder only case. In conclusion, when the number of particles is set to 100 for low computational burden, the number of particles over 5 is required for better performance.

For further verification of suggested algorithm, the simulation condition is changed. A difference in simulation is an angle of the vision sensor. Figure 12 (left) illustrates attached angle of vision sensor. The angle is changed from 0 to 180 degree.

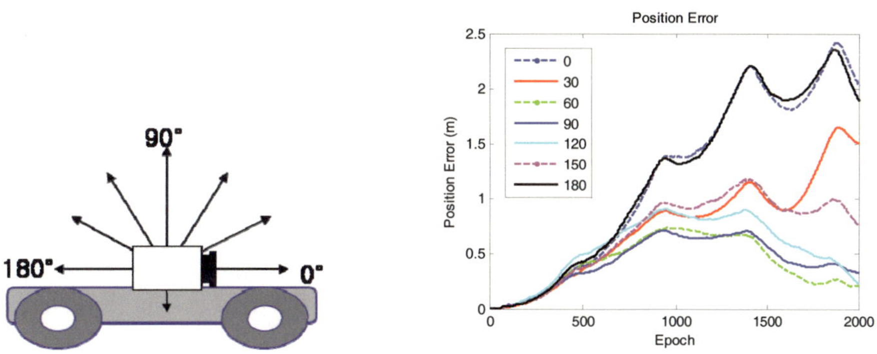


Fig. 12. Attached angle of a vision sensor and position error through the changing of attached angle

Figure 12 (right) shows the position error through the changing of the angle. The error is below 1m when the angle is changed from 60 to 120, whereas the other cases have larger errors.

In other words, if the vision sensor looks toward the top direction (90 degree), the feature point position of the horizontal direction is detected and measured more precisely. So the quality of measurement is better than the other cases. For this reason, the horizontal error can be decreased. Since the angle is far from 90 degree, the quality of measurements on horizontal direction becomes poor. That affects the position and heading error.

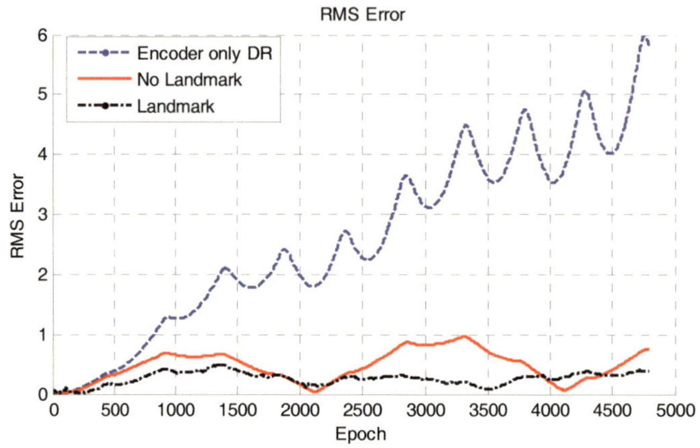


Fig. 13. Position error in case of using the landmark

If there are some points which are known positions with previous information, the navigation performance can be increased. In this chapter, such a point is called the 'Landmark'. The landmark has precise position information, so it is used for accurate measurements when it is observed. In the integration filter, feature point estimation step will be skipped because the position is already known. In Figure 13, landmark aids position estimation performance to be increased compare with general case (without landmark). The landmark aiding case has 0.2567m RMS error (mean), which is smaller than 0.4656m in general cases. This means that precise measurements (landmark) prevent a divergence of estimation error sand improve the estimation performance in mobile robot navigation.

## 6. Conclusion

In this chapter, a vision based SLAM and encoder integrated system is presented for mobile robot navigation. By considering the nonlinear measurement model and feature point availability around the trajectory, a distributed particle filter approach is applied. Simulation results demonstrate the performance of the implemented mobile robot. Further results confirm that the estimation performance largely depends on the number of feature points and particles, which will be mutually associated while implementing the embedded navigation computer. It also depends on the attached angle of a vision sensor and the landmark.

## 7. References

Carlo Tomasi and Takeo Kanade, "Detection and Tracking of Point Features", *Technical Report CMU-CS-91-132*, April 1991

Edward Rosten and Tom Drummond, "Fusing Points and Lines for High Performance Tracking", *IEEE Int. Conf. Computer Vision*, October 2005, pp.1508-1511.

Hugh Durrant-Whyte and Tim Bailey, "Simultaneous Localization and Mapping", *IEEE Robotics & Automation Magazine*, June 2006, pp.99-117

J. Leonard and H. F. Durrant-Whyte, "Simultaneous map building and Localization for an autonomous mobile robot," *IEEE Int. Wkshp. Intell. Robots Syst.*, Osaka, Japan, vol. 3, November 1991, pp. 1442-1447.

Jing Wu and Hong Zhang, "Camera Sensor Model for Visual SLAM", *IEEE. Conf, Computer and Robot Vision*, May 2007, pp.149-156

Konstantinos G. Derpanis, " The Harris Corner Detector", October 2004.

Philip J. Schneider and David H. Eberly, *Geometric Tools for Computer Graphics*, Morgan Kaufmann Publishers, 2003, pp.400-433

Rafael C. Gonzalez, *Digital Image Processing using MATALB*, Prentice Hall, 2004, pp.26-27

Ristic, Arulman and Gordon, *Beyond the Kalnman Filter*, Artech House, 2004, pp.35-48

S. M. Smith and J. M. Brady, "SUSAN - a new approach to low level image processing," *International Journal of Computer Vision*, May 1997.pp.45-78.

Sheng Fu, Hui-ying Liu, Lu-fang Gao and Yu-xian Gai, "SLAM for Mobile Robots Using Laser Range Finder and Monocular Vision", *IEEE Int, Conf. Mechatronics and Machine Vision in Practice*, December 2007, pp.91-96

Sukhan Lee and Seongsoo Lee, "Recursive Particle Filter with geometric Constraints for SLAM", *IEEE Int. Conf, Multisensor Fusion and Integration for Intelligent Systems*, 2006, pp.359-401

9

# Omnidirectional vision based topological navigation

Toon Goedemé and Luc Van Gool
*Lessius Campus De Nayer Technical University, VISICS, Catholic University of Leuven*
*Belgium*

## 1. Introduction

In this work we present a novel system for autonomous mobile robot navigation. With only an omnidirectional camera as sensor, this system is able to build automatically and robust accurate topologically organised environment maps of a complex, natural environment. It can localise itself using that map at each moment, including both at startup (kidnapped robot) or using knowledge of former localisations. The topological nature of the map is similar to the intuitive maps humans use, is memory-efficient and enables fast and simple path planning towards a specified goal. We developed a real-time visual servoing technique to steer the system along the computed path.
The key technology making this all possible is the novel *fast wide baseline feature matching*, which yields an efficient description of the scene, with a focus on man-made environments.

### 1.1 Application

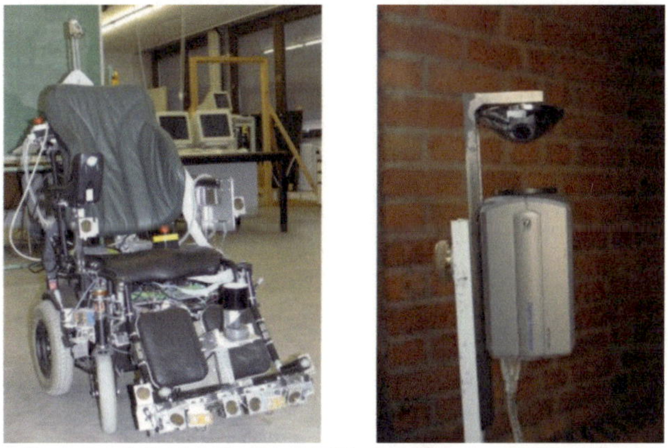

Fig. 1. Left: the robotic wheelchair platform. Right: the omnidirectional camera, composed by a colour camera and an hyperbolic mirror.

This chapter describes a total navigation solution for mobile robots. It enables a mobile robot to efficiently localise itself and navigate in a large man-made environment, which can be

indoor, outdoor or a combination of both. For instance, the inside of a house, an entire university campus or even a small city lie in the possibilities.
Because of reliability problems of other sensors like e.g. GPS, why we aim at a *vision-only* solution to navigation. Vision is, in comparison with these other sensors, much more informative. Moreover, cameras are quite compact and increasingly cheap. We observe also that many biological species, in particular migratory birds, use mainly their visual sensors for navigation. We chose to use an omnidirectional camera as visual sensor, because of its wide field of view and thus rich information content of the images acquired with. For the time being, we added a range sensing device for obstacle detection, but this is to be replaced by an omnidirectional vision range estimator under development.
Our method works with *natural* environments. That means that the environment does not have to be modified for navigation in any way. Indeed, adding artificial markers to every room in a house or to an entire city doesn't seem feasible nor desirable.
In contrast to classical navigation methods, we chose a *topological* representation of the environment, rather than a metrical one, because of its resemblance to the intuitive system humans use for navigation, its flexibility, wide usability, memory-efficiency and ease for map building and path planning.
The targeted application of this research is the visual guidance of electric wheelchairs for severely disabled people. More in particular, the target group are people not able to give detailed steering commands to navigate around in their homes and local city neighbourhoods. If it is possible for them to perform complicated navigational tasks by only giving simple commands, their autonomy can be greatly enhanced. For most of them such an increase of mobility and independence from other people is very welcome.
Our test platform and camera are shown in fig. 1.

### 1.2 Method overview

An overview of the navigation method presented is given in fig. 2. The system can be subdivided in three parts: map building, localisation and locomotion.
The map building stage has to be gone through only once, to train the system in a new environment. The mobile system is lead through all parts of the environment, while it takes images at a constant rate (in our set-up one per second). Later, this large set of omnidirectional images is automatically analysed and converted into a topological map of the environment, which is stored in the system's memory and will be used when the system is actually in use.
The next stage is localisation. When the system is powered up somewhere in the environment, it takes a new image with its camera. This image is rapidly compared with all the images in the environment map, and an hypothesis is formed about the present location of the mobile robot. This hypothesis is refined using Bayes' rule as soon as the robot starts to move and new images come in.
When the present location of the robot is known and a goal position is communicated by the user to the robot, a path can be planned towards that goal using the map. The planned route is specified as a sequence of training images, serving as a reference for what the robot should subsequently see if on course. This path is executed by means of a visual servoing algorithm: each time a visual homing procedure is executed towards the location where the next path image is taken.

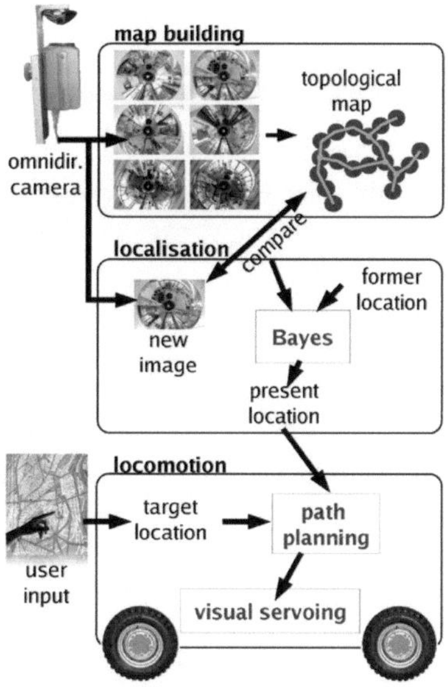


Fig. 2. Overview of the navigation method

The remainder of this chapter is organised as follows. The next section gives an overview of the related work. In section 3, our core image analysis and matching technique is explained: fast wide baseline matching. The sections thereafter describe the different stages of our approach. Section 4 discusses the map building process, section 5 explains the localisation method, section 6 describes the path planning, and section 7 details the visual servoing algorithm. We conclude with an overview of experimental results (section 8) and a conclusion (section 9).

## 2. Related Work

### 2.1 Image comparison

A good image comparison method is of utmost importance in a vision-based navigation approach. Global methods compute a measure using all the pixels of the entire image. Although these methods are fast, they cannot cope with e.g. occlusions and severe viewpoint changes. On the other hand, techniques that work at a local scale, extracting and recognising *local features*, can be made robust to these effects. The traditional disadvantage of these local techniques is time complexity. In our approach, we combine novel global and local approaches resulting in fast and accurate image comparison.

#### 2.1.1 Global techniques

Many researchers use global image comparison techniques. Straightforward *global methods* like histogram-based matching, used by (Ulrich & Nourbakhsh, 2000) don't seem distinctive enough for our application. Another popular technique is the use of an eigenspace decomposition of the training images (Jogan & Leonardis, 1999), which yields a compact database. However, these methods proved not useful in general situations because they are not robust enough against occlusions and illumination changes. That is why (Bischof et al., 2001) developed a PCA-based image comparison that is robust against partial occlusions, respectively varying illumination.

### 2.1.2 Local techniques

A solution to be able to cope with partial occlusions is comparing local regions in the images. The big question is how to detect these local features, also known as visual landmarks.

A simple solution to do this is by adding artificial markers to strategically chosen places in the world. In this project we use *natural landmarks*, extracted from the scene itself, without modifications. Moreover, the extraction of these landmarks must be automatic and robust against changes in viewpoint and illumination to ensure the detection of these landmarks under as many circumstances as possible.

Many researchers proposed algorithms for natural landmark detection. Mostly, local regions are defined around interest points in the images. The characterisation of these local regions with descriptor vectors enables the regions to be compared across images. Differences between approaches lie in the way in which interest points, local image regions, and descriptor vectors are extracted. An early example is the work of (Schmid et al., 1997), where geometric invariance was still under image rotations only. Scaling was handled by using circular regions of several sizes. (Lowe, 1999) extended these ideas to real scale-invariance. More general affine invariance has been achieved in the work of Baumberg (Baumberg, 2000°), Tuytelaars & Van Gool (Tuytelaars et al., 1999; Tuytelaars & Van Gool, 2000), Matas (Matas et al.,2002), and Mikolajczyk & Schmid (Mikolajczyk & Schmid, 2002).

Although these methods are capable to find high quality correspondences, most of them are too slow to use in a real-time mobile robot algorithm. That is why we propose a much faster alternative, as explained in section 3.

## 2.2 Map structure

Many researchers proposed different ways to represent the environment perceived by vision sensors. We can order all possible map organisations by metrical detail: from dense 3D over sparse 3D to topological maps. We believe that the outer topological end of this spectrum offers the top opportunities.

One approach is building dense 3D models out of the incoming visual data (Pollefeys et al., 2004; Nistér et al., 2004). Such approach has some disadvantages. It is computationally and memory demanding, and fails to model planar and less-textured parts of the environment such as walls. Nevertheless, these structures are omnipresent in our application, and collisions need to be avoided.

One way to reduce the computational burden is to make abstraction of the visual data. Instead of modelling a dense 3D model containing billions of voxels, a sparse 3D model is built containing only special features, called *visual landmarks*.

Examples of researchers solving the navigation problem with sparse 3D maps of natural landmarks are (Se et al., 2001) and (Davison, 2003). They position natural features in a metrical frame, which is as big as the entire mapped environment. Although less than the dense 3D variant, these methods are still computationally demanding for large environments since their complexity is quadratic in the number of features in the model.

Also, for larger models the metric error accumulates, so that feature positions are drifting away.
As a matter of fact, the need for explicit 3D maps in navigation is questionable. One step further in the abstraction of environment information is the introduction of topological maps. The psychological experiments of (Bülthoff et al., 1998) show that people rely more on a topological map than a metrical one for their navigation. In these topological maps, locally places are described as a configuration of natural landmarks. These places form the nodes of the graph-like map, and are interconnected by traversable paths. Other researchers (Vale & Ribeiro, 2003; Ulrich & Nourbakhsh, 2000; Kosecká & Yang, 2004) also chose for topological maps, mainly because they scale better to real-world applications than metrical, deterministic representations, given the complexity of unstructured environments. Other advantages are the ease of path planning in such a map and the absence of drift.

### 2.3 Toplogical map building

Vale (Vale & Ribeiro, 2003) developed a clustering-based method for automatic building of a topological environment map out of a set of images. Unfortunately, the latter method is only suited for image comparison techniques which are a metric function, and does not give correct results if *self-similarities* are present in the environment, i.e. places that are different but look similar.
Very popular are various *probabilistic* approaches of the topological map building problem. (Ranganathan et al., 2005) for instance use Bayesian inference to find the topological structure that explains best a set of panoramic observations, while (Shatkay & Kaelbling, 1997) fit hidden Markov models to the data. If the state transition model of this HMM is extended with robot action data, the latter can be modeled using a partially observable Markov decision process or POMDP, as in (Koenig & Simmons, 1996; Tapus & Siegwart, 2005). (Zivkovic et al., 2005) solve the map building problem using graph cuts.
In contrast to these global topology fitting approaches, an alternative way is detecting *loop closings*. During a ride through the environment, sensor data is recorded. Because it is known that the driven path is traversable, an initial topological representation consists of one long edge between start and end node. Now, extra links are created where a certain place is revisited, i.e. an equivalent sensor reading occurs twice in the sequence. This is called a loop closing. A correct topological map results if all loop closing links are added.
Also in loop closing, probabilistic methods are introduced to cope with the uncertainty of link hypotheses and avoid links at self-similarities. (Chen & Wang, 2005), for instance, use Bayesian inference. (Beevers & Huang, 2005) recently introduced Dempster-Shafer probability theory into loop closing, which has the advantage that ignorance can be modelled and no prior knowledge is needed. Their approach is promising, but limited to simple sensors and environments. In this chapter, we present a new framework for loop closing using rich visual sensors in natural complex environments, which is also based on Dempster-Shafer mathematics but uses it differently.

### 2.4 Visual Servoing

As explained in section 2, the execution of a path using such a topological environment map boils down to a series of visual servoing operations between places defined by images.
(Cartwright & Collett, 1987) proposed the so-called bearing-only 'snapshot' model, inspired

by the visual homing behaviour of insects such as bees and ants. Their proposed algorithm consists of the construction of a home vector, computed as the average of landmark displacement vectors. (Franz et al., 1998) analysed the computational foundations of this method and derived its error and convergence properties. They conclude that every visual homing method based solely on bearing angles of landmarks like this one, inevitably depends on basic assumptions such as equal landmark distances, isotropic landmark distribution or the availability of an external compass reference. Unfortunately, because none of these assumptions generally hold in our targeted application we propose an alternative approach.

If both image dimensions are taken into account, not limiting the available information to the bearing angle, the most obvious choice is working via epipolar geometry estimation (e.g. Tuytelaars et al., 1999; Basri et al., 1993). Unfortunately, in many cases this problem is ill conditioned. A workaround for planar scenes is presented by (Sagüés et al., 2005), who opted for the estimation of homographies. (Svoboda et al., 1998) proved that motion estimation with omnidirectional images is much better conditioned compared to perspective cameras. That is why we chose a method based on omnidirectional epipolar geometry. Other work in this field is the research of (Mariottini et al., 2005), who split the homing procedure in a rotation phase and a translation phase, which can not be used in our application because of the non-smooth robot motion.

## 3. Fast wide baseline matching

The novel technique we use for image comparison is *fast wide baseline matching*. This key technique enables extraction of natural landmarks and image comparison for our map building, localisation and visual servoing algorithms.

We use a combination of two different kinds of wide baseline features, namely a rotation reduced and colour enhanced form of Lowe's *SIFT* features (Lowe, 1999), and the *invariant column segments* we developed (Goedemé et al., 2004). These techniques extract local regions in each image, and describe these regions with a vector of measures which are invariant to image deformations and illumination changes. Across different images, similar regions can be found by comparing these descriptors. This makes it possible to find correspondences between images taken from very different positions, or under different lighting conditions. The crux of the matter is that the extraction of these regions can be done beforehand on each image separately, rather than during the matching. Database images can be processed off-line, so that the images themselves do not have to be available at the time of matching with another image.

### 3.1 Camera motion constraint

The camera we use is a catadioptric system, consisting of an upward looking camera with a hyperboloidal mirror mounted above it. The result is a field of view of 360° in horizontal direction and more than 180° in vertical direction. The disadvantage is that these images contain severe distortions, as seen for instance in fig. 4.

We presume the robot to move on one horizontal plane. The optical axis of the camera is oriented vertically. In other words, allowed movements consist of translations in the plane and rotation around a vertical axis. Figure 3 shows an illustration on this.

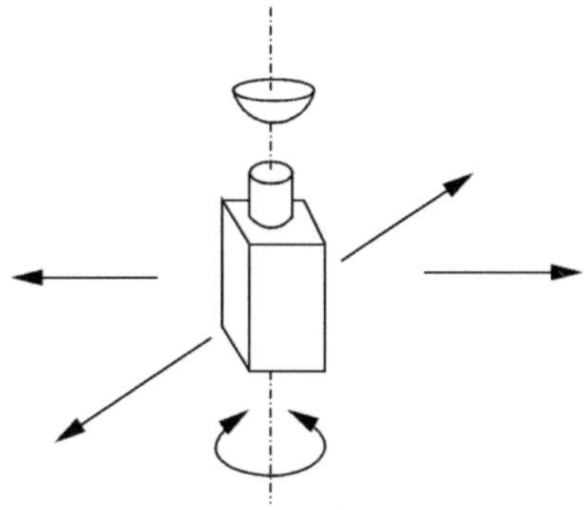

Fig. 3. Illustration of the allowed movements of the camera.

### 3.2 Rotation reduced and colour enhanced SIFT

David Lowe presented the *Scale Invariant Feature Transform* (Lowe, 1999), which finds interest points around local extrema in a scale-space of difference-of-Gaussian (DoG) images. The latter tend to correspond to blobs which contrast with their background. A dominant gradient orientation and scale factor define an image patch around each interest point so that a local image descriptor can be found as a histogram of normalised gradient orientations. SIFT features are invariant to rotation and scaling, and robust to other transformations.

A reduced form of these SIFT features for use on mobile robots is proposed by (Ledwich & Williams, 2004). They used the fact that rotational invariance is not needed for a camera with a motion constraint as in fig. 3. Elimination of the rotational normalisation and rotational part of the descriptor yields a somewhat less complex feature extraction and more robust feature matching performance.

Because the original SIFT algorithm works on greyscale images, some mismatches occur at similar objects in different colours. That is why we propose an outlier filtering stage using a colour descriptor of the feature patch based on global colour moments, introduced by Mindru *et al.*. We chose three colour descriptors: $C_{RB}$, $C_{RG}$ and $C_{GB}$, with

$$C_{PQ} = \frac{\int PQ\,\mathrm{d}\Omega \int \mathrm{d}\Omega}{\int P\,\mathrm{d}\Omega \int Q\,\mathrm{d}\Omega}, \qquad (1)$$

where $P,Q \in \{R,G,B\}$, i.e. the red, green, and blue colour bands, centralised around their means. After matching, the correspondences with Euclidean distance between the colour description vectors above a fixed threshold are discarded.

We tested these algorithms on the image pair in fig. 5. With the original SIFT algorithm, the first 13 matches are correct. Using our rotation reduced and colour enhanced algorithm, we see that up to 25 correct matches are found without including erroneous ones.

### 3.3 Invariant column segments

We developed wide baseline features which are specially suited for mobile robot navigation. There we exploited the special camera motion and the fact that man-made environments contain many vertical structures. Examples are walls, doors, and furniture. These don't have to be planar, so cylindrical elements like pillars do comply too. Vertical lines in the world always project to radial lines in the omnidirectional image for the constrained camera motions.

Here, these new wide baseline features are described for the use on omnidirectional images. More details and how they can be used on perspective camera images are described in (Goedemé et al, 2004). The extraction process of the wide baseline features starts as illustrated in figure 4. We stress that every step is invariant to changes in viewpoint and illumination. Along every line through the centre of the original image (left), we look for points having a local maximum gradient value (centre). Every consecutive pair of gradient maxima along the line defines the begin and end of a new invariant column segment (right).

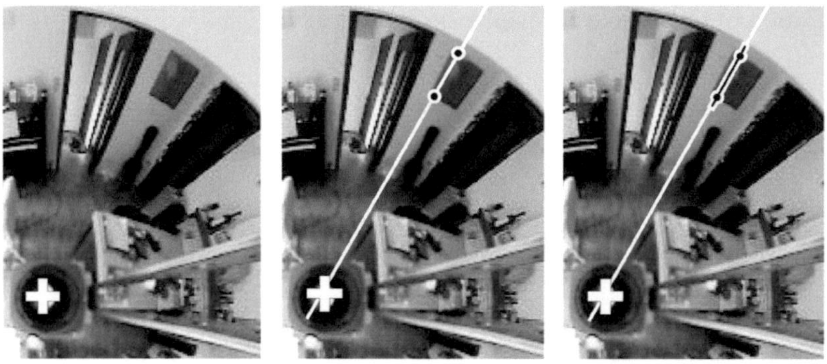

Fig. 4. Illustration of the invariant column segment extraction algorithm: (left) part of the original image, the white cross identifies the projection centre, (centre) local maxima of the gradient for one radius, (right) one pair of maxima defines a column segment.

We characterise the extracted column segments with a descriptor that holds information about colour and intensity properties of the segment. This 10-element vector includes:

- Three *colour invariants*. To include colour information in the descriptor vector, we compute the colour invariants, based on generalised colour moments (equation 1), over the column segment. To include information about the close neighbourhood of the segment, the line segment is expanded on both sides with a constant fraction of the segment length (in our experiments 0.2). Figure 4 (right) shows this.
- Seven *intensity invariants*. To characterise the intensity profile along the column segment, the best features to use are those obtained through the Karhunen-Lòeve transform (PCA). But because all the data is not known beforehand this is not practical. As is well known, the Fourier coefficients can sometimes offer a close approximation of the KL coefficients. In our method, because it is computationally less intensive and gives real output values, we choose to use the seven first coefficients of the *discrete cosine transform* (DCT), instead of Fourier.

In many cases there are horizontally constant elements in the scene. This leads to many very resembling column segments next to each other. To avoid matching over and over again very similar line segments, we first do a clustering of the line segments in each image. As a clustering measure we use the Mahalanobis distance of the descriptor vectors, extended with the horizontal distance between the line segments. In each cluster a prototype segment is chosen for use in the matching procedure.

### 3.4 Matching

These two kinds of local wide baseline features are very suited to quickly find correspondences between two widely separated images. A correspondence pair is established if for both features the other is the closest to it in the feature space, for the entire data set. Also, this match must be at least a fixed ratio better than the second best match. To be able to cope with different ranges of the elements of the descriptor vectors, distances are computed using the Mahalanobis measure (where we assume the cross-correlations to be zero):

$$d_{ij} = \sqrt{\sum_k \frac{(x_{ik} - x_{jk})^2}{\sigma_k^2}} \quad (2)$$

To speed up the matching, a Kd-tree of the reference image data is built.

Fig. 5 shows the matching results on a pair of omnidirectional images. As seen in these examples, the SIFT features and the column segments are complementary, which pleads for the combined use of the two. The computing time required to extract features in two 320x240 images and find correspondences between them is about 800 ms for the enhanced SIFT features and only 300 ms for the vertical column segments (on a 800 MHz laptop). Typically 30 to 50 correspondences are found.

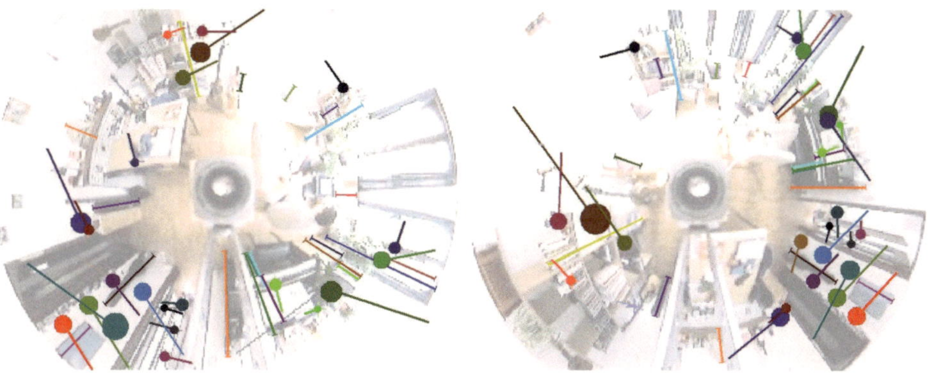

Fig. 5. A pair of omnidirectional images with colour-coded corresponding column segments (radial lines) and SIFT features (circles with tail).

For the description of a feature only the descriptors are used in the end, and not the underlying pixel data. As a result, the memory requirements for storing the reference images of entire environments can be kept limited.

## 4. Map Building

The navigation approach proposed is able to automatically construct a topological world representation out of a sequence of training images. During a training tour through the entire environment, omnidirectional images are taken at regular time intervals. The order of the training images is known. Section 4.1 describes the map structure targeted. In section 4.2, the image comparison technique based on fast wide baseline features is described which is used by the actual map building algorithm, presented in section 4.3.

### 4.1 Topological Maps

To be of use in the following parts of the navigation method, the topological map must describe all 'places' in the environment and the possible connections between these places. The topology of the world, being the maze of streets in a city or the structure of a house, must be reflected in the world model. The question remains what exactly is meant with such a *place* and how to delimit it. In a building, a place can be defined as a separate room. But then, what to do with long corridors, and outdoors with city streets?

That is why we define a place with regard to the needs of the localisation and locomotion algorithms. To be able to get a sufficiently detailed localisation output, the sampling of places must be dense enough. For the locomotion algorithm, which performs visual homing between two places at a time, the distance between these places must be not too big to ensure errorless motion. On the other hand, a compact topological map with fewer places requires less memory and enables faster localisation and path planning.

We discuss the image comparison method used in the map building algorithm before deciding on this place definition, as this comparison will lie at its basis.

### 4.2 Image Comparison Measure

The main goal of this section is to determine for each arbitrary pair of images a certain similarity measure, which tells how visually similar the two images are. Our image comparison approach consists of two levels, a global and a local comparison of the images. We first compare two images with a coarse but fast global technique. After that, a relatively slower comparison with more precision based on local features only has to be carried out on the survivours of the first stage.

#### 4.2.1 Global colour similarity measure

To achieve a fast global image similarity measure between two images, we compute the same moments we used for the local features (equation 1) over the entire image. These moments are invariant to illumination changes, i.e. offset and scaling in each colour band. The Euclidean distance between two sets of these image colour descriptors gives a visual dissimilarity measure between two images.

With this dissimilarity measure, we can clearly see for instance the difference of images taken in different rooms. Because images taken in the same room but at different positions have approximately the same colour scheme, a second dissimilarity measure based on local features is needed to distinguish them.

#### 4.2.2 Local measure based on matches

First, we search for feature matches between the two images, using the techniques described in section 3. The dissimilarity measure is taken to be inversely proportional to the *number of matches*, relative to the average number of features found in the images. Also the difference in relative configuration of the matches is taken into account. Therefore, we first compute a global angular alignment of the images by computing the average angle difference of the matches. The dissimilarity measure $D_m$ is now also made proportional to the average angle difference of the features after this global alignment:

$$D_m = \frac{1}{N} \cdot \frac{n_1 + n_2}{2} \cdot \frac{\sum |\theta_i|}{N} \quad (3)$$
$$= \frac{(n_1 + n_2) \sum |\theta_i|}{2N^2}, \quad (4)$$

,where $N$ corresponds to the number of matches found, $n_i$ the number of extracted features in image $i$, and $\theta$ the angle difference for one match after global alignment.

#### 4.2.3 Combined dissimilarity measure

We combine these two measures: only those pairs of images who have a colour dissimilarity under a predefined threshold are candidates for computing a matching dissimilarity.
This combined *visual* distance between two images is related to the physical distance between the corresponding viewpoints, but is certainly not a linear measure for it. As a matter of fact, the disparity and appearance difference of features is also related to the distance of the corresponding natural landmark to the cameras. Therefore, in large spaces (halls, market squares), a certain visual distance will be corresponding to a much larger physical distance, compared to the same visual distance between two images in a small space.
With this visual distance, the place definition problem can be addressed on the basis of a constant visual distance between places instead of a constant physical distance.

### 4.3 Map Building Algorithm

We apply the mathematical theory of Demster and Shafer (Dempster, 1967; Shafer, 1976) on the topological map building problem posed. Out of a series of omnidirectional images, acquired at constant rate during a tour through the environment. Firstly, these images are clustered into *places*. Then, loop closing hypotheses are formulated between visually similar places of which evidence is collected using Dempster-Shafer theory. Once decisions are made about these hypotheses, the correct topology of the world is known.
This technique makes it possible to cope with *self-similar* environments. Places that look alike but are different will more likely get their link hypothesis rejected.

#### 4.3.1 Image clustering

The dots in the sketch of figure 6 denote places where images are taken. Because they were taken at constant time intervals and the robot did not drive at a constant speed, they are not evenly spread. We perform agglomerative clustering with complete linkage based on the combined visual distance on all the images, yielding the ellipse shaped clusters in fig. 6. The black line shows the exploration path as driven by the robot.

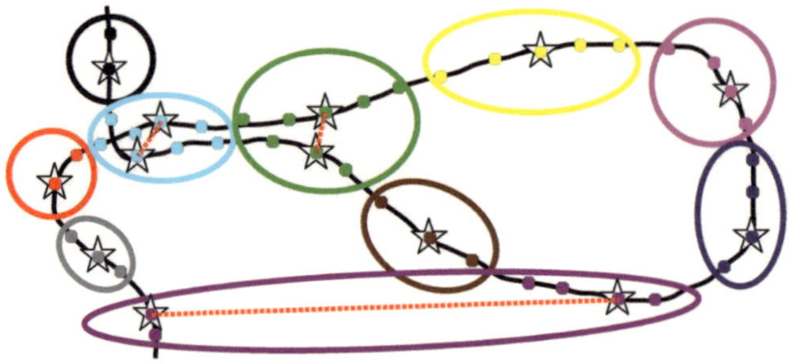

Fig. 6. Example for the image clustering and hypothesis formulation algorithms. Dots are image positions, black is exploration path, clusters are visualised with ellipses, prototypes of (sub)clusters with a star. Hypotheses are denoted by a dotted red line.

#### 4.3.2 Hypothesis formulation

As can be seen in the lower part of fig. 6, not all image groups nicely cover one distinct place. This is due to *self-similarities*, or distinct places in the environment that are different but look alike and thus yield a small visual distance between them.

For each of the clusters, we can define one or more subclusters. Images within one cluster which are linked by exploration path connections are grouped together. For each of these *subclusters* a prototype image is chosen as the *medoid*[1] based on the visual distance, denoted as a star in the figure.

For each pair of these subclusters within the same cluster, we define a *loop closing hypothesis H*, which states that if *H=true*, the two subclusters describe the same physical place and must be merged together. We will use Dempster-Shafer theory to collect evidence about each of these hypotheses.

#### 4.3.3 Dempster-Shafer evidence collection

For each of the hypotheses defined in the previous step, a decision must be made if it was correct or wrong. Figure 7 illustrates four possibilities for one hypothesis. We observe that a hypothesis has more chance to be true if there are more hypotheses in the neighbourhood, like in case *a* and *b*. If no neighbouring hypotheses are present (*c*,*d*), no more evidence can be found and no decision can be made based on this data.

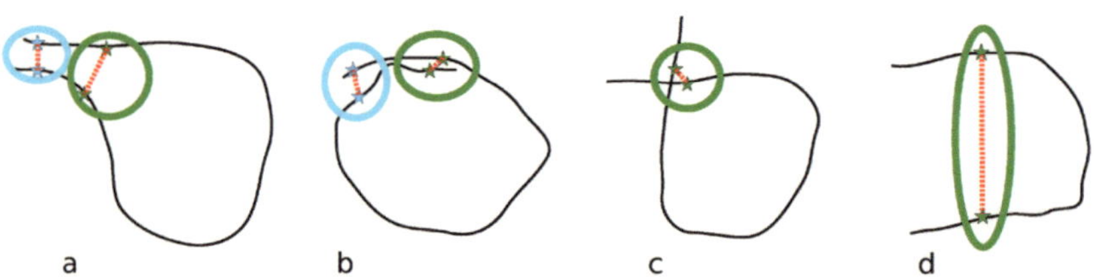


Fig. 7. Four topological possibilities for one hypothesis.

We conclude that for a certain hypothesis, a neighbouring hypothesis adds evidence to it. It

---

[1]The medoid of a cluster is computed analogous to the centroid, but using the median instead of the average.

is clear that, the further away this neighbour is from the hypothesis, the less certain the given evidence is. We chose to model this subjective uncertainty by means of the ignorance notion in Dempster-Shafer theory. That is why we define an *ignorance function* $\xi$ containing the distance between two hypotheses $H_a$ and $H_b$:

$$\xi(H_a, H_b) = \begin{cases} 1 - \sin\left(\frac{d_H(H_a,H_b)\pi}{2d_{th}}\right) & (d_H \leq d_{th}) \\ 0 & (d_H > d_{th}) \end{cases} \quad (8)$$

where $d_{th}$ is a distance threshold and $d_H$ is the sum of the distances between the two pairs of prototypes of both hypotheses, measured in number of exploration images.
To gather *aleatory* evidence, we look at the visual similarity of both subcluster prototypes, normalised by the standard deviation of the intra-subcluster visual similarities. The visual similarity is the inverse of the visual distance, defined in equation 4.
Each neighbouring hypothesis $H_b$ yields the following set of Dempster-Shafer masses, to be combined with the masses of the hypothesis $H_a$ itself:

$$\begin{aligned} m(\{\emptyset\}) &= 0 \\ m(\{H_a\}) &= s_V(H_b)\xi(H_a, H_b) \\ m(\{\neg H_a\}) &= (1 - s_V(H_b))\xi(H_a, H_b) \\ m(\{H_a, \neg H_a\}) &= 1 - \xi(H_a, H_b) \end{aligned} \quad (9)$$

Hypothesis masses are initialised with the visual similarity of its subcluster prototypes and an initial ignorance value (0.25 in our experiments), which models its influenceability by neighbours.

### 4.4 Hypothesis decision

After combination of each hypothesis's mass set with the evidence given by neighbouring hypotheses (up to a maximum distance $d_{th}$), a decision must be made if this hypothesis was correct and thus if the subclusters must be united into one place or not.
Unfortunately, as stated above, only positive evidence can be collected, because we can not gather more information about totally isolated hypotheses (like *c* and *d* in fig. 7). This is not too bad, because of different reasons. Firstly, the chance for correct, but isolated hypothesis (case *c*) is low in typical cases. Also, adding erroneous loop closings (*c* and *d*) will yield an incorrect topological map, whereas leaving them out will keep the map useful for navigation, but a bit less complete. Of course, new data about these places can be aqcuired later, during navigation.
It is important to remind oneself that the computed Dempster-Shafer masses can not directly be interpreted as probabilities. That is why we compute the support and plausibility of each hypothesis after evidence collection. Because these values define a confidence interval for the real probability, a hypothesis can be accepted if the lower bound (the support) is greater than a threshold.
After this decision, a final topological map can be built. Subclusters connected with accepted hypotheses are merged into one place, and a new medoid is computed as prototype of it. For hypotheses that are not accepted, two distinct places should be constructed.

## 5. Localisation

When the system has learnt a topological map of an environment, this map can be used for a variety of navigational tasks, firstly *localisation*. For each arbitrary new position in the known environment, the system can find out *where* it is. The output of this localisation algorithm is a *location*, which is—opposed to other methods like GPS—not expressed as a metric coordinate, but as one of the topological places defined earlier in the formerly explained map building stage.

The training set doesn't need to cover every imaginable position in the environment. A relatively sparse coverage is sufficient to localise every possible position. That is because the image comparison method we developed is based on wide baseline techniques and hence can recognise scene items from substantially different viewpoints.

Actually, two localisation modes exist. When starting up the system, there is no *a priori* information on the location. Every location is equally probable. This is called *global localisation*, alias the *kidnapped robot problem*. Traditionally, this is known to be a hard problem in robot localisation. In contrast, if there is knowledge about a former localisation not too long ago, the locations in the proximity of that former location have a higher probability than others further away. This is called *location updating*.

We propose a system that is able to cope with both localisation modes. A probabilistic approach is taken. Instead of making a hard decision about the location, a probability value is given to each location at each time instant. The Bayesian approach we follow is explained in the next subsection.

### 5.1 Bayesian Filtering

Define $x \in X$ a place of the topological map. $Z$ is the collection of all omnidirectional images $z$, so that $z(x)$ corresponds to the training observation at place $x$. At a certain time instant $t$, the system acquires a new image $z_t$. The goal of the localisation algorithm is to reveal the place $x_t$ where this image was taken.

We define the *Belief function* $Bel(x,t)$ as the probability of being at place $x$ at time $t$, given all previous observations. So,

$$Bel(x,t) = P(x_t|z_t, z_{t-1}, \ldots, z_0), \qquad (10)$$

for all $x \in X$. In the *kidnapped robot* case, there is no knowledge about previous observations hence $Bel(x,t_0)$ is initialised equal for all $x$.

Using Bayes' rule, we find:

$$Bel(x,t) = \frac{P(z_t|x_t, z_{t-1}, \ldots, z_0)P(x_t|z_{t-1}, \ldots, z_0)}{P(z_t|z_{t-1}, \ldots, z_0)}. \qquad (11)$$

Because the denominator of this fraction is not dependent on $x$, we replace it by the normalising constant $\eta$. If we know the current location of the system, we assume that future locations do not depend on past locations. This property is called the *Markov Assumption*. Using it, together with the probabilistic sum rule, equation 11 yields:

$$Bel(x,t) = \eta\, P(z_t|x_t) \sum_{x_{t-1} \in X} [P(x_t|x_{t-1}) Bel(x, t-1)] \qquad (12)$$

This allows us to calculate the belief recursively based on two variables: the *next state density* or *motion model* $P(x_t \mid x_{t-1})$ and the *sensor model* $P(z_t \mid x_{t-1})$.

### 5.2 Motion Model

The motion model $P(x_t \mid x_{t-1})$ explicits the probability of a transition from one place $x_{t-1}$ to another $x_t$. It seems logical to assume that a transition in one time instant between places that are far from each other is less probable than between places close to each other. We model this effect with a Gaussian:

$$P(x_t|x_{t-1}) = \frac{1}{\beta_x} e^{-dist(x_{t-1},x_t)/\sigma_x{}^2}. \qquad (13)$$

In this equation, the function $dist(x_1,x_2)$ corresponds to a measurement of the distance between the two places. We approximate it as the minimum number of place transitions needed to go from $x_1$ to $x_2$ on the topological map, computed with the Dijkstra algorithm (Dijkstra, 1959). In equation 13, $\beta_x$ is a normalisation constant, and $\sigma_x^2$ is the variance of the distances, measured on the map data. Once the topological map is known, the complete motion model can be computed off-line for usage during localisation.

### 5.3 Sensor Model

The entity $P(z_t \mid x_{t-1})$, called the sensor model, is the probability of acquiring a certain observation $z_t$ if the location $x_{t-1}$ is known. This is related to the visual dissimilarity of that observation and the training observation at location . The probability of acquiring an image at a certain place that differs much from the training image taken at that place has a low probability. We model this sensor model also by a Gaussian:

$$P(z_t|x_t) = \frac{1}{\beta_z} e^{-diss(z_t,z(x_t))/\sigma_z}. \qquad (14)$$

This time, the function $diss(z_1,z_2)$ refers to the visual dissimilarity explained in section 4. Unlike the motion model, the sensor model cannot be computed beforehand. It depends on the newly incoming query image data. Every location update step the visual dissimilarities of the query image with many database images must be computed. This validates our efforts to make the computation of the visual dissimilarity measure as fast as possible.

## 6. Path planning

With the method of the previous section, at each time instant the most probable location of the robot can be found, from which a path to a goal can be determined. How the user of the system, for instance the wheelchair patient, gives the instruction to go towards a certain goal is highly dependent on the situation. For every disabled person, for instance, an individual interface must be designed adapted to his/her possibilities.

We assume a certain goal is expressed as a certain place of the topological map, e.g. as a voice command <<*Kitchen!* >>. From the present pose, computed by the localisation algorithm, a path can be easily found towards it using Dijkstra's algorithm (Dijkstra, 1959). This path is expressed as a series of topological places which are traversed.

## 7. Visual servoing

The algorithm described in this section makes the robot move along a path, computed by the previous section. Such a path is given as a sparse set of prototype images of places. The physical distance between two consecutive path images is variable (1 to 5 metres in our tests), but the visual distance is constant, such that there are enough local feature matches as needed by this algorithm.
It is easy to see that following such a sparse visual path boils down to a succession of *visual homing* operations. First, the robot is driven towards the place where the first image on the path is taken. When arrived, it is driven towards the next path image, and so on. Because a smooth path is desired for the application, the motion must be continuous without stops at path image positions.
We tackle this problem by estimating locally the spatial structure of the wide baseline features using epipolar geometry. Hence, at this point we bring in some 3D information. This may seem at odds with our topological approach, but the depth maps are very sparse and only calculated locally so that errors are kept local, don't suffer from error build-up, and are efficient to compute.

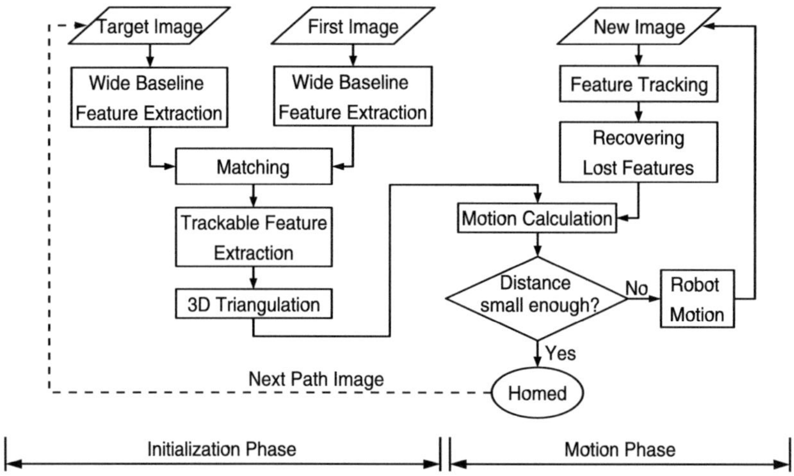


Fig. 8. Flowchart of the proposed algorithm for visual servoing along a path.

Fig. 8 offers an overview of the proposed method. Each of the *visual homing* operations is performed in two phases, an initialisation phase (section 7.1) and an iterated motion (section 7.2) phase.

### 7.1 Initialisation phase

From each position within the reach of the next path image (the *target* image), a visual homing procedure can be started. Our approach first establishes wide baseline local feature correspondences between the present and the target image, as described in section 3. That information is used to compute the epipolar geometry, which enables us to construct a local map containing the feature world positions, and to compute the initial homing vector.

### 7.1.1 Epipolar geometry estimation

Our calibrated single-viewpoint omnidirectional camera is composed of a hyperbolic mirror and a perspective camera. As imaging model, we use the model proposed by (Svoboda & Pajdla, 1998). This enables the computation of the epipolar geometry based on 8 point correspondences. In (Svoboda, 1999), Svoboda describes a way to robustly estimate the *essential matrix* $E$, when there are outliers in the correspondence set. The essential matrix is the equivalent of the fundamental matrix in the case of known internal camera calibration, . Svoboda's so-called *generate-and-select* algorithm to estimate $E$ is based on repeatedly solving an overdetermined system built from the correspondences that have a low *'outlierness'* and evaluating the *quality measure* of the resulting essential matrix. Because our tests with this method did not yield satisfactory results, we implemented an alternative method based on the well-known Random Sample Consensus (RANSAC (Fischler & Bolles, 1981)) paradigm.

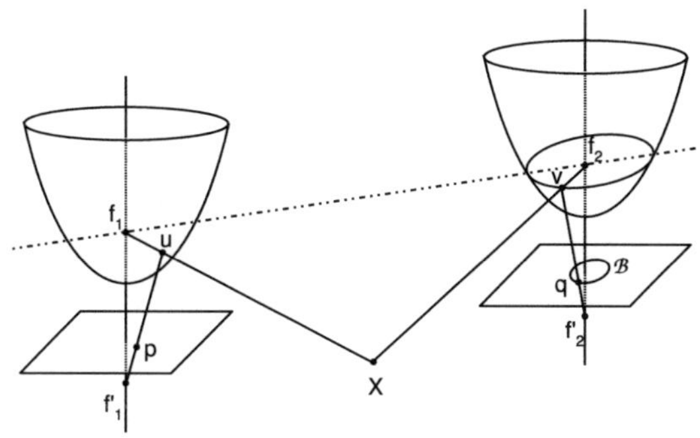


Fig. 9. Projection model for a pair of omnidirectional images.

The set-up is sketched in fig. 9. One visual feature with world coordinates **X** is projected via point **u** on the first mirror to point **p** in the image plane of the first camera. In the second camera, the mirror point is called **v** and the image plane point **q**. For each of the correspondences, the mirror points **u** and **v** can be computed as

$$\mathbf{u} = \mathcal{F}(K^{-1}\mathbf{p})K^{-1}\mathbf{p} + \mathbf{t}_C, \qquad (15)$$

with $\boldsymbol{t}_C=[0,0,-2e]^T$ and

$$\mathcal{F}(\mathbf{x}) = \frac{b^2(ex_1 + a\|\mathbf{x}\|)}{b^2x_1^2 - a^2x_2^2 - a^2x_3^2}. \qquad (16)$$

In these equations $K$ is the internal calibration matrix of the camera, and $a$, $b$ and $e$ are the parameters of the hyperbolic mirror.

If $E$ is the *essential matrix*, for all correspondences $\boldsymbol{v}^T E \boldsymbol{u}=\boldsymbol{0}$. This yields for each correspondence pair one linear equation in the coefficients of $E=[e_{ij}]$.

For each random sample of 8 correspondences, an $E$ matrix can be calculated. This is repeatedly done and for each $E$ matrix candidate the inliers are counted. A correspondence is regarded an inlier if the second image point **q** lies within a predefined distance from the epipolar *ellipse*, defined by the first image point **q**. This epipolar ellipse $B$ with equation $\boldsymbol{x}^T B \boldsymbol{x}=\boldsymbol{0}$ is computed with $B$=

$$\begin{bmatrix} -4t^2a^2e^2 + r^2b^4 & rsb^4 & rtb^2(-2e^2 + b^2) \\ rsb^4 & -4t^2a^2e^2 + s^2b^4 & stb^2(-2e^2 + b^2) \\ rtb^2(-2e^2 + b^2) & stb^2(-2e^2 + b^2) & t^2b^4 \end{bmatrix} \quad (17)$$

From the one essential matrix $E$ with the maximal number of inliers the motion between the cameras can be computed using the SVD based method proposed by (Hartley,1992). If more than one $E$-matrix is found with the same maximum number of inliers, the one is chosen with the best (i.e. smallest) quality measure , where is the $i$th singular value of the matrix $E$. Out of this relative camera motion, a first estimate of the homing vector is derived. During the motion phase this homing vector is refined.

### 7.1.2 Local feature map estimation

In order to start up the succession of tracking iterations, an estimate of the local map must be made. In our approach the local feature map contains the 3D world positions of the visual features, centred at the starting position of the visual homing operation. These 3D positions are easily computed by triangulation.

We only use two images, the first and the target image, for this triangulation. This has two reasons. Firstly, these two have the widest baseline and therefore triangulation is best conditioned. Our wide baseline matches between these two images are also more plentiful and less influenced by noise than the tracked features.

## 7.2 Motion phase

Then, the robot is put into motion in the direction of the homing vector and an image sequence is recorded. We rely on lower-level collision detection, obstacle avoidance and trajectory planning algorithms to drive safely (Demeester et al., 2003). In each new incoming image the visual features are tracked. Robustness to tracking errors (caused by e.g. occlusions) is achieved by reprojecting lost features from their 3D positions back into the image. These tracking results enable the calculation of the present location and from that the homing vector towards which the robot is steered.

When the (relative) distance to the target is small enough, the entire homing procedure is repeated with the next image on the sparse visual path as target. If the path ends, the robot is stopped at a position close to the position where the last path image was taken. This yields a smooth trajectory along a sparsely defined visual path.

### 7.2.1 Feature tracking

The corresponding features found between the first image and the target image in the previous step, also have to be found in the incoming images during driving. This can be done very reliably performing every time wide baseline matching with the first or target image, or both. Although our methods are relatively fast this is still too time-consuming for a driving robot.

Because the incoming images are part of a smooth continuous sequence, a better solution is *tracking*. In the image sequence, visual features move only a little from one image to the next, which enables to find the new feature position in a small search space.

A widely used tracker is the KLT tracker of (Shi & Tomasi, 1994). KLT starts by identifying interest points (corners), which then are tracked in a series of images. The basic principle of

KLT is that the definition of corners to be tracked is exactly the one that guarantees optimal tracking. A point is selected if the matrix

$$\begin{bmatrix} g_x^2 & g_x g_y \\ g_x g_y & g_y^2 \end{bmatrix}, \qquad (18)$$

containing the partial derivatives of the image intensity function over an *N*×*N* neighbourhood, has large eigenvalues. Tracking is then based on a Newton-Raphson style minimisation procedure using a purely translational model. This algorithm works surprisingly fast: we were able to track 100 feature points at 10 frames per second in 320×240 images on a 1 GHz laptop.
Because the well trackable points are not necessarily coinciding with the anchor points of the wide baseline features to be tracked, the best trackable point in a small window around such an anchor point is selected. In the assumption of local planarity we can always find back the corresponding point in the target image via the relative reference system offered by the wide baseline feature.

### 7.2.2 Recovering lost features

The main advantage of working with this calibrated system is that we can recover features that were lost during tracking. This avoids the problem of losing all features by the end of the homing manoeuvre, a weakness of our previous approach (Goedemé et al., 2005). This feature recovery technique is inspired by the work of (Davison, 2003), but is faster because we do not work with probability ellipses.
In the initialisation phase, all features are described by a local intensity histogram, so that they can be recognised after being lost during tracking. Each time a feature is successfully tracked, this histogram is updated.
When tracking, some features are lost due to invisibility because of e.g. occlusion. Because our local map contains the 3D positions of each feature, and the last robot position in that map is known, we can reproject the 3D feature in the image. Svoboda shows that the world point $\mathbf{X}_C$ (i.e. the point $\boldsymbol{X}$ expressed in the camera reference frame) is projected on point $\boldsymbol{p}$ in the image:

$$\mathbf{p} = \frac{K}{2e}(\lambda \mathbf{X}_C - \mathbf{t}_C), \qquad (19)$$

wherein λ is the largest solution of

$$\lambda = \frac{b^2(-e)\mathbf{X}_{C3} \pm a\|\mathbf{X}_C\|}{b^2\mathbf{X}_{C3}^2 - a^2\mathbf{X}_{C1}^2 - a^2\mathbf{X}_{C2}^2}. \qquad (20)$$

Based on the histogram descriptor, all trackable features in a window around the reprojected point **p** are compared to the original feature. When the histogram distance is under a fixed threshold, the feature is found back and tracked further in the next steps.

### 7.2.3 Motion computation

When in a new image the feature positions are computed by tracking or backprojection, the camera position (and thus the robot position) in the general coordinate system can be found based on these measurements.

It is shown that the position of a camera can be computed when for three points the 3D positions and the image coordinates are known. This problem is know as the *three point perspective pose estimation problem*. An overview of the proposed algorithms to solve it is given by (Haralick et al., 1994). We chose the method of Grunert, and adapted it for our omnidirectional case.
Also in this part of the algorithm we use RANSAC to obtain a robust estimation of the camera position. Repeatedly the inliers belonging to the motion computed on a three-point sample are counted, and the motion with the greatest number of inliers is kept.

#### 7.2.4 Robot motion

In the subsections above, it is explained how the position and orientation of the target can be extracted from the computed epipolar geometry. Together with the present pose results of the last subsection, a homing vector can easily be computed. This command is communicated to the locomotion subsystem. When the homing is towards the last image in a path, also the relative distance and the target orientation w.r.t. the present orientation is given, so that the locomotion subsystem can steer the robot to stop at the desired position. This is needed for e.g. docking at a table.

## 8. Experiments

### 8.1 Test platform

We have implemented the proposed algorithm, using our modified electric wheelchair 'Sharioto'. A picture of it is shown in the left of fig. 1. It is a standard electric wheelchair that has been equipped with an omnidirectional vision sensor (consisting of a Sony firewire colour camera and a Neovision hyperbolic mirror, right in fig. 1). The image processing is performed on a 1 GHz laptop.

### 8.2 Map building

The wheelchair was guided around in a large environment, while taking images. The environment was a large part of our office floor, containing both indoor and outdoor locations. This experiment yielded a database of 545 colour images with a resolution of 320×240 pixels. The total distance travelled was approximately 450 *m*. During a second run 123 images were recorded to test the localisation. A map and some of these images are shown in fig. 10.

After place clustering with a fixed place size threshold (in our experiments 0.5), this resulted in a set of 53 clusters. Using the Dempster-Shafer based evidence collection, 6 of 41 link hypotheses were rejected, as shown in fig. 10. Fig. 11 shows the resulting 59 place prototypes along with the accepted interconnections.

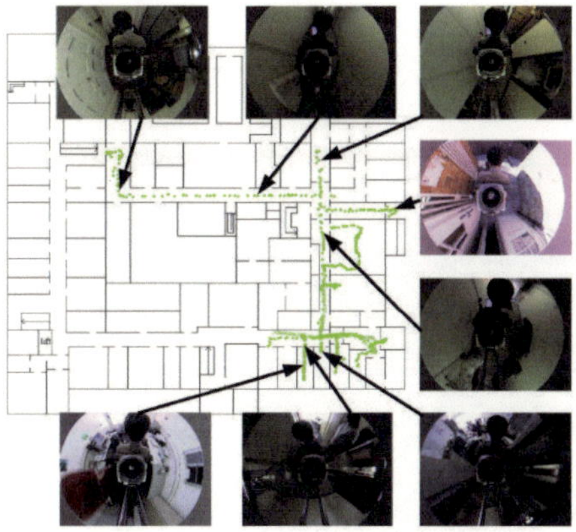

Fig. 10. A map of the test environment with image positions and some of the images.

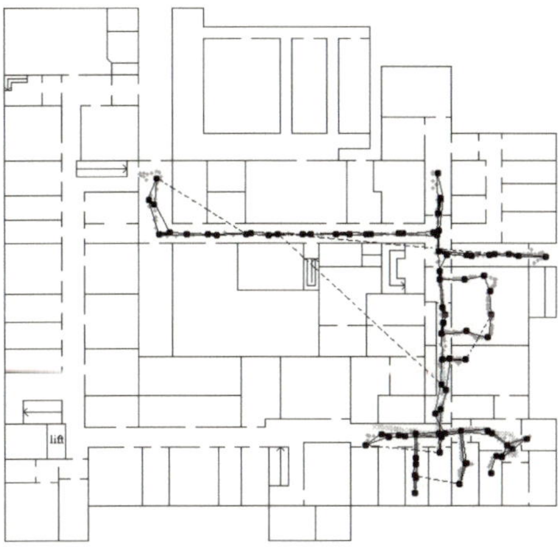
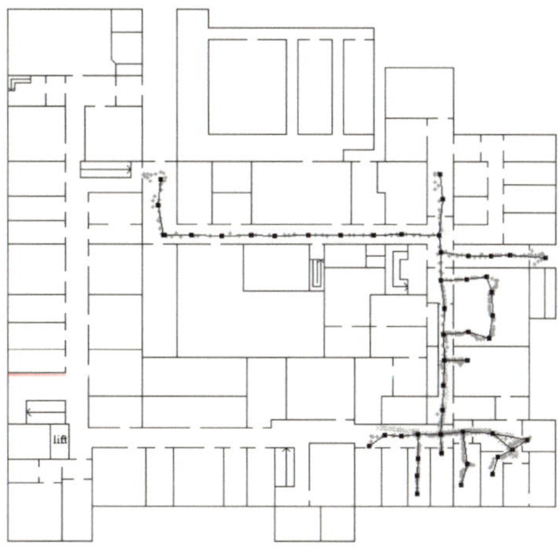

Fig. 11. (left) and 12. (right): Left: topological loop closing, accepted hypotheses are shown in thick black lines, rejected in dashed thin black lines. Right: the resulting topological map, locations of the place prototypes with interconnections.

Instead of keeping all the images in memory, the database is now reduced to only the descriptor sets of each prototype image. In our experiment, the memory needed for the database was reduced from 275 MB to 1.68 MB.

### 8.3 Localisation

From this map, the motion model is computed offline. Now, for the separate test set, the accuracy of the localisation algorithm is tested. A typical experiment is illustrated in fig. 13.

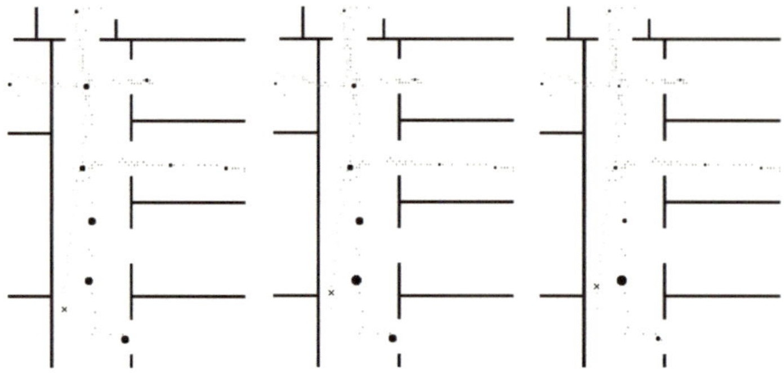

Fig. 13. Three belief update cycles in a typical localisation experiment. The black x denotes the location of the new image. Place prototypes with a higher belief value are visualised as larger black circles.

In total, for 78% of the trials the maximum of the belief function was located at the closest place at the first iteration, after the second and third belief update this percentage raised to 89% and 97%.

### 8.4 Visual servoing

#### 8.4.1 Initialisation phase

During the initialisation phase of one visual homing step, correspondences between the present and target image are found and the epipolar geometry is computed. This is shown in fig. 14.

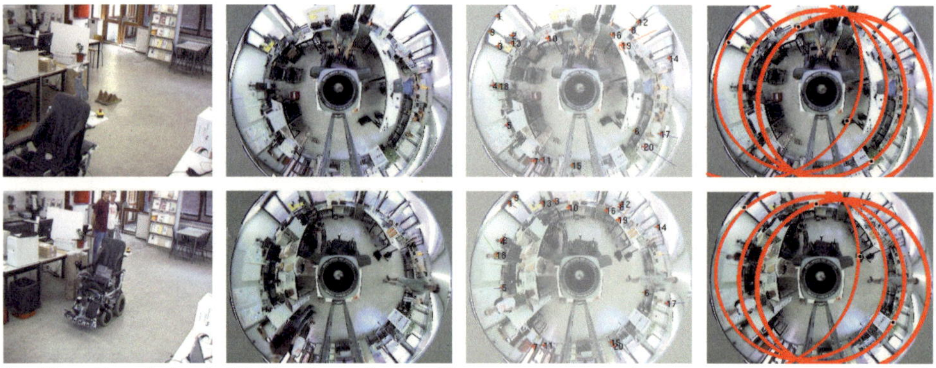

Fig. 14. Results of the initialisation phase. Top row: target, bottom row: start. From left to right, the robot position, omnidirectional image, visual correspondences and epipolar geometry are shown.

To test the correctness of the initial homing vector, we took images with the robot positioned at a grid with a cell size of 1 meter. The resulting homing vectors towards one of these images (taken at (6,3)) are shown in fig. 15. This proves the fact that even if the images are situated more than 6 metres apart the algorithm works thanks to the use of *wide baseline* correspondences.

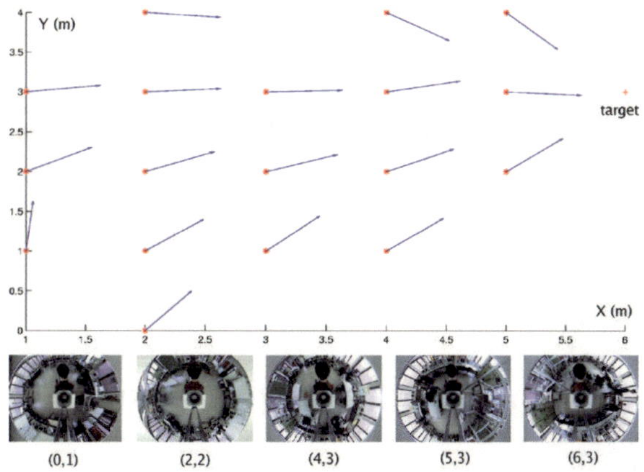


Fig. 15. Homing vectors from 1-meter-grid positions and some of the images.

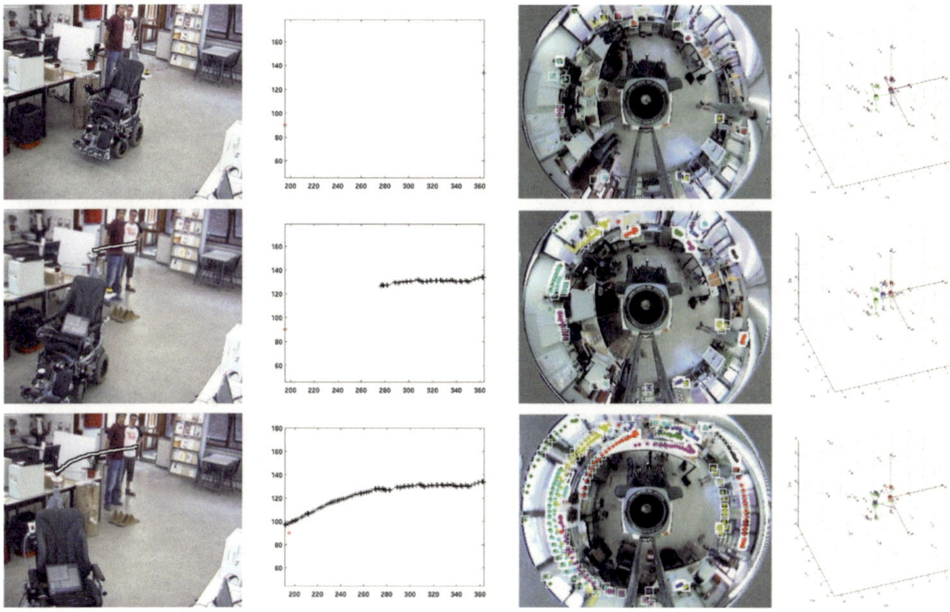

Fig. 16. Three snapshots during the motion phase: in the beginning (left), half (centre) and at the end (right) of the homing motion. The first row shows the external camera image with tracked robot position. The second row shows the computed world robot positions [cm]. The third row shows the colour-coded feature tracks. The bottom row shows the sparse 3D feature map (encircled features are not lost).

### 8.4.2 Motion phase

We present a typical experiment in fig. 16. During the motion, the top of the camera system was tracked in a video sequence from a fixed camera. This video sequence, along with the homography computed from some images taken with the robot at reference positions, permits calculation of metrical robot position ground truth data.

Repeated similar experiments showed an average homing accuracy of 11 cm, with a standard deviation of 5 cm, after a homing distance of around 3 m.

### 8.4.3 Timing results

The algorithm runs surprisingly fast on the rather slow hardware we used: the initialisation for a new target lasts only 958 ms, while afterwards every 387 ms a new homing vector is computed. For a wheelchair driving at a cautious speed, it is possible to keep on driving while initialising a new target. This resulted in a smooth trajectory without stops or sudden velocity changes.

## 9. Conclusion

This chapter describes and demonstrates a novel approach for a service robot to navigate autonomously in a large, natural complex environment. The only sensor is an omnidirectional colour camera. As environment representation, a topological map is chosen. This is more flexible and less memory demanding than metric 3D maps. Moreover, it does not show error build-up and enables fast path planning. As natural landmarks, we use two kinds of fast wide baseline features which we developed and adapted for this task. Because these features can be recognised even if the viewpoint is substantially different, a limited number of images suffice to describe a large environment.

Experiments show that our system is able to build autonomously a map of a natural environment it drives through. The localisation ability, with and without knowledge of previous locations, is demonstrated. With this map, a path towards each desired location can be computed efficiently. Experiments with a robotic wheelchair show the feasibility of executing such a path as a succession of visual servoing steps.

## 10. References

Baumberg, A. (2000). Reliable feature matching across widely separated views, *Computer Vision and Pattern Recognition*, pp. 774-781, Hilton Head, South Carolina.

Basri, R.; Rivlin, E. & Shimshoni, I. (1998), Visual homing: Surfing on the epipoles, *in IEEE International Conference on Computer Vision ICCV'98*, pp. 863-869, Bombay.

Beevers, K. & Huang W.(2005). Loop Closing in Topological Maps, *ICRA*, Barcelona, Spain.

Bischof, H.; Wildenauer, H. & Leonardis, A. (2001). Illumination insensitive eigenspaces, *In Proc. ICCV01*, pages 233-238, IEEE Computer Society, Vancouver, Canada.

Bülthoff, H.H.; van Veen, H.; Distler, H.K. & Braun, S.J. (1998), Navigating through a virtual city: Using virtual reality technology to study human action and perception, *Future Generation Computer Systems*, 14, pp. 231-242.

Cartwright, B. & Collett, T. (1987). *Landmark Maps for Honeybees*, Biological Cybernetics, 57, pp. 85-93.

Chen Ch. & Wang, H. (2005). Appearance-based topological Bayesian inference for loop-closing detection in cross-country environment, *IROS*, pp. 322-327, Edmonton.

Davison, A. (2003). Real-time simultaneous localisation and mapping with a single camera, *Intl. Conf. on Computer Vision*, Nice, France.

Dempster, A. P. (1967). *Upper and Lower Probabilities Induced by a Multivalued Mapping*, The Annals of Statistics (28), pp. 325–339, 1967.

Demeester, E.; Nuttin, M.; Vanhooydonck, D. & Van Brussel, H. (2003). Fine Motion Planning for Shared Wheelchair Control: Requirements and Preliminary Experiments, *International Conference on Advanced Robotics*, pp. 1278-1283, Coimbra, Portugal.

Dijkstra, E. W. (1959). *A note on two problems in connection with graphs*, Numerische Mathematik, 1: 269-271, 1959.

Fischler, M. & Bolles, R. (1981). *Random Sample Consensus: A Paradigm for Model Fitting with Applications to Image Analysis*, Comm. of the ACM, Vol 24, pp 381-395, 1981.

Franz, M.; Schölkopf, B.; Mallot, H. & Bülthoff, H. (1998). *Where did I take that snapshot? Scene-based homing by image matching*, Biological Cybernetics, 79, pp. 191-202, 1998.

Goedemé, T.; Tuytelaars, T. & Van Gool, L. (2004). Fast Wide Baseline Matching with Constrained Camera Position, *Computer Vision and Pattern Recognition*, Washington, DC, pp. 24-29.

Goedemé, T.; Tuytelaars, T.; Vanacker, G.; Nuttin, M. & Van Gool, L. (2005). Feature Based Omnidirectional Sparse Visual Path Following, *IEEE/RSJ International Conference on Intelligent Robots and Systems, IROS 2005*, Edmonton, Canada.

Haralick, R.; Lee, C.; Ottenberg, K. & Nölle, M. (1994). *Review and analysis of Solutions of the Three Point Perspective Pose Estimation Problem*, International Journal of Computer Vision, 13, 3, pp. 331-356, 1994.

Hartley, R. (1992). Estimation of relative camera positions for uncalibrated cameras, *2nd European Conference on Computer Vision*, pp. 579-587, Springer-Verlag, LNCS 588.

Jogan, M. & Leonardis, A. (1999). Panoramic eigenimages for spatial localisation, *In Proceedings 8th International Conference on Computer Analysis of Images and Patterns(1689)*, Solina, Franc and Leonardis, Ales, Eds., pages 558-567, Ljubljana.

Koenig, S. & Simmons, R. (1996). Unsupervised learning of probabilistic models for robot navigation, *Proceedings of ICRA, 1996.*

Kosecká, J. & Yang, X. (2004). Global Localization and Relative Pose Estimation Based on Scale-Invariant Features, *ICPR* (4), pp. 319-322, 2004.

Ledwich, L. & Williams, S. (2004). Reduced SIFT Features For Image Retrieval and Indoor Localisation, *Australasian Conf. on Robotics and Automation ACRA*, Canberra, Australia.

Lowe, D. (1999). Object Recognition from Local Scale-Invariant Features, *International Conference on Computer Vision*, pp. 1150-1157, Corfu, Greece.

Matas, J.; Chum, O.; Urban, M. & Pajdla, T. (2002). Robust wide baseline stereo from maximally stable extremal regions, *British Machine Vision Conference*, Cardiff, Wales, pp. 384-396.

Mariottini, G.; Alunno, E.; Piazzi, J. & Prattichizzo, D. (2005). Epipole-Based Visual Servoing with Central Catadioptric Camera, *IEEE ICRA05*, Barcelona, Spain.

Mikolajczyk, K. & Schmid, C. (2002). An affine invariant interest point detector, *ECCV*, vol. 1, 128–142, Copenhagen, Denmark.

Mindru, F.; Moons, T. & Van Gool, L. (1999). Recognizing color patters irrespective of viewpoint and illumination, *Computer Vision and Pattern Recognition*, vol. 1, pp. 368-373.

Nistér, D.; Naroditsky, O. & Bergen, J. (2004). Visual Odometry, *Conference on Computer Vision and Pattern Recognition*, Washington, DC.

Nuttin, M., Demeester, E.; Vanhooydonck, D. & Van Brussel, H. (2001). *Shared autonomy for wheelchair control: Attempts to assess the user's autonomy*, in Autonome Mobile Systeme, pp. 127-133, 2001.

Pollefeys, M.; Van Gool, L.; Vergauwen, M.; Verbiest, F.;. Cornelis, K; Tops, J.; Koch, R. (2004). Visual modeling with a hand-held camera, *International Journal of Computer Vision* 59(3), 207-232, 2004.

Ranganathan, A.; Menegatti E. & Dellaert, F., (2005). *Bayesian Inference in the Space of Topological Maps*, IEEE Transactions on Robotics, 2005.

Sagüés, C. & Guerrero, J. (2005). Visual correction for mobile robot homing, *Robotics and Autonomous Systems*, Vol. 50, no. 1, pp 41-49, 2005.

Schmid, C.; Mohr, R.; Bauckhage, C. (1997). *Local Grey-value Invariants for Image Retrieval,* International Journal on Pattern Analysis an Machine Intelligence, Vol. 19, no. 5, pp. 872-877, 1997.

Se, S.; Lowe, D. & Little, J. (2001) Local and Global Localization for Mobile Robots using Visual Landmarks, *In proceedings of the IEEE/RSJ International Conference on Intelligent Robots and Systems (IROS '01),* Hawaii, USA, 2001.

Shafer, G. (1976). *A Mathematical Theory of Evidence,* Princeton University Press, 1976.

Shatkay, H. & Kaelbling, L. P. (1997) Learning Topological Maps with Weak Local Odometric Information, *IJCAI* (2), pp. 920-929, 1997.

Shi, J. & Tomasi, C. (1994) Good Features to Track, *Computer Vision and Pattern Recognition,* Seattle, pp. 593-600, 1994.

Svoboda, T.; Pajdla, T. and Hlavá, V. (1998) Motion Estimation Using Panoramic Cameras, *Conf. on Intelligent Vehicles,* Stuttgart, pp. 335-340, 1998.

Svoboda, T. (1999) *Central Panoramic Cameras, Design, Geometry, Egomotion,* PhD Thesis, Czech Technical University, 1999.

Tapus, A. & Siegwart, R. (2005) Incremental Robot Mapping with Fingerprints of Places, *IROS*, Edmonton, Canada.

Tuytelaars, T.; Van Gool, L.; D'haene, L. & Koch, R. (1999) Matching of Affinely Invariant Regions for Visual Servoing, *Intl. Conf. on Robotics and Automation,* pp. 1601-1606, 1999.

Tuytelaars, T. & Van Gool, L. (2000) Wide baseline stereo based on local, affinely invariant regions, *British Machine Vision Conference,* pp. 412-422, Bristol, UK.

Ulrich, I. & Nourbakhsh, I.(2000)Appearance-Based Place Recognition for Topological Localisation, *IEEE International Conference on Robotics and Automatisation,* pp. 1023-1029, San Francisco, CA, April 2000,

Vale, A.; Ribeiro, M. I. (2003) Environment Mapping as a Topological Representation, *Proceedings of the 11th International Conference on Advanced Robotics - ICAR2003,* Universidade de Coimbra, Portugal, June 30 - July 1- 3, 2003.

Zivkovic, Z.; Bakker, B. & Kröse, B. (2005), Hierarchical Map Building Using Visual Landmarks and Geometric Constraints, in proceedings of the International conference on Intelligent Robots and Systems (IROS), pp. 7-12, Edmonton, Canada.

10

# Neural Networks Based Navigation and Control of a Mobile Robot in a Partially Known Environment

Diana D. Tsankova
*Technical University – Sofia, Branch Plovdiv*
*Bulgaria*

## 1. Introduction

Developing mobile robots able to move autonomously and staying operational in unexplored environments is one of the most important aims of intelligent robotics research. Navigation (here in the sense: collision-free goal following behaviour) depends on the amount of a priori available information. It could be assumed that this information comes in the following three categories: (I) complete information about the robot's workspace is available, (II) a priori knowledge about the environment is not available and the robot would have to rely on its sensors at execution time to obtain the information needed to accomplish the task, (III) partial knowledge of the environment is available and sensors are used at the execution time to acquire additional information. The robot may interweave planning and execution monitoring activities, but the more incomplete the prior knowledge, the less important the role of global planning.

Most of the global path-planning methods, which assume a static and completely known to the robot environment, are considered to belong to or to be variations of a few basic approaches: roadmap, cell decomposition, and potential field methods (Latombe, 1991; Choset et al., 2005). However, the dynamic and complex features of robot environments make their accurate modelling and predicting fundamentally impossible. Hence, during path following a local re-planning is needed in order to make the robot avoid collisions with unknown obstacles. By combining global and local path-planning methods mobile robots are able to operate in dynamic environments (Yahja et al., 2000). Each of the path-planning methods has its own advantages and drawbacks and is more or less proper to be used in a particular situation. How to select the most suitable approach and how to use local path-planning in collaboration with the global one in a practical example, related to a partially known or fast-changing environment - these are still open research problems.

Navigation relies on the tracking and regulation capabilities of the feedback control design (the lower control level of the robot). Standard approaches to non-holonomic control design often deal only with the kinematic steering system, but good performance cannot be obtained without taking into account the vehicle dynamics. A stable control algorithm that considers the complete vehicle dynamics has been developed using back stepping kinematics into dynamics and the results have been reported in the literature (Fierro & Lewis, 1995; Zulli et al., 1995). According to this approach the dynamics of the vehicle has to be completely known. However, in many practical cases, exact knowledge of the mobile robot dynamics is unattainable, e.g., friction is very difficult to model by conventional techniques. A solution to this problem requires implementation of robust adaptive control methods combining conventional approaches with new learning techniques (Gomi & Kawato, 1990; Gomi & Kawato, 1993), and these techniques have to be suitable for the real time applications.

This research work treats problems that belong to the third of the categories, mentioned above. The proposed integrated three-component system for navigation and control is similar to that, reported in Ref. (Topalov et al., 1998b), but all the three parts are implemented as neural networks here and the overall performance of the system is improved. The global path-planning algorithm seeks the shortest collision-free path by minimizing a cost function using neural penalties for obstacle collisions (Meng & Picton, 1992). The local obstacle avoidance controller is implemented as Braitenberg's No.3c vehicle modified by an artificial emotion mechanism. The emotional intervention significantly improves the performance of the vehicle (Mochida, 1995; Tsankova, 2009). Finally the path planner and local obstacle avoidance controller are integrated with a nonlinear trajectory tracking controller, whose control algorithm is based on a dynamical extension that makes possible the integration of a kinematic controller and a radial basis function (RBF) net based torque controller into an adaptive feedback controller using on-line *feedback-error-learning* scheme. The use of the recursive least squares technique based on Givens QR decomposition (Rosipal et al., 1998) makes the RBF net's weights tuning faster than the learning procedures based on genetic algorithms (Topalov et al., 1997; Topalov et al., 1998a) or backpropagation (Topalov et al., 1998b). The effectiveness of the proposed navigation and control strategies, is confirmed in MATLAB simulations by a collision-free goal following task, presented in different types of environment (including static and dynamic obstacles, a priori known and unknown obstacles).

## 2. The Task, the Robot and the General Design

The basic effort in this work is directed toward developing a system integrating three control devices (algorithms), which implements a collision-free trajectory tracking behaviour in a partially known environment. The devices (a path planner, a local obstacle avoidance controller and a trajectory tracking controller) are briefly described below.

### 2.1 General Design Algorithm

The overall structure of the design algorithm is shown in Fig.1. The design algorithm uses three different models of the mobile robot - the geometric, kinematic, and dynamic models. The path planner needs only the geometric model of the robot and the information about the a priori known static obstacles, which are a part of the „partially known" environment. It produces the global collision-free path between the given start and goal positions, which furthermore is transformed by an appropriate interface, linking the path-planning and trajectory tracking devices, into time-indexed sequence of points - reference trajectory. Based on the knowledge of the three models, the trajectory tracking controller is able to find the appropriate commands (torques) for the wheel actuators, so that the robot follows the desired trajectory. If the robot, tracking the trajectory, encounters an obstacle that has been unknown during the path-planning procedure, then the local obstacle avoidance controller is switched to turn the robot aside from the trajectory and thus to overcome the obstacle. Reactive or behaviour-based control architectures are very appropriate for this purpose. It is evident that the obstacle avoidance controller needs to know the geometric model of the robot and needs to collect information about the a priori unknown static and/or dynamic obstacles.

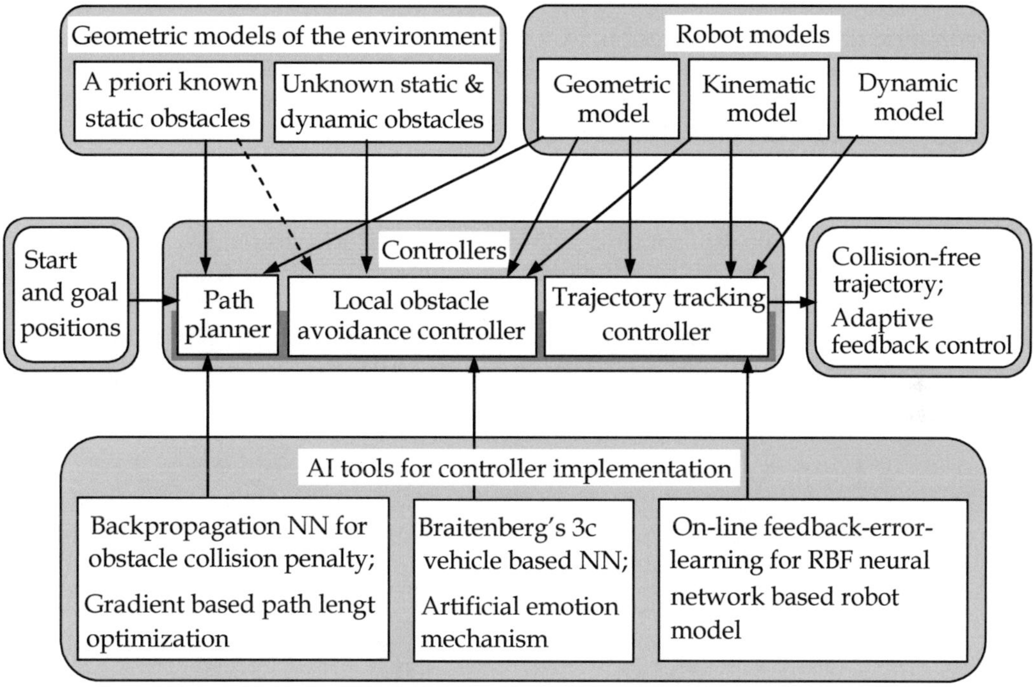


Fig. 1. Structure of the design algorithm

The three types of algorithms, included in the general design, make use of neural networks (NNs) as a powerful tool of artificial intelligence (AI). The path planner generates a path, which is optimal according to a criterion, which guarantees „minimum path length and no collisions with obstacles“ (Meng & Picton, 1992). An appropriate gradient based method is used for minimization of the path length together with the penalties for collisions with a priori known obstacles. The penalties are produced using an approximation of obstacle-oriented repulsive potential function, made by a feed-forward neural network, trained by error backpropagation algorithm.

The local obstacle avoidance controller includes the Braitenberg's No.3c architecture (Braitenberg, 1984), implemented as a sum of two NNs: one for an obstacle avoidance behaviour and another for a goal following one. The goal is represented by the corresponding reference trajectory points. To improve the performance, a special type of neural network known as „artificial emotion mechanism“ (Mochida et al., 1995) is applied to modulate the weights of the goal-oriented NN.

The trajectory tracking controller represents an adaptive feedback controller using the on-line *feedback-error-learning* scheme (Gomi & Kawato, 1990; Gomi & Kawato, 1993) based on an RBF net based model. The tracking behaviour has to be adaptable and robust to changes in the robot dynamics and/or in the external conditions. The contribution to this controller (in its „model" part) is the use of the recursive least squares technique based on Givens QR decomposition (Rosipal et al., 1998), that makes the RBF net's weights tuning relatively fast and easy.

### 2.2 Geometric and Kinematic models of the robot

Consider a mobile robot with two driving wheels, mounted on the same axis, and a front free wheel (Fig.2a). The mobile robot is equipped with two kinds of detectors: obstacle detectors and a goal detector. The obstacle detectors are installed in five directions, as shown in Fig.2b. They can detect the existence of obstacles in their directions (sectors $S_i$, $i = 1,2,...,5$), and the detecting range of sensors is assumed to be equal to the diameter of the robot. The obstacles are physical existing structures, but the goal is a virtual one - it corresponds to the current reference trajectory point. The goal detector is a software detector, which has to calculate the distance and orientation to this „goal" with respect to the mobile base, using information for the reference point and the robot position and heading. After that the simulated goal detector can recognize the direction of the goal at any position of the obstacle detectors.

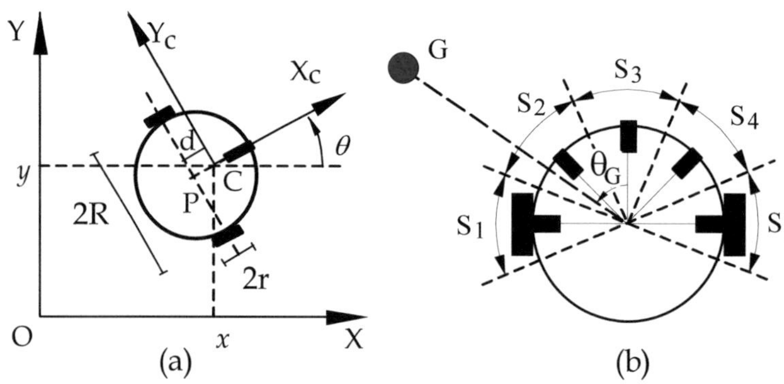


Fig. 2. Geometric model of a non-holonomic mobile robot

The motion and orientation of the robot are achieved by independent actuators (DC motors). The position of the robot in an inertial Cartesian frame $\{O, X, Y\}$ (Fig.2a) is completely specified by the *posture* $\mathbf{q} = (x, y, \theta)^T$ where $(x, y)$ and $\theta$ are the coordinates of the reference point $C$, and the orientation of the mobile basis $\{C, X_C, Y_C\}$ with respect to the inertial basis, respectively. The motion of the mobile robot is controlled by its *linear velocity* $\upsilon$ and *angular velocity* $\omega$, which are also functions of time. The kinematics of the vehicle is defined by Jacobian matrix $\mathbf{J}$, which transforms velocities $\mathbf{v} = (\upsilon,\ \omega)^T$ expressed in mobile basis into velocities $\dot{\mathbf{q}}$ expressed in a Cartesian one (Fierro & Lewis, 1995):

$$\begin{bmatrix} \dot{x} \\ \dot{y} \\ \dot{\theta} \end{bmatrix} = \dot{\mathbf{q}} = \mathbf{J}\mathbf{v} = \begin{bmatrix} \cos\theta & -d\sin\theta \\ \sin\theta & d\cos\theta \\ 0 & 1 \end{bmatrix} \begin{bmatrix} \upsilon \\ \omega \end{bmatrix}, \tag{1}$$

where $|\upsilon| \leq \upsilon_{\max}$ and $|\omega| \leq \omega_{\max}$. $\upsilon_{\max}$ and $\omega_{\max}$ are the maximum linear and angular velocities of the mobile robot. System (1) is called the steering system of the vehicle.

### 2.3 Dynamic model of the robot

The dynamical equations of the mobile robot system having an $n$-dimensional configuration space with generalized coordinates $\mathbf{q} = (q_1, q_2, \ldots, q_n)^{\mathrm{T}}$ and $m$ constraints can be described by (Fierro and Lewis, 1995):

$$\mathbf{M}(\mathbf{q})\ddot{\mathbf{q}} + \mathbf{V}_{\mathrm{m}}(\mathbf{q}, \dot{\mathbf{q}})\dot{\mathbf{q}} + \mathbf{F}(\dot{\mathbf{q}}) + \mathbf{G}(\mathbf{q}) + \boldsymbol{\tau}_{\mathrm{d}} = \mathbf{B}(\mathbf{q})\boldsymbol{\tau} - \mathbf{A}^{\mathrm{T}}(\mathbf{q})\lambda, \tag{2}$$

where $\mathbf{M}(\mathbf{q}) \in R^{n\times n}$ is a symmetric positive definite inertia matrix, $\mathbf{V}_{\mathrm{m}}(\mathbf{q}, \dot{\mathbf{q}}) \in R^{n\times n}$ is the centripetal and coriolis matrix, $\mathbf{F}(\dot{\mathbf{q}}) \in R^{n}$ denotes the surface friction, $\mathbf{G}(\mathbf{q}) \in R^{n}$ is the gravitational vector, $\boldsymbol{\tau}_{\mathrm{d}} \in R^{n}$ denotes bounded unknown disturbances including unstructured unmodelled dynamics, $\mathbf{B}(\mathbf{q}) \in R^{n\times r}$ is the input transformation matrix, $\boldsymbol{\tau} \in R^{r}$ is the input vector, $\mathbf{A}(\mathbf{q}) \in R^{m\times n}$ is the matrix associated with the constraints, and $\lambda \in R^{m}$ is the vector of constraint forces. The dynamical equations of the mobile base presented in Fig.2 can be expressed in the matrix form (2) where

$$\mathbf{M}(\mathbf{q}) = \begin{bmatrix} m & 0 & md\sin\theta \\ 0 & m & -md\cos\theta \\ md\sin\theta & -md\cos\theta & I \end{bmatrix}, \qquad \mathbf{V}(\mathbf{q}, \dot{\mathbf{q}}) = \begin{bmatrix} md\dot{\theta}^2\cos\theta \\ md\dot{\theta}^2\sin\theta \\ 0 \end{bmatrix},$$

$$\mathbf{B}(\mathbf{q}) = \frac{1}{r}\begin{bmatrix} \cos\theta & \cos\theta \\ \sin\theta & \sin\theta \\ R & -R \end{bmatrix}, \qquad \mathbf{G}(\mathbf{q}) = 0, \qquad \boldsymbol{\tau} = \begin{bmatrix} \tau_r \\ \tau_l \end{bmatrix}, \qquad \mathbf{A}^{\mathrm{T}}(\mathbf{q}) = \begin{bmatrix} -\sin\theta \\ \cos\theta \\ -d \end{bmatrix}, \tag{3}$$

$$\lambda = -m(\dot{x}\cos\theta + \dot{y}\sin\theta)\dot{\theta}.$$

## 3. The Path Planner

The collision-free path-planning can be stated as: Given an object with an initial position, a desire goal position, and a set of obstacles; the problem is to find a continuous path from the initial position to the goal position, which avoids colliding with obstacles along it. The path-planning procedure proposed in this work is based on the theoretical work of Meng and Picton (1992). The path for the object is represented by a set of $N$ via points. The path

finding algorithm is equivalent to optimizing a cost function, defined in terms of the total path length and the collision penalty, by moving the via points in the direction that minimizes the cost function. A two-layer log-sigmoid/log-sigmoid backpropagation neural network is used to produce the collision penalty. The environment area is divided into 2D grid cells and a binary value is assigned to each grid cell to indicate an obstacle presence: "0" means that the cell is fully unoccupied, and "1" - the cell is occupied. To take into account the robot's geometry the obstacles could be modified by growing its size isotropically by the robot's radius plus a small tolerance.
The $x$, $y$ coordinates of the cells' centres and the corresponding assigned binary values are used as learning patterns for the 2-input/1-output neural network. The output of the network represents the collision penalty for the current position $(x, y)$ of the object. The collision penalty of a path is defined as the sum of individual collision penalties of all the via points. The energy function for collision is defined as:

$$E_C = \sum_{i=1}^{N} C_i \,, \tag{4}$$

where $C_i$ is the collision penalty for the $i^{th}$ via point. The energy function for the path length is defined as the sum of squares of all the segments' lengths connecting the via points:

$$E_L = \sum_{i=1}^{N} L_i^{\,2} = \sum_{i=1}^{N} [(x_i - x_{i-1})^2 + (y_i - y_{i-1})^2]\,, \tag{5}$$

The total energy is

$$E = E_C + E_L \,. \tag{6}$$

The equation (6) is modified by multiplying both energy functions with positive weight coefficients $k_C$ and $k_L$ to express our preference for the collision criterion or for the path length criterion, respectively (Topalov et al., 1997):

$$E = k_C E_C + k_L E_L \,. \tag{7}$$

The dynamical equation for a via point is chosen to make time derivative of the energy be negative along the trajectory, because the low energy implies less collisions and a shorter path. The time derivative of $E$ is

$$\frac{dE}{dt} = \frac{d}{dt}(k_C E_C + k_L E_L) = \sum_{i=1}^{N} (k_L \frac{\partial L_i^2}{\partial x_i} + k_C \frac{\partial C_i}{\partial x_i}) \frac{dx_i}{dt} + \sum_{i=1}^{N} (k_L \frac{\partial L_i^2}{\partial y_i} + k_C \frac{\partial C_i}{\partial y_i}) \frac{dy_i}{dt} \,. \tag{8}$$

Let

$$\frac{dx_i}{dt} = -(k_L \frac{\partial L_i^2}{\partial x_i} + k_C \frac{\partial C_i}{\partial x_i}), \qquad \frac{dy_i}{dt} = -(k_L \frac{\partial L_i^2}{\partial y_i} + k_C \frac{\partial C_i}{\partial y_i}), \tag{9}$$

then

$$\frac{dE}{dt} = -\sum_{i=1}^{N}[(\frac{dx_i}{dt})^2 + (\frac{dy_i}{dt})^2] < 0. \quad (10)$$

$dE/dt = 0$ if and only if $dx_i/dt = 0$ and $dy_i/dt = 0$. Therefore, all via points move along trajectories, decreasing the energy, and finally reach equilibrium positions. In (9)

$$\frac{\partial C_i}{\partial x_i} = f'(I2)\sum_{j=1}^{S} W2_j f'(I1_j) W1^x{}_j, \qquad \frac{\partial C_i}{\partial y_i} = f'(I2)\sum_{j=1}^{S} W2_j f'(I1_j) W1^y{}_j, \quad (11)$$

where $C_i$ is the output of the network, $I2$ is the input of the output layer neuron, $I1_j$ is the input of the j^th^ hidden layer neuron, $S$ is the number of hidden layer neurons, $W1^x{}_j$ is the weight coefficient of the input " $x$ " with respect to j^th^ hidden layer neuron, $W2_j$ is the weight coefficient of j^th^ hidden layer neuron's output with respect to the output layer neuron. From (9) and (11), the dynamical equations for $x_i$ and $y_i$ are derived as:

$$\begin{aligned} \frac{dx_i}{dt} &= -[k_L(2x_i - x_{i-1} - x_{i+1}) + k_C f'(I2)\sum_{j=1}^{S} W2_j f'(I1_j) W1^x{}_j] \\ \frac{dy_i}{dt} &= -[k_L(2y_i - y_{i-1} - y_{i+1}) + k_C f'(I2)\sum_{j=1}^{S} W2_j f'(I1_j) W1^y{}_j]. \end{aligned} \quad (12)$$

## 4. The Local Obstacle Avoidance Controller

Following the trajectory that is based on the path-planning procedure, described in the previous section, the mobile robot would not have an ability to avoid unexpected local obstacles (obstacles whose location or geometry is different from those in the robot's model of the workspace). A solution to this problem is the introduction of a sensor-based local-obstacle avoidance controller. It operates in cooperation with the trajectory tracking controller as it switches and functions *only when* the sensors detect an obstacle. Based on sensor readings the local obstacle avoidance controller calculates the target velocities of the robot and passes them to trajectory tracking controller. The last produces torques for the robot's motion control. As it has been mentioned in Section 2.1, the sensor-based controller consists of Braitenberg's No.3c architecture as an action selection mechanism and an artificial emotion mechanism as a superstructure over it. The reason for this choice together with the description of both the components and the functions of the controller are presented below.

### 4.1 Artificial emotion mechanism

Emotion is a key element of adaptive behaviour, increasing the possibility of survival of living organisms. According to J. LeDoux (1996) the *amygdala*, being deep in the brain's center, makes the memory emotional. It is responsible for the emotions, especially for the most fundamental one - the fear. Sensor information obtained by receptors is firstly gathered at *thalamus*, and then forks into cerebral *cortex* and amygdala. In cerebral cortex, fined-grained information processing is carried out. The signals coming from the thalamus are fast and crude, reaching the amygdala before signals from the cortex, but providing only general information about the incoming stimulus. The signals from the cortex are slow and refined, providing detailed information about the stimulus. The coarse information processing accomplished in amygdala just evaluates whether the current situation is pleasant or not. It requires less computing time compared with the one in the cortex. This coarse but fast computation in the emotional system is indispensable for self preservation since living organisms have to cope with a continually changing world. The pathways that connect the amygdala with the cortex ("the thinking brain") are not symmetrical - the connections from the cortex to the amygdala are considerably weaker than those from the amygdala to the cortex. The amygdala is in a much better position to influence the cortex. Due to the above characteristics, it can be considered that the emotional system regulates activities in the cerebral cortex feedforwardly (Mochida et al., 1995).

In the computational model of the amygdala proposed by Mochida et al. (1995), the emotion of robots is divided into two states: *pleasantness* and *unpleasantness*, represented by a state variable called *frustration*. The neural network representation of this model is shown in Fig.3. Using sensory inputs the level of frustration is formulated as (Mochida et al., 1995):

$$f_{k+1} = \xi_1 \sum_{i=1}^{n} W_i S_i + (1 + \xi_2 \sum_{i=1}^{n} W_i S_i - b) f_k \, , \tag{13}$$

where $f_k$ represents the frustration level of the agent at the moment $k$, $\xi_1$ and $\xi_2$ are coefficients, $W_i$ denotes the weight parameter with respect to the obstacle detector $S_i$, and $b$ is the threshold, which determines the patience for unpleasantness; $n$ is the number of equipped obstacle detectors; $S_i, i = 1,2,...,5$ is the continuous signal from the obstacle detector, which is presented in Fig. 4, and represents the level of danger derived from the distance between the agent and the detected obstacle; In (13), the first and the second terms on the right hand side denote the frustration levels caused by the direct stimulation of the agent and the recently experienced relationship between the agent and the situation, respectively. The regulation output $\gamma = \gamma(f)$ of the emotional mechanism is determined here as:

$$\gamma = -\frac{1 - e^{-2(f + b_\gamma)}}{1 + e^{-2(f + b_\gamma)}} \, , \tag{14}$$

where $f \leftarrow f_k$ is taken from (13), $b_\gamma$ is a bias, and $\gamma \in [-1; 1]$ (Fig.3).

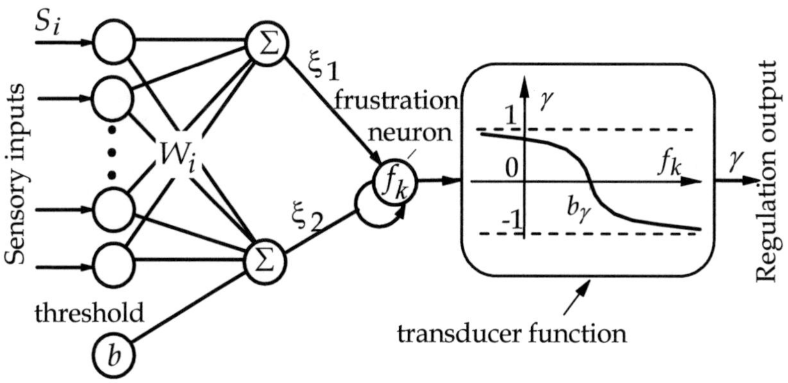


Fig. 3. Model of amygdala (Mochida et al., 1995)

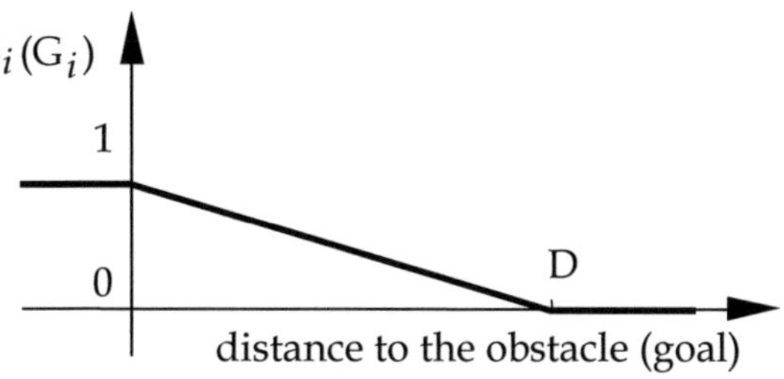


D is: the diameter of the robot (obstacle detector) or
the diagonal of the rectangular working field (goal detector)

Fig. 4. Normalized sensor readings of the obstacle/goal detectors

### 4.2 Emotionally influenced Braitenberg's vehicle No.3c

Before incorporating the artificial emotion mechanism into the local sensor-based controller, the innate action selection mechanism has to be determined. It is realized in the form of Braitenberg's architecture No.3c (Braitenberg, 1984) containing two neural networks, NN1 and NN2, for obstacle avoidance and goal following behaviours, respectively. The goal, as it has been mentioned in Section 2.1, is a virtual target, which corresponds to the current reference trajectory point. These networks directly connect sensors with the motor neurons of each wheel as shown in Fig. 5. Each of the connections has a positive (excitatory) or negative (inhibitory) weight. Depending on the weights of the connections, these structures can display repulsive (Fig. 5a) or attractive (Fig. 5b) behaviours. The robot motor control is determined by simple summation of the output signals of the corresponding motor neurons. The two neural networks, NN1 and NN2, simultaneously take part in a control action, that represents their consensus (Fig. 6a). In Fig. 6a $S$ and $G$ are sensor readings of the obstacle and goal detectors, respectively (as it has been stated in Fig. 4 and in Section 2.2); $\mathbf{v}_S = (v_{S,right}, v_{S,left})^{\mathrm{T}}$ and $\mathbf{v}_G = (v_{G,right}, v_{G,left})^{\mathrm{T}}$ are vectors with target linear velocities towards the motors of the right and left wheels in the obstacle avoidance and goal following networks, respectively. Each wheel is assumed to rotate according to the total activation level of the corresponding motor neuron.

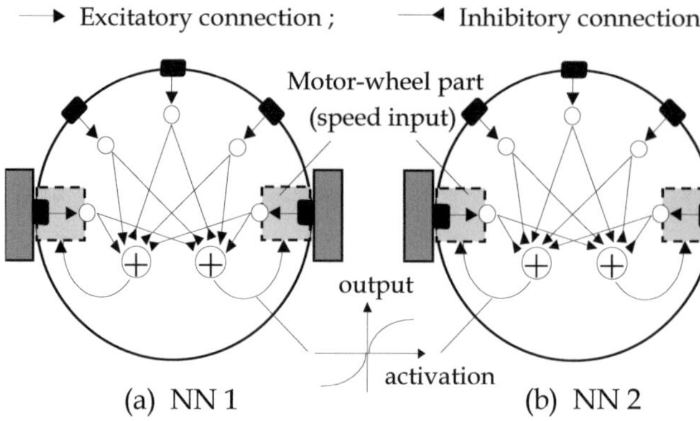


Fig. 5. Neural network architecture for: (a) obstacle avoidance behaviour - and (b) goal following behaviour (Tsankova & Topalov, 1999)

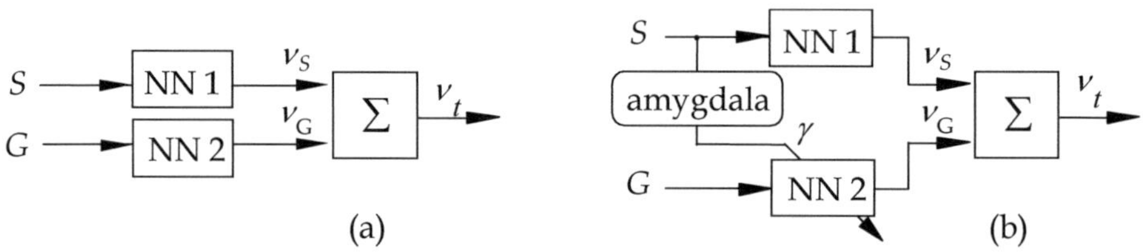


Fig. 6. Braitenberg's architecture No.3c: (a) independently acting, and (b) emotionally influenced

To incorporate the artificial amygdala into Braitenberg's architecture No.3c the outputs of the motor neurons in the goal following network have to be generated by multiplying together the weight of the network connections and the obtained regulation output $\gamma$ (Fig.6b). The artificial emotion (fear) mechanism influences on NN2 feed-forwardly by modulating the weights of connections with $\gamma$, taking positive and negative values. Thus, if obstacles are available, especially if they are critically disposed at the agent's movement to the goal, the goal following behaviour is manipulated: from decreasing the influence of the goal, through forgeting it, to the appearance of a mirror (false) goal, instead of the real one. In case of absence of obstacles the artificial amygdala does not influence on the action selection mechanism, because the weights of NN2 are multiplied by $\gamma = 1$.

## 5. The Adaptive Trajectory Tracking Controller

### 5.1 The trajectory tracking controller – task, scheme, and control laws

The trajectory tracking problem for non-holonomic vehicles is posed as follows (Fierro and Lewis, 1995): Given a reference cart

$$\dot{x}_r = \upsilon_r \cos\theta_r, \quad \dot{y}_r = \upsilon_r \sin\theta_r, \quad \dot{\theta}_r = \omega_r, \quad q_r = [x_r \quad y_r \quad \theta_r]^T, \quad v_r = [\upsilon_r \quad \omega_r]^T, \tag{15}$$

with $\upsilon_r > 0$ for all $t$, find a smooth velocity control $\upsilon_t$ such that $\lim_{t\to\infty}(q_r - q) = 0$, and compute the torque input $\tau(t)$ for (2) by some means such that $\nu \to \nu_t$ as $t \to \infty$.

The proposed structure for the mobile robot's tracking control system is presented in Fig. 7. It is assumed that the solution to steering system tracking problem as found in Kanayama et al.(1990) is available. This is denoted as $\nu_t$. The tracking error posture $\mathbf{e}(t) = (e_1(t), e_2(t), e_3(t))^{\mathrm{T}}$ is expressed in the basis of a frame linked to the mobile platform (Fierro and Lewis, 1995)

$$\mathbf{e} = \mathbf{T}_e(\mathbf{q}_r - \mathbf{q}),$$

$$\begin{bmatrix} e_1 \\ e_2 \\ e_3 \end{bmatrix} = \begin{bmatrix} \cos\theta & \sin\theta & 0 \\ -\sin\theta & \cos\theta & 0 \\ 0 & 0 & 1 \end{bmatrix} \begin{bmatrix} x_r - x \\ y_r - y \\ \theta_r - \theta \end{bmatrix}. \quad (16)$$

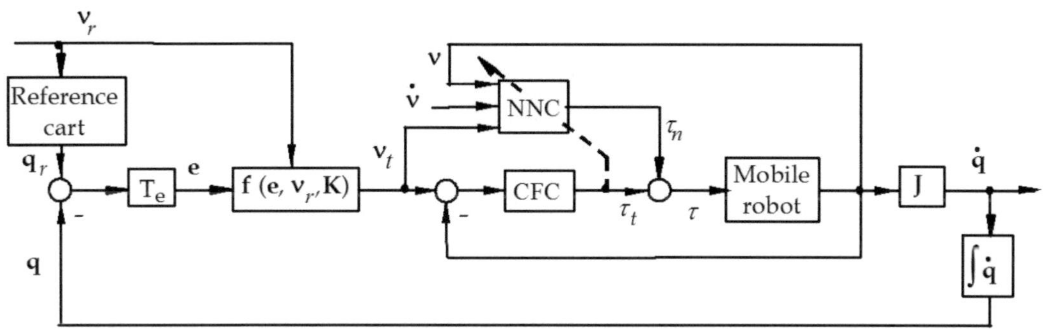


Fig. 7. Block diagram of neural network based tracking control of mobile robot.

The auxiliary velocity control input (target velocity vector $\mathbf{v}_t(t)$) that achieves tracking for (1) is given by (Kanayama et al., 1990):

$$\mathbf{v}_t = \mathbf{f}(\mathbf{e}, \mathbf{v}_r, \mathbf{K}) = \begin{bmatrix} \upsilon_r \cos e_3 + k_1 e_1 \\ \omega_r + k_2 \upsilon_r e_2 + k_3 \upsilon_r \sin e_3 \end{bmatrix}, \quad (17)$$

where $k_1$, $k_2$ and $k_3$ are positive constants, and $\upsilon_r > 0$.

Given the desired velocity $\mathbf{v}_t(t) \in R^{n-m}$, the auxiliary velocity tracking error can be defined as

$$\mathbf{e}_t = \mathbf{v}_t - \mathbf{v}. \quad (18)$$

In the proposed velocity control scheme, no preliminary knowledge of the cart dynamics is assumed. The velocity control input vector $\mathbf{v}_t$ is transformed into desired left and right wheel velocities and compared to the current wheel velocities of the mobile robot. Each wheel velocity is under independent control. In accordance with the *feedback-error-learning* approach (Gomi & Kawato, 1990; Gomi & Kawato, 1993), a conventional PD feedback controller (CFC) is used both as an ordinary feedback controller to guarantee global asymptotic stability during the learning period, and as a reference model for the responses of the controlled object. To obtain the desired response during tracking-error convergence movement by compensating for the nonlinear object dynamics, the RBF neural network is trained to become an adaptive nonlinear controller (NNC). The output of the CFC is fed to the NNC as the error signal used for learning.

There are two NNCs with identical structure for each robot's wheel velocity. The structure of an NNC is presented in Fig. 8. It has three inputs ($\upsilon^{left/right}$, $\dot{\upsilon}^{left/right}$, $\upsilon_t^{left/right}$) corresponding to the actual wheel velocity and acceleration, and the desired wheel velocity of the robot's left or right wheel, respectively. The output ($\tau_n^{left/right}$) of the network is computed using on-line RBF training procedure, described in Section 5.2. As output learning patterns the values of the signal $\tau^{left/right}$, generated with contribution of the conventional PD controller ($\tau_t^{left/right}$) (Fig. 7), are used.

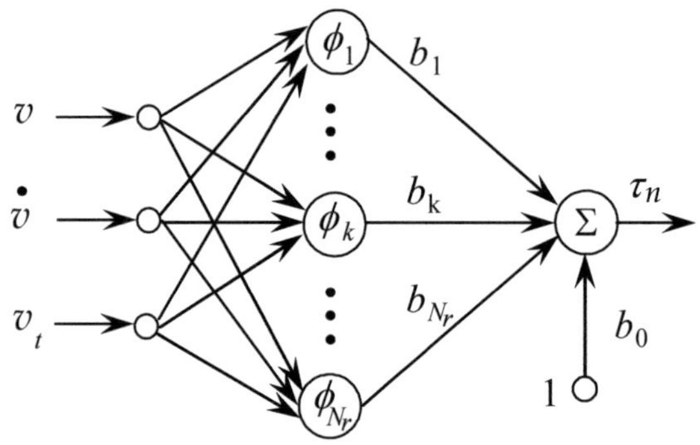


Fig. 8. RBF neural network structure. There are two networks with identical structure for each wheel – left and right. The names of the variables and parameters have to be considered with the label "left" or "right" (e.g., $\upsilon^{left}$ / $\upsilon^{right}$), respectively.

## 5.2 RBF network

In this Section, the approach of Rosipal et al. (1998) to on-line constructing a resource-allocating radial basis function network is presented. It exploits weights adaptation using recursive least squares technique based on Givens QR decomposition.

Consider a two-layered RBF network, which implements a mapping $\hat{y}: R^n \rightarrow R$ according to

$$\hat{y} = b_0 + \sum_{i=1}^{N_r} b_i \phi_i (\|\mathbf{x} - \mathbf{c}_i\| / \sigma_i), \tag{19}$$

where $\mathbf{x} \in R^n$ is an input vector, $\phi_i(.)$ is the transfer function (Gaussian function), $\sigma_i$ is the $i$-th center width, $\|.\|$ denotes the Euclidean norm, $b_i \in R$ are the weights, $\mathbf{c}_i \in R^n$ represent the positions of RBF centers, and $N_r$ is the number of centers.

**Growing phase.** The network starts to learn with no centers. The condition for allocating a new center at discrete time $j$ exploits two criteria, proposed by Platt (1991). The first criterion is based on the prediction error $|e_{RBF}(j)|=|y(j)-\hat{y}(j)|$ ($y(j)$ is the desired output), which is compared with the critical value $\varepsilon$. The second criterion is satisfied if the Euclidean distance of input $\mathbf{x}(j)$ from the nearest center $\mathbf{c}_{nearest}$ is greater than the critical scale resolution $\delta(j)$. The learning starts with the largest scale of resolution, i.e. $\delta(0)=\delta_{\max}$ and $\delta$ is shrunk during the time steps by a decay constant $T_d$ until it reaches the smallest value $\delta_{\min}$. If both criteria are satisfied, a new center is set as

$$\mathbf{c}_{new}(j)=\mathbf{x}(j), \tag{20}$$

with width

$$\sigma_{new}=k_{RBF}\left\|\mathbf{x}(j)-\mathbf{c}_{nearest}(j)\right\|^2, \tag{21}$$

where $k_{RBF}$ is a constant, and corresponding output weight

$$b_{new}(j)=e_{RBF}(j). \tag{22}$$

LMS gradient descent is used to update the positions of the centers

$$\Delta\mathbf{c}_i(j)=2\frac{\alpha}{\sigma_i^2}(\mathbf{x}(j)-\mathbf{c}_i(j))\phi_i(\mathbf{x}(j))e_{RBF}(j)b_i(j), \tag{23}$$

where $\alpha>0$ is the learning factor.
The steps, describing the network learning strategy, are given in Table 1 (Rosipal et al., 1998; 1997).

| |
|---|
| I. Initialization: $\delta(0)=\delta_{\max}$, $b_0(0)=y(0)$, $N_r=0$, set $\varepsilon$<br>II. At each iteration $j$, do:<br>1. Present a new input-output pair $(\mathbf{x}(j), y(j))$;<br>2. Evaluate output of network $\hat{y}(j)=b_0(j)+\sum_{i=1}^{N_r} b_i(j)\phi_i(\mathbf{x}(j))$;<br>3. Compute error $e_{RBF}(j)=y(j)-\hat{y}(j)$;<br>4. Find distance to nearest center $d=\min_{1\le i\le N_r}\lVert \mathbf{c}_i(j)-\mathbf{x}(j)\rVert$;<br>5. If $\lvert e_{RBF}(j)\rvert>\varepsilon$ and $d>\delta(j)$,<br>then allocate new center: $N_r=N_r+1$, $\mathbf{c}_{N_r}(j)=\mathbf{x}(j)$, $b_{N_r}(j)=e_{RBF}(j)$,<br>$\sigma_{N_r}=k_{RBF}d^2$;<br>Otherwise,<br>update $(b_0(j), \{b_i(j), \mathbf{c}_i(j)\}_{i=1}^{N_r})$.<br>6. If $\delta(j)>\delta_{\min}$,<br>then update $\delta(j+1)=\delta(j)\exp(-1/T_d)$;<br>Otherwise,<br>$\delta(j+1)=\delta(j)$. |

Table 1. The RBF network learning strategy

After adding a new center, another center addition is forbidden and during the next $T_W$ time steps only adaptation of the network parameters is allowed.

The on-line adaptation of output-layer weights $\mathbf{b}=[b_0, b_1, ..., b_{N_r}]^{\mathrm{T}}$ can be formulated as a problem of finding a weights vector $\mathbf{b}$, which minimizes a performance criterion, e.g.

$$V_t(\mathbf{b})=\sum_{j=0}^{t} w_t(j)\, e_{RBF}^2(j, \mathbf{b}(j)), \tag{24}$$

where $e_{RBF}(j, \mathbf{b}(j))$ is the error signal at time $j$ defined as $e_{RBF}(j, \mathbf{b}(j))=y(j)-\hat{y}(j, \mathbf{b}(j))$, and $w_t(j)$ is a window function reflecting time-varying significance of past and recent error signal:

$$w_t(j)=\lambda^{t-j}, \tag{25}$$

where $0<\lambda<1$ is a forgetting factor.

From the weights adaptation point of view, the RBF network can be considered as a special case of linear regression model

$$\mathbf{y}(t)=\mathbf{\Phi}(t)\mathbf{b}(t)+\mathbf{e}(t), \tag{26}$$

where $\mathbf{y}(t)=[y(1), y(2), ..., y(t)]^{\mathrm{T}}$, $\mathbf{e}_{RBF}(t)=[e_{RBF}(1), e_{RBF}(2), ..., e_{RBF}(t)]^{\mathrm{T}}$, $\mathbf{\Phi}(t)$ is $t\times N_{r+1}$

matrix whose transpose $\mathbf{\Phi}^{\mathrm{T}}(t) = [\phi(1), \phi(2),..., \phi(t)]$, in which $\phi(j) = [1, \phi_1(j), ..., \phi_{N_r}(j)]^{\mathrm{T}}$ is a $(N_{r+1} + 1) \times 1$ vector of the hidden-layer outputs (and bias term) at time $j$.
In accordance with the criterion (24), weights $\mathbf{b}(t)$ can be determined by a recursive least squares technique based on Givens QR decomposition. The estimation of the vector $\mathbf{b}(t)$ is based on the solution of the following system:

$$\mathbf{\Theta}(t)\hat{\mathbf{b}}(t) = \mathbf{q}(t), \tag{27}$$

where $\mathbf{\Theta}(t)$ is an upper triangular square matrix and $\mathbf{q}(t)$ is a vector. The solution of (27) consists of the steps presented in Table 2 (Rosipal et al., 1998). In Table 2 when a new center is allocated at time $j$, the forgetting factor $\lambda(j)$ is re-initialized to $\lambda(0)$.

**Pruning phase.** The above algorithm does not eliminate centers whose contribution to performance accuracy of the network is becoming negligible. Therefore, the updated network has to be checked to prune the redundant hidden units that have minimal contribution to the network output for $M$ consecutive observations in the sliding data window. In this work the pruning strategy of Salahshoor and Jafari (2007) is used. For this purpose, the contributions of all the available hidden units ($i = 1,...,N_r$) are first calculated on the network output as follows:

$$o_i(\mathbf{x}) = b_i\phi_i(\mathbf{x}), \qquad i = 1,...,N_r. \tag{28}$$

The highest absolute value of these contribution values is:

$$o_{\max}(\mathbf{x}) = \max_i |\, o_i(\mathbf{x}) \,|. \tag{29}$$

The contribution values on the network output are normalized to the highest absolute value:

$$r_i = |\, o_i / o_{\max} \,|, \qquad i = 1,...,N_r. \tag{30}$$

If the normalized contribution value of the $i$-th hidden unit on the network output ($r_i$) falls below a threshold ($r_{\text{threshold}}$) for $M$ consecutive observations, the $i$-th hidden unit is regarded as a redundant unit and is pruned (Salahshoor & Jafari, 2007).

1. Initialize matrix $\mathbf{\Theta}(.)$ and vector $\mathbf{q}(.)$

- At time $k=0$ set $\mathbf{\Theta}(0)=\eta$ and $\mathbf{q}(0)=0$;

- If a new center $N_r$ is allocated at time $k \neq 0$, increase dimensions of $\mathbf{\Theta}(k)$ and $\mathbf{q}(k)$ to $(N_r+1)\times(N_r+1)$ and $(N_r+1)$, respectively, and set diagonal elements $\mathbf{\Theta}_{N_r+1,N_r+1}(k)=\eta / e_{RBF}(k)$ and $\mathbf{q}_{N_r+1}(k)=\eta$ (where $\eta$ is a small positive number);

2. At each time $j \neq k$, update $\lambda(j)$ via $\lambda(j)=\lambda_0\lambda(j-1)+1-\lambda_0$, the appropriate values for $\lambda_0$ and $\lambda(0)$ are just less than one;

3. At time $j \neq k$, form the following $(N_r+2)\times(N_r+2)$ matrix;

$$\Omega(j)=\left[\begin{array}{c|c} \lambda^{1/2}\mathbf{\Theta}(j-1) & \lambda^{1/2}\mathbf{q}(j-1) \\ \hline \phi^{\mathrm{T}}(j) & y(j) \end{array}\right]$$

4. At time $j \neq k$, perform a sequence of elementary Givens rotations (GR) to transform the matrix $\Omega(j)$ to an upper triangular matrix;

$$\Omega(j)=\left[\begin{array}{c|c} \lambda^{1/2}\mathbf{\Theta}(j-1) & \lambda^{1/2}\mathbf{q}(j-1) \\ \hline \phi^{\mathrm{T}}(j) & y(j) \end{array}\right] \overset{\mathrm{GR}}{\Rightarrow} \left[\begin{array}{c|c} \mathbf{\Theta}(j) & \mathbf{q}(j) \\ \hline 0 & * \end{array}\right],$$

where 0 is $1\times(N_r+1)$ zeros vector and $*$ is a *don't care* variable;

5. At time $j \neq k$, compute $\hat{\mathbf{b}}(j)=\mathbf{\Theta}^{-1}(j)\mathbf{q}(j)$.

Table 2. The recursive technique for determining $\mathbf{b}(t)$

### 5.3 The trajectory tracking controller in collaboration with the local obstacle avoidance one

The overall integrated system for navigation and control of the mobile robot is shown in Fig.9. The path planner generates a path as a sequence of points (piece-wise linear path). Under assumption for a rectilinear movement with a constant reference velocity $\upsilon_r$ between each two points, discrete time step $T_0$ and the sequence of path points, the box $\mathrm{T_q}$ calculates the reference posture $\mathbf{q}_r$. When sensors detect an obstacle, the local obstacle avoidance controller switches on. The NN1 and the amygdala obtain inputs from obstacle detectors $\mathbf{S}$, and the NN2 – from a goal detector $\mathbf{G}$. Here, the goal detector is a virtual detector that uses the current reference point $(x_r, y_r)$ seen with respect to the mobile base as a virtual goal. The box $\mathrm{T_G}$ derives the first two components of the vector $\mathbf{e}$ ($\mathrm{T_G}:\ \mathbf{G}\subset\mathbf{e}$) and transforms them into a shape, appropriate for the NN2 inputs. In the presence of obstacles the local obstacle avoidance controller replaces the kinematic controller, proposed by Kanayama et al. (1995) and generates the target velocities $\mathbf{v}_t$.

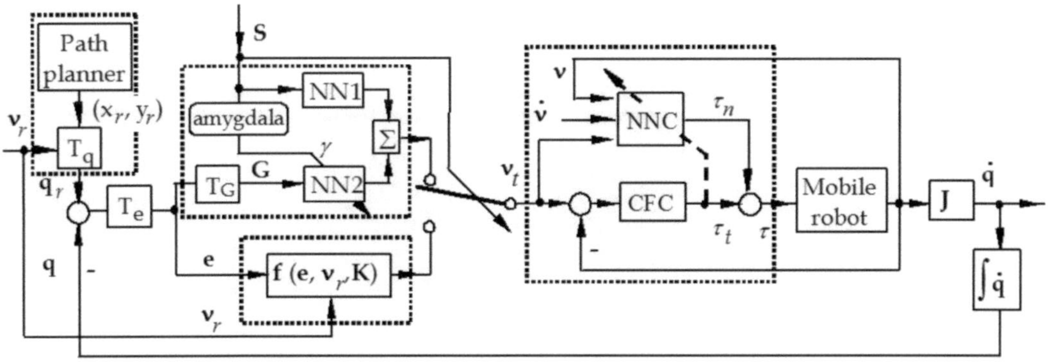


Fig. 9. Block diagram of the overall integrated system for navigation and control of mobile robot.

## 6. Simulation Results and Discussions

To illustrate the performance of the proposed system integrating the three control devices, they (and the overall system) were simulated in MATLAB environment and tested on several examples. In accordance with Fig. 2 the following parameters of the geometric model were chosen: $R = 0.036\ \text{m}$, $d = 0.02\ \text{m}$. The mass of the mobile robot was set to be $m = 0.5\ \text{kg}$. All the rest parameters used in simulations are presented in the corresponding sections below, where they are at first hand employed.

### 6.1 Path-planning in a previously known environment

The workspace consisted of two a priori known obstacles. The neural network of the path planner was trained with the error backpropagation algorithm with momentum and adaptive learning rate. The number of neurons in the hidden layer was $25$. The workspace was divided into $40 \times 40$ grid cells and the neural network used 1600 training samples. After training the neural network was used to produce collision penalties needed to solve the equations (12) and thus to find a path described by a set of $N = 25$ via points (without initial and target points). The virtual sampling time was $T_0 = 0.02\ \text{s}$. The initial position was $(0, 2)$, and the goal position was $(2, 0)$. The initial path, needed for solution of the equations (12) was chosen as a straight line from the initial position to the goal position (it is shown in Fig.10b as a dashed line). The weight coefficients in (12) were chosen to be $k_L = 1$ and $k_C = 1$. Fig. 10 shows the approximation performance of the neural network ((a) the surface; (b) the dotted area). In the same figure, case (b), the piece-wise liner path obtained by the described above algorithm is presented by the bold line with 'o'-marks.

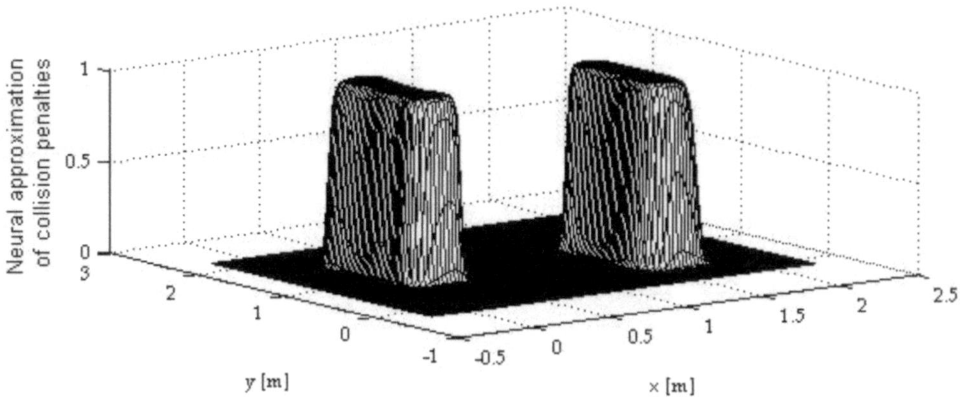


(a)

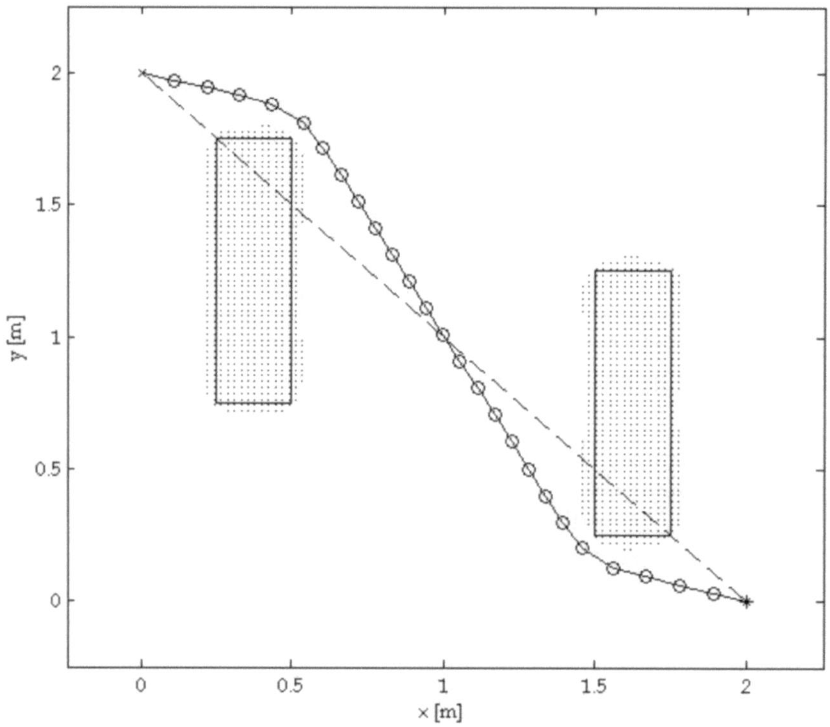


(b)

Fig. 10. Results from the path-planning procedure: (a) Surface of collision penalties generated by a neural network, trained to recognize a priori known obstacles; (b) Collision-free path obtained by the path-planning algorithm.

### 6.2 Performance of the sensor-based local obstacle avoidance controller

In accordance with the neural network structure depicted in Fig.5, a matrix of weights for obstacle avoidance behaviour is shown in Table 3 (Tsankova & Topalov, 1999). To give the robot a basic motivation to go forward when there are no obstacles, a bias was added to the tan-sigmoid threshold function of the motor speed neurons. The matrix of weights for goal following behaviour is shown in Table 4. The biases of the two output neurons were chosen with different signs one compared to another and much smaller than those in Table 3. These biases make the robot turn when the goal is not detected by the sensor in sectors $S_i, i = 1,2,...,5$, i.e., it is behind the robot.

| Sensor: | S1 | S2 | S3 | S4 | S5 | Bias |
|---|---|---|---|---|---|---|
| Motor Left: | 0.2 | 0.3 | -0.4 | -0.72 | -0.33 | 0.3 |
| Motor Right: | -0.38 | -0.77 | -0.4 | 0.3 | 0.2 | 0.3 |

Table 3. Matrix of weights for neural network NN 1

| Sensor: | S1 | S2 | S3 | S4 | S5 | Bias |
|---|---|---|---|---|---|---|
| Motor Left: | -0.2 | -0.3 | 0.35 | 0.72 | 0.33 | 0.1 |
| Motor Right: | 0.38 | 0.77 | 0.35 | -0.32 | -0.2 | -0.1 |

Table 4. Matrix of weights for neural network NN 2

The Braitenberg's vehicle No.3c, equipped with the neural networks described above, was simulated in MATLAB environment and tested on a few examples. The results are shown in Fig.11, where the symbols $*$, $\times$ and $\circ$ denote the goal position, the initial and final position of the agent, respectively. The robot orientation with respect to the inertial basis is signed as $\theta$. For the sake of simplicity (of simulations) the obstacles were chosen to be with circular shape. The obstacle avoidance behaviour derived by NN1 (Fig.5a) working alone is illustrated in Fig.11a, and the goal following behaviour obtained by NN2 (Fig.5b) operating unassisted - in Fig.11b. When the two NNs co-operated by simple summation of their outputs (Fig.6a) the resultant behaviour was a collision-free goal following behaviour (Fig.11c). It was evident that the performance of the Braitenberg's vehicle No.3c was not very good, as in some cases the robot did not reach the goal.

In accordance with the choice made in Section 4, the performance of this Braitenberg architecture was improved by emotional intervention on it. The parameters of the amygdala's model, selected as appropriate in Ref. (Tsankova, 2009), were used in this work, i.e.: $\xi_1 = 0.003$, $\xi_2 = 0.3$, $b = 0.12$, $b_\gamma = 5$ and $\mathbf{W} = (0.5 \;\; 0.75 \;\; 2.5 \;\; 0.75 \;\; 0.5)^{\mathrm{T}}$. The frustration signal was limited in accordance with its definition domain $f \in [f_{\min}; f_{\max}]$, where $f_{\min} = 0$ and $f_{\max} = 18$. The simulation results with the emotionally affected Braitenberg's vehicle No.3c are shown in Fig.11d. It can be seen that the artificial amygdala assists the innate controller to negotiate some dead-end situations successfully.

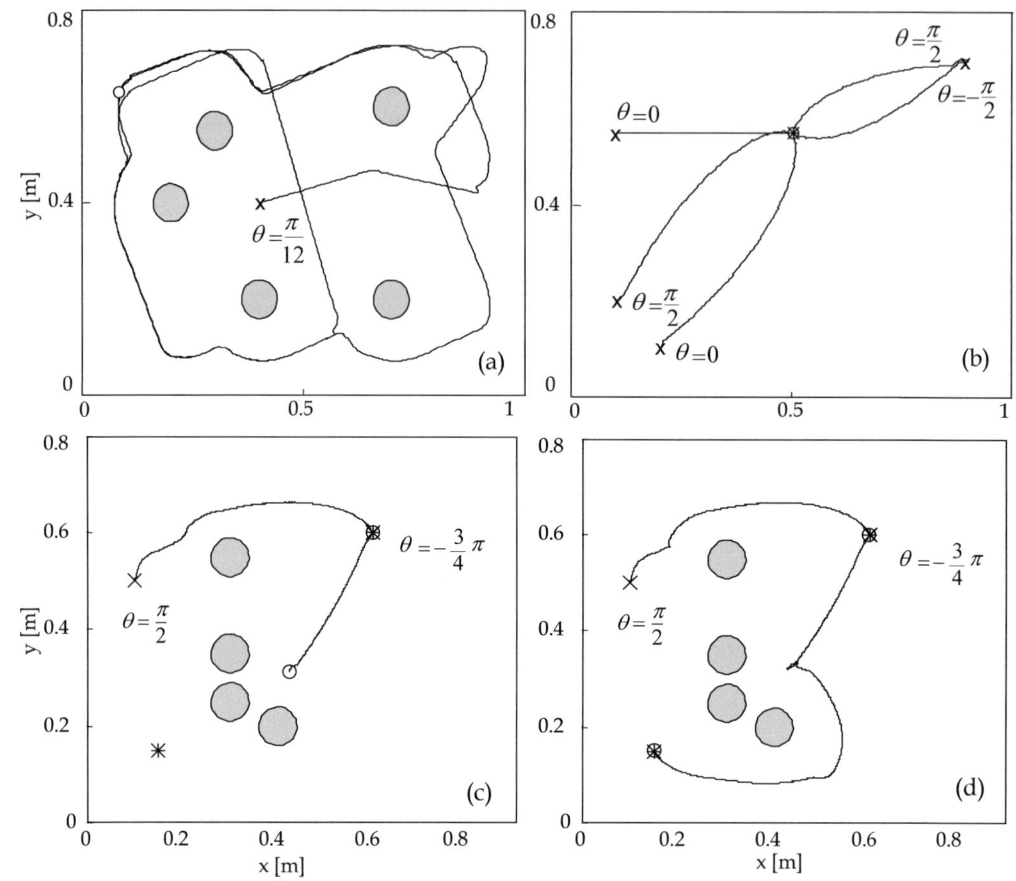


Fig. 11. Simulations with Braitenberg's vehicle No.3c emotionally affected and unaffected: (a) NN1 alone - obstacle avoidance behaviour; (b) NN2 alone – goal following behaviour; (c) NN1+NN2 together – collision-free goal following behaviour; and (d) emotionally affected vehicle.

### 6.3 The tracking performance and adaptability of the on-line feedback-error-learning controller

The trajectory tracking controller takes into account the operational issues of the previous two devices – the path planner and local obstacle avoidance controller. In order to handle non-smooth reference paths without causing the robot's slippage, the target velocities $(\upsilon_t, \omega_t)$ were limited by constants $(\hat{\upsilon}_t, \hat{\omega}_t)$. The values of these constants were adopted to be $\hat{\upsilon}_t = \upsilon_{\max} = 0.4\,\mathrm{m/s}$ and $\hat{\omega}_t = \omega_{\max} = 1.57\,\mathrm{rad/s}$. The posture controller gains were chosen to be as in Ref. (Kanayama et al., 1995): $k_1 = 10/\mathrm{s}$, $k_2 = 64/\mathrm{m}^2$ and $k_3 = 16/\mathrm{m}$. The robot's sampling time was set to $T_0 = 0.005\,\mathrm{s}$. The following parameters for the PD controller were used: $k_{\mathrm{P}} = 1$ and $k_{\mathrm{D}} = 0.5$. The reference linear velocity was set to $\upsilon_r = 0.25\,\mathrm{m/s}$, and reference angular velocity – to $\omega_r = 0\,\mathrm{rad/s}$. In accordance with Section 5.2 the following

RBF network parameters were used: $\delta_{max} = 0.1$, $\delta_{min} = 0.005$, $\lambda(0) = 0.9$, $\lambda_0 = 0.99$, $\varepsilon = 0.005$, $\eta = 5.10^{-5}$, $\alpha = 0.001$, $k_{RBF} = 7$, $T_d = 2$, $T_W = 10$, $M = 10$, $r_{threshold} = 0.1$.
Fig. 12a shows simulation results for $\Delta\theta$ discontinuous jumps ($\Delta\theta = \pi/4$, $\pi/2$, and $\Delta\theta = 3\pi/4$) carried out without the velocity limiter. The actual trajectories are marked with solid lines, while the reference trajectories - with dashed lines. Fig. 12b shows simulation results for the same directional discontinuities with the velocity limitation. The mobile robot tracked the reference trajectory, but its responses under limitation of velocities were slower than those without a limitation.

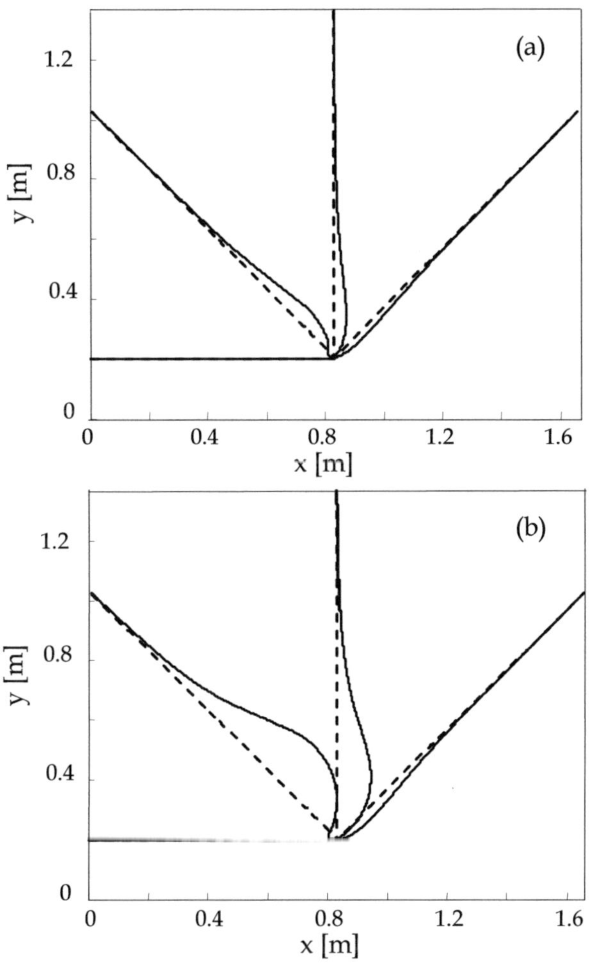


Fig. 12. Tracking under directional discontinuities: (a) without the limiter; (b) with the limiter

In order to demonstrate the adaptation capability of the control scheme to execute desired motions in the face of unforeseen parameter variations, an example when the robot picked up a load during the trajectory tracking task was carried out. During tracking a reference trajectory the mobile robot has to pick up an unknown load, which is considered as an increase in the robot's mass from 0.5 kg to 0.75 kg. The performance of the considered control scheme under these conditions is depicted in Fig.13, where the actual trajectory is marked with a solid line while the reference trajectory - with a dashed line. The position where the robot undergoes change of its mass is marked with a black point. The tracking performance of the mobile robot before and after the change of mass is almost the same. This is due to the adaptive capabilities of the tracking controller using RBF network with on-line feedback-error-learning.

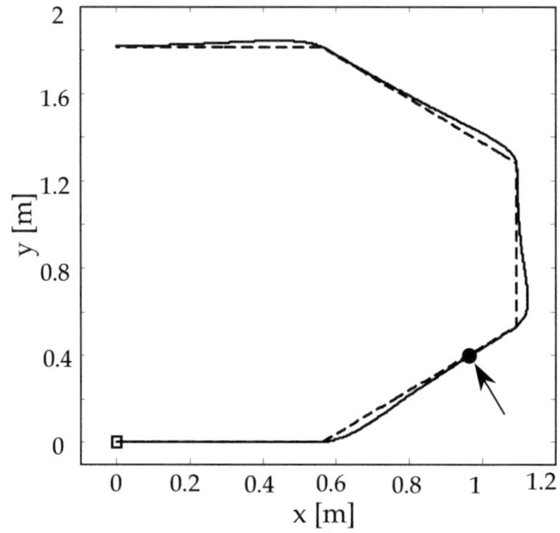


Fig. 13. Trajectory tracking with change of the robot's mass.

### 6.4 The robot, equipped with the overall integrated system: simulations and discussions

This Section treats the performance of the robot supplied with the overall integrated system tested by simulations in a partially known environment. Tracking the reference trajectory, the robot encountered accidentally appearing obstacles. Two types of obstacles were presented in the experiments: static (non-movable) and dynamic (movable) obstacles. The static obstacles were of both small and large size. Here, it is assumed that the small size is a size, commensurable with the robot size. In the simulations the small obstacles were with circular shape and a radius $R_{obst} = 0.04$ m . The large obstacles were compositions of a certain number of identical small obstacles. In Fig. 14a it is shown how the robot escapes a small obstacle, accidentally encountered during the movement along the reference trajectory. When the sensors no longer detect the obstacle, the robot returns to the trajectory tracking behaviour. However, going round the obstacle slows down the robot and the next reference point (which is a time-indexed one) turns out to be quite remote from it. In Fig.14 a few pairs of points with equal time indexes on the current and reference trajectory are marked with signs "+" and "★", respectively, to illustrate this delay. Depending on the maximum linear velocity and the location of a priori known obstacles with respect to the actual and the reference positions of the robot after avoiding the local obstacle, the trajectory tracking task may finish successfully or not. The probability for tracking without success increases when the robot encounters obstacles with bigger sizes

(Fig.14c). This drawback could be sold by an on-line modification of the time indexes of the reference trajectory points after the detour of a local obstacle by applying the following rule:

***Rule 1: If*** (there is not a detected obstacle at time k) **&** (there was an obstacle detected at time k-1) ***then*** find the reference point $(x_r^{\min}, y_r^{\min})$ nearest to the current robot position and set its time index to be $k$. Recalculate the time indexes of the rest of the points (to the end) of the reference trajectory and start trajectory tracking behaviour with $(x_r^{\min}, y_r^{\min})$ at time index $k$.

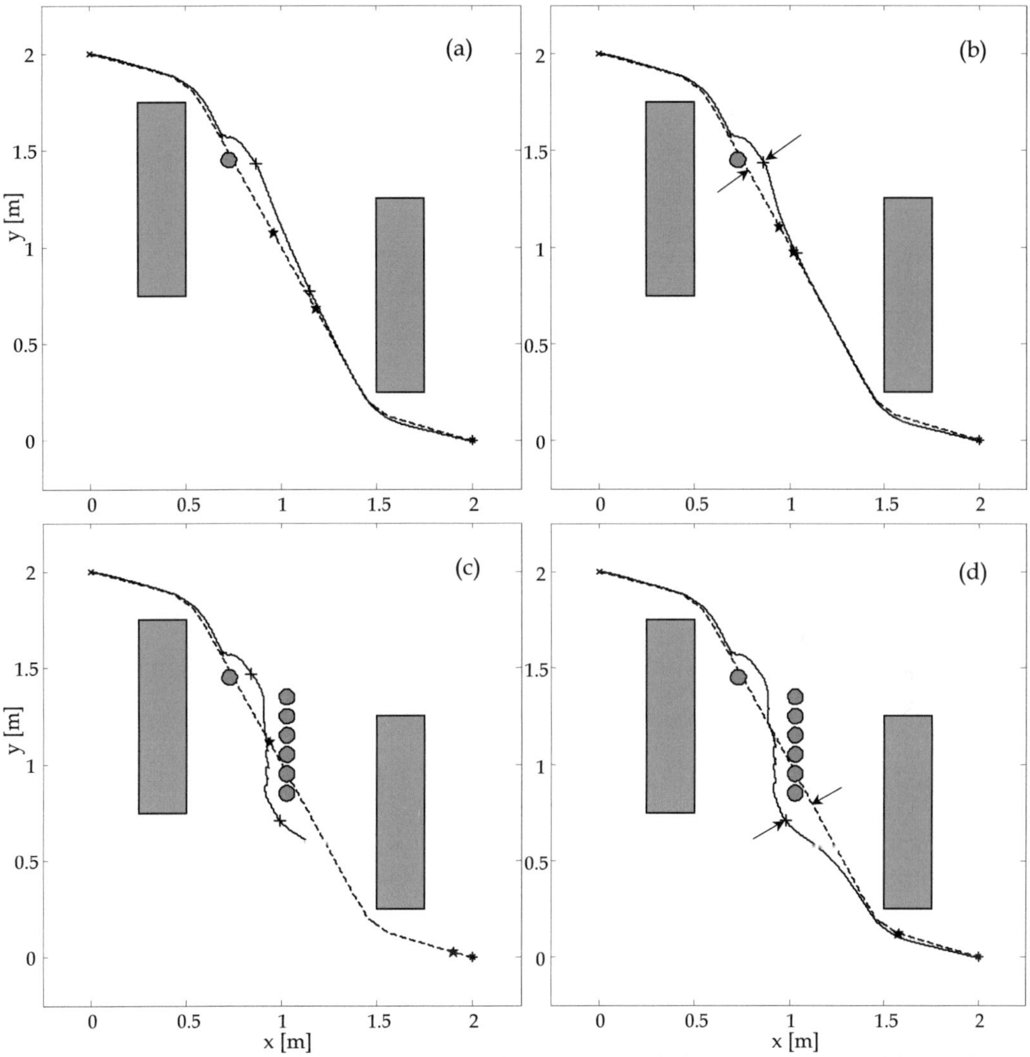


Fig. 14. Trajectory tracking and local obstacle avoidance behaviour: (a) and (c) - ordinary algorithm; (b) and (d) - modified by *Rule 1* algorithm

The contribution of this rule is demonstrated in Fig.14b and Fig.14d, where the two arrows show: (I) the current trajectory point where Rule 1 is switched on, and (II) the recalculated starting reference point $(x_r^{\min}, y_r^{\min})$ at time index $k$.

The proposed three-component system for a robot navigation and control copes with dynamic obstacles as well as with static ones. In Fig.15 tracking the reference trajectory, the robot successfully passes each other with two movable obstacles. They move rectilinearly with constant velocity $\upsilon_{\text{obst}} = 0.15\ \text{m/s}$.

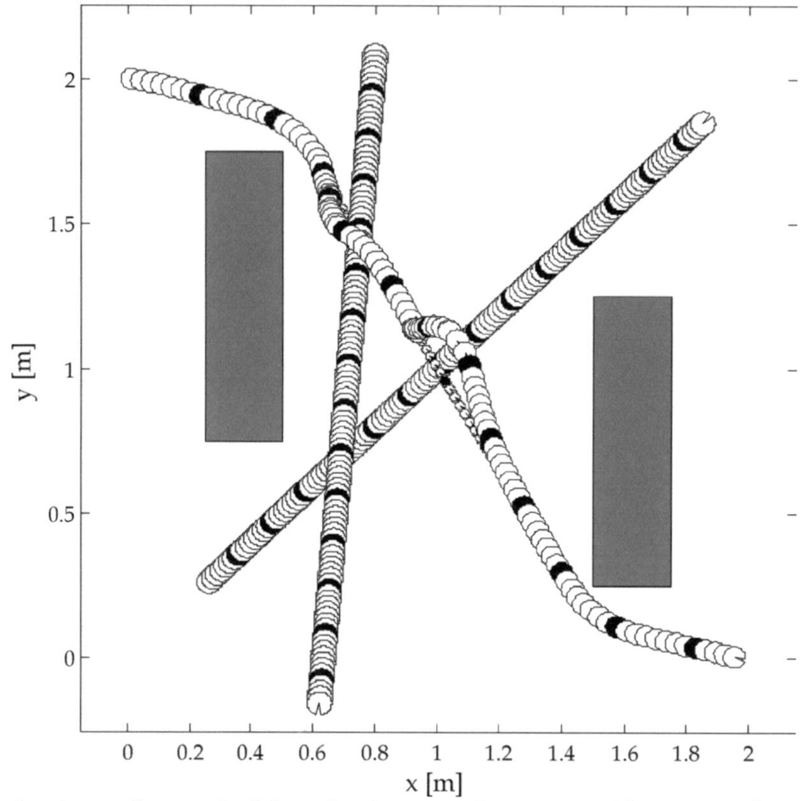


Fig. 15. Navigation and control of the robot in an environment with moving obstacles

The advantages and drawbacks of each of the devices in the integrated control system reflect on the overall performance of the system. Hence the improvement of each of them separately matters to the integrated system. The planner not only generates the collision-free path, but optimizes it in the sense of the criterion "the shortest path". However, for the sake of the gradient nature of the planning algorithm, it suffers from a significant drawback – the trapping in local minima of the potential function, which could frustrate the planning. The emotional intervention on the Braitenberg's architecture No.3c significantly improves it and makes it very proper as a sensor-based local obstacle avoidance controller. It has been reported in the literature (Tsankova, 2008; Tsankova, 2009) that the performance could be improved once more if an artificial immune network is used as innate action selection mechanism, but in compromise with the computation complexity. The trajectory tracking controller, designed using the approach proposed by Gomi and Kawato (1993), ensures the adaptability of the robot to the uncertainties or changes in its dynamics. Here the contribution to this controller is the use of RBF NNC and especially the learning algorithm. The advantage of using weights adaptation by a recursive least squares technique based on Givens QR decomposition is a faster and easier solution of the problem than the cases, when: (1) a genetic algorithm has been used for fuzzy-neural network learning (Topalov et al., 1997); (2) a NN with backpropagation algorithm has been applied for the same purpose (Topalov et al., 1998).

## 7. Conclusions

A complete integrated system for navigation and control of a mobile robot in a partially known environment is developed. The neural network in the path-planning algorithm can automatically build up the collision penalty function (as an obstacles' potential function approximation). The training procedure uses the information of the environment, no matter how many obstacles there are and how they look like. This makes the algorithm effective, but because of its gradient character, it meets the local minima problem. The local obstacle avoidance controller performs very well under accidentally encountered obstacles. Probably this is due to the property of the artificial amygdala to be the short term memory for the 'fear' from the encountered obstacles. The trajectory tracking controller is highly adaptable to uncertaities and changes in the robot dynamics. The recursive least squares technique using Givens QR decomposition is a comparatively fast and easy way for training the RBF net's weights. Future work will include the development of a modification of the path-planning algorithm, which aims at filling the local minima along the path to calculate and thus to ensure reaching the goal reliably. This modification, as well as the overall integrated control system, will be further verified on a real mobile robot.

## Acknowledgements

The paper presents research and development, supported by Ministry of Education and Science in Bulgaria (National Scientific Fund) under the Research Project BY-TH-108/2005.

## 8. References

Braitenberg, V. (1984). *Vehicles: Experiments in Synthetic Psychology*, MIT Press, Cambridge, MA

Choset, H.; Lynch, K.M.; Hutchinson, S.; Kantor, G.; Burgard, W.; Kavraki, L.E. & Thrun, S. (2005). *Principles of Robot Motion: Theory, Algorithms, and Implementations*, MIT Press, ISBN: 0262033275, Boston

Fierro, R. & Lewis, F.L. (1995). Control of a nonholonomic mobile robot: backstepping kinematics into dynamics. *Proceedings of the 34th Conference on Decision and Control*, pp. 3805-3810

Gomi, H. & Kawato, M. (1990). Learning control for a closed loop system using feedback-error-learning. *Proceedings of the 29th Conference on Decision and Control*, pp. 3289-3294

Gomi, H. & Kawato, M. (1993). Neural network control of a closed-loop system using feedback-error-learning. *Neural Networks*, Vol. 6, pp. 933-946

Kanayama, Y.; Kimura, Y.; Miyazaki, F. & Noguchi, T. (1990). A stable tracking control method for an autonomous mobile robot. *Proceedings of IEEE International Conference on Robotics and Automation*, Vol.1, pp. 384-389

Latombe, J.C. (1991). *Robot Motion Planning*, Kluwer Academic Publishers, ISBN: 0792391292, Boston

LeDoux, J. (1996). *The Emotional Brain*, Simon & Schuster, New York

Meng, H. & P.D. Picton (1992). A neural network for collision-free path-planning. In: *Artificial Neural Networks*, Aleksander, I. & Taylor, J. (Ed.), Vol.2, pp. 591-594

Mochida, T.; Ishiguro, A.; Aoki, T. & Uchikawa, Y. (1995). Behaviour arbitration for autonomous mobile robots using emotion mechanisms. *Proceedings of IROS'95*, Vol.1, pp. 516-521

Platt, C.J. (1991). A resource-allocating network for function interpolation. *Neural Computation*, Vol.3, No.2, pp. 213-225

Rosipal, R.; Koska, M. & Farkas, I. (1997). Chaos time-series prediction using resource-allocating RBF networks. *Proceedings of Conference Measurement '97*, Smolenice, Slovakia

Rosipal R.; Koska, M. & Farkas, I. (1998). Prediction of chaotic time-series with a resource-allocating RBF network. *Neural Processing Letters*, Vol.7, No.3, pp. 185-197

Salahshoor, K. & Jafari, M.R. (2007). On-line identification of non-linear systems using adaptive RBF-based neural networks, *International Journal of Information Science and Technology*, Vol.5, No.2, pp. 100-121

Topalov, A.; Tsankova, D.; Petrov, M. & Proychev, T. (1997). Intelligent motion planning and control for a simulated mobile robot, *Proc. of the 2nd IFAC Workshop on New Trends in Design of Control Systems*, pp. 338-343, Smolenice, Slovac Republic, 1997

Topalov, A.V.; Kim, J.-H. & Proychev, T.P. (1998a). Fuzzy-net control of non-holonomic mobile robot using evolutionary feedback-error-learning, *Robotics and Autonomous Systems*, Vol.23, pp. 187-200, Elsevier

Topalov, A.; Tsankova, D.; Petrov, M. & Proychev, T. (1998b). Intelligent sensor-based navigation and control of mobile robot in a partially known environment, *Proc. of the 3rd IFAC Symposium on Intelligent Autonomous Vehicles*, pp. 439-444, Madrid, 1998

Tsankova, D. (2008). Emotional intervention on stigmergy based foraging behaviour of immune network driven mobile robots, Chapter 27, pp. 517-542, In: *Frontiers in Evolutionary Robotics*, Iba, H. (Ed.), I-tech Education and Publishing, Vienna, Austria, www.i-technoline.com, ISBN 978-3-902613-19-6

Tsankova, D.D. (2009). Emotional intervention on an action selection mechanism based on artificial immune networks for navigation of autonomous agents, In: *Adaptive Behavior*, Vol.17, No.2, pp. 135-152, June 2009, Sage Press

Tsankova, D.D. & Topalov, A.V. (1999). Behaviour arbitration for autonomous mobile robots using immune networks. *Proceedings of the 1st IFAC Workshop on Multi-Agent Systems in Production (MAS'99)*, pp. 25-30, Vienna, Austria, 2-4 December, 1999

Yahja, A.; Singh, S. & Stentz, A. (2000). An efficient on-line path planner for outdoor mobile robots. *Robotics and Autonomous Systems*, Vol.32, pp. 129-143, Elsevier Science.

Zulli, R.; Fierro, R.; Conte, G. & Lewis, F.L. (1995). Motion planning and control for non-holonomic mobile robots. *Proc. IEEE Int. Symp. on Intelligent Control*, pp. 551-557

11

# Navigation Planning with Human-Like Approach

Yasar Ayaz*, Atsushi Konno*, Khalid Munawar**,
Teppei Tsujita* and Masaru Uchiyama*
**Tohoku University, **National University of Sciences and Technology (NUST)*
**Japan, **Pakistan*

## 1. Introduction

Moving about in everyday human environments such as homes and offices would require a robot to cope with terrains not ordinarily traversable by robots on wheels. These may include stairs and stepping stones etc. This is because our homes and offices are custom designed for biped walkers i.e. humans. The usage of legs rather than wheels for mobility was thus an inevitable evolution in the history of mobile robotics. Giving legs to a robot instead of wheels gives it a lot more than just resemblance to a human being. Unlike ordinary mobile robots, humanoids have the ability to cross obstacles by stepping over or upon them. This ability is left unexploited if the algorithms used for ordinary mobile robot navigation among obstacles are employed for humanoids too.

Various approaches have been adopted in order to address this problem in the past. (McGhee & Iswandhi, 1979) developed a method that divides the terrain into 'permissible' and 'impermissible' stepping positions. Keeping in view the direction in which the ultimate goal position is located, the robot selects the next foothold from amongst the permissible ones in the immediately reachable footholds while taking care of postural stability constraints. While this method targeted a general application to legged robotics, (Yagi & Lumelsky, 1999) presented another one that deals specifically with humanoid robots. Depending upon the distance from the obstacle nearest to the robot in the direction of motion (presumably the direction in which the goal position is located) the robot adjusts the length of its steps until it reaches the obstacle. Now, if the obstacle is small in size, it is overcome by stepping over or upon it whereas if it is too tall to be overcome in this way, the robot starts sidestepping until it moves clear of the obstacle. Obviously, whether to sidestep left or right is also a pre-programmed decision. These and other such localized reactive approaches have the tendency to lead the robot into a local loop or a deadlock in which case the robot would have to be backtracked in order to follow an alternate path.

(Kuffner et al., 2001) argued that since such reactive algorithms failed to consider complexities occurring in the path at a later stage before opting to take it, they ended up stuck in local loops and deadlocks. In order to tackle this problem they presented a footstep planning algorithm based upon game theory that takes into account global positioning of obstacles in the environment. This technique has been tested on H6 (Kuffner et al., 2001), H7

(Kuffner et al., 2003), Asimo Honda (Chestnutt et al., 2005; Michel et al., 2005) and HRP-2 (Michel et al., 2006) humanoid robots with some improvements. The algorithm uses a discrete set of predefined stepping locations in the robot's immediate vicinity for either leg. Also predefined are intermediate postures that the robot assumes while moving its feet between any two of these stepping locations. Selecting a set of these possible stepping locations, each of which leads to a similar set of descendants, a tree is spread out from the initial footstep position. Now, after pruning from the tree those branches that do not end up with a step in the destination area, a two-levelled polygon-polygon collision search is conducted in order to identify trajectories that result in collision with the obstacles in the environment. Once these are discarded too, a greedy search is conducted in order to hunt out the best among the paths generated. This strategy cannot distinguish between the paths without obstacles and those with some obstacles that can be stepped over due to which the robot always needs to consider high stepping profiles in every step. In addition, because of generation of a large search tree of possible stepping locations, greedy search has to be applied instead of an exhaustive one. On the other hand, a human in such a situation would start by searching for an obstacle-free path directly connecting initial and goal locations. If found, it would simply be adopted. Whereas some options would be considered once an obstacle-free direct path is not found. Again, an obstacle can be dodged from the right, or dodged from the left, or is crossed by stepping over.

Our algorithm starts by checking if it is possible for the robot to go directly from the initial to the final location. If so, such a path is adopted. If not, trajectories are formed in order to dodge the hindering obstacle from the right and the left. In addition, keeping in view the dimensional constraints, obstacles are classified into three types. The smallest ones that can be stepped over: type-1. For these an additional trajectory considering stepping over is also formed. The larger ones that cannot be stepped over but the robot can pass right next to them: type-2. And the largest ones that might collide with the robot's arms (sticking out wider than its feet) if the robot passes right next to them: type-3. The robot keeps a larger distance from obstacles of type-3 as compared to those of type-2 in order to avoid collision. In this way, the algorithm starts by considering the shortest path and then expands its alternatives on either side spreading out considering the obstacle type and proximity. In addition, by identifying obstacles once encountered, it avoids getting stuck into the local loops and deadlocks. It is noticeable here that branches formed from every node in the search tree can be a maximum of 3, and for several straight line segments each node leads to only one child. This reduces the branching factor making a smallest possible search tree which in turn enables us to apply the exhaustive search for seeking the best possible path. It can immediately be noticed that this search, though exhaustive, is computationally cheap.

This article explains our proposed method in detail. In section 2 we examine some related research. Section 3 introduces the concept of human-like footstep planning and highlights salient differences with the existing approaches. The footstep planner is explained in detail in section 4 along with simulation results using a model of HRP-2 humanoid robot. Section 5 states the conclusion and some ideas for future work.

## 2. Game Theory Based Footstep Planning

Kuffner's algorithm for footstep planning among obstacles for biped robots (Kuffner et al., 2001; Kuffner et al., 2005) is a global motion planning strategy based on using game theory for finding alternate paths in an obstacle cluttered environment by considering a discrete set of heuristically chosen statically stable footstep locations for a humanoid robot and identifying the better path via cost heuristic implementation using greedy search. The assumptions made by the algorithm are:

(i). The environment floor is flat and cluttered with non-moving obstacles of known position and height,

(ii). A discrete set of feasible, statically-stable footstep placement positions and associated stepping trajectories are pre-computed,

(iii). Only the floor surface is currently allowed for foot placement (not obstacle surfaces).

A static stability analysis is carried out for the stepping motion of the humanoid robot as a result of which a continuous region is identified where the robot can place its lifted foot (while balanced on one leg) without losing stability. Since it is computationally extensive to check every single point in this region for a footstep placement possibility in every step, a discrete set of 'feasible' footstep locations is heuristically identified in this region. Standing balanced on one foot, the robot would alternatively consider placing its foot at each of these locations and from each stepping possibility more nodes are generated in a similar manner. Foot transition from each location to the next is carried out via heuristically identified statically stable intermediate postures $Q_{right}$ and $Q_{left}$ (for standing balanced on right and left foot respectively) in order to limit the number of possible transition trajectories considered in every step. This is followed by best path identification through greedy search by employing a heuristic cost function which attempts to approximate an exhaustive search while applying a greedy one since exhaustive search itself would be very time consumptive given the large search tree of possible stepping locations the algorithm generates (Kuffner et al., 2001; Kuffner et al., 2005).

The algorithm was simulated on a model of H6 humanoid robot (Kuffner et al., 2001; Kuffner et al., 2005). Considering 15 discrete footstep placement options in each step in a room with 20 obstacles, it formed search trees of 6,700 and 830,000 nodes in (Kuffner et al., 2001) and (Kuffner et al., 2005) respectively. Best path was traced out using greedy (Kuffner et al., 2001) and A* (Kuffner et al., 2005) search strategies with total processing time of 12 s on 800 MHz Pentium-II and 4 s on 1.6 GHz Pentium-IV running Linux respectively.

The main drawback of this algorithm is high computational complexity because of generation of a large search tree of possible stepping locations which reflects in terms of time consumption. Moreover, a closer look reveals that since this algorithm is not reactive in nature, it does not distinguish between steps taken to cross over obstacles and those taken ordinarily. Since the same intermediate postures as in ordinary steps ($Q_{right}$ and $Q_{left}$) have to provide the robot with trajectories capable of stepping over obstacles too, they involve lifting of the foot to a high position as if stepping over obstacles in every step. Yet, since the

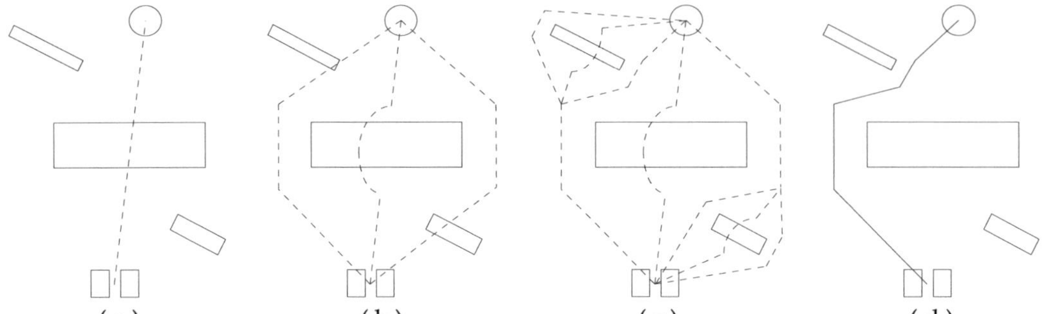

Fig. 1. An illustration of the human-like footstep planning strategy.

tree of step options in game theory based footstep planning strategy is spawned without paying heed to presence or absence of obstacles in the environment, stepping over is merely a result of obstacles fortunately found lying between present and next footstep locations without causing a collision during the stepping trajectory. In addition, when it comes to searching for the best path amongst those traced out by the algorithm, an exhaustive search cannot be applied because of the large tree of possible stepping locations generated (Kuffner et al., 2001; Kuffner et al., 2005). A greedy search, however, might not hunt out the best solution under many circumstances (Cormen et al., 1994). In order to tackle some of these problems a tiered planning strategy has been introduced that splits the planner into high, middle and low level layers in order to traverse different terrain types (Chestnutt & Kuffner, 2004).

## 3. Human-Like Footstep Planning

### 3.1 Graph Expansion Strategy

As humans we know that, unlike localized reactive navigation strategies, paths adopted by us to traverse obstacle cluttered terrains are planned while taking into account global positioning of obstacles around us. Also, humans do not conduct navigation planning by random spreading of footstep placement possibilities throughout the terrain while ignoring the relative position of obstacles in the environment and later evaluating which of these would result in collision as is done by the game theory based approaches. Quite differently, humans start by first searching for the path directly connecting the start and destination points. If found available, this path is simply taken without considering alternatives. On the other hand, if the direct path is found hindered by an obstacle, we consider paths dodging it as well as stepping over it with the better one of these being adopted eventually.

This is precisely the approach we seek to adopt. In our method the planner starts by looking for the presence of obstacles in the path directly connecting the initial and goal locations. If available, this path is taken without heed to alternate possibilities. If unavailable, paths are generated from both right and left of the nearest obstacle hindering the direct passage. In addition, if the obstacle is small enough, a path is also planned by stepping over it. If another obstacle is found hindering a generated alternate path, paths from right, left and over this obstacle are also analyzed in a similar fashion. In this way the graph evolves from simple to complex paths.

Fig. 1 elaborates upon graph expansion in a simple scenario. Intended destination is marked by a circle. Checking the direct path from initial to final position (Fig. 1 (a)), an obstacle is found hindering the way. Pivot points are chosen near the corners of this obstacle in order to find paths to dodge it and a path is also planned by stepping over (Fig. 1 (b)). Even sections of these alternate paths formed are found hindered by other obstacles, following which, paths are designed by circumvention and stepping over (Fig. 1 (c)). Based on heuristic cost assignment (depending upon step complexity in case of humanoids), exhaustive search for best path gives us the desired solution (Fig. 1 (d)).

Keeping in view the dimensional constraints, obstacles are classified into three types. The smallest ones that can be stepped over: type 1. For these a trajectory considering stepping over is also formed. The larger ones that cannot be stepped over but the robot can pass right next to them: type 2. And the largest ones that might collide with the robot's arms (sticking out wider than its feet) if it passes right next to them: type 3. The robot keeps a larger distance from obstacles of type 3 as compared to those of type 2 in order to avoid collision. In this way, the algorithm starts by considering the shortest path and then expands its alternatives on either side, spreading out considering the obstacle type and proximity. In addition, by identifying obstacles once encountered, it avoids getting stuck into the local loops and deadlocks. It is noticeable here that branches formed from every node in the search tree can be a maximum of three, and for several straight line segments each node leads to only one child. This reduces the branching factor making a smallest possible search tree which in turn enables us to apply exhaustive search for seeking the best possible path. It can immediately be noticed that this search, though exhaustive, is computationally cheap.

### 3.2 Comparison with Visibility Graph

Due to the fact that our algorithm uses straight line footstep sequences to approach pivot points (located near obstacle edges) on which turning is carried out, some readers may confuse our method with the well-known visibility graph approach used for mobile robot motion planning (Latombe, 1991). Therefore, it is important to highlight salient differences between the two approaches.

Of course one clear difference is that we also plan paths for crossing obstacles by stepping over, which the visibility graph algorithm does not. However, more fundamental differences in the graph, as well as in the way that it is expanded, reveal themselves upon a keener examination. For the analysis given in this subsection, we limit our graph expansion methodology to only dodging obstacles by circumvention in order to be able to readily make comparison with the visibility graph approach.

#### 3.2.1 Graph Planning

A visibility graph is a graph of *intervisible* locations. Each obstacle corner in the environment is included in the graph and represents a vertex (or pivot point) about which turning is possible. A path is a visible (i.e. collision free) connection between any two obstacle corners.

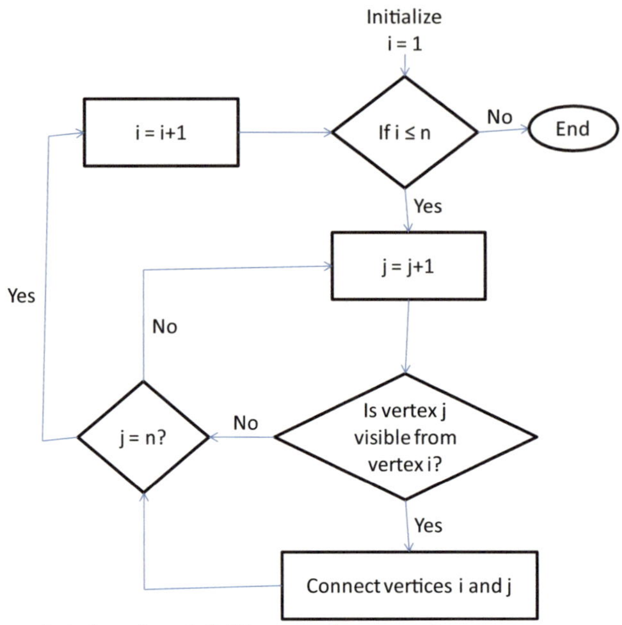


Fig. 2. A flowchart explaining the visibility graph expansion process.

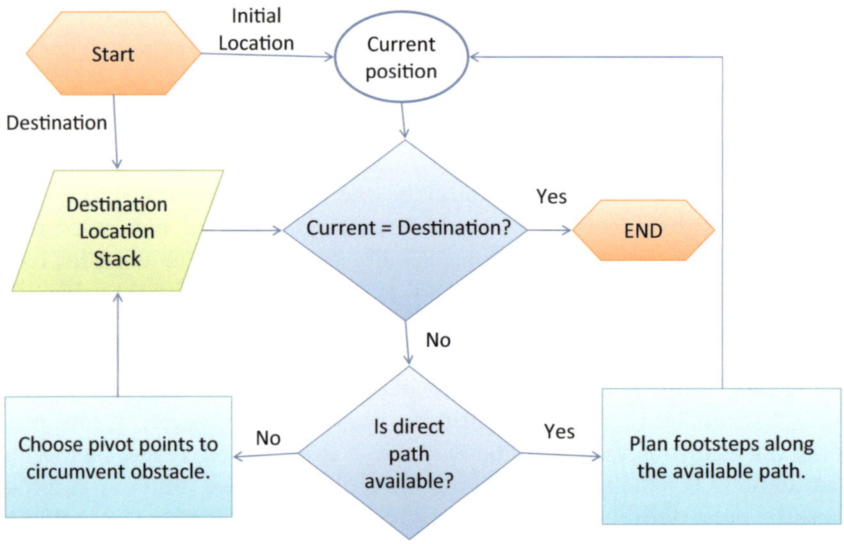


Fig. 3. A flowchart describing the graph expansion process using our planning strategy. Here, the graph expansion is limited to path planning by circumvention only.

If two obstacle corners can see each other (i.e. if a collision free direct path is possible in order to travel between them), they are connected by a possible path.

Fig. 2 illustrates the visibility graph expansion process with the help of a flowchart. The way to expand a visibility graph is to simply include all the obstacle corners into the graph as vertices (or pivot points) and interconnecting them by all possible collision free direct paths.

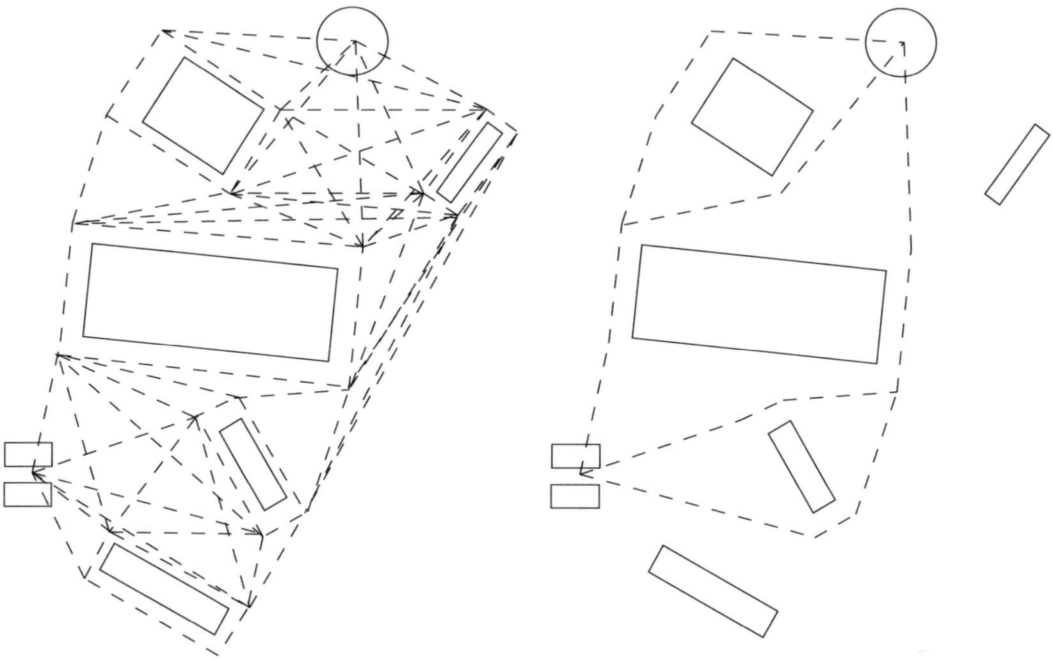

(a) The visibility graph expanded. (b) Graph expanded using our approach.

Fig. 4. A comparison between the visibility graph and the graph expanded using our method.

On the contrary, our approach is evolutionary and the graph is expanded from simpler to more complex paths. First, only the direct path interconnecting the start and destination points is checked for the presence of obstacles. If it is available, it is simply taken and no other paths are generated. If it is not available, paths are generated from the right and left to dodge the nearest obstacle hindering the direct path. If any of these alternate paths are also found hindered by an obstacle, alternate paths are generated to dodge this obstacle from right and left too in a similar manner. A flow diagram of graph expansion using our approach is given at Fig. 3. Please note that in the flow diagram at Fig. 3 as well as in the above description we are not referring to the paths formed by stepping over obstacles in order to restrict graph expansion to planning by circumvention only for the sake of comparison with visibility graph.

The difference between the two algorithms is very clear in terms of the manner of graph expansion. A visibility graph is directionless and simply includes all the obstacle edges into the graph. Our algorithm, on the other hand, is of an evolutionary nature. It expands the graph keeping in view the direction of motion and increases the number of vertices or paths only if a direct path to an intermediate or ultimate destination point is found hindered by an obstacle.

The graph expansion approach in the two algorithms, therefore, is significantly different.

### 3.2.2 Graph Output

It is insightful to look at the outputs produced by both the algorithms in order to observe their comparative uniqueness.

Fig. 4 shows a robot in an obstacle cluttered environment with the destination marked by a circle. Fig. 4 (a) shows a visibility graph expanded in this scenario whereas Fig. 4 (b) shows the graph expanded using our approach. As it can clearly be seen, our algorithm produces a much simpler graph comprising only the meaningful paths. This is because of the evolutionary approach inspired by human navigation that we adopt for path planning.

## 4. Planner Design and Simulation

### 4.1 Kinematics and Stability Analysis

For initial simulation work, our algorithm inherits assumptions (i) and (iii) from the game theory based method described in section 2. Also, as in assumption (ii) we also employ a static stability based criteria for stepping motions in the simulation phase.

Selection of footstep locations for ordinary walking is done according to the intended direction of motion. Right or left foot respectively is used to take the first step depending upon whether the destination is located to the right or left of the robot's current position. For ordinary steps a half-sitting posture is used. Maximum possible step-length (1.9 cm approximately for HRP-2) is used in each step. Final step when reaching the destination is shortened to step exactly within the destination area.

Backward steps are not considered while planning since back-stepping is one of the very situations a footstep planner is primarily meant to avoid (Ayaz et al., 2006; Ayaz et al., 2007). Thus, kinematically reachable area in the forward direction only (highlighted in Fig. 5 for the right foot) is used for stepping.

Fig. 5. Stepping area for normal half sitting posture.

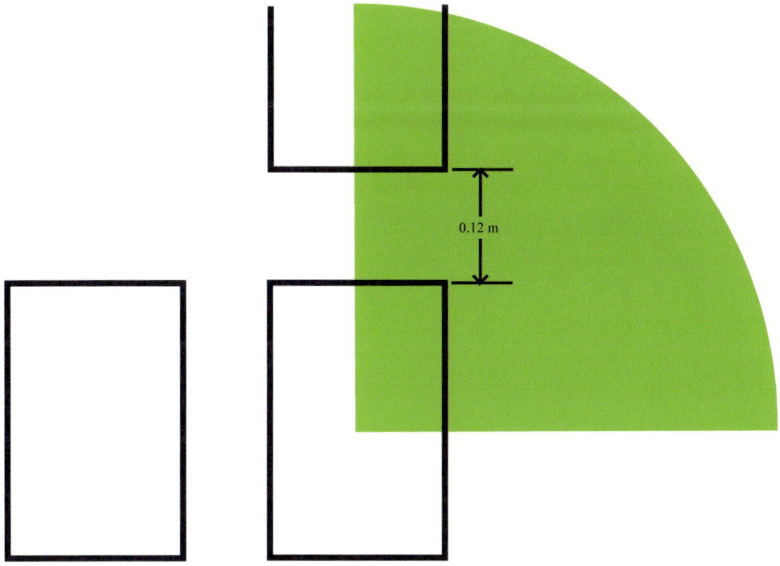


Fig. 6. Creating space between current and next stepping locations by lowering the waist.

As it can be seen from Fig. 5, even stretching the foot to its maximum is not sufficient enough to create space between current and next step positions. In other words, this posture for normal stepping cannot be used to step over obstacles as well. Thus for stepping over obstacles, the knees of the robot are bent further and the body is lowered in order in order to increase the possible step-length. In this way a gallery of 11.9 cm is created between current and next stepping locations in which the obstacle could be located while it is stepped over (illustrated in Fig. 6). The trajectory of the foot used is similar to that employed for our earlier work on Saika-3 humanoid robot (Ayaz et al, 2007). The height of HRP-2's ankle joint is approximately 11 cm. Obstacles of height ≤ 10 cm and depth ≤ 11 cm are thus considered possible to be stepped over for the sake of this simulation work. It is to be borne in mind that here the dimensional constraints are determined neither while accounting for dynamic stability of the robot nor are limiting in terms of the maximum obstacle height possible to be overcome using HRP-2.

### 4.2 Obstacle Classification

In human environments, obstacles are of a variety of shapes and sizes. Our algorithm requires identification of a box-like boundary around each obstacle such that the obstacle of arbitrary shape is entirely contained inside it. This box-like boundary must be a four cornered shape when viewed from the top but there is no restriction on it being a rectangle or a parallelogram as long as it is a four cornered shape in top view. However, stepping over is only attempted in case the width and height of the boundary that binds the obstacle satisfy the constraints described in Section 4.1. Our algorithm at this stage considers the entire area inside this boundary, in which the obstacle of arbitrary shape is contained, as an obstacle. These obstacles are classified into three types:

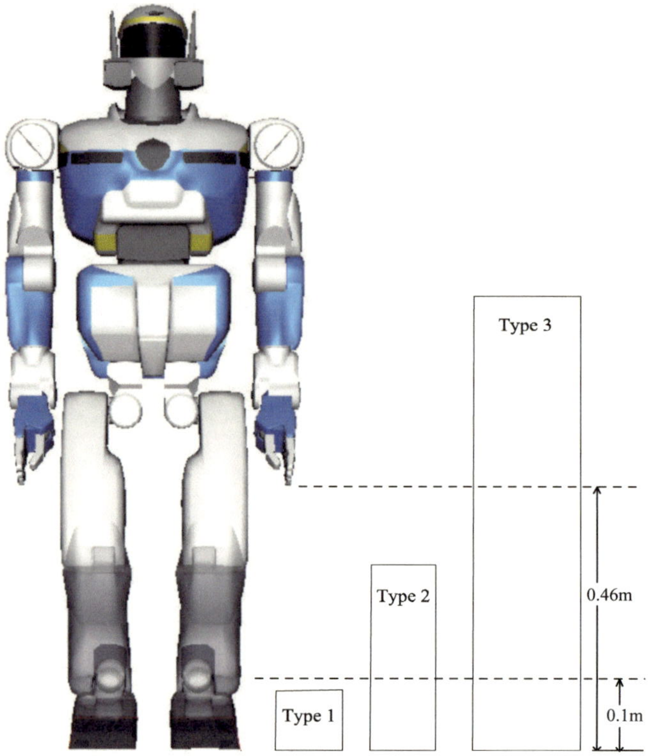


Fig. 7. Obstacles classified into types on the basis of their heights.

| | |
|---|---|
| Type-1:<br>Height ≤ 0.1 m | Can be crossed by stepping over if depth ≤ 0.11 m. |
| Type-2:<br>0.1 m < Height < 0.46 m | Cannot be crossed by stepping over and cannot collide with the robot's arms. |
| Type-3:<br>Height ≥ 0.46 m | Cannot be crossed by stepping over and while dodging we have to make sure the robot's arms do not collide with it. |

### 4.3 Collision Detection

Extending lines from the current body-corners of the robot to the to-be body-corner locations at the goal position, a gallery is formed. Using two of these boundary-lines and more drawn between corresponding current and destined points, we form a mesh. If any of these line segments is found intersecting a side of an obstacle (marked by a line between two of its neighbouring corners), we say that a collision has been detected. Then, based upon the distance from the intersection point and the angles made from current position, near and far sides and left and right corners of the obstacle are identified respectively.

The collision detection strategy can be understood more easily using Fig. 8. Using the angle of the intended line of motion, an imaginary set of initial footstep locations is generated at

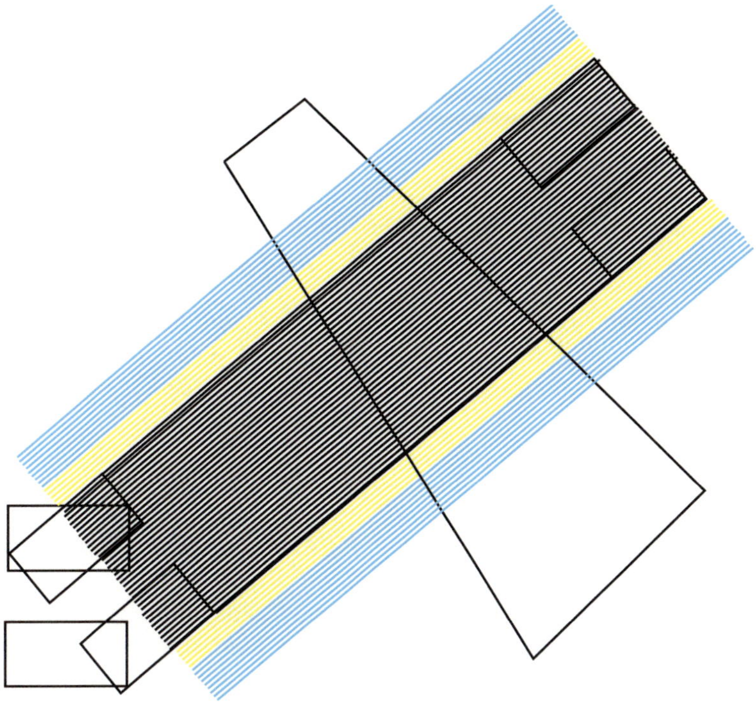

Fig. 8. Mesh of lines drawn for collision detection / prediction.

the current position which indicates the initial position of the robot after it will orientate itself towards the destination. In this direction of motion, another set of imaginary footstep locations is generated at the destination point, which marks the locations at which the robot will place its feet upon reaching the destination. Joining the outer boundaries of these initial and final imaginary footstep locations, a gallery is formed. Inside this gallery, a mesh of equidistant line segments joining corresponding points at the initial and final locations is generated. Each of these lines is checked for intersection with each of the walls of all obstacles in the environment (which are also marked by line segments when looked at from the top). If an intersection is found, a collision is said to have been detected and the robot seeks to form trajectories in order to overcome this obstacle in the path. This type of collision detection is performed not only between the starting point and the ultimate destination, but between every two points which mark the initial and final locations of every intermediate path the robot considers traversing.

From Fig. 8 it can also be seen that a part of the mesh of lines extends beyond the outer boundary of the footholds (drawn in blue and green). These are drawn by joining the outer ends of the arms of the robot (which stick out wider than its feet) at initial and goal locations. Only if an obstacle of type-3 is found to be colliding with a line in the blue part of the mesh, a collision is said to have been detected. The green part, however, checks for both type-2 and 3 obstacles but is insensitive to obstacles of type-1.

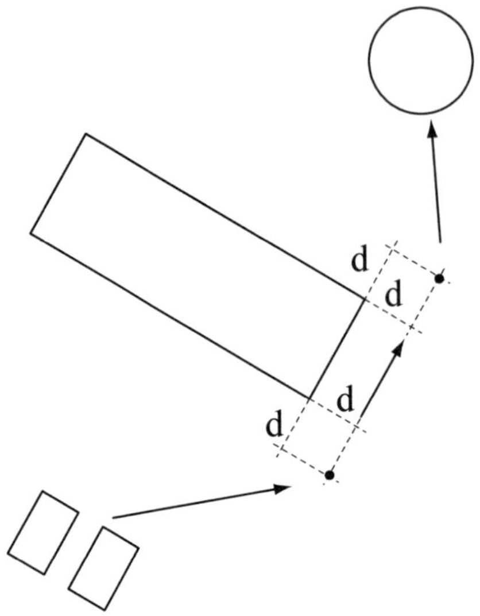


Fig. 9. Chosing pivots to dodge an obstacle.

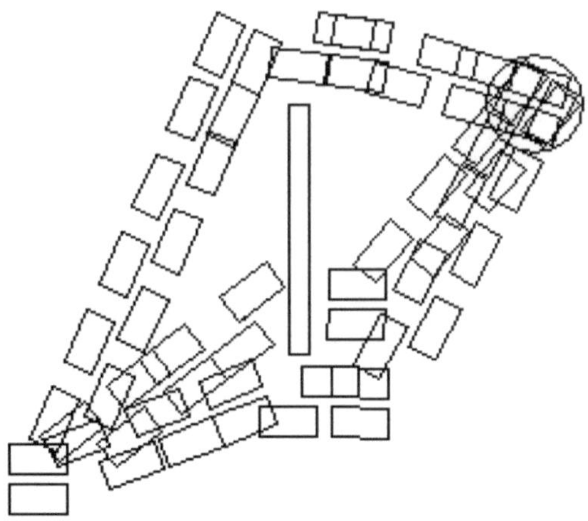

Fig. 10. Overcoming an obstacle of type-1.

For simulation results presented in this chapter, the collision checking mesh used comprises 31 lines with a distance of less than 2 cm between every two neighbouring line segments.

### 4.4 Overcoming an Obstacle

To dodge an obstacle from a side we choose pivot points near the obstacle corners as shown in Fig. 9.

The distance 'd' along the extended side is chosen such that no part of the robot's body collides with the obstacle as the robot stands in double support state with its body centre at the pivot point. For instance, in case of type-1 obstacles, this distance is equal to half the length of the diagonal of the rectangular support polygon formed when the robot stands with both feet next to each other. This is because the outer most edges of the feet are the points closest to the obstacle with which there might be a chance of collision. 'd' can thus be different for each obstacle type but not for obstacles of one group.

As explained in section 4.3, to step over an obstacle, the robot balances itself on both legs one step short of the closest step-location near the obstacle. Next, based on rightward or leftward tilt of the obstacle in relevance with the robot's trajectory, it places forward left or right foot respectively and balances itself on it. Using enhanced stepping motion, the robot now takes a step forward with its movable leg, making sure that the extended foot lands with its back-side parallel to the obstacle's away-side. Fig. 10 displays possible trajectories generated to overcome an obstacle of type-1 by circumvention and stepping over.

### 4.5 Local Loops and Deadlocks

Each obstacle in the surroundings is allotted an identification number. The planner keeps a history of the obstacles it has overcome in a path as it plans footsteps. Whenever an obstacle

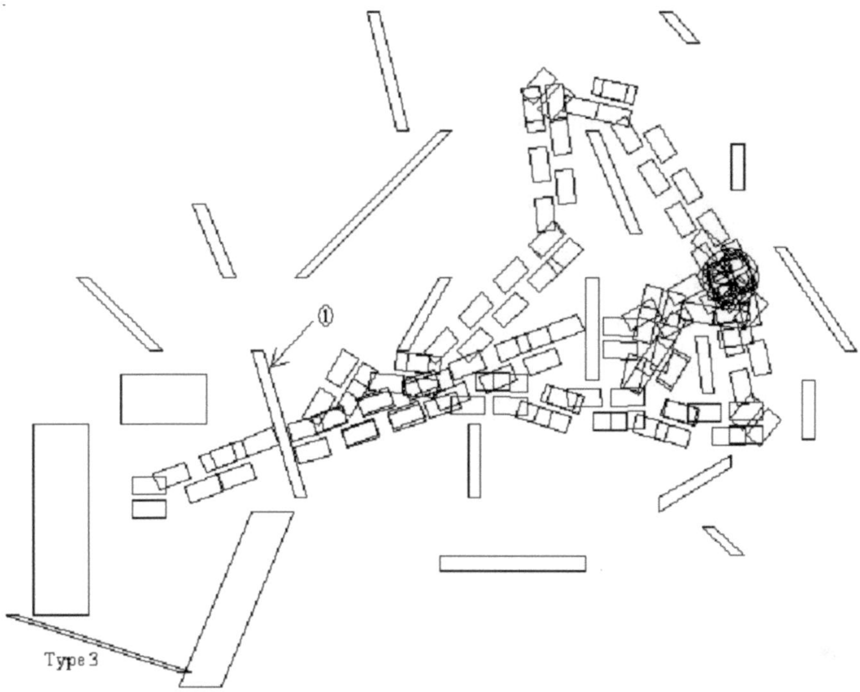


Fig. 11. Avoiding local loops and deadlocks while crossing obstacle '1' of type-1.

occurs twice, it indicates a local loop or deadlock since attempting to overcome it again is bound to bring the planner back to the same position again and again. Such a trajectory is immediately abandoned and pruned from the search tree.

One such situation where the robot encounters a local loop and deadlocks while trying to dodge the obstacle labelled '1' from both right and left sides is shown in Fig. 11. For instance, when trying to dodge the obstacle labelled '1' in Fig. 11 from the right, the robot chooses pivot points to pass from its right side as elaborated upon in section 4.4. However, this path is obstructed by another obstacle on the robot's right. To dodge this newly encountered obstacle, once again the robot forms trajectories from its right and left. The one attempted to pass from left finds the obstacle labelled '1' in the way again. Since this obstacle is present in the history as one that has already been attempted to be dodged, the trajectory for dodging the newly encountered obstacle from the left is discarded as a situation where a deadlock is encountered. The trajectory formed to dodge it from the right, however, finds another obstacle (indicated as a type-3 obstacle in Fig. 11). Attempting to dodge this type-3 obstacle from the left results in a deadlock just as in the previous case whereas the one attempted from its right encounters yet another obstacle. This process is repeated twice more until, in an effort to dodge from the right the obstacle to the left of the obstacle labelled '1', the obstacle labelled '1' is encountered again. This trajectory, therefore, is also abandoned and a local loop is said to have been encountered.

A similar situation occurs while trying to dodge the obstacle labelled '1' in Fig. 11 from its left side. Thus the only trajectory possible to overcome this obstacle which is free of local loops and deadlocks, is the one formed by stepping over. As we can see, the planner only plans the successful trajectory, avoiding getting stuck in local loops and deadlocks.

### 4.6 Cost Assignment

A heuristic cost based on the order of complexity of each stepping motion is given at Table 1. These costs are assigned to foot placements as they are planned.

| Step type | Description | Cost |
|---|---|---|
| Straight | Step placed straight in forward direction | 1 |
| Turning | Step placed to change orientation of the robot | 2 |
| Extended | Step in which foot turning is combined with extension | 3 |
| Enhanced | Step taken for stepping over obstacles | 4 |

Table 1. Heuristic costs assigned to different step types

### 4.7 Intermediate Paths

If, in order to proceed towards a pivot point while dodging an obstacle from its side, another obstacle is encountered along the way, alternate paths based upon the obstacle type are formed. All these alternate paths converge at the pivot point, meaning that they will all have similar descendants. Thus, the number of these intermediate alternate paths is multiplied by the number of descendent paths and added to the total number of possible alternate paths. Thus, evaluating cost of each intermediate path and keeping only the best during alternate path creation reduces the number of paths to only useful ones.

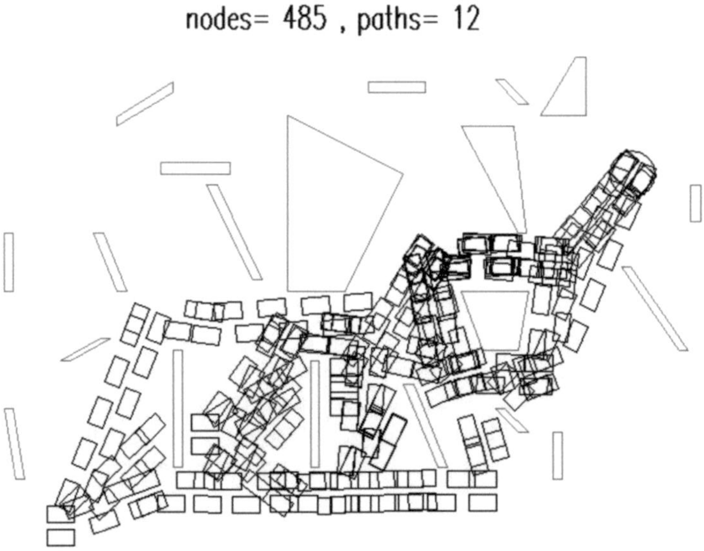


Fig. 12. Various paths for reaching the destination planned for HRP-2 humanoid robot.

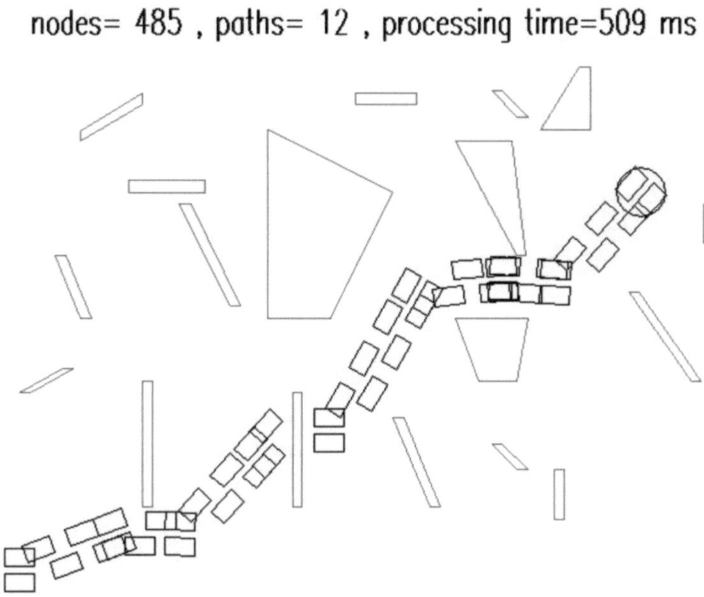


Fig. 13. Best path determined using depth first exhaustive search.

### 4.8 Search for Best Path

In order to identify the best one of the paths formed by our algorithm, we employ a depth first exhaustive search since greedy or A* search would not filter out for us the best path in every scenario (Cormen, 1994).

### 4.9 Simulation Results

Figs. 12 and 13 show results of a simulation of HRP-2 planning footsteps in a room with 20 obstacles. We see that a graph of only 485 nodes is formed consisting of 12 paths all reaching the destination. The whole process takes only 509 ms on a 2.4 GHz Pentium DUO (2 GB RAM) running Windows Vista. A comparison with previous techniques reviewed in section 2 shows reduction in the number of nodes as well as in processing time even though the algorithm employs exhaustive search for hunting out best path which guarantees optimality of the chosen path among those traced out by the algorithm.

Some more simulation results are given in Figs. 14 and 15. We see that the fastest result is obtained in Fig. 14. This corresponds to the lowest number of nodes as well as paths in the graph which also reduces the time taken in search for best path. Fig. 15 shows the trajectories formed around the obstacle of type-3 (drawn with red). It is readily observed that the robot keeps a greater distance with this obstacle than with obstacles of other types.

## 5. Conclusion

Our algorithm successfully demonstrates a novel global reactive footstep planning strategy with a human-like approach. Incremental graph expansion from simpler to more complex

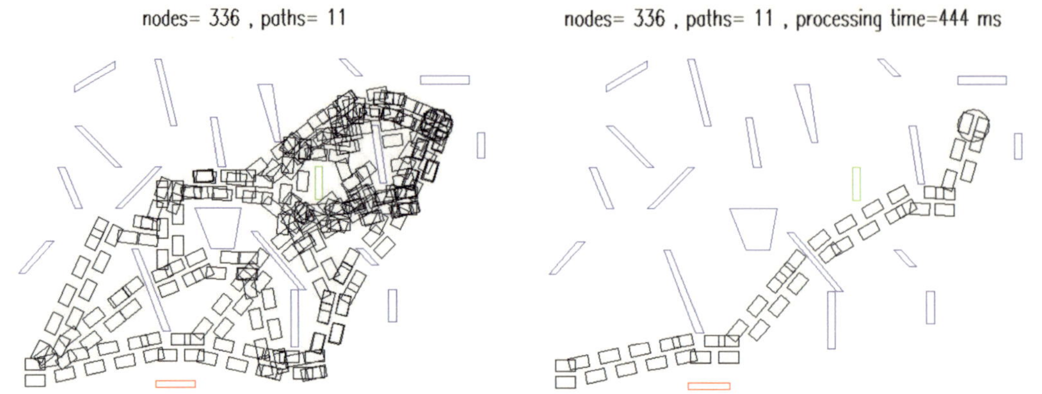


(a) Alternate paths. (b) Best path.
Fig. 14. Result 2: Nodes formed 336, paths 11, time taken 444 ms.

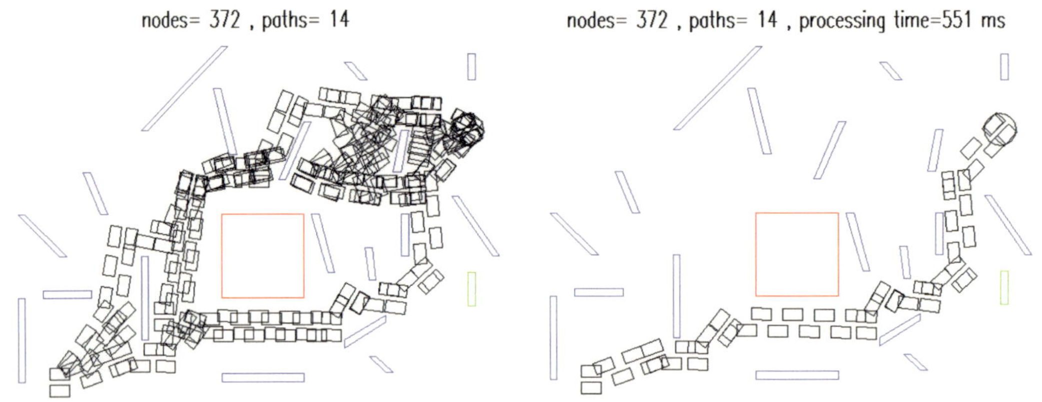


(a) Alternate paths. (b) Best path.
Fig. 15. Result 3: Nodes formed 372, paths 14, time taken 551 ms.

paths ensures formation of a simpler and more useful graph as compared to that formed by approaches such as the visibility graph. The trajectory generated is more energy-efficient since the robot does not have to lift its foot to a high location in every step as in case of game theory based approaches. The algorithm is considerably fast and reduces computational complexity by minimizing the number of alternate steps considered after planning each step. However, basing the cost of each step on energy or time optimization criteria instead of just the complexity level of the stepping motion can further improve the performance of the algorithm. Future work avenues include hardware implementation of the complete footstep planner as well as footstep planning in the presence of moving obstacles.

## 6. References


Ayaz, Y.; Munawar, K.; Malik, M.B. ; Konno, A. & Uchiyama, M. (2006). Human-Like Approach to Footstep Planning Among Obstacles for Humanoid Robots, Proc. of IEEE/RSJ Int. Conf. on Intelligent Robots and Systems (IROS), pp. 5490-5495

Ayaz, Y.; Munawar, K.; Malik, M.B. ; Konno, A. & Uchiyama, M. (2007). Human-Like Approach to Footstep Planning Among Obstacles for Humanoid Robots, International Journal of Humanoid Robotics, World Scientific Publishing Company, Vol. 4, No. 1, pp. 125-149

Ayaz, Y.; Munawar, K.; Malik, M.B. ; Konno, A. & Uchiyama, M. (2007). A Human-Like Approach to Footstep Planning, Humanoid Robots: Human-Like Machines, I-Tech Education & Publishing, pp. 295-314

Chestnutt, J. & Kuffner, J.J. (2004). A Tiered Planning Strategy for Biped Navigation, Proceedings of IEEE Int. Conf. on Humanoid Robotics (ICHR)

Chestnutt, J.; Lau, M.; Cheung, G.; Kuffner, J.J.; Hodgins, J. & Kanade, T. (2005). Footstep Planning for the Honda Asimo Humanoid, Proceedings of IEEE Int. Conf. on Robotics & Automation (ICRA), pp. 629-634

Cormen, T.H.; Leiserson, C.E. & Rivest, R.L. (1994). Introduction to Algorithms, McGraw-Hill Book Co.

Guan, Y.; Yokoi, K.; Neo, E.S. & Tanie, K. (2004). Feasibility of Humanoid Robots Stepping over Obstacles, Proceedings of IEEE/RSJ Int. Conf. on Intelligent Robots and Systems (IROS), pp. 130-135

Guan, Y.; Neo, E.S.; Yokoi, K. & Tanie, K. (2005). Motion Planning for Humanoid Robots Stepping over Obstacles, Proceedings of IEEE/RSJ Int. Conf. on Intelligent Robots and Systems (IROS), pp. 364-370

Guan, Y.; Neo, E.S.; Yokoi, K. & Tanie, K. (2006). Stepping Over Obstacles With Humanoid Robots. IEEE Trans. on Robotics, Vol. 22, No. 5, pp. 958-973

Huang, Q.; Yokoi, K.; Kajita, S.; Kaneko, K.; Arai, H.; Koyachi, N. & Tanie, K. (2001). Planning Walking Patterns for a Biped Robot. IEEE Trans. on Robotics, Vol. 17, No. 3, pp. 280-289

Kajita, S.; Kanehiro, F.; Kaneko, K.; Fujiwara, K.; Harada, K.; Yokoi, K. & Hirukawa, H. (2003). Biped Walking Pattern Generation by using Preview Control of Zero-Moment Point, Proceedings of IEEE Int. Conf. on Robotics & Automation (ICRA), pp. 1620-1626

Kajita, S.; Morisawa, M.; Harada, K.; Kaneko, K.; Kanehiro, F.; Fujiwara, K. & Hirukawa, H. (2006). Biped Walking Pattern Generator allowing Auxiliary ZMP Control, Proceedings of IEEE/RSJ Int. Conf. on Intelligent Robots and Systems (IROS), pp. 2993-2999

Kaneko, K.; Kanehiro, F.; Kajita, S.; Hirukawa, H.; Kawasaki, T.; Hirata, M. ; Akachi, K. & Isozumi, T. (2004). Humanoid Robot HRP-2, Proceedings of IEEE Int. Conf. on Robotics & Automation (ICRA), pp. 1083-1090

Konno, A.; Kato, N.; Shirata, S.; Furuta, T. & Uchiyama, M. (2000). Development of a Light-Weight Biped Humanoid Robot, Proceedings of IEEE/RSJ Int. Conf. on Intelligent Robots and Systems (IROS), pp. 1565-1570

Kuffner, J.J.; Nishiwaki, K.; Kagami, S.; Inaba, M. & Inoue, H. (2001). Footstep Planning Among Obstacles for Biped Robots, Proceedings of IEEE/RSJ Int. Conf. on Intelligent Robots and Systems (IROS), pp. 500-505

Kuffner, J.J.; Nishiwaki, K.; Kagami, S.; Inaba, M. & Inoue, H. (2003). Online Footstep Planning for Humanoid Robots, Proceedings of IEEE/RSJ Int. Conf. on Robotics and Automation (ICRA), pp. 932-937

Kuffner, J.J.; Nishiwaki, K.; Kagami, S.; Inaba, M. & Inoue, H. (2005). Motion Planning for Humanoid Robots, Robotics Research, Paolo Dario and Raja Chatila (Eds.), Springer Tracts in Advanced Robotics, Vol.15, pp. 365-374

Latombe, J.C. (1991). Robot Motion Planning, Kluwer Academic Publishers, Boston, Massachusetts, United States of America

McGhee, R.B. & Iswandhi, G.I. (1979). Adaptive Locomotion of a Multi-legged Robot over Rough Terrain, IEEE Trans. on Systems, Man and Cybernetics, Vol. SMC-9, No. 4, pp. 176-182

Michel, P.; Chestnutt, J.; Kuffner, J.J. & Kanade, T. (2005). Vision-Guided Humanoid Footstep Planning for Dynamic Environments, Proceedings of IEEE/RAS Int. Conf. on Humanoid Robotics (ICHR), pp. 13-18

Michel, P.; Chestnutt, J.; Kagami, S.; Nishiwaki, K.; Kuffner, J.J. & Kanade, T. (2006). Online Environment Reconstruction for Biped Navigation, Proceedings of IEEE Int. Conf. on Robotics & Automation (ICRA), pp. 3089-3094

Tsujita, T.; Tsuchie, T.; Myojin, T.; Konno, A. & Uchiyama, M. (2007). Development of an Impact Motion Generation Support System, Proceedings of JSME Conf. on Robotics and Mechatronics (RoboMech), ID: 1A1-B01 (in Japanese)

Verrelst, B.; Stasse, O.; Yokoi, K. & Vanderborght, B. (2006). Dynamically Stepping Over Obstacles by the Humanoid Robot HRP-2, Proceedings of IEEE Int. Conf. on Humanoid Robotics (ICHR), pp. 117-123

Vukobratović, M. & Borovac, B. (2004). Zero Moment Point---35 Years of Its Life, International Journal of Humanoid Robotics, World Scientific Publishing Company, Vol. 1, No. 1, pp. 157-173

Yagi, M. & Lumelsky, V. (1999). Biped Robot Locomotion in Scenes with Unknown Obstacles, Proceedings of IEEE Int. Conf. on Robotics & Automation (ICRA), pp. 375-380

# 12

# Approaches to door identification for robot navigation

E. Jauregi, E. Lazkano and B. Sierra
*University of Basque Country,*
*Donostia*

## 1. Introduction

Even though great technical progress has been made in the area of mobile robotics, some fundamental control problems, such as autonomous navigation, remain unresolved. Many animals have shown that they are very good at navigating but autonomous navigation is still a complicated task for engineered robots. Therefore, research efforts have aimed at incorporating biologically inspired strategies into robot navigation models (Mallot and Franz, 2000; Trullier et al., 1997). Among these, *Local navigation strategies* allow the agent to choose actions based only on its current sensory input, while *way finding strategies* are responsible for driving the agent to goals out of the agent's perceptual range and require the recognition of different places and correlations among them. Acquiring the appropriate set of landmarks remains a challenge.

Many navigation tasks can be fulfilled by point to point navigation, door identification and door crossing (Li et al., 2004). Indoor semi-structured environments are full of corridors that connect different offices and laboratories where doors give access to many of those locations that are defined as goals for the robot. Hence, endowing the robot with the door identification ability would undoubtedly increase the navigating capabilities of the robot.

Doors in indoor environments are not necessarily uniform; door panels can be of different texture and colour and handles can vary in shape. Specific methods can be designed for each type of door. However, feature-based methods are more easily applicable to different object recognition tasks they are computationally expensive and therefore less attractive to be applied for real-time problems such as for mobile robot navigation. This chapter compares several approaches developed for door identification based on handle recognition, from specific methods based on colour segmentation techniques to more general ones that focus on feature extraction methods, like SIFT (Lowe, 2004) and (U)SURF (Bay et al., 2008). A new two-step multiclassifier that combines region detection and feature extraction is also presented. The objective of the approach is to extract the most relevant part of the image, the subimage that with high probability should contain the most interesting region of the image, and limit the application of the feature extraction techniques to that region. The developed algorithm improves the relevance of the extracted features, it reduces the superfluous keypoints to be compared at the same time that increases the efficiency by improving accuracy and reducing the computational time.

The developed approaches are first evaluated off-line and tested afterwards in a real robot/environment system using a a Peoplebot robot from *Mobilerobots*. The door recogni-

tion module is integrated in a behaviour-based control architecture (Brooks, 1986) that allows the robot to show a door-knocking behaviour.
The chapter is structured as follows: next section reviews the literature about door identification. Section 3 presents the first approach taken to solve the task: a three-stage procedure that relies mainly on colour segmentation. Section 4 summarises some feature extraction methods and shows the results obtained for the door recognition problem. Section 5 presents a new two-step algorithm that combines region detection and feature extraction that incorporate the applicability of the feature extraction methods to the performance of the environment specific ones. The obtained results are commented in section 6. Experiments performed on a real robot/environment system are described in section 7. Finally, conclusions and further work are outlined in section 8.

## 2. The door identification task

Several references can be found that tackle the problem of door identification. In (Kragic et al., 2002), where doors are located in a map, rectangular handles are searched for manipulation purposes using cue integration by consensus. However, most of the references we found are vision-based approaches (Seo et al., 2005) and focus on edge detection. Muñoz-Salinas et al. (2005) present a visual door detection system that is based on the Canny edge detector and Hough transform to extract line segments from images. Another vision-based system for detection and traversal of doors is presented in (Eberset et al., 2000). Door structures are extracted from images using a line based filtering parallel method, and an active tracking of detected door line segments is used to drive the robot through the door.
A different proposal for vision-based door traversing behaviour can be found in (Seo et al., 2005). Here, the PCA (Principal Component Analysis) pattern finding method is applied to the images obtained from the camera for door recognition. (Stella et al., 1996) presents SAURO, a robotic autonomous system oriented to transportation tasks in indoor environments. SAURO uses a vision-based self-localisation subsystem that allows the robot to determine its location to verify that the planned path is correctly followed. A door identification and crossing approach is also presented in (Monasterio et al., 2002), where a neural network based classification method was used for both, the recognition and crossing steps. More recently, in (Lazkano et al., 2007) a Bayesian Network based classifier was used to perform the door crossing task. Doors are assumed to be open, the opening is identified, and doors crossed using sonar sensor information.
But navigating in narrow corridors makes it difficult to identify doors by line extraction due to the inappropriate viewpoint restrictions imposed by the limited distance to the walls. The goal of the work presented in this chapter is to try to identify the doors by recognising the handles, extracting the necessary local features needed for robust identification. In the experiments here described the aim is to identify doors by means of the door handles in an environment with two type of handles: circular ones, located at doors with pladour (a type of laminated surface) and consequently, completely textureless blades; and rectangular ones located at wooden door blades. The problem is approached from two angles. The first one relies on the segmentation of the door blade surfaces using colour segmenters for each type of door. The second approach is based on feature extraction methods and aimed at finding a more general algorithm easily applicable to other types of doors. Both approaches are vision based and are tackled from a classification problem perspective: given an image, the goal is to determine whether the image contains or not a door handle.

Although there are tasks where images may not be restricted to a single object identification, and thus could contain more than one region to be detected and extracted, that is not the case in the problem described in this chapter, where it is assumed that doors contained a single handle.

## 3. Segmentation based identification

Monochrome image segmentation algorithms are based on two basic properties of grey levels: discontinuities and similarities. Discontinuity based segmentation means partitioning the image by detecting abrupt changes in grey levels. On the other hand, similarity based segmentation pretends to group regions with similar properties (Gonzalez and Woods, 1993).
*Discontinuity detection* is performed by looking for three different types of discontinuities: isolated points, lines and edges. Points and lines are not too frequent for most practical applications. On the contrary, edges represent the frontiers among regions of different grey level properties and are the most widely applied segmentation technique. But discontinuity detection methods rarely characterise a border in a unique manner and therefore, additional local or global methods are needed to join pixels associated to the edges.
Among the methods that apply *similarity detection*, *thresholding* is probably the most important technique. But it is known that variant illumination affects the result of the segmentation. Thereby, several methods allow to change the threshold levels making it adaptive. On the other hand, *region growing* techniques consist of grouping pixels or regions into bigger regions. Alternatively, *region splitting* techniques start with an arbitrary number of regions in an image and try to group or divide them into regions that comply with certain properties.
It is noticeable the large number of grey level segmentation techniques. On the contrary, colour image segmentation requires more information about the objects to be identified. Colour image segmentation extracts one or more connected regions from an image and the extracted regions must satisfy certain uniformity criteria based on characteristics derived from the spectral components (Skarbek and Koschan, 1994).
Some colour segmentation techniques are based on mathematical models and some are algorithmic approximations. Basically, there are *histogram based* techniques, which try to identify peaks on the colour space and then use these extrema to classify pixels; *Clustering* techniques that group pixels according to the obtained cluster representatives; and *Fuzzy Clustering techniques* that use fuzzy functions to decide the membership of every pixel in each of the defined cluster.

### 3.1 The door identification process

In order to identify images containing a door handle a three-stage algorithm has been designed. This three-stage process proceeds as follows:

1. Region detection: this step is designed to approximate the handle area, if any, in the image. The most intuitive approach to detect circular handles seems the extraction of circular edges. The *Circular Hough Transform* is an appropriate method to achieve this objective. But handles can also be non circular. A more general method is needed as *blob* extraction also known as region detection or labelling.
2. Colour segmentation: using the position information of the previous step, the surroundings of the candidate object are segmented in order to see if the candidate handle is well surrounded by the appropriate door blade.

3. Statistical validation: measures of the segmented image are afterwards needed to classify an image as containing or not a handle.

### 3.1.1 Region detection

As mentioned before, some handles are assumed to be circular in shape. Therefore, circle detection can be performed to locate handles in the images taken by the robot. Although many circle extraction methods have been developed, probably the most well-known algorithm is the Circle Hough Transform (CHT). Moreover, Hough transform methods have shown to be robust enough to deal with noisy images (Ayala-Ramírez et al., 2006).

The use of the Hough Transform to detect circles was first outlined by Duda and Hart (1972) and then, Kimme et al. (1975) gave probably the first known application of the Hough transform to detect circles in real images. Later on, Yuen et al. (1990) investigated five circle detection methods which are based on variations of the Hough Transform. One of those methods is the *Two stage Hough Transform* and it is implemented in the OpenCV vision library (http://www.intel.com/research/opencv) used in the experiments described later on.

The circle finding function can identify a huge number of circles depending on the image background. From the returned list of circles, only the most probable one is considered. However, due to the local navigation strategies of the robot the images will be obtained within the same distance range and therefore, it is possible to know in advance the approximate radius of the handles. In this manner only the identified circumferences with a radius that lies within a known range would be considered as handle candidates.

The CHT detects circular shapes; hence, an alternative method is needed in order to first approximate the non circular handles. The method can be generalised by scanning the image for continuous connected regions or *blobs*. A blob (binary large object) is an area of touching pixels with the same logical state. Blob extraction, also known as region detection or labelling, is an image segmentation technique that categorises the pixels in an image as belonging to one of many discrete regions. The process consists of scanning and numbering any new regions that are encountered, but also merging old regions when they prove to be connected on a lower row. Therefore, the image is scanned and every pixel is individually labelled with an identifier which signifies the region to which it belongs (see (Horn, 1986) for more details).

Blob detection is generally performed on the resulting binary image from a thresholding step. Instead, we apply the SUSAN (Smallest Univalue Segment Assimilating Nucleus) edge detector (Smith and Brady, 1997), a more stable and faster operator.

Again, the blob extraction process can give many different regions for a single image. In order for a blob to be confirmed as a candidate, the result of the blob detection process should be filtered and false positives should be discarded. Different filtering techniques can be used. For instance, in (Ye and Zhong, 2007) location-related pixel information is used for blob discrimination where the aim is to count persons in images taken by a surveillance system. A similar approach is used in our proposal where blobs that are not consistent with the defined size restrictions are discarded. The size restrictions depend on the distance the images are taken from.

### 3.1.2 Colour segmentation

The region extraction step is not reliable enough, neither to confirm, nor to reject, the presence of a handle in an image. Therefore, it needs to be endowed with some complementary processes in order to improve its performance. The approach used here is to use colour information around the location of the detected shape for door recognition. The objective of

this second step is to segment the pixels belonging to the door blades in an image. Candidate handles not surrounded by the proper blades should help to reject false positives.
Within the robot environment, a circular handle is always associated to a pladour door and rectangular ones are on wooden doors presenting different tonalities according to lighting conditions –e.g. presence or absence of electric or natural lighting.

#### 3.1.2.1 Segmenting pladour door blades

A supervised machine learning approach has been selected for this goal. To build the classifier we chose *Oc1* (Oblique Classifier 1) (Murthy et al., 1994), a decision tree induction system well suited for applications where the instances have numeric feature values. *Oc1* builds decision trees containing linear combinations of one or more attributes at each internal node. The trees built in this way partition the instance space with both oblique and axis-parallel hyperplanes. Images taken by the robot are represented in RGB colour space and thereby, each pixel is a three component vector, each component taking a value that ranges from 0 to 255.
In every classification problem, a training set is required to get a model to be later used when a new instance is presented to the model. To get the training set, we firstly constructed a database of positive instances (those associated to pladour doors), as well as negative instances (those not associated to pladour doors). The size of these databases was about two million pixels, obtained from about sixty images taken from the robot camera in a corridor. From these databases we extracted, 80,000 pixels randomly, 40,000 of them labelled as *pladour* and the remaining 40,000 as *not pladour*. Then, these 80,000 pixels were presented to the *Oc1* tree generation procedure, to get the decision tree used in the segmentation of the images taken by the robot camera. The obtained performance after applying 10 fold crossvalidation to this database was 93.61%. Fig. 1 shows several examples of this segmentation process. The first row contains images with handles, although the variant lighting conditions affect the pladour colour recognition process and therefore, the segmentation process. Notice for example the upper right corner of the fourth image. The bottom row shows images without handles. Original images contain the found circles detected by the Hough transform, in spite of the radius exceeds the preset threshold, for sake of clarity. It can be observed that the segmentation of these images does not imply the existence of any handle.

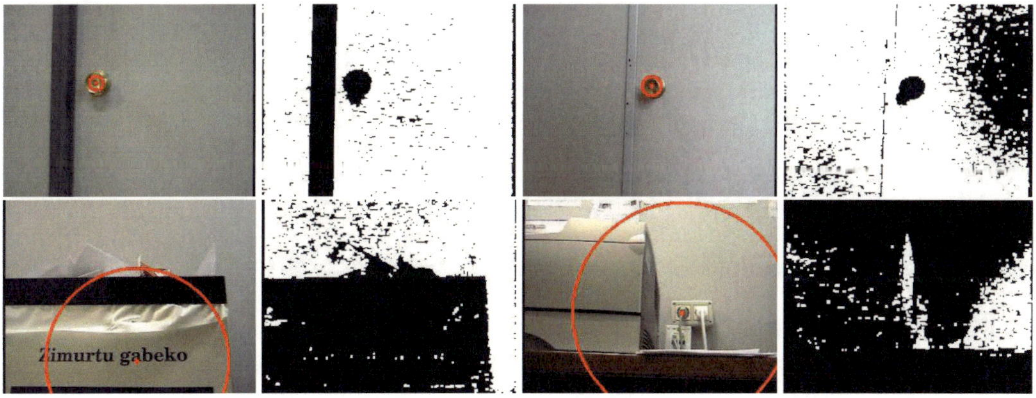


Fig. 1. Pladour segmentation examples

#### 3.1.2.2 Segmenting wooden door blades

The goal now is to classify pixels belonging to wooden surfaces. To segment wooden surfaces a specific thresholding algorithm has been designed. First of all, the following reference values are selected from an image containing uniquely wooden coloured pixels using the normalised RGB colour space:

- $R_{min}$: minimum value of the red channel.
- $G_{min}$: minimum value of the green channel.
- $B_{min}$: minimum value of the blue channel.
- $Rdiff_{min}$: value of the minimum difference among the red and the maximum of the green and blue.
- $Gdiff_{min}$: value of the minimum difference among the green and the maximum of the red and blue.
- $Bdiff_{min}$: value of the minimum difference among the blue and the maximum of the green and red.

Then, given a new image, a pixel will be labelled as belonging to a wooden surface according to the algorithm described in figure 2. The process is performed using the normalised RGB colour space.

```
wood_segmenter(R, G, B,
               Rmin, Rdiffmin, Gmin, Gdiffmin, Bmin, Bdiffmin)
{
  Rdiff = R - max(G,B)
  Gdiff = G - max(R,B)
  Bdiff = B - max(R,G)
  if (R > Rmin && G > Gmin && B > Bmin &&
      Rdiff > Rdiffmin &&
      Gdiff > Gdiffmin &&
      Bdiff > Bdiffmin)
      return 1;
  return 0;
}
```

Fig. 2. Wooden surface segmentation algorithm

Figure 3 shows some examples of this segmentation process.

### 3.1.3 Statistics for decision taking

So far we are provided with two procedures to recognise the handle and its surroundings. However, both procedures are prone to errors due to noise, lighting conditions and other objects in the environment (printers, dustbins, tables, and so on). We cannot relay in the accuracy of any of them separately. The aim of this third step is to analyse segmented pixels of the candidate handle and those surrounding the circle in order to definitively confirm or reject the handle candidate. The segmentation process yields a black-and-white image, where white

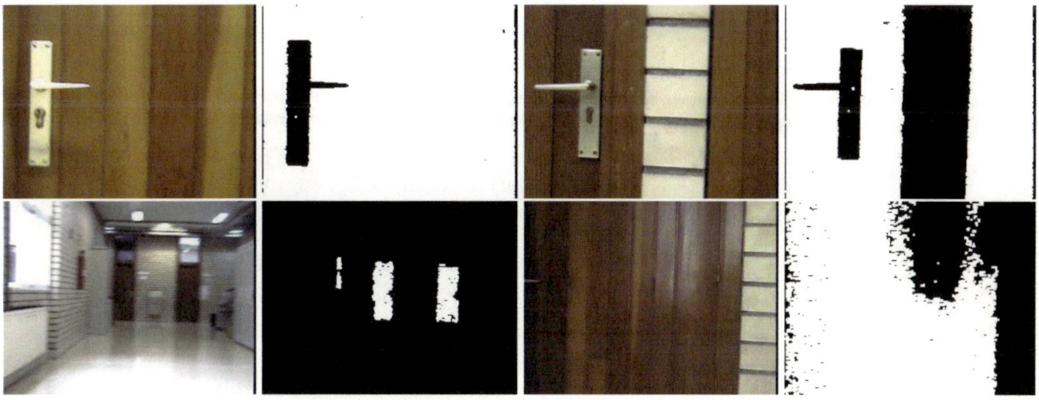

Fig. 3. Segmentation of wooden surfaces

points are those classified as *door blade*, and black ones are those classified as not belonging to the *door blade*. To analyse the surroundings of the prospective handle, and using the information about the detected region, the centre $(x_0, y_0)$ is obtained and then, the following values are calculated (see Figure 4):

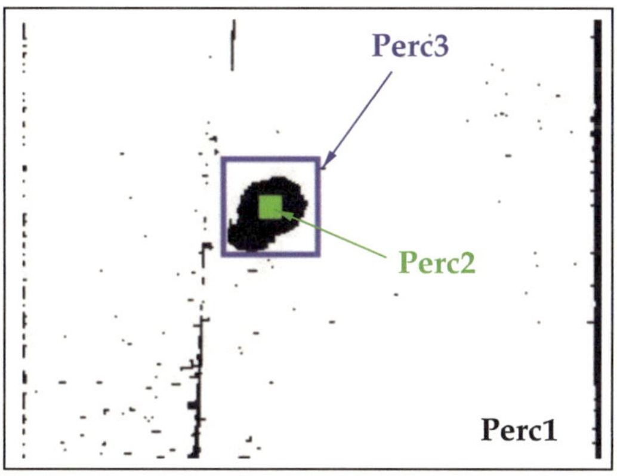


Fig. 4. Zones from which *Perc2* and *Perc3* are computed; *Perc1* is computed over the whole image

- *Perc1*: Percentage of white points in the whole image. When the robot is close to a door, a huge percentage of the pixels in its visual field are likely to be segmented as white. So, the bigger *Perc1*, the more likely the robot is in front of a pladour door.
- *Perc2*: The pixels around the centre should belong to the handle, not to the pladour class. This value represents the percentage of black points in the $5 \times 5$ grid centred at $(x_0, y_0)$. Therefore, the bigger *Perc2*, the more likely the robot is in front of a handle.
- *Perc3*: When the procedure that returns the region has really located a handle, the farther surroundings of the centre are expected to be segmented as white, as they do not fall in the handle, but in the door blade. We define:

- $S1$: set of points in the squared area centred at $(x_0, y_0)$ and of size length $2 \times r$.
- $S2$: set of points in the squared area centred at $(x_0, y_0)$ and of size length $2 \times (r + d)$, where $d$ represents the desired shift from circle perimeter.

Hence, $S = S_2 - S_1$ defines the closest surroundings of the handle and *Perc3* represents the percentage of white points (*pladour*) in $S$. Again, the bigger *Perc3*, the more likely the robot is in front of a handle.

The CHT gives the radius of each detected circle, together with some more information, whereas using region labelling $r$ is defined as the maximum value between the width and the height of the obtained region.
The combination of these percentages (weighted mean) give us a measure of confidence ($cl_{handle}$) of being in front of a door and that the region recognition procedure has really recognised a circular handle.

### 3.2 Results

At this stage of the experimentation, the handle recognition was performed for each class of handle, using separate databases for each one.
All the images (learning and testing DBs) were taken while the robot followed the corridors using its local navigation strategies and therefore, they were taken at distances at which the robot is allowed to approach the walls. Both databases contained almost equal positive and negative cases.
The databases characteristics were the following:

1. For circular handles, the database contained about 3000 entries
2. For rectangular handles, the size was bigger with about 5000 images

The experiments were executed for a maximum allowed radius of 20 pixels ($r_{max} = 20$), a shift of 15 pixels from the circle perimeter ($d = 15$), and the confidence level of being in front of a handle should rise above 0.6 to confirm the door ($cl_{handle} > cl_{TH} = 0.6$).
In order to evaluate the improvement introduced by the three-stage process, the classification accuracy was computed for the sequence and compared with the accuracy that would be obtained if only region detection was used.
However, accuracy is considered a fairly crude score that does not give much information about the performance of a categoriser. The F1 measure, also known as F-score or F-measure, is a measure of a test's accuracy and can be viewed as a weighted average of the precision and recall (see equation 1), where an F1 score reaches its best value at 1 and worst score at 0. F1 measure combines both precision and recall into a single metric and favour a balanced performance of the two metrics (Chen, 1996).

$$F1 = \frac{2 \times Precision \times Recall}{Precision + Recall} \tag{1}$$

Table 1 shows the obtained classification accuracies and the corresponding F1 measure values. These results show how both performance measures, the classification accuracy and the F1 value are highly increased using the three stage approach for both types of handles. Surprisingly, the blob detector shows to be more appropriate for the circular handles. The robot's viewpoint affects the shape of the handle in the sense that the circular handle distorts and often does not look like a circle but more like an ellipsoid making it more difficult to be identified by the CHT.

| Method | Circular |  | Rectangular |  |
|---|---|---|---|---|
|  | Acc. | F1 | Acc. | F1 |
| CHT | 30.6 | 0.319 | – | – |
| CHT 3-stage | 85.1 | 0.685 | – | – |
| Blobs | 91.1 | 0.867 | 61.0 | 0.36 |
| Blobs 3-stage | 96.7 | 0.946 | 93.6 | 0.72 |

Table 1. Experimental results

The door identification process should be performed in real time by the robot. Hence, the time needed to compute the identification must be analysed. Table 2 shows the time requirements of the developed three-stage algorithm.

| Method | Circular | Rectangular |
|---|---|---|
| CHT 3-stage | 0.043 | – |
| Blobs 3-stage | 0.050 | 0.057 |

Table 2. Computational payload (s) for processing an image applying the 3-stage procedure

## 4. Feature extraction based identification

The approach described above showed to be too specific to the robot's particular environment and not easily generalisable. Although the region detector is robust enough, the segmentation processes are specific and new ones should be developed in order to detect handles located in different surfaces.

Feature extraction in images is an important issue in mobile robotics, as it helps the robot to understand its environment and fulfil its objectives. There are several applications where it is mandatory to recognise objects or scenes: image retrieval (Ledwich and Williams, 2004), mobile robot localisation (Gil et al., 2005);(Tamimi et al., 2006) and SLAM (Se et al., 2002), physical sign recognition (Roduner and Rohs, 2006) and automatic guidance of vehicles (Primdahl et al., 2005). Although different sensors can be used, vision turns out to be the most appropriate one due to the rich information that can be extracted from images. However, object identification becomes a complex task specially because of varying environmental conditions and changes in object scale and camera viewpoint.

Objects can be identified in images extracting local image descriptors. These descriptors should be distinctive and invariant to image transformations. The idea is to detect image regions co-variant to a class of transformations, which are then used as support regions to compute invariant descriptors (Mikolajczyk and Schmid, 2005). While complete invariance has yet to be achieved, features which are robustly resilient to most image transforms have been proposed by D. G. Lowe (Lowe, 2004).

Several invariant feature extraction techniques exist in the literature. SIFT (*Scale-invariant feature transform*) (Lowe, 2004) attempts to extract distinctive features from an image in order to find correspondences among different images or to identify objects. SIFT keypoints are nowadays the most common descriptors used. The features extracted by the algorithm are invariant to scale and rotation, and partially invariant to changes in illumination.

**SIFT** features are extracted according to the following procedure:

1. Detect scale-space extrema, searching over all scales and image locations. Potential interest points invariant to scale and orientation are efficiently computed using a DoG (differential of Gaussian) function.
2. Localise keypoints, detecting local extrema and removing low contrast points or candidates located in edges.
3. Assign orientations to the candidate keypoints based on local image gradient directions.

After keypoints are localised, for each keypoint a descriptor is computed by calculating a histogram of local oriented gradients around the interest point and storing the bins in a 128 dimensional vector. These descriptors can be then compared with stored ones in a reference database for object recognition purposes. But SIFT suffers from high computational payload and for on-line applications, each one of the three steps (detection of local extrema, keypoint description computation and keypoint matching) should be computed faster.
**SURF** (Bay et al., 2006) descriptors are computed in two steps. First, a reproducible orientation is found based on the information of a circular region around the interest point. This is performed using Haar-wavelet responses in the $x$ and $y$ directions at the scale where the interest point was detected. The dominant orientation then is estimated by calculating the sum of all responses within a sliding window. Next, the region is split up into smaller square subregions and some simple features are computed (weighted Haar wavelet responses in both directions, sum of the absolute values of the responses). This yields a descriptor of length 64, half size of the original SIFT descriptor and hence, offers a less expensive computationally matching process.
The upright version of SURF, named **USURF**, skips the first step of the descriptor computation process, resulting in a faster version. USURF is proposed for those cases where rotation invariance is not mandatory.
In all the three methods, the matching process requires a matching criteria and a reference database where the keypoints of the interesting objects must be stored.

### 4.1 Extracting features from door images

As mentioned previously, in our robot's environment circular and rectangular handles are distinguished. Circular ones are located into pladour blades. These have the advantage that almost every keypoint is located at the door handle and only a few of them appear at the handle surroundings (see Figure 5(a)). On the other hand, rectangular handles are located into wooden door blades that are not textureless and therefore, keypoints appear outside the handle (see Figure 5(b)).
In order to see the adequateness of the feature extraction techniques for the handle identification task, the same databases used for the previous experiments were used, albeit a few extra images where taken to use as reference objects (reference DBs). It must be noted that the test cases were collected in a different environment from where reference cases were taken. Therefore, the test databases did not contain images of the handles in the reference database. The databases characteristics were the following:

1. For circular handles, the testing database contained about 3000 entries and the reference database contained 32 images.
2. For rectangular handles, the testing database contained about 5000 images and the reference database contained 19 images.

The keypoint matching criteria used in these experiments is the 1-NN, the same proposed in (Lowe, 2004).

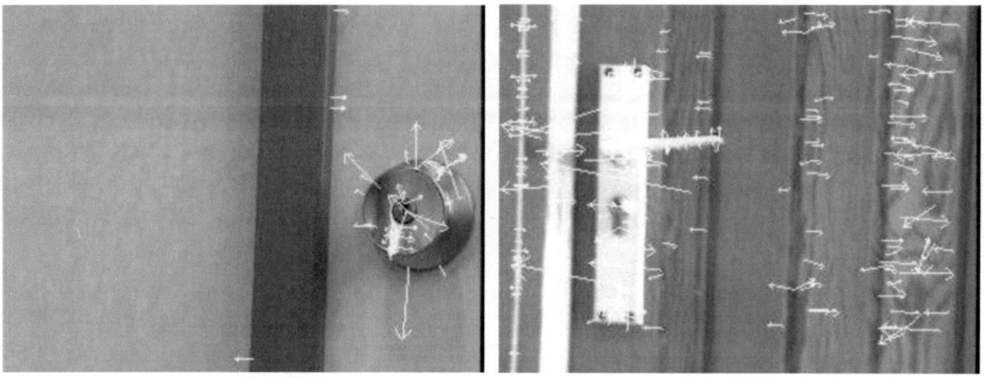

(a) Circular Handle (b) Rectangular Handle

Fig. 5. SIFT keypoints in 320 × 240 images

### 4.2 Results

Table 3 shows the obtained results. Again, accuracy and F1 measure are calculated. USURF outperforms SIFT in both, accuracy and F1 measure in the case of circular handles, but its performance degrades for rectangular ones.

| | Circular | | Rectangular | |
|---|---|---|---|---|
| | Acc. | F1 | Acc. | F1 |
| SIFT | 62.39 | 0.592 | **55.41** | **0.362** |
| SURF | 72.5 | 0.691 | 42.34 | 0.324 |
| USURF | **72.5** | **0.691** | 47.0 | 0.344 |

Table 3. Performance of the keypoint extraction methods

These values are disappointing. The feature extraction approach does not give reasonable performance values to be applied on the robot for navigation purposes. Comparing these results with the ones shown in table 1, both the classification accuracy and the F1 measure are much lower that the values obtained using the three-stage procedure previously described.

## 5. Combining region extraction and feature extraction

This section presents a new two-step algorithm that aims at reducing the superfluous keypoints obtained by methods like SIFT and (U)SURF by firstly extracting the region of the image where the object or objects to be identified are likely to be located.

As mentioned before, instead of computing the invariant features of the whole image, the approach presented here aims to reduce the size of the image to be processed by extracting the portion of the image that, will have a high probability of containing crucial features. Here on, we will call that portion the region of interest (ROI) of the image.

Several methods can be used for extracting the ROI but the experiments performed while developing the segmentation based door identifier showed that blob extraction gives robust results for the handle identification goal.

Once the most interesting blob is located, blob's length and width values are used to find its centre and afterwards, a square subimage is extracted. The size of the square is determined by the maximum value between the length and width of the candidate blob. The subimage is then scaled to a fixed size to obtain the portion of the image with a high probability of containing the object of interest, i.e a ROI. Then, the keypoint extraction and matching procedure is performed to the extracted ROI. Figure 6 summarises the process.

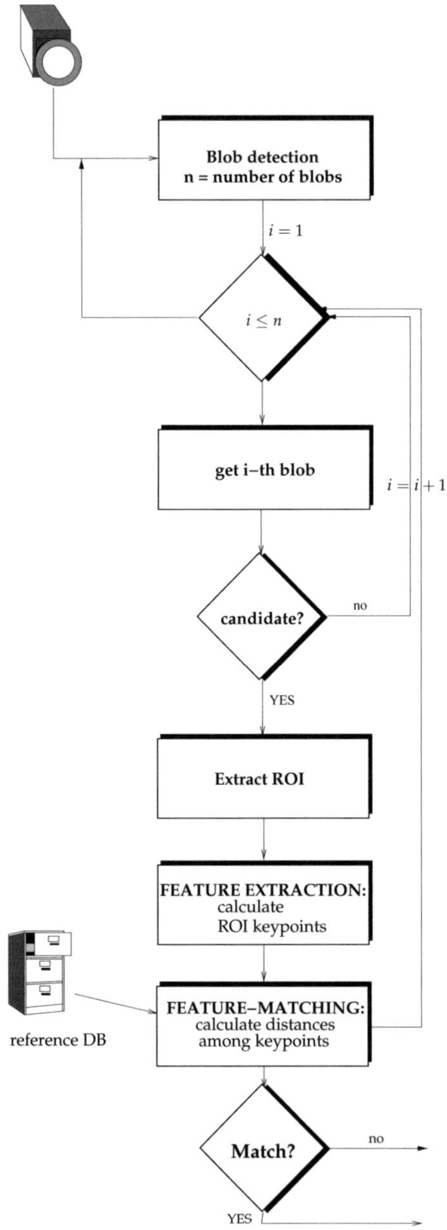


Fig. 6. Flow diagram of the new algorithm

Although it is not the case for the problem stated in this paper, the algorithm is drawn in a general form where more than one object can be searched in an image and hence, more than one ROI should be extracted and afterwards, processed for identification purposes.

### 5.1 Application to handle identification: results

Figure 7 shows examples of the result of extracting the ROI for both handle types. Obviously, the rectangular handle identification problem is more difficult due to the non symmetrical property of the handle itself and the featured blades where they are located.

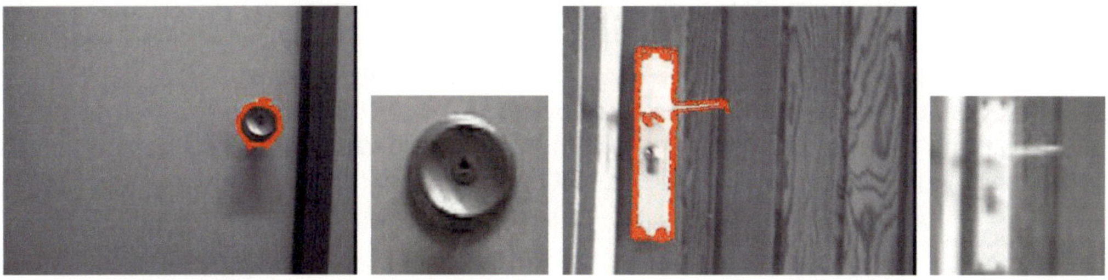

(a) Circular handle: blob (b) Circular handle: ROI (c) Rectangular handle: blob (d) Rectangular handle: ROI

Fig. 7. Blob extraction and ROI scaling

Table 4-a) shows the improvement obtained applying the method proposed in the paper. Again, USURF seems to be the most adequate feature extraction method for circular handles, whereas SIFT is clearly the best approach to be applied to the obtained ROI when non circular handles must be identified. The accuracy is increased in almost 30% for SIFT and about $20-22\%$ for SURF and USURF respectively for the case of circular handles. The improvement is better for rectangular handles, with an increase of 35% for SIFT and about 40% for SURF. Notice that this improvement is also reflected in the F1 measure, proving the adequateness of the methodology.

| | Circular | | | Rectangular | | |
|---|---|---|---|---|---|---|
| | acc. | size | F1 | acc. | size | F1 |
| SIFT | 91.82 | 150 | 0.863 | **91.52** | 80 | **0.669** |
| SURF | 92.94 | 100 | 0.890 | 87.77 | 40 | 40.6 |
| USURF | **94.35** | 150 | **0.911** | 87.77 | 40 | 0.406 |

a) Region labelling + feature extraction

| | Circular | | | Rectangular | | |
|---|---|---|---|---|---|---|
| | acc. | size | F1 | acc. | size | F1 |
| SIFT | 94.48 | 240 | 0.909 | **92.67** | 80 | **0.699** |
| SURF | 95.70 | 80 | 0.931 | 89.87 | 100 | 66.84 |
| USURF | **96.09** | 150 | **0.936** | 89.71 | 100 | 0.666 |

b) Region labelling + feature extraction + MR

Table 4. Best results

As mentioned in (Pfeifer and Bongard, 2006), vision may take advantage of the physical interaction of the agent with its environment. Taking into account the robot's morphology and the environmental niche, more specifically, the height at which the camera is mounted on the robot, and the height at which the handles are located on the doors, the handles should always appear at a specific height on the image. The improvement introduced by this *morphological restriction* (MR) is also showed in Table 4-b).

## 6. Discussion

Figures 8 and 9 show the evolution of the performance, for circular and rectangular handles. Both, the classification accuracy and the F1 measure are plotted when increasing the ROI size. These figures correspond to the proposed two-step method together with the morphological restriction previously mentioned. It can be appreciated how USURF is the best approach for circular handles and, on the contrary, SIFT improves speeded-up variants when rectangular handles need to be detected, even for larger ROI sizes.

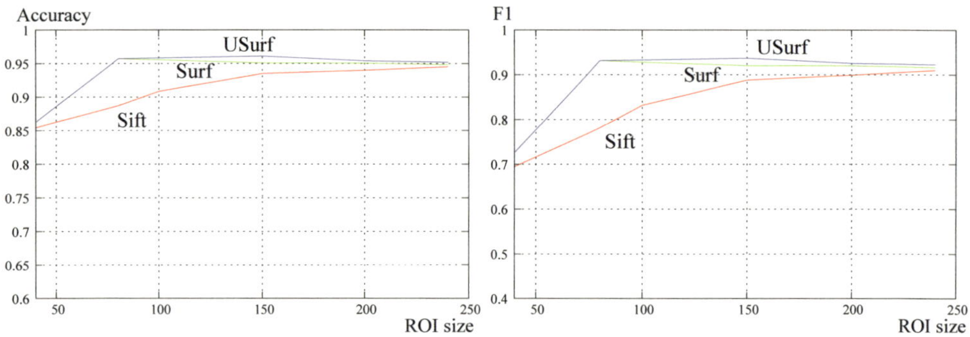


(a) Accuracy (b) F1

Fig. 8. Performance on circular handles while varying the ROI size

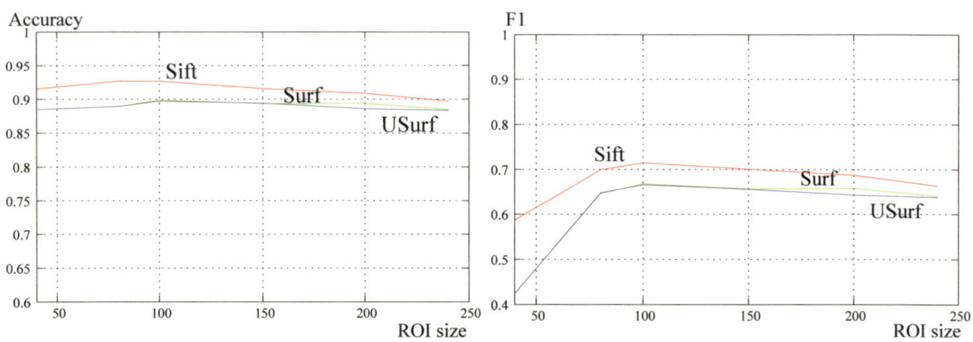


(a) Accuracy (b) F1

Fig. 9. Performance on rectangular handles while varying the ROI size

Performances are comparable to those obtained with the three-stage approach.
Table 5 reflects how the average number of keypoints increases together with the ROI size. These values were calculated using the reference databases used in each experimental trial

and hence, are only tentative values. Although the number of keypoints increases together with the ROI size, stable keypoints are found on small image sizes. Increasing the scale of the ROI gives a higher number of keypoints but the repeatability of these new keypoints does not seem good.

| Ref. DB | 40 | 80 | 100 | 150 | 200 | 320 × 240 |
|---|---|---|---|---|---|---|
| Circular | 10.09 | 21.43 | 41.5 | 74.03 | 105.53 | 49.28 |
| Rectangular | 5.95 | 23.47 | 38.79 | 72.10 | 97.78 | 139.89 |

Table 5. Average number of keypoints

Table 6 shows the average time needed to process a single image using SIFT (blob location, keypoint extraction and matching against the reference database). Note that the time needed is sensibly higher for larger ROI sizes due to the larger number of keypoints that occur in those images. Moreover, the computational requirements for processing a single image is larger than the 20 fps attainable with the three-stage process; the same frequency can be reached only using ROI size of 40.

| Database | 40 | 80 | 100 | 150 | 200 | No ROI |
|---|---|---|---|---|---|---|
| Circular | 0.06 | 0.18 | 0.22 | 0.46 | 0.75 | 0.31 |
| Rectangular | 0.06 | 0.13 | 0.17 | 0.42 | 0.58 | 0.89 |

Table 6. Computational payload (s) for processing an image using SIFT

To summarise, although the off-line experimental step showed a degraded accuracy for the ROI of size 40 extracted using SIFT for circular handles, the short time needed to compute the identification and the better performance of SIFT for the rectangular handle identification problem that proved more difficult, makes it more appealing for the real time problem stated in this paper.

## 7. Experiments with the robot

*Tartalo* is a PeopleBot robot from MobileRobots, provided with a Canon VCC5 monocular PTZ vision system, a Sick Laser, several sonars and bumpers and some other sensors. Player-Stage (Gerkey et al., 2003) is used to communicate with the different devices and the software to implement the control architecture is SORGIN (Astigarraga et al., 2003), a specially designed framework that facilitates behaviour definition. To evaluate the robustness of the handle identification system developed, it has been integrated in a behaviour-based control architecture that allows the robot to travel across corridors without bumping into obstacles. When the robot finds a door, it stops, turns to face the door and knocks it with its bumpers a couple of times asking for the door to be opened and waiting for someone to open it. If after a certain time the door is still closed, Tartalo turns back to face the corridor and carries on, looking for a new handle. On the contrary, if someone opens the door the robot detects the opening with its laser and activates a door crossing behaviour module that allows it to enter the room. All the computation is carried out in its on-board Pentium (1.6GHz).

The overall control architecture is composed of nine threads that communicate and activate/deactivate each other when appropriate (see Figure 10). These nine threads can be grouped as follows:

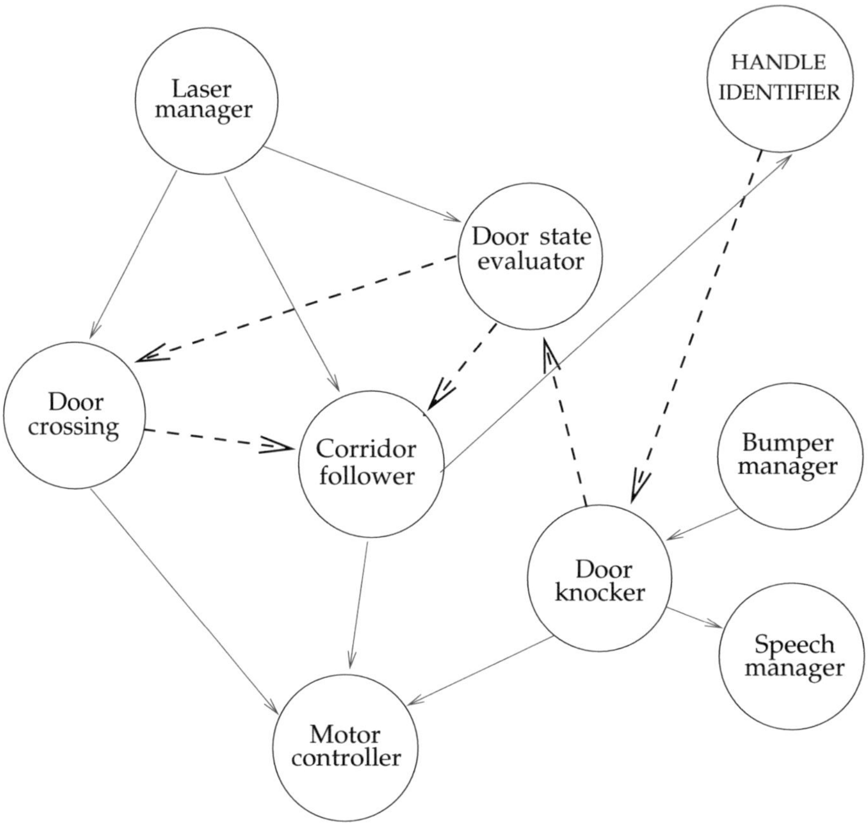


Fig. 10. Behaviour modules and communication

- Four device managers: these modules are not behaviour modules themselves, but they are needed to get/send up-to-date data from/to physical devices. These managers are needed for laser, bumper, speech and motor control.
- Three local navigation strategies: *corridor following*, *door crossing* and *door knocker* behaviours.

  The *corridor following* behaviour represents the main local navigation strategy and its aim is to balance free space on its sides, is such a way that corridors are travelled in smooth trajectories, while avoiding obstacles at the same time.
- Two landmark identification subsystems: *handle identification* and *door state evaluator* subsystems. This last module calculates the difference between the reading taken just after the robot stops for the waiting period and the current reading, and decides if the door has been opened according to that difference.

Dashed lines in Fig.10 represent behaviour activation/deactivation links, while thin links are data communication connections among modules. Only most relevant connections are drawn for sake of clarity.
As mentioned previously, experiments within the real robot/environment system were performed using a ROI size of 40 and applying the SIFT feature extraction method. Also, to make the behaviour more robust, instead of relying on a single image classification, the robot will base its decision upon the sum of the descriptor matches accumulated for the last five consecutive images. Experiments were carried out in three different environments.

### 7.1 Environment 1: circular handles

Figure 11 shows the environment together with the evolution of the sum of the matching keypoints over time. The horizontal line represents the value at which the threshold was fixed. The 18 doors present in the environment were properly identified and no false positives occurred.

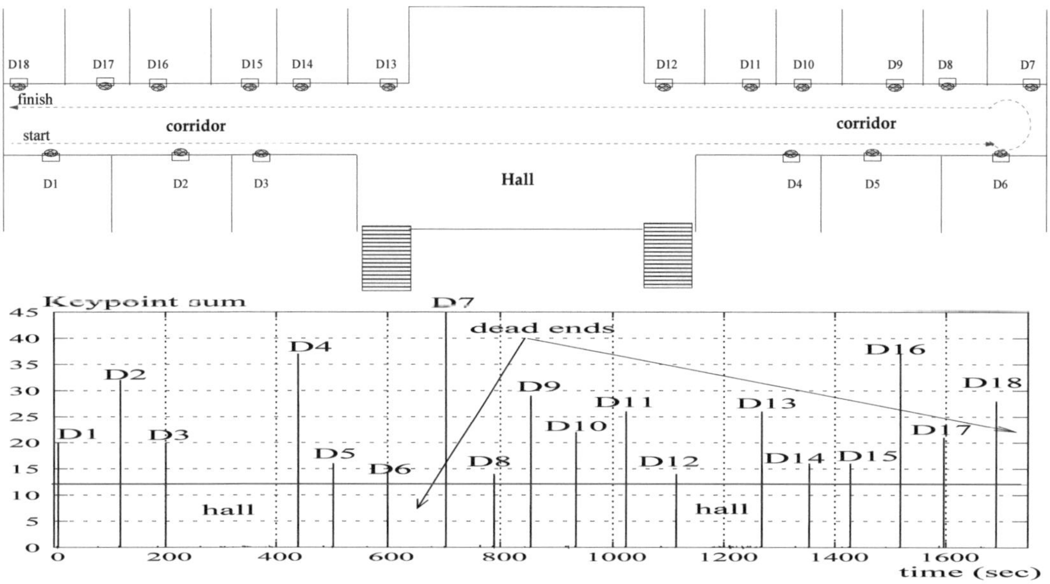


Fig. 11. First floor results

### 7.2 Environment 2: rectangular handles

Second floor of the building, lower corridor. 39 rectangular handles were consecutively to be identified. Figure 12 shows the original environment and the evolution of the keypoint sum over time.
The robot started on the left side of the corridor, with its camera pointing to its right and travelled all the way along the corridor successfully fulfilling the sequence of 6+7+7 handle identification, and then turned at the dead end. In its way back, only one of the handles of the central sector of the corridor, the one marked with a circle in the upper image of figure 12, was not recognised (6+6+6) and again, no false positives were given. Hence, a success ratio of 0.97 was achieved.

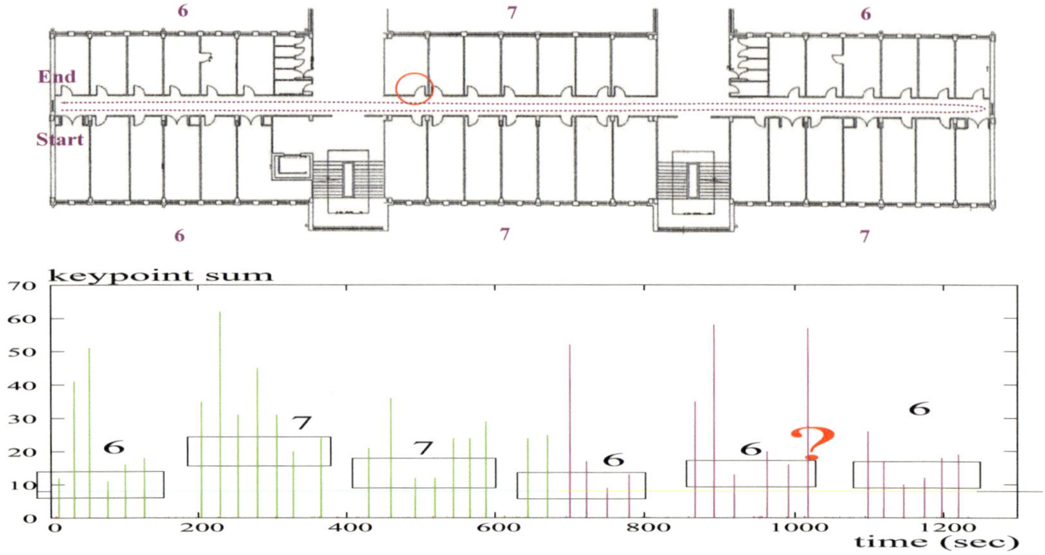


Fig. 12. Second floor results

### 7.3 Environment 3: mixed identification

Mixed handle identification over time. Experiments were performed on a part of the third floor of the faculty (figure 13). Three circular handles and three rectangular handles were to be identified. The robot was left running for three rounds in the corridor.

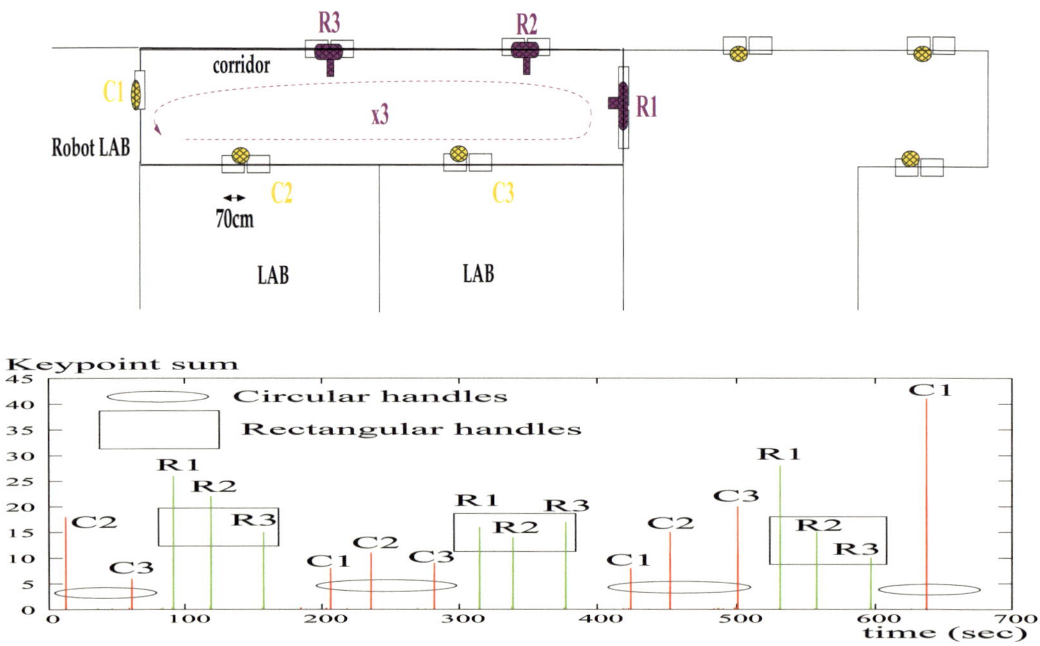


Fig. 13. Mixed environment

In each round, three circular handles and three rectangular ones were positively identified and no false positive was given by the system. The total amount of handles correctly recognised was 18.

Summing up, during the experiments performed within the real robot/environment system, 74 handles were identified, achieving a success of 98.66% and not a false positive occurred.

## 8. Conclusions and further work

The chapter summarises several approaches aimed at identifying doors for robot navigation purposes. Environmental specific methods showed a good performance. However, the segmentation based approach is not easily applicable to other environments since new segmenters need to be developed. Feature extraction methods were easy to apply because they only required building a reference database containing the objects to be identified. However, the classification performance and the computational payload needed to process a single image make these methods unsuitable to be used during robot navigation. A new two step algorithm was presented based on feature extraction that aimed at improving the extracted features to reduce the superfluous keypoints to be compared at the same time that it increased its efficiency by improving accuracy and reducing the computational time.

ROI extraction improves handle identification procedure and depending on the ROI size, the computational time to classify an image can be considerably reduced. The system showed a very low tendency to give false positives while providing a robust identification. The authors' opinion is that the presented approach is appropriate for situations where the background is so varied that its characteristics are either irrelevant for object identification or just add noise to the recognition process.

The area of mobile robotics needs general methods for several reasons. The use of algorithms specifically designed to work in a concrete environment makes the comparison among methods difficult as well as the application of the approach in different robot/environment systems. The developed system outperforms the performances obtained without extracting the ROIs, and experiments carried out in a real robot-environment system showed the adequateness of the approach. Opposite to the segmentation based method, the two-step method can easily be generalised to other types of handles. The same blob extraction method could be applied to obtain the region of interest, adapting the reference database to contain the correct reference keypoints and, adjusting the size restrictions applied to the region detector results –which depend on the distance the images are taken from by the robot.

Currently, the aim is to extend the proposed method to other applications such as face recognition and road signal identification. Still, the keypoint matching criteria has to be analysed more deeply. More sophisticated and efficient algorithms remain to be tested and the performance of different distance measures still needs to be studied. Obviously, the proposed method is opened to any other feature extraction method and to any improvements that could be made.

**Aknowledgements**

This work has been supported by the Basque Country Government under Research Team Financiation, and by the ROBAUCO Ministery project.

## 9. References


Astigarraga, A., Lazkano, E., Rañó, I., Sierra, B., and Zarautz, I. (2003). SORGIN: a software framework for behavior control implementation. In *CSCS14*, volume 1, pages 243–248.

Ayala-Ramírez, V., Garcia-Capulin, C., Perez-Garcia, A., and Sanchez-Yanez, R. R. (2006). Circle detection on images using genetic algorithms. *Pattern Recognition Letters*, 27(6):652–657.

Bay, H., Ess, A., Tuytelaars, T., and Gool, L. V. (2008). SURF: Speeded up robust features. *Computer Vision and Image Understanding*, 110(3):346–359.

Bay, H., Tuytelaars, T., and Gool, L. V. (2006). SURF: Speeded up robust features. In *Proceedings of the 9th European Conference on Computer Vision*.

Brooks, R. A. (1986). A robust layered control system for a mobile robot. *IEEE Journal of robotics and automation*, RA–26:14–23.

Chen, S.-H. (1996). Elements of information theory : Tomas m. cover and joy a. thomas, (john wiley & sons, new york, ny, 1991). *Journal of Economic Dynamics and Control*, 20(5):819–824.

Duda, R. and Hart, P. E. (1972). Use of Hough transform to detect lines and curves in pictures. *Communications of the ACM*, 15(1):11–15.

Eberset, C., Andersson, M., and Christensen, H. I. (2000). Vision-based door-traversal for autonomous mobile robots. In *Proceedings of the IEEE/RSJ International Conference on Intelligent Robots and Systems*, pages 620–625.

Gerkey, B. P., Vaughan, R. T., and Howard, A. (2003). The Player/Stage project: tools for multi-robot and distributed sensor systems. In *Proc. of the International Conference on Advanced Robotics (ICAR)*, pages 317–323.

Gil, A., Reinoso, O., Vicente, A., Fernández, C., and Payá, L. (2005). Monte Carlo localization using SIFT features. *Lecture Notes in Computer Science*, 3522:623–630.

Gonzalez, R. C. and Woods, R. E. (1993). *Digital image processing*. Addison Wesley.

Horn, B. K. P. (1986). *Robot Vision*. MIT Press.

Kimme, C., Ballard, D., and Sklansky, J. (1975). Finding circles by array accumulators. *Communications of the ACM*, 18(2):120–122.

Kragic, D., Petersson, L., and Christensen, H. I. (2002). Visually guided manipulation tasks. *Robotics and Autonomous Systems*, 40(2-3):193–203.

Lazkano, E., Sierra, B., Astigarraga, A., and Martínez-Otzeta, J. M. (2007). On the use of bayesian networks to develop behavior for mobile robots. *Robotics and Autonomous Systems*, 55(3):253–265.

Ledwich, L. and Williams, S. (2004). Reduced sift features for image retrieval and indoor localisation. In *Australian Conference on Robotics and Automation*.

Li, W., Christensen, H. I., and Orebäck, A. (2004). An architecture for indoor navigation. In *Proceedings of the IEEE International Conference on Robotics and Automation*, pages 1783–1788.

Lowe, D. G. (2004). Distinctive image features from scale-invariant keypoints. *International Journal of Computer Vision*, 60(2):91–100.

Mallot, H. A. and Franz, M. A. (2000). Biomimetic robot navigation. *Robotics and Autonomous System*, 30:133–153.

Mikolajczyk, K. and Schmid, C. (2005). A performance evaluation of local descriptors. *IEEE Transactions on Pattern Analysis and Machine Intelligence*, 27(10):1615–1630.

Monasterio, I., Lazkano, E., Rañó, I., and Sierra, B. (2002). Learning to traverse doors using visual information. *Mathematics and Computers in Simulation*, 60:347–356.

Muñoz-Salinas, R., Aguirre, E., and García-Silvente, M. (2005). Detection of doors using a genetic visual fuzzy system for mobile robots. Technical report, University of Granada.

Murthy, S. K., Kasif, S., and Salzberg, S. (1994). A system for induction of oblique decision trees. *Journal of Artificial Intelligence Research*, 2:1–33. ftp://blaze.cs.jhu.edu/pub/ocl.

Pfeifer, R. and Bongard, J. (2006). *How the body shapes the way we think. A new view of intelligence.* MIT Press.

Primdahl, K., Katz, I., Feinstein, O., Mok, Y., Dahlkamp, H., Stavens, D., Montemerlo, M., and Thrun, S. (2005). Change detection from multiple camera images extended to non-stationary cameras. In *Proceedings of Field and Service Robotics*, Port Douglas, Australia.

Roduner, C. and Rohs, M. (2006). Practical issues in physical sign recognition with mobile devices. In Strang, T., Cahill, V., and Quigley, A., editors, *Pervasive 2006 Workshop Proceedings (Workshop on Pervasive Mobile Interaction Devices, PERMID 2006)*, pages 297–304, Dublin, Ireland.

Se, S., Lowe, D. G., and Little, J. (2002). Mobile robot localization and mapping with uncertainty using scale-invariant visual landmarks. *International Journal of Robotics Research*, 21(9):735–758.

Seo, M. W., Kim, Y. J., and Lim, M. T. (2005). *LNAI*, chapter Door Traversing for a Vision Based Mobile Robot using PCA, pages 525–531. Springer-Verlag.

Skarbek, W. and Koschan, A. (1994). Colour image segmentation – a survey. Technical report, Polish Academy of Sciences.

Smith, S. M. and Brady, J. M. (1997). SUSAN: a new approach for low level image processing. *International Journal of Computer Vision*, 23(1):45–78.

Stella, E., Cirelli, G., Attolico, G., and Distante, A. (1996). Sauro: An autonomous mobile vehicle for indoor environment. In *MELECON*, volume 2, pages 1145–1150.

Tamimi, H., Halawani, A., Burkhardt, H., and Zell, A. (2006). Appearance-based localization of mobile robots using local integral invariants. In *In Proc. of the 9th International Conference on Intelligent Autonomous Systems (IAS-9)*, pages 181–188, Tokyo, Japan.

Trullier, O., Wiener, S. I., Berthoz, A., and Meyer, J. A. (1997). Biologically-based artificial navigation systems: review and prospects. *Progress in Neurobiology*, 51:483–544.

Ye, W. and Zhong, Z. (2007). Robust people counting in crowded environment. In *Proceedings of the 2007 IEEE International Conference on Robotics and Biomimetics*, pages 1133–1137.

Yuen, H. K., Princen, J., Illingworth, J., and Kittler, J. (1990). Comparative study of Hough transform methods for circle finding. *Image and Vision Computing*, 8(1):71–77.

# 13

# Path Planning and Execution for Planetary Exploration Rovers based on 3D Mapping

Andres Mora, Keiji Nagatani and Kazuya Yoshida
*Space Robotics Laboratory, Department of Aerospace Engineering, Tohoku University*
*Japan*

## 1. Introduction

The unmanned exploration of the Moon has steadily increased in the past years due to the renewed interest in creating a permanent human settlement on our natural satellite. After detailed remote sensing from orbiters, the focus is now shifting onto autonomous landing and surface locomotion by robotic devices on and around a specific area of highest interest. Especially, robotic exploration is a precursor for future human settlements on the Moon and in-situ resource utilization (ISRU) is one important aspect in the selection process of candidate locations for such settlements. In the early 1960's, Watson (Watson et al., 1961) estimated that it would be possible to find deposits of ice caps at the bottom of craters located at the Moon's polar regions. He argued that the shadows produced on these craters due to the very small deviation of the Moon's equatorial plane position with respect to the Sun create an environment that would present the proper conditions to retain such deposits.
Robotic exploration of these regions becomes a vital initial step on the way to build a permanent base of operations for humans to live on extended periods. The navigation of a vehicle on the Moon is presented with a series of issues such as the trafficability over the lunar soil (called "regolith"), the irregularity of the topography and the poor illumination at the Polar Regions, especially at the craters. In this chapter, attention is centered on the irregularity of the topography and the poor illumination issues. Given that we assume an environment with a very low angle of illumination which results in large shadowed areas, we propose the use of Light Detection and Ranging (LIDAR) systems to perceive the surroundings of the vehicle. These systems are not impaired by the lighting conditions assumed and have been successfully utilized in several outdoor mobile robotic applications (Langer et al., 2000),(Morales et al., 2008),(Skrzypczński, 2008).
The features of the topography of the surface of the Moon may translate as obstacles for the navigation of the vehicle and also may limit the visual range of the exteroceptive sensors at a given location. These obstacles cause an "occluding" effect on the readings of the sensors therefore decreasing the size of the perceivable area.
Various researchers have approached the occlusion problem presented on indoor and outdoor autonomous navigation of mobile robotic systems. In (Dupuis et al., 2005),(Rekleitis et al., 2009), an irregular triangulated mesh is created based on a LIDAR's sensory data, and then "filtered" in order to find "shadows" or occluding obstacles and eliminate them from the map. A description of the problems that occluding obstacles may present in teleoperated navigation is presented in (Kunii & Kubota, 2006). In (Heckman et al., 2007), a method to classify different voxels based on its location with respect to an occluding obstacle is given, whereas

(Kunii & Kubota, 2006) presents a method to determine the position of a planetary rover by matching a set of 3D maps. An algorithm to calculate the shortest path between two given points within a completely known workspace implementing an art gallery method which selects guards and then connect them is presented in (Danner & Kavraki, 2000).
In this chapter, we present a path planning and navigation system based on various components introduced as follows. The perception of the environment surrounding a mobile robot through LIDAR technology. The recognition of occluded areas in a local map, the selection of a position within the boundaries of the environment known to the vehicle which maximizes the information gained once the vehicle reaches it, a generated path from the initial location of the vehicle to the selected goal and the control procedure implemented to navigate through the environment while following the generated local path.

## 2. Robotic Planetary Exploration

In the last decade the number of robotic space missions whose main goal is the exploration of our Moon and other planets has increased. These missions can be divided into three main phases: *Orbital exploration*, in which an orbiter conducts remote sensing around a target body in order to acquire "global" characteristics of the body such as terrain, geomagnetic, internal constituent data, etc. The second phase is called *Surface exploration with a robotic probe*, where a robotic probe is deployed on the targeted planet to conduct more investigations regarding the surface/internal characteristics of the surveyed planet. In this phase a mobile robotic probe provides a wider range of locations than that of a lander from where different scientific tests can be carried out. The final phase is known as *Sample return*. In this phase of the mission, the samples gathered by the robotic probe are delivered back to the Earth for further analysis. In this chapter, we focus on the second phase of this mission categorization.
Due to the broader working field that a mobile robotic probe (or rover) provides to the numerous scientific instruments onboard, this particular type of probe has been implemented successfully in various planetary exploration missions. The first rover to ever land on and traverse the surface of another celestial body was the Soviet Union's Lunokhod 1 followed by the Lunokhod 2 which reached our Moon and and combined covered up to 47[km] of its surface. Their main tasks were to obtain geological data as well as detailed panoramic views of the surface of the Moon. The first space exploration rover to successfully reach another planet was NASA's Mars Path Finder (Sojourner). The Sojourner was a six-wheeled vehicle weighting 10.6[kg] that traveled about 500[m] from its lander. During its operation, it sent 550 photographs of scenery and terrain close-ups and it analyzed the chemical properties of sixteen locations near the lander. The most recent sucessfully rover exploration missions have been those carried out by the Mars Exploration Rovers (MER) Spirit and Opportunity. These twin mobile robots landed on January 2004, and up to January 2009 they were still engaging their ongoing robotic missions of exploring the Martian surface and geology. Similarly to the Sojourner, the MERs are six-wheeled vehicles but much larger in size than their predecesor with a weight of 180[kg].

### 2.1 The Moon's robotic exploration

Most of the scientific experiments performed by robotic missions (satellites, landers and rovers) on the Moon have dealt with mapping its surface, understanding its soil components and several other geological characteristics. We present here some of the characteristics of our natural satellite and the issues a mobile robotic probe would encounter there.

Figure 1. Mobile robotic testbed implemented: El Dorado-II

The distance between the Earth and our natural satellite is 384,403[km], being this the average center-to-center distance due to the change produced by the orbit of the Moon around the Earth. The diameter of the Moon is 3,474[km], its mass is approximately $7.3477 \times 10^{22}$[kg] and it has a gravity at the equator of 1.622[m/s$^2$] which allows for a relatively easy escape velocity of 2.38[km/s]. One of the features of the Moon the naked eye is able to look at is its craters.
The comets and interplanetary materials that continuously hit the lunar surface are believed to carry a certain amount of water to the lunar surface. Upon impact against the regolith (dusty, upper layer of the surface), the water would then split into its constituent elements which subsequently, would be liberated to space. One theory (Watson et al., 1961) states, however, that due to the small 1.5 [deg] deviation of the axial tilt of the Moon's spin axis to the ecliptic plane, craters located near or at the poles never receive sunlight thus creating a permanent "shadow" over large areas of these polar craters. Inside these shadows, extremely low temperature remain constant allowing the otherwise escaping water constituent elements to be trapped at the bottom of the craters.

## 3. Proposed Scenario

Water and mineral prospecting missions on the surface of the Moon and inside its polar craters are one the interest of major national space agencies around the world. In order to accomplish such task, an initial approach is to send an unmanned robotic mission to our natural satellite and upon arrival initiate its reconnaissance operation. This paper assumes a mission on which a mobile robotic vehicle traverses an area starting at a base close to the landing site and that extends to the locations of interest. The Japanese SELenological and ENgineer Explorer (SELENE) was launched in September 2007 having as main mission objectives: (1) gather information about the lunar geological history and (2) obtain the topographic data required to generate a complete representation of its surface (Kato et al., 2007). The data obtained from the lunar surface can be treated in such manner that a Digital Elevation Map (DEM) can be generated. SELENE's payload allows it to obtain terrain data with a resolution of 10[m] per pixel and a height detection of objects with a minimum height of 5[m]. This map could work as a global map and it would correspond to the large areas surrounding the mobile robot.
The illumination conditions in the vecinity of the Moon's polar craters and at its inner walls is the main reason why the authors assume the usage of a Light Detection and Raging (LIDAR)

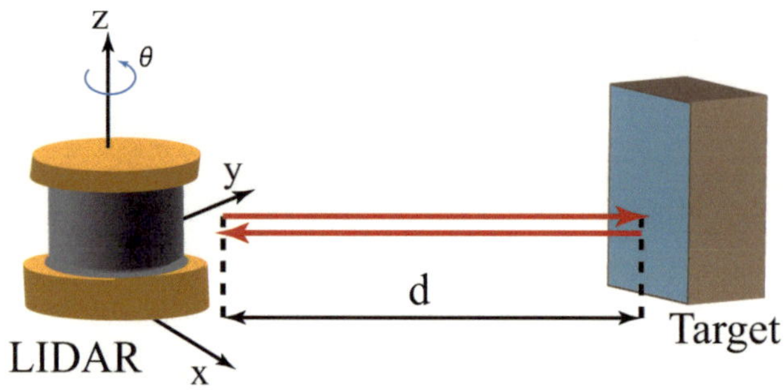


Fig. 2. Time of Flight Concept of laser range sensors. Disparity is avoided by the coaxial characteristic of the emitted and received laser beam.

sensory system on board of the vehicle to gather the information required to build a local map corresponding to the smaller, more detailed areas immediately surrounding the vehicle. The range of compact sensory systems typically goes from 5[m] to up 80[m] and with angular resolutions as small as of 0.25[deg].
In this research, we also assume a wheeled-rover as the mobile robot platform that will navigate from a lunar base to the polar lunar crater. The robotic system used in this research as a model for simulation purposes is a four wheeled-rover called "El Dorado II" and it is shown in Fig. 1. This rover is 0.75[m] long (measured between front and rear axles), 0.56[m] wide (measured from the outer side of its left wheels to the outer side of its right wheels). It is 0.50[m] high without any equipment on it and 0.70[m] when it has a laser range sensor mounted on it.

## 4. Mapping based on 3D sensing

A laser range sensor is a device that can obtain range measurements through the use of one of two methods: the "Time-of-Flight" (TOF) principle and the determination of the phase-shift between an emitted continuous wave and its reflection (Adams, 1999) on a target's surface.

### 4.1 LIDAR principle

The laser range sensor used through out this research, a Hokuyo Top-URG UTM-30LX/LN (Hokuyo, 2008), has being designed under the Time-of-Flight principle. With this method, a laser beam is fired towards an opaque object in order to determine the distance separating the device and the targeted object. It then measures the time taken by the pulse to be reflected off the target and return to the sender. Through the use of elementary physics, the distance, $d$, covered in a round-trip is determined by multiplying the velocity, $v$, of the energy wave by the time, $t$, required to travel that distance, $d = v \times t$.
The measured time is representative of travelling twice the separation distance (round-trip), and must therefore be reduced by half to result in the actual range to the target (Everett, 1995). With this in mind, one can deduce that the velocity, $v$, is equal to the speed of light, $c$, which is considered to be approximately $3 \times 10^8$[m/s]. The time that it takes to reach the measured point is then $d = \frac{c \times t}{2}$.
In order to take advantage of this principle and to obtain measurements with resolutions in the order of millimeters, it is necessary for the sensor to have a timing circuitry capable of picosecond accuracy. Although nowadays these circuits have reduced their price, various

Fig. 3. Scene taken from an outdoors location using a laser range sensor.

other methods have been designed as a way around to overcome this financial issue. A review of the methods that can accomplish such level of accuracy is presented in (Kalisz, 2004).
Laser range sensors are used extensively in 3D object recognition, 3D object modelling, and a wide variety of computer vision related fields. The principle of this technology, which constitutes the heart of the so-called TOF 3D scanners, is shown in Fig. 2, where the inertial coordinate frame is inside the scanner. Laser range sensors offer high-precision scanning abilities, with either single-face or 360[deg] scanning modes.
Sensing systems based on pairs of cameras, or also known as stereo cameras, suffer of a problem caused due to its geometric configuration: disparity. The problem of disparity is produced by the distance separating the two cameras on the $x$-axis. This distance between the cameras generates some overlapping of the same object being registered by both cameras which can be surmounted by a number of calibration algorithms (Lucas & Kanade, 1981),(Barnea & Silverman, 1972), (Moravec, 1979), (Sanger, 1988) which eliminate the disparity from one camera's registered object to the other. These procedures are well known in the field of robotics and are of common use, nevertheless they are time and computation power consuming. Laser range sensors eliminate the disparity problem inherent in the stereo cameras systems by keeping the transmitted and received beams coaxial, as can be seen from Fig. 2.
The values a laser range sensor outputs are sets of locii of points given in Polar coordinates, $(\theta,\phi,\rho)$. Each scan performed by the sensor is a plane defined by each of those coordinates. That is, a plane with an initial value of $\theta$, $\theta_{init}$, equal to 0[deg], is defined by the initial value of $\phi$ (for example, $\phi_{init} = 0$[deg] and its corresponding value of $\rho$) and it ends at the final value of $\phi$ (for example $\phi_{final} = 180$[deg]) with its corresponding $\rho$. The sensing information of a given area will be completed when the final value of $\theta$ (for example $\theta_{fin} = 180$[deg]) has been reached.
For computations using these locus of points at each plane, it is necessary to transform the polar coordinates of each point $(\theta,\phi,\rho)$, into its corresponding set of cartesian coordinates $(x,y,z)$. This transformation is done for every point sensed by the device in order to form what is known as a "cloud of points".

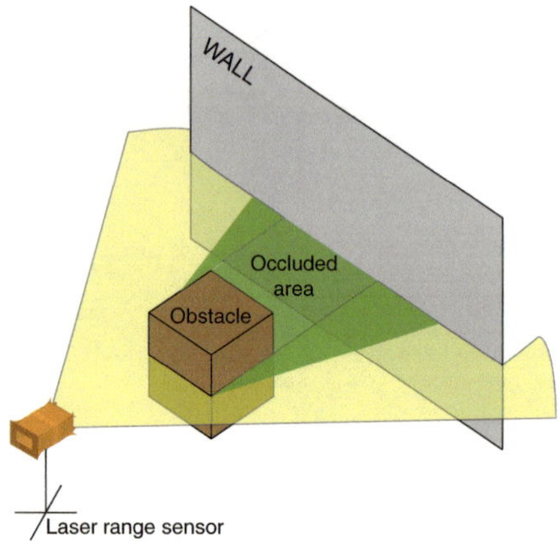


Fig. 4. Occlusion effect conceptual diagram.

After this coordinate transformation, the generated cloud of points can be represented through a graphical user interface (GUI) which can allow a human user to observe the information of the environment retrieved by the sensor. Depending on the range of the sensor, large, very detailed areas can be mapped with a single scanning procedure. An example of such cloud points representation can be seen in Fig. 3. A cloud of points is one of the multiple representations that can be produced from the raw sensor's data, other representations include rectangular meshes, quad-tree representations and triangular grids.

### 4.2 Occlusion problem

Exteroceptive sensors such as LIDAR-based devices and stereo cameras are capable to retrieve the distance that separates them from an object within its field of view. However, these sensors are not able to "look" behind the objects they perceive. If for example, a laser range sensor is placed in front of a wall and an object moves from one end of the wall to the other, the sensor would register the object as it passes but it would also loose information about the area of the wall immediately behind the object. The effect that the passing object produces onto the laser range sensor while it registers the distance to the wall is known as *occlusion*.

Consider the static environment presented in Fig. 4. The laser range sensor is placed facing a wall and an object is located between the sensor and the wall. Assuming a single plane sweeping action, when the sensor's laser beam reaches the object a discontinuity from the wall reading spans until the occluding object has been completely registered. The occlusion caused by the object can be thought as the shadow the object projects onto the wall once a source of light is set upon the object. When a mobile robot constructs maps in a cluttered environment or in large irregular outdoor environments, the occlusion effect must be considered since the environment information lost in the occluded areas of the map may present hazards for the vehicle such as undetected holes, rocks, etc. A procedure proposed in this chapter to alleviate the occlusion effect in a given map is discussed in Subsection 5.2.

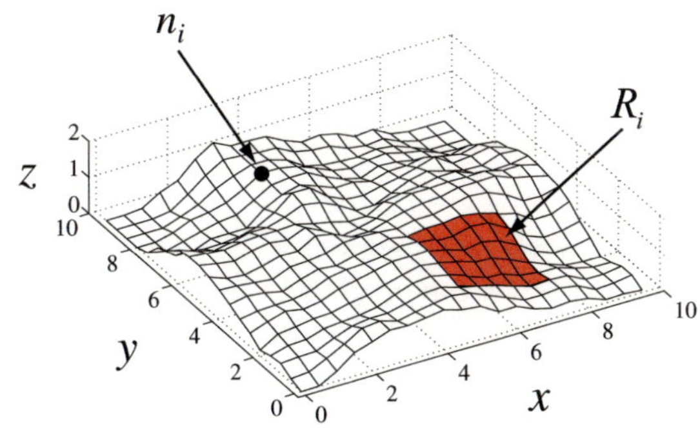


Fig. 5. A Digital Elevation Map as a discrete representation of a cloud of points.

### 4.3 Digital elevation map

The map representation implemented in this research is a commonly used method to display a cloud of points as a set of heights, $z$, correspondent to a coordinate pair $(x,y)$. This representation is known as a Digital Elevation Map (DEM) as previously introduced in Section 3. A digital elevation map can be considered as a representation of the topography of any given surface in digital format, that is, by coordinates and numerical descriptions of the value of every $z$ correspondent to a pair $(x,y)$. A DEM can be defined by Equation 1;

$$\begin{cases} x = l_{grid} \cdot i & (for\ i = 0,1,2,...,X_{max} - 1) \\ y = l_{grid} \cdot j & (for\ j = 0,1,2,...,Y_{max} - 1) \\ z = (x_i, y_j) \end{cases} \quad (1)$$

In Equation 1, $X_{max}$ and $Y_{max}$ define the size of the grid, $l_{grid}$ defines the length of the smallest possible grid, and the pair coordinate $(x_i,\ y_j)$ defines the value of $z$ for every iteration $i$ and $j$. An example of such representation is given in Fig. 5.

Although ranging measuring systems based on the TOF principle have high-precision measurement characteristics, some potential errors can originate due to various factors, such as (Everett, 1995), (Borenstein et al., 1996):

- Uncertainties in determining the exact time of arrival of the reflected pulse
- Inaccuracies in the timing circuitry used to measure the round-trip time-of-flight
- Interaction of the incident wave with the target surface

these and other sources of errors in the ranging measurement of the laser range sensors have been the motivation to investigate additional characteristics of such sensors (Pascoal et al., 2008), (Ye & Borenstein, 2002) and also to find a model that can describe the likelihood that a given measurement obtained from these devices can be relied as true and not faulty measurements. This analysis is not within the scope of this chapter and shall not be considered. All measurements obtained from the implemented laser range sensor are assumed to be accurate.

## 5. Map Representations

In Section 3, the assumption of having two representations of the surface of the Moon with different levels of detail was introduced. One of them having a wider or *global* coverage of a given area with a lower resolution and the other one covering a smaller or *local* area but with higher resolution.

The wider representation will be referred hereinafter as *Global DEM* or *GDEM*. Based on the imagery information obtained from a lunar orbiter mission such as SELENE, a GDEM can be constructed. This GDEM can be generated with a 10[m] × 10 [m] cell resolution and a 500 [m] × 500 [m] grid area and can be used as an input for a path planner that will generate a path to the specific position where the rover should go and perform the desired science tasks intended during the mission. This path planner is called "Global Path Planner"and will determine a candidate path to the specified goal, however, since the generated path is based on the information provided by the GDEM, the global path can only be used as a "guideline" since the resolution of the GDEM is not detailed enough for a rover with the dimensions presented earlier.

Once the rover leaves the lander, the vehicle will map the environment surrounding it and it will then traverse the targeted area. The map representing this area is called *Local DEM* or *LDEM*. Depending on the LIDAR system used, the resolution of the LDEM can be increased, in this chapter a cell resolution of 10[cm] × 10[cm] and a grid area of 20[m] × 20[m] is used. Throughout the use of this representation, a local path that considers various issues such as obstacle avoidance, visibility and distance to the global goal can be devised. This local path would be generated by what the authors called the "Local Path Planner". This local path planner will refine the path outlined by the global path planner by getting a better understanding of the local environment given by the local map. The destination where the robot will intend to move to (within its known local map) will be the position from where the vehicle would avoid occluded areas as much as possible in the next sensing or scanning position.

Although the GDEM is a an important element in this research, the forthcoming sections will focus on the LDEM, how it is obtained, its characteristics and the relationship it holds with the calculation of the next sensing position planning.

### 5.1 Local DEM

The LDEM's main purpose is to determine the workspace of the vehicle by recognizing its boundaries, help the vehicle perceive the presence of obstacles within the map and by doing it so provide the information required to calculate the free space and obstacle's space at a given map. Having determined a connected free space, the proposed planning strategy can be calculated within that specific region.

#### 5.1.1 Characterization of the Local DEM

Let $n_i$ represent a terrain elevation node of the LDEM as presented in Fig. 5. $i$ indicates the ubication of the node of the node $n_i$ in the LDEM and takes values from 0 to $N$, where $N$ is equal to the total number of nodes in the LDEM. Define the map space non-occluded $\chi_{noc}$ as the set of nodes $n_i$ of the LDEM that are non-occluded or visible with respect to the location of the laser range sensor and the map space occluded $\chi_{occ}$ as the set of nodes $n_i$ of the LDEM that are occluded or invisible with respect to the location of the laser range sensor. The union of the sets $\chi_{noc}$ and $\chi_{occ}$, $\{\chi_{noc} \cup \chi_{occ}\}$, contains all the nodes $n_i$ within a single LDEM. This union is named the map working space $\chi_{wspace}$ and can be defined as the semi-circle with a

radial boundary equivalent to the maximum range measurement of the LIDAR as depicted in Fig. 8.
$\chi_{noc}$ is formed by two subsets called the map space free $\chi_{free}$ and the map space obstacle $\chi_{obs}$. $\chi_{free}$ is defined as that subset of nodes $n_i$ of the $\chi_{wspace}$ that are not part of the subset of nodes contained in $\chi_{occ}$ and that are not part of the subset of nodes contained in the map space obstacle $\chi_{obs}$ as given by Equation 2;

$$\chi_{free} = \{n_i : n_i \in \chi_{wspace} | n_i \notin \chi_{obs} \text{ and } n_i \notin \chi_{occ}\}, 0 \leq i \leq N \quad (2)$$

$\chi_{obs}$ is defined as the set that contains all nodes $n_i$ of the $\chi_{wspace}$ as presented by Equation 3 and that present a problem to the trafficability of the vehicle.

$$\chi_{obs} = \left\{n_i : n_i \in \chi_{wspace} | n_i \notin \chi_{free} \text{ and } n_i \notin \chi_{occ}\right\}, 0 \leq i \leq N \quad (3)$$

Having defined the subsets $\chi_{free}$ and $\chi_{obs}$, the map space occluded $\chi_{occ}$ can be defined as Equation 4.

$$\chi_{occ} = \left\{n_i : n_i \in \chi_{wspace} | n_i \notin \chi_{free} \text{ and } n_i \notin \chi_{obs}\right\}, 0 \leq i \leq N \quad (4)$$

The relationship between the set $\chi_{wspace}$ and its three subsets of nodes $\chi_{free}$, $\chi_{obs}$ and $\chi_{occ}$ is given by Equation 5.

$$\chi_{wspace} = \chi_{free} \cup \chi_{obs} \cup \chi_{occ} \quad (5)$$

Finally, it can be noted that there is no intersection between any of the three subsets of the $\chi_{wspace}$ as given by $\chi_{free} \cap \chi_{obs} = \emptyset$, $\chi_{free} \cap \chi_{occ} = \emptyset$ and $\chi_{free} \cap \chi_{occ} = \emptyset$.

### 5.2 Visibility index

The visibility of an area of interest viewed from a node $n_i$, can be determined by the calculation of an index named by the authors the *visibility index*. This index returns the number of nodes that are within the field of view of the laser range sensor. The calculation of this index is given by Equation 6;

$$Vis = \frac{|\chi_{noc}|}{|\chi_{wspace}|} \quad (6)$$

where $Vis$ indicates the visibility index within the analyzed given LDEM, $|\chi_{noc}|$ is the cardinality or number of elements (nodes $n_i$) inside of $\chi_{noc}$ and $|\chi_{wspace}|$ is the cardinality or number of elements (nodes $n_i$) contained by the set $\chi_{wspace}$. In order to classify a given node as an element of the set given by $\chi_{free}$ or as an element of the set $\chi_{occ}$ and to quantify the number of nodes corresponding to each set, the author presents the following procedure Algorithm 1.
In line 2 a single virtual laser beam for each angle of the area to be scanned is traced in the form of a vector $\vec{L}_i$. Once this vector has been defined, in line 3 a unitary vector $\vec{u}_i$ is calculated based on the laser beam's vector $\vec{L}_i$. This unitary vector will have a magnitude equivalent to that of the resolution of the LDEM $d_{Res}$. Later on, the initial values of the variables of the number of segments $j$ within each $\vec{L}_i$ and the maximum value of the slope $S_{max}$ at any given iteration are set. The maximum number of segments per laser beam $\vec{L}_i$, is then calculated. The initial value of a vector that moves along the laser beam's vector $\vec{L}_i$ in equal displacements of $\vec{u}_i$ is set to be where the laser range sensor is located, in line 6.
The algorithm then calculates $\vec{r}_{end_j}$. This vector represents the location of a node $n_i$ within the $\chi_{wspace}$. Following this calculation, the height of the studied node within the actual $\vec{L}_i$ represented by $\vec{r}_{end_j}$ is compared to the height of the node where the laser range sensor is

located $\vec{p}_{org}$. If the slope of the current position is higher than the maximum slope so far, the node is determined to be a non-occluded node or an element of the set $\chi_{noc}$ and the maximum slope value is changed to the current position's one. Otherwise it is classified as an occluded node or an element of the set $\chi_{occ}$. The procedure is repeated until the final position of the virtual laser beam is reached and until all the perceivable area has been covered. Once the perceivable area has been covered, the visibility index is calculated and its value returned.

**Algorithm 1** Calculate the visibility index, *Vis*, in a given LDEM

**Input:** LDEM, nodes $n_i \in \chi_{wspace}$
**Output:** Vis index

**while** $i = 0 < i_{max}$ **do**
  Define $\vec{L}_i$: $\vec{p}_{org} \rightarrow \vec{p}_{end_i}$
  Calculate $\vec{u}_i = \frac{\vec{L}_i}{||\vec{L}_i||} \cdot d_{Res}$
  Set $j = 0, S_{max} = 0.0$
  Calculate $j_{max} = \frac{||\vec{L}_i||}{d_{Res}}$
  $\vec{r}_{org} \Leftarrow \vec{p}_{org}$
  **while** $j \leq j_{max}$ **do**
    $\vec{r}_{end_j} = \vec{r}_{org_j} + \vec{u}_i$
    Calculate $S_j$ between the heights $\vec{r}_{end_j.z}$ and $\vec{p}_{org.z}$
    **if** $S_j > S_{max}$ **then**
      $\vec{r}_{end_j} \in \chi_{noc}$
      $S_{max} \Leftarrow S_j$
    **else**
      $\vec{r}_{end_j} \in \chi_{occ}$
    **end if**
    $j = j + 1$
  **end while**
  $i = i + 1$
  $\theta_i = \theta_{i-1} + \theta_{res}$
**end while**
Calculate $Vis = \frac{|\chi_{noc}|}{|\chi_{wspace}|}$
**return** *Vis*

A candidate path, can now be generated by the Local path planner within this map space free $\chi_{free}$ at the set $\chi_{wspace}$ (Ishigami et al., 2007). It only remains necessary to select an appropriate location that serves as a goal within the LDEM. The goal at every LDEM is in reality a subgoal of the final path which has as it final goal the destination initially selected for the global path planner.

In order to select each of these subgoals, the visibility index along with other parameters that will be explained later in this chapter are considered. Due to the occlusion effect of the natural obstacles present in the topography the mobile robot traverses, a considerable amount of the environment's characteristics is lost and the parameters utilized to select a goal at a given LDEM may not be used.

## 6. Fractal-based Terrain Generation

Euclidean geometry has been the traditional method to describe man-made objects, it is based on the characteristics of the size or scale of the object and can be calculated through algebraic formulas. A fractal in the other hand, is a fragmented geometrical shape too irregular to be analyzed through traditional Euclidean geometry, it can be split into smaller "self-similar" components independently of the scale, and can be generated by a recursive algorithm. The term "fractal" was first coined by the mathematician Benoit MandelBrot in 1975 (Mandelbrot, 1983) who elaborated one of the nowadays well known fractal sets, the Mandelbrot set.
One of the fundamental characteristics of fractals is the concept of *self-similarity* or *scaling* which is closely related to the Euclidean notion of dimension (Barnsley et al., 1998). This concept is based on the fact that a $D$-dimensional self-similar object can be divided into "N" smaller copies of itself each of which is scaled down by a factor $r$ as shown by Equation 7;

$$r = \frac{1}{N^{1/D}}$$
$$N = \frac{1}{r^D} \tag{7}$$

and the fractal dimension of a $N$-parts self-similar object scaled by the same ratio as in Equation 7 is given by Equation 8

$$D_f = \frac{\log(N)}{\log(1/r)} \tag{8}$$

and it differs from the Euclidean dimension in the fact that it does not need to be an integer.

### 6.1 Natural terrain generation

An effective mathematical model to simulate natural terrain with topographical accidents such as those of the lunar soil or Mars is the *fractional Brownian motion* or *fBM*. This model is an extension of the random walk or Brownian motion which was initially proposed by R. Brown and further studied by A. Einstein (Mazo, 2002) after observing under the microscope the random movement of particules suspended in a fluid. Brownian motion can be informally defined as the displacement of a particle in one time interval that is independent of the displacement of the particle during another time interval (Sharma & Vishwamittar, 2005).

#### 6.1.1 Terrain generation based on $H$ and $\sigma$

A number of methods can be implemented in order to generate a fractional Brownian motion-based natural terrain. In this research the authors have used the spectral synthesis approach also known as the Fourier filtering method to produce such terrain (Barnsley et al., 1998). A fBm-based curve can be obtain through an appropiate selection of the spectral density $S_X(f)$ of a random process $X(t)$. If a random process $X(t)$ has a spectral density proportional to $\frac{1}{f^\beta}$ as shown in Equation 9,

$$S_X(f) \propto \frac{1}{f^\beta} \tag{9}$$

where $\beta$ is known as the spectral exponent, $S_X(f)$ will represent a fBM curve with a $\beta = 2H + 1$. The scaling behavior exhibited by such curve is characterized by a parameter called the "Hurst Parameter", $H$, after E. H. Hurst who introduced it when studying the statistics of the annual water level fluctuations of the Nile river. The Hurst parameter can take values between $0 \leq H \leq 1$ and it determines the "roughness" of the curve. When $H$ tends to 0 the function's curve becomes rougher, conversely, when $H$ gets closer to 1 the curve smoothens. Having $1 < \beta < 3$ will produce a fBm curve with a fractal dimension as given by Equation 10 for a Euclidean dimension $E = 1$:

$$D_f = 2 - H = \frac{5 - \beta}{2} \tag{10}$$

In a general form that can be extended to higher dimensions, the spectral density $S_X(f)$ for a random process $X(t)$, can be expressed as Equation 11,

$$S(u_i) \propto \frac{1}{\left(\sum_{i=1}^{E} u_i^2\right)^{\frac{2H+E}{2}}} \tag{11}$$

where $u_i$ represents every coordinate axis $(x,y,z)$, $H$ is the Hurst parameter taking values as $0 < H < 1$ as previously presented and the relationship between $E$ and $H$ in the Euclidean space is given by $D_f = E + 1 - H = E + \frac{3-\beta}{2}$. Based on this general form, Equation 11 can be written to represent a uni-dimensional space fBM curve equivalent to Equation 9 as described by Equation 12,

$$S(u) \propto \frac{1}{u^{2H+1}} \tag{12}$$

where a value $H = 1/2$ is called Brownian Noise, $H = 0$ describes the $1/f$ noise and $H = -1/2$ is known as white noise. The general equation for a two-dimensional spectral density is then provided by Equation 13,

$$S(u,v) \propto \frac{1}{(u^2 + v^2)^{H+1}} \tag{13}$$

To build a natural terrain bounded in a DEM the following procedure is performed. Assume that the heights contained in a plane are known and constrained by a two-dimensional grid represented mathematically by a matrix with the corresponding coordinates $x$ and $y$ of every height $Z(x,y)$ as its indexes. The spectral density of this collection of heights can be described by $S_Z(u,v)$ as Equation 14 shows

$$S_Z(u,v) = \frac{\sigma^2}{(u^2 + v^2)^{H+1}} \tag{14}$$

In Equation 14, $\sigma$ is a proportional constant representing the amplitude coefficient of a given terrain. In other words, $\sigma$ represents the behavior of the heights $Z(x,y)$ within the natural terrain to be generated. Based on the characteristic line of the Fourier transform, $\sigma$ can be obtained from the values of the height $Z(x,y)$. If we call $F_Z(u,v)$ the Fourier transform of $Z(x,y)$, the power spectral density $S_Z(u,v)$ becomes Equation 15,

$$S_Z(u,v) = \frac{1}{N\Delta} |F_Z(u,v)|^2 \tag{15}$$

where $NxN$ is the number of nodes into which the terrain was discretized, $\Delta$ is the grid's spacing between nodes which is in accordance to $S_X(f) = \frac{|X(f)|}{\Delta f}$ which is an equivalent definition of the spectral density given by Equation 9. From Equation 14 and Equation 15, the $F_Z(u,v)$ of $Z(x,y)$ can found as presented by Equation 16,

$$|F_Z(u,v)| = \frac{\sigma\sqrt{(N\Delta)}}{(u^2 + v^2)^{\frac{H+1}{2}}} \tag{16}$$

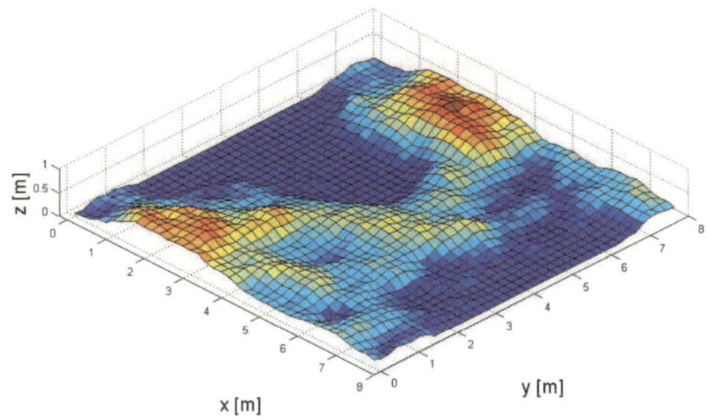


Figure 6. A natural terrain generated with the implemented fBM model.

The parameters required by the implemented model to generate a simulated natural terrain are $\sigma, H, N$ and $\Delta$, where $N$ and $\Delta$ are kept constant in order to maintain the same constrained area on every terrain generation. Therefore, the parameters that will determine the roughness and amplitude (height) of the simulated natural terrain will be determined by $\sigma$ and $H$. Using this characterizing parameters, its possible to compute $|F_Z(u,v)|$ as shown in Equation 16. Now that the Fourier transform is known based on the characterizing parameters, it is possible to calculate $Z(x,y)$ by finding its inverse Discrete Fourier transform. It should be noticed that $Z(x,y)$ will have white noise added due to this computational process. In order to avoid the issues this added noise may introduce in the calculation, random values can be utilized. The inverse Fourier transform of $F_Z(u,v)$ can be determined as shown in Equation 17,

$$Z(\hat{x},\hat{y}) = \frac{1}{(N\Delta)^2} \sum_{\hat{u}=0}^{N-1} \sum_{\hat{v}=0}^{N-1} F_Z(\hat{u},\hat{v}) e^{2i\pi(\hat{u}\hat{x}+\hat{v}\hat{y})/N} \tag{17}$$

where the variables that have a "hat" such as $\hat{}$ are in the interval spanning from 0 to $N-1$ and the sampling cycle $\Delta$ can be considered as Equation 18:

$$u = \left(\frac{\hat{u}}{N\Delta}\right), x = \hat{x}\Delta \tag{18}$$

Then since $Z(x,y)$ has real numbers, the conjugate of $F_Z(\hat{u},\hat{v})$ can be found as described by Equation 19,

$$F_Z(\hat{u},\hat{v}) = F_Z^*(N-\hat{u},N-\hat{v}) \tag{19}$$

Now if one combines Equation 16 and Equation 18 in order to avoid the indefinition caused by $u = v = 0$, we can obtain the result given by Equation 20:

$$|F_Z(u,v)| = \frac{\sigma(N\Delta)^{H+\frac{3}{2}}}{(\hat{u}^2+\hat{v}^2)^{\frac{H+1}{2}}} \tag{20}$$

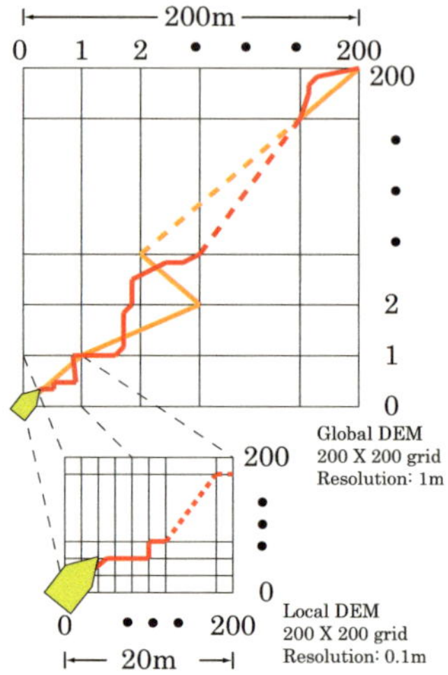


Fig. 7. Description of the path planning strategy based on the Global and Local DEMs

which can be introduced in Equation 17 to obtain the desired natural terrain model. Fig. 6 shows an example of a natural terrain that can be obtained through the implementation of this procedure.

## 7. Path Planning and Execution

Several techniques have been designed and utilized in the robotics field to generate paths for a robot to move from its initial position to the desired destination. In this section, the authors describe the two different techniques that have been implemented: a "global path planner" and a "local path planner". The first method determines a path based on the information gathered by an orbiter and represented in the GDEM. The local path planner, in the other hand, is based on the topographic information of the local environment gathered by the laser range sensor. As shown by Fig. 7, the global path planner will serve as an "guideline" the mobile robot would follow. As the vehicle navigates the environment, its knowledge of the terrain increases and the originally determined global path is reconfigured.

### 7.1 Global path planner

In this section, a procedure that calculates a candidate path based on the information provided by the GDEM is explained. The path planner's objective function is to find a candidate path that can be composed of three criteria indices: terrain roughness, path length and inclination. This path planner takes into account the features of the terrain to be traversed, however, it does not include an index that weights the visibility from the actual position of the mobile robot Ishigami et al. (2006).

The path planner used in this research assumes a perfect, namely without uncertainties, knowledge of the terrain map and the DEM obtained from it. The terrain map is represented by a DEM, which is defined as a series of elevations at a grid's node $n_i$ in $(x_i, y_i, z_i)$, where $i$ is the index of the node. This DEM-based terrain map can be measured by an orbiter flying around the Moon or celestial body of interest.

The terrain roughness index aims to define the traversability over an uneven terrain. The planner avoids defining the path over a too rough zone that the vehicle would not be able to overcome. The terrain roughness index $B_i$ is given as a standard deviation of the terrain elevation over a projection region (described in Fig. 5) of the robot $R_i$ Iagnemma & Dubowsky (2004) as shown by Equation 21 :

$$B_i = \sqrt{\frac{1}{n}\sum_{R_i}(z(R_i) - \bar{z}(R_i))^2} \tag{21}$$

where, $n$ represents the number of the node inside the region $R_i$ and $\bar{Z}(R_i)$ denotes an average elevation inside the same region. The rougher the terrain is, the larger $B_i$ becomes. The projection region $R_i$ includes the set of terrain elevation points inside the region surrounded by the wheels of the mobile robot.
This index aims to define the shortest path from the initial position of the rover to a desired destination. The path length index $L_i$ between adjacent nodes is calculated by Equation 22:

$$L_i = |n_i - n_j| = \sqrt{(x_i - x_j)^2 + (y_i - y_j)^2 + (z_i - z_j)^2} \tag{22}$$

If the nodes $n_i$ are not adjacent to the nodes $n_i$, $L_i$ gets a large value.
When the robot is climbing up a hill or traversing the slope of a crater, the risk of slippage or tipping over increases. The index of the terrain inclination aims to mitigate such risks. The terrain inclination angles are divided into two axis related to the robot body coordinates. An inclination angle around $x$-axis of the robot coordinates is denoted by $\theta_x$, while one in the $y$-axis is $\theta_y$. The indices $\Theta_{xi}$ and $\Theta_{yi}$, associated with each terrain inclination are respectively determined by the average inclination at the region $R_i$ as presented by Equation 23:

$$\Theta_{xi} = \bar{\theta}_x(R_i) \tag{23}$$

$$\Theta_{yi} = \bar{\theta}_y(R_i) \tag{24}$$

The above criteria indices are weighted in order to define the $C(p)$ economical function to minimize, that generates a candidate path $p$.

$$C(p) = \sum_{i=p}(W_B N_B B_i + W_L N_L L_i + W_{\theta_x} N_{\theta_x} \Theta_{xi} + W_{\theta_y} N_{\theta_y} \Theta_{yi}) \tag{25}$$

where, $W_B$, $W_L$, $W_{\theta_x}$ and $W_{\theta_y}$ are the weighting factors to give specific priorities between the terrain roughness, path length, and terrain inclinations. Note that $W_{\theta_x}$ or $W_{\theta_y}$ respectively take large enough values when the index $\Theta_{xi}$ or $\Theta_{yi}$ exceed threshold angles $\theta_{x_{max}}$ and $\theta_{y_{max}}$. $N_B$, $N_L$, $N_{\theta_x}$ and $N_{\theta_y}$ are constants to normalize each corresponding indices and eliminate the dimensions. The path $p$ consists of a series of neighboring nodes, $p = \{n_{start}, ..., n_i, ..., n_{goal}\}$. A small index means that the robot is less affected by the criteria. For example the smaller $W_B$ is, the less the planner will try to avoid rocks and boulders.
Therefore the smallest the sum of the weighted indices is, the optimal as less hazardous the path is, and supposing that the objective function is a hypothetical distance function, the path planning problem is considered as a shortest path search. And considering that the minimum objective function derives the "shortest" path $p_s$, the following equation can be defined by Equation 26:

$$min\ C(p) = C(p_s) \tag{26}$$

where $p_S$ is derived by the commonly used Dijkstra algorithm LaValle (2006).

### 7.2 Local path planner

Once a global destination has been determined and the global path calculated, the procedure to generate a local path follows. After having introduced the methods to characterize the terrain and determine the occluded areas in Sections 4- 5, it is possible to calculate the destination at every LDEM to where the vehicle should go to. This position is called the "next sensing position".

#### 7.2.1 Next sensing position

The next sensing position in this paper is defined as the position within the map space free $\chi_{free}$, at a given LDEM, from where it is estimated that the LIDAR sensory system will cover a larger non-occluded area on the next scanning procedure.

Consider Fig. 8 as a conceptual example. Assume that the robot knows the coordinates of the global goal and that this goal is behind an obstacle Fig. 8(a). Once the information from the LIDAR sensory system has been acquired and the LDEM has been created, the visibility from the initial position of the vehicle can be determined by the algorithm presented in Subsection 5.2. This algorithm will classify the nodes occupied by the viewable edges of the obstacle as non-occluded and the area behind it in the other hand, will be determined to be occluded. We can then complete the characterization of the LDEM and determine its $\chi_{free}$. In Fig. 8(b) a node representing a new location is evaluated and its $\chi_{free}$ compared to other visited locations such as the one described in Fig. 8(c). The estimated location for the next sensing position will sense the same viewable edges of the obstacle detected previously, but due to its new orientation towards the global goal and its position in the $\chi_{free}$, it will obtain the largest non-occluded area from the sampled positions as presented in Fig. 8(d). This estimation procedure is based on the evaluation of various parameters of a node contained in $\chi_{free}$. These parameters are: the visibility index *Vis* presented in Subsection 5.2, the Euclidean distance $L_{n_i}$ separating the inspected node and the global goal as given by Equation 27,

$$L_{n_i} = \sqrt{(x_g - x_{n_i})^2 + (y_g - y_{n_i})^2 + (z_g - z_{n_i})^2} \tag{27}$$

where $(x_g, y_g, z_g)$ denote the coordinates of the global destination and $(x_{n_i}, y_{n_i}, z_{n_i})$ represent the coordinates of the currently inspected node. The third parameter is the value *obs* a node is given to if it is a node element of $\chi_{obs}$. The evaluation function combining these parameters is presented in Equation 28;

$$E_{n_i} = W_{occ} \cdot Vis_{n_i} + W_L \cdot L_{n_i} + obs_{n_i} \tag{28}$$

where $E_{n_i}$ is the value of the evaluation function at node $n_i$ and $W_{occ}$ and $W_L$ represent the weighting factors that give specific priorities between the size of the occluded area in the next sensing position and distance from the inspected node $n_i$ to the global goal. The parameter $obs_{n_i}$ works as a penalty index, since it takes large values if the inspected node $n_i$ is contained in $\chi_{obs}$. It should be noted that if a higher value is set to a given weighting factor, a higher priority will be given to that parameter. For example, the higher the weighting factor $W_{occ}$ is, the more importance to the amount of non-occluded area in the next sensing procedure will be. In the same way, the higher the weighting factor $W_L$ is, the more importance to a position closer to the global goal will be given. In our implementation, the node $n_i$ with the evaluation function having the highest value from the sample taken will be selected as the next sensing position. This next sensing position is consider as optimal with respect to the priorities given to the weighting factors.

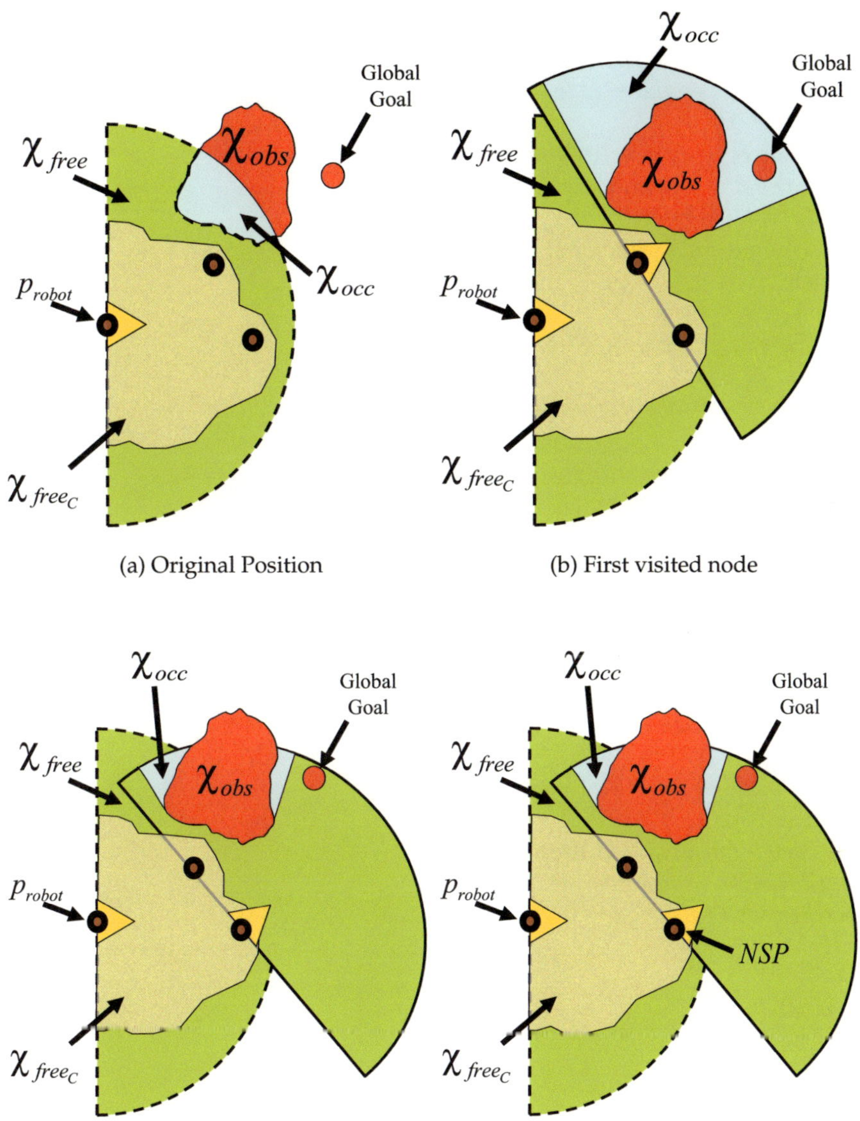


(a) Original Position (b) First visited node

(c) Second visited node (d) NSP selected node

Fig. 8. Conceptual representation of an estimated next sensing position (NSP) for the mobile robot.

**Algorithm 2** Next Sensing Position in a given LDEM

**Input:** Nodes contained in a LDEM's $\chi_{free}$
**Input:** Number of seed nodes for the Hill Climbing algorithm, *numSeeds*
**Output:** Next sensing position within a LDEM, *nextSenPos*

**for** $i = 0$ to *numSeeds* **do**
  Select a random node *rndmNode* from $\chi_{free}$
  Set $currentNode \leftarrow rndmNode$
  **while** TRUE **do**
    Calculate $Nghbrs = neighbors(currentNode)$
    Set $nxtEval = -\infty$ and $nxtNode = NULL$
    **for** $\forall n_i \in Nghbrs$ **do**
      Calculate distance to goal $L_{n_i}$ from $n_i$
      Determine the visibility index $Vis_{n_i}$
      **if** $n_i \in \chi_{obs}$ **then**
        Set $obs_{n_i} = arbitrary\ low\ value$
      **else**
        Set $obs_{n_i} = arbitrary\ high\ value$
      **end if**
      Calculate $E_{n_i}$
      **if** $E_{n_i} > nxtEval$ **then**
        $nxtNode \leftarrow n_i$
        $nxtEval \leftarrow E_{n_i}$
        $tmpEval \leftarrow nxtEval$
      **end if**
    **end for**
    **if** $nxtEval \leq E_{n_i}$ **then**
      return nxtNode
    **end if**
    $currentNode \leftarrow nxtNode$
  **end while**
  **if** $tmpEval > tmpEvalMax$ **then**
    $tmpEvalMax \leftarrow tmpEval$
    $nextSenPos \leftarrow nxtNode$
  **end if**
**end for**
**return** *nextSenPos*

Since the number of nodes in the $\chi_{free}$ can be large, the computation of Equation 28 for every node inside the $\chi_{free}$ is time-consuming. Therefore, the authors utilized a hill-climbing search within the $\chi_{free}$. To avoid the local minima issue that can arise by the use of this algorithm, several random nodes were used as initial nodes. The Algorithm 2 describes in pseudocode the procedure to obtain the next sensing position.

A set of results obtained from a numerical simulation are depicted in Fig. 9. The global goal is represented by a red sphere and the local goal determined by the next sensing position is shown as a purple sphere in Fig. 9. Various local goal candidates that were considered by the algorithm are depicted as blue spheres. Fig. 9 shows the results obtained by the Algo-

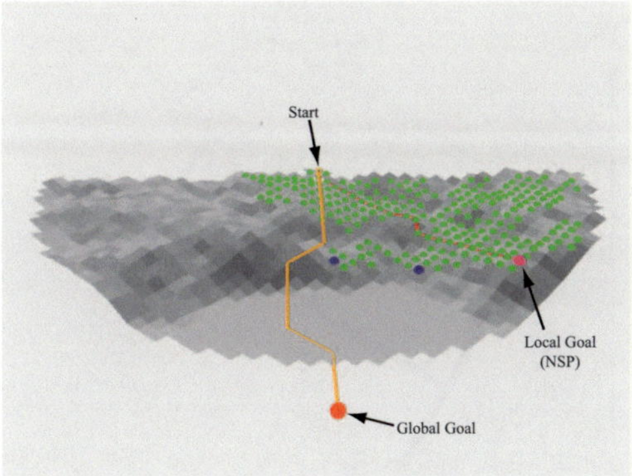


Figure 9. Visual description of the next sensing position algorithm

rithm 2 after 5 iterations. It is possible however, that the same node may be selected multiple times among the 5 iterations as is the case presented in Fig. 9. After the pre-setted number of iterations is reached the node with the highest evaluation function among these nodes is determined to be the next sensing position.

Once the next sensing position has been obtained, the destination where the mobile robot would go to is passed on to the implemented local path planner. The local path planner utilized in this research is called the Distance Transform and it was introduced by (Jarvis, 1994). This algorithm finds the optimal collision-free path between the initial and final destination of the vehicle based on a given known stationary environment map. Its output is a two-dimensional array of exact cell decomposition. In our implementation, each cell is equivalent to a node of the LDEM and each cell has a value that represents a collision-free distance cost to reach the goal.

The local goal, or the subglobal goal, is initially given an arbitrary low distance cost, i.e. zero, while other nodes are given much higher values. Notice that the inverse procedure works equally well. The algorithm then visits every node and assigns to each of them a different distance cost based on the propagating distance costs of the surrounding nodes. The optimal path is found by using the steepest descent or ascent track from the actual location of the vehicle in the $\chi_{free}$. The generated path has no risk of falling into the local minima problem since every node has been assigned a different distance cost at the propagating phase of the algorithm.

The distance transform is a robust algorithm that can be efficiently implemented either with an a priori knowledge of the $\chi_{obs}$ or without it. In our proposed scenario, we have a completely known global enviroment with a low resolution and a partially known local environment with a high resolution.

## 8. Path following strategy

Having defined a path as a set of way points and the mobile robot being able to determine its actual position based on the features the laser range sensor registers in the environment surrounding the rover, it is necessary to adopt a control strategy that assists the rover to comply as close as possible with the path it was commanded to follow. This control strategy aims to

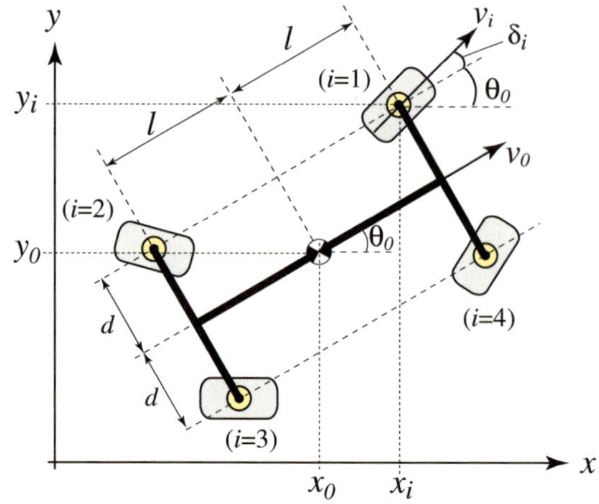


Fig. 10. Kinematic model of a four-wheeled vehicle

correct the steering and driving maneuvers generated by the vehicle while traversing an area in order to reach its destination.

In this section the nonholonomic kinematic constraints that govern the vehicle, and the control method known as "following the carrot" that implements them to minimize the steering and driving errors are presented.

### 8.1 Nonholonomic kinematic model and constraints

In the nonholonomic kinematic model of a slipless vehicle shown in Fig. 10, the following assumptions are considered: the distance between wheels are strictly fixed, the steering axle of each wheel is perpendicular to the terrain surface and the vehicle does not consist of any flexible parts. In Fig. 10, the subscript $i$ denotes the identification of the wheel, $(x_0, y_0, \theta_0)$ the position and orientation of the center of gravity of the vehicle, $(x_i, y_i)$ the position of each wheel, $v_i$ the linear velocity of each wheel, $l$ longitudinal distance from the centroid of the vehicle to the front or rear wheel and $d$ the lateral distance from the centroid of the vehicle to the left or right wheel.

The nonholonomic constraints of the implemented mobile robot are expressed by Equation 29 and Equation 29,

$$\dot{x}_0 \sin\phi_0 - \dot{y}_0 \cos\phi_0 = 0 \tag{29}$$

$$\dot{x}_i \sin\phi_i - \dot{y}_i \cos\phi_i = 0 \tag{30}$$

where, $\phi_0 = \theta_0$, and $\phi_i = \theta_0 + \delta_i$, having $\delta_i$ as the wheel steering angle. By looking at the configuration between every wheel and the center of the gravity of the vehicle, the geometric constraints can be given by

$$\left.\begin{array}{l} x_1 = x_0 + l\cos\theta_0 - d\sin\theta_0 \\ x_2 = x_0 - l\cos\theta_0 - d\sin\theta_0 \\ x_3 = x_0 - l\cos\theta_0 + d\sin\theta_0 \\ x_4 = x_0 + l\cos\theta_0 + d\sin\theta_0 \end{array}\right\} \tag{31}$$

$$\left.\begin{array}{l} y_1 = y_0 + l\sin\theta_0 + d\cos\theta_0 \\ y_2 = y_0 - l\sin\theta_0 + d\cos\theta_0 \\ y_3 = y_0 - l\sin\theta_0 - d\cos\theta_0 \\ y_4 = y_0 + l\sin\theta_0 - d\cos\theta_0 \end{array}\right\} \tag{32}$$

which can be rewritten as $x_i = x_0 + X_i$ and $y_i = y_0 + Y_i$.

### 8.2 Steering and driving manuevers

Having defined the nonholonomic constraints of the implemented mobile robot, it is possible to determine the steering angle of each wheel $\delta_i$ and the driving manuevers of the vehicle. By transforming the nonholonomic constraints given by Equation 30, the steering angle can be obtained as follows,

$$\delta_i = \tan^{-1}(\dot{y}_i/\dot{x}_i) - \theta_0 \tag{33}$$

Then, by substituting Equation 31 and Equation 32 into Equation 33, $\delta_i$ is derived by the following equation:

$$\delta_i = \tan^{-1}\left(\frac{v_0\sin\theta_0 - \dot{Y}_i(\dot{\theta}_0)}{v_0\cos\theta_0 - \dot{X}_i(\dot{\theta}_0)}\right) - \theta_0 \tag{34}$$

where, $v_0$ represents the linear velocity of vehicle and $\dot{\theta}_0$ is the turning angular velocity of vehicle. The driving maneuver of the vehicle can be achieved by the manipulating the wheel's angular velocity $\omega_i$. The relationship between $\omega_i$ and wheel's linear velocity $v_i$ can be written as:

$$\omega_i = v_i\cos\delta_i/r \tag{35}$$

where, $r$ is wheel radius. Here, $v_i$ can be expressed by $\dot{x}_i$ or $\dot{y}_i$ as:

$$v_i = \dot{x}_i/\cos\phi_i = \dot{y}_i/\sin\phi_i \tag{36}$$

Substituting Equation 31 or Equation 32 into the above equations, the angular velocity of the wheel $\omega_i$ can be finally derived as shown by Equation 37:

$$\omega_i = \begin{cases} (v_0\cos\theta_0 + \dot{X}_i(\dot{\theta}_0))\cos\delta_i/r\cos\phi_i & (\theta_0 \le \pi/4) \\ (v_0\sin\theta_0 + \dot{Y}_i(\dot{\theta}_0))\cos\delta_i/r\sin\phi_i & (\theta_0 \ge \pi/4) \end{cases} \tag{37}$$

### 8.3 Follow the carrot

Once a local path has been generated from the actual location of the vehicle to the optimal position found by the algorithm proposed on Subsection 7.2.1 the robot will execute the path by following its set of configurations. The method used in this research to track the path provided by the path planner is called "follow the carrot" and is illustrated in Fig. 11(a). This approach calculates the desired heading $\phi_{carrot}$ based on the current position of the mobile robot and the local path. This approach draws a line equivalent to the radius $r$ of a circle, from the center of the vehicle's coordinate system to the desired path. The carrot point, or goal point, is then defined to be the point on the path that intersects the generated circle and that has an orientation given by $\phi_{err}$. The most important parameter of this method is the orientation error, defined to be the angle between the current vehicle heading and the line drawn to the carrot point. A proportional control law can then aim at minimizing the orientation error. The magnitude of a turn is decided by Equation 38

$$\varphi = K_p\,\phi_{err} \tag{38}$$

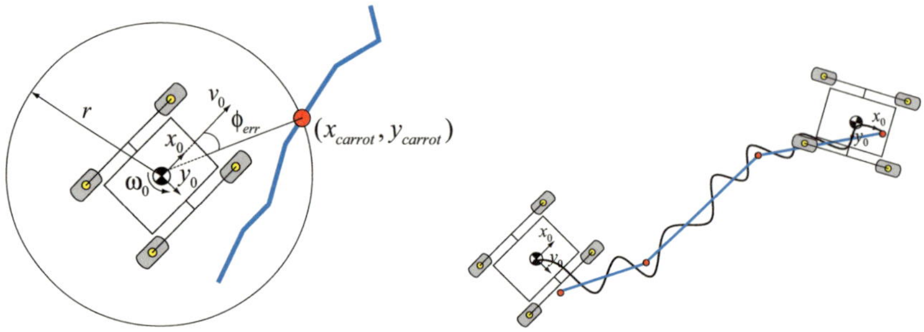


(a) Path following concept (b) Oscillating behavior when using small circles

Fig. 11. Conceptual detail of the implemented path following strategy.

where $K_p$ is the proportional gain and $\phi_{err}$ is the orientation error defined by

$$\phi_{err} = \phi_{carrot} - \phi_{heading} \tag{39}$$

In Equation 39, $\phi_{heading} = \delta_i$ and the desired carrot angle $\phi_{carrot} = \theta_d$, where $\theta_d$ represents the desired steering orientation angle of each the wheels of the vehicle. With this in mind Equation 39 can be rewritten as Equation 40;

$$\theta_{err} = \theta_d - \delta_i \tag{40}$$

The mobile robot can move towards the carrot by controlling the turning angular velocity $\dot{\theta_d}$. Thus by substituting $\theta_0 = \theta_d$ and $v_0 = v_d$ into Equation 37, the desired angular velocity of each wheel $\omega_{di}$ can be achieved.
One of the most important characteristics of this approach is its simplicity in terms of implementation, alas, it has a couple of major drawbacks (Helmick et al., 2004),(Kelly, 1994). First, the vehicle has a tendency to naturally cut corners. This happens because the vehicle immediately tries to turn towards each new carrot point. Another drawback is that the vehicle could oscillate about the path, particularly in the case of a small "lookahead distance" (smaller circles) or at higher speeds as presented in Fig. 11(b). In our implementation we solve these issues by having a constant linear speed of $v_d = 11$[cm/s] which allows the vehicle to change its heading without loosing track of the path.
A large lookahead distance will tend to filter out small features of the path, but results in a smooth motion of the vehicle. A small value results in large heading changes of the vehicle for small path errors (which is extremely inefficient), but results in an overall smaller path following error. A lookahead distance that balances the tradeoff between getting closer to the path and decreasing the magnitude of the heading error $\phi_{err}$ is then determined. Under nominal conditions, the vehicle path error will always be smaller than the lookahead distance. If this is not the case, then the radius is grown until an intersection does occur.

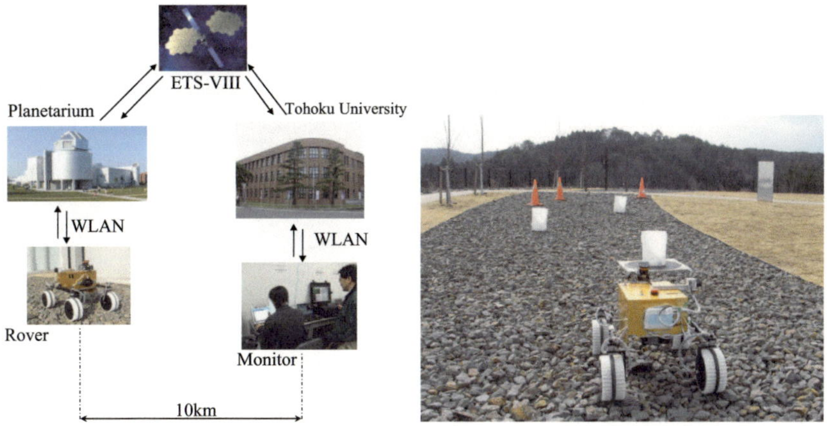


(a) (b)

Fig. 12. Setup of the experiments realized at Sendai city's planetarium.

## 9. Experimental Results

In order to validate the procedures presented in the previous sections of this chapter, a set of experiments that recreated the occluding effect on the laser range sensor implemented in our mobile robotic testbed were carried out. This set of experiments were conducted between Tohoku University and the Planetarium of the city, both located in Sendai, Japan but separated by a distance of approximately 10 [km]. The communication between the human monitors at Tohoku University and the mobile robotic platform at the Planetarium relied on the Japanese Engineering Test Satellite VIII (ETS-VIII). The communication setup implemented between both ends is presented in Fig. 12(a). The communication link provided by the ETS-VIII can be set to three possible bandwidth specifications and data package loss rates in order to simulate the communication conditions on a possible lunar mission. The experiments consisted on the navigation of our mobile robotic platform through an environment where various obstacles were located in such a way that the field of view of the laser range sensor would be limited due to the characteristics of the environment as can be seen in Fig. 12(b). This location at the Planetarium has an area of 20[m] $\times$ 20[m] and is covered by loose rocks that simulate the possible conditions the vehicle may encounter on the lunar surface. During the navigation, the vehicle acquires the information of its environment and through the procedures presented in this chapter, determined the next sensing position, the local path to that position and finally it executed the generated local path. The process was then repeated until the global destination was achieved.
The comparison between the generated global path, the succession of the generated local paths that reconfigured the global path and the actual path followed by the vehicle (represented by the odometry obtained from its proprioceptive sensors) are described in Fig. 13(a). The orientation profile the vehicle realized through out the navigation experiment is depicted by Fig. 13(b).

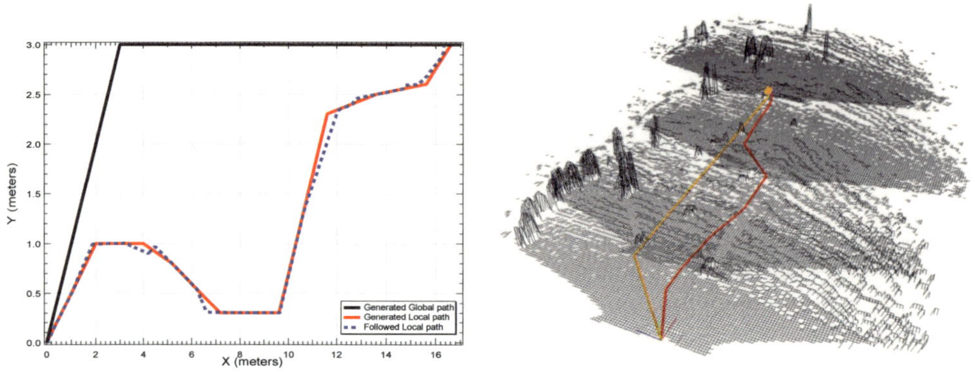


(a) (b)

Fig. 13. Experimental results: generated and followed paths of the vehicle.

As it can be seen from Fig. 13(a), the vehicle initially follows the global path but it soon steers towards the left in order to avoid the obstacle on the right and when it reaches its local destination it rotates around its $z$-axis in order to cover a bigger area in the next mapping procedure. Instead of following the path suggested by the global path, the vehicle continuously moves towards a position that will enable it maximize the number of nodes it can see from the determined next sensing position.

## 10. Concluding Remarks

In this chapter, the combination of two path planning strategies, a global path planner and a local path planner, was discussed for a mobile robot (rover) to explorer in a lunar surface environment with poor illumination and rich obstacles. The global path planner would use the representation of the elevation data called the *Global DEM* (digital elevation map) gathered by a satellite mission orbiting the targeted celestial body to calculate a general path to the desired destination. The local path planner in the other hand, uses the representation of the elevation data called *Local DEM* obtained by its laser range sensor mounted on the rover to compute a path that would reconfigure the global path initially determined.

An analysis of the LDEM, or local map, was made to classify the different areas according to the occlusion effect due to the presence of obstacles. This analysis allowed the quantification of the occluded and non-occluded regions within the area that can be covered by a laser range sensor. The algorithm that quantifies the non-occluded areas that can be perceived by the sensory system from the position where the rover is located within a local map was addressed. This quantification is given in the form of an index called the visibility index which is later incorporated in the local path planner through a method called the next sensing position. The next sensing position concept and its algorithm were explained thoroughly, and numerical simulation results were presented. The local path planner evaluates the terrain and finds an optimal path in terms of distance to the local goal obtained by the next sensing position algorithm.

Fractal geometry was introduced to generate a discrete natural terrain representation which can describe topographic features similar to those of the lunar surface. Particularly, a method based on the fractional Brownian motion approach was implemented to validate through numerical simulations the concepts presented in this chapter. Experiments that utilized the Japanese Engineering Test Satellite VIII (ETS-VIII) in order to simulate the conditions of a mobile robotic lunar exploration mission were carried out. In these experiments, the vehicle navigated through an irregular, rocky terrain by perceiving its environment, determining the next sensing position that maximizes the non-occluded region within each local map and executing the local path generated.

## 11. References


Adams, M. D. (1999). Sensor modelling, design and data processing for autonomous navigation, *World Scientific* .

Barnea, D. I. & Silverman, H. F. (1972). A class of algorithms for fast digital image registration, *IEEE Transactions on Computers* pp. 179–186.

Barnsley, M. F., Devaney, R. L., Mandelbrot, B. B., H. O. Peitgen, D. S. & Voss, R. F. (1998). The science of fractal images, *Springer-Verlag* .

Borenstein, J., Everett, H. R. & Feng, L. (1996). Navigating mobile robots, *A. K. Peters, Wellesley, MA* .

Danner, T. & Kavraki, L. E. (2000). Randomized planning for short inspection paths, *Proc of the 2000 IEEE Int. Conf. on Robotics and Automation* .

Dupuis, E., Allard, P., Bakambu, J., Lamarche, T., Zhu, W. & Rekleitis, I. (2005). Towards autonomous long-range navigation, *Proceedings 8th International Symposium on Artificial Intelligence, Robotics and Automation in Space* .

Everett, H. R. (1995). Sensor for mobile robots, theory and application, *A. K. Peters* .

Heckman, N., Lalonde, J., Vandapei, N. & Hebert, M. (2007). Potential negative obstacle detection by occlusion labeling, *Proceedings IEEE International Conference on Intelligent Robotics and Systems* .

Helmick, D. M., Cheng, Y., Clouse, D. S., Matthies, L. H. & Roumeliotis, S. I. (2004). Path following using visual odometry for a mars rover in high-slip environments, *Proc of the IEEE International Aerospace Conference* .

Hokuyo (2008). Scanning laser range finder utm-30lx/ln specification, *Hokuyo Automatic Specification Report* . Hokuyo Automatic Co. LTD.

Iagnemma, K. & Dubowsky, S. (2004). Mobile robots on rough terrain, *Springer Tracts in Advanced Robotics* .

Ishigami, G., Nagatani, K. & Yoshida, K. (2006). Path following control with slip compensation on loose soil for exploration rover, *Proc of the 2006 IEEE International Conference on Intelligent Robots and Systems* .

Ishigami, G., Nagatani, K. & Yoshida, K. (2007). Path planning for planetary exploration rovers and its evaluation based on wheel slip dynamics, *Proc of the IEEE Int. Conf. on Robotics and Automation* .

Jarvis, R. A. (1994). On distance transform based collision-free path planning for robot navigation in known, unknown and time-varying environments, *Advanced Mobile Robots*, World Scientific Publishing Co. Pty. Lt, pp. 3–31.

Kalisz, J. (2004). Review of the methods for time interval measurements with picosecond resolution, *Metrologia* **41**: 17–32.

Kato, M., Takizawa, Y., Sasaki, S. & Team, S. P. (2007). The selene mission: Present status and science goals, *Procedings 38th Lunar and Planetary Science Conference* .

Kelly, A. (1994). A feedforward control approach to the local navigation problem for autonomous vehicles, *Robotics Institute Technical Report, CMU-RI-TR-94-17* .

Kunii, Y. & Kubota, T. (2006). Human machine cooperative tele-drive by path compensation for long range traversability, *Proceedings IEEE/RSJ International Conference on Intelligent Robots and Systems* .

Langer, D., Mettenleiter, M., Hart, F. & Frohlich, C. (2000). Imaging laser scanners for 3-d modeling and surveying applications, *Proc. of the 2000 IEEE International Conference on Robotics and Automation (ICRA 2000)*, San Francisco, CA, U.S.A.

LaValle, S. (2006). Planning algorithms, *Cambridge University Press* .

Lucas, B. D. & Kanade, T. (1981). An iterative image registration technique with an application to stereo vision, *Proc. of Imaging Understanding Workshop* .

Mandelbrot, B. B. (1983). The fractal geometry of nature, *W. H. Freeman* .

Mazo, R. M. (2002). Brownian motion fluctuations, dynamics, and applications, *Oxford University Press* .

Morales, Y., Takeuchi, E., Carballo, A., Tokunaga, W., Kuniyoshi, H., Aburadani, A., Hirosawa, A., Nagasaka, Y., Suzuki, Y. & Tsubouchi, T. (2008). Imaging laser scanners for 3-d modeling and surveying applications, *Proc. of the 2008 IEEE/RSJ International Conference on Intelligent Robots and Systems*, Nice, France.

Moravec, H. P. (1979). Visual mapping by a robot rover, *Sixth International Joint Conference on Artificial Intelligence* .

Pascoal, J., Marques, L. & de Almeida, A. T. (2008). Assessment of laser range finders in risky environments, *Proc. of the 2008 IEEE/RSJ International Conference on Intelligent Robots and Systems* .

Rekleitis, I., Bedwani, J.-L. & Dupuis, E. (2009). Autonomous planetary exploration using lidar data, *Proceedings of the IEEE International Conference on Robotics and Automation* .

Sanger, T. D. (1988). Stereo disparity computation using gabor filters, *Journal of Biological Cybernetics* **59**(6): 405–418.

Sharma, S. & Vishwamittar (2005). Brownian motion problem: Random walk and beyond, *Resonance* **10**(8): 49–66.

Skrzypcznski, P. (2008). How to recognize and remove qualitative errors in time-of-flight laser range measurements, *Proc. of the 2008 IEEE/RSJ International Conference on Intelligent Robots and Systems*, Nice, France.

Watson, K., Murray, B. & Brown, H. (1961). On the possible presence of ice on the moon, *Journal of Geophysical Research* **Vol. 66**: 1598–1600.

Ye, C. & Borenstein, J. (2002). Characterization of a 2-d laser scanner for mobile robot obstacle negotiation, *Proc. of the 2002 IEEE International Conference on Robotics and Automation (ICRA 2002)*, Washington, D.C., U.S.A.

14

# A Decentralised Software Process Approach For Real time Navigation of Service Robots

S. Veera Ragavan and Velappa Ganapathy
*Monash University, Sunway Campus,*
*Malaysia*

## 1. Introduction

Convergence of Communication, Computing and Control is considered by many as the future of Information Technology. In the same breadth, the recent advances in Key Technology such as Telematics, M2M, Smart Wireless Communication Devices (SWCD) and Sensor Networks (WSN) from Telecommunications' domain; Artificial Intelligence (AI), Sensor Fusion , Behavior Fusion and Hybrid algorithms (ANFIS, GA) from the Robotics Domain; Open Source Systems (OSS), Geographical Information Systems ( GIS), Service Oriented Architecture (SOA) and Session Initiated Protocol (SIP) from the Information Technology Domains necessitates a paradigm shift in approaches to building applications for future Robots.

Early single function robotic systems were designed as control systems with a clear feedback model. A Sensor generates feedback, and any deviation from the expected feedback generated by the Sensors updates a control signal to minimize the error over time. As complexity of modern day Service Robots grew and the robots needed to perform multiple functions, the perception-action loop was extended linearly to include a planning component resulting in the sense-plan-act paradigm. Obviously it does not perform very well in dynamic and unpredictable environments: the sensors and real world models are usually inadequate as the actions are not always a direct consequence of perception.

The hybrid deliberate/reactive approach has proven very successful, practical and robust in a large number of implementations, and it appears that there is general agreement among the research community that this is the best type of architecture for Service Robots. However, some types of modules are hard to force into any particular layer. In a general framework, it is imperative that no special architecture is preferred for enforcement and a good support for builders of the hybrid deliberate/reactive architecture is important so that the framework supports parallel execution of behaviours. This is precisely where the proposed architecture scores above the other architectures.

The major problems for robotics today lie, not in the hardware but on the software side. There are plenty of well functioning and robust algorithms developed by competent researchers readily available. Each new implementation would provide significant gains in the performance and capabilities, but it will be lost due to non portability and reuse issues.

While the lowest layer or reactive layer has to be embedded on the robot controller due to the obvious fact that this layer requires the highest response and lowest CPU time, in our approach the Middleware layer helps us to switch based on deliberative approaches from the repository of allowable robot behaviours. What was essentially an traditionally AI Rule Based Behaviour Switching now graduates to a Location Based Behaviour Switching based on a host of Soft computing techniques .

Most of the commercial Service Robotics industries even today, employ at best the "earlier" or traditional approaches to software reuse, which are built on the paradigm of a set of libraries containing many small building blocks. Some of the more advanced establishments use object-oriented frameworks resulting in a generic design that can be instantiated for each object system constructed. While the Object oriented framework ideally captures the elements common to a family of related systems, it is essentially a large design pattern capturing bulk of the system functionality in one specific kind of object system, which is maintained as a single entity. Each software system using framework is only an instantiation of that framework. Traditionally, control of generalised Service Robots required well-defined activities tightly coupled across different hierarchies on single platform i.e. monolithic control architecture.

In a distributed system or multiprocessor, Middleware allocates system resources, giving requests to the operating systems on the individual processors to implement those decisions. One of the key differences between Middleware and Software Libraries are that Middleware manages resources dynamically. In a uniprocessor, the operating system manages the resources on the processor (for example, the CPU itself, the devices, etc.) and Software Libraries perform computational tasks based on those allocations.

By Process we mean a system element that is independent, and can be freely deployed and versioned. This approach loosely couples the various layers into process components that are well defined entities that can be replaced or made redundant without affecting the rest of the systems. It is shown here how they can be developed and tested separately and integrated later building on the Middleware Framework to provide a systematic approach to developing software that would be easy to integrate, manage, and reuse

The field of robotics relies heavily on various technologies such as Mechatronics, computing systems, and wireless communication. Given the fast growing technological progress in these fields, robots can cater to a wide range of applications. However real world integration and application development for such a distributed system composed of many robotic modules and networked robotic devices is very difficult. Therefore, Middleware services provide a novel approach offering many possibilities and drastically enhancing the application development for robots.(Mohamed, Al-Jaroodi et al. 2008)

Middleware is prevalent in IT-oriented server systems than in Embedded Systems. Several Embedded Middleware Systems have been developed based on IT standards but with heavy footprints. Middleware architectures are slowly and steadily evolving to support embedded operations. A Telematics device Middleware and the GPRS link enable configuring and reconfiguring the devices as many times as required and this is effectively used to change the control program and actions remotely to use the robot for different services without having to make any major changes to the design, hardware or software.

This paper discusses our domain, existing architectures, component and processes execution techniques and the approach we took to integrate these to form a distributed decentralised web enabled service that is robust and safe-fail. The resulting architecture is simple, and can support a wide range of trade-offs that can be manipulated easily at run-time based on control of processes rather than control of discrete actions. The current approach is a loosely coupled integration of different process technologies and computational mechanisms.

## 2. The Service Robot Domain

The Services area is just emerging as the future of field application and Services computing has been recognized as a new foundational discipline of the modern services industry(Zhang 2008). Even though the number of Service Robots currently in operation is small, the number of people working in the service field shows a constant growth rate. Consequently, there is an enormous potential for the expansion of this sector. Due to often strict requirements with respect to safety (personal and functional) associated with robot autonomy and navigation in partially or entirely unknown environments (coupled with the convergence issues) the use of robots for carrying out service tasks has been very limited. Only a fraction of existing services which incorporate handling, transportation and working can be automated. A successful design of future Service Robots will be based upon detailed knowledge of available technologies and methodologies to design handling devices or mobile platforms, peripheral devices, and organizational schemes which account for a flexible, fault-tolerant and user friendly human-machine interaction. (Schraft 1994)

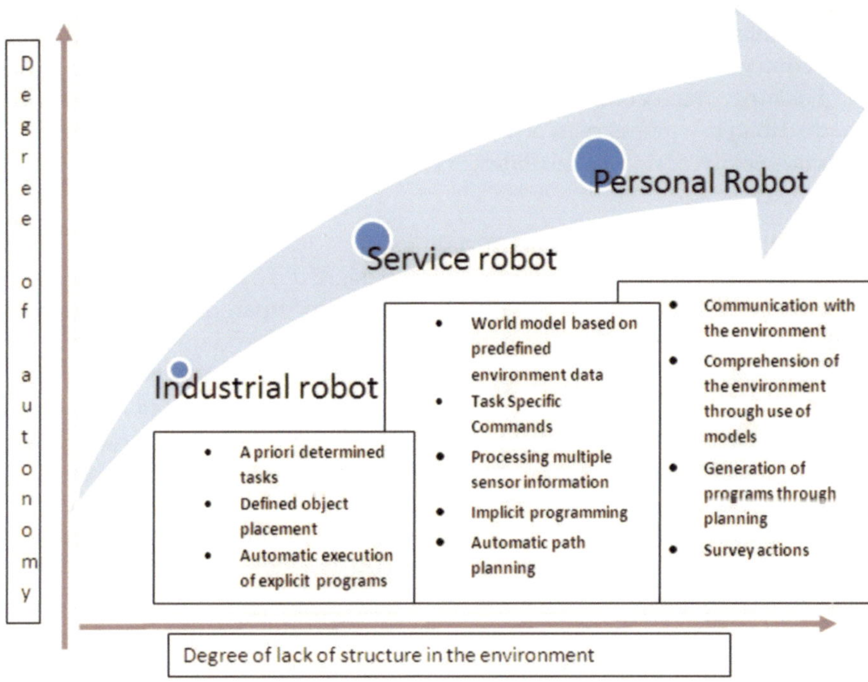


Fig. 1. Characteristics of Industrial Service and Personal Robots [based on (Schraft 1994)]

Such a Design philosophy calls for architecture that must also provide a structure that improves the coherence and modularity of the system, while supporting large scale robotic system development which include considerations of real-time response and a layered

decomposition of the system to distinguish the granularity of the modules responsibilities. Service Robots must have highly competent reactive mechanisms to be safe, flexible and easy to use. At the same time, planning and sequencing are useful to reduce the repetition and taxation on the user for direction. (Kawamura, Pack et al. 1995).
The uncertainties from the environment, the complexities of software/hardware interactions, and the variability of the robotic hardware make the task of developing robotic software complex, hard, and costly. Hence, it has become increasingly important to leverage robotic developments across projects and platforms.(Nesnas, Wright et al. 2003)
In spite of an explosion of technology and methods, the Service Robots are still not complex and in their early stages of development. Many researchers specialize in one or more areas/topics, which usually involve development of algorithms. However, in order to test the competence on a real robot, a complete system is needed involving a process based approach. Many of these are required to run in parallel and need to communicate both synchronously and asynchronously. It has to also accommodate changing application requirements, incorporate new technology, interoperate in heterogeneous environments, and maintain viability in changing environments. This puts a tremendous burden on the developer if he or she has to build everything from scratch and hence a delay in "Market ready" products. Hence, it has become increasingly important to develop Service Robots on General Platforms and Frameworks. (Ragavan and Ganapathy 2007).
We present a Novel Decentralised Architecture for Navigation and Control of Service Robots based on control of processes rather than control of discrete actions. The current approach is a loosely coupled integration of different process technologies and computational mechanisms. It is our firm contention that a well designed software architectural framework is necessary to effectively leverage microcontrollers (Read Service Robots), wireless networks (read Telematics, distributed wireless networks) and process orchestration (read service) to address problems of complexity, scale and reliability of networked Service Robots

### a. Layered Architecture and Hybrid approaches

Early robotic systems for single functions were designed as control systems with a clear feedback model. A Sensor generates feedback, which is compared to the expected feedback derived from a model of the system. Any deviation is used to update the control signal so as to minimize the error over time. As complexity grew and the robots needed to perform more than one function, the perception-action loop was extended to have a planning component. This was a natural linear extension beyond traditional control towards modern day Service Robots. This resulted in a hierarchical system having an elaborate model of the world, using sensors to update this model, and to draw conclusions based on the updated model. Obviously it does not perform very well in dynamic and unpredictable environments as the sensors and real world models are usually inadequate. That the actions are not a direct consequence of perception is perhaps the reason why it is also called the sense-plan-act paradigm.
Reactive approaches are often capable of autonomously exploring new regions in the environment and, as there is no fixed plan, they are generally able to respond rapidly to any changes that may occur in the operating environment. Moreover, they are more tolerant to uncertainties in sensor measurements and the errors. Robots that were running reactive behaviour based systems performed very well, also in changing environments. However, the purely reactive scheme is not capable of performing complex tasks. A software

architecture based on purely reactive approach is usually monolithic and requires rewriting of control software for even small changes in the task, or environment.

On the other hand deliberative navigation methods generally assume that the obstacles in the environment in which a robot moves are known in terms of their physical location and dimensions. The navigation task is then to plan a path that is both collision free and satisfies certain optimization criteria. The classical deliberative approach to navigation is based entirely on planning and on explicit symbolic models of the world exhausts the computation resources all along the way (Brooks 1991). Even more, it does not seem to operate successfully in a dynamic changing world. It has difficulties in dealing with sensors' errors as well. The models it uses are not realistic; it appears that the world is too complicated to be presented completely. Attempts to create a complete model that includes all the essential knowledge needed to deal with the uncertainties and surprises of the real world, became enormously big and the planning too expensive in time and computer resources. Hence, it has become increasingly important to leverage upon Hybrid Approaches to robotic developments across projects and platforms.

#### b. Hybrid Approaches

A hybrid approach, combining low-level reactive behaviours with higher level deliberation and reasoning, has since then been common among researchers (Arkin 1990; Cattoni, Di Caro et al. 1994). The hybrid systems are usually modelled as having three layers as shown in Figure 1; one deliberative, one reactive and one middle layer (Gat 1992) and this approach for a long time now remains vastly unchallenged.

There is also a sound architectural rationale for having exactly three major components and not just because three is an aesthetically pleasing number. It has to do with the role of internal state and with ability to organize algorithms according to whether they contain no state, contain state reflecting memories about the past, or contain state reflecting predictions about the future. Abstractions are then used to isolate aspects of reality that can be tracked or predicted reliably, and to ignore other aspects (Erann 1998).

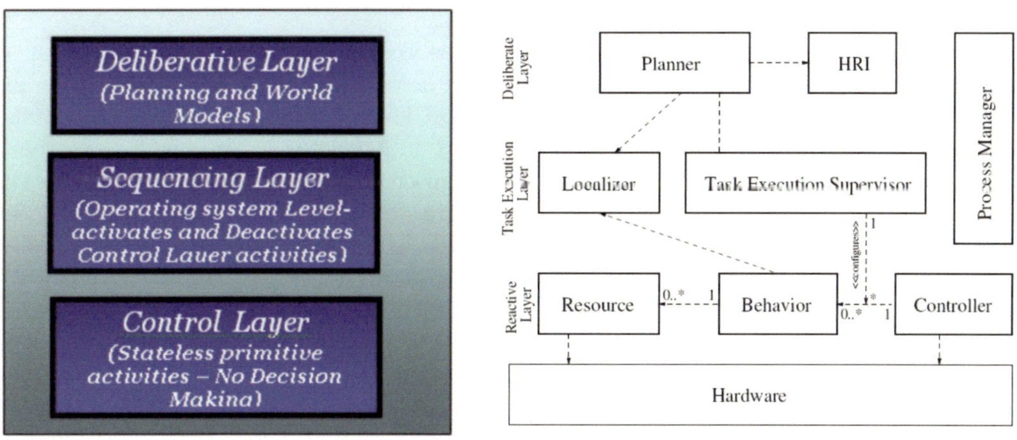


Fig. 2. Three Layer architectures - ATLANTIS (Gat 1992) and BERRA (Lindstrom, Oreback et al. 2000)

The Lowest Layer or control layer is mainly reactive with no decision making and the process computations at this layer use the least amount of CPU time and are tightly coupled to the Sensor-Actuator layer. The Middle level or grey layer bridges the gap between the two layers (Cattoni, Di Caro et al. 1994) and is usually either a Rule Based Layer or a Finite State Machine deciding which set of behaviours should be running. It also acts as a supervisory layer and catches failures of the reactive layer and then executes deliberative plans. The highest level of activities like planning and world modelling are done at this level and these are the areas that need significant computing resources. The advances in distributed computing techniques and communication infrastructure are leveraged in the proposed architecture to offer a decentralized control system.

## 3. Decentralised Hybrid Software approach

The hybrid deliberate/reactive approach has proven very successful, practical and robust in a large number of implementations, and it appears that there is a general agreement among the research community that this is the best type of architecture for Service Robots. However, some types of modules are hard to force into any particular layer. In a general framework, it is imperative that no special architecture is preferred for enforcement and a good support for builders of the hybrid deliberate/reactive architecture is important so that the framework supports parallel execution of behaviours. This is precisely where this proposed architecture scores above the other architectures.

The major problems for robotics today lie, not in the hardware but on the software side. There are plenty of well functioning and robust algorithms developed by competent researchers readily available(Ibrahim and Fernandes 2004). Each new implementation would provide significant gains in the performance and capabilities but it will be lost due to non portability and reuse issues.

While the lowest layer or reactive layer has to be embedded on the robot controller due to the obvious fact that this layer requires the highest response and lowest CPU time, the Middleware layer helps us to switch from the repository of allowable robot behaviours. What was essentially an AI Rule Based Behaviour Switching now graduates to a Location Based Behaviour Switching in the current architecture.

In contrast to the "earlier" or traditional approaches to software reuse, which are built on the paradigm of a set of libraries containing many small building blocks, object-oriented frameworks allow the highest common abstraction level between a number of similar systems to be captured in terms of general concepts and structures. The result is a generic design that can be instantiated for each object system constructed(Lewandowski 1998). The Object oriented framework(Bagchi and Kawamura 1992) , (Miller and Lennox 1990) (Chochon 1993) is ideally suited for capturing the elements common to a family of related systems. In this sense, the framework is essentially a large design pattern capturing the essence of one specific kind of object system. The bulk of the system functionality is captured in the framework, which is maintained as a single entity. Each software system using framework is an instantiation of that framework (Lewandowski 1998). Object and component frameworks can also be seen as a special breed of object-oriented systems — they are extensible, semi-finished pieces of software. Completing the semi-finished software leads to different software pieces, typically specific applications, that share the same core.

Though frameworks have been developed for a wide range of domains, they use common construction principles(Marcus, Wolfgang et al. 2000)

## 4. Overview of Software approaches in Robotics

Overview of some relevant software systems (implemented architectures) can be found in (Mohamed, Al-Jaroodi et al. 2008) (Utz, Sablatnog et al. 2002) . Examples of existing systems are AuRA(Arkin 1990; Arkin and Balch 1997), Task Control Architecture(Simmons 1994), Saphira, Teambots, Smartsoft, Mobility, Player/stage , MIRO(Utz, Sablatnog et al. 2002), LICA (Hu, Brady et al. 1995), ORCA (Oreback and Christensen 2003), BERRA (Lindstrom, Oreback et al. 2000), PeLote (Ruangpayoongsak, Roth et al. 2005) and Loosely coupled Layered architecture for Robot Autonomy CLARAty.(Volpe, Nesnas et al. 2001; Urmson, Simmons et al. 2003)

## 5. Middleware approach

In a distributed system or multiprocessor, Middleware allocates system resources, giving requests to the operating systems on the individual processors to implement those decisions. One of the key differences between Middleware and Software Libraries are that Middleware manages resources dynamically. In a uniprocessor, the operating system manages the resources on the processor (for example, the CPU itself, the devices, etc.) and Software Libraries perform computational tasks based on those allocations. Real world integration and application development for such a distributed system composed of many robotic modules and networked robotic devices becomes very difficult without Middleware. Middleware is prevalent in IT-oriented server systems than in Embedded Systems. Several embedded Middleware systems have been developed based on IT standards but with heavy footprints. Middleware architectures are slowly and steadily evolving to support embedded operations (Schmidt 2002). Middleware is software infrastructure that has been used to successfully integrate and manage software for complex distributed systems (Baliga, Graham et al. 2004; Baliga, Graham et al. 2004). Middleware is generally constructed to provide communication between application software and processes in P2P, client-Server or Publish - Subscribe models. Most Middleware address a particular domain such as Web Services, RTOS etc and define simple and uniform architectures for developing applications in that domain. Standard mechanisms for defining software interfaces and functionalities encourage the development of well-defined and reusable software. The Middleware concepts introduced make implementing the control systems more flexible so that they can be dynamically reconfigured with ease and can be upgraded or adapted in a flexible manner. An appropriate Middleware would allow software components to be integrated easily and provide standard functionalities such as support for robustness and fault tolerance, which can be easily reused in most applications.

Therefore, Middleware Services provide a novel approach offering many possibilities and drastically enhancing the application development for robots. We present a novel decentralized architecture as shown in Figure 2 for navigating and controlling a Service Robot based on control of processes rather than control of discrete actions.

By Process we mean a system element that is independent, and can be freely deployed and versioned. This approach loosely couples the various layers into process components that are well defined entities that can be replaced or made redundant without affecting the rest of the systems. It is shown here how they can be developed and tested separately and integrated later building on the Middleware Framework to provide a systematic approach to developing software that would be easy to integrate, manage, and reuse.

At the core of the framework lies a Smart Wireless Communication device (SWCD) operating on Middleware. The SWCD Telematics framework based design accommodates three dimensions of variability namely:

- varying operating systems,
- varying hardware platforms, and
- varying implementation languages.

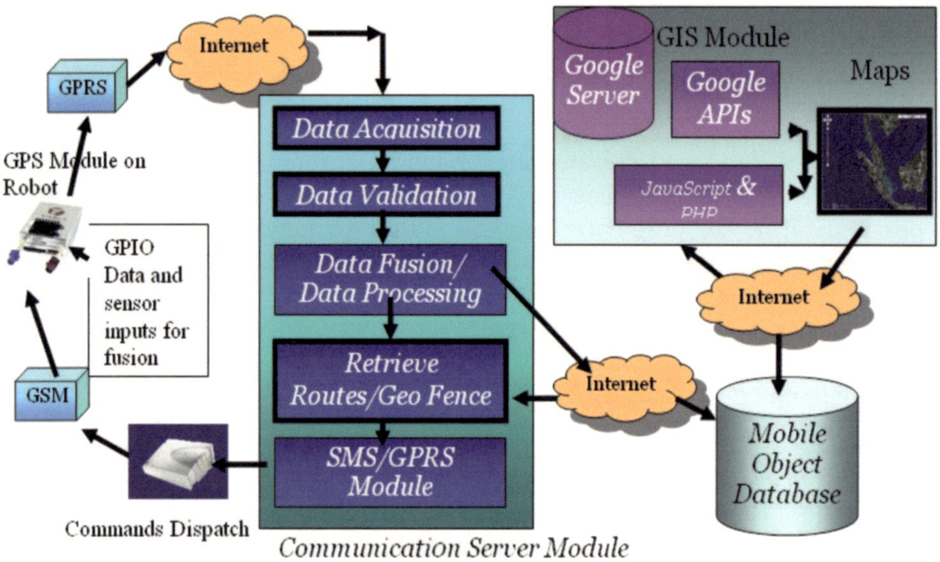


Fig. 3. SWCD based Decentralized System Architecture

Since the framework does a mapping from a constant high-level API to a set of native APIs of different operating systems and hardware platforms, only few interactions are sufficiently complex to warrant dynamic modeling. This static management of variables (as opposed to run-time dependencies) is sufficient since OS, platform, and language dependencies can all be resolved at build time. This still allows for later variability with respect to the features used in an application since the latter will only include those portions of the framework that are actually required and included in a given hardware configuration.(Marco 2003)

## 6. Implementation, Integration and Experimental Test setup

The implementation of the distributed software engineering concepts introduced in the earlier section permits our Service Robot application to be dynamically reconfigured with ease and to be upgraded or adapted in a flexible manner using the P2P, Client Server and Producer-Consumer models.

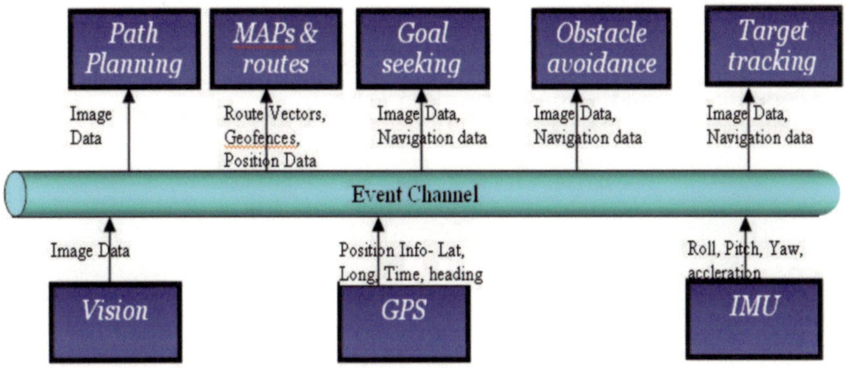


Fig. 4. Producer Consumer Middleware Model

The robot communicates to communication server (GPRS Data Acquisition Module, Data Validation and Command Dispatch Modules located in different physical locations through a secure web connection) in a P2P mode using GPRS over TCP/IP. We utilized Falcom StepIII Telematic modules (STEPPIII 2009) with Middleware (PFAL commands) to dynamically configure and process the sensor modules like GPS units, distance sensors and video camera. The communication between the Telematics terminal and the robot GPIO's (General Purpose Input and Outputs) is shown in the Figure 4-6.

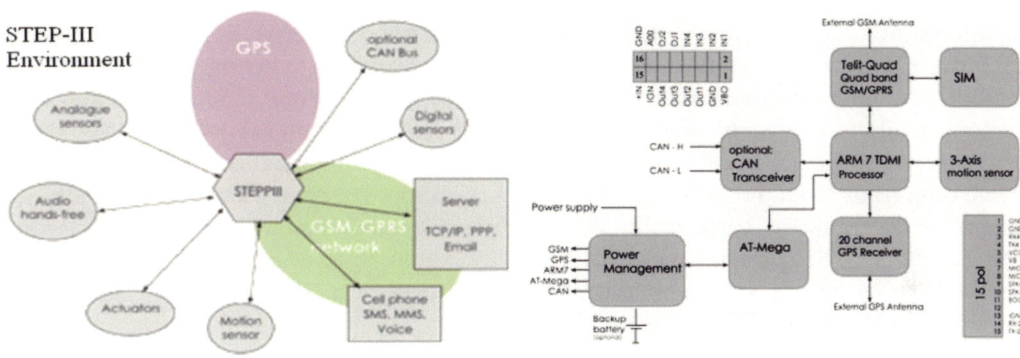


Fig. 5. Telematic unit – Function Blocks in Falcom - Step III.

The Robot communicates with the sensors through the event channels in the Publish – Subscribe mode through the GPIO's. The control components are software modules that perform the tasks of path planning, goal seeking, obstacle avoidance target tracking and localisation. The sensor components consist of device drivers and hardware. The sensors used here are GPS, vision systems and IMU and there exists a loose coupling through the Middleware providing an abstract communication channel referred to in the Figure 4 as Event Channel. The sensors register with the event channel as Publishers of data (e.g. camera as image data and GPS as position data) and Process components (e.g. Obstacle Avoidance and Target Tracking) are Subscribers. Subscribers get the data from whatever is available from the Publishers and new Publishers can be added at will.

## 7. Component Decomposition

The data flow diagram of the communication server is shown in the picture below. Many SWCD's can communicate to the communication server.

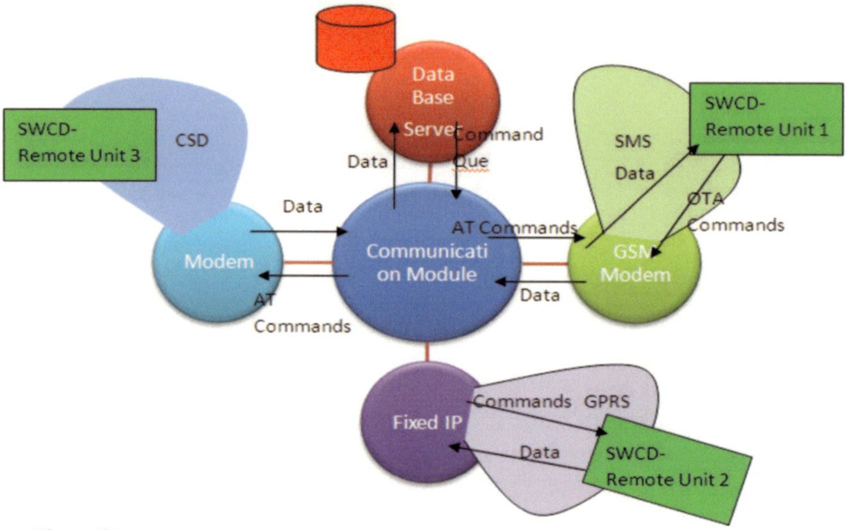


Fig. 6. Data Flow diagrams

The way-point and alert management processes are not shown separately in the picture as they form a part of the communications module. In addition to acting as a gateway for incoming and outgoing data communications, the communications module also acts as a monitor for alerts. Events monitored on the server side like way-points, SWCD not-reporting events, Geofence alerts, etc are handled by the communications module based on the live data feeds from the SWCD and hence, this component is an always-on module. As seen in the diagram, the communications module communicates with the SWCD's using either TCP over GPRS (preferred) or it falls back to SMS (over GSM modem). In case history data needs to be read from a remote unit in the absence of a GPRS network, the module does a data call to the SWCD via the modem.

## 8. GPRS Data Acquisition Module

A heterogeneous asynchronous communication process spread over four process layers and two physical layers is at the core of this design process as shown in Figure 7. The Communication Server Module was developed based on Client/Server architecture to acquire the GPS data over a GPRS network using a TCP/IP connection. In regions of poor GSM coverage the module switches to SMS for command transfer. GPS devices running on GSM/GPRS SIM cards were configured as clients to stream positional information to a Test Communication Server which had to be located external to the university network due to Static IP/Firewall restrictions. The units streamed data directly from GPS devices to an external communication server which performed the processes of data acquisition and validation functions before passing on the data for Data Fusion.

The remaining process of the Communication Services such as data and alert processing, command and alert service responses and configuration despatches was spread over a remote system within the university campus through the web. Therefore, data was collected externally, and the client application was used to stream the bulk data to the Server for processing.

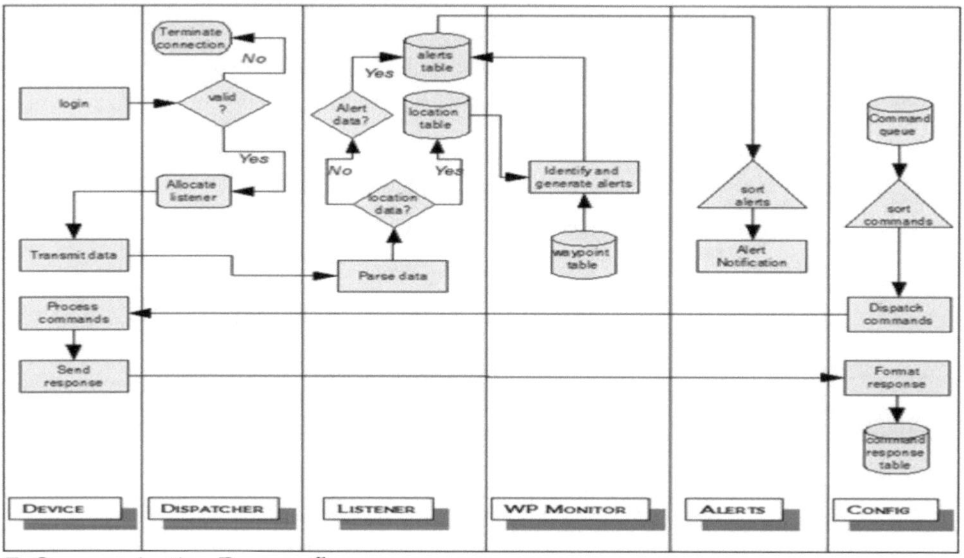


Fig. 7. Communication Process flow

Data received at the server was validated prior to structuring based on the starting characters of the string ($GPRMC) and by checking whether the third element in the string represents character 'A', which implies the validity of the string. Once validated, each string was structured using Sensor Bridge components.

| ID | AsText(LATLON) | TIME | DATE | SPEED |
|---|---|---|---|---|
| 1 | POINT(3.07004 101.59712) | 05:57:59 | 8:9:2005 | 31 |
| 2 | POINT(3.06996 101.59712) | 05:58:09 | 8:9:2005 | 0 |
| 3 | POINT(3.06896 101.59708) | 05:58:15 | 10:9:2005 | 20 |
| 4 | POINT(3.06896 101.59708) | 05:58:24 | 10:9:2005 | 48 |
| 5 | POINT(3.06896 101.59708) | 05:58:24 | 10:9:2005 | 48 |

Fig. 8. Spatial Table containing GPS data

Structuring Data and Data fusion is done using Sensor Bridge. Sensor Bridge is a component suite developed for Visual Studio 2005. It can incorporate data from different types of sensors and actuators. It consists of a hierarchy of class structures designed to manipulate raw sensor data received. Using Sensor Bridge, data received from different GPS devices were structured into separate series tables created from Sensor Bridge Components. Thereafter, the useful information such as latitude, longitude, time, date and speed were extracted to be stored for further processing. The inertial measurement readings obtained through the GPIO of the GPS modules are also streamed along with the position info to provide for near real time location awareness.

## 9. Mobile Object Database and Database Management

A Mobile Object Database (MOD) system was developed using MySQL as the data repository to store the processed GPS data. MySQL is an open source database management system, noted mostly for its speed, reliability and flexibility. Furthermore, MySQL incorporates spatial extensions under the specification of the Open GIS Consortium (OGC), which is an organization that groups many other organizations that prescribe standards for GIS data processing. The Mobile Object Database for this system was developed adhering to the structure of the geometry types proposed by the OGC. Figure 8 depicts the structure of a spatial table created for a GPS device using geometry type 'POINT' to store latitude and longitude values.

| DATE | LAT | LON | SPEED |
|---|---|---|---|
| 8:9:2006 | 3.084960 | 101.592280 | 89.000000 |
| 8:9:2006 | 3.084480 | 101.594440 | 90.000000 |
| 8:9:2006 | 3.083240 | 101.601320 | 85.000000 |
| 8:9:2006 | 3.083440 | 101.600040 | 81.000000 |
| 8:9:2006 | 3.083280 | 101.602120 | 81.000000 |
| 8:9:2006 | 3.084000 | 101.612320 | 35.000000 |

Fig. 9. Filtered and structured Position Information

A database connection between the MIS process and MySQL was established using a .NET connector component provided by MySQL. Data processed and structured into series tables using Sensor Bridge components were written into separate spatial tables to manage and store Geo-fences and Route patterns that have either been acquired through reactive navigation or through route patterns marked on the GIS system as shown in Figure 9, which can be re-transmitted to the Mobile robot through GPRS in order to navigate objects successively.

## 10. Transmission of Routes and Virtual Boundaries

Geo-Fences are virtual boundaries the robot is supposed to remain within to reach the goal and /or keep away to avoid dangers. Geo-Fences and the route vectors saved in the MOD are retrieved from the application to be transmitted back to the corresponding Mobile devices as shown in Figure 10.

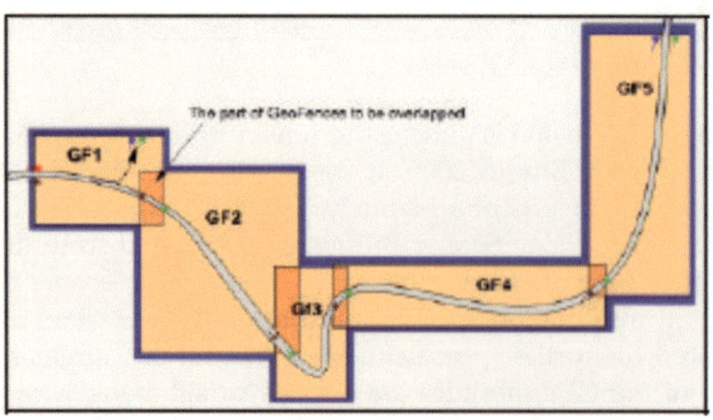


Fig. 10. Route planning and Geo-fence configuration "Over The air"

The Geo-Fences can also be created from the Maps like Google Earth and the boundary information can be sent to the Robot through the Telematic unit. PFAL (Device Middleware) commands were used to configure predefined routes and virtual boundaries in the GPS devices. For route configuration, a virtual boundary was created within a 30m radius for each waypoint in the route as required by the PFAL commands. Figure 10 depicts the configuration of a route to be transmitted to GPS devices

## 11. Over The Air (OTA) Configuration and GPIO Enabling and Disabling

The Device Middleware and the GPRS link enable configuring and reconfiguring the devices as many times as required and this is effectively used to change the control program and actions remotely to use the robot for different services without having to make any major changes to the design, hardware or software.

The feature diagram of a Smart Telematics Platform at various levels of abstraction as applied to a functional Product design is shown below. Middleware framework provides the maximum flexibility as the underlying SWCD hardware Platform is variable with respect to its assortment of components.

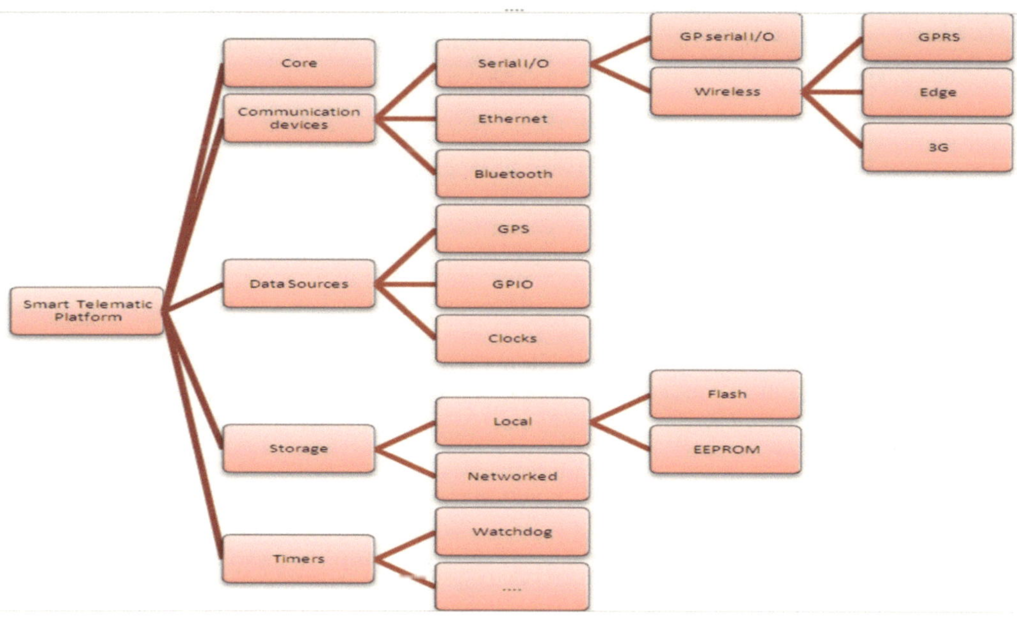


Fig. 11. Variable assortment of components form the variable SWCD hardware Platform

Sample configuration scripts sent over the air to configure, activate and connect to the device is shown below in Figure 12.

```
$PFAL,CNF.Set,PPP.AUTOPING=0,36000000
$PFAL,CNF.Set,PPP.PASSWORD=
$PFAL,CNF.Set,PPP.USERNAME=
$PFAL,CNF.Set,PROT.AREA=0
$PFAL,CNF.Set,PROT.BIN=0
$PFAL,CNF.Set,PROT.GGA=0
$PFAL,CNF.Set,PROT.GLL=0
$PFAL,CNF.Set,PROT.GSA=0
$PFAL,CNF.Set,PROT.GSM=10
$PFAL,CNF.Set,PROT.GSV=0
$PFAL,CNF.Set,PROT.IOP=0
$PFAL,CNF.Set,PROT.RMC=10
$PFAL,CNF.Set,PROT.START.BIN=$!!
$PFAL,CNF.Set,PROT.VTG=0
$PFAL,CNF.Set,TCP.CLIENT.ALTERNATIVE=0,217.1
$PFAL,CNF.Get,TCP.CLIENT.CONNECT=1,80.237.15
$PFAL,CNF.Set,TCP.CLIENT.DNS.TIMEOUT=86400
$PFAL,CNF.Set,TCP.CLIENT.LOGIN=1
$PFAL,CNF.Set,TCP.CLIENT.PING=1,300000
$PFAL,CNF.Set,TCP.CLIENT.SENDMODE=0
$PFAL,CNF.Set,TCP.CLIENT.TIMEOUT=300000,3000
$PFAL,CNF.Set,TCP.SMTP.CONNECT=0,<mailserver
$PFAL,CNF.Set,TCP.SMTP.FROM=<valid mailadres
$PFAL,CNF.Set,TCP.SMTP.LOGIN="<domain>",180,
```

Fig. 12. Middleware configuration scripts sent over GPRS and SMS

The GPIO's data requests can be enabled or disabled "on the fly" through Over The Air Commands to optimise on the performance of the mobile device that is resource and energy starved. Video camera for instance which consumes huge amounts of battery power can be switched ON/OFF or at optimised rates at any time based on events or by the remote operator. Figure 13 shows the Middleware commands that are sent through GPRS or SMS to dynamically configure, enable or reconfigure the GPIO's.

```
$PFAL,CNF.Set,AL14=IO.IN.e0=redge:TCP.Client.Send,100,"INPUT1:HIGH"
$PFAL,CNF.Set,AL15=IO.IN.e0=fedge:TCP.Client.Send,100,"INPUT1:LOW"
$PFAL,CNF.Set,AL16=IO.IN.e1=redge:TCP.Client.Send,100,"INPUT2:HIGH"
$PFAL,CNF.Set,AL17=IO.IN.e1=fedge:TCP.Client.Send,100,"INPUT2:LOW"
$PFAL,CNF.Set,AL18=IO.IN.e2=redge:TCP.Client.Send,100,"INPUT3:HIGH"
$PFAL,CNF.Set,AL19=IO.IN.e2=fedge:TCP.Client.Send,100,"INPUT3:LOW"
$PFAL,CNF.Set,AL20=IO.IN.e3=redge:TCP.Client.Send,100,"INPUT4:HIGH"
$PFAL,CNF.Set,AL21=IO.IN.e3=fedge:TCP.Client.Send,100,"INPUT4:LOW"
$PFAL,CNF.Set,AL22=IO.IN.e6=fedge:TCP.Client.Send,100,"BATTERY:LOW"
$PFAL,CNF.Set,AL23=IO.IN.e7=redge:GPS.Geofence.Park.Remove
$PFAL,CNF.Set,AL24=IO.IN.e7=fedge:GPS.Geofence.Park.Set
```

Fig. 13. GPIO's Enabling and Disabling Events "Over The Air"

The approach to the design of the Middleware Framework appears to be through the mapping of elements of the SWCD platform's feature model (see fig. 11) to the classes and packages. Atomic features have generally been turned into classes and composite features into packages. Additional packages and classes have been created for elements not covered by the hardware feature model but for which abstractions needed to be provided, such as utilities for handling concurrency etc. Classes have been interrelated using refinement/ abstraction, implementation, and generic dependency relations. Relations have been defined in such a fashion that hardware/OS dependencies have been kept within as few classes as possible, reducing the effort posed by adding support for a new platform. Common base and dedicated interface classes have been used to capture commonalities among devices of a kind, such as stream- or message-based communications devices/classes.

## 12. Geographic Information System (GIS)

A GIS system is a technique that manipulates, integrates and maps geographical information based on positional coordinates. (Latitude / Longitude). Many GIS software applications have been developed and integrated to precisely plot and display positional information, such as ArcGIS, MapPoint, Google Maps etc.

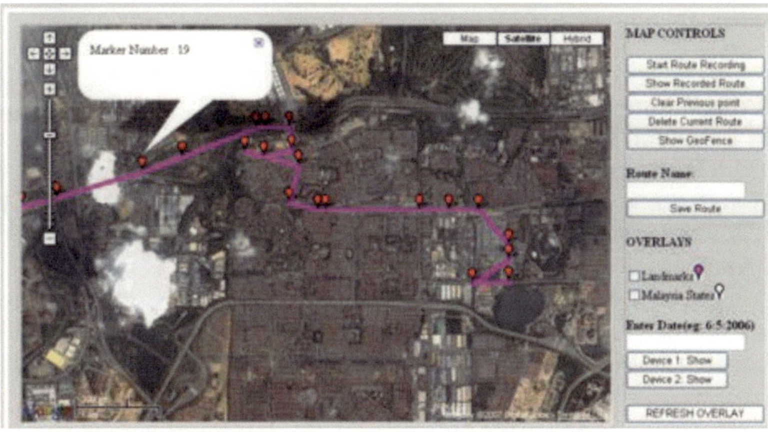


Fig. 14. GIS system used for Planning.

Companion papers (A.U. Alahakone 2007; Ragavan and Ganapathy 2007) describe the GIS systems that was developed along this work as a separate processes to remotely track the mobile object/robot on a web page embedding Google Maps APIs. Google APIs are freely available and accessible on the internet, and provide satellite maps as well as street maps. GIS systems incorporate several layers, each providing a different set of information to represent positional data. Google APIs provides the ability to embed many types of layers to enhance the quality of the data representation of the GIS system. This service as shown in Figure 14 is used to plan the path, create Geo-fence specifications and waypoint locations. Like wise options are available for the robot to be controlled and commanded over the web as a Tele-robot if need be.

## 13. Results and Discussion

In spite of a host of integration issues, software integration of the process modules was seamless. It has been established that dynamic heterogeneous systems can be evolved such as:- an embedded RTOS controller of the robot communicating through a GPRS mobile network, through a Windows based communication server that is a client/server application over TCP/IP, communicating the raw data acquired over a secure connection to be structured, processed, validated and fused at a remotely located server running on a different version of a Windows system across a Firewall, to be further stored in an Open Source Spatial MOD Database System running on MySQL, for further reporting and tracking of the robot movements in near Real-Time over a Web based GIS Tracking System continuously and accurately on a Map and send the information stored or created such as a Geofence or Route all the way back to the robot. The entire operation is completed with a round trip time of a few hundreds of milliseconds. Lastly it should be noted that our hybrid approach has considerably evolved over time based on lessons learnt in real-time and in distributed systems.

As a test example the Path Planning and Navigation System for Service Robots was successfully developed and deployed. It has also been established that Middleware can be used reliably to integrate disparate systems and processes and help in smooth evolution of Complex Dynamical Systems.

## 14. Conclusion

Modules and Components have been developed for an Asynchronous, Loosely Coupled to Uncoupled, Process Based, and Safe-Fail System as discussed and reported in this paper. The processes have been successfully deployed across hybrid and heterogeneous platforms from dedicated RTOS processors on the robot to distributed and disparate server machines connected through the World Wide Web. It has also been established that Middleware can be used reliably to integrate disparate systems and processes and helps in smooth evolution of Complex Dynamical Systems.

Having demonstrated how these strategies can be successfully implemented using the distributed networked software infrastructure such as Middleware, Webware and Hardware, a major challenge lies as future work in understanding how to make the most of it especially,

- understanding the tradeoffs between knowledge representations that are process based reactive, deliberative or hybrid and
- how to reduce the risk by managing software related failures in network controlled systems.
- Secure and fast methods for OTA download.

## Acknowledgements

The authors wish to thank Monash University Sunway Campus for providing grants to purchase GPS modules, sensors, software and equipment necessary for the conduct of these experiments. The authors also wish to thank Transnoble Sdn Bhd, Kuala Lumpur for field support, streamed data, data acquisition and test communication server hosting.

## 15. References

A.U. Alahakone, S. V. Ragavan. (2007). Geographic Information System for Path Planning and Navigation of Mobile Objects. Conference on Innovative Technolgoies in Intelligent Systems & Industrial Applications, Kuala Lumpur , Malaysia. Kuala Lumpur , Malaysia. Proceedings of: Pages 6.

Arkin, R. C. (1990). "Integrating behavioral, perceptual, and world knowledge in reactive navigation." Robotics and Autonomous Systems 6(1-2): 105-22.

Arkin, R. C. and T. Balch (1997). "AuRA: principles and practice in review." Journal of Experimental and Theoretical Artificial Intelligence 9(2-3): 175-89.

Bagchi, S. and K. Kawamura (1992). An Architecture Of A Distributed Object-oriented Robotic System. Intelligent Robots and Systems, 1992., Proceedings of the 1992 IEEE/RSJ International Conference on.

Baliga, G., S. Graham, et al. (2004). Etherware: Domainware for wireless control networks, Vienna, Austria, IEEE Computer Society.

Baliga, G., S. Graham, et al. (2004). "Service continuity in networked control using etherware." IEEE Distributed Systems Online 5(9).

Brooks, R. A. (1991). Intelligence without reason, Sydney, Aust, Morgan Kaufmann Publ Inc.

Cattoni, R., G. Di Caro, et al. (1994). Bridging the gap between planning and reactivity: a layered architecture for autonomous indoor navigation, New York, NY, USA, IEEE.

Chochon, H. (1993). Object-oriented design of mobile robot control systems Springer Berlin/ Heidelberg.
Erann, G. (1998). Three-layer architectures. Artificial intelligence and mobile robots: case studies of successful robot systems, MIT Press: 195-210.
Gat, E. (1992). Integrating Planning and Reacting in a Heterogeneous Asynchronous Architecture for Controlling Real-World Mobile Robots. PROCEEDINGS OF THE NATIONAL CONFERENCE ON ARTIFICIAL INTELLIGENCE, JOHN WILEY & SONS LTD. NUMBER 10, : 809
Hu, H., J. M. Brady, et al. (1995). LICAs: a modular architecture for intelligent control of mobile robots, Los Alamitos, CA, USA, IEEE Comput. Soc. Press.
Ibrahim, M. Y. and A. Fernandes (2004). Study on mobile robot navigation techniques. Industrial Technology, 2004. IEEE ICIT '04. 2004 IEEE International Conference on.
Kawamura, K., R. T. Pack, et al. (1995). Design philosophy for service robots. Systems, Man and Cybernetics, 1995. Intelligent Systems for the 21st Century., IEEE International Conference on.
Lewandowski, S. M. (1998). "Frameworks for Component-Based Client/Server Computing." ACM Computing Surveys 30(1): 3-27.
Lindstrom, M., A. Oreback, et al. (2000). BERRA: a research architecture for service robots, Piscataway, NJ, USA, IEEE.
Marco, G. (2003). A Flexible Object-Oriented Software Architecture for Smart Wireless Communication Devices. Proceedings of the conference on Design, Automation and Test in Europe: Designers' Forum - Volume 2, IEEE Computer Society.
Marcus, F., P. Wolfgang, et al. (2000). The UML Profile for Framework Architectures, Addison-Wesley Longman Publishing Co., Inc.
Miller, D. J. and R. C. Lennox (1990). An object-oriented environment for robot system architectures, Los Alamitos, CA, USA, IEEE Comput. Soc. Press.
Mohamed, N., J. Al-Jaroodi, et al. (2008). Middleware for Robotics: A Survey. Robotics, Automation and Mechatronics, 2008 IEEE Conference on.
Nesnas, I. A. D., A. Wright, et al. (2003). CLARAty and Challenges of Developing Interoperable Robotic Software, Las Vegas, NV, United states, Institute of Electrical and Electronics Engineers Inc.
Oreback, A. and H. I. Christensen (2003). "Evaluation of architectures for mobile robotics." Autonomous Robots 14(1): 33-49.
Ragavan, S. V. and V. Ganapathy (2007). A General Telematics Framework for Autonomous Service Robots. Automation Science and Engineering, 2007. CASE 2007. IEEE International Conference on.
Ruangpayoongsak, N., H. Roth, et al. (2005). Mobile robots for search and rescue, Kobe, Japan, Institute of Electrical and Electronics Engineers Computer Society.
Schmidt, D. C. (2002). "Middleware for real-time and embedded systems." Communications of the ACM 45(6): 43-8.
Schraft, R. D. (1994). "Mechatronics and robotics for service applications." IEEE Robotics and; Automation Magazine 1(4): 31-5.
Simmons, R. G. (1994). "Structured control for autonomous robots." IEEE Transactions on Robotics and Automation 10(1): 34-43.

STEPPIII, F. (2009). "Falcom product website
http://www.falcom.de/fileadmin/downloads/documentation/STEPPIII/STEPPIII_flyer_v1.0.0_pre_web.pdf." last accessed 31.08.2009.

Urmson, C., R. Simmons, et al. (2003). A generic framework for robotic navigation. Aerospace Conference, 2003. Proceedings. 2003 IEEE.

Utz, H., S. Sablatnog, et al. (2002). "Miro - middleware for mobile robot applications." Robotics and Automation, IEEE Transactions on 18(4): 493-497.

Volpe, R., I. Nesnas, et al. (2001). The CLARAty architecture for robotic autonomy, Piscataway, NJ, USA, IEEE.

Zhang, L.-J. (2008). "Introduction to the IEEE transactions on services computing." IEEE Transactions on Services Computing **1**(1): 2-4.

# 15

# Multi-robot collective path finding in dynamic environments

Carlos Astengo-Noguez, Gildardo Sanchez-Ante*,
José Ramón Calzada and Ricardo Sisnett-Hernández*
*Tecnológico de Monterrey, Campus Monterrey, *Campus Guadalajara*
*México*

## 1. Introduction

Planning the motion of a rigid or articulated object among static obstacles is known as the basic motion planning problem. In its simplest definition, it is a purely geometric problem that only considers avoiding collision with static obstacles, given that the robot is the only object that is able to move in the environment. This problem has been extensively studied in the last two decades, and we do know that the problem itself can be very challenging. For instance, it has been demonstrated that planning the motion of a set of polyhedral objects forming an articulated structure in an environment with static polyhedral obstacles is PSPACE-hard (Reif, 1979). That means that although some elegant complete algorithms exist –the most efficient complete and general algorithm for basic motion planning has a $O(2^n)$ complexity (Canny, 1988)-- their prohibitive computational cost has motivated the search for efficient algorithms that can run in significantly less time despite the fact that they are not complete.

In many practical problems the environments is more complex than in the basic motion planning problem. For instance, it might be the case that there are objects that are supposed to be moved by the robot, or the objects might not be rigid but deformable, or there could be objects moving in the environment whose trajectory might not be known in advance by the robot. All those cases can be considered as extensions of the basic motion planning problem. In this chapter we are interested in analyzing problems where several robots are moving in a shared workspace. This topic is raising interest in the community not only because many researchers argue that we have covered very well the simplest cases by designing efficient algorithms for them, but also because some applications are already pushing the envelop towards the automatic generation of behaviors for agents in dynamic and uncertain environments.

The chapter is structured as follows: first, we will introduce some concepts and notation commonly used in motion planning. Then, a description of some methods for motion planning in single-robot environments is offered. The last part of the chapter is about methods for multi-robot motion planning.

## 2. Concepts and Notation

A *configuration* of a robot is a list of parameters that uniquely defines the placement of the robot in its workspace (Lozano-Perez 1979; Lozano-Perez, 1983). For example, the configuration of a rigid body is its position and orientation. A configuration $q$ is free if the object does not collide with the obstacles in the environment or with itself when placed at $q$. For any given robot, there are multiple ways of defining a configuration. The selected definition may affect the geometry of some sets, like the free space, but not their connectivity, which is what matters in our case. The set of all configurations forms the *configuration space* C. The C space is then the union of all the configurations of the robot, some of them free of collision ($C_{free}$) and some others in collision ($C_{obs}$). $C = C_{free} \cup C_{obst.}$ For a polygonal rigid body $R$ that translates in a 2-D polygonal space, the configuration space can be explicitly computed by taking the Minkowsky difference of $R$ and the obstacles. Intuitively, that means that we "grow" the obstacles by the shape of $R$ and then $R$ becomes a point in this space. This transformation implies that finding the path for a robot means constructing a one-dimensional curve in this space, regardless the fact that the robot itself can be located in a three-dimensional space and may have many degrees of freedom (dof). It is important to mention that the configuration space will have as many dimensions as degrees of freedom may have the robot.

There is a distinction between path planning and motion planning. A *path* is a continuous curve on the configuration space. It is represented by a continuous function that maps some path parameter, usually taken to be in the unit interval [0, 1], to a curve in $C_{free}$ . The choice of unit interval is arbitrary; any parameterization would suffice. The solution to the path planning problem is a continuous function $c \in C^0$ such that
$c : [0, 1] \rightarrow C$ where $c(0) = q_{start}$, $c(1) = q_{target}$ and $c(s) \in C_{free}\ \forall\ s \in [0, 1]$.

When the path is parameterized by time *t*, then *c(t)* is a trajectory, and velocities and accelerations can be computed by taking the first and second derivatives with respect to time. This means that c should be at least twice-differentiable. Finding a feasible trajectory is called trajectory planning or motion planning.

*Navigation* is the problem of finding a collision-free motion for an agent-system from one configuration (or state) to another. The agent could be a videogame avatar, a mobile robot, or something else. Localization is the problem of using a map to interpret sensor data to determine the configuration of the agent. *Mapping* is the problem of exploring and sensing an unknown environment to construct a representation that is useful for navigation or localization. Localization and mapping can be combined.

## 3. Single-Robot Motion Planning Algorithms

Most of the work in motion planning has been done considering the case of a single robot moving in an environment populated with static obstacles. The problem can then be stated as finding a collision-free path from any given starting position to a goal or desired location for the robot. In some cases, a function to measure cost is introduced, so that the algorithm is able to search for the optimal path (i.e., minimum cost).

In general, we could classify the algorithms as complete or incomplete. On the one hand, a **complete** algorithm is one that either finds a solution or proves that it does not exist. Some authors call these algorithms *exact*. Algorithms under this classification are usually computationally expensive and, by consequence, impractical for many important instances of the problem. That is the case of the algorithm of Canny (Canny, 1987). Canny's algorithm is the most efficient general path planning algorithm known to date, with a time complexity of $O(2^n)$. The method involves powerful techniques from real algebraic geometry (Canny, 1987; Canny, 1988), but is nevertheless exponential in $n$.

On the other hand, **incomplete** algorithms are not able to offer such guarantee. When a complete algorithm is impractical, despite its elegance, what computer scientists would like to have is an algorithm that satisfies another notion of completeness, such as resolution completeness or probabilistic completeness. In the first case, we say that an algorithm is *resolution complete* if its accuracy arbitrarily improves when the resolution is increased, and it becomes an exact algorithm at the limit where the resolution approaches the continuum. Some of the cell decomposition methods are resolution complete, for instance (Zhu, 1990; Kondo, 1991). For the second case, we say that an algorithm is probabilistically complete if the probability of finding a solution (if one exists) approaches 1, as the running time increases. Algorithms such as those described in (Barraquand, 1990; Kavraki, 1996; Hsu, 1997) are probabilistically complete.

One of the main drawbacks of resolution complete methods is that the number of points required for the discretization of the configuration space grows exponentially in the number of degrees of freedom. Conversely, in a probabilistic complete method, we would need to guarantee an adequate coverage of the configuration space in order to have a high probability of finding a path whenever it exists.

Some well known examples of motion planning algorithms are:

**Cell Decomposition**: Based on the idea of decomposing the $C_{free}$ space into convex regions (either regular or not) and using that discretization to build a representation of the connectivity of $C_{free}$ (usually a graph), and then searching for the path in the graph. Cell decomposition is an approach that sounds simple at first sight, but whose complexity grows quickly with the number of degrees of freedom of the robot. Also, since it is conceived to work on the $C$ space, it requires the computation of $C_{free}$, which may take very long time, depending on the dofs of the robot. It is an approach that, in general, can only be applied in very simple environments, making it unsuitable for real problems where obstacles may have complex geometries and the robot may have many degrees of freedom (Latombe, 1991).

**Potential Fields**: Khatib introduced this approach, in which the idea is to consider the point robot in the configuration space as a charged particle under the influence of an artificial potential field in a way that the particle is "pushed" away from the obstacles and "attracted" towards the goal. The vectorial sum of those forces defines the motion of the robot to a new location and then the potentials are computed again and the whole loop is repeated. The approach originally was intended to be used as an on-line process for mobile robots

equipped with sensors such as sonars. It has proven to be an approach useful for real robots moving on a plane, i.e., 2 or 3 degrees of freedom (Khatib, 1983).
The main drawback of this approach is the tendency it has to get trapped in local minima (since it relies on a greedy search algorithm). Also, unless the C space is pre-computed, it is not possible to guarantee optimal paths or even the completeness of the algorithm. In practice, the main problem is usually the definition of "good" functions to represent attractive and repulsive forces.

**Roadmaps**: In this approach, the idea is to represent the connectivity of the $C_{free}$ space by a one-dimensional curve, called roadmap. Once this curve is constructed, it is used to search for paths connecting some given initial and goal configurations. There are different ways of constructing the roadmap, such as: visibility graphs, Voronoi Diagrams, Silhouettes, etc. The main problem with the roadmaps approach is that computing such curve in high-dimensional spaces is almost prohibitive in terms of computational time, making the approach not useful for environments involving many degrees of freedom.

More recently, a new family of methods have been proposed, based on the idea of sampling the configuration space instead of actually computing it. The approach relies on recent results for algorithms that can enable the computation of collision checks in very short amounts of time, such as object oriented bounding boxes (OBB) (ref), axis aligned bounding boxes (AABB), Spheres and other options (Lin, 1998).

The methods based on this idea are called Probabilistic Roadmaps (PRM), and we will describe them in more detail in the next section.

**The PRM Planning Approach**

Sampling-based motion planning is a well-known concept, (see, for instance (Donald, 1987)) that was originally used to deal with some difficulties encountered while implementing complete planners. PRM planning consists of sampling the configuration space at random and testing the sampled points, as well as connections between them, for collision.

The obstacle regions in configuration space make explicit the constraints on the possible motions of a robot. These constraints derive from the interaction between the geometric shapes of the robot and the obstacles in the workspace. Small and geometrically simple obstacles in the workspace may yield complex and large obstacle regions in the configuration space C.

There are two major issues in computing an explicit geometric representation of the obstacle regions in C. First, C has as many dimensions as the robot system has dofs. Second, the shape of $C_{free}$ may be complex even when both the robot and the obstacles have simple shapes.

For those reasons, computing an explicit geometric representation of $C_{free}$ is not possible in practice (even with much greater computational power than is available today). On the other hand, efficient collision-detection techniques have existed for some years, which can determine quickly whether an arbitrary robot configuration is collision-free, or not.

The existence of fast collision-checking techniques has led to the idea of probing the configuration space at random. One may pick many sampled configurations and test them for collision. Thus, the collision-free samples form a discrete approximation of the free space. This is the basic idea underlying probabilistic roadmaps (Kavraki, 1994).

There are two main classes of PRM planners: multi-query and single-query planners. Historically, multi-query planners were developed first (Kavraki, 1994; Kavraki, 1996; Svestka, 1997; Amato, 1998b}. Single-query planners are more recent (Hsu, 1997; Hsu, 2000; Kuffner, 2000; LaValle, 2001), but they have significant advantages over multi-query planners. Additionally, mixed planners have also been proposed in (Amato, 1998; Bohlin, 2000; Nielsen, 2000; Song, 2001). Basically, their goal is to distribute the time over a pre-computation and a query phase.

**Multi-Query Planners**

A multi-query PRM planner operates in two phases. It first pre-computes a probabilistic roadmap *R*. Then it uses *R* to answer an arbitrary number of queries, each defined by a pair of configurations. Each query must be made in the same robot-obstacle environment for which *R* was computed. The pre-computation of *R* may be rather expensive, but it is "amortized" over the several queries that are subsequently made (Kavraki, 1994; Kavraki, 1996). Processing one query is usually extremely fast.

*Building the roadmap*

A roadmap *R* is pre-computed by repeatedly sampling the configuration space *C* at random. Each sample is tested for collision, and the collision-free samples are retained as *milestones*. Then, the planner connects pairs of milestones that are not too far apart by simple paths and retains those which test collision-free as *local paths*. The milestones and local paths form a network over $C_{free}$, which is called the *probabilistic roadmap*. It is stored as an undirected graph, which usually has a large number of vertices (typically, several thousands to a few millions) (Kavraki, 1994; Kavraki, 1996).

More formally, the algorithm is as follows (*d* is the metric function in *C*):

BUILD_ROADMAP

1 $R \leftarrow$empty graph
2 Repeat until *s* milestones have been generated
    3 Pick a configuration *q* uniformly at random in *C*
    4 If *q* is collision-free then add *q* as a new vertex (milestone) of *R*
    5 For each pair of milestones *q, q'* such that $d(q,q') \leq \rho$
    6 If the line segment joining *q* and *q'* tests collision-free then
      add it as an edge (local path) of *R*
7 Return *R*

The algorithm consists of two independent loops. The first loop (lines 2-4) adds milestones to the roadmap. The parameter $s$ defines the number of milestones to be generated. It is often called the *size* of the roadmap. The second loop (lines 5-7) establishes the local paths. Local paths are created only between milestones that are closer apart than some predefined distance ρ. Indeed, there are $O(s^2)$ pairs of milestones and testing all segments would be too costly. Moreover, if two milestones are far apart, the segment joining them is unlikely to be collision-free, and, if it is collision-free, then the above algorithm is likely to connect the two milestones by a sequence of local paths through intermediate milestones. Figure 1 illustrates the process.

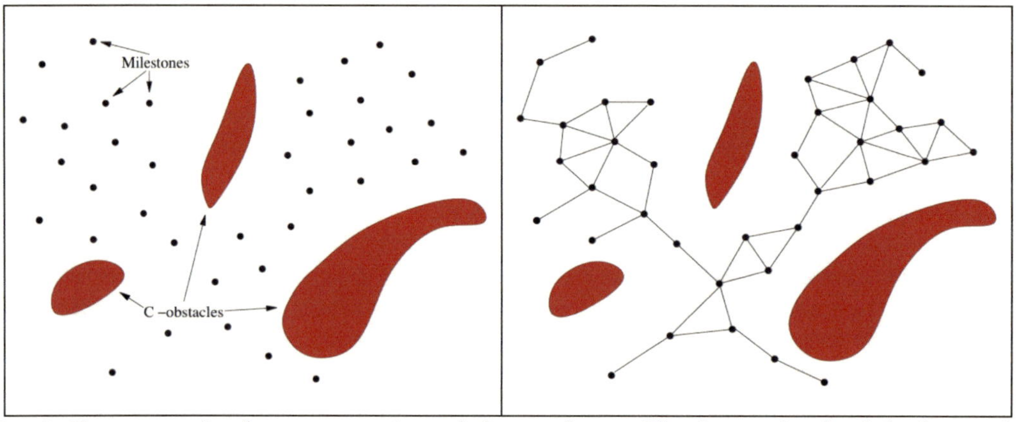


Fig. 1. Two steps in the computation of the roadmap. The image in the left shows the sampled milestones. The image on the right shows the graph built after calling the local planner.

*Querying the roadmap*

A query is defined by two configurations, $q_i$ and and $q_g$. To answer the query, the planner first attempts to connect each of these configurations to a milestone of $R$ by a local path. If the two connections succeed, then the planner searches $R$ for a sequence of local paths connecting $q_i$ and $q_g$. Such a sequence, if one is found, constitutes a free path for the robot.

The algorithm is as follows:

QUERY_ROADMAP($q_i$,$q_g$)

1 Repeat for all $q \in R$ such that $d(q, q_i) \leq \rho$
    2 If the line segment joining $q_i$ and $q$ tests collision-free then connect $q_i$ to $q$ and exit loop
3 If $q_i$ have not been connected to a milestone of $R$ then return failure
4 Repeat for all $q \in R$ such that $d(q, q_g) \leq \rho$
    5 If the line segment joining $q_g$ and $q$ tests collision-free then connect $q_g$ to $q$ and exit loop
6 If $q_g$ have not been connected to a milestone of $R$ then return failure
7 Search $R$ for a path connecting $q_i$ to $q_g$
8 If a path has been found, then return this path, else return *no path*

The algorithm returns *failure* if it fails to connect $q_i$ or $q_g$ to the roadmap. It returns *no path* if it fails to find a path connecting $q_i$ to $q_g$ after they have been successfully connected to the roadmap. The possible interpretations of these two outcomes will be discussed below.

*Probabilistic completeness*

There are two cases where QUERY_ROADMAP does not return a path: (1) if it fails to connect $q_i$ or $q_g$ to the roadmap, and (2) if it fails to find a path in the roadmap. Clearly, it is desirable that the first case happens as rarely as possible. In the second case, the algorithm's output is *no path*. This output may be correct, as it is possible that $q_i$ and $q_g$ belong to two distinct components of the free space $C_{free}$. But it may also be incorrect: $q_i$ or $q_g$ may belong to the same component of $C_{free}$, but the roadmap $R$ may have more than one component lying in it. It is desirable that the planner rarely returns "*no path*" incorrectly.

**Single-Query Planners**

Multi-query PRM planners are appropriate when the pre-computation of a roadmap can be amortized over a rather large number of queries performed in the same environment. However, in practice, the number of queries in a given environment is rather small, as one or several objects are often moved, deleted, or added between two queries. Single-query PRM planners are a better solution in those cases.

A single-query PRM planner computes a new probabilistic roadmap for each query (Hsu, 1997; Hsu, 2000; Kuffner, 2000). While multi-query planners must use a sampling strategy that covers well the whole free space, in order to later successfully deal with any query, a single-query planner applies a more focused strategy aimed at exploring the smallest portion of free space needed to find a solution path. More precisely, it takes advantage that it knows the two query configurations to explore restricted subsets of the components of $C_{free}$ that are reachable from these configurations. This is done either by growing one tree of milestones rooted at one query configuration, until a connection is found with the other query configuration (*single-directional* search), or by growing two trees concurrently, respectively rooted at one of the two query configurations, until a connection is found between the two trees (*bi-directional* search) (Hsu, 2000). In both cases, milestones are iteratively added to the roadmap. Each new milestone $m'$ is selected in a neighbourhood of a milestone $m$ already installed in a tree and is connected to $m$ by a local path (hence, $m'$ becomes a child of $m$). Bi-directional planners are usually more efficient than single-directional ones (Amato, 1998; Hsu, 1998; Hsu, 1999; Hsu, 2000). Fig. 2 shows an example of a single-query planner in process.

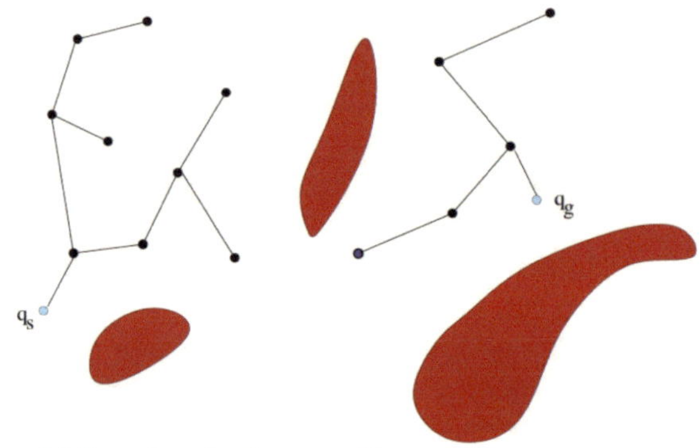


Fig. 2. A single-query, bidirectional planner. Two trees are grown rooted at the start and goal configurations.

A single-query planner is allowed to generate a maximal number $s$ of milestones. Either it outputs a free path between the query configurations, or it indicates that it has not found a path after generating $s$ milestones. The second output occurs when the two query configurations lie in two distinct connected components of $C_{free}$. It may also occur when a solution path exists, but the planner did not find one because $s$ was set too small.

In (Hsu, 2000), it is shown that if $C_{free}$ is expansive, then the probability that a slightly idealized version of a single-query PRM planner fails to find a path when one exists goes to 0 exponentially in $s$. This property defines the probabilistic completeness of a single-query PRM planner. The proof in (Hsu, 2000) requires that the milestones generated by the planner eventually ``diffuse'' through the components of $F$ reachable from the query configurations.

**Intermediate Planners**

There are several planners which posses characteristics of both multi-query and single-query planners. Below we briefly sketch some of them.

In (Bohlin, 2000) a Lazy PRM is described. The algorithm is similar to the original PRM (Kavraki, 1994) in the sense that the aim is to find the shortest path in a roadmap constructed by randomly distributed configurations. Nevertheless, in this approach, instead of constructing a roadmap of feasible paths, a roadmap of paths *assumed* to be feasible is build. The idea is to lazily evaluate the feasibility of the roadmap as planning queries are processed. In other words, a number of uniformly distributed points form nodes in a roadmap, and the connections between nodes being sufficiently close form the edges on the roadmap. Neither nodes nor edges are validated until a possible solution path is found. At that moment, both edges and nodes are checked for collision. If a collision is found, the corresponding node/edge is removed and the search process is re-started.

In (Nielsen, 2000) a Fuzzy PRM planner is presented. In such approach, the process is started in the query phase, and if the roadmap does not contain a possible solution path, it enters to the learning phase, adding milestones and edges. The milestones are always collision-free configurations, while in the case of the edges, they are annotated by a probability. This probability is an estimate of the chance that the edge is actually feasible. The query phase is split into three steps: update, search and upgrade. The update step adds nodes to the graph, starting with the query ones. In the search step, the most probable path is found from start to goal configurations on the graph. The upgrade step is used to do the actual verification of the path. As a result of the application of this step, the probabilities on the edges are upgraded.

In (Song, 2001) a "Customizable" PRM planner is described. In the learning step a coarse roadmap is constructed by performing only approximate validation of nodes and edges. In the query step, the roadmap is validated and refined only in the area of interest for the query. Moreover, it is ``customized" in accordance with any specified query preferences (e.g., maintaining certain clearance from the obstacles).

In (Vallejo, 2001) an adaptive framework for single-query (or *single-shot*) planning is presented. In this approach, two trees are constructed, rooted on the start and goal configurations. In each iteration, it is attempted to generate a path that connects both query configurations. To do this, all potential query pairs with one configuration in each tree, and all the algorithms in the bank are evaluated, and it is selected the query pair and algorithm combination that is most likely to make a connection. The approach assumes that several planners are available.

## 4. Multiple-Robot Motion Planning Algorithms

There are two established approaches to multi-robot motion planning: centralized and decoupled (Latombe, 1991). So far, the prevalent approach has been decoupled planning. In most cases, centralized planning has been beyond the practical capabilities of existing planning techniques, as it requires searching configuration spaces with many dimensions. Instead, decoupled planning breaks the original planning problem into several more tractable sub-problems. Despite the fact that decoupled planning is known to be inherently incomplete - that is, it is not guaranteed to find a solution whenever one exists - it has been assumed that the loss of completeness is relatively small in most practical cases and worth the gain in computational time.

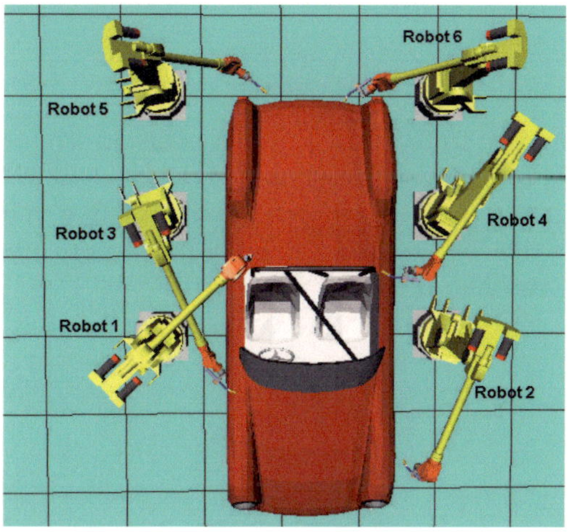


Fig. 3. Model of six-robot spot-welding station.

**Centralized planning**

It consists of considering all the robots involved in the problem as if they were forming a single multi-limb robot, by encoding all their dofs in a single "composite" configuration space C and searching that space for a collision-free path between the initial composite configuration and the goal one. In such case, the "total" configuration space is given by the combination of the configuration spaces of all the robots: $C = C_1 \times C_2 \times ... \times C_p$, where $p$ is the number of robots and $C_i$ is the configuration space of the $i$-th robot ($i \in [1,p]$). Thus, the number of dimensions of C is equal to the total number of dofs of the robots. In the example of Fig. 3, where each robot has 6 dofs, the composite configuration space has 36 dimensions.

Let $\tau$: $s \in [0,1] \leftarrow \tau(s) \in F$ be a path in the free subset $F$ of $C$. The projection $\tau_i$ of $\tau$ into the subspace $C_i$ is the path to be followed by the $i$-th robot. For each $s \in [0,1]$, $\tau(s)$ is of the form $(\tau_1(s), \tau_2(s), ... , \tau_p(s))$, which describes the configurations of the $p$ robots at a single point along the path $\tau$. Hence, a collision-free path in $F$, if one exists, not only describes the individual path to be followed by each robot, but also how the robots are to be coordinated.

In principle, any sufficiently general path-planning algorithm can be used to implement centralized planning. This only requires applying this algorithm to the composite space C. However, in the past, centralized planning has not been considered practical because it usually leads to searching large-dimensional configuration spaces that are beyond the practical capabilities of existing planning techniques. Most proposed centralized planners have been based on ad-hoc and incomplete heuristics, for example potential field techniques, which are too unreliable to be widely useful (Tournassoud, 1986; Barraquand, 1990; Barraquand, 1991; Barraquand, 1992). Complete centralized planning algorithms have only been proposed for very simple robotic systems, e.g., the coordination of discs among polygonal obstacles (Schwartz, 1983). The complexity of the algorithm described in that work is $O(n^3)$ for two discs, and $O(n^{13})$ for three discs.

Centralized approaches have the advantage that, at least in theory, they allow for complete planners.

**Decoupled planning**

This is a two-phase approach. In the first phase, a collision-free path is generated for each robot by considering only the obstacles in the environment and ignoring the other robots. In the second phase, called *velocity tuning*, the relative velocities of the robots along their respective paths are selected to avoid collision among them (Kant86, Odonnell89, Alami98,Aronov99}.

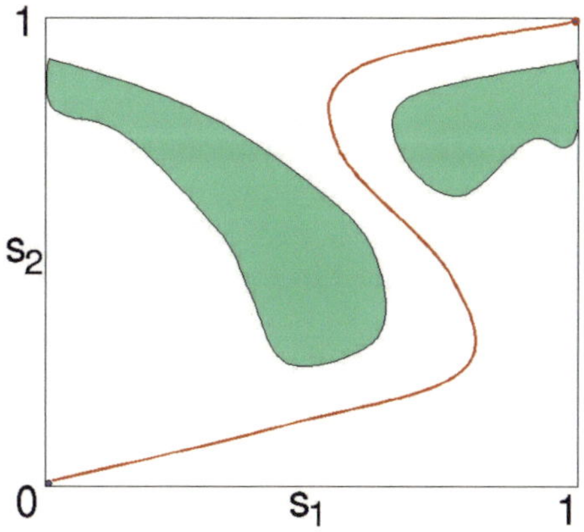


Fig. 4. Coordination space for two robots.

Velocity tuning consists of searching a coordination space. Consider two robots, and let $\tau_1$ and $\tau_2$ be the two paths (one for each robot) generated in the first phase of decoupled planning. By forcing the robots to move along these paths, we actually reduce the number of dofs of each robot to 1, hence the dimension of their composite configuration space -- now called the *coordination space* -- to 2 (O'Donnell, 1989).

Let each path $\tau_i$ (i = 1,2) be parameterized by some $s_i \in [0,1]$. The set $P = [0,1] \times [0,1]$ represents the coordination space of the two robots (see Fig. 4). Each point $(s_1,s_2) \in P$ defines a placement of the two robots at their respective configurations $\tau_1(s_1)$ and $\tau_2(s_2)$. This point is collision-free if at this placement the two robots do not collide with each other. (Collisions with obstacles in the environment were already taken care of during the generation of $\tau_1$ and $\tau_2$). A path joining the point (0,0) -- where both robots are at their respective initial configurations -- to the point (1,1) -- where they are at their goal configurations -- in the collision-free subset of *P* defines a valid coordination of the two robots along $\tau_1$ and $\tau_2$; it determines the relative velocities of the robots along their respective paths. Note that this path may not be monotonic along any of the dimensions of *P*. If it is not non-monotonic along $s_i$ (i = 1 or 2), then for a while the *i*-th robot will move backward along $\tau_i$. Such motion may be required to provide maneuvering space to the second robot. An unfortunate choice of $\tau_1$ and $\tau_2$ in the first phase of decoupled planning may lead the points (0,0) and (1,1) to lie in two distinct connected component of the free subset of *P*.

If there are $p > 2$ robots, one may coordinate all the robots by generating a collision-free path in the *p*-dimensional space *P* where the *i*-th axis encodes the parameter $s_i$ of the path of the *i*-th robot, from the point (0,...,0) to the point (1,...,1). We term this approach to velocity tuning *global coordination.* An alternative, *pairwise coordination,* consists of planning *p*-1 paths in a series of *p*-1 two-dimensional spaces $P_2,...,P_p$. The axes of $P_2$ encode the parameters $s_1$ and $s_2$ along the paths of the 1st and 2nd robots, and a collision-free path $\tau_{1,2}$: $s_{1,2} \in [0,1] \rightarrow \tau_{1,2}(s_{1,2}) \in P_2$ defines a valid coordination of these two robots. One axis of $P_3$ encodes the parameter $s_3$ along the path of the 3rd robot, while the other axis encodes the parameter $s_{1,2}$ along the coordinated path of the 1st and 2nd robots. Hence, each point in $P_3$ determines a placement of the first three robots, and a collision-free path in $P_3$ defines a valid coordination of these robots.

Decoupled planning leads to searching lower-dimensional spaces than centralized planning. But, it is inherently incomplete, even if the core planning algorithms used in the first and second phases are complete. Velocity tuning may fail because the paths generated in the first planning phase cannot be coordinated without collision between robots, while this coordination would have been possible if other paths had been selected. A decoupled planner based on global coordination is less incomplete than one based on pairwise coordination, since a specific path selected in the path space $P_i$ may result into a space $P_{i+1}$ with no collision-free path between (0,...,0) and (1,...,1). Nevertheless, in the past, pairwise coordination has been more widely used than global coordination, since it only requires planning in two-dimensional spaces. In theory, when velocity tuning fails, the planner could backtrack and generate new robot paths. But this option has rarely been used. Indeed, it is difficult to extract from a failure the information that can be used to generate new paths. Moreover, backtracking quickly increases the planner's running time.

An alternative to velocity tuning, called *prioritized planning,* is proposed in (Erdmann, 1986). It consists of processing the robots in some predefined order and planning the path of each robot by treating the robots whose paths have already been planned as moving obstacles of known trajectories. A problem with this approach is finding a good way of defining the priorities for the robots, as this assignment affects the likelihood of finding the solution.

In (Sánchez, 2002), we describe the use of the SBL planner to implement both the centralized and the decoupled approaches. Interestingly, SBL can be invoked at each stage of decoupled planning, not only to plan individual paths of robots, but also to coordinate these paths (velocity tuning). We give experimental results obtained for the model of a 6-robot spot-welding station shown in Fig. 3, comparing the relative performance and reliability of centralized planning, decoupled planning with global coordination, and decoupled planning with pairwise coordination (these terms will be precisely defined below). The results reveal that, in the context of multi-robot spot welding, which requires rather tight robot coordination, decoupled planning is too un-reliable to be practical. This is an important observation, since it invalidates the assumption that the loss of completeness in adopting decoupled planning is not very significant in practice and indicates that centralized planning is a more desirable approach. By no means, however, does this imply that decoupled planning is useless. First, it may be reasonably reliable for other applications where interactions among robots are less constraining. Second, there are distributed-robot systems where centralized planning is not possible because no robot or processor knows the global state of the system or the goals of all robots. Finally, even in cases where decoupled planning is possible but unreliable, a decoupled planner may still have some utility if it receives interactive hints from a human user.

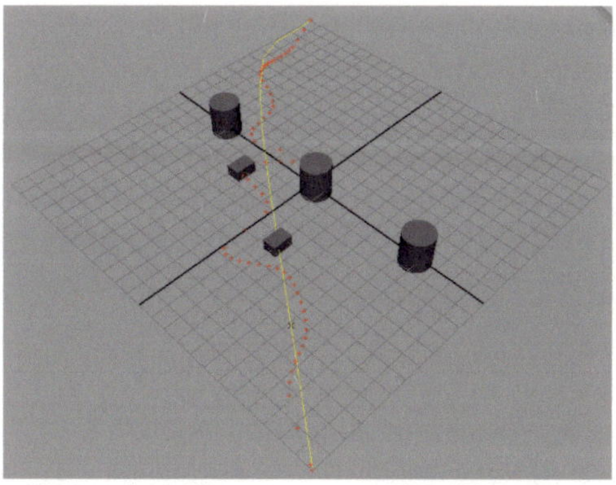

Fig. 5. Snapshot of a path obtained in Maya™ using SBL Planner

We have recently programmed SBL inside Autodesk Maya software, using Python. In Fig 5. it is shown the path obtained by SBL (red) and the optimized path (yellow). In general, computing such paths take less than a second for uncluttered 3D environments and a robot with 2-3 dofs.

## 5. Multiple-Robot Planning as Multi-Agent Systems

A somewhat different way of dealing with the coordination of multiple robots is based on the idea of the robots forming a multi-agent system. A multi-agent system (MAS) is a system composed of multiple interacting intelligent agents. Multi-agent systems can be used to solve problems which are difficult or impossible for an individual agent to solve.
The agents in a multi-agent system have several important characteristics (Shoham, 2008):

- *Autonomy*: the agents are at least partially autonomous.
- *Local views*: no agent has a full global view of the system, or the system is too complex for an agent to make practical use of such knowledge.

* *Decentralization*: there is no designated controlling agent (or the system is effectively reduced to a monolithic system).

Typically multi-agent systems research refers to software agents. However, the agents in a multi-agent system could equally well be robots, humans or human teams. A multi-agent system may contain combined human-agent teams.

Multi-agent systems can manifest self-organization and complex behaviors even when the individual strategies of all their agents are simple.

Agents are assumed to operate in a planar ($R^2$) or three dimensional ($R^3$) environment or (vectorial) space, called workspace *W*. This workspace will often contain obstacles; let $WO_i$ be the i-th obstacle. Motion planning, however, does not usually occur in the workspace. Instead, it occurs in the configuration space (also called C-space).

The game industry is demanding algorithms that allow multiple agents to plan for non-colliding routes on congested-dynamical environments (Pottinger, 1999), such problems also appear in flock traffic navigation (FTN) based on negotiation (Astengo, 2007).

FTN uses A* for path planning and, only at intersections, flocks use the Dresner and Stone reservation algorithm (Dresner, 2005). We intend to improve this algorithm using cooperative strategies.

**Cooperative Pathfiding**

In cooperative pathfinding each agent is assumed to have full knowledge of all other agents and their planned routes (Silver, 2005). The complementary problem is called "non-cooperative pathfinding", where the agents have no knowledge of each other's plans and must predict their future movements. There is also another approach called "antagonist pathfinding" where agents try to reach their own goals while preventing other agents from reaching theirs.

FTN based on negotiation uses a decoupled approach called local repair A*: each agent searches for a route to the destination using A*, ignoring all other agents except for its current neighbors. It is in this neighborhood that negotiation takes place and the flock is created. The agents then follow their route (according to the bone-structure) until a collision is imminent.

It is clear that collisions will happen at the intersections, so there are, with this approach, two possible solutions:

1. The Dresner and Stone reservation method (Dresner, 2004; 2005; 2006; 2007).
2. The Zelinsky (Zelinsky, 1992) brute force algorithm: whenever an agent or a flock is about to move into an occupied position it instead recalculates the remainder of its route.

The implementation of each method depends on the information and the time that the agent or flock has at that particular moment. The Zelinsky algorithm usually suffers from cycles and other severe breakdowns in scenarios where bottlenecks are present (Pottinger, 1999; Zelinksy, 2992).

The Dresner and Stone reservation model was developed for individual agents that can accelerate or decelerate according to the reservation agenda. In simulations performed it was shown that, at the beginning, it works properly, but as time moves on the agents are eventually stopped.

**Cooperative Pathfinding with A***

The task is decoupled into a series of single agent searches. The individual searches are performed in three-dimensional space-time and takes the planned routes of other agents into account. A wait move is included in the agent's action set to enable it to remain stationary. After each agent's route is calculated the states along the route are marked into a reservation table. Entries in the reservation table are considered impassable and are avoided during searches by subsequent agents.

The reservation table represents the agents' shared knowledge about each other's planned routes. The reservation table is a three-dimensional grid: two spatial dimensions and one time dimension. Each cell that is intersected by the agent's planned route is marked as impassable for precisely the duration of the intersection, thus preventing any other agent from planning a colliding route. Only a small proportion of grid locations will be touched, and so the grid can be efficiently implemented as a hash table (Silver, 2005).

**New Cooperative Pathfinding Algorithm**

Using the Cooperative A* a D* algorithms as a starting point, we can propose a new algorithm that can use similar techniques to plan paths in a dynamic environment in which several agents exist.

As in Cooperative A*, a reservation table is used to store the agents' planned paths and the time at which they will occupy certain regions in space. Assuming a three-dimensional space, we need a four-dimensional table that allows us to reference a specific point in space at a certain point in time. Any dynamic elements of the environment and their movement need to be included in the table as well.

Agents rely on a visibility index to determine how far ahead in time they can detect potential collisions with other agents or objects. In a fully cooperative environment where all agents have complete access to the planning information of other agents, this index is equal to the total amount of time required for all agents to move through their planned paths.

Once the visibility index has been established, all agents will be added to a priority queue, where their priority will be the delay the agent has suffered due to adjustments made to its path in order to avoid collisions. At first, all agents start with a priority value of zero.

During execution, each agent will be removed from the queue and then, assuming an initial planned path that was the result of an A* algorithm, will attempt to make a reservation in the table that covers the different points in space-time that correspond to the path of the agent within the visibility index previously established.

If the reservation is successful, then we update the agent's position according to its planned path and velocity. If not, then we need to roll back any entries made to the table within the current reservation for this particular agent and detect at which point in time the collision would have occurred.

Once, and if, we have set the collision time, we can use it to calculate a lower velocity that would allow us to avoid it. Agents can optionally set a speed threshold which would prompt the agent to calculate a new path, using A*, from its current position to its desired position in case the new speed would fall below it. In any case, the delay caused by the modifications is calculated and then stored. This is the value that determines the agent's priority when making a reservation, which ensures that agents do not receive preferential treatment.

```
COOPERATIVE_PF()
Until all agents have reached their goals do
   For each agent in priority queue do
      Remove agent from queue
         For t=0 to t=v
            Calculate position occupied at t according to current path
            Attempt to make reservation
            if reservation not successful
               Undo previous reservations
                set collision Time
                set collision Agent
                break;
            end
            if collision Time = -1 do
               update agent position according to its velocity and path
            else
                calculate speed decrease required to avoid collision
                if speed after decrease < speed threshold
                   calculate new path from goal to current position

          calculate time difference between modified time and speed and
          original time and speed.

         Use that value to re-insert the agent into priority queue
         Store it in agent.total-delay
      end
   Insert all agents into queue using their total delays as priority values.
END COOPERATIVE_PF
```

Once all agents have been removed from the priority queue, they are inserted back in if they haven't reached their goals yet.

**New Cooperative Pathfinding Algorithm Applied to Flock Traffic Navigation**

We will explain the classic FTN based on Negotiation algorithm and then propose an improvement to avoid collisions at intersections.

**Flock Traffic Navigation**

Flock Traffic Navigation (FTN) based on negotiation is a new approach for solving traffic congestion problems in big cities (Astengo, 2007). In FTN, vehicles can navigate automatically in groups called flocks allowing the coordination of intersections at the flock level, instead of at the individual vehicle level, making it simpler and far safer. To handle flock formation, coordination mechanisms are issued from multi-agent systems.

The mechanism to negotiate (Astengo, 2006) starts with an agent who wants to reach its destination. The agent knows an a priori estimation of the travel time that takes to reach its goal from its actual position if he travels alone. In order to win a speed bonus (social bonus) he must agree with other agents (neighbors) to travel together at least for a while. The point in which the two agents agree to get together is called the meeting point and the point where the two agents will separate is called the splitting point. Together they form the so-called "bone" structure diagram.

Individual reasoning plays the main role in this approach. Each agent must compare its a priori travel time estimation versus the new travel time estimation based on the bone-diagram and the social bonus and then make a rational decision. Decision will be made according to whether they are in Nash equilibrium (there is no incentive for either of them to choose another neighbor agent over the agreed one) or if they are in a Pareto Set (Wooldridge, 2002).

If both agents are in Nash equilibrium, they can travel together as partners and can be benefited with the social bonus. In this moment a new virtual agent is created in order to negotiate with future candidates. Agents in a Pareto Set can be added to this "bone" diagram if their addition benefits them without affecting the original partners negatively. Simulations indicate that flock navigation of autonomous vehicles could substantially save time to users and let traffic flow faster (Astengo, 2007).
Until now, Agents make their own path planning according to A* and only at intersections use the Dresner-Stone Algorithm.

**A Collision Detection Flock Traffic Navigation Algorithm**

FTN based on Negotiation now are decoupled in two main parts:

```
OFFLINE Algorithm
  For each Agent do
   plan using A* and find
        an optimal path traveling  and
        an arrival time estimation.

REAL Time Algorithm
  For each spirit flock do
     A reservation δ-time units forward according to its path
     If reservation = TRUE
     Follow previous calculated A*-path
     else
     Conflict Module
```

The critical issue is the conflict Module that can be changed according to the social rules. Here we present a conflict module according to the rules in (Astengo 2006).

```
Conflict Module
Rules
  Priority 1: Larger Flock goes first
              Then tile is marked as obstacle.
  Priority 2: If Size of Flocks are equal
              compare delay-table
              delayed Flock goes first
              Then tile is marked as obstacle
  Priority 3: If none previous priority is accomplished
              Use Stone-Dresner Algorithm
```

**Application to Continuous Domains**

When applied to computer games, path finding usually takes place on top of one of two representations: A grid structure that wholly describes the traversable game world or a waypoint graph that samples the continuous space over which game agents can move.

The first variant can usually be seen in Real-Time Strategy Games (RTS) and bi-dimensional role-playing games (RPG). The cooperative path finding algorithm in dynamic environments is a perfect fit for these representations and allows game agents to react realistically to the presence of other agents and unforeseen obstacles. Other genres, however, require the simulation of a continuous space updated in fixed time-steps.

Applying a grid-like structure to such spaces can be prohibitively expensive. The algorithm can be modified to work in continuous domains by replacing the reservation table with an analysis of world geometry.

The algorithm consists of an update function, responsible for advancing the state of each agent by a single time-step. The agents are stored in a priority queue that uses each agent's total delay as its key. The function calculates the time elapsed between the last and the current call and uses this value to update the agents. The agents are updated by first performing a collision-detection test between the Minkowsky sum of the agent's path and an assigned bounding volume and each of the Minkowsky sums of the other agents' predicted paths and their bounding volumes. If a collision is detected, the agent will try to adjust its speed to avoid the intersection point at the intersection time. If this value falls under a specified threshold, the agent will, instead, attempt to calculate a different path.

The predicted paths are calculated using only the other agent's current position, orientation, and speed. These predictions are also limited by the forecasting index, which indicates how far in time are agents willing to predict.

This approach allows the path-planning operation to be distributed across frame updates and the individual update steps for each agent cause no side-effects, making them trivially parallelizable. The algorithm can be further optimized by pruning the collision-detection search by using spatial partitioning schemes such as kd-Trees and by limiting the path used to calculate the Minkowsky sums with the forecasting index.

The update function is presented below:

```
INPUTS: agents[0...N], forecastIdx[0...N], agentPositions[0...N], agentSpeeds[0...N], speedBound,
agentOrientations[0...N], agentDelays[0...N], agentPaths[0...N], boundingVolumes[0...N]

  WHILE agents NOT EMPTY
    a <- agents[0]
    removeFromQueue(a)
    currentPath <- minkowski(agentPaths[a], boundingVolumes[a], forecastIdx[a])
      FOR o IN 0...N
        otherPath <- minkowski(predictPath(agentPositions[o], agentSpeeds[o],
                   agentOrientations[o], forecastIdx[a]), boundingVolumes[o], forecastIdx[a])
        IF intersects(currentPath, otherPath) THEN
          slowdown <- calculateSlowdown(agentPaths[a], agentSpeeds[a], intersectionPoint,
          intersectionTime)
          IF slowdown < speedBound THEN
            IF isOtherPathAvailable(a) THEN
              previousPath <- agentPaths[a]
              agentPaths[a] <- calculateNewPath(a)
              delay <- calculatePathDelay(agentPaths[a], previousPath)
              agentDelays[a] <- agentDelays[a] + delay
              addToQueue(a, agentDelays[a])
          ELSE
            delay <- calculateSpeedDelay(agentSpeeds[a], slowdown)
            agentSpeeds[a] <- slowdown
            agentDelays[a] <- agentDelays[a] + delay
            addToQueue(a, agentDelays[a])
          END
        ELSE
          agentPositions[a] <- updatePosition(agentPaths[a], agentSpeeds[a])
        END
      END
    END
    agents <- buildPriorityQueue(agentDelays)
```

## 5. Concluding Remarks and future work

Pathfinding is a critical element of AI in many modern applications like multiple mobile robots, game industry and flock traffic navigation based on negotiation (FTN).

We develop a new algorithm capable of planning paths for multiple agents on partially known and changing environments inspired by cooperative A* and D*.

From a distributed approach (Decoupled) our collective pathfinding in dynamic environments algorithm decomposes the task of individual plan into weakly-dependent problems for each agent. Each agent can search greedily for a path according to its destination, given the current state of all other agents. Then based on a space-time search space each agent attempt to make a reservation on $(x,y,t,\delta)$ where $x,y$ are in the Euclidean space, t is a time measure and $\delta$ is a forward planning-vision measure (forecasting index).

Because these kinds of algorithms are problem dependent we developed a modification of our collective pathfinding in dynamic environments algorithm in the FTN context. Taking care that in FTN the main issue is that it is based on negotiation a conflict-solver module that has the social rules within.

Evidently in FTN we will not in general obtain a globally optimal path from the individual agent perspective but it is a at least a better plan compared with traveling alone ( the worst scenario is if an agent can't find Nash or Pareto partners in the whole path so, it becomes a 1-individual flock).

It was shown that the algorithm can, with relatively few modifications, work on continuous domains updated in fixed time-steps, such as those used by most 3D computer games. The shift from using a reservation table to analyzing world geometry allows the work to be cleanly distributed amongst the agents. This creates a clear separation of concerns and the lack of side-effects makes it trivially parallelizable.

## 6. References


R. Alami and F. Ingrand and S. Qutub (1998), A Scheme for Coordinating Multi-Robot Planning Activities and Plans Execution, *Proc. 13th European Conf. on Artificial Intelligence ECAI 98*, pp. 617-621.

N. M. Amato, C. Jones & D. Vallejo. (1998), An Adaptive Framework for "Single Shot" Motion Planning, *Technical Report 98-025*, Texas A&M University.

N. M. Amato, O. B. Bayazit, L. K. Dale, C. Jones & D. Vallejo. (1998), OBPRM: An Obstacle-Based PRM for 3D Workspaces, Ed. P. K. Agarwal et al., *Robotics: The Algorithmic Perspective*, pp. 155-168.

B. Aronov, M. de Berg, A. F. Van der Stappen, P. Svestka & J. Vleugels. (1999), Motion Planning for Multiple Robots, *Discrete and Computational Geometry*, Vol. 22, pp. 505-525.

C. Astengo-Noguez and G. Sánchez-Ante G. (2007), Collective Methods on Flock Traffic Navigation Based on Negotiation. *Proc. 6th Mexican International Conference on Artificial Intelligence*, Springer.

C. Astengo-Noguez and R. Brena (2006) Flock Traffic Navigation Based on Negotiation. *Proc. of 3rd International Conference on Autonomous Robots and Agents (ICARA 2006)* . Palmerston North, New Zealand, pp. 381-384.

J. Barraquand & J. C. Latombe. (1990), A Monte-Carlo Algorithm for Path Planning with Many Degrees of Freedom, *Proc. IEEE Int. Conf. Rob. & Autom.*, pp. 1712-1717.

J. Barraquand & J. C. Latombe. (1991), Robot Motion Planning: A Distributed Representation Approach, *Int. J. of Robotics Research*, Vol. 10, Nr. 6, pp. 628-649.

J. Barraquand, B. Langlois & J. C. Latombe. (1992), Numerical Potential Field Techniques for Robot Path Planning, *IEEE Trans. on Systems, Man and Cybernetics*, Vol. 22, Nr. 2, pp. 224-241.

R. Bohlin & L. E. Kavraki. (2000), Path planning using lazy PRM, *Proc. IEEE Int. Conf. Rob. & Autom.*, pp. 521-528.

J. F. Canny & J. Reif. (1987), New Lower Bound Techniques for Robot Motion Planning Problems}, *Proc. 28th Symp. Foundations of Computer Science*, pp. 49-60.

J. F. Canny, (1988), *The Complexity of Robot Motion Planning,* MIT Press, Cambridge, MA.

B.R. Donald. (1987), A Search Algorithm for Motion Planning with Six Degrees of Freedom, *Artificial Intelligence,* Vol. 31, Nr. 3, pp. 295-353.

K. Dresner and P. Stone (2004), Multiagent Traffic Management: A Reservation-Based Intersection Control Mechanism. *Proc. Third International Joint Conference On Autonomous Agents and Multiagent Systems,* pp. 530-537.

K. Dresner and P. Stone (2005), Multiagent Traffic Management: An Improved Intersection Control Mechanism. *Proc. Fourth International Joint Conference on Autonomous Agents and Multiagent Systems (AAMAS 05).* pp. 471-477.

K. Dresner and P. Stone (2006), Human-Usable and Emergency Vehicle–Aware Control Policies for Autonomous Intersection. Hakodate, Japan. *Proc. Fourth Workshop on Agents in Traffic and Transportation (ATT 06).* pp. 17-25.

K. Dresner and P. Stone (2007) Sharing the Road: Autonomous Vehicles Meet Human Drivers. Sharing the Road: Autonomous Vehicles Meet Human Drivers. Hyderabad, India. *Proc. Twentieth International Joint Conference on Artificial Intelligence (IJCAI 07).* pp. 1263-1268.

D. Hsu, J. C. Latombe & R. Motwani. (1997), Path Planning in Expansive Configuration Spaces, Proc. *IEEE Int. Conf. Rob. & Autom.,* pp. 2719-2726.

D. Hsu. (2000), Randomized Single-Query Motion Planning in Expansive Spaces, *PhD Thesis,* Stanford University, Stanford, CA, USA.

L. E. Kavraki. (1994), *Random Networks in Configuration Space for Fast Path Planning,* PhD Thesis, Stanford University, Stanford, CA, USA

L. E. Kavraki, P. Svestka, J. C. Latombe & M. H. Overmars. (1996), Probabilistic Roadmaps for Path Planning in High-Dimensional Configuration Spaces, *IEEE Trans. on Robotics and Automation,* Vol. 12, Nr. 4, pp. 566-580

K. Kondo. (1991), Motion Planning with six degrees of freedom by multistrategic, bidirectional heuristic free space enumeration, *IEEE Trans. on Robotics and Automation,* Vol. 7, Nr. 3, pp. 267-277.

J. J. Kuffner, Jr., & S. M. LaValle. (2000), RRT-Connect: An Efficient Approach to Single-Query Path Planning, *Proc. IEEE Int. Conf. Rob. & Autom.*

J. C. Latombe. (1991), *Robot Motion Planning,* Kluwer Academic Publishers, Boston, MA.

S. M. LaValle & J.J. Kuffner, Jr. (2001), Randomized Kinodynamic Planning, Int. J. on Robotics Research, Vol. 20, Nr. 5, pp. 378-400.

M. C. Lin & S. Gottschalk. (1998) Collision Detection Between Geometric Models: A Survey, *IMA Conference on Mathematics of Surfaces.*

T. Lozano-Perez & M. A. Wesley (1979), An algorithm for planning collision-free paths among polyhedral obstacles, *Communications of the ACM,* Vol. 22, Nr. 10, pp. 560-570.

T. Lozano-Perez. (1983), Spatial Planning: A configuration Space Approach, *IEEE Trans. on Computers,* Vol. 32, Nr. 2, pp. 108-120.

C. Nielsen & L. Kavraki. (2000), A Two level Fuzzy PRM for Manipulation Planning, *Proc. IEEE/RSJ Int. Conf. on Intelligent Robots and Systems.*

P. A. O'Donnell & T. Lozano-Perez. (1989), Deadlock-Free and Collision-Free Coordination of Two Robot Manipulators, *Proc. IEEE Int. Conf. Rob. & Autom.,* pp. 484-489.

D. C. Pottinger (1999), Coordinated Unit Movement . *Game Developer Magazine,* Vol. 3.

J.H. Reif (1979), Complexity of the Mover's Problem and Generalizations, *Proc. 20th Symp. on the Foundations of Computer Science*, pp. 421-427.

G. Sánchez-Ante & J. C. Latombe. (2002), On Delaying Collision-Checking in PRM Planning -- Application to Multi-Robot Coordination, *Int. J. of Robotics Research.*

J. T. Schwartz & M. Sharir (1983), On the Piano Movers' Problem I: The Case of a Two-Dimensional Rigid Polygonal Body Moving Amidst Polygonal Barriers, *Communications Pure and Applied Mathematics*, Vol. 36, pp. 345-398.

Yoav Shoham and Kevin Leyton-Brown. (2008), *Multiagent Systems: Algorithmic, Game-Theoretic, and Logical Foundations*, Cambridge University Press.

D. Silver (2005), Cooperative Pathfinding, *American Association for Artificial Intelligence*, 2005.

G. Song, S. Miller & N. Amato. (2000), Customizing PRM Roadmaps at Query Time, *Proc. IEEE Int. Conf. Rob. & Autom.*

P. Svestka. (1997), *Robot Motion Planning using Probabilistic Roadmaps*, Ph.D. Thesis, Utrecht University, The Netherland.

P. Tournassoud. (1986), A strategy for obstacle avoidance and its applications to multi-robot systems, *Proc. IEEE Int. Conf. Rob. & Autom.*, pp. 1224-1229.

D. Vallejo, I. Remmler & N. M. Amato. (2001), An Adaptive Framework for "Single Shot" Motion Planning: A Self-Tuning System for Rigid and Articulated Robots, *Proc. IEEE Int. Conf. Rob. & Autom.*

M. Wooldridge (2002), *Introduction to Multiagent Systems.* John Wiley and Sons.

A. A. Zelinsky (1992), Mobile Robot Exploration Algorithm, *IEEE Transactions on Robotics and Automation*, Vol. 8.

D. Zhu & J. C. Latombe. (1990), Constraint reformulation in a hierarchical path planner, *Proc. IEEE Int. Conf. Rob. & Autom.*, pp. 1918-1923.

16

# Navigation for mobile autonomous robots and their formations: An application of spatial reasoning induced from rough mereological geometry

Lech Polkowski and Pawel Osmialowski
*Polish - Japanese Institute of Information Technology*
*Poland*

## 1. Introduction

This Chapter is intended as a sequel to our Chapter 21 in the book on "Mobile Robots Motion Planning. New Challenges" by this Publisher. It is now commonly accepted that problems of planning and navigation are inseparable: "Most recent contribution to the field combine effective algorithms tested on significant problems, along with some formal guarantees of performance" (J.-C. Latombe in Foreword to "Principles of Robot Motion. Theory, Algorithms and Implementations" by H. Choset et al.). Therefore, as with the former Chapter, we provide in this Chapter planning mechanisms along with navigation tests and a theoretical analysis of underlying constructs. We extend our scope of analysis by considering formations of mobile autonomous robots. We introduce a definition of a robot formation, based on the spatial relation of betweenness and we give a treatment of planning and navigation problems for robot formations. In our investigations into problems of multi-robot planning and navigation, we apply rough mereological theory of spatial reasoning. This theory is briefly recalled in this Chapter for completeness sake. The software system Player/Stage is employed as means of simulation and visualization of robot trajectories to chosen goals. To this end, it has been provided with SQL functions rendering predicates of rough mereological geometry. Robotics of autonomous mobile robots presents the most intricate field for applications of techniques of artificial intelligence, decision making and cognitive methods. Among the basic problems in this area are planning and navigation problems and we are concerned with them both in their mutual bond.The planning and navigation problem for mobile robots is addressed from many angles and a multitude of approaches and techniques have emerged; it suffices to mention a division of planners according to assumptions about robot equipment and abilities as well theoretical principles used in planner construction, from simple bug-type algorithms though potential functions and potential field based strategies to roadmaps constructed by exploiting visibility in configuration spaces, metric-based ideas like Voronoi diagrams and graphs, cell decompositions of various types, and probabilistic (sampling) planners allowing for incremental space exploration by building trees of configuration points like EST-trees or

RRT-trees, see (Choset et al., 2005) for an excellent account of these approaches.Path planning methods, according to (Latombe, 1991) can be divided into *centralized*, in which case planning considers all robots in a team, or *decoupled*, when path is planned for each robot independently. Another division of path planning methods consists in *local* vs. *global* approach; in the local method, planning is provided for some neighborhood of a robot, whereas in the global approach, the whole environment is taken into consideration. Path planning consists in finding controls which secure the desired path of a robot toward a goal with, e.g., obstacle avoidance. As with a single robot, the path planning problem arises for teams of robots. In particular, centralized methods for single robots are extended to teams of robots see, e.g., (Švestka&Overmars, 1998) where such planning is applied with help of relational structures called super-graphs on which admissible paths are searched for. This approach however assumes that the situation is static, i.e., no changes in the environment happen during plan execution. The assumption of environment stability is not valid in real world situations and the reactive approach (Arkin, 1998) to planning considers simple, sensor - actuator coupling schemes expressible as low-level behaviors (Urdiales et al., 2006) ; in these schemes, potential field methods, vector field histograms, dynamic window approach are used (Urdiales et al., 2006). Some reactive methods use dynamic variants of search algorithms like A*, e.g., D* (Brumitt et al., 2001).

From among those methods, we choose to adopt the method of potential field, see sect.5. In classical setting, the potential field is built as the sum of two components: repulsive, induced by obstacles, and attractive, induced by goals. The field force is defined as the gradient of the repulsive, respectively, attractive, potential, see (Choset et al., 2005). In either case, the potential is defined with the use of a metric, in analogy to classical physical examples of a potential field like Coulomb or gravitational fields. Our approach is different: the potential field is constructed by means of a chosen rough inclusion - the primitive predicate of rough mereology, see sect.5. A robot is driven to the goal by following areas of increasing density of the field as shown in sect.5. The problem for a single robot is presented fully in (Polkowski&Osmialowski, 2009) where mereological potential fields have been constructed and applied in planning of paths and robot navigation.

Problems of cooperative mobile robotics are even more demanding as they require an accounting for group behavior of many autonomous mobile robots. There is the increasing need for making use of such teams in practical problems of performing complex tasks inaccessible for a single robot (like pushing large objects, rescue operations, assembling); there is also a theoretical interest in research on aspects of their behavior: cooperative mechanisms, leadership, conflict resolution, consensus making, many of which belong as well in biology and environmental studies, see, e.g., (Balch&Arkin, 1998; Brumitt et al., 2001; Chen&Luh, 1998; Leonard&Fiorelli, 2001; Shao et al., 2005) and also a discussion in (Cao et al., 1997). These motifs have propelled research in direction of multi-robot planning.

Cooperative behavior is perceived by many authors, see, e.g., (Cao et al., 2005) and references therein, as a specialization of collective behavior having the tint of achieving jointly some goal. The goal may mean an economic advantage, reaching a specified position, learning jointly a feature or a category of objects, etc., etc. Main directions of research in this area of schemes for cooperative mobile robotics, include, as distinguished in the literature, see, e.g., (Cao et al., 2005; Kramer&Scheutz, 2007), a study on group architectures, conflicts of resources, motivations for cooperation, learning of a cooperative behavior, spatiotemporal aspects: path planning, moving to formations, pattern generation.

In this work, which extends our earlier results (Osmialowski, 2009; Polkowski&Osmialowski, 2008a), we are concerned with the last aspect, i.e., moving to formations and path planning in order to make robots into a given formation. Alongside, we are concerned with problems of repairing formations and navigating formations in static environments. We study the problem of path planning in order to make robots in a team into a formation. We apply as a theoretical framework for our approach, a qualitative theory of spatial reasoning as provided by Rough Mereology, see, e.g., (Polkowski&Osmialowski, 2008). In this framework, we give a definition of a formation by a team of robots, by means of a rough mereological betweenness relation among them, see (Osmialowski, 2009; Polkowski & Osmialowski, 2008). In the same framework, we study the problem of path planning for moving into a given formation. We propose some procedures to this end. We model our robots on Roomba[1] robots by iRobot(R), i.e., we assume our robots to be planar disk-shaped objects. We use Player/Stage system, see (Osmialowski, 2007; 2009; http://playerstage.sourceforge.net) , as a tool for simulation and visualization.

## 2. On formations of autonomous mobile robots

A study of the concept of a robot formation was initially based on a perception of animal behavior like herding, swarming, flocking or schooling. In this respect, a few principles emerged, see, e.g., (Balch&Arkin, 1998), keeping all animals within a certain distance from one another (e.g., to ensure mutual visibility), moving away when the distance becomes too close (to avoid congestion, collision, or resource conflict), adapting own movement to movement of neighbors (e.g., velocity of motion), orienting oneself on a leader.

From those observations a geometric approach to formations has been derived: a formally simplest approach (Balch&Arkin, 1998), uses referencing techniques; reference is made either to the team center or to the team leader, or to a specified neighbor in a coordinate system given by the position of the team center or the leader along with the orientation given by the nearest navigation point; positions are determined, e.g., with the help of GPS or dead reckoning. Another method for forming a geometric formation relies on a direct usage of a metric, say *rho*, see, e.g., (Chen&Luh,1998; Sugihara&Suzuki,1990): given a threshold $\delta$, and a parameter $\varepsilon$, for each robot $r$ in a team, its farthest neighbor $r1$ and the nearest neighbor $r2$, if $\rho(r, r1) > \delta$ then $r$ moves toward $r1$, if $\rho(r, r1) < \delta - \varepsilon$ then $r$ moves away from $r1$, if $\delta - \varepsilon < \rho(r, r1) < \delta$ then $r$ moves away from $r2$. By this method, robots are arranged on a circle. Some methods rely on the potential field technique (Leonard& Fiorelli, 2001); in this approach, the potential of the field is defined dependent on the distance among robots in the team in order to keep distances among them as prescribed. In addition, also the technique of a virtual leader is involved to keep robots in a team at a prescribed distance from their current leaders; in some approaches the relation the leader - the follower is expressed by means of control laws in a given coordinate system (Das et al., 2002; Shao et al., 2005), with execution of movement controlled by an omnidirectional camera.

1 Roomba is the trademark of iRobot Inc.

It seems desirable to propose an approach which in principle would be metric independent and which would take into account only relative positions of robots one to another. In this work we propose a definition of a formation which is based on spatial predicates defined within rough mereological theory of spatial reasoning, see, e.g., (Polkowski, 2001; Polkowski&Osmialowski, 2008; 2008a).

## 3. Qualitative spatial reasoning: a nutshell reminder

In this Section, we recall elements of spatial theory induced in the rough mereological framework which have already been presented extensively elsewhere, in particular in (Polkowski&Osmialowski, 2008), Ch. 21. Qualitative Spatial Reasoning emerged on basis of an idea by (A. N. Whitehead, 1929; Leonard & Goodman, 1940) of an extension, dual to a notion of part in mereology theory (Lesniewski,1916;1982), reformulated as a theory of Connection (Leonard&Goodman, 1940; Clarke, 1981). Qualitative Spatial Reasoning is a basic ingredient in a variety of problems in mobile robotics, see, e.g., (Kuipers&Byun, 1987). Spatial reasoning which deals with objects like solids, regions etc., by necessity refers to and relies on mereological theories of concepts based on the opposition part - whole (Gotts et al., 1996). Mereological ideas have been early applied toward axiomatization of geometry of solids, see (De Laguna, 1922; Tarski,1929).

Mereological theories rely either on the notion of a part (Lesniewski, 1916; 1982), or on the notion of objects being connected (Clarke, 1981; Gotts et al., 1996). Our approach to spatial reasoning is developed within the paradigm of rough mereology. Rough mereology, see, e.g., (Polkowski, 2003; 2004; 2008), is based on the predicate of being a part to a degree, called a *rough inclusion* and thus it is a natural extension of mereology based on part relation, as proposed by (Lesniewski, 1916; 1982). A rough inclusion, cf., (Polkowski, 2008), is a ternary relation $\mu$ such that for any pair of objects $u$, $v$ and real $r$ the formula $\mu(u, v, r)$ means that *u is a part of v to a degree of r* where $r \in [0,1]$.

In our applications to spatial reasoning, objects will be regions in Euclidean spaces, notably rectangles, in particular squares, or discs in 2-dimensional space, and the rough inclusion applied will predominantly be the one defined by the equation,

$$\mu^0(u, v, r) \text{ if and only if } \frac{|u \cap v|}{|u|} \geq r \qquad (1)$$

where $|u|$ is the area of the region $u$.

On the basis of a given rough inclusion $\mu$, we can introduce predicates of a certain geometry of regions in low-dimensional spaces. Points in such geometries are recovered usually be means of the technique proposed by Alfred Tarski of ultrafilters of regions, see (Tarski, 1929).

## 4. Mereogeometry: a geometry of regions

We are interested in introducing into the mereological world defined by $\mu^0$ a geometry in whose terms it will be possible to express spatial relations among objects; a usage for this geometry will be found in navigation and control tasks of multi-agent mobile robotics.

### 4.1 A notion of a quasi-distance

We first introduce a notion of a quasi-distance $\kappa$ in our rough mereological universe by letting,

$$\kappa(u, v) = \min\{\mathrm{argmax}r\mu^0(u, v, r), \mathrm{argmax}s\mu^0(v, u, s)\} \tag{2}$$

Observe that mereological distance differs essentially from the standard distance: the closer are objects, the greater is the value of $\kappa$ : $\kappa(u, v) = 1$ means $u = v$, whereas $\kappa(u, v) = 0$ means disjointness in the sense of $\mu^0$ of $u$ and $v$ regardless of the Euclidean distance between them.

### 4.2 Nearness and Betweenness: Van Benthem's variant

We apply the distance $\kappa$ to define in our context the predicate $N$ of nearness proposed in (van Benthem, 1983),

$$N(z, u, v) \iff (\kappa(z, u) = r, \kappa(u, v) = s \Rightarrow s < r) \tag{3}$$

Here, nearness means that $z$ is closer to $u$ than $v$ is to $u$.

We make an essential use of the betweenness predicate $T_B$ proposed by van Benthem [3], in analogy to the Tarski betweenness (Tarski,1959) on the basis of the nearness predicate,

$$T_B(z, u, v) \iff [\text{for all } w \ (z \text{ is } w \text{ or } N(z, u, w) \text{ or } N(z, v, w))] \tag{4}$$

**Example 1**. We consider a context in which objects are rectangles positioned regularly, i.e., having edges parallel to axes in $R^2$. The measure $\mu$ is $\mu^0$ of (1). In this setting, given two disjoint rectangles $C$, $D$, the only object between $C$ and $D$ in the sense of the predicate $TB$ is the extent $ext(C, D)$ of $C$, $D$, i.e., the minimal rectangle containing the union $C \cup D$. As linear stretching or contracting along an axis does not change the area relations, it is sufficient to consider two unit squares $A$, $B$ of which $A$ has (0, 0) as one of vertices whereas $B$ has $(a, b)$ with $a, b > 1$ as the lower left vertex (both squares are regularly positioned). Then the distance $\kappa$ between the extent $ext(A, B)$ and either of $A$, $B$ is $\frac{1}{(a+1)(b+1)}$. For a rectangle $R$: $[0, x] \times [0, y]$ with $x \in (a, a+1)$, $y \in (b, b+1)$, we have that $\kappa(R, A) = \frac{(x-a)(y-b)}{xy} = \kappa(R, B)$. For $\varnothing(x, y) = \frac{(x-a)(y-b)}{xy}$, we find that

$\frac{\partial \phi}{\partial x} = \frac{a}{x^2} \cdot (1 - \frac{b}{y}) > 0$, and, similarly, $\frac{\partial \phi}{\partial y} > 0$, i.e., *ø* is increasing in *x*, *y* reaching the maximum when *R* becomes the extent of *A*, *B*. An analogous reasoning takes care of the case when *R* has some (*c*, *d*) with *c*, *d* > 0 as the lower left vertex.

The betweenness predicate allows for definitions of various *patterns Pt*. For instance, we define a *line pattern*. We let,

$$Pt(u, v, z) \iff z \text{ is } TB(u, v) \text{ or } u \text{ is } TB(z, v) \text{ or } v \text{ is } TB(u, z) \quad (5)$$

We will say that a finite sequence *u1, u2, ..., un* of objects belong in a line segment whenever *Pt(ui, ui+1, ui+2)* for *i* = 1, ..., n - 2; formally, we introduce the functor *Line* of finite arity defined via

$$Line(u1, u2, ..., un) \iff \text{for all } i < n - 1 : Pt(ui, ui+1, ui+2) \quad (6)$$

**Example 2**. With reference to Example 1, rectangles *C*, *D* and their extent *ext*(*C*, *D*) form a line segment.

## 5. Mereological potential fields

As mentioned in sect.2, the technique of potential fields, see (Krogh, 1984; Khatib, 1986) for seminal ideas, cf., (Choset et al., 2005; Latombe, 1991), well-known from planning in case of a single robot, has been extended to the case of robot teams. An example of this approach is given in (Leonard&Fiorelli, 2001), where robots in a team are organized around a set of beacons called *leaders*, and are subjected to repulsive and attractive forces induced by potential fields generated for pairs of robots and pairs of the form robot–leader in such a way as to prevent too close distance among robots and to keep them along leaders.

In our case, we apply the idea already exposed, see (Osmialowski, 2009; Osmialowski& Polkowski,2009; Polkowski&Osmialowski, 2008a), of building a potential field from the rough inclusion $\mu^0$. Our path planner accepts target point coordinates and provides a list of *waypoints* from a given robot position to the goal. It takes as an input a map of static obstacles that a robot should avoid while approaching the target point. A robot and a target should both lay within the area delimited by surrounding static obstacles that form borders of the robot environment. There can be other static obstacles within the area, all marked on the provided map. After the path is proposed, a robot is lead through the path until it reaches given target. If a robot cannot move towards the target position for some longer time (e.g., it keeps on hitting an other robot reaching its target or some unknown non-static obstacle), a new path is proposed.We tested our planner by running simulations in which we have had a model of Roomba robot, see (Tribelhorn&Dodds, 2007) traveling inside an artificially created environment. Real Roomba robots are disc-shaped and therefore easy to model, but they do not provide many useful sensor devices (except bumpers which we were using to implement lower-level reaction to hitting unexpected obstacles). Also, odometry of

Roomba robots is unreliable (loc.cit) hence we assume that simulated robots are equipped with a global positioning system.Right after the target position is given, our planner builds the mereological potential field filled with squared areas each of the same size. The field is delimited by environment's borders. Only space free of obstacles is filled.
The algorithm for building the potential field is the following.

***SQUARE_FILL_ALGORITHM***
Structure: a queue Q
1. Add to the queue Q, x and y coordinates of a given goal together with 0 as current distance from current squared area to the next neighboring area (so they will be part of each other to the maximal degree). Also put clockwise as current direction of exploration. These are initial values.
2. Spin in the main loop until there are no more elements in the queue Q:
2.1. Extract x, y, current distance and current direction of exploration from the beginning of queue Q.
2.2. Check if there is any other squared area already present in potential field to which the distance from current x and y coordinates is equal or shorter than current distance. If so, skip taken element and run new main loop turn.
2.3. Form new squared area with current x and y as the coordinates of the centroid of this new area. Check if there are any common part with any static obstacle within this new squared area. If so, skip taken element and run new main loop turn.
2.4. Add new squared area to the potential field.
2.5. Increase current distance by 0.01.
2.6. Add eight neighbor areas to the queue Q (for each area add these data: x and y coordinates, current distance and direction of exploration opposite to current); if direction is clockwise neighbors are: left, left-up, up, right-up, right, right-down, down, left-down; if direction is anti-clockwise neighbors are: left-down, down, right-down, right, right-up, up, left-up, left.
2.7. Run new main loop turn.

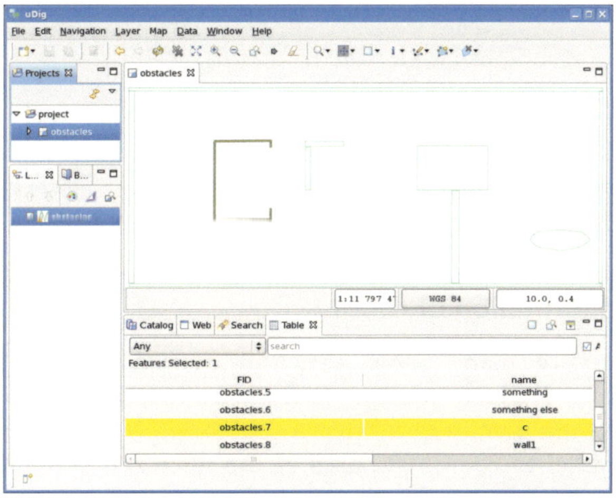


Fig. 1. Map of our artificial world edited by the *uDig* application (created and maintained by Refractions Research). The map consists of number of layers whose can be edited individually; on the figure we can see how solid obstacles are situated within *obstacles* layer

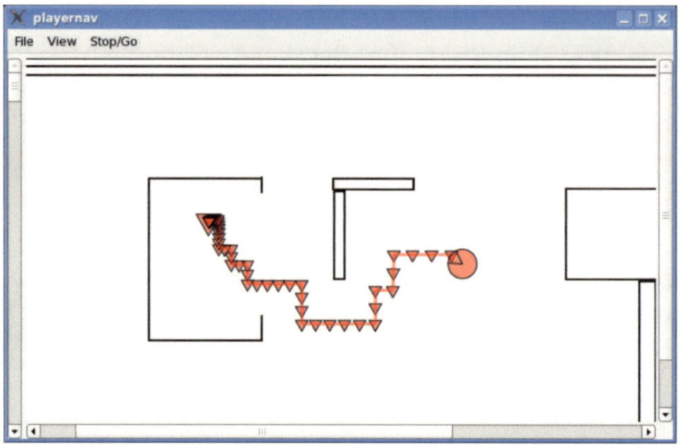


Fig. 2. The *playernav* program can be used to indicate the goal position for given robot and to show trajectory computed by underlying planer. In his situation, our mereological planner computed the trajectory.

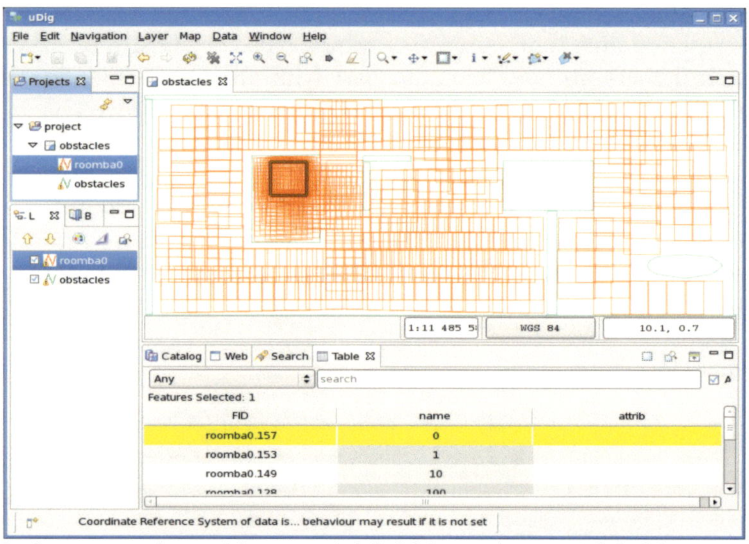


Fig. 3. *Obstacles* layer together with potential field layer (potential field generated for given goal is stored as another map layer, here called *roomba0*). Observe increasing density towards the goal.

## 6. A definition of a formation of robots

We propose a theory of many robot structures, based on the predicates of rough mereological geometry, of which foremost is the predicate $T_B$ of betweenness. A Roomba robot is a disc-shaped robot and due to this we model it as the square circumscribing the robot with edges parallel to coordinate axes of the reference system. This allows for the extent of two given robots to be always oriented as a regular rectangle, i.e., with edges parallel to coordinate axes. In particular, this feature allows for translational and rotational invariance of extents, more generally under affine transformations of the plane.

**Definition 1.** *We say that a robot B is between robots A and C, in symbols (between B A C), in case the rectangle ext(B) is contained in the extent of rectangles ext(A), ext(C), i.e.,* $\mu^0(ext(B), ext(ext(A), ext(C)), 1)$.

This allows as well for a generalization to the notion of *partial betweenness* which models in a more precise manner spatial relations among *A, B, C* (we say in this case that robot *B* is between robots *A* and *C* to a degree of at least *r*): in symbols,

$$(\text{between-deg } r\ B\ A\ C) \tag{7}$$

if and only if

$$\mu^0(ext(B), ext[ext(A), ext(C)], r) \tag{8}$$

We now give the central definition in this work: the definition of a formation. By a formation, we mean a set of robots along with a structure imposed on it as a set of spatial relations among robots.

**Definition 2.** *For a team of robots, T(r1, r2, …, rn) = {r1, r2; …, rn}, an ideal formation IF on T(r1, r2, …, rn) is a betweenness relation (between …) on the set T(r1, r2, …, rn) of robots.*

In practice, ideal formations will be given as a list of expressions of the form,

$$(\text{between } r0\ r1\ r2) \tag{9}$$

indicating that the object *r0* is between *r1; r2*, for all such triples, along with a list of expressions of the form,

$$(\text{not-between } r0\ r1\ r2) \tag{10}$$

indicating triples which are not in the given betweenness relation.

To account for dynamic nature of the real world, in which due to sensory perception inadequacies, dynamic nature of the environment, etc., etc., we allow for some deviations from ideal formations by allowing that the robot which is between two neighbors can be between them to a degree in the sense of (7).

This leads to the notion of a real formation.

**Definition 3.** *For a team of robots, T(r1, r2, …, rn) = {r1, r2, …, rn}, a real formation RF on T(r1, r2, …, rn) is a betweenness to degree relation (between-deg …) on the set T(r1, r2, …, rn) of robots.*

In practice, real formations will be given as a list of expressions of the form,

$$(\text{between-deg } \delta\ r0\ r1\ r2) \tag{11}$$

indicating that the object $r0$ is to degree of $\delta$ in the extent of $r1$, $r2$, for all triples in the relation (between-deg …), along with a list of expressions of the form,

$$\text{(not-between } r0\ r1\ r2) \tag{12}$$

indicating triples which are not in the given betweenness relation.

In Fig. 4, we sketch some cases of instances of relations (between-deg $\delta$ $r0$ $r1$ $r2$ ).

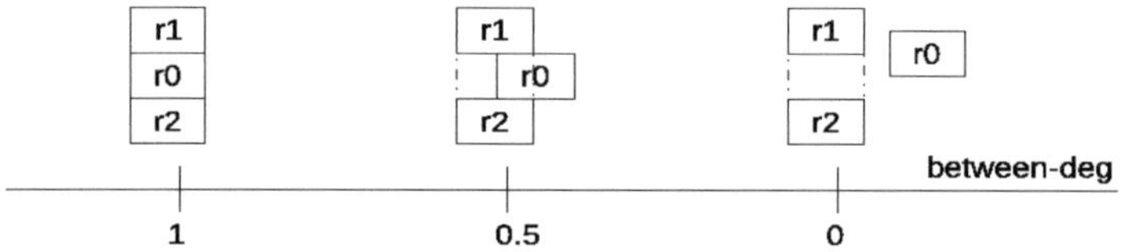


Fig. 4. Object $r_0$ is in extent of $r_1$ and $r_2$ to degree $\delta$.

## 7. On complexity of formation description

Description of formations, as proposed in Def. 1, 2 of sect. 6, can be a list of relation instances of large cardinality, cf., Examples 3 and 4, below. The problem can be posed of finding a minimal set of instances wholly describing a given formation. It turns out (Polkowski&Osmialowski,2008) that the problem is intractable. We have

**Proposition 1.** *The problem of finding a minimum size description of a given formation is NP-hard.*

## 8. Implementation in Player/Stage software system

Player/Stage is an Open-Source software system designed for many UNIX-compatible platforms, widely used in robotics laboratories (Kramer&Scheutz, 2007; Osmialowski, 2007; http://playerstage.sourceforge.net) . Main two parts are Player - message passing server (with bunch of drivers for many robotics devices, extendable by plug-ins) and Stage - a plug{in for Player's bunch of drivers which simulates existence of real robotics devices that operate in the simulated 2D world.

Player/Stage offers client-server architecture. Many clients can connect to one Player server, where clients are programs (robot controllers) written by a user who connects to Player client-side API. Player itself uses drivers to communicate with devices and in this activity it does not make distinction between real and simulated hardware. It gives the user means for testing programmed robot controller in both real and simulated world.

Among all Player drivers that communicate with devices (real or simulated), there are drivers not intended for controlling hardware, instead those drivers offer many facilities for sensor data manipulation, for example, camera image compression, retro-reflective detection of cylindrical markers in laser scans, path planning. One of the new features added to Player version 2.1 is the PostGIS[2] driver: it connects to PostgreSQL database in order to obtain and/or update stored vector map layers.

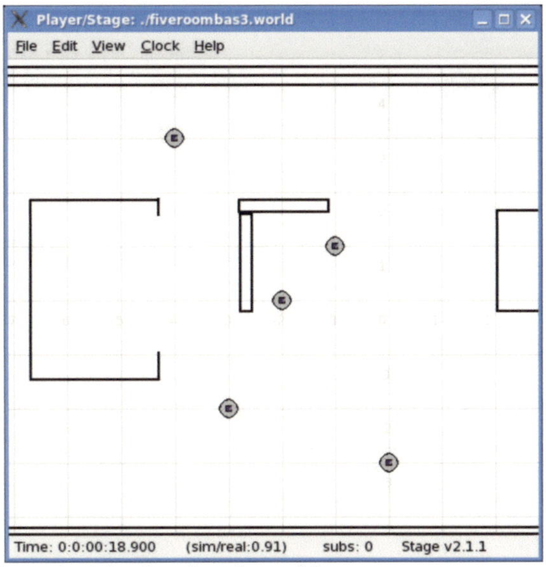


Fig. 5. Five Roomba robots inside simulated world

PostGIS itself is an extension to the PostgreSQL object-relational database system which allows GIS (Geographics Information Systems) objects to be stored in the database. It also offers new SQL functions for spatial reasoning. Maps which are to be stored in SQL database can be created and edited by graphical tools like *uDig* or by C/C++ programs written using GEOS library of GIS functions. PostGIS, uDig and GEOS library are projects maintained by Refractions Research. A map can have many named layers, and for each layer a table in SQL database is created. We can assume that the layer named *obstacles* consists of objects which a robot cannot walk through. Other layers can be created in which we can divide robot's workspace into areas with an assigned attribute which for example tells whether a given area is occupied by an obstacle or not. During our experimentations, we have created a plug-in for Players bunch of drivers which constantly tracks changes of position of every robot and updates *obstacles* layer so robots are marked as obstacles. As a result, the map stored in SQL database is kept always up to date. This feature is also useful in multi-agent environments: at any time a robot controller can send a query to SQL database server regarding every other robot position.

### 8.1. SQL queries representing rough mereogeometric predicates

A roboticist can write a robot controller using Player client-side API which obtains information about current situation through the *vectormap* interface. Additionally, to write such a program, PostgreSQL client-side API can be used in order to open direct connection to the database server on which our mereogeometry SQL functions are stored together with

---

[2] PostGis is an intelectual property of Refractions Research, Inc.

map database. These functions can be called using this connection and results are sent back to the calling program. This gives robot controller program ability to perform spatial reasoning based on rough mereology. Using PostGIS SQL extensions we have created our mereogeometry SQL functions, see (Ladanyi, 1997). Rough mereological distance is defined with help of the following SQL function *meredist*.

```
CREATE FUNCTION meredist(object1 geometry, object2 geometry)
RETURNS DOUBLE PRECISION AS
$$
        SELECT min(degrees.degree) FROM
                ((SELECT
                        ST Area(ST Intersection(extent($1), extent($2)))
                        / ST Area(extent($1))
                        AS degree)
                UNION (SELECT
                        ST Area(ST Intersection(extent($1), extent($2)))
                        / ST Area(extent($2))
                        AS degree))
                AS degrees;
$$ LANGUAGE SQL STABLE;
```

Having mereological distance function we can derive nearness predicate:

```
CREATE FUNCTION merenear(obj geometry, o1 geometry, o2 geometry)
RETURNS BOOLEAN AS
$$
        SELECT meredist($1, $2) > meredist($3, $2)
$$ LANGUAGE SQL STABLE;
```

The equi-distance can be derived as such:

```
CREATE FUNCTION mereequ(obj geometry, o1 geometry, o2 geometry)
RETURNS BOOLEAN AS
$$
        SELECT (NOT merenear($1, $2, $3))
                AND (NOT merenear($1, $3, $2));
$$ LANGUAGE SQL STABLE;
```

Our implementation of the betweenness predicate makes use of a function that produces an object which is an extent of given two objects:

```
CREATE FUNCTION mereextent(object1 geometry, object2 geometry)
RETURNS geometry AS
$$
        SELECT GeomFromWKB(AsBinary(extent(objects.geom))) FROM
                ((SELECT $1 AS geom)
```

```
                    UNION (SELECT $2 AS geom))
                    AS objects;
$$ LANGUAGE SQL STABLE;
```

The betweenness predicate is defined as follows:

```
CREATE FUNCTION merebetb(obj geometry, o1 geometry, o2 geometry)
RETURNS BOOLEAN AS
$$
        SELECT
                meredist($1, $2) = 1
                OR meredist($1, $3) = 1
                OR
                        (meredist($1, $2) > 0
                        AND meredist($1, $3) > 0
                        AND meredist(mereextent($2, $3),
                                mereextent(mereextent($1, $2), $3)) = 1);
$$ LANGUAGE SQL STABLE;
```

Using the betweenness predicate we can check if three objects form a pattern:

```
CREATE FUNCTION merepattern(object1 geometry, object2 geometry, object3 geometry)
RETURNS BOOLEAN AS
$$
        SELECT merebetb($3, $2, $1)
                OR merebetb($1, $3, $2)
                OR merebetb($2, $1, $3);
$$ LANGUAGE SQL STABLE;
```

Also having pattern predicate we can check if four objects form a line:

```
CREATE FUNCTION mereisline4(obj1 geometry, obj2 geometry, obj3 geometry, obj4
geometry)
RETURNS BOOLEAN AS
$$
        SELECT merepattern($1, $2, $3) AND merepattern($2, $3, $4);
$$ LANGUAGE SQL STABLE;
```

To figure out if a set of objects form a line an aggregate can be used:

```
CREATE FUNCTION mereisline_state(state array geometry[4], input data geometry)
RETURNS geometry[4] AS
$$
        SELECT ARRAY[$1[2], $1[3], $2, result.object]
                FROM (SELECT CASE
                WHEN $1[4] IS NOT NULL
                        THEN $1[4]
```

```
                WHEN $1[3] IS NULL
                    THEN NULL
                WHEN ($1[2] IS NULL) AND (meredist($1[3], $2) > 0)
                    THEN NULL
                WHEN ($1[2] IS NULL) AND (meredist($1[3], $2) = 0)
                    THEN $2
                WHEN ($1[1] IS NULL) AND merepattern($1[2], $1[3], $2)
                    THEN NULL
                WHEN ($1[1] IS NULL) AND (NOT merepattern($1[2], $1[3], $2))
                    THEN $2
                WHEN merepattern($1[1], $1[2], $1[3]) AND merepattern($1[2], $1[3], $2)
                    THEN NULL
                ELSE $2
        END AS object)
AS result;
$$ LANGUAGE SQL STABLE;

CREATE FUNCTION mereisline_final(state array geometry[4])
RETURNS BOOLEAN AS
$$
        SELECT ($1[4] IS NULL)
                AND ($1[3] IS NOT NULL)
                AND ($1[2] IS NOT NULL);
$$ LANGUAGE SQL STABLE;

CREATE AGGREGATE mereisline
(
SFUNC = mereisline_state,
BASETYPE = geometry,
STYPE = geometry[],
FINALFUNC = mereisline_final,
INITCOND = 'fg'
);
```

For our convenience we have derived betweenness predicate in more general form:

```
CREATE FUNCTION merebet(object geometry, object1 geometry, object2 geometry)
RETURNS BOOLEAN AS
$$
        SELECT (
                ST Area(ST Intersection(extent($1), mereextent($2, $3)))
                / ST Area(extent($1))
                ) = 1.0;
$$ LANGUAGE SQL STABLE;
```

### 8.2. A driver for Player server to maintain formations

We have created a plug-in driver (written in C++ programming language) for Player server that keeps on tracking all robots in a team in order to make sure their positions form desired formation. If formation is malformed, our driver tries to repair it by moving robots to their proper positions within the formation. Also our driver is responsible for processing incoming orders: position commands which are dispatched to formation leader (selected member of a team) and geometry queries which are replied with information about current formation extent size and its global position. As such, our driver can be considered as a finite state machine which by default is constantly switching between two states: *process orders* and *formation integrity check*. If formation integrity check fails it switches to *repair formation* state.

Formation integrity check is done according to a given description. As pointed earlier, description of formation is a list of s-expressions (LISP-style symbolic expressions). To parse those descriptions efficiently we have used *sfs-exp* programming library written by Matthew Sottile (sfsexp). Each relation between robots in given description is checked and if related robots positions do not fulfill requirements, error value is incremented. Also while traversing through a description, overall error value is computed in order to figure out what could be the maximum error value for the given description. Finally, error value derived from each robot position is divided by computed overall error value which gives the normalized formation fitness value between 0 (all requirements were fulfilled) and 1 (none of requirements were fulfilled). If the fitness value is below some threshold (typically 0.2), then we can conclude that robots are in their desired positions.

Typical formation description may look like below.

**Example 3.**

```
(cross
        (set
                (max-dist 0.25 roomba0 (between roomba0 roomba1 roomba2))
                (max-dist 0.25 roomba0 (between roomba0 roomba3 roomba4))
                (not-between roomba1 roomba3 roomba4)
                (not-between roomba2 roomba3 roomba4)
                (not-between roomba3 roomba1 roomba2)
                (not-between roomba4 roomba1 roomba2)
        )
)
```

This is a description of a formation of five Roomba robots arranged in a cross shape. The *max-dist* relation is used to bound formation in space by keeping all robots close one to another.

Example below describes diamond shape formed by team of eight Roomba robots.

**Example 4.**

*(diamond*

*(set*

*(max-dist 0.11 roomba1 (between roomba1 roomba0 roomba2))*
*(max-dist 0.11 roomba3 (between roomba3 roomba1 roomba4))*
*(max-dist 0.11 roomba5 (between roomba5 roomba4 roomba6))*
*(max-dist 0.11 roomba7 (between roomba7 roomba0 roomba6))*
*(between roomba1 roomba0 roomba2)*
*(between roomba1 roomba0 roomba3)*
*(between roomba1 roomba2 roomba7)*
*(between roomba1 roomba3 roomba7)*
*(between roomba3 roomba2 roomba4)*
*(between roomba3 roomba2 roomba5)*
*(between roomba3 roomba1 roomba5)*
*(between roomba3 roomba1 roomba4)*
*(between roomba5 roomba4 roomba6)*
*(between roomba5 roomba4 roomba7)*
*(between roomba5 roomba3 roomba7)*
*(between roomba5 roomba3 roomba6)*
*(between roomba7 roomba0 roomba6)*
*(between roomba7 roomba0 roomba5)*
*(between roomba7 roomba1 roomba5)*
*(between roomba7 roomba1 roomba6)*
*(not-between roomba1 roomba0 roomba4)*
*(not-between roomba1 roomba2 roomba6)*
*(not-between roomba1 roomba2 roomba3)*
*(not-between roomba3 roomba0 roomba4)*
*(not-between roomba3 roomba2 roomba6)*
*(not-between roomba3 roomba1 roomba2)*
*(not-between roomba5 roomba6 roomba7)*
*(not-between roomba5 roomba2 roomba6)*
*(not-between roomba5 roomba0 roomba4)*
*(not-between roomba7 roomba5 roomba6)*
*(not-between roomba7 roomba2 roomba6)*
*(not-between roomba7 roomba0 roomba4)*
*(not-between roomba0 roomba1 roomba5)*
*(not-between roomba0 roomba3 roomba7)*
*(not-between roomba2 roomba1 roomba5)*
*(not-between roomba2 roomba3 roomba7)*
*(not-between roomba4 roomba1 roomba5)*
*(not-between roomba4 roomba3 roomba7)*
*(not-between roomba6 roomba1 roomba5)*
*(not-between roomba6 roomba3 roomba7)*

*)*

*)*

If formation is malformed our driver can try to repair it. A run-time parameter of the driver indicates which one of three methods should be used to move robots into their desired positions within the formation.

#### 8.2.1. Three methods of formation repairing

We propose three methods for restoring a team to its prescribed formation shape. The first method is behavioral and does not use any planning. The second one is decoupled as planning is made for each robot separately, and global as all robots are taken into consideration at the same time. The third method is decoupled and global, and in addition is behavioral, as all robots move simultaneously.

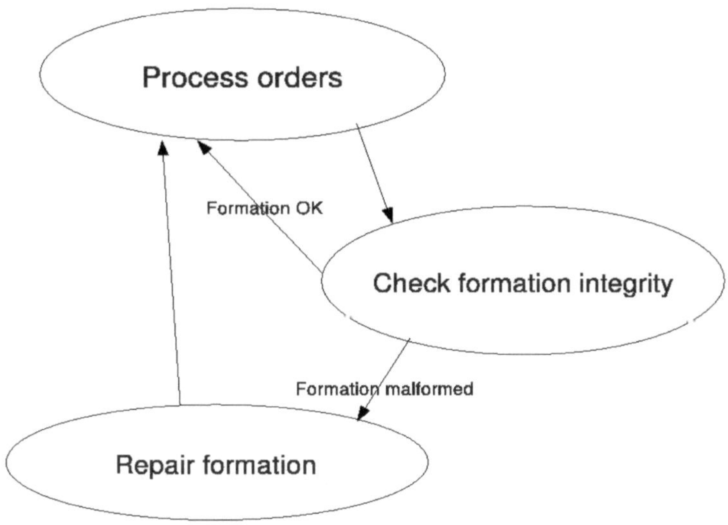


Fig. 6. States of our formation keeping driver for Player server

**Method 1. Pure behavioral.** Each robot (except a selected leader) moves to the goal position. Whenever collision is detected (on the robot bumper device), robot goes back for a while then turns left or right for a while and from this new situation, it tries again to go towards goal position. Due to the nature of this method, formation repair process is time-consuming (reactions to possible collisions take additional time) and may be even impossible. Formation is repaired relatively to one selected member of a team called a leader (therefore this selected member sticks in place while all other robot moves to their positions). If formation is not repaired after some grace time, a next member of a team is selected to be the new leader (therefore this new selected member sticks in place while all other robot moves which changes whole situation). If there are no members left to be new leaders, this method signals that the formation shape is impossible to be restored.

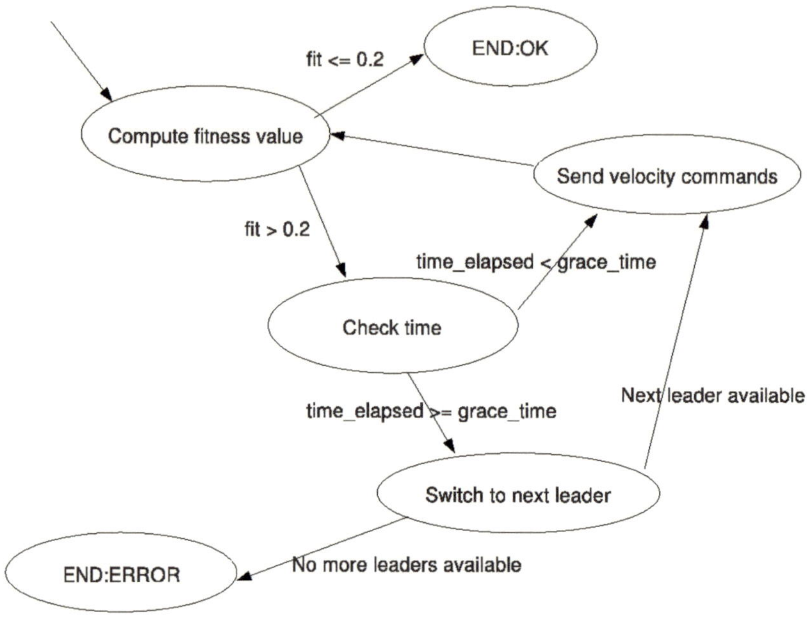


Fig. 7. Pure behavioral method of repairing the formation

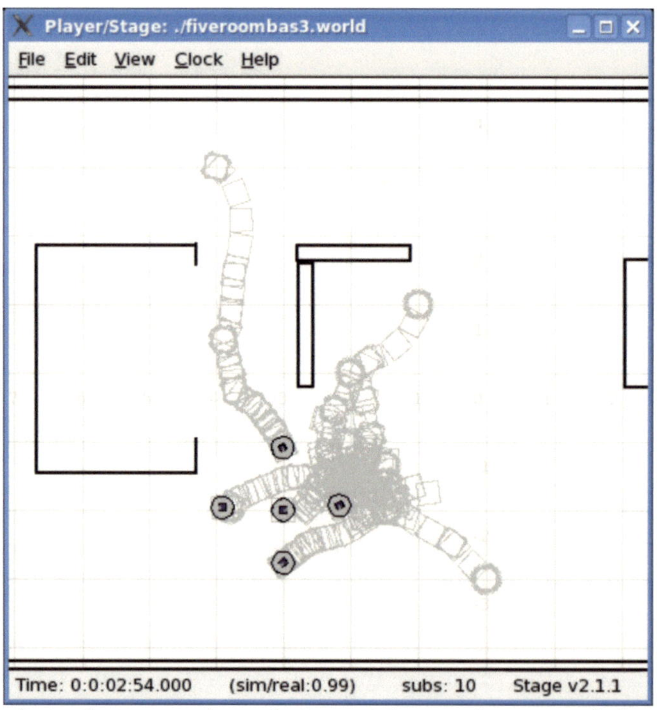


Fig. 8.Trails of robots that moved to their positions using pure behavioral method of repairing the formation

**Method 2. One robot at a time**. The procedure is repeated for each robot in a team: a path is planned by using any available planner (e.g., *wavefront* planner shipped with Player, or *mereonavigator* planner created during our experimentations (Osmialowski, 2009)); then, a robot moves to the goal position. This is the most reliable method, however it is time too consuming for bigger teams.

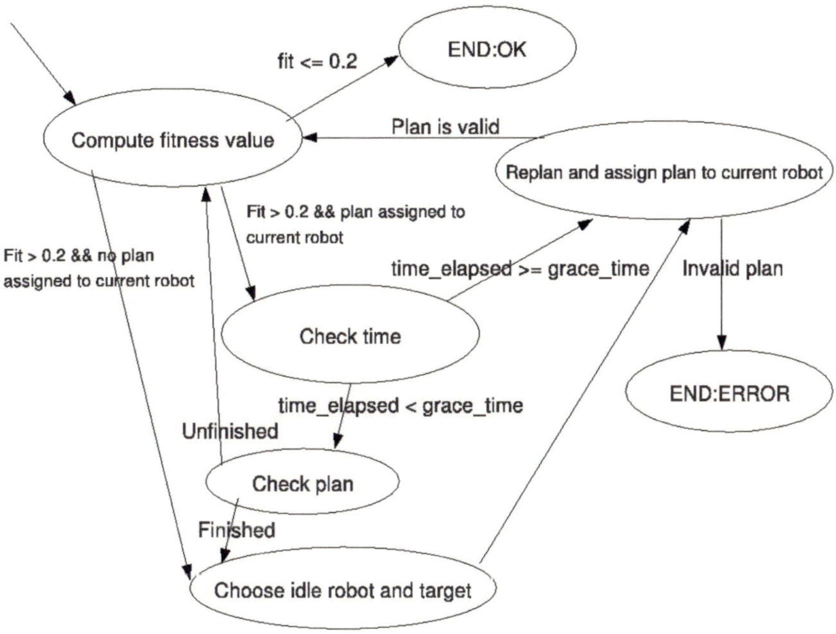


Fig. 9. *One robot at a time* method for repairing the formation

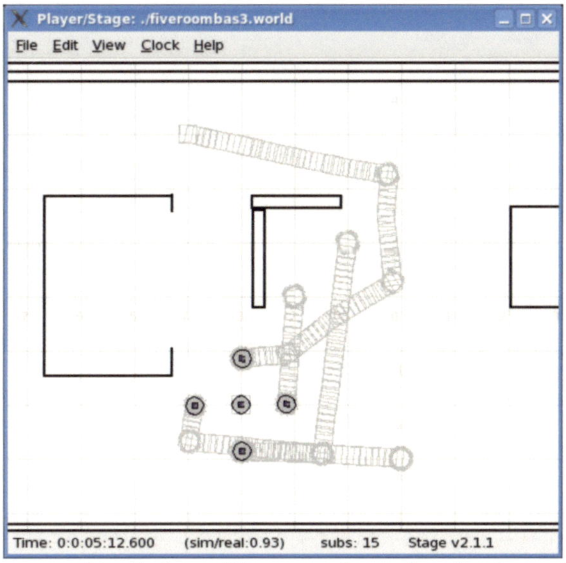


Fig. 10. Trails of robots moved to their positions on planned paths (*one robot at a time* method)

**Method 3. All robots at a time**. Paths are planned and executed for all robots simultaneously. Whenever collision occurs during plan execution, lower level behavior causes involved robots to go back for a while, turn left or right for a while and new paths for those robots are planned. This is the fastest method and despite the fact that it is not collision aware, it is reliable enough.

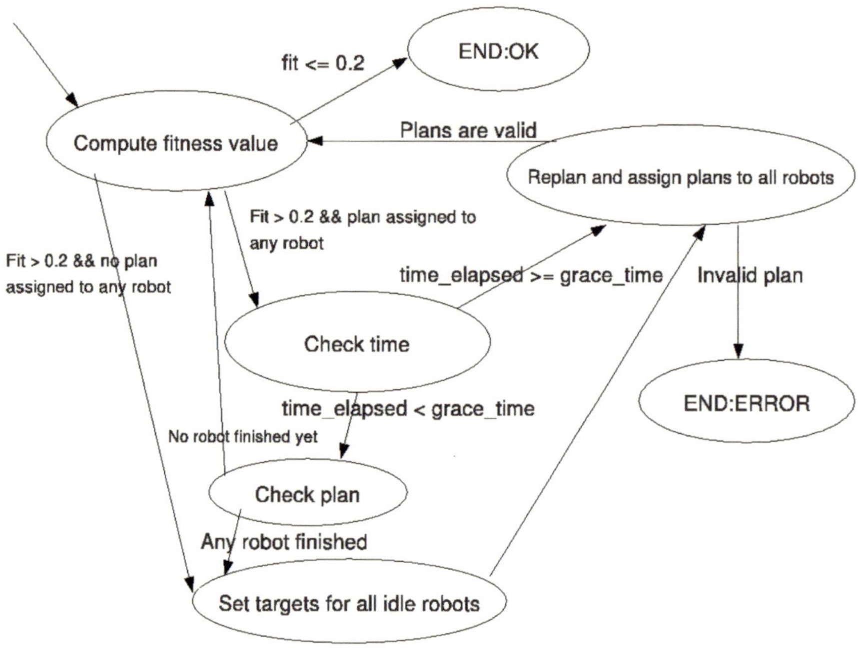


Fig. 11. All robots at a time method of repairing the formation

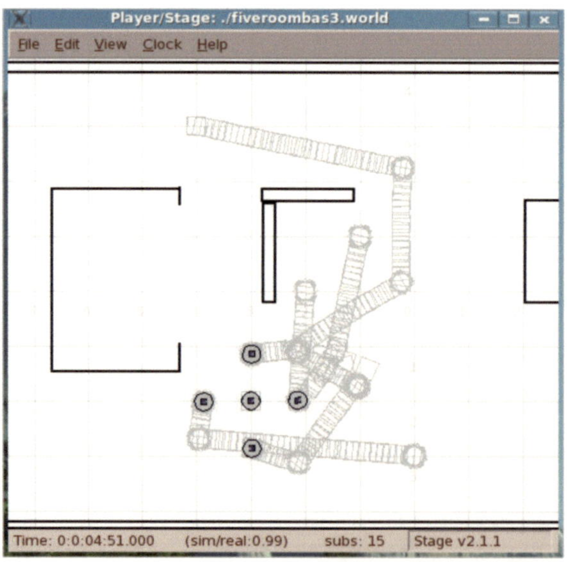


Fig. 12. Trails of robots moved to their positions on planned paths (*all robots at a time* method)

## 9. Navigation by obstacles with robot formations. Formation changing and repairing

The final stage of planning is in checking its soundness by navigating robots in an environment with obstacles. We show results of navigating with a team of robots in the initial formation of cross-shape in a crowded environment, see Fig. 13. In order to bypass a narrow avenue between an obstacle and the border of the environment, the formation changes to a line, and after bypassing it can use repairing to restore to the initial formation (if it is required), see Figs.14-18. The initial cross-shaped formation is shown in Fig. 13 along with obstacles in the environment.

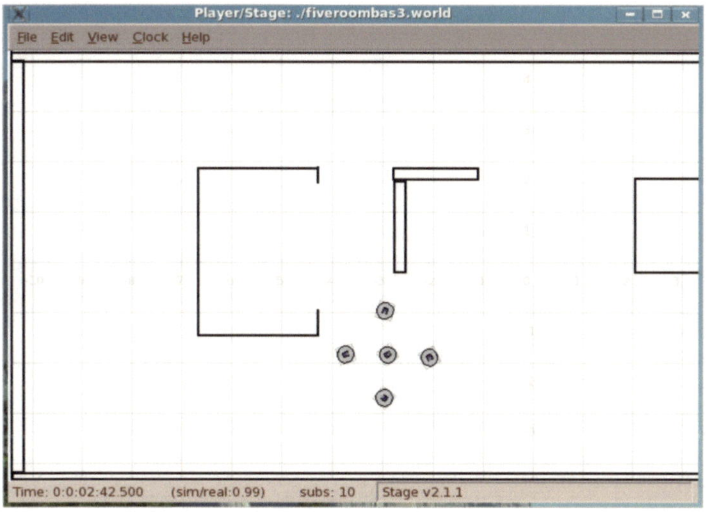


Fig. 13. Initial formation of robots and the obstacle map

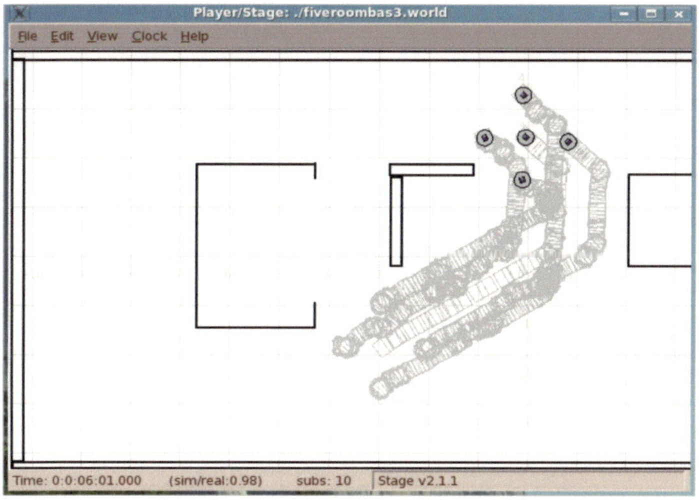


Fig. 14. Trails of robots moved to their positions on the cross formation

Reaching the target requires passing by a narrow passage between the border and the rightmost obstacle. To carry out this task, robots in the formation are bound to change the initial formation. They try the line formation, see Figs. 14-15.

However, making the line formation at the entrance to narrow passage is coupled with some difficulties: when the strategy *all robots at a time* is applied, robots at the lower part of the formation perceive robots at the upper part as obstacles and wait until the latter move into passage, which blocks whole routine as it is assumed that from the start each robot has a plan until it does reach goal or until it does collide with another robot.

To avoid such blocking of activity the behavior *wander* was added, see clouded area in Figs. 15, 16, which permitted robots to wander until they find that they are able to plan their paths into the line. It can be observed that this wandering consumes some extra time. When the strategy *one robot at a time* is applied it is important to carefully select the order in which robots are moved: the robots that have a clear pass to their target positions should go first.

Surprisingly, *pure behavioral* strategy showed good performance in managing with these difficulties, however, (as we expected) when this strategy is applied, it is time -consuming to reshape the formation. After the line was formed and robots passed through the passage, see Figs. 17-18, the line formation could be restored to the initial cross-shape, if necessary, with the help of a strategy for repairing formations of section 8.2.1. The results presented in Figs. 14-19 have been witnessing that our approach has proved its usefulness and validity: in quite complicated obstacle-ridden environments, robots are able to reach the goal. The important feature of this approach is the invariance of the notion of formation with respect to metric relations among robots: as no metric constraint bounds robots, they are able to disperse when facing an obstacle with the only requirement being to keep spatial relationships as set by the betweenness relation imposed upon them.

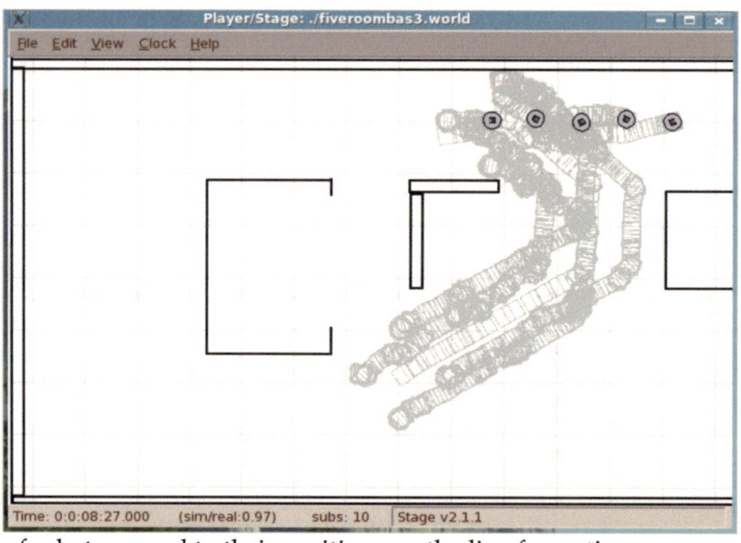


Fig. 15. Trails of robots moved to their positions on the line formation

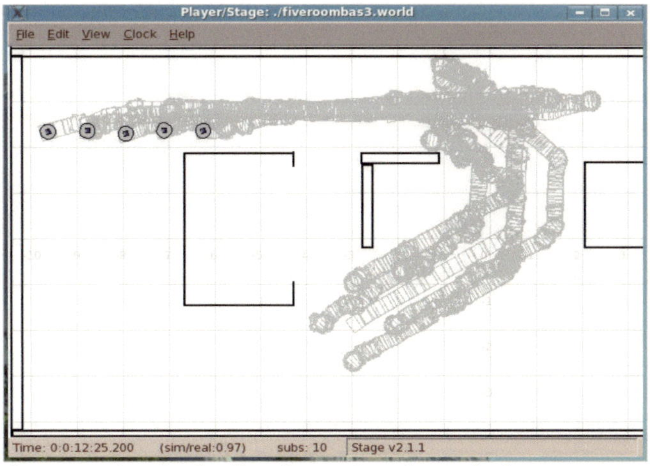

Fig. 16. Trails of robots moving in the line formation through the narrow passage

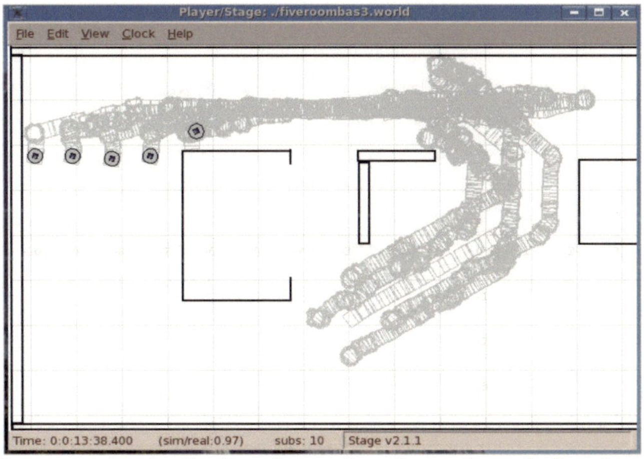

Fig. 17.Trails of robots moving in the line formation through and after the passage

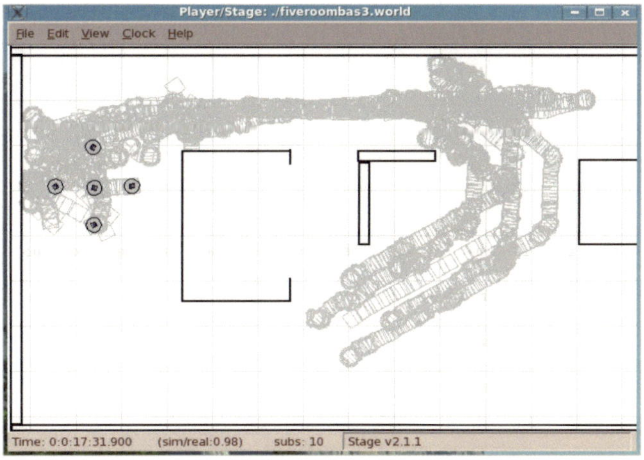

Fig. 18. Yet another formation change: back to the cross formation

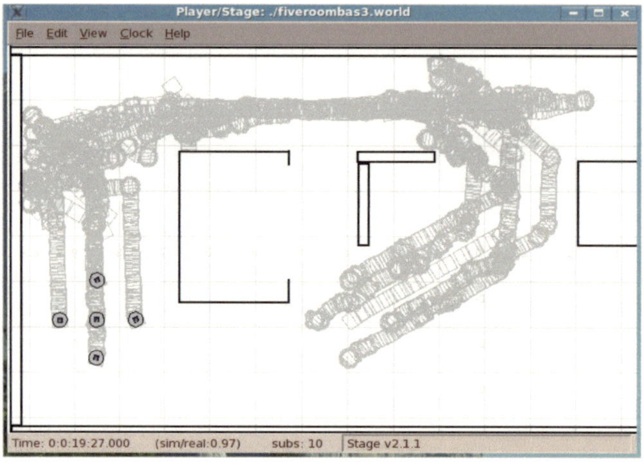


Fig. 19. Trails of robots in the cross formation in the free workspace after the passage

## 10. Conclusions and future research

We have proposed a precise formal definition of a formation and we have presented a Player driver for making formations according to our definition. Our definition of a formation is based on a set of rough mereological predicates which altogether define a geometry of the space. The definition of a formation is independent of a metric on the space and it is invariant under affine transformations. We have examined three methods of formation restoring, based on a reactive (behavioral) model as well as on decoupled way of planning. We have performed simulations in Player/Stage system of planning paths for formations with formation change. The results show the validity of the approach. Further research will be directed at improving the effectiveness of execution by studying divisions into sub-formations and merging sub-formations into formations as well as extending the results to dynamic environments.

## 11. References


Arkin, R.C. (1998). *Behavior-Based Robotics*, MIT Press, ISBN 0-262-01165-4 , Cambridge MA

Balch, T. & Arkin, R.C. (1998). Behavior-based formation control for multi-robot teams. *IEEE Transactions on Robotics and Automation*, 14(6), 926-939, ISSN 1042-296X

vanBenthem, J. (1983). *The Logic of Time*, Reidel, ISBN 9-027-71421-5, Dordrecht

Brumitt, B.; Stentz, A.; Hebert, M. & CMU UGV Group. (2001). Autonomous driving with concurrent goals and multiple vehicles: Mission planning and architecture. *Autonomous Robots*, 11, 103-115, ISSN 0929-5593

Uny Cao, Y.; Fukunaga, A.S. & Kahng, A.B. (1997). Cooperative mobile robotics: Antecedents and directions. *Autonomous Robots*, 4, 7-27, ISSN 0929-5593

Chen, Q. & Luh, J. Y. S. (1998). Coordination and control of a group of small mobile robots, *Proceedings of IEEE Intern. Conference on Robotics and Automation*, pp. 2315-2320, ISBN 0-7803-4300-X, Leuven, May 1998, IEEE Press

Choset, H.; Lynch, K.M.; Hutchinson, S.; Kantor, G.; Burgard, W.; Kavraki, L.E.& Thrun, S. (2005). *Principles of Robot Motion. Theory, Algorithms, and Implementations,* MIT Press, ISBN 0-262-03327-5, Cambridge MA

Clarke, B. L. (1981). A calculus of individuals based on connection. *Notre Dame Journal of Formal Logic,* 22(2), 204-218, ISSN 0029-4527

Das, A.; Fierro, R.; Kumar, V.; Ostrovski, J.P.; Spletzer, J. & Taylor, C.J. A vision-based formation control framework. *IEEE Transactions on Robotics and Automation,* 18(5), 813-825, ISSN 1042-296X

Gotts, N.M.; Gooday, J.A. & Cohn, A.G. (1996). A connection based approach to common—sense topological description and reasoning. *The Monist* , 79(1), 51-75

Khatib, O. (1986). Real-time obstacle avoidance for manipulators and mobile robots. *International Journal of Robotic Research* , 5, 90-98, ISSN 0278-3649

Kramer, J. Scheutz, M. (2007). Development environments for autonomous mobile robots: A survey. *Autonomous Robots,* 22, 101-132, ISSN 0929-5593

Krogh, B. (1984). A generalized potential field approach to obstacle avoidance control. *SME-I Technical paper MS84-484,* Society of Manufacturing Engineers, Dearborn MI

Kuipers, B.J. & Byun, Y.T. (1987). A qualitative approach to robot exploration and map learning, *Proceedings of the IEEE Workshop on Spatial Reasoning and Multi-Sensor Fusion,* pp. 390-404, St. Charles IL, October 1987, Morgan Kaufmann, San Mateo CA

Ladanyi, H. (1997). SQL Unleashed, Sams Publishing, ISBN 0-672-31133-X

De Laguna, T. (1922). Point, line, surface as sets of solids. *J. Philosophy,* 19, 449-461, ISSN 0022-362-X

Latombe, J. (1991). *Robot Motion Planning,* Kluwer, ISBN 0-792-39206-X, Boston

Ehrich Leonard, N. & Fiorelli, E. (2001). Virtual leaders, artificial potentials and coordinated control of groups, *Proceedings of the 40th IEEE Conference on Decision and Control,* ISBN 0-780-37061-9, pp. 2968-2973, Orlando Fla, December 2001, IEEE Press

Leonard, H. & Goodman, N. (1940). The calculus of individuals and its uses. *The Journal of Symbolic Logic,* 5, 45-55, ISSN 0022-4812

Lesniewski, S. (1916). *O Podstawach Ogolnej Teorii Mnogosci* (*On Foundations of General Theory of Sets,* in Polish), The Polish Scientific Circle in Moscow, Moscow

Lesniewski, S. (1982). On the foundations of mathematics, *Topoi* 2, 7-52, ISSN 0167-7411

Osmialowski, P. (2007). Player and Stage at PJIIT Robotics Laboratory, *Journal of Automation, Mobile Robotics and Intelligent Systems,* 2, 21-28, ISSN 1897-8649

Osmialowski, P. (2009). On path planning for mobile robots: Introducing the mereological potential field method in the framework of mereological spatial reasoning, *Journal of Automation, Mobile Robotics and Intelligent Systems,* 3(2), 24-33, ISSN 1897-8649

Osmialowski, P. & Polkowski, L. (2009). Spatial reasoning based on rough mereology: path planning problem for autonomous mobile robots, *Transactions on Rough Sets. Lecture Notes in Computer Science,* in print, ISSN 0302-9743, Springer Verlag, Berlin

Player/Stage: Available at http://playerstage.sourceforge.net

Polkowski,L. (2001). On connection synthesis via rough mereology, *Fundamenta Informaticae,* 46, 83-96, ISSN 0169-2968

Polkowski, L. (2008). A unified approach to granulation of knowledge and granular computing based on rough mereology: A survey, In: *Handbook of Granular Computing,* Pedrycz, W.; Skowron, A. & Kreinovich, V., (Eds.), 375-400, John Wiley and Sons, ISBN 987-0-470-03554-2, Chichester UK

Polkowski, L. & Osmialowski, P. (2008) Spatial reasoning with applications to mobile robotics, In: *Mobile Robots Motion Planning. New Challenges,* Xing-Jian Jing, (Ed.), 43-55, I-Tech Education and Publishing KG, , ISBN 978-953-7619-01-5, Vienna

Polkowski, L. & Osmialowski, P. (2008). A framework for multiagent mobile robotics: Spatial reasoning based on rough mereology in Player/stage system, *Lecture Notes in Artificial Intelligence,* 5306, 142-149, Springer Verlag, ISSN 0302-9743, Berlin

P. Ramsey, PostGIS Manual, in: postgis.pdf file downloaded from Refractions Research home page.

J. Shao, G. ; Xie, J. Yu & Wang, L. (2005). Leader-following formation control of multiple mobile robots, In: *Proceedings of the 2005 IEEE Intern. Symposium on Intelligent Control*, pp. 808-813, ISBN 0-780-38936-0, Limassol, June 2005, IEEE Press

*sfsexp*: Available at http://sexpr.sourceforge.net

Sugihara, K. & Suzuki, I. (1990). Distributed motion coordination of multiple mobile robots, In: *Proceedings 5th IEEE Intern. Symposium on Intelligent Control*, pp. 138-143, ISBN 9-991-32943-9, Philadelphia PA, Sept. 1990, IEEE Press

Švestka, P. & Overmars, M.H. (1998). Coordinated path planning for multiple robots, *Robotics and Autonomous Systems,* 23, 125-152, ISSN 0921-8890

Tarski, A. (1929) Les fondements de la géométrie des corps, In: *Supplement to Annales de la Sociéte Polonaise de Mathématique*, 29-33, Krakow, Poland

Tarski, A. (1959). What is elementary geometry?, In: *The Axiomatic Method with Special Reference to Geometry and Physic*s, Henkin, L.; Suppes, P. & Tarski, A., (Eds)., 16-29, North-Holland, Amsterdam

Tribelhorn, B. & Dodds, Z. (2007). Evaluating the Roomba: A low-cost, ubiquitous platform for robotics research and education, In: *2007 IEEE International Conference on Robotics and Automation, ICRA 2007,* pp. 1393-1399, ISBN 1-4244-0601-3, April 2007, Roma, Italy, IEEE Press

Urdiales, C.; Perez, E.J.; Vasquez-Salceda, J.; Sanchez-Marrμe, M. & Sandoval, F. (2006). A purely reactive navigation scheme for dynamic environments using Case-Based Reasoning, *Autonomous Robots* , 21, 65-78, ISSN 0929-5593

Whitehead, A. N. (1979). *Process and Reality. An Essay in Cosmology* 2nd. ed., The Free Press, ISBN 0-029-34570-7, New York NY

# An Artificial Protection Field Approach For Reactive Obstacle Avoidance in Mobile Robots

Victor Ayala-Ramirez, Jose A. Gasca-Martinez, Rigoberto Lopez-Padilla and Raul E. Sanchez-Yanez
*Universidad de Guanajuato DICIS*
*Mexico*

## 1. Introduction

Mobile robots using topological navigation require of being able to react to dynamical obstacles in their environment. Reactive obstacle avoidance is also an essential capability needed by a mobile robot evolving in a cluttered dynamic environment. Scenarios like offices where people and robots share a common workspace are difficult to be modeled by static maps. In such scenarios, robots need to avoid people and obstacles when executing any other task involving its motion.

Another application for reactive obstacle avoidance arises when the robot is building a map of an unknown environment. The robot will need to react to different events during its exploring task. For example, the robot needs to be able to avoid a wall not previously known or to evade another moving object in its neighborhood.

In the past, several researchers have proposed reactive navigation methods. Some examples of these methods are: The artificial potential field (Khatib, 1986) and the elastic band approach (Quinlan & Khatib, 1993; Lamiraux & Laumond, 2004) proposed originally both by Khatib; the vector field histogram (Borenstein & Koren, 1990), the dynamic window approach proposed by (Fox et al., 1997) and the nearness diagram, recently proposed by (Minguez & Montano, 2004).

Our approach is to build an artificial protection field around the robot and to survey it by fusing laser range finder and odometry measurements. In this work, an approach to reactive obstacle avoidance for service robots is proposed. We use the concept of an artificial protection field along a robot pre-planned path. The artificial protection field is a dynamic geometrical neighborhood of the robot and a set of situation assessment rules that determine if the robot needs to evade an object not present in its map when its path was planned. This combination results in a safety zone where no other object can be present when the robot is executing a motion primitive; a zone where the robot needs to recalculate its path; and some other zones where the object can perform successfully its navigation task, even if obstacles are detected near the robot path. During the execution of a motion primitive, dynamical obstacles are detected by using a laser range finder. If the obstacles detected in the neighborhood of the robot path enter the artificial protection field of the robot, reactive behaviours are launched to recalculate the path online in order to avoid collisions with them. Our method has been tested in an experimental setup both in a simulation platform and in the real robot in a qualitative manner. The robot has demonstrated successful evolution on these tests for static and dynamic scenarios.

The structure of the chapter will be as follows: Firstly, we are going to review recent approaches in reactive navigation for robots. We will also review their usefulness in topological navigation approaches. A second section will describe in detail what are the specific features of the proposed artificial protection field approach and a critical comparison of its characteristics against some other algorithms for obstacle evasion in the recent literature. Test protocols used to validate our approach will be inspected in detail in a subsequent section. We will show results in a custom-developed robotic simulation platform for our robot. We will also show experimental tests that have been implemented on a Pioneer P3-AT mobile robot named XidooBot under several scenarios. Our main findings will then be discussed and we finish our proposal with a section giving our conclusions and perspectives of future work.

## 2. Reactive Navigation Methods

Robot navigation using reactive methods has been studied extensively. Here we present main approaches and recent methods proposed in literature.

One of the first approaches used for reactive navigation is the potential field (PF) approach. (Khatib, 1986) proposed the generation of an artificial potential field that repels the robot from the obstacles and that attracts it toward its goal. Main problem of this method is the emergence of isopotential regions because of the potential selection for the environment elements; that traps the robot in local minima regions, impeding it to attain its goal. This drawback limits the applicability of the method in complex environments. Recently (Antich & Ortiz, 2005) have proposed to define a behaviour based navigation function that combined with the PF approach can partially overcome main drawbacks of the PF approach. PF methods are global planning methods so they can be used also for motion planning.

A variation of the PF methods is known as the elastic band (EB) approach. EB methods propose to deform an a-priori computed path when obstacles not considered at planning time are detected during the execution of a give trajectory. Some examples of this approach are the works by (Quinlan & Khatib, 1993) for manipulator robots and (Lamiraux & Laumond, 2004; Lamiraux et al., 2004) for mobile robots. As said before, EB methods require a pre-planned path, so they can not be used for exploration tasks.

The vector field histogram (VFH) is a reactive navigation method proposed by (Borenstein & Koren, 1991). The main idea is to represent the free space surrounding the current position of a robot using an occupancy grid. A polar histogram is created and the robot selects the direction with the maximal cell count of free space as a preferential orientation for its motion. This method is essentially a local navigation method to avoid obstacles.

(Fox et al., 1997) have originally proposed the dynamic window (DW) approach to avoid collision with obstacles in a reactive way. In this method, the dynamic restrictions of differential and synchro-drive steered robots are taken into account to generate arc motion primitives that avoid intruding elements. The optimization of the motion primitives find the optimal values for the translational and rotational velocities with respect to the current target heading, robot velocity and clearance.

A more recent approach is presented by (Minguez & Montano, 2004; Minguez, 2005; Vikenmark & Minguez, 2006) as the nearness diagram (ND) approach. This method models objects and free space in the proximity of the robot. The robot recognizes its situation with respect to the task to be done. The robot takes then consequent actions (motion laws) according to the assessment. Recently, (Li et al., 2006) have proposed the hybridization of this technique with the DW approach. Main contribution of this improvement is to increase the speed of the mobile robot even in troublesome scenarios.

Some other methods have also been proposed recently for reactive navigation. Some of them use fuzzy logic to control the reactive motion of the robot. Some examples of this approach are the works by (Mester, 2008) and (Larson et al., 2005). Some other consider also the identification of the behaviours associated to dynamic obstacles by using Bayesian approaches, as for example, (Lopez-Martinez & Fraichard, 2008) and (Laugier et al., 2008). However, they are not very related with our approach even if they are alternatives for reactive robot motion.

## 3. The Artificial Protection Field Approach

Our approach is based in the concurrent execution of two tasks: the execution of a navigation command and the obstacle detection task. The navigation commands implement the planned path in a static and known environment configuration. We define the robot pose by using the $(x,y,\theta)$ coordinates. The motion primitives link two poses by using advance and rotate primitives. Nevertheless, each time an obstacle is detected the motion command is stopped and a re-planning process is spawned. In the following, we give the details of the above procedure implemented in a mobile robot platform.

### 3.1 Obstacle Detection

In any reactive navigation method, the robot needs to acquire information about its surroundings in order to detect obstacles. In our robot, obstacles are detected by using the laser range finder (LRF) sensor. The LRF measures acquired by the mobile robot are classified to determine a safety condition for it. In particular, they are classified according to its closeness to a protection zone around the robot. We call such safety zone an artificial protection field (APF).

### 3.2 Artificial Protection Field

The APF is defined in terms of three restrictions:

Firstly, we consider a region where the robot can freely execute the motion commands without re-planning its path, namely the minimal obstacle free space $E$. The shape of the boundary of this region can be arbitrarily defined by using a polar defined function $r(\theta)$. The value of this function $r_\theta$ is taken as the maximal distance being free of obstacles at an orientation $\theta$ with respect to a robot-centered coordinate system. That is,

$$E = \left\{ \theta, r) \middle| 0 \le r \le r_\theta, r_\theta = r(\theta), \theta \in \left[ -\pi/2, \pi/2 \right] \right\} \tag{1}$$

In our particular application, we have used a half-ellipse boundary referenced in the robot as shown in Figure 1.

The second restriction considers the distance of the current position of the robot with respect to the goal of the motion primitive, named $d_g$. That is, we do not want to react to an obstacle in $E$ which is farther than the goal. The zone satisfying this restriction is represented by $G$ in Figure 1, and it is shown in yellow. Given that we take a Euclidean distance metric, the shape of $E$ is a half circle centred on the robot reference point, that is

$$G = \left\{ \theta, r) \middle| 0 \le r \le d_g, \theta \in \left[ -\pi/2, \pi/2 \right] \right\} \tag{2}$$

Finally, the critical space $C$ (see Figure 1) is a safety zone to avoid collisions of the robot with obstacles. If we take a critical distance $d_c$, region $C$ is defined as follows:

$$C = \left\{ \theta, r) \middle| 0 \le r \le d_c, \theta \in \left[ -\pi/2, \pi/2 \right] \right\} \tag{3}$$

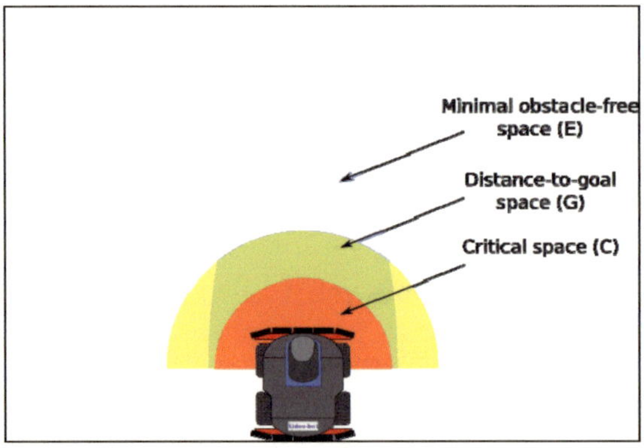


Fig. 1. Description of the interest regions around a mobile robot.

### 3.3. Situation Assessment

Reactive behaviours of the robot are launched after recognizing a danger context. To recognize which behaviour has to be taken by the robot, we use the LRF measures as input information. As usual, the sensor provides us with a range distance to the next obstacle at a pre-defined set of orientations. We process this set of measurements to infer the free zone L in front of the current position of the mobile robot. We assess then the situation that the robot is facing. Our goal is to classify the situational context of the robot as a class of the following list of exhaustive states: Obstacle free situation (labelled as $O_f$), a low risk situation (labelled as $O_l$), a medium risk situation ($O_m$) and a high risk situation ($O_h$).

**Obstacle Free Situation** $O_f$

This is the ideal situation for the robot because it has not detected an obstacle in E. That situation lets the robot to continue with the execution of the current motion primitive. Formally this is represented by:

$$\bar{L} \cap E = \varnothing \rightarrow O_f \,. \tag{4}$$

**Low Risk Situation** $O_l$

When an object invades the minimal free space of the robot, we consider the event as a low risk situation if the obstacle is farther than the goal position, i.e.,

$$\bar{L} \cap E \cap \bar{G} \neq \varnothing \rightarrow O_l \,. \tag{5}$$

**Medium Risk Situation** $O_m$

The robot is facing a medium risk situation if the following three conditions are fulfilled:

- There is an object invading the minimal free space.
- The target position for the motion primitive is closer than the current position of the robot.
- The obstacle does not enter into the critical space $C$.

Formally, that implies,

$$\bar{L} \cap E \cap G \cap \bar{C} \neq \varnothing \rightarrow O_m \,. \tag{6}$$

**High Risk Situation** $O_h$

A high risk situation is the event where the robot has detected an object invading ist critical space. In a formal way,

$$\bar{L} \cap C \neq \varnothing \rightarrow O_h \,. \tag{7}$$

Figure 2 shows how obstacles are detected and used to assess the robot situation.

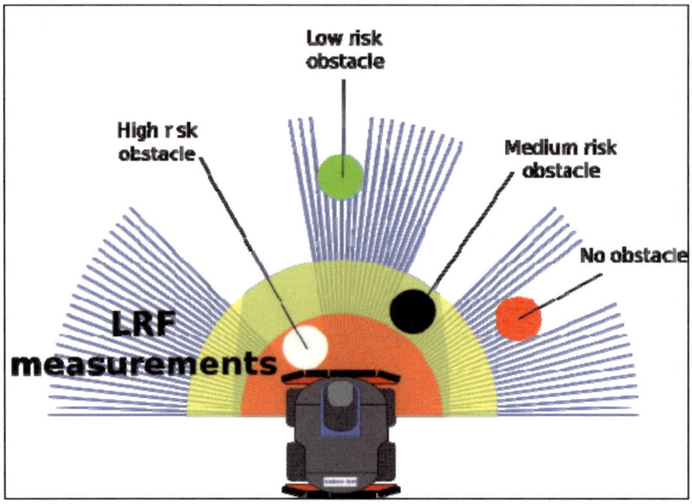


Fig. 2. Situation assessment in a service robot using the APF.

### 3.4. Reactive Behaviour

During the execution of a motion command, the robot is polling its sensors in order to assess the navigation situation. When a no risk ($O_f$) or low risk ($O_l$) situation is detected, the motion primitive continues its execution. A medium risk situation $O_m$ causes the robot to execute the following actions: 1) Stop motion execution, 2) Re-plan its path to get an obstacle free trajectory, and 3) Execute the modified path. If a high risk situation $O_h$ is recognized, the service robot interrupts inmediately ist motion to avoid collision with the intruding object. An activity diagram of the actions taken as a reaction to the situation assessment is shown in Figure 3.

### 3.5 Path Re-Planning

If the robot recognizes a medium risk situation, it stops execution of the current motion primitive. It also launches a path re-planning process that uses LRF measurements as input information. The purpose of this process is to find an intermediate goal $s_g^*$ in the free space perceived by the robot and that minimizes some metric related to the path.

To do so, we analyse some points in the free-space L (see Section 3.3) detected by the LRF. The feasible intermediate target goal $S_g$ consists of all the points in a given set of radial distances from the current position of the robots that belong also to L. Another characteristic that all points in $S_g$ have is that an inspection windos W centered on them presents no collision with any obstacle. The inspection window extendes from an angle $\theta - \Delta\theta$ to an angle $\theta + \Delta\theta$, where $\theta$ is the angular coordinate of the point in $S_g$ being tested with respect to the robot-centered coordinate system, and $\Delta\theta$ is the width of the inspection window.

All target goals in $S_g$ are then attainable. We carry out an optimization process over all these points and we choose the optimal one according to the objective function being optimized.

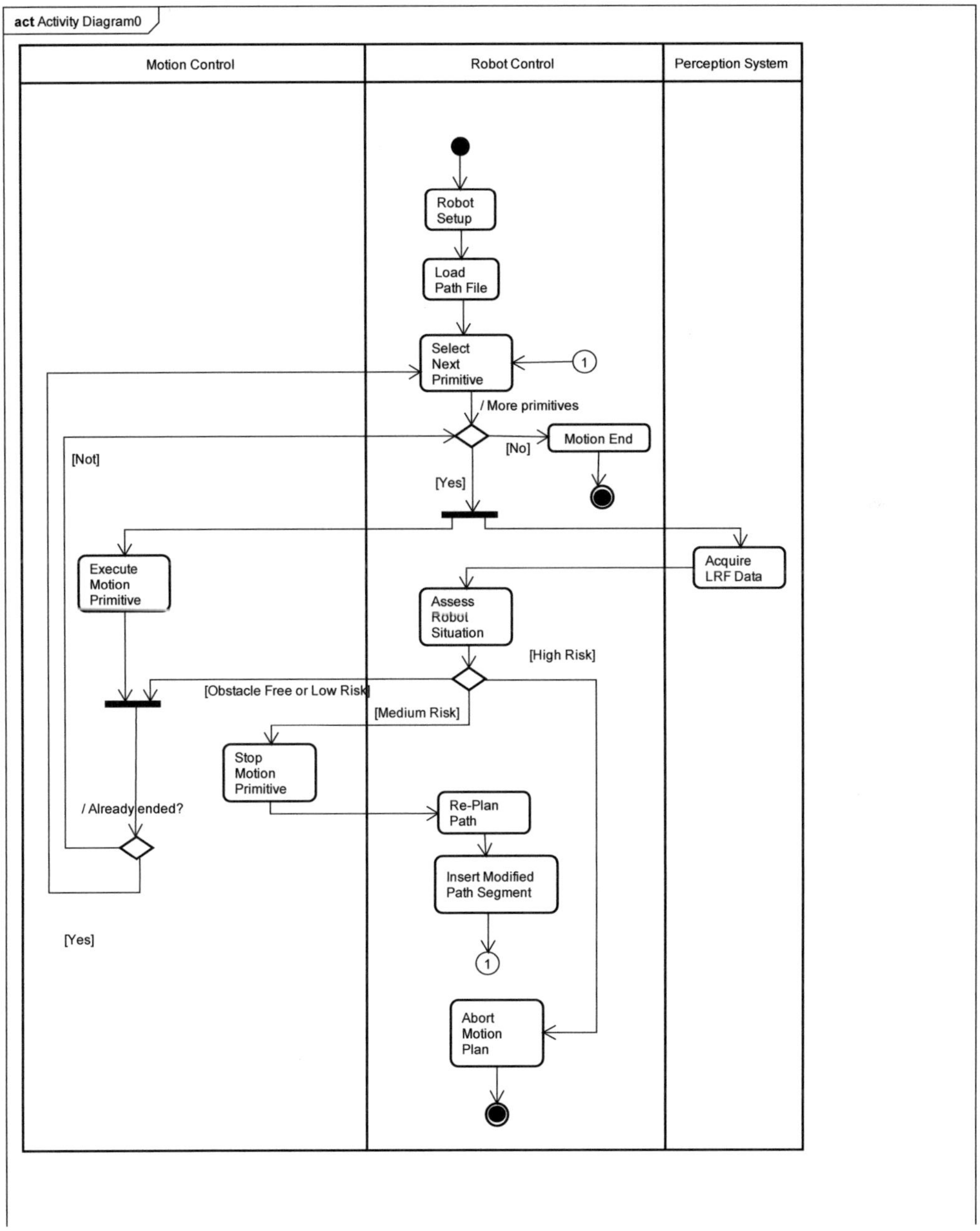


Fig. 3. Activity diagram for the reactive behaviour of the service robot.

Let us consider, for example that our objective is to minimize closeness to the target position $s_t$ in the interrupted motion command. Then, for all $s_k$ in $S_g$:

$$f(s_k) = \|s_t - s_k\|. \quad (8)$$

and then

$$s_g^* = s_k \Leftrightarrow f(s_k) = \min_{s_k \in S_g} \|s_t - s_k\|. \quad (9)$$

where for example, $\|\bullet\|$ could be an Euclidean norm.

## 4. Test and Results

We have run tests firstly in a custom-developed simulation platform. The goal of these experiments was to define the best parameters for the implementation on a real P3-AT robotics platform named XidooBot. We have then executed motion scripts that tested exhaustively all the contexts that could arise in the reactive navigation task.

The first set of tests has been performed in a custom-developed simulation environment. A scenario with multiple objects is shown in Figure 4. Here, we can observe the re-planning of a new trajectory once the robot finds an obstacle in its APF. After that, an obstacle-free trajectory is executed until it reaches its goal. The half-elliptical safety zone at the end of the trajectory is specifically indicated in the Figure.

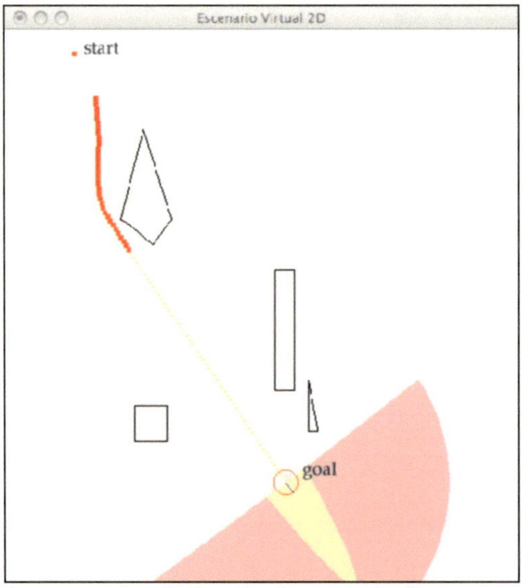


Fig. 4. Simulation of a script composed of a single straight-line segment.

In Figure 5, the simulation of a more complex trajectory is given. Here, point A is an intermediate goal and the final one is point B. We observe a reactive behaviour zone in the trajectory between the start point and point A. From A to B, its necessary beginning with a continuous re-planning of the trajectory until the goal is attained executing a straight trajectory.

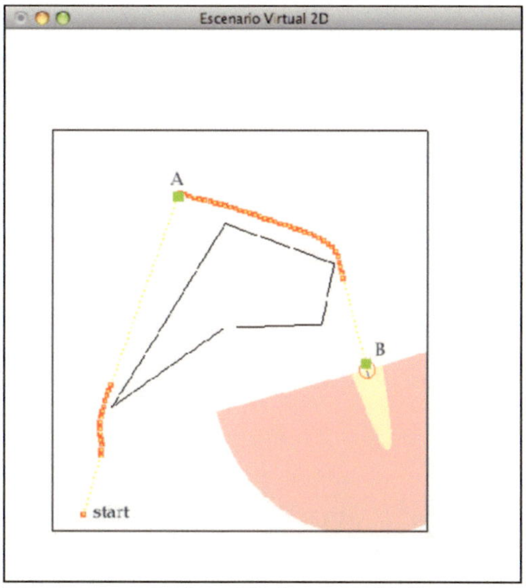


Fig. 5. Execution of a script composed of multiple straight-line segments.

A second phase of tests was run in the real platform for robotics experiments. We have also setup scenario configurations that cover all the cases foreseen in the APF approach. The reactive navigation system has been tested qualitatively. The test procedure includes the execution of a set of paths in an environment where obstacles were added dynamically during the motion commands execution. Figure 6 shows a successful motion command execution because no obstacle was detected along the execution of the motion primitives. Even if an obstacle is near the target position of the robot, path is not re-planned because the object does not interfere the robot motion.

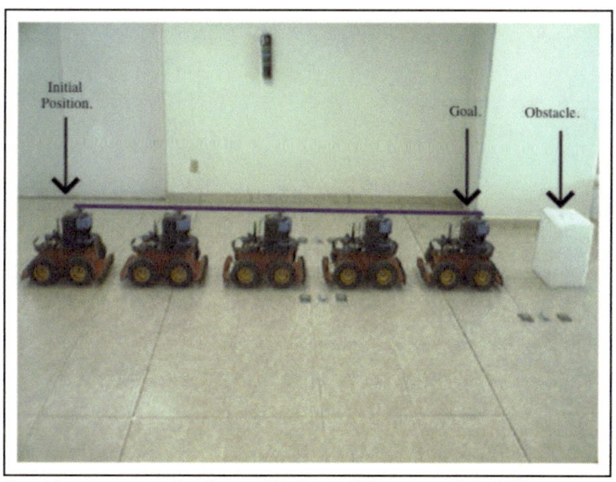


Fig. 6. A reactive navigation scenario (case I) for a Pioneer P3-AT.

A second example is shown in Figure 7. When the motion execution is launched, the robot detects a medium risk situation, so it re-plans its trajectory. We can see in Figure 7 what are the sub-goals chosen by the robot in order to reach the originally planned target position.

A more complex environment is shown in Figure 8. We have several obstacles near the planned path (shown in blue). In the figure, we can see the executed path (shown in red) after reacting to the presence of obstacles.

A closed path execution is shown in Figure 9. Our reactive strategy is applied when during the execution of a path segment an obstacle appears. We show the planned path in blue and the modified executed path in red. As we can see, the robot passes by the control points of the motion primitive sequences without problems.

Fig. 7. A reactive navigation scenario (case II) for a Pioneer P3-AT.

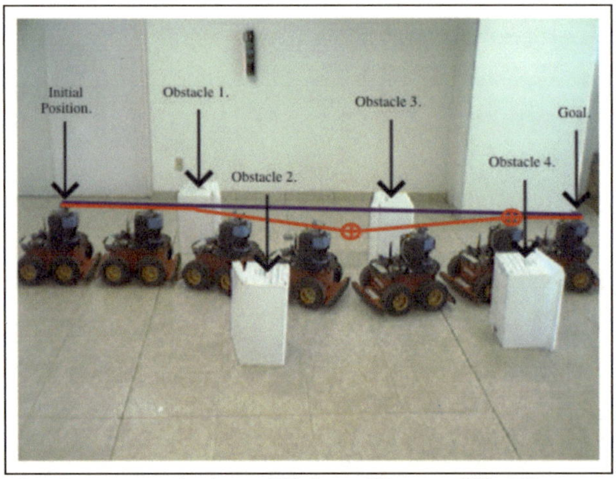


Fig. 8. A reactive navigation scenario (case III) for a Pioneer P3-AT.

Fig. 9. A reactive navigation scenario (case IV) for a Pioneer P3-AT.

## 5. Conclusion and Perspectives

In this work, we have presented a reactive obstacle avoidance approach based on a situated-activity paradigm. Our system has demonstrated to be robust in qualitative tests developed on a dynamically changing environment. We have shown tests in a custom-developed simulation platform and in the real robot. Our approach is based in the risk assessment made by the robot by using an artificial protection field to avoid moving obstacles.

In the near future, we will evaluate quantitative performance of the method by analyzing pose errors after the re-planning step both in simulation and in the real robotic platform XidooBot. Fusion of several sensors information will be done in order to improve the robustness of this approach. This method will be integrated with a topological navigation system.

## 6. References

Antich, J. & Ortiz A. (2005), Extending the potential fields approach to avoid trapping situations, Proc. Int. Conf. on Intelligent Robots and Systems, 2005 (IROS 2005), pp. 1386 1391, August, 2005, IEEE Press.

Borenstein, J. & Koren, Y. (1990), Real-time obstacle avoidance for fast mobile robots in cluttered environments, Proc. of the 1990 IEEE Int. Conf. on Robotics and Automation (ICRA1990), pp. 572–577, Cincinnati, Ohio, May, 1990, IEEE Press.

Fox, D.; Burgard, W. & Thrun, S. (1997), The dynamic window approach to collision avoidance, *IEEE Robotics & Automation Magazine*, Vol. 4, No.1 (March, 1997), pp. 23–33.

Khatib, O. (1986), Real-time obstacle avoidance for manipulators and mobile robots, *International Journal of Robotics Research*, Vol. 5, No. 1 (1986), pp. 90–98.

Koren, Y. & Borenstein, J. (1991), Potential field methods and their inherent limitations for mobile robot navigation, Proc. IEEE Int. Conf. Robotics and Automation, April, 1991, pp.1398-1404.

Lamiraux, F.; Bonnafous, D. & Lefebvre, O. (2004), Reactive path deformation for nonholonomic mobile robots, *IEEE Trans. on Robotics and Automation*, Vol. 20, (Dec. 2004), pp. 967–977.

Lamiraux, F. & Laumond, J.-P. (2004), Reactive obstacle avoidance and trajectory optimization for non-holonomic systems: two problems, one solution, Proc. World Automation Congress 2004, vol. 15, pp. 473–478, June, 2004.

Larsson, J.; Broxvall, M. & Saffiotti, A. (2008), Laser Based Intersection Detection for Reactive Navigation in an Underground Mine, 4th IEEE Int. Conf. on Intelligent Systems, 2008 (IS '08), Vol. 1, pp. 2-45 -2-50, Nice, France, September 2008.

Laugier, C.; Vasquez-Govea, D.A.; Yguel, M.; Fraichard, T. & Aycard, O. (2008), Geometric and Bayesian models for safe navigation in dynamic environments, *Intelligent Service Robotics*, Vol. 1, No. 1 (2008), pp. 51-72.

Li, G.; Wu, G. & Wei, W. (2006), ND-DWA: A Reactive Method for Collision Avoidance in Troublesome Scenarios, Proc. of the 6th World Congress on Intelligent Control and Automation, 2006 (WCICA 2006), Volume 2, pp. 9307–9311, 2006.

Martinez-Gomez, L. & Fraichard, T. (2008), An efficient and generic 2D Inevitable Collision State Checker, Proc. Int. Conf. on Intelligent Robots and Systems, IROS 2008, Sept. 2008, pp. 234–241.

Mester, G. (2009), Obstacle-slope avoidance and velocity control of wheeled mobile robots using fuzzy reasoning, Proc. Int. Conf. on Intelligent Engineering Systems (INES 2009), April, 2009, pp. 245–249.

Minguez, J. & Montano, L. (2004), Nearness Diagram (ND) Navigation: Collision Avoidance in Troublesome Scenarios, *IEEE Trans. On Robotics and Automation*, Vol. 20, No. 1 (Feb. 2004), pp. 45-59.

Minguez, J. (2005), Integration of planning and reactive obstacle avoidance in autonomous sensor-based navigation, Proc. Int. Conf. on Intelligent Robots and Systems, (IROS 2005), pp. 2486–2492, August 2005.

Quinlan, S. & Khatib, O. (1993), Elastic bands: Connecting path planning and control. Proc. of IEEE Int. Conf. on Robotics and Automation, Vol. 2, pp. 802–807, May 1993.

Vikenmark, D. & Minguez, J. (2006), Reactive obstacle avoidance for mobile robots that operate in confined 3D workspaces, Proc. IEEE Mediterranean Electrotechnical Conference, MELECON 2006, pp. 1246–1251, May 2006.

# 18

# Hierarchical action control technique based on prediction time for autonomous omni-directional mobile robots

Masaki Takahashi, Yoshimasa Tada, Takafumi Suzuki, and Kazuo Yoshida
*Keio University*
*Japan*

## 1. Introduction

Recently, various essential technologies of an autonomous mobile robot such as a self localization scheme, an environmental map formation and path planning, learning algorithm and communication are developed in the area of robot. In addition, a variety of service robots which offers service with the actual environment with other moving objects, including people are proposed and developed(B. Graf, 2004; Asoh,2001; DeSouza,2002; Thrun,1999; R. Bischoff). A variety of tasks are required for such a service robot, but here we will focus on problems related to moving, which is the most fundamental and important of tasks. In the environment include humans, safe and efficient movement should be required. To deal with this problem, there have been a variety of mobile control techniques for the autonomous mobile robot. The visibility graph-like method and the Voronoi diagram method are the technique to design the pass which can reach a goal without colliding with obstacles on the basis of the advance knowledge of environment. However, it is not easy to work effectively with an actual environment where the unknown static or moving obstacles exist, because the complete knowledge regarding environment is needed. On the one hand, sensor based approach and reactive technique are the technique where recognizes the information of environment with the sensor and then decides behaviour in real time. Therefore it is the technique which is suited for actual environment.

This study proposes a hierarchical action control method for autonomous omni-directional mobile robot to achieve a smooth obstacle avoidance ensuring safety in the presence of moving obstacles including humans. The hierarchical control method considers various time scales for actions such as goal path planning, obstacle avoidance within the recognizable range, and emergency avoidance to deal with unexpected events. In the proposed method, several modules for each action are composed in parallel.

The vertical axis is the time scale in the control system. In the lowest module, the robot can move to the goal safely and efficiently by planning from the environment information which is obtained in advance. In the higher module, the robot moves more safely by using the estimated information about the obstacles to avoid them. By integrating the output of each module comprehensively, it is possible to realize a safe and efficient movement according to

the situation. In this paper, as an example of the proposed moving control technique, we present a method based on virtual potential fields (Khatib, 1986; Brooks, 1986; Ge, 2002).
First, the module which generates the potential field based on each prediction time is formed hierarchically. Second, the virtual force which is derived from the respective potential fields is synthesized. Third, the velocity command is decided on the basis of the resultant force.
To verify the effectiveness of the proposed technique, some experiments using the real apparatus of an autonomous omni-directional robot have been carried out. Moreover, experimental results in an hospital environment are shown.

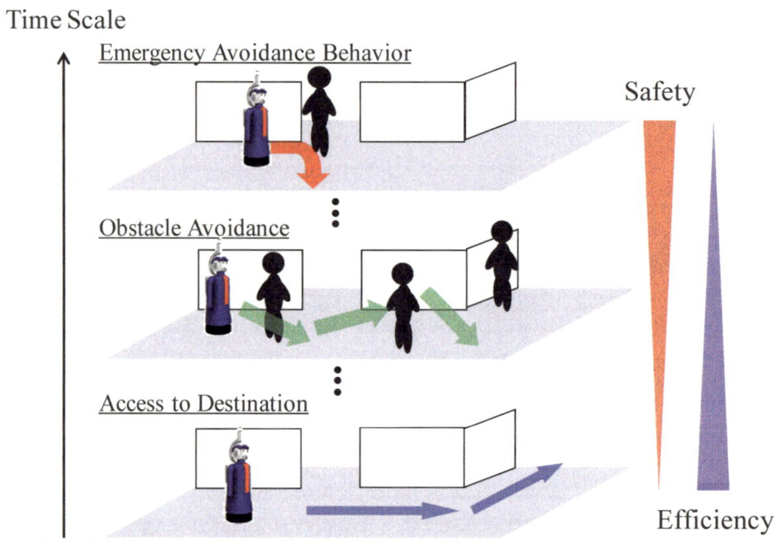

Fig. 1. Problem Establishment for Action of Service Robot

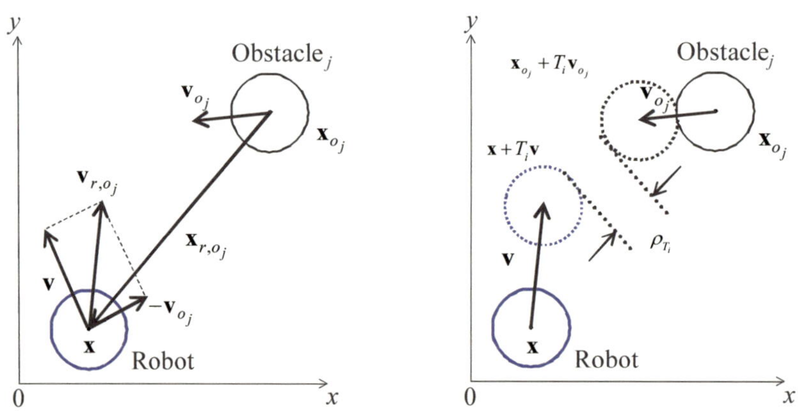

Fig. 2. World coordinate system and predicted shortest distance.

# 2. Important Hierarchical Action Control Method Considering Prediction Time of its Actions

## 2.1 Nomenclature

| Symbol | Quantity |
|---|---|
| $T_i$ | Prediction time |
| $\rho_g$ | Distance between the robot and the goal |
| $\rho_{T_i}$ | Predicted shortest distance between the robot and the obstacle |
| $\rho_0$ | Minimum of repulsive potential |
| $\mathbf{x}$ | Position vector of robot |
| $\mathbf{x}_g$ | Position vector of the goal |
| $\mathbf{x}_{O_j}$ | Position vector of object $O_j$ |
| $\mathbf{x}_{r,O_j}$ | Position vector of the obstacle $O_j$ relative to the robot |
| $\mathbf{v}$ | Velocity vector of robot |
| $\mathbf{v}_{O_j}$ | Velocity vector of object $O_j$ |
| $\mathbf{v}_{r,O_j}$ | Velocity vector of the obstacle $O_j$ relative to the robot |
| $U_{O_j}^{T_i}$ | Virtual potential about object $O_j$ on each time scale |
| $\mathbf{F}_{O_j}^{T_i}$ | Virtual force vector from $U_{O_j}^{T_i}$ |
| $i$ | Index of each time scale |
| $j$ | Index of object |
| $x$ | $x$-axis |
| $y$ | $y$-axis |

Table 1. Nomenclature

## 2.2 Concept

In this study, we address the issues concerning robots that are expected to move within general buildings, such as public spaces, offices, hospitals, or homes, with other moving objects, including people. A variety of tasks are required for such a service robot, but here we will focus on problems related to moving, which is the most fundamental and important of tasks. As the service robot comes in direct contact with humans, we must attach importance to safety to ensure that the robot will not harm humans, but the time it requires for its movements must be minimized as much as possible. It is desirable for the robot to reach its goal without colliding with obstacles, including humans. However, this type of environment includes events that are unpredictable at the design phase, and it is difficult to respond to various circumstances using a scenario-based method. Thus, we propose a moving control method that generates actions corresponding to the circumstances, by embedding a fundamental problem-solving method for moving into the control system.

We propose a hierarchical moving control method that considers the differences in time scale among actions to achieve both safe and effective movement; e.g., movement of the robot over the shortest distance can be realized based on prediction of the movements of objects, such as those that have a relatively large time scale, including structures in the environment that do not change, or moving obstacles that only move in a constant and predictable direction. However, in a dynamic environment with moving objects such as humans, spontaneous events may occur that cannot be predicted, and thus it is desirable to use very short-term predictions to realize safe movement. To realize movement over such short distances and safely, decision making regarding actions considering time scale differences is required. The hierarchical control method considers various time scales for actions in problems with various time scales, such as goal path planning, obstacle avoidance within the recognizable range, and emergency avoidance to avoid spontaneous events. The module is a potential function with the time scale as a design parameter, and generates a potential field.

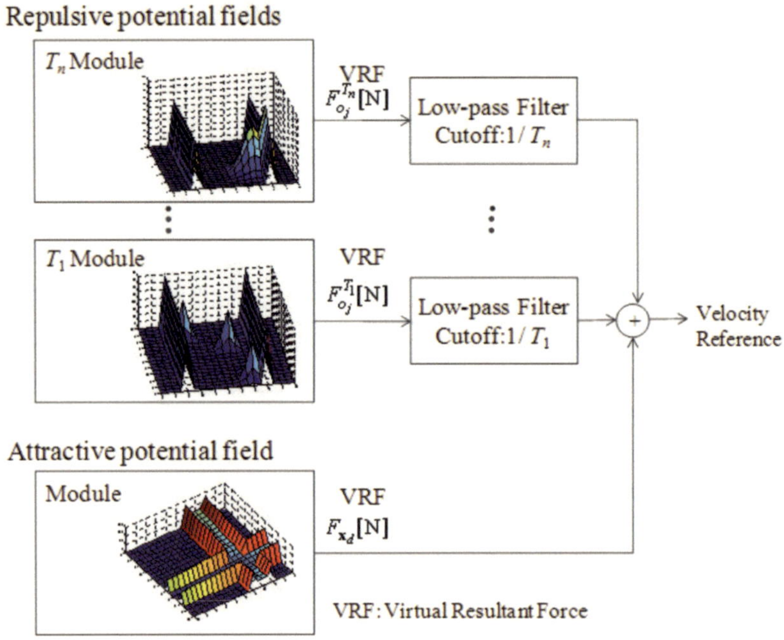


Fig. 3. Output of each module and integration.

### 2.3 Design Method

The module is a potential function with the prediction time as a parameter, and generates a potential field for each problem and virtual force on the robot is calculated. In this study, the proposed potential function was designed based on the repulsive potential reported by Khatib (Khatib, 1986).

$$U = U_{\mathbf{x}_g} + U_o \tag{1}$$

$$U_{\mathbf{x}_g} = k_a \rho_g \tag{2}$$

$$U_o = \sum U_{o_j}^{T_i} \tag{3}$$

$$U_{o_j}^{T_i} = \begin{cases} \eta T_i \left( \dfrac{1}{\rho_{T_i}} - \dfrac{1}{\left\| T_i \mathbf{v}_{r,o_j} \right\| + \rho_0} \right)^{\frac{1}{T_i}} & , \rho_{T_i} \le \left\| T_i \mathbf{v}_{r,o_j} \right\| + \rho_0 \\ 0 & , \rho_{T_i} > \left\| T_i \mathbf{v}_{r,o_j} \right\| + \rho_0 \end{cases} \tag{4}$$

where $\mathbf{x}_g$ is the goal position and $U_{\mathbf{x}_g}$ is an attractive potential field. In the proposed method, a repulsive potential function in consideration with prediction time $T_i$ is used.
$\rho_{T_i}$ is the predicted shortest distance between the robot and the obstacle $O_j$ at time scale $T_i$. This is determined by calculating the predicted positions of the robot and the obstacles after time $T_i$ using the current velocity, and by calculating the shortest distances in these positional relationships.
$1/\left(\left\| T_i \mathbf{v}_{r,o_j} \right\| + \rho_0\right)$ in Eq. (4) is a parameter that determines the range of the potential field in the repulsive potential of Khatib. $\rho_0$ is the limit distance of the potential field influence and $\eta$ is a constant gain. By considering the relative velocity, the response of the robot is expected to change even if the relative position of the obstacle to the robot is the same. For example, when the relative velocity is large, the robot reacts more quickly.
The exponential part $1/T_i$ is the part of the equation that determines the priority level of the output of each module. When the robot faces an emergency situation, the time scale is small, and the output of the module with a small time scale is large. Thereby, the robot acts more rapidly.
A force for the position **x** of the robot is derived from the following equation.

$$\mathbf{F}(\mathbf{x}) = -\frac{\partial U}{\partial \mathbf{x}} \tag{5}$$

where $\dfrac{\partial U}{\partial \mathbf{x}}$ denotes the partial derivation vector of the total virtual potential U. From Eqs. (2) and (5), the attractive force allowing the position x of the robot to reach the goal position $\mathbf{x}_g$ is as follows:

$$\mathbf{F}_{\mathbf{x}_g} = -k_a \frac{\partial \rho_g}{\partial \mathbf{x}} \tag{6}$$

From Eqs. (4) and (5), the repulsive force to the obstacle $O_j$ are as follows:

$$\mathbf{F}_{o_j}^{T_i} = \begin{cases} \eta \left( \dfrac{1}{\rho_{T_i}} - \dfrac{1}{\left\| T_i \mathbf{v}_{r,o_j} \right\| + \rho_0} \right)^{\frac{1}{T_i}-1} \dfrac{1}{\rho_{T_i}^2} \dfrac{\partial \rho_{T_i}}{\partial \mathbf{x}} & , \rho_{T_i} \le \left\| T_i \mathbf{v}_{r,o_j} \right\| + \rho_0 \\ 0 & , \rho_{T_i} > \left\| T_i \mathbf{v}_{r,o_j} \right\| + \rho_0 \end{cases} \tag{7}$$

The command vector **F** of the robot is derived from the following equation.

$$\mathbf{F} = \mathbf{F}_{\mathbf{x}_g} + \mathbf{F}_o \tag{8}$$

When combining the virtual force derived from a potential field which is generated at each module, we consider the robot as a point mass. The velocity command with the same magnitude and direction is determined by combining the forces to the robot. In addition, the potential approach has a vibration problem caused by the magnitude of velocity and roughness of the control period. Thus, in the method, a low pass filter on each element of the virtual force output in each module is used to suppress such vibration as shown in Fig.3. It was confirmed that safe and effective motion is possible even in a situation where movement to the destination, avoiding moving obstacles, and emergency avoidance all coexist. In the simulations, each low pass filter uses the reciprocal of each prediction time as a cut-off frequency.

Fig. 4. Omni wheeled platform.

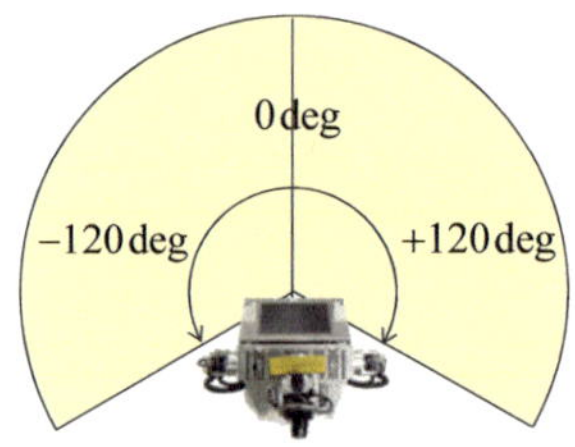


Fig. 5. The range of LRF.

## 3. Simulation Results

### 3.1 Simulation Environments

To verify the usefulness of the proposed method, several simulations were carried out. The method of the Khatib was used as a comparative method. In this study, several conditions based on the real robot are assumed.

1. The robot has an omni-directional wheel as shown in Fig.4 and thereby can move in all directions.
2. In order to recognize the environment, the robot has a laser range finder (LRF) which the measuring range is 4.0 m at the angle of the 240 degrees as shown in Fig.5.
3. The velocity of the obstacle can be estimated based on the information from LRF.

The proposed control system used in this study consists of three modules. The lowest module generates the attractive potential field and the repulsive potential field to the wall. The second module is the middle time scale $T_1$ module to avoid collision with the obstacles. The third module is the short time scale $T_2$ module to avoid collision with the sudden appearing obstacles. From the parameter studies, $T_1$=1.2 s and $T_2$=0.5 s are decided.

To verify the performance of the proposed method against two moving obstacles in different conditions, the simulation was carried out. Figure 6 shows the situation of the simulations. The initial position of the robot is (0 m, 0 m). The obstacle 1 moves straight forward to the robot from the initial position (0.2 m, 6.0 m). The obstacle 2 starts the movement from right to left as shown in Fig.4 when the robot comes close to the corner. The robot cannot recognize the obstacle 2 until just before. Therefore, the robot should deal with the situation instantaneously. The velocity of the robot and two obstacles are 1.3 m/s.

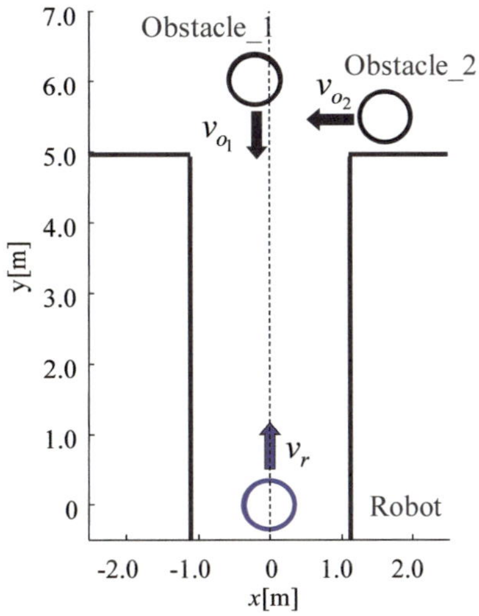


Fig. 6. Simulation Environment.

### 3.2 Simulation Results

Figure 7 shows the simulation results. Figures 7(a) and (b) show the trajectories of the robot and two obstacles using the Khatib and the proposed methods respectively. Figure 7(c) shows the time history of the module activations.

From the result in Fig. 7(a), it was confirmed that the robot with the Khatib method can reach the goal without colliding with two moving obstacles. However, the robot moves backward to avoid the obstacle 1 and then starts the movement to the goal after the obstacle 1 passes over. In the case of the obstacle 2, the robot moves in the direction of movement of the obstacle 2 because the predicted information of the obstacle 2 is not used. Thereby, the arrival time to the goal is longer than our method.

On the other hand, it was confirmed in Fig. 7(b) that the robot can reach the goal earlier than the other method without colliding with two moving obstacles. The robot moves in the right direction in advance to avoid the obstacle 1 by considering the predicted information and thereby the efficient collision avoidance is performed. In the case of the obstacle 2, when the robot recognizes the obstacle 2, by generating the repulsive potential fields based on the estimated information of the obstacle 2, the robot stops until the obstacle 2 passes over, and then starts the movement to the goal. From Fig. 7(c), the robot can move without colliding with two moving obstacle 2 by acting on the $T_2$ module simultaneously with the $T_1$ module at about 6.0 s. From the results, it was confirmed that the robot with the proposed method can realize the safe and efficient movement according to the situation.

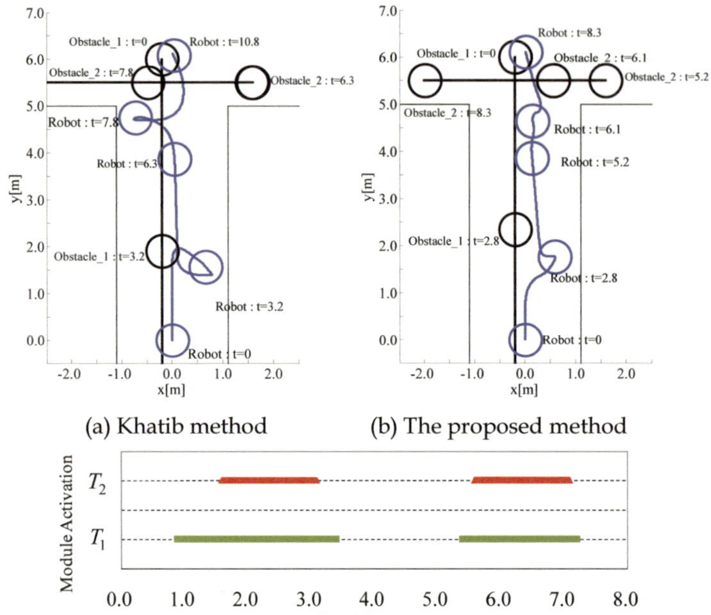


(a) Khatib method (b) The proposed method

(c) Time History of the Module Activations

Fig. 7. Simulation Result.

## 4. Experimental Results

### 4.1 Experimental Environment

To verify the effectiveness of the proposed method in the actual situation, the experiments using the real robot were carried out. The robot size is 0.55 m × 0.75 m × 1.25 m and the weight of the robot is about 60 kg. In order to recognize environment, though the stereo camera and the stemma camera, the laser range finder and the ultrasonic sensor are loaded, in this research the robot recognizes environment making use of only the laser range finder. The velocity limit of the robot is 0.5 m/s and the acceleration limit is 1.0 m/s$^2$.

Figure 8 shows the experimental environment to verify the effectiveness of the proposed method to a static obstacle. The initial position of the robot is (0 m, 0 m). The obstacle size is 0.20 m × 0.33 m × 0.50 m and its initial position is (-0.5 m, 3.0 m).

Figure 11 shows the experimental environment to verify the effectiveness of the proposed method to a moving obstacle. In this case, the moving obstacle appears through the blind corner at the velocity of 0.5 m/s when the robot comes close to the corner.

### 4.1 Experimental Results

Figures 9 (a), (b) and (c) show the trajectory of the robot by using the Khatib ( $\rho_0 = 0.8$ , $\eta = 0.064$ ), the Khatib ( $\rho_0 = 1.5$ , $\eta = 0.064$ ) and the proposed method respectively.

From the result in Fig. 9(a), it was confirmed that the robot comes close to the static obstacle because the repulsive potential fields for the obstacle is small. Fig. 9 (b) shows that the robot does not approach to the obstacle because the influence of the obstacle is large. In addition, because the robot receives the influence of repulsive forces from the wall, the robot is the stable state before it reaches the goal. On the other hand, it was confirmed in Fig. 9(c) that the robot with the proposed method can reach the goal earlier than other methods without colliding with the moving obstacle.

Figures 12 (a) and (b) show the trajectories of the robot and the moving obstacle by using the Khatib and the proposed method respectively. Figure 12(c) shows the time history of the module activations in the proposed method. Figure 12(a) shows the same result as the simulation. The robot can reach the goal without colliding with the obstacle. However, the robot moves in the direction of movement of the obstacle because the predicted information of the obstacle is not used. This movement of the robot is shown in Fig. 13. Thereby, the arrival time to the goal is longer than our method.

From the results in Figs. 12 (b) and 14, it was confirmed that when the robot recognizes the moving obstacle, the robot stops on the moment until the obstacle passes over, and then starts the movement to the goal. As shown in Fig.12(c), the robot can move without colliding with the moving obstacle by acting on the $T_1$ and $T_2$ modules at the same time at about 5.0 s.

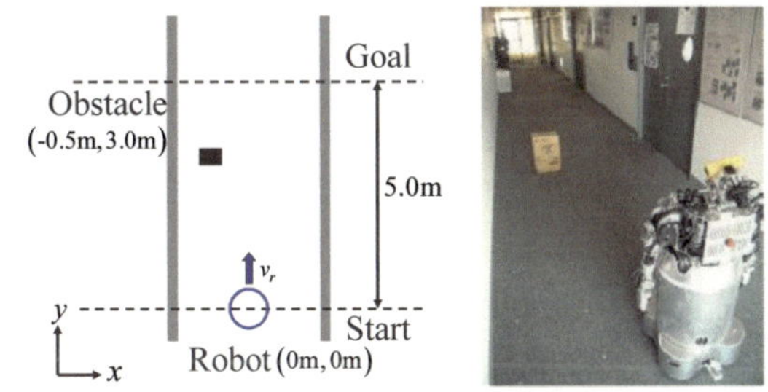


Fig. 8. Experimental Environment.

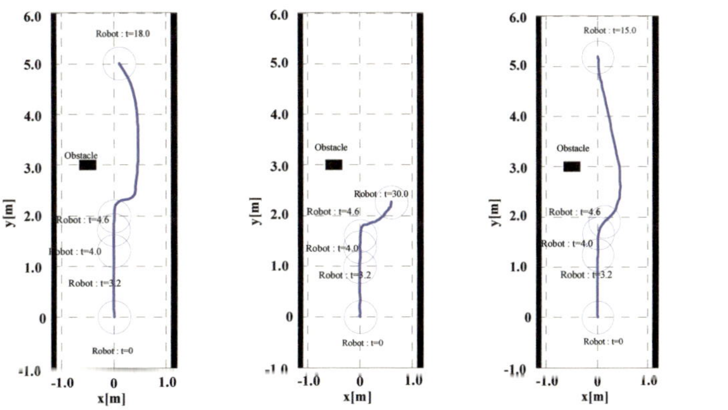


(a) $\rho_0 = 0.8, \eta = 0.064$ (b) $\rho_0 = 1.5, \eta = 0.064$ (c) The proposed method

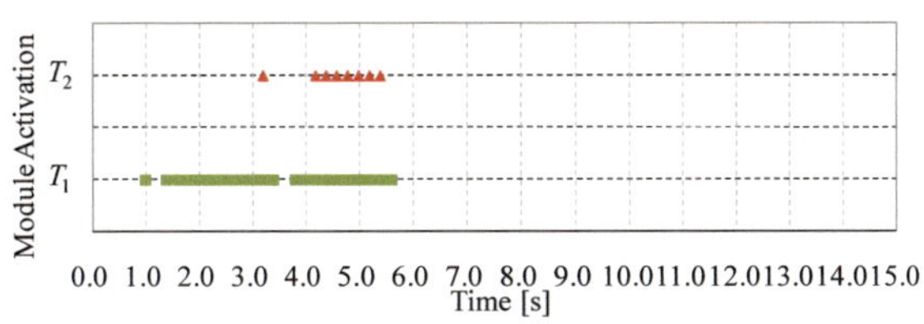


(c) Time History of the Module Activations

Fig. 9. Experimental Result.

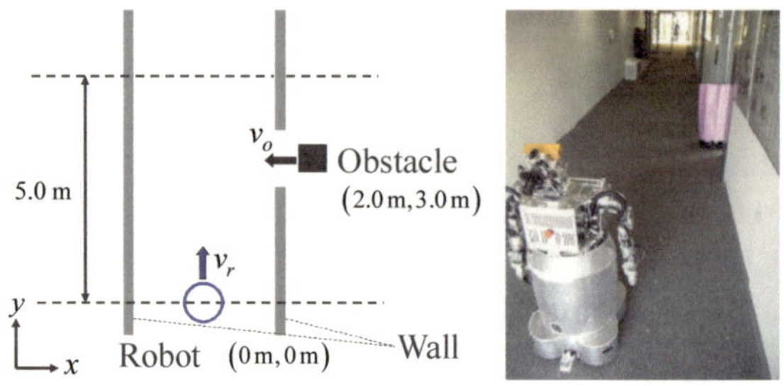


Fig. 10. Experimental Environment.

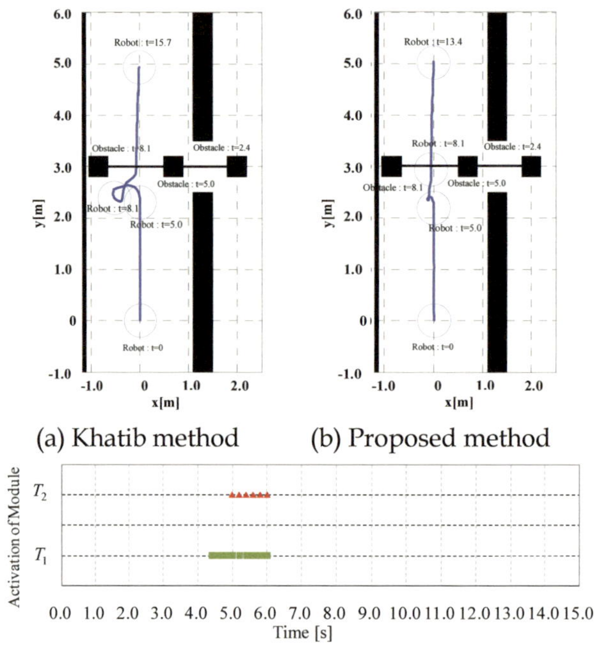


(a) Khatib method (b) Proposed method

(c) Time History of the Module Activations in the proposed method

Fig. 11. Experimental Result.

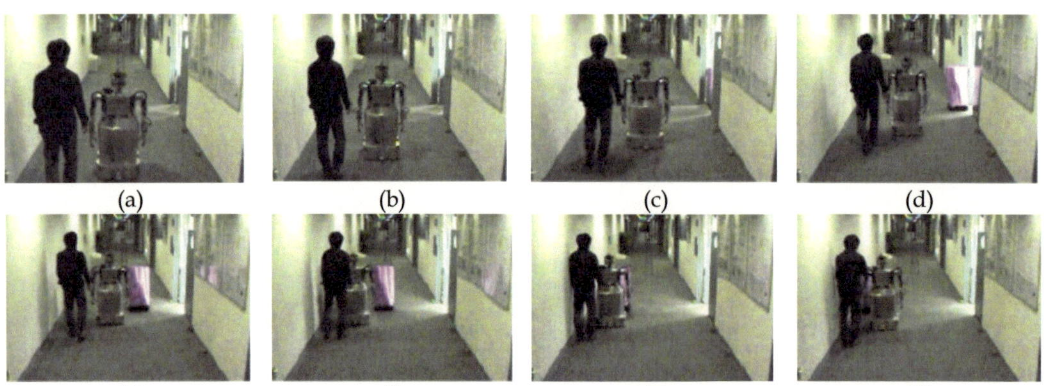

(a) (b) (c) (d)

(e) (f) (g) (h)

Fig. 12. Experimental Result.

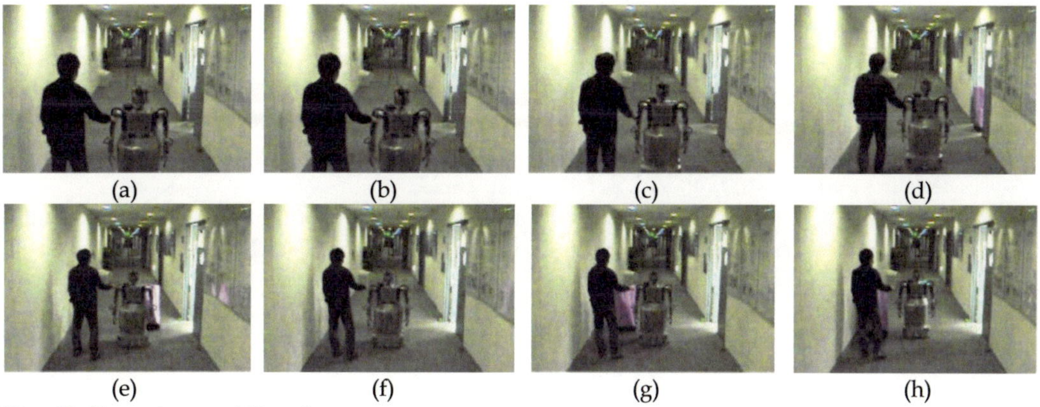


Fig. 13. Experimental Result.

## 5. Conclusions

This study proposed the hierarchical action control method for an autonomous omni-directional mobile robot to realize safe and effective movement. In the method, the module with different prediction time processes in parallel, and the command velocity to the robot is decided by integrating them. As for each module, the selection condition is different according to relative position and velocity between the robot and the obstacle.

This study proposed the design procedure of the proposed method based on the virtual potential method. From the results of the numerical simulations and the experiments, it was confirmed that the robot can reach the goal efficiently without colliding with both the static and the moving obstacles by using the estimated information of them. Moreover, the robot can deal with the emergency situation adequately by acting on several modules simultaneously according to the situation. From the results, the effectiveness of the proposed method in dynamic environment was confirmed.

## Acknowledgements

This work was partly supported by a project budget for "Development Project for a Common Basis of Next-Generation Robots" from the New Energy and Industrial Technology Development Organization (NEDO) in Japan.

## 6. References

Graf, B., Hans, M. and Schraft, R. D. (2004). Mobile Robot Assistant, *IEEE Robotics and Automation*, Vol.11, No.2, pp. 67-77.

Asoh, H. et al. (2001). Jiji-2: An Office Robot That Communicates and Learns, *IEEE Intelligent Systems*, Vol.16, No.5, pp. 46-55.

DeSouza, G. N. and Kak, A. C. (2002). Vision for Mobile Robot Navigation: A Survey, *IEEE Trans. on Pattern Analysis and Machine Intelligence*, Vol.24, No.2, pp. 237-267.

Thrun, S. et al (1999), MINERVA: A Second-Generation Museum Tour-Guide Robot, *Proc. IEEE Int. Conf. on Robotics and Automation*, pp.1999-2005.

Bischoff, R. and Graefe, V. (2004), HERMES - A Versatile Personal Robotic Assistant, *Proc. IEEE,* Vol.92, No.11, pp.1759-1779.

Khatib, O. (1986), Real-time obstacle avoidance for manipulators and mobile robots, *International Journal of Robotics Research*, Vol. 5, No. 1, pp. 90-98.

Brooks, R. A. (1986), A robust layered control system for a mobile robot, *IEEE Journal of Robotics and Automation*, Vol. 2, No. 1, pp. 14-23.

Ge, S.S., and Cui, Y. J. (2002), Dynamic motion planning for mobile robots using potential field method, Autonomous Robots, Vol. 13, No. 3, pp. 207-222.

# 19

# Stable Switching Control of Wheeled Mobile Robots

Juan Marcos Toibero, Flavio Roberti, Fernando Auat Cheein,
Carlos Soria and Ricardo Carelli
*Universidad Nacional de San Juan*
*Argentina*

## 1. Introduction

This Chapter addresses the wheeled autonomous mobile robots navigation problem starting from very simple solutions which are used to solve more complex tasks. This way, it is considered a hybrid (discrete-time/continuous-time) approach, composed by several individual continuous controllers, each of them solving a particular navigation problem, but when are considered as a whole constitute a new approach for more demanding navigation challenges. The discrete part of this system is compound by a high order supervisor, which gives the logic switching. In particular, this Chapter deals with Switching Systems, which are hybrid system characterized by the fact that only a unique controller could be active at any time. This kind of control architecture presents some benefits as for example the modularity and the possibility to analyze the whole system behavior, allowing the designer to conclude stability. It also allows, conversely, decomposing a given task in simpler subtasks which could be treated separately in order to facilitate the design process.
The interest on asymptotic stability is due to our interest on the control system analysis. For this reason, several well-known concepts such as Lyapunov stability and common Lyapunov functions (Liberzon, 2003) will be treated along this Chapter, in order to prove stability for the individual controllers as well as for the switching controller. So, it will be shown under a control point of view the design and analysis of stable switching controllers and two applications examples to wheeled robots navigation: firstly, developing a switching controller for reactive obstacle avoidance (Toibero et al., 2009), and secondly, its applications at the local navigation level in a simultaneous localization and map building (SLAM) task (Auat Cheeín, 2009).
Robotic autonomy is directly related to the capability the robots have to perform given task without continuous human guidance. Arkin (1998) proposed a way to deal with this autonomy, by decomposing a task into individual behaviors such as e.g. move-to-goal, find-next-point, return-to-start, wander and halt. Next, these behaviors may be selected or fused according to a desired performance. Within the several available behaviours, the interest is on a key feature for robot autonomy: obstacle detection, identification and avoidance.

In this context, there are two main approaches: path planning controllers or reactive controllers. In path planning controllers, the obstacle is to be detected and avoided by defining a new secure path. Instead of that, reactive controllers only react in response to the sensory inputs without major computational effort. In the following, a wall-following (WF) controller already designed is considered as the basis of our obstacle avoider controller. A WF task is characterized by maintaining a constant distance from the robot to a wall (or to an object) and can be properly combined with other tasks in order to obtain higher degrees of autonomy for the mobile platform, e.g. the skill Go-to-goal avoiding obstacles in (Boada et al., 2003). Typically a WF controller has to recognize the distance and orientation errors between the robot and the wall. These two control states can be measured or estimated using sonar, laser rangefinder or information extracted from a video camera. Besides, WF controllers can be easily adapted for navigation along corridors by setting a desired distance relative to the centre of the corridor. Regarding the environment surrounding the robot, a WF controller becomes useful for reactive navigation in unknown environments. In this context, it can be used as an obstacle avoider by considering the obstacle as part of a wall to be followed. Therefore, it is relevant to design controllers for walls with several kinds of contours. The inclusion of such situations leads to a new control problem: to deal with discontinuities on the wall contour. To solve this problem most of the papers in the literature use fuzzy-logic: (Braunstingl & Ezkerra, 1995), (Wang & Liu, 2004), (Ando, 1996), or switching controllers: (Fazli & Kleeman, 2005), (Borenstein & Koren, 1989), (Zhang et al., 2004). The WF problem, as an important component of current robotic research, has been widely considered. In the early work in (van Turennout & Hounderd , 1992) a WF controller was used to avoid obstacles and to follow unknown walls. Then (Ando, 1996) proposes a way to extend the capabilities of this behavior but as a path planning problem, without describing the controller. In (Braunstingl & Ezkerra, 1995) the contour following problem is treated by using a fuzzy logic controller. More recently (Bicho, 2000) has used neural networks in order to estimate the relative orientation of the robot and the wall. In (Zhang et al., 2004) a complete switching controller that allows the robot to track sharp discontinuous trajectories is presented. This last switching approach includes a stability demonstration at the switching times but the controllers need a path to be tracked and the paper does not consider obstacle avoidance while tracking the trajectory. Here, the WF problem is addressed by using a laser rangefinder and odometry in fully unknown environments. Initially, it will be assumed that the wall's contour is smoothly varying and it is proposed an expression relating the contour variations with the saturation of control actions which is useful to design a saturation-free controller. This first part includes the simpler case of straight walls. Then, the contour-following (CF) problem is treated as an extended approach of the WF problem when considering discontinuities such as corners. For this CF controller three individual subsystems were considered, one characterizing each behavior: re-orientation, wall-following and circle-performer. The stability at the switching times is considered and discussed.

In addition to the use of WF in move-to-goal and obstacle avoidance behaviors considered by several authors, other complementary applications of a CF controller are the mapping of interior environments (Auat Cheein, 2009), (Edlinger & von Puttkamer, 1994), (Fazli & Kleeman, 2005) in surveillance tasks and within teleoperation contexts (Wang & Liu, 2004) using a similar behavior for telecommanding a mobile robot.

The remaining of this Chapter is organized as follows. Section 2 presents the mobile robot kinematcis. Section 3 proposes a WF controller to maintain this desired distance for straight and smooth varying walls based on laser and odometric information only. Next, in Section 4 some limitations of this approach are shown; and a switching scheme that allows the fully contour-following of the object is presented, including several experimental results in Section 5. As a final example application, in Section 6 it is shown the incorporation of this controller to a SLAM task showing also additional experimental results.

## 2. Mobile Robot

The unicycle-type wheeled mobile robot is to be considered in this Chapter. Let us set the specific notation, with reference to Figure 1, the global or task coordinate system is denoted by [W] and the coordinate system attached to the robot by [R]. The robot states variables are $x$, $y$ and $\theta$, where $\theta$ denotes the heading of the vehicle relative to the $^{W}X$-axis of the world coordinate system. Vector $[x \quad y \quad \theta]^T$ defines the posture of the vehicle. The kinematics of the robot can be modeled by the well known kinematics model.

$$\dot{\chi} = \begin{bmatrix} \dot{x} \\ \dot{y} \\ \dot{\theta} \end{bmatrix} = f(\chi, u) = \begin{bmatrix} cos(\theta) & 0 \\ sin(\theta) & 0 \\ 0 & 1 \end{bmatrix} \begin{bmatrix} v \\ \omega \end{bmatrix} \tag{1}$$

The robot is equipped with a laser radar sensor. As it can be seen in Figure 1, some lateral beams are used to estimate the object contour angle, whereas all beams are used to define a safety-zone, which purpose is to detect possible collisions between the robot and the object. The selection of the side of the object to follow is based on the initial conditions. Firstly the robot computes the object orientation and the distance to the object for its right side and also for its left side, then, the robot will take the nearest side to follow and will keep following this side. Each beam is identified by its angle, e.g. for the beam at 0°, the notation *d000* is used to denote the distance measured by this beam. From here on, to simplify the notation, it will be assumed that the robot follows only the wall at the right side of the robot.

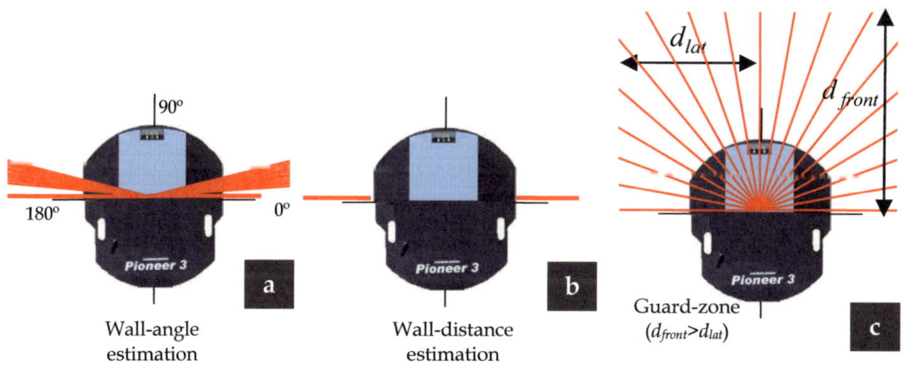


Fig. 1. Laser radar

## 3. Standard Wall-Following Control

A WF controller will be referred as *standard* in this Chapter, when do not allow to follow discontinuous walls. It could be said that allows the robot to follow continuous (straight) and smooth varying contours. This appreciation is based on its stability proof. Two control states must be defined: the robot-to-wall angle in Section3.1; and the robot-to-wall distance in Section3.2.

### 3.1 Wall Angle Computation

The wall angle estimation is performed by using only two beams: one perpendicular to the heading of the robot, and the other may be selected depending on the desired robot-wall distance. In order to obtain a more reliable value, it is considered a group of beams at each side of the robot, thus resulting ten different angle values. Each value is obtained according to

$$\theta_{wall}^{10} = atan\left(\frac{-\Delta y}{-\Delta x}\right) = atan\left(\frac{y_{010} - y_{000}}{x_{010} - x_{000}}\right). \quad (2)$$

These values are then fused together using a decentralized Kalman filter (Brawn & Hwang, 1997) which returns an improved estimation for the wall angle $\theta_{wall}$ with respect to the world coordinate system. In Toibero et al. (2009) details and derivations of this filter can be found.

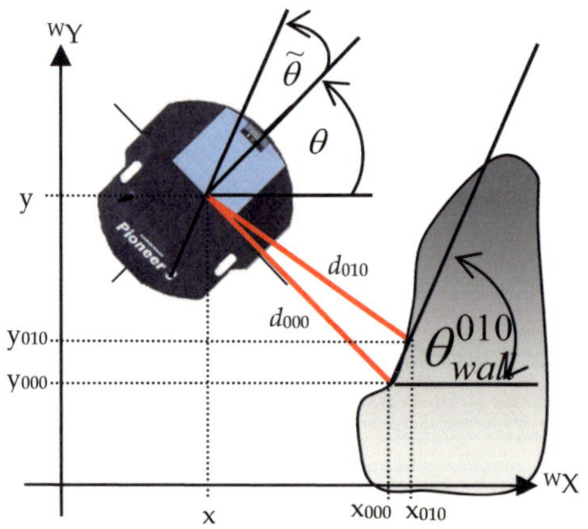


Fig. 2. Wall angle computation. Resulting in the value $\theta_{wall}^{010}$ of the wall-angle related to robot posture x, y and θ in the world coordinate system for the beam at 10°

### 3.2 Robot-wall distance computation

The wall-distance computation to a wall at the right side of the robot is obtained by

$$d_{wall} = d_{000} \cos(\tilde{\theta}) \quad (3)$$

where $\tilde{\theta}$ is the difference between the heading angle of the robot and the wall angle. This angle is properly defined in the next paragraphs. It can be obtained a similar expression for the distance to a wall at the left side of the robot by replacing $d_{000}$ with $d_{180}$. Note that according with (3) it is assumed that the distance from the robot to the wall is considered for null orientation error.

### 3.2.1 Controller description and analysis

This controller renders a control law for wall-following based on the laser sensorial information and the odometry of the robot. The reference for this controller is the desired distance from the robot to the wall $d_{des}$. As can be seen in Figure 3, the control errors are defined (Toibero et al., 2006b), as follows

$$\tilde{d} = d_{des} - d_{wall} \tag{4.a}$$
$$\tilde{\theta} = \theta_{wall} - \theta \tag{4.b}$$

Equation (4.a) states the distance error between the actual distance to the wall and the desired distance to the wall $d_{des}$ while equation (4.b) states the orientation error between the estimated wall angle and the heading angle of the robot in the world coordinate system [W].

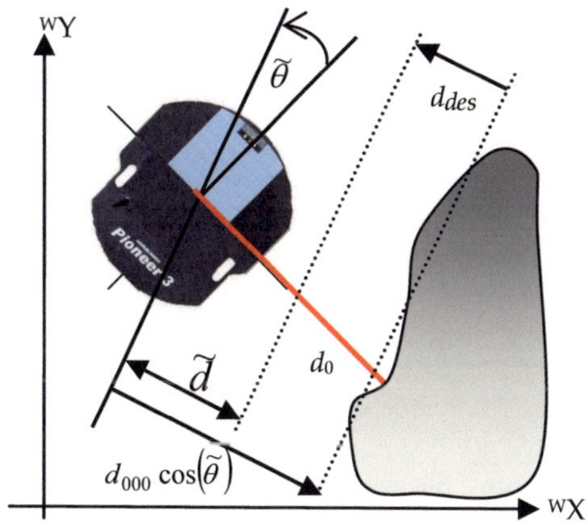


Fig. 3. Wall-following controller scheme. Distance error calculation. Note that the value for the distance error could be negative.

The time variations of the control errors are given by

$$\dot{\tilde{d}} = -v\, sin(\tilde{\theta}) \tag{5.a}$$
$$\dot{\tilde{\theta}} = \dot{\theta}_{wall} - \omega \tag{5.b}$$

where, $\dot{\theta}_{wall}$ is the derivative of the reference for the wall orientation. This derivative is null for a straight wall. Now, taking into consideration the control system analysis, the following - globally positive definite and radially unbounded - Lyapunov candidate function is proposed

$$V = \frac{1}{2}\tilde{\theta}^2 + \int_0^{\tilde{d}} k_{\tilde{d}}(\xi)\, \xi\, d\xi, \tag{6}$$

where $k_{\tilde{d}}(\tilde{d})$ was selected as in (Kelly & Carelli, 1996) and $\xi$ is a dumb variable:

$$k_{\tilde{d}} = \frac{k_1}{1+|\tilde{d}|} \geq 0. \tag{7}$$

The time derivative of the Lyapunov candidate function is

$$\dot{V} = \tilde{\theta}\dot{\tilde{\theta}} + k_{\tilde{d}}\, \tilde{d}\, \dot{\tilde{d}} \tag{8}$$

Then, the proposed control actions are

$$v = v_{max} \geq 0 \tag{9.a}$$
$$\omega = \dot{\theta}_{wall} + K_{\tilde{\theta}}\, tanh\big(k_{\tilde{\theta}}\, \tilde{\theta}\big) + k_{\tilde{d}}\, \tilde{d}\, v \frac{sin(\tilde{\theta})}{\tilde{\theta}} \tag{9.b}$$

where $v_{max}$ is the desired linear velocity; $K_{\tilde{\theta}} \geq 0$ is a positive definite design constant, and $k_{\tilde{\theta}}$ is a design constant that indicates the slope of the *tanh* function which can be adjusted with $K_{\tilde{\theta}}$ in order to bound the control action to its saturation values. By substituting (9) in (8) the derivative of the Lyapunov candidate function becomes

$$\dot{V} = -K_{\tilde{\theta}}\, \tilde{\theta}\, tanh(k_{\tilde{\theta}}\, \tilde{\theta}) \tag{10}$$

Equation (10) states that the control system is Lyapunov stable but not asymptotically stable. To prove the asymptotic stability the autonomous nature of the closed-loop system is considered. The closed-loop equations can be obtained by replacing the control actions in (5)

$$\dot{\tilde{d}} = -v_{max}\, sin\,(\tilde{\theta}) \tag{11.a}$$
$$\dot{\tilde{\theta}} = -K_{\tilde{\theta}}\, \tilde{\theta}\, tanh\big(k_{\tilde{\theta}}\, \tilde{\theta}\big) - k_{\tilde{d}}\, \tilde{d}\, v_{max}\, \frac{sin(\tilde{\theta})}{\tilde{\theta}} \tag{11.b}$$

By applying the Krasovskii-LaSalle theorem (Vidyasagar, 1993) in the Ω region, with

$$\Omega = \left\{\begin{bmatrix}\tilde{d}\\ \tilde{\theta}\end{bmatrix} : \dot{V}(\tilde{d}, \tilde{\theta}) = 0\right\} \Rightarrow \left\{\begin{bmatrix}\tilde{d}\\ \tilde{\theta}\end{bmatrix} = \begin{bmatrix}\tilde{d}\\ 0\end{bmatrix}\right\} \tag{12}$$

the only invariant is $\tilde{d} = 0$. Therefore, by invoking LaSalle theorem, it can be concluded that the origin of the state space is globally uniformly asymptotically stable.

### 3.2.2 Experimental results

This controller was proved in a real office environment with a Pioneer3-DX mobile robot. The robot length is about 330millimeters. The values of the design parameters were set to: $k_1 = 0.025$, $K_{\tilde{\theta}} = 11.5$, $k_{\tilde{\theta}} = 0.3$ and $d_{des} = 500$millimeters. In order to show the performance of the proposed controller, three different experiments were carried out:

*i)* The robot follows a straight wall.
*ii)* The robot follows a wall with smoothly varying contour.
*iii)* The robot follows an almost-circular contour.

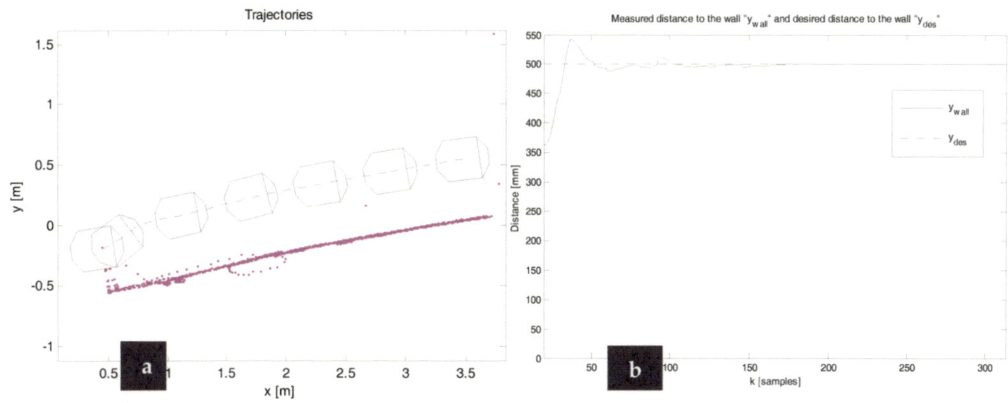


Fig. 4. a) Trajectory described by the robot following a wall with straight contour. The contour of the wall was reconstructed by using laser information. b) Distance between the robot and the straight wall. The linear velocity for this experiment was set to 200millimetres per second.

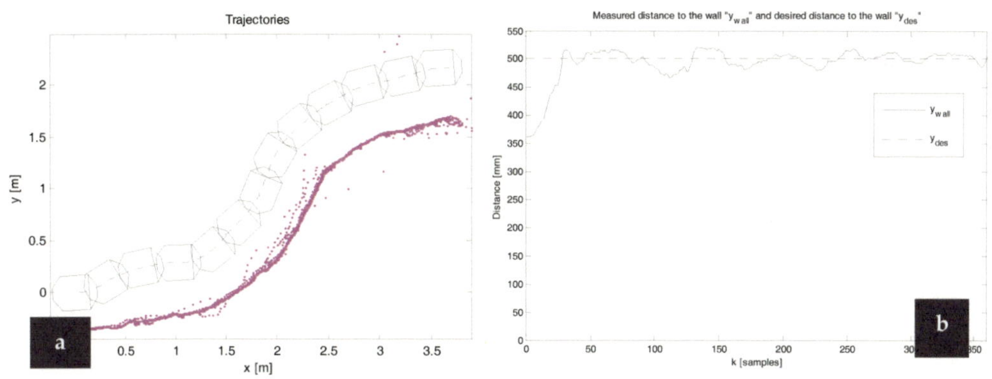


Fig. 5.a) Trajectory described by the robot following a wall of variable contour. As can be seen, the wall presents variations on its contour and on its texture. b) Robot wall distance in the variable contour experiment. The error is less than 25millimetres - a small value when compared to the size of the robot.

From Figures 4 to 6, a good performance of the proposed controller can be concluded. For all the experimental situations the errors are not significant. But it can be noted the impossibility of the controller to follow discontinuous walls' contours. These results inspire the development of the Contour-following controller of the next Section.

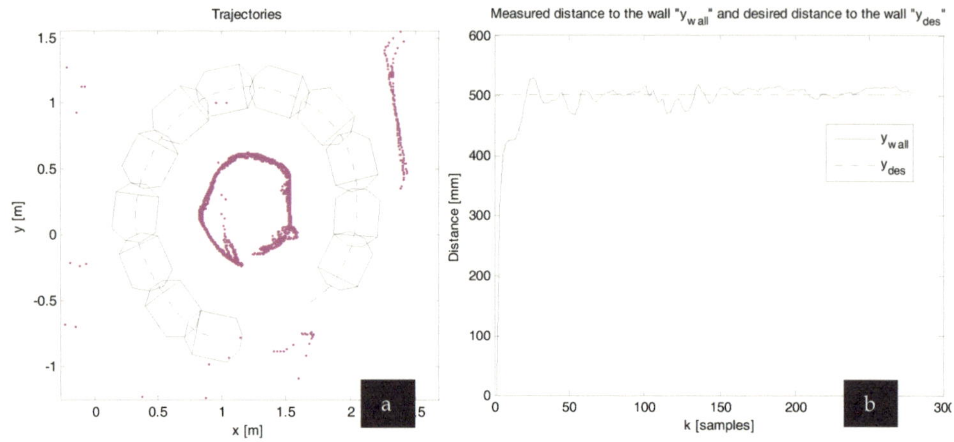


Fig. 6. a) Robot following an almost-circular contour. The linear velocity for this experiment was set to 200millimetres per second. b) Robot wall distance in the circular path experiment. The error is less than 25millimetres.

## 4. Contour-Following Controller

A stable switching between several controllers is a proper way to improve performance or to achieve control objectives that are difficult or impossible to consider under continuous nonlinear control. The proposed switching controller is designed to follow the contour of objects which have at least a size comparable with that of the robot (Toibero et al., 2006a). The basic idea is to use the wall-following controller described in Section 2. The controller is designed within the context of stability theory for switching systems. The WF controller acting alone shows a good performance when following a wall, but there are two cases which it cannot deal with:

*i)* Abrupt variations in the contour of the object which makes it difficult to estimate the actual object's angle. In this case (Figure 7.a) the distance measured by the laser beams varies suddenly.

*ii)* Abrupt variations in the contour due to an interior corner that causes the robot to crash against the object (Figure 7.b).

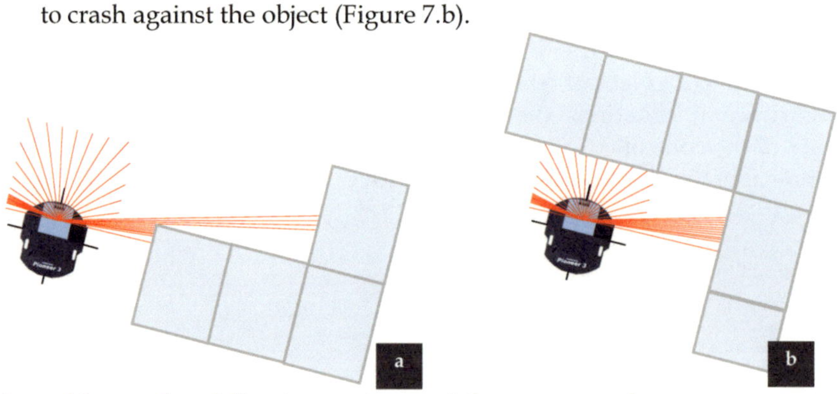


Fig. 7. Main problems when following contours: a) the presence of an open corner and b) the presence of an interior corner

The first step is given by the proper detection of these situations, and fortunatelly, both can be detected by using the laser radar. For example, the first case can be detected by computing the variance of the length of the lateral beams (note that these lateral beams estimates the wall angle). So, this variance would have a value close to zero when the robot is following a wall, but it will become larger when this contour is lost. Therefore, a threshold value can be defined in order to generate a *loss of wall* event. On the other hand, for the second case, collisions can be detected as an invasion of the robot safety-zone. If the length of any beam lies within this predefined safety-zone, then a *possible collision* event is generated.

At this point, we aggregate two new individual subsystems, one to deal with each situation, thus building a switching system.

### 4.1 Handling the case of missing the wall to follow

In this case, the robot has lost the wall to follow. This behavior is activated when non-consistent measurements of the control states $\tilde{\theta}$ and $\tilde{d}$ are obtained. The proposed solution is to describe a circular path of radius R on the floor (Figure 8) until one of the following conditions is fulfilled:

*i)* The wall can be followed again using the wall-following behavior provided that $\tilde{d} = \tilde{d}_{lost}$ (Figure 8.a and Figure 8.b); being $\tilde{d}_{lost}$ the last value of $\tilde{d}$ just before switching. This way, it could be assumed that $\dot{\tilde{d}} = 0$ considering the value of $\tilde{d}$ does not change at the beginning and at the end of this behavior.

*ii)* A possible collision between the robot and the wall is detected (Figure 8.c.), as considered in Section 3.2.

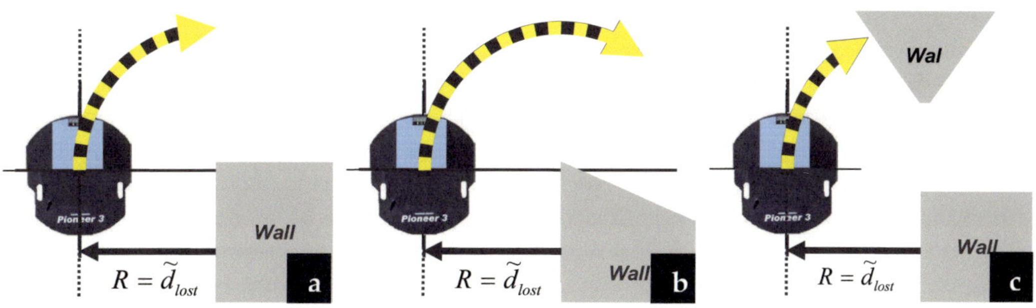


Fig. 8. Handling *"loss of wall"* situations

In order to be able to guarantee stability, a controller must be included, and this controller should share the same control states which the standart WF controller had. For this reason,we included a simple controller to describe a circular path of radius $R > 0$, based on the previous wall-following controller, though the distance error is not considered. Therefore, considering the following errors as in (4)

$$\tilde{\theta} = \theta_{circ} - \theta \tag{13}$$

where $\theta_{circ}$ is the heading reference to perform the desired circular path

$$\theta_{circ} = \pm\frac{v_{max}}{R}\, t \tag{14}$$

Here, the plus/minus sign of (14) denotes that the reference $\theta_{circ}$ will generate a left or right turn. That is, the robot will turn left (right) if it is following an object at its left (right) side. Then, replacing (14) into (13) and considering its time derivative, we obtain

$$\dot{\tilde{\theta}} = \dot{\theta}_{circ} - \omega = \pm\frac{v_{max}}{R} - \omega. \tag{15}$$

In order to analyze the stability of this control system, consider the following Lyapunov candidate function and its time derivative

$$V_{\tilde{\theta}} = \frac{\tilde{\theta}^2}{2} \tag{16}$$

$$\dot{V}_{\tilde{\theta}} = \tilde{\theta}\,\dot{\tilde{\theta}} = \tilde{\theta}\left(\pm\frac{v_{max}}{R} - \omega\right). \tag{17}$$

Now, to achieve the control objective, the following control actions are proposed

$$v = v_{max} \tag{18.a}$$

$$\omega = \pm\frac{v_{max}}{R} + K_{\tilde{\theta}}\tanh\left(k_{\tilde{\theta}}\tilde{\theta}\right), \qquad K_{\tilde{\theta}} > 0, k_{\tilde{\theta}} > 0. \tag{18.b}$$

Finally, by replacing (18) in (17), the asymptotic stability of the control system can immediately be proved. That is, this controller guarantees that the robot will perform a circular path by acting only on the robot angular velocity.

### 4.2 Handling possible collisions

This behavior is activated when an obstacle appears in front of the robot at a distance $d_{impact} \leq d_{front}$, being $d_{impact}$ the smallest distance to the obstacle measured inside the robot safety-area, which is also characterized by and angle $d_{impact}$ on the laser range finder framework. Also, $d_{front}$ (defined in Section2) is selected to be equal to the actual robot wall distance $d_{wall}$ and $d_{lat} < d_{front}$. The objective of this behavior is to avoid possible collisions by rotating the robot until a free-path condition (characterized by an empty safety-area) is again achieved, and $\tilde{d}_{k+T} = \tilde{d}_k$, where $k$ is the obstacle detection instant, and $T$ is the time during this behavior was active until switching to the wall-following behavior. Under the above mentioned conditions, this behavior will always achieve a free- path condition satisfying $\tilde{d}_{k+T} = \tilde{d}_k$ (see Figure 9). Analogously to the behavior described in Section4.1, it could be assumed that $\dot{\tilde{d}} = 0$, considering that $\tilde{d}$ does not change at the beginning and at the end of this behavior.

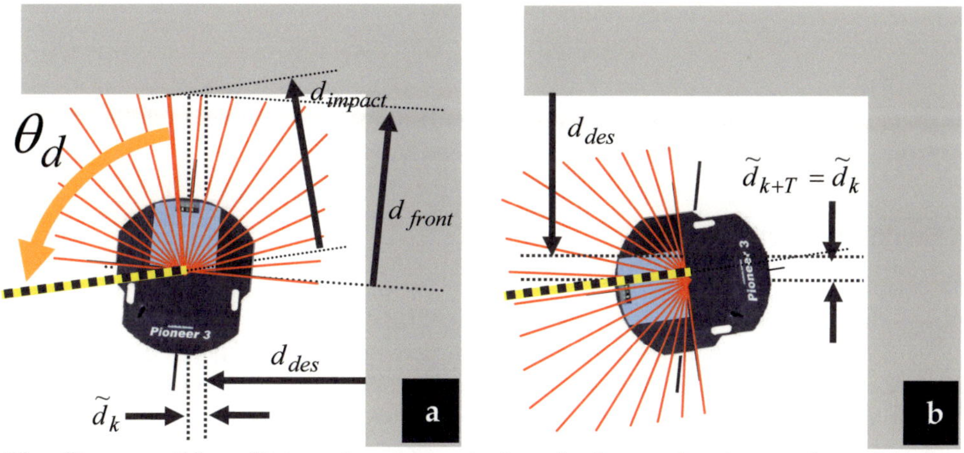


Fig. 9. Handling possible collision situations, a) obstacle detected at instant $k$, and b) free-path condition achieved with $\tilde{d}_{k+T} \leq \tilde{d}_k$ at instant $k + T$

To this aim, it is proposed a controller to allow the robot to positioning itself at a desired orientation angle $\theta_d$. The orientation error between the heading angle of the robot and the desired orientation is defined as shown in Figure 10

$$\tilde{\theta} = \theta_d - \theta, \tag{19}$$

where $\theta_d = \theta_{impact} \pm 90^{\circ}$ is a constant value that is updated so as to attain an open area.

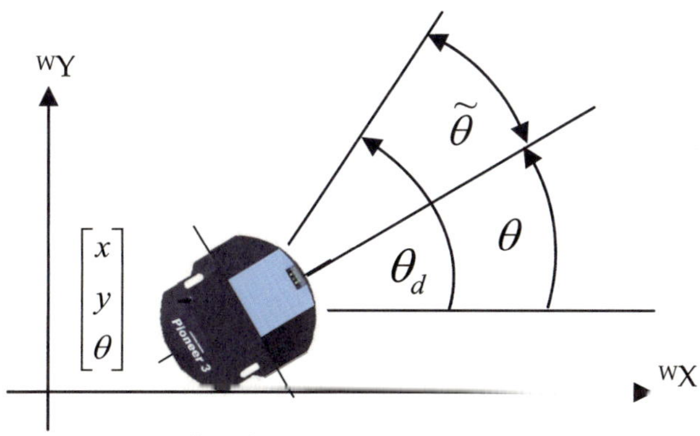


Fig. 10. Angular position controller description

Then, the time derivative of (19) is,

$$\dot{\tilde{\theta}} = -\omega \tag{20}$$

The asymptotic stability analysis of this controller can be proved by considering again (16) along with the following control actions:

$$\nu = 0 \tag{21.a}$$
$$\omega = K_{\tilde{\theta}} \tanh\left(k_{\theta}\, \tilde{\theta}\right), \qquad K_{\tilde{\theta}} > 0 \tag{21.b}$$

As done in the previous section, by replacing (21.b) into (20), the asymptotic stability of this control system, that is $\tilde{\theta}(t) \rightarrow 0$ as $t \rightarrow \infty$, can easily be proved.

## 5. Switching System

Once each part was designed it is necessary to desing the supervisor logics. Figure 11 presets the final block diagram, which includes three behaviors: wall-following, orientation and rotation (circular path performer). The switching signal $\sigma$ generated by the supervisor, takes one of three possible values: a) $\sigma = 0$ if the controller for object-following is active, b) $\sigma = 1$ if the orientation controller is active and c) $\sigma = 2$ if the controller to perform a circular path is active.

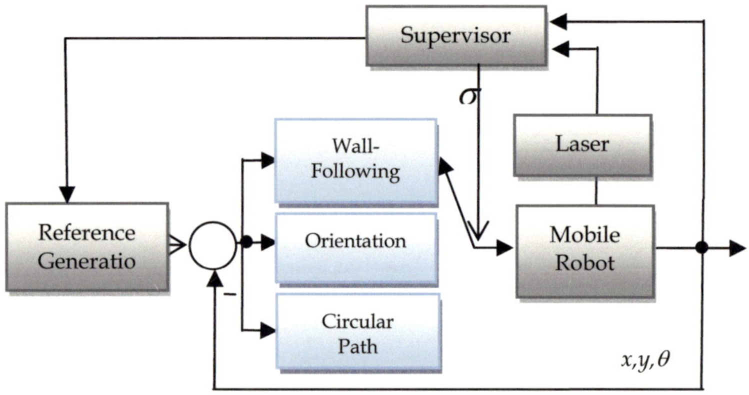


Fig. 11. Block diagram of the Contour-Following controller

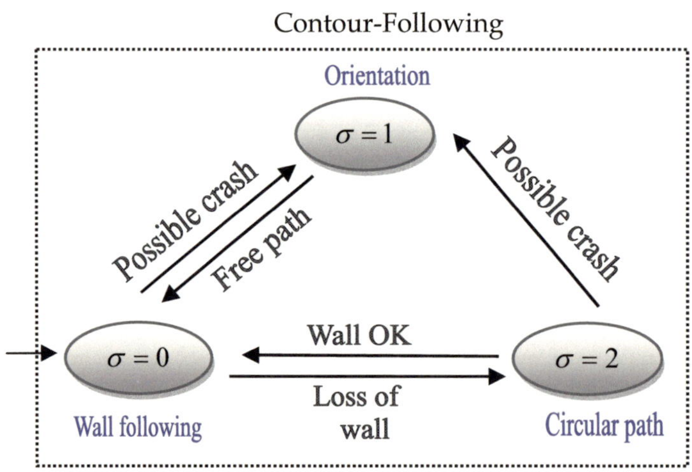


Fig. 12 Supervisor logic

Transitions between these discrete states are ruled by the logic described in Figure 12 (Toibero et al., 2006a). Where, as mentioned before, the "Possible crash" condition is detected as an invasion to the guard-zone shown in Figure 1.c and the "Loss of wall" condition depends on the variance of the length of the laser beams on the side of the robot. It is easy to see that these transitions do not depends on the control states values, but on data provided by the laser range finder. This *a priori* unpredictable data will define the value for the switching signal $\sigma$.

### 5.1 Stability Analysis

Given a family of systems, it is desired that the switched system be stable for every switching signal $\sigma$. This is known as stability under arbitrarily switching and has been the subject of several studies (Vu & Liberzon, 2005), (Liberzon, 2003) and (Mancilla-Aguilar, 2000). In fact, common Lyapunov functions are considered in order to prove stability for arbitrary switching. Finding a common Lyapunov function could be a difficult task, but if such function is found, the restrictions over the switching signal disappear, allowing the stable switching between the involved controllers. So, a basic fact that is used in this Section is that the existence of a Common Lyapunov Function with suitable properties guarantees uniform stability, as stated in (Liberzon & Morse, 1999). Let us define a Common Lyapunov Function

*Definition I* (Liberzon, 2003)
Given a positive definite continuously differentiable function $V: \mathbb{R}^n \to \mathbb{R}$, then, it is said that V is a common Lyapunov function for the family of systems $\dot{\chi} = f_p(\chi), p \in P$ ,if there exists a positive definite continuous function $W: \mathbb{R}^n \to \mathbb{R}$ such that

$$\frac{\partial V}{\partial \chi} f_p(\chi) \leq -W(\chi) \qquad \forall \chi, \forall p \in P \tag{22}$$

Where $p$ is some index set. Now, the following theorem can be stated

*Theorem I* (Liberzon, 2003)
If all systems in the family $\dot{\chi} = f_p(\chi), p \in P$ share a radially unbounded common Lyapunov function, then the switched system $\dot{x} = f_\sigma(x)$ is globally uniformely asymptoticaly stable (GUAS).

The main point in this well-known theorem is that the rate of decrease of $V$ along solutions, given by (22), is not affected by switching, hence asymptotic stability is uniform with respect to $\sigma$. Now, the purpose is to find a Common Lyapunov Function for the three continuous controllers. Then, it is first necessary to show that the closed-loop systems corresponding to each controller share a common equilibrium point at the origin. From Sections 3, 4.1 and 4.2, the closed-loop equations are

$$\dot{\chi} = f_0(x) = \begin{bmatrix} \dot{\tilde{\theta}} \\ \dot{\tilde{d}} \end{bmatrix} = \begin{bmatrix} -K_{\tilde{\theta}}(\tilde{\theta}) \tanh(k_{\tilde{\theta}}\tilde{\theta}) - k_{\tilde{d}}(\tilde{d})\tilde{d}\, v_{max} \frac{\sin(\tilde{\theta})}{\tilde{\theta}} \\ -v_{max} \sin(\tilde{\theta}) \end{bmatrix} \tag{23}$$

$$\dot{\chi} = f_1(x) = \begin{bmatrix} \dot{\tilde{\theta}} \\ \dot{\tilde{d}} \end{bmatrix} = \begin{bmatrix} -K_{\tilde{\theta}}(\tilde{\theta}) \tanh(k_{\tilde{\theta}}\tilde{\theta}) \\ 0 \end{bmatrix} \tag{24}$$

$$\dot{\chi} = f_2(x) = \begin{bmatrix} \dot{\tilde{\theta}} \\ \dot{\tilde{d}} \end{bmatrix} = \begin{bmatrix} -K_{\tilde{\theta}}(\tilde{\theta}) \tanh(k_{\tilde{\theta}}\tilde{\theta}) \\ 0 \end{bmatrix} \tag{25}$$

It is clear that the origin is a common equilibrium point for the involved controllers. Then, from (6) and (16), a common Lyapunov function for the switching among these controllers is given by (6). Therefore, it can be concluded that the switching control system composed by the three subsystems described in the previous sections is stable for any switching signal $\sigma$, because every behavior is at least stable considering the common Lyapunov function (6). However, the proposed application is composed by an asymptotically stable main behavior (wall-following) and two complementary stable behaviors (orientation and circular-path performer). As it was shown along this paper, the complementary behaviors are active only during special situations and the system always returns to the main behavior. Therefore, the GUAS property for the overall switched system can be concluded, provided that the wall following behavior is asymptoticaly stable (AS).

## 6. Experimental results

The switching contour-follower controller described in this Chapter has been implemented in a Pioneer III mobile robot navigating through a typical office environment at 150millimetres per second. In the following figures it can be seen the trajectory described by the robot when following the interior contour of the Institute of Automatics (INAUT). As mentioned before, the office contour was reconstructed by using the laser sensorial information.

The first experiment (Figure 13) shows a typical situation when following discontinuous contours. The robot follows the outline of a rectangular box at 400millimetres. From this picture it can be appreciated the good performance of the controller when switching between the wall-following and the circular path controller, in this specially discontinuous contour.

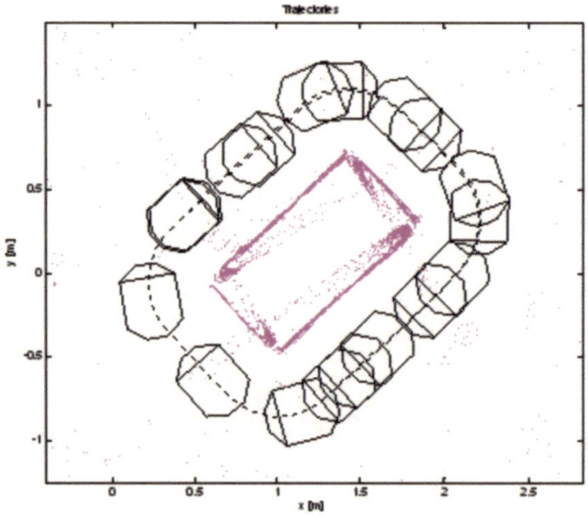


Fig. 13. Robot following a rectangular contour

Figure 14 depicts the performance of the control system when unknown obstacles appear in front of the robot: in this case first avoiding a human being and finally due to a block in the corridor, the robot returns to keep following the corridor the opposite side.

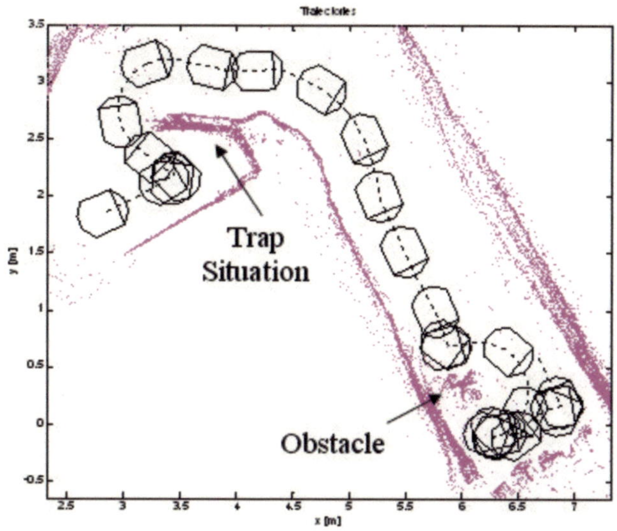


Fig. 14. Robot moving along a corridor avoiding obstacles

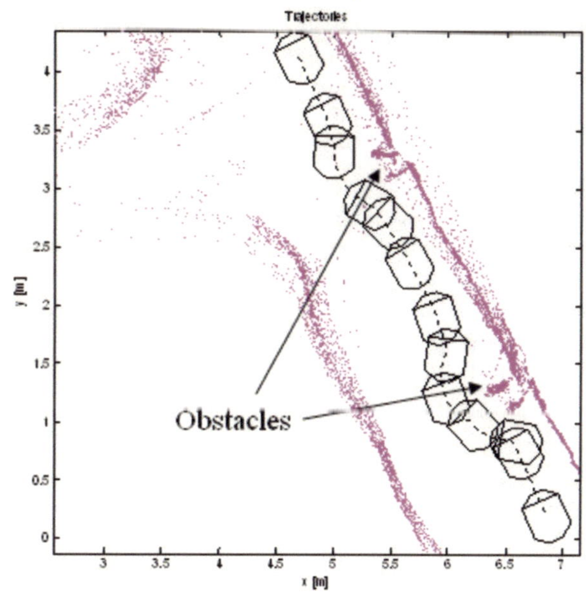


Fig. 15. Robot moving along a corridor avoiding obstacles

The controller constants were set to: $v_{max}$=150millimeters per second, $k_1 = 0.02$, $K_{\tilde{\theta}} = 0.25$ and $k_{\tilde{\theta}} = 5$. The sample time was of 100milliseconds. The desired distance to the object was selected as 0.38meters, the selection of this value was based on the size of the doors, and larger values for $d_{wall}$ do not allow the robot to pass across them.

## 7. Mobile robot SLAM algorithm combined with a stable switching controller

Once treated the obstacle avoidance problem, its inclusion to a new system is considered in this Section. In the introductory section, some paragraph dedicated to possible applications to this algorithm mentioned mapping of unknown environments. In fact, simultaneous localization and map building (SLAM) is a challenging problem in mobile robotics that has attracted the interest of an increasing number of researchers in the last decade (Thrun et al., 2005), (Briechle & Hanebeck, 2004). Self-localization of mobile robots is obviously a fundamental issue in autonomous navigation: a mobile robot must be able to estimate its position and orientation (pose) within a map of the environment it is navigating within (Pfister et al., 2003), (Garulli et al., 2005). However, in many applications of practical relevance (like exploration tasks or operations in hostile environments), a map is not available or it is highly uncertain. Therefore, in such cases the robot must use the measurements provided by its sensory equipment to estimate a map of the environment and, at the same time, to localize itself within the map. Several techniques have been proposed to solve the SLAM problem (Thrun et al., 2005), (Durrant-Whyte & Bailey, 2006a), (Durrant-Whyte & Bailey, 2006b).

The SLAM algorithm implemented in this final Section consists on a sequential Extended Kalman Filter (EKF) feature-based SLAM. Prior to the experimental results, it is necessary to devote some paragraphs to its functioning algorithms for completness. This algorithm fuses corners (convex and concave) and lines of the environment in the SLAM system state. Corners are defined in a Cartesian coordinate system whereas lines are defined in the polar system. In addition to the SLAM system state, a secondary map is maintained. This secondary map stores the information on the endpoints of each segment associated with a line of the environment (representing walls). The secondary map and the SLAM system state are updated simultaneously. Once a new feature is added to the SLAM system state, it is also added to the secondary map. The feature extraction method allows the rejection of moving agents of the environment. Equations (26) and (27) show the SLAM system state and its covariance. As it can be seen, the SLAM system state has the robot's pose estimation ($\chi$) and the parameters that define the features of the environment. The covariance matrix has covariance of the robot's pose estimation ($P_{vv}$), the covariance of the features' parameters ($P_{mm}$) and the corresponding cross correlations. The elements of the SLAM system state are attached to a global coordinate system determine at the SLAM's first execution (Durrant-Whyte & Bailey, 2006a).

$$\varsigma(k) = \begin{bmatrix} \chi(k) \\ \varsigma_1(k) \\ \vdots \\ \varsigma_n(k) \end{bmatrix}_G \tag{26}$$

$$P(k) = \begin{bmatrix} P_{vv}(k) & P_{vm}(k) \\ P_{vm}^T(k) & P_{mm}(k) \end{bmatrix} \tag{27}$$

The combination of the SLAM algorithm with a strategy for exploration or navigation inside the environment is known as Active SLAM and has been a key poblem in the implementation of autonomous mobile robots. The integration of SLAM algorithms with control strategies to govern the motion of a mobile robot and the ability of selecting feasible destinations on its own will endow the vehicle with full autonomy (Liu et al., 2008), (Liu et al., 2007).

The combination of control strategies with the SLAM algorithm has been addressed from two significantly different points of view. Whereas the first one considers how the control is used to reduce errors during the estimation process (Durrant-Whyte & Bailey, 2006a), (Durrant-Whyte & Bailey, 2006b) the second one concerns exploration techniques providing the best map from the reconstruction perspective (Andrade-Cetto & Sanfeliu, 2006). Despite the duality between regulation and estimation, whatever the control strategy is implemented, it will not be guaranteed that, in general, the mobile robot will follow a specific trajectory inside the environment. In many applications, the control signal is not considered as an input of the SLAM algorithm, and, instead, odometry measurements of the robot are used (Guivant & Nebot, 2001), (Durrant-Whyte & Bailey, 2006a), (Durrant-Whyte & Bailey, 2006b). Thus, most of the associated implementations focus on the low-level, basic control-reactive behavior, leaving the motion planning and control as a secondary algorithm. Thus, albeit restricted to a local reference frame attached to the robot, active exploration strategies for indoor environments are proposed in (Andrade-Cetto & Sanfeliu, 2006), (Liu et al., 2008). As an example, a boundary exploration problem is proposed in (Xi, 2008). In this case, the robot has to reach the best point determined in the boundary of its local point of view. From a global reference perspective, these implementations have a random behavior inside the environment. To solve the lack of global planning, some implementations have included algorithms for searching optimal path based on the information acquired of the environment. These algorithms usually require the map to be gridded and, accordingly, they compute a feasible path to a possible destination (closure of the loop or global boundary points) without specifying the control law implemented on the mobile robot. Despite of the advances made so far, the integration of control strategies based on the SLAM system state (map and vehicle) to guide the robot inside an unknown environment from a local and a global reference frame following a pre-established plan is not quite studied or implemented in real time.

In this work, the robot starts at an unknown position inside an unknown environment and by an active exploration, searches for boundary points from a local reference frame attached to the robot. From now on, these boundary points will be designated by local uncertainty points. Once a local uncertainty point is determined, a trajectory connecting this point and the robot's current pose is generated. The trajectory is continued by one resulting from the execution of a switching adaptive controller (De la Cruz et al., 2008) articulated with an avoiding obstacle algorithm to prevent the collision with moving agents. Once a neighborhood of the local uncertainty point is reached by the robot, the vehicle searches for a new boundary point. This process is repeated until no additional local uncertainty point can be determined. Once this stage is reached - mainly due to the fact that the environment is occupied with already mapped features -, the robot searches for global destination regions. The global destination regions are represented by global uncertainty points. These points are determined by using the Gaussian distribution of each random variable involved in the EKF-SLAM system state. Thus, according to the geometrical information given by the secondary map of the environment, the entire map is circumscribed by four virtual segments that determine the limits of the known environment. Then, a set of points contained inside the limits of the virtual features is generated by a Monte Carlo experiment, and those that are not navigable are rejected. The probability attached to each one of the remaining point is estimated based on the concept of sum of Gaussians, which yields an estimate of the occupational probability of each point. Those points whose probability values do not fall near zero (free space point) or near one (non-empty point) will be considered as a global uncertainty point. This kind of points can be attached to the boundary of the map - unexplored region - or to badly mapped features. Once a global uncertainty point is determined -according to the needs of the navigation purposes - a hybrid contour-following control (Toibero et al., 2007) is implemented to drive the robot to that point. Once a neighborhood of the global uncertainty point is reached, the exploration switches to the mode of searching for local uncertainty points in a local reference frame. The entire navigation system stops when no global uncertainty points are found, what will happen when the map is complete. During the entire navigation or exploration phase, the SLAM algorithm continues been executed Additional information about this topic can be found in (Auat Cheein, 2009). Figure 16 shows the general architecture of the SLAM algorithm with the non-reactive behavior controllers (the adaptive switching controller and the hybrid contour following) implemented on the mobile robot.

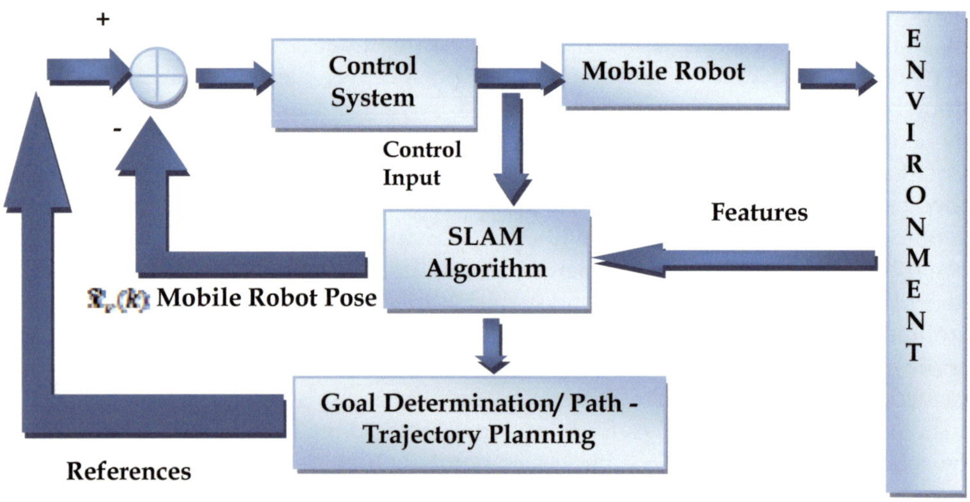


Fig. 16. General SLAM - Control system architecture where the control uses SLAM system state information for planning purposes

Figure 17 shows the real time experimentation of the SLAM algorithm combined with the two control strategies. The experimentation was carried out at the facilities of the Instituto de Automatica. As it can be seen, the map obtained by the SLAM was built consistently. The local controller (adaptive switching control) and the global controller (hybrid contour following) have allowed the entire navigation of the mobile robot within the Instituto de Automatica.

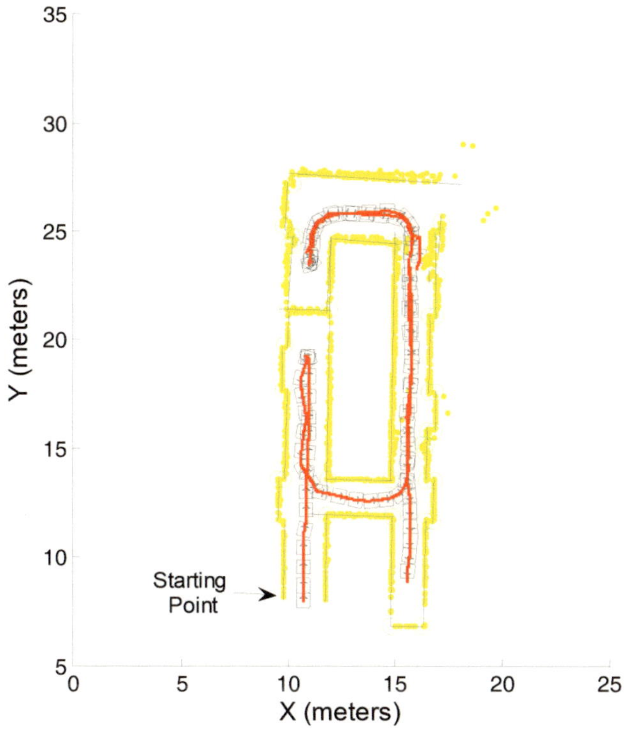


Fig. 17. Map reconstruction of the Institute of Automatics

In Figure 17, the green circles are the corners detected in the environment, the solid dark lines are the segments associated with walls whereas red crosses are the beginning and ending points of such segments; the red points are the path travelled by the mobile robot and the yellow points are raw laser data.

## 8. Conclusions

In this Chapter it has been presented a switched countour-following controller, which allows a wheeled mobile robot to follow discontinuous walls' contours. This controller has been developed by considering a standard (stable) wall-following controller and aggregating two complementary (also stable) controllers. One risponsible for avoid collisions between the robot and the object which is being followed, and the other responsible for find a lost contour of this same object. This new swiching controller proved to be stable with respect to its swiching signal, guaranteing that the robot will be able to stay at a desired robot-to-object distance, and with the same object orientation.
Next, this controllers was included into a SLAM algorithm, in order to deal with the exploration and map construction of unknown environments, exposing the modularity capability of this kind of control architecture.

In designing the control system, the asymptotic stability of the individual controllers as well the asymptotic stability at the switching times (for the switching controller) were considered and proved.

Several experimental results in a Pioneer III mobile robot with odometry and laser radar sensor have been included; showing the good performance of the proposed control strategy in a laboratory setting, in the first attemps, and later, in a large scale setting for the SLAM experiment.

## Acknowledgments

Authors thank to the National Council of Scientific and Technical Research of Argentina (CONICET) for partially supporting this research.

## 9. References

Ando, Y. & S. Yuta. (1996). A reactive wall-following algorithm and its behaviors of an autonomous mobile robot with sonar ring. Journal of Robotics and Mechatronics, Vol. 8(1), pp.33-39.

Andrade-Cetto, J. & Sanfeliu, A. (2006). Environment Learning for Indoors Mobile Robots, ISBN: 978-3-540-32795-0, Series: Springer Tracts in Advanced Robotics , Vol. 23, Springer

Arkin, R.C. (1998). Behavior-based Robotics, MIT Press: Cambridge, M A.

Auat Cheein, F. A. (2009). Simultaneous Localization and Mapping of a Mobile Robot based on Uncertainty Zones Navigation, PhD Thesis, ISBN: 978-987-05-6727-1, INAUT: National University of San Juan

Bailey, T.; Nieto, J.; Guivant, J.; Stevens, M. & Nebot, E. (2006). Consistency of the EKF-SLAM algorithm, in IEEE/RSJ International Conference on Intelligent Robots and Systems (IROS), pp. 3562–3568, ISBN: 1-4244-0259-X, Beijing, China

Bicho, E. (2000). Detecting, representing and following walls based on low-level distance sensors, Proc. of the 2nd Int. Symposium on Neural Computation.

Boada, M.J.L.; Rodriguez, F.J.; Barber, R. & Salichs, M.A. (2003). A control system based on reactive skills for autonomous mobile robots, IEEE Int. Conf. on Advanced Robotics.

Borenstein, J. & Koren, Y. (1989). Real-time obstacle avoidance for fast mobile robots, IEEE Trans. on Systems, Man and Cybernetics. Vol. 19(5), pp.1179-1187

Braunstingl, R. and Ezkerra, J.M. (1995). Fuzzy logic wall following of a mobile robot based on the concept of general perception, Int. Conf. on Advanced Robotics, pp. 367-376

Brawn, R. & Hwang, P. (1997). Introduction to Random Signals and Applied Kalman Filtering 3rd ed., John Wiley & Sons: New York, pp. 371-375

Briechle, K. & Hanebeck, U. D. (2004). Localization of mobile robot using relative bearing measurements, IEEE Transactions on Robotics and Automation, vol. 20, no. 1, pp. 36–44, ISSN: 1042-296X

De la Cruz, C.; Carelli, R. & Bastos, T. F. (2008). Switching Adaptive Control of Mobile Robots, in: International Symposium on Industrial Electronics, pp. 835-840, ISBN: 978-1-4244-1665-3, June 30 2008-July 2, Cambridge, UK

Durrant-Whyte, H. & Bailey, T. (2006a). Simultaneous localization and mapping (SLAM): part I essential algorithms. IEEE Robotics and Automation Magazine, vol. 13, pp. 99-108, ISSN: 1070-9932

Durrant-Whyte, H. & Bailey, T. (2006b). Simultaneous localization and mapping (SLAM): part II state of the art, IEEE Robotics and Automation Magazine. Vol.13, No. 3, pp. 108-117, Sept., ISSN: 1070-9932

Edlinger, T. & von Puttkamer, E. (1994). Exploration of an indoor-environment by an autonomous mobile robot, IEEE/RSJ/GI Int. Conf. on Intelligent Robots and Systems 2, pp.1278-1284

Fazli, S. & Kleeman, L. (2005). Wall following and obstacle avoidance results from a multi-dsp sonar ring on a mobile robot, In Proc. of the IEEE Int. Conf. on Mechatronics & Automation, 5, pp.432-437

Garulli, A.; Giannitrapani, A.; Rosi, A. & Vicino, A. (2005). Mobile robot SLAM for line-based environment representation, Proceedings of the 44th IEEE Conference on Decision and Control, pp. 2041- 2046, ISBN: 0-7803-9567-0, December 2005

Guivant, J.E. & Nebot, E.M. (2001). Optimization of the simultaneous localization and map-building algorithm for real-time implementation, IEEE Transactions on Robotics and Automation, vol. 17, pp. 242–257, ISSN: 1042-296X

Kelly, R. & Carelli, R. (1996). A class of nonlinear PD-type for robot manipulator, Journal of Robotic Systems, 13(12), pp.793-802

Liberzon, D. & Morse, A.S. (1999). Basic problems in stability and design of switched systems, IEEE Control Systems Magazine, 19(5):pp.59-70

Liberzon, D. (2003). Switching in Systems and Control, Birkhauser

Liu, Y. & Sun, F. (2007). A solution to active simultaneous localization and mapping problem based on optimal control, in: Proceedings of IEEE International Conference on Mechatronics and Automation, pp. 314-319, ISBN: 978-1-4244-0828-3, 5-8 August, Harbin, China, 2007

Liu, Y; Dong, J. & Sun, F. (2008). An Efficient Navigation Strategy for Mobile Robots with Uncertainty Estimation, in: Proc. of the World Congress on Intelligent Control and Automation, pp. 5174-5179, ISBN: 978-1-4244-2113-8, 25-27 June, Chongquing, China

Mancilla-Aguilar, J. L. (2000). A condition for the stability of Switched Nonlinear Systems, IEEE Trans. on Automatic Control, 45, pp.2077-2079

Pfister, A. T.; Roumeliotis, S. I. & Burdick, J. W. (2003). Weighted line fitting algorithms for mobile robot map building and efficient data representation, in Proceedings of the 2003 IEEE International Conference on Robotics and Automation, pp. 1304–1311, ISBN: 0780377362, Taiwan, September 2003

Thrun, S.; Burgard, W. & Fox, D. (2005). Probabilistic Robotics, ISBN: 978-0-262-20162-9, MIT Press, Cambridge.

Toibero, J.M.; Carelli, R. & Kuchen, B. (2006a). Stable Switching Contour-Following Controller for Wheeled Mobile Robots, IEEE Int. Conf. on Robotics and Automation

Toibero, J.M.; Carelli, R. & Kuchen, B. (2006b). Wall-following stable control for wheeled mobile robots, Proc. of the 8th Int. IFAC Symposium on Robot Control

Toibero, J.M.; Carelli, R. & Kuchen, B. (2007). Switching control of mobile robots for autonomous navigation in unknown environments, in: IEEE International Conference on Robotics and Automation, pp. 1974-1979, ISBN: 1-4244-0601-3, 10-14 April, Rome, Italy

Toibero, J.M.; Roberti, F. & Carelli, R. (2009). Stable Switching Contour-Following Control of Wheeled Mobile Robots. ROBOTICA, 27, 1-12

van Turennout, P. & Hounderd, G. (1992). Following a wall with a mobile robot using ultrasonic sensors, In Proc. of the 1992 IEEE/RSJ Int. Conf. on Intelligent Robots and Systems 2, pp.1451-1456

Vidyasagar, M. (1993). Nonlinear Systems Analysis 2nd edition, Prentice Hall, New Jersey

Vu, L. & Liberzon, D. (2005). Common Lyapunov Functions for families of commuting nonlinear systems, Systems & control letters, 54: pp.405-416

Wang, M. & Liu, J. (2004). Autonomous robot navigation using fuzzy logic controller. Proc. of the Int. Conf. on Machine Learning and Cybernetics, pp.26-29, 2004

Xi, B.; Guo, R.; Sun, F. & Huang, Y. (2008). Simulation Research for Active Simultaneous Localization and Mapping Based on Extended Kalman Filter, in: Proc. of the IEEE International Conference on Automation and Logistics, pp. 2443-2448, ISBN: 978-1-4244-2502-0, 1-3 Sept, Quingdao, China.

Zhang, Z.; Sarkar, N. & Yun, X. (2004). Supervisory control of a mobile robot for agile motion coordination, IEEE Int. Conf. on Robotics and Automation, Vol.3, pp.2196-2203

# 20

# PFC Fuzzy Decision-Making Control and Its Application to Car-Like Mobile Vehicle

You-gen Chen[1], Seiji Yasunobu[2], Wei-hua Gui[1],
Ren-yong Wei[1] and Zhi-yong Li[1]
[1]*Central South University, China*
[2]*University of Tsukuba, Japan*

## 1. Introduction

Human have a remarkable capability to perform a wide variety of physical and mental tasks without any measurements and any computations. Underlying this capability is brain's crucial ability to manipulate perceptions and remarkable capability to operate on and reason with, perception-based information which are intrinsically vague, imprecise and fuzzy. Human's action decision (Fig. 1) is based on multi-targets and can respond flexibly under different situations just based on information which are intrinsically vague, imprecise and fuzzy. The best alternative target is selected in real time based on experiences by predicting and evaluating the object's states with taking dynamic environment information into account.
From the process of human decision, it is clear that the multi-targets-based methodology results in their dynamic soft decision and flexibility rather than single-target-based conventional method. The image of multi-targets and its utilization is denoted in Fig. 2, when disturbances make the best target $r_3$ in multi-target set with membership 1.0 become unavailable, system selects the sub-optimal target $r_2$ as control target automatically to respond to situation change in environment.
For a real control system, the fact is that there usually exist plural targets (all possible and reachable sub-targets) with their membership values (reflecting the target how good or bad) for the object to select at all time. Thus, it is possible to provide multi-targets for controlled object rather than single target like the conventional method. This provides a quite important premise for mobile body to act like human and adapt to changing situation flexibly.
Fuzzy logic which firstly introduced by Lotfi Zadeh provides a theoretic foundation for combining multi-targets with fuzzy set to form a target set with membership values reflecting satisfaction degree (how good or bad) of its elements. The target set is defined as "soft target" in this chapter.
Decision making problems are described as applying approximate reasoning and incomplete or uncertain information to find a fuzzy set of decision alternatives and choose the best one from possible alternatives (Bellman & Zadeh, 1970). Fuzzy control presents a formal methodology for representing, manipulating, and implementing a human's heuristic knowledge about how to control a system, which make it possible to incorporate the knowledge into a fuzzy controller that emulates the decision-making process of the human. Especially,

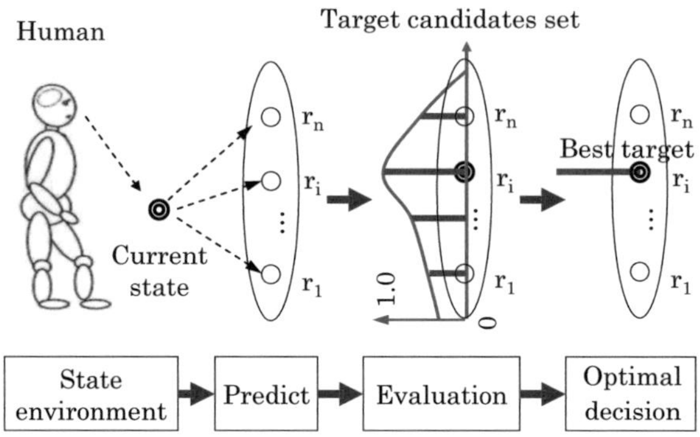


Fig. 1. Decision process of human with fuzzy multi-targets

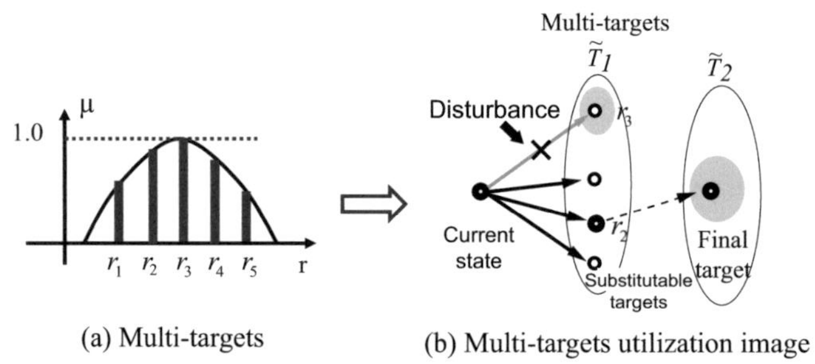


Fig. 2. Image of multi-targets and its utilization

predictive fuzzy control (PFC) method proposed by Prof. Yasunobu provides a quite potential means to realize human's prediction, evaluation and optimal decision-making process to select the best target element from a soft target set by using fuzzy multiple criteria decision making.

In many control systems, the situations always change with the constraints or disturbances. Responding flexibly to the dynamic situation changes in the external world like human is very necessary but difficult for an autonomous control system (Barraquand et al., 1992; Jing, 2005). The motion control of mobile vehicle with dynamic constraints (static and dynamic obstacles) is still a difficult and challenging problem because the surrounding situations are not qualified in static, knowledge is only partial and the execution is often associated with uncertainty.

Therefore, responding flexibly to the situation changes like human for mobile vehicle is considered in this chapter. We proposed a soft-target-based fuzzy decision-making control method to realize flexible autonomous operation like human for motion control of mobile vehicle in a narrow space with change situations to verify the validity of the proposed method. The soft target including many target elements with different satisfaction grade is a multi-target set defined with fuzzy logic. Fuzzy decision-making is realized by predictive fuzzy control to emulate human's prediction, evaluation and optimal decision-making process. The experiment system is mainly composed of a control PC, a CCD camera, and a reconstructed vehicle from a RC car. Simulation and experiment results demonstrate the validity and flexibility of the proposed soft-target-based fuzzy decision-making control method. Collision-free

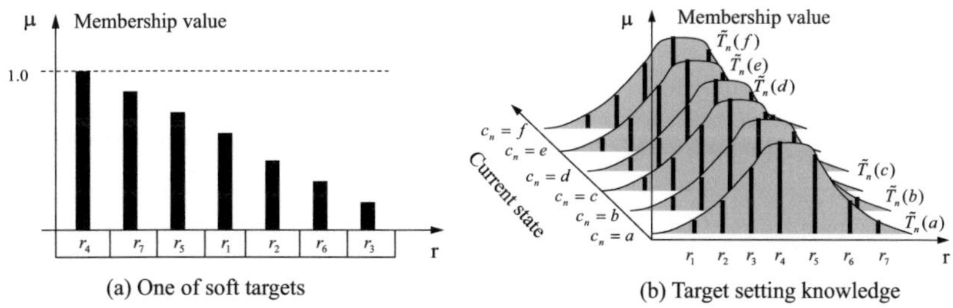


Fig. 3. Definition of soft target

and low cost motion control is achieved, and realized flexible adaptation to changing situations like human.

## 2. Soft-Target-Based PFC Fuzzy Decision Making Control

### 2.1 Definition of Soft Target

Fuzzy logic provides a mathematical strength to capture the uncertainties associated with human cognitive processes, such as thinking and reasoning. It gives us an inference morphology that enables approximate human reasoning capabilities to be applied to knowledge-based systems. Fuzzy set is a means of representing and manipulating data that was not precise, but rather fuzzy.

Fuzzy logic guarantees a theoretic foundation for combining multi-targets with fuzzy set to form a target set with membership values reflecting satisfaction degree (how good or bad) of its elements (Ordonez & Zumberge, 1997). The target set is defined as "soft target" in this chapter (Chen & Yasunobu, 2007b).

Soft target is defined as a target set and is converted into target setting knowledge by soft computing. It is constructed from all available target candidates by combining with fuzzy logic based on the final target and constraint information, and can be expressed as a control target set defined by fuzzy set, which includes many alternative candidates. Each candidate has its membership value defined as satisfaction grade in [0, 1].

It is denoted as Fig. 3 (a), and can be expressed by the membership function of enumeration type in a discrete space (Chen & Yasunobu, 2006).

The total set of the target is assumed as $R$. Soft target $\widetilde{T_n}$ assumed to be a control target set can be defined by the following expression in state $c_n$ of the object.

$$\widetilde{T_n} = \int_R \mu_{\widetilde{T_n}}(r_i)/r_i \qquad r_i \in R \tag{1}$$

Here, $\mu_{\widetilde{T_n}}(r_i)$ is the membership value of alternative $r_i$ in $\widetilde{T_n}$ corresponding with state $c_n$, and the integral denotes the union of the fuzzy singletons $\mu_{\widetilde{T_n}}(r_i)/r_i$, $r_i \in R$.

As shown in Fig. 3 (b), target setting knowledge can be expressed as set clusters which correspond with different states. According to different "current" state $c_n$(a ~ f), the soft target candidates set is $\widetilde{T_n}$(a ~ f) respectively. Once the current soft target is set, it is possible for the system to select the best alternative candidate instruction corresponding with one of the substitutable target element $r_i$ by decision-making. This is repeated until final target or mission achievement.

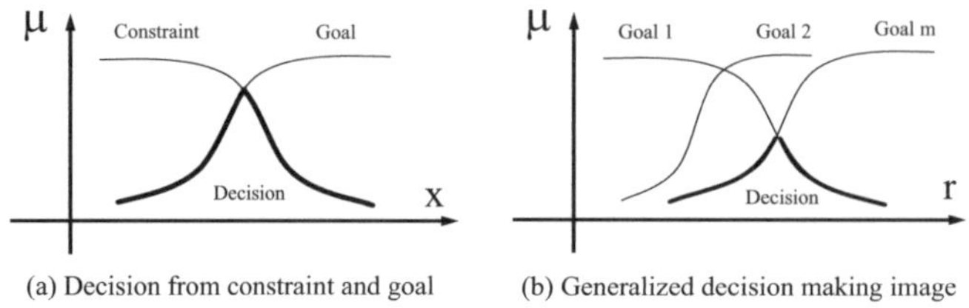


(a) Decision from constraint and goal (b) Generalized decision making image

Fig. 4. Decision-making from fuzzy goal and constraint

By using soft target, it is possible to construct an intelligent controller based on multi-targets for a system with dynamic constraints to achieve a flexible operation.

### 2.2 PFC Fuzzy Decision-Making

Decision making problems are described as applying approximate reasoning and incomplete or uncertain information to find a fuzzy set of decision alternatives and choose the best one from possible alternatives. Multi-criteria decision making (MCDM) involves determining the optimal alternative among multiple, conflicting, and interactive criteria, such control goals and external constraints (Kim & Cho, 2006).

#### 2.2.1 Fuzzy Decision-Making

Fuzzy goals and fuzzy constraints can be defined precisely as fuzzy sets in the space of alternatives. A fuzzy decision, then, may be viewed as as intersection of the given goals and constraints. A maximizing decision is defined as a point in the space of alternatives as which the membership function of a fuzzy decision attains its maximum value.
As shown in Fig. 4 (a), assume that we are given a fuzzy goal $G$ and a fuzzy constraint $C$ in a space of alternatives $X$. $G$ and $C$ combine to form a decision, $D$, which is a fuzzy set resulting from intersection of $G$ and $C$.

$$D = G \cap C \tag{2}$$

The membership function of the intersection is given by

$$\mu_D = \mu_G \wedge \mu_C \tag{3}$$

Then, a maximizing decision can be made from the acquired fuzzy decision set by choosing the one with maximal $\mu$. Fuzzy multi-criteria decision making (FMCDM) which combines MCDM with fuzzy logic provides a promising theoretical framework for alternatives or candidates selection decision.

#### 2.2.2 Predictive Fuzzy Control

Predictive fuzzy control is an intelligent control method based on human control strategy. As shown in Fig. 5, firstly, a series of control instruction candidates are prepared based on expert's experience. Next, the future state of controlled object is predicted by using all the instruction candidates in parallel. Then the future state is evaluated by fuzzy inference using expert's knowledge described by fuzzy rules. Lastly, the operation instruction candidate with the highest evaluation value is selected as the current control instruction. Consequently, an

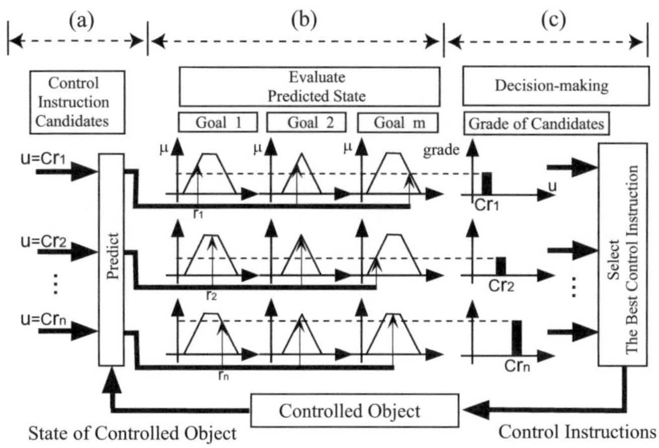


Fig. 5. Predictive fuzzy control

intelligent controller with a similar mechanism to predictive control based on expert's knowledge and experiences can be achieved with PFC.

Obviously, PFC has the three main characteristics of prediction, evaluation and decision in human's action decision. PFC method provides a potential means to realize human's prediction, evaluation and optimal decision-making process to select the best target element from a series of control instruction candidates by using FMCDM.

### 2.2.3 Fuzzy Decision-Making by PFC

MCDM is not only an efffective means for management decision but also valid for action decision for mobile body although there are not much applications until now. As described in the previous section, it should be noted that PFC can be regarded as a typical realization method of FMCDM with integrating predictive control, fuzzy logic and decision-making theories.

Let's consider the example in Fig. 4 (a) again, as a constraint can be regarded as a "goal" which satisfies the constraint, it is possible to convert all constraints to goals and expand them to m goals like Fig. 4 (b). Assume in a discrete decision space, let $R = \{r_1, r_2, \cdots, r_n\}$ be the set of decision alternatives, $\tilde{G}_j$ be the fuzzy sets representing control goal $j$. When the attainment of the goal $\tilde{G}_j$ by alternative $r_i$ can be expressed by the degree of membership $\mu_{\tilde{G}_j}(r_i)$, combining with the PFC architecture in Fig. 5, the decision set $\tilde{D}$ can be expressed as follow,

$$\tilde{D}(r_i) = \tilde{G}_1(r_i) \cap \cdots \cap \tilde{G}_j(r_i) \cap \cdots \cap \tilde{G}_m(r_i) \quad (i = 1, 2, \cdots, n; j = 1, 2, \cdots, m) \tag{4}$$

and correspondingly its membership expression is

$$\mu_{\tilde{D}}(r_i) = \mu_{\tilde{G}_1}(r_i) \wedge \cdots \wedge \mu_{\tilde{G}_j}(r_i) \wedge \cdots \wedge \mu_{\tilde{G}_m}(r_i) \tag{5}$$

More generally, we may express $\mu_{\tilde{D}}$ as below,

$$\mu_{\tilde{D}}(r_i) = \sum_{j=1}^{m} \alpha_j \cdot \mu_{\tilde{G}_j}(r_i) \qquad \sum_{j=1}^{m} \alpha_j = 1 \tag{6}$$

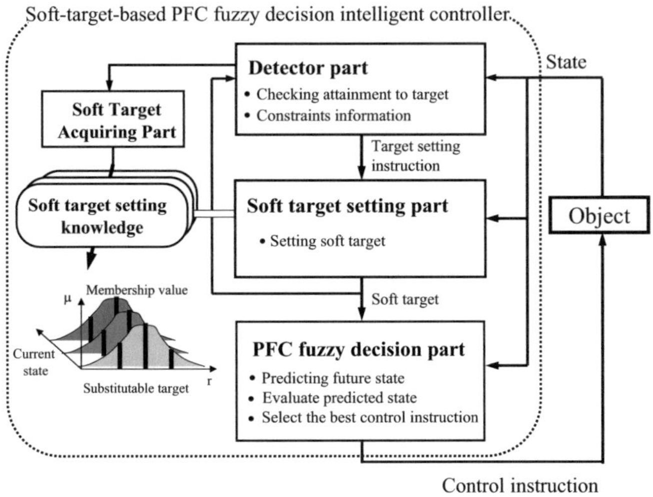


Fig. 6. Outline of soft-target-based fuzzy decision system

where weight coefficient $\alpha_j$ can be chosen in such a way to reflect the relative importance of each control goal $\tilde{G}_j$. It is consistent with fact since each control goal usually has different weight in a real system. Naturally, the $r_i$ with the highest membership degree will be selected as the optimal alternative. In short, a fuzzy decision-making may be completely realized by PFC.

### 2.3 Soft Target Based PFC Fuzzy Decision

The control instruction candidates of a conventional predictive fuzzy controller are usually acquired from experiences or try and error. They are limit and difficult to cover the whole control instructions domain. While if we get the control instruction candidates from the all current available target elements (acquired soft targets), the control instructions will more reasonable and have wider domain. This make the soft-target-based PFC fuzzy decision-making is greatly different from a conventional PFC.

The constructed PFC fuzzy decision-making system's configuration based on soft target is outlined as Fig. 6. It is composed of four parts: state detecting, soft target learning, soft target setting and fuzzy decision-making.

#### 2.3.1 Soft Target Learning Part

For any sampling time, it is necessary to acquire all available sub-targets and their membership values based on the current state and environment information to prepare the soft target setting knowledge. It can be learned by final target based method or reinforcement learning method.

#### 2.3.2 Detector Part

This part is used to detect the state variables and obstacles information, and judge the attainment degree to target (either sub-targets or final target) and the contact degree to obstacles. When it detected the previous sub-target was achieved or the vehicle can't move forward

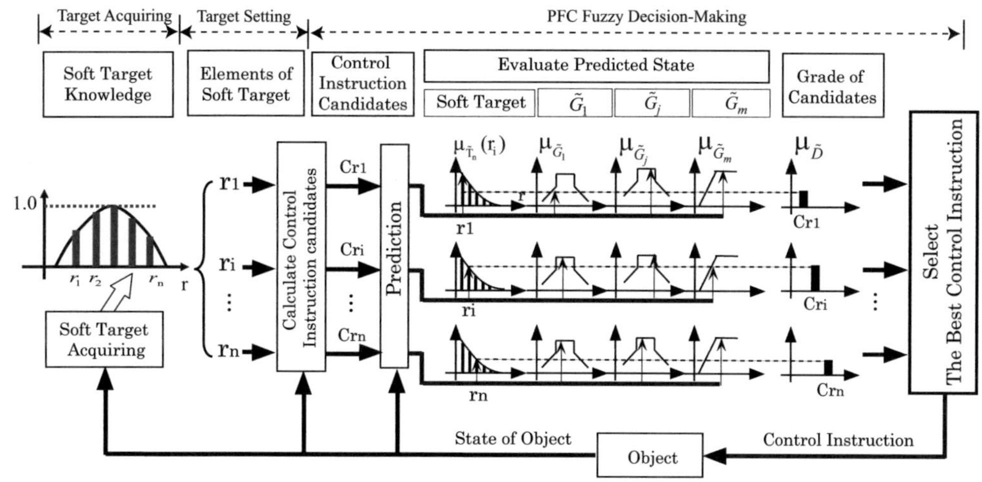


Fig. 7. PFC fuzzy decision-making with soft target

(or backward) anymore because of the influence of obstacles, target setting instruction is outputted to the soft target setting part.

### 2.3.3 Soft Target Setting Part

When target setting instruction is received, the soft target setting part carries out on-line learning for all possible candidates to calculate their membership values based object's current state and around obstacle information. It provides all possible candidates to fuzzy decision-making part to calculate control instruction candidates.

### 2.3.4 Fuzzy Decision-Making Part

Fuzzy decision is made by PFC as following process just like human's decision. It is easier to understand by combining Fig. 7. Firstly, each element of soft target is assumed as the control target, and the operation instruction candidate to each target is calculated. Next, the future state of controlled object is predicted by using all the operation instruction candidates in parallel. Then the future state is evaluated by fuzzy inference, and the evaluation value of the operation instruction candidate is calculated by equation (5) or (6). Lastly, the operation instruction candidate with the highest evaluation value is selected and given to the object as a control instruction.

These operations are repeated in the whole control process until achieving the final target. Thus, the fuzzy decision based on soft target is realized.

## 2.4 Summary

Soft target expand the conventional single target to a fuzzy target set by using fuzzy logic. PFC fuzzy decision-making provides a potential means to realize human's prediction, evaluation and optimal decision-making process to select the best target element from a series of control instruction candidates by using FMCDM. Soft-target-based PFC fuzzy decision-making control makes the control instructions more reasonable and have wider domain, which makes it greatly different from a conventional PFC. It is possible to solve the motion control problem

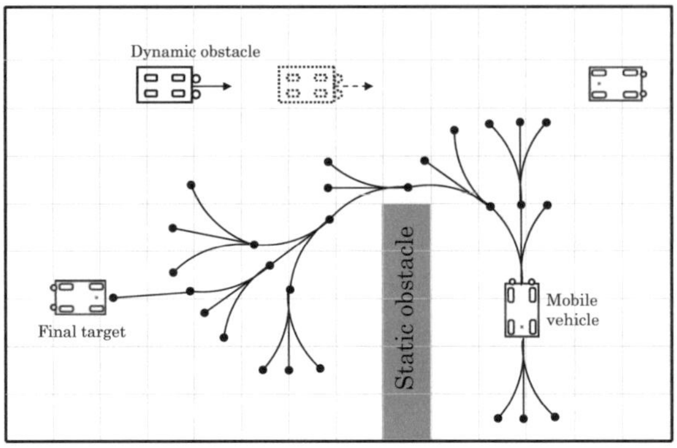


Fig. 8. Problem description

of mobile body with dynamic external constraints to achieve a flexible operation and self-adaptation to change situations.

## 3. Application to Car-Like Mobile Vehicle

The motion control problem of mobile vehicle can be typically formulated as planning a path between two specified locations, which is collision-free and satisfies certain optimization criterias. Autonomous moving in dynamic environment is still a very challenging problem because the surrounding situations are not qualified in static, knowledge is only partial and the execution is often associated with uncertainty. The car parking problem in a static environment has been studied by Prof. Yasunobu. An intelligent controller based on fuzzy target has been proposed. It solved the parking problem in a fixed space without moving obstacles. However, because the target is acquired off-line for a fixed parking lot, when the final goal or space changed, soft target had to been explored once more. It is difficult to respond to the dynamic environment such as moving obstacles, arbitrary placement of static obstacles or discretional initial position of the vehicle.

### 3.1 Problem Description

Four-wheeled car-like vehicle's motion control is very difficult because it has a strong internal constraint — minimal turn radius, which results in higher dependence of external environment and the demand of more well-rounded control knowledge. Let's consider a mobile vehicle moving in a narrow space in indoor environment with static and dynamic obstacles as denoted in Fig. 8 in which the dots means possible target candidates and the forking solid line means one moving route to one of the available multi-target candidates corresponding to the current state and obstacle information. The final target is able to be set as we want. The static obstacles can be placed at any position with arbitrary shape, and the vehicle can start at arbitrary initial position and orientation. In order to achieve a collision-free and low cost motion, the moving action from initial position to final target had to be designed online. Because of the nonholonomic characteristic and the impact of obstacles, it is necessary to find appropriate sub-targets corresponding to each current state and constraint information until arrive at the final target (Chen & Yasunobu, 2008).

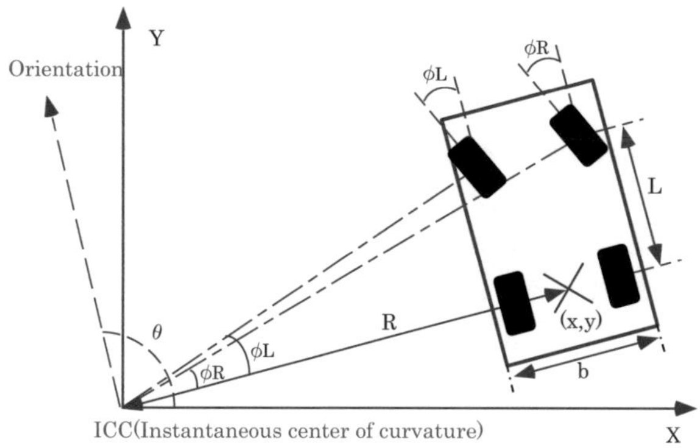


Fig. 9. Model of Ackerman steering mobile vehicle

### 3.2 Four-Wheeled Mobile Vehicle Model

Fig. 9 is the model of a car-like vehicle with Ackerman Steering. Where, $(x, y)$ is the Cartesian location of its rear wheels' center, $\theta$ is the heading angle between the body axis and the horizontal axis, $\phi_L$ and $\phi_R$ are relative steering angle of left and right wheel respectively, and $\phi = (\phi_L + \phi_R)/2$ represents the steering angle with respect to the car body ($|\phi| \leq \phi_{\max}$). $L$ is the wheelbase (longitudinal wheel separation). $b$ is the width of car (lateral wheel separation). $R$ is turning radius which is the distance from instantaneous center of curvature (ICC) to centerline of the vehicle. This system has 2 degrees of nonholonomy since the constraints on the system arise by allowing the wheels to roll and spin, but not slip. Thus, the Pfaffian constraints on the mobile vehicle become:

$$\begin{aligned} \sin(\theta+\phi)\dot{x} - \cos(\theta+\phi)\dot{y} - L\cos\phi\cdot\dot{\theta} = 0 \\ \sin\theta\cdot\dot{x} - \cos\theta\cdot\dot{y} = 0 \end{aligned} \tag{7}$$

Choosing $u_1 = v\cos\phi$ and $u_2 = \dot{\phi}$ as inputs yields:

$$\dot{q} = \begin{pmatrix} \dot{x} \\ \dot{y} \\ \dot{\theta} \\ \dot{\phi} \end{pmatrix} = \begin{pmatrix} \cos\theta \\ \sin\theta \\ \frac{\tan\phi}{L} \\ 0 \end{pmatrix} u_1 + \begin{pmatrix} 0 \\ 0 \\ 0 \\ 1 \end{pmatrix} u_2 \tag{8}$$

where, $v$ is the driving speed, $u_1$ corresponds to the translational velocity of the driving wheels and $u_2$ corresponds to the angular velocity of the steering wheels. Obviously, (8) is a so-called driftless nonlinear system with 2 inputs $(v, \phi)$ and 3 outputs $(x, y, \theta)$ constrained by $R_{\min} = L/\tan\phi_{\max}$. Where, $R_{min}$ is the minimal turn radius, $\phi_{max}$ is the maximal steering angle.

The parameters of the four-wheeled vehicle (assumed as about the same size of an RC car) are width of the car $b = 18\ cm$, wheelbase $L = 25.6\ cm$, minimal outside turning radius $R_{min} = 36.5\ cm$, and the moving speed $v = 10\ cm/s$ in both ahead and back. The map is set as Fig. 10, and the static obstacle is placed with different mode such as in the parking side, in opposite side and so on. The moving obstacle car with the same size of the controlled vehicle

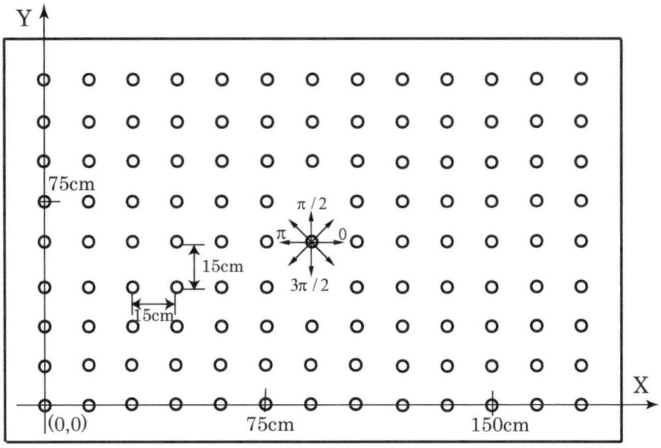


Fig. 10. Discretized moving space for four-wheeled vehicle

moves at a speed of 10 $cm/s$ from left to right with steering angle 0. Final target and initial position can be set arbitrarily.

### 3.3 Soft Target Acquisition

In order to acquire the target setting knowledge for the current state, the 180 $cm \times 120$ $cm$ space is described by occupancy grid maps with 15 $cm$ interval as showed in Fig. 10 in which each small circle denotes a target location $(x, y)$ of vehicle (total $13 \times 9 = 117$ points). And the orientation $\theta$ is divided into eight azimuths $(0, 0.25\pi, 0.5\pi, 0.75\pi, \pi, 1.25\pi, 1.5\pi, 1.75\pi)$. Thus, the space results $117 \times 8 = 936$ target candidates. For arbitrary state, we can approximate it to the nearest grid location and orientation. So it is possible to obtain all possible targets corresponding to the current state and obstacles information to form a state-action table with its action evaluation value named membership value in this study. This state-action table is defined as the soft target setting knowledge for the vehicle. It is acquired based on the final target, current state and obstacle information in real-time. The acquiring process is to find all possible sub-targets that can reach the final target directly (Chen & Yasunobu, 2007a).

To obtain the evaluation value of each target, it is supposed that the vehicle moves from an arbitrary position $r_i = (x_i, y_i, \theta_i)$ in the space to achieve the final target $r_{final} = (x_{final}, y_{final}, \theta_{final})$ controlled by cascade fuzzy control method as showed in Figure 11. In which, the current target orientation $\theta_T$ is fuzzy inferred from deflection $e_X$ of current position $X_t$ and target $X_T$, then, operation steering angle $\phi$ is fuzzy inferred from error $e_\theta$ of the target direction $\theta_T$ and the current body direction $\theta_t$. The membership functions used for evaluating $e_X$ and $e_\theta$ are denoted in Figure 12, where NB, NS, ZO, PS and PB are defined fuzzy linguistics variables corresponding with "Negative Big", "Negative Small", "Zero", "Positive Small" and "Positive Big" respectively. And the $1^{st}$ stage and $2^{nd}$ stage fuzzy inference models are denoted by Table 1 and Table 2 respectively.

The evaluation value $\mu_{\widetilde{T_n}}(r_i)$ is calculated according to the following cost functions.

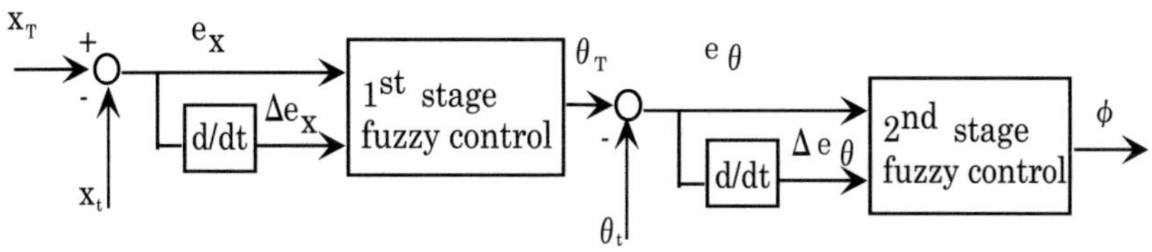


Fig. 11. Cascade fuzzy controller

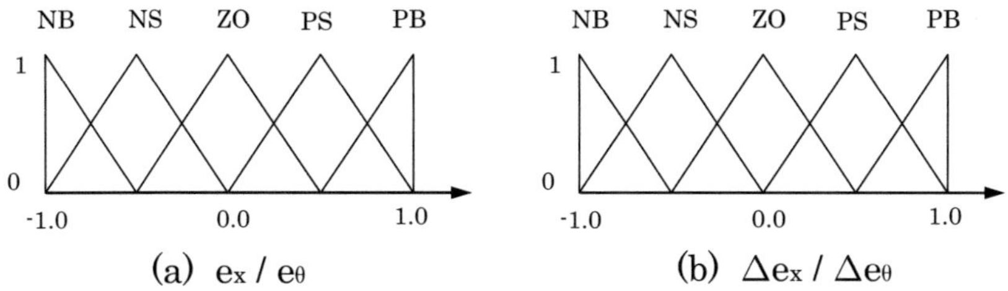


Fig. 12. Membership functions of cascade fuzzy controller

| $\theta_T(rad)$ | | $e_X(m)$ | | | | |
|---|---|---|---|---|---|---|
| | | NB | NS | ZO | PS | PB |
| $\Delta e_X$ | NB | -2.5916 | -2.4738 | -2.356 | -2.2382 | -2.1204 |
| | NS | -1.413 | -1.2958 | -1.178 | -1.0602 | -0.9424 |
| | ZO | -0.2356 | -0.1178 | 0.0 | 0.1178 | 0.2356 |
| | PS | 0.9424 | 1.0602 | 1.178 | 1.2958 | 1.4136 |
| | PB | 2.1204 | 2.2382 | 2.356 | 2.4738 | 2.5916 |

Table 1. $1^{st}$ stage fuzzy inference model

| $\phi(rad)$ | | $e_\theta(rad)$ | | | | |
|---|---|---|---|---|---|---|
| | | NB | NS | ZO | PS | PB |
| $\Delta e_\theta$ | NB | 7.0 | 4.5 | 2.0 | -0.5 | -3.0 |
| | NS | 6.0 | 3.5 | 1.0 | -1.5 | -4.0 |
| | ZO | 5.0 | 2.5 | 0.0 | -2.5 | -5.0 |
| | PS | 4.0 | 1.5 | -1.0 | -3.5 | -6.0 |
| | PB | 3.0 | 0.5 | -2.0 | -4.5 | -7.0 |

Table 2. $2^{nd}$ stage fuzzy inference model

$$
\begin{aligned}
\mu_{\widetilde{T_n}}(r_i) &= \mu_{time}(r_i) \wedge \mu_{ope}(r_i) \wedge \mu_{err}(r_i) \\
\mu_{time}(r_i) &= (t_{max} - t)/t_{max} \in [0,1] \\
\mu_{ope}(r_i) &= 1.0 - \alpha \sum_{t=0}^{time} |ope(t)| \in [0,1] \\
\mu_{err}(r_i) &= \mu_{dx}(x) \wedge \mu_{dy}(y) \wedge \mu_{d\theta}(\theta) \in [0,1]
\end{aligned}
\tag{9}
$$

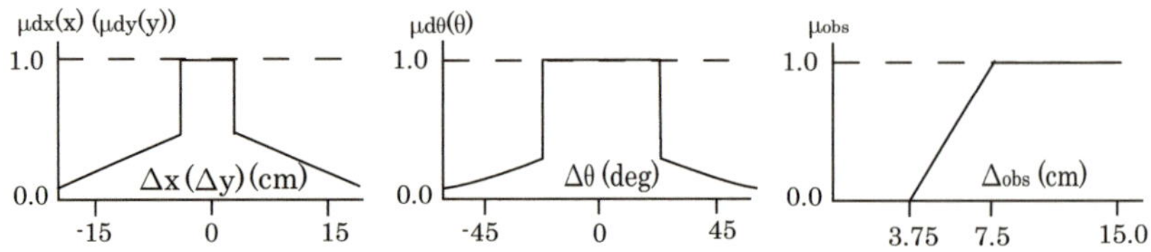


Fig. 13. Error evaluation membership functions

| Number | Sub-target position | Membersip value $\mu$ |
|---|---|---|
| 1 | (15cm, 15cm, -0.75$\pi$) | 0.660 |
| 2 | (15cm, 30cm, -0.75$\pi$) | 0.596 |
| 3 | (15cm, 45cm, -0.75$\pi$) | 0.576 |
| ⋮ | ⋮ | ⋮ |
| 42 | (60cm, 105cm, -1.0$\pi$) | 0.546 |
| 43 | (75cm, 0cm, 0.5$\pi$) | 0.999 |
| 44 | (75cm, 15cm, 0.5$\pi$) | 0.980 |
| ⋮ | ⋮ | ⋮ |
| 134 | (165cm, 90cm, 0.25$\pi$) | 0.497 |
| 135 | (165cm, 90cm, 0.5$\pi$) | 0.464 |
| 136 | (165cm, 105cm, 0.25$\pi$) | 0.475 |

Table 3. Soft target for final target $(75\ cm, 0\ cm, 0.5\pi)$

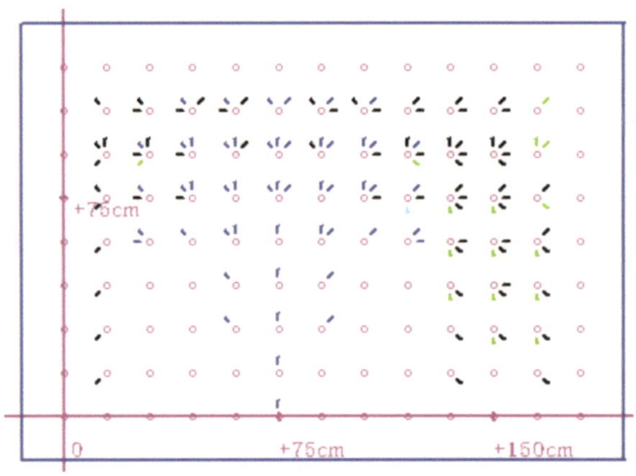


Fig. 14. Distribution map of soft target elements for final target $(75\ cm, 0\ cm, 0.5\pi)$ with $\mu \geq 0.1$

Where, $\mu_{time}(r_i)$ is evaluation of moving time, $\mu_{ope}(r_i)$ is evaluation of steering amount, $\mu_{err}(r_i)$ is evaluation of arrival grade to final target. $t_{max}$ is the maximal limit time for a moving action, $t$ is the consumption time till arriving at the final target or being unable to move

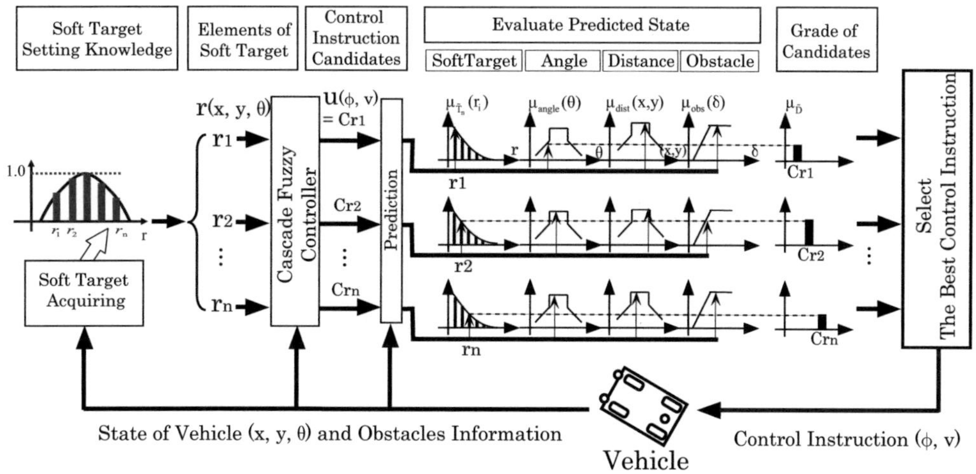


Fig. 15. PFC fuzzy decision with soft target for vehicle

anymore, $\sum_{t=0}^{time} |ope(t)|$ is the total steering amount, $\alpha$ is coefficient of it. $\mu_{dx}(x)$, $\mu_{dy}(y)$, $\mu_{d\theta}(\theta)$ are error evaluations of current position $(x, y, \theta)$ to final target respectively whose error evaluation membership functions are shown in Figure 13. Here, $t_{max}$ is set as 250 $s$, and $\alpha$ is set as 1.5. They are estimated in experiences from the possible longest running time and maximal steering change amount. The less the consumption time or total steering amount or error evaluations to final target, the higher the evaluation value of the alternative target. For those that are unable to reach the final target, the membership values are set as 0. For the first time, the soft target is acquired in this space without considering any obstacle, after then, it is acquired in an area near the vehicle current position (around 60 $cm$ range) to reduce the computation expenses. If there is no available target for the current state, system selects the one acquired at the first time as the soft target set (Chen & Yasunobu, 2009).

Based on the vehicle's kinematics model (8) and the cost evaluation function (9), we can obtain each available sub-target and its membership value which presents its satisfactory degree. Table 3 lists the acquired soft target set for final target $(75\ cm, 0\ cm, 0.5\pi)$ without considering any obstacle from an arbitrary initial position in the 936 target candidates. There are 136 possible candidates in which $(75\ cm, 0\ cm, 0.5\pi)$ has the highest evaluation value 0.999. Figure 14 denotes the candidates who have the membership value no less than 0.1. Here, blue cords denote the targets with membership value greater than 0.7, black means the membership value is greater than 0.5, green means greater than 0.3, and magenta means greater than 0.1.

### 3.4 PFC Fuzzy Decision with Soft Target

The constructed PFC fuzzy decision-making control system based on soft target for the four-wheeled car-like vehicle is denoted as Figure 15. Based on the vehicle's posture and environment information, state detector part decides whether giving target setting instruction or not by three flags. One is target achievement judgment flag which reflects whether the vehicle reached current target or not (judgment terms: $|\Delta x| < 3.75\ cm, |\Delta y| < 3.75\ cm$, and $|\Delta\theta| < 0.5\ rad$). The other two are moving flags which reflect whether the vehicle is able to move forward and backward respectively with judgment term $|\Delta obs| < 3.75\ cm$). When it

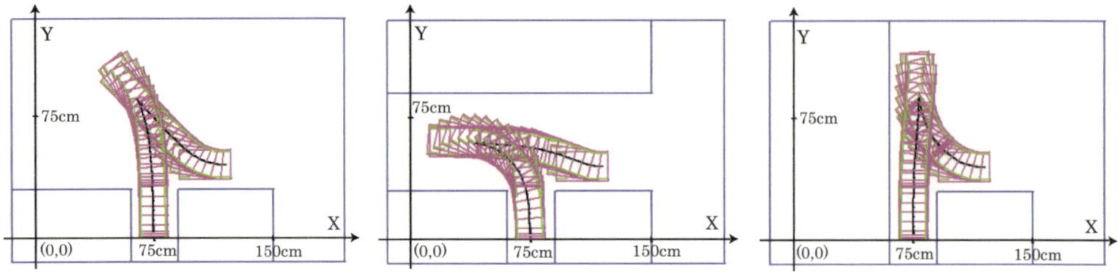


(a) Obstacles in parking side (b) Obstacles in parking and oppsite side (c) Obstacles in parking and left side

Fig. 16. Parking simulation trajectories with different placement of obstacles

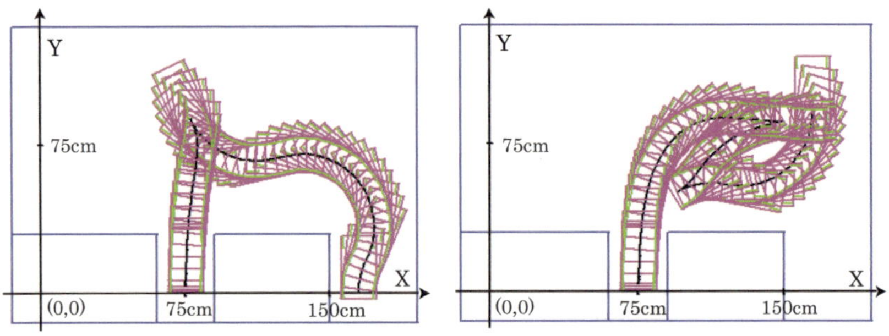


(a) Initial position (165 cm, 0 cm, $\pi/2$) (b) Initial position (165 cm, 90 cm, $\pi/2$)

Fig. 17. Simulation trajectories with different initial position

detected the previous sub-target was achieved or the vehicle can't move forward (or backward) anymore because of the influence of obstacles, target setting instruction is outputted to the soft target setting part. If it is necessary to reset target, soft target for the current state is obtained in target acquiring part to obtain all possible candidates.
For each candidate $r_i$ in $\widetilde{T_n}$, the control instruction $C_{r_i}$ is calculated by the cascade fuzzy control (Huang & Yasunobu, 1999). And the future pose $(x_{t+1}, y_{t+1}, \theta_{t+1})$ of vehicle is predicted for each instruction candidate $C_{r_i}$ by the kinematics model (8). Then multipurpose fuzzy evaluation is conducted for angle deflection, distance deflection and the minimal distance to obstacles. It calculates the evaluation values of all candidates and makes decision to select the one with the highest evaluation value as the control target by the following equation.

$$\mu_{\tilde{D}} = \mu_{\tilde{T}_n}(r_i) \wedge \mu_{angle}(\theta) \wedge \mu_{dist}(x, y) \wedge \mu_{obs}(\delta) \quad (10)$$

The evaluation value of the operation instruction candidate which results moving in the opposite direction is reduced a half to make the vehicle select others' target candidates with higher evaluation to avoid trapping into dead loop of local minima.

### 3.5 Simulation

In order to confirm the validity of the constructed control system based on soft target, we carried out parking simulations with different placement of obstacles, arbitrary setting of initial position and final target, and autonomous running simulation with dynamic obstacle as well as static obstacles using the vehicle's parameters listed in 3.2.

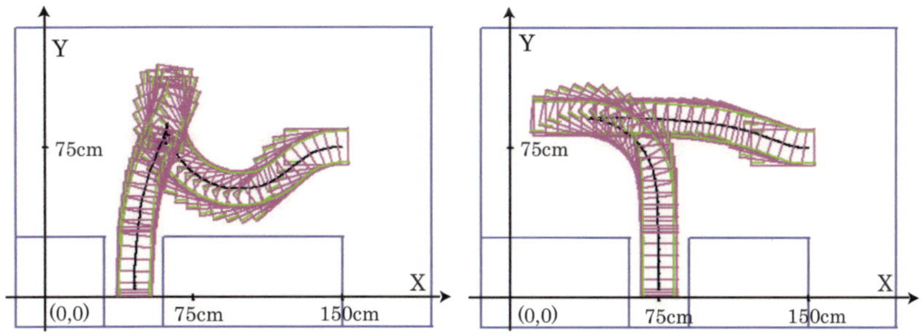


(a) Final target (45 cm, 0 cm, $\pi/2$) (b) Final target (75 cm, 90 cm, $\pi/2$)

Fig. 18. Simulation trajectories with different final target

The coordinate of parking target point is set as $(75\ cm, 0\ cm, 0.5\pi)$. Parking simulation is conducted in three kinds of suppositions of different obstacles placement. One is some parking lots in parking side are occupied, the other is parking side and opposite side are occupied, and another is parking side and its left side are occupied. The initial position for them is set as $(120\ cm, 45\ cm, \pi)$. Results are shown as Fig. 16. When parking lots in parking side are occupied, vehicle selects $(60\ cm, 90\ cm, 0.75\pi)$ with the membership value $\mu = 0.787$ as the best sub-target by acquisition and evaluation, and moves to final target as denoted in Fig. 16 (a). When parking lots in parking side and opposite side are occupied, vehicle selects $(30\ cm, 60\ cm, \pi)$ with $\mu = 0.718$ as the best via-point, and achieves final target as shown in Fig. 16 (b). When parking lots in parking side and left side are occupied, vehicle selects $(90\ cm, 60\ cm, 0.5\pi)$ $(\mu = 0.854)$ and $(75\ cm, 90\ cm, 0.5\pi)$ $(\mu = 0.879)$ as sub-targets in turn, and moves to final target as Fig. 16 (c).
To confirm the control method's flexibility further, simulations with same final target but different initial position and same initial position but different final target were executed too.
Fig. 17 denotes the running trajectories for same final target $(75\ cm, 0\ cm, 0.5\pi)$ from different initial position $(165\ cm, 0\ cm, 0.5\pi)$ and $(165\ cm, 90\ cm, 0.5\pi)$. For $(165\ cm, 0\ cm, 0.5\pi)$, the vehicle achieved final target smoothly by selecting soft target element $(165\ cm, 60\ cm, 0.75\pi)$ and $(60\ cm, 90\ cm, 0.75\pi)$. But for $(165\ cm, 90\ cm, 0.5\pi)$, the vehicle couldn't achieve final target by just once reverse. It firstly selected target $(90\ cm, 60\ cm, 0.25\pi)$ to try to approach final target, but potential collision was predicted, so the vehicle reversed and select $(150\ cm, 90\ cm, 0\pi)$ as sub-target to achieve a better via-points. Then it reversed again and reached final target successfully.
Fig. 18 shows the motion trajectories for different final target $(75\ cm, 0\ cm, 0.5\pi)$ and $(45\ cm, 0\ cm, 0.5\pi)$ from the same initial position $(150\ cm, 75\ cm, \pi)$. For final target $(75\ cm, 0\ cm, 0.5\pi)$, the vehicle moved to left to approach the selected sub-target $(30\ cm, 90\ cm, \pi)$, and then reversed and arrived at final target quickly. But for $(45\ cm, 0\ cm, 0.5\pi)$, because there is no enough space for the vehicle to reverse and park, membership value of $(30\ cm, 90\ cm, \pi)$ is decreased as it is close to the left wall. It is not the optimal one anymore. So the vehicle selected new optimum sub-target element $(30\ cm, 90\ cm, \pi)$ by fuzzy decision-making and achieved final target $(45\ cm, 0\ cm, 0.5\pi)$ smoothly too.
From these results, it is confirmed that the vehicle controlled by this method can avoid the obstacles flexibly, and select the path with the lowest cost to achieve the task. And it can

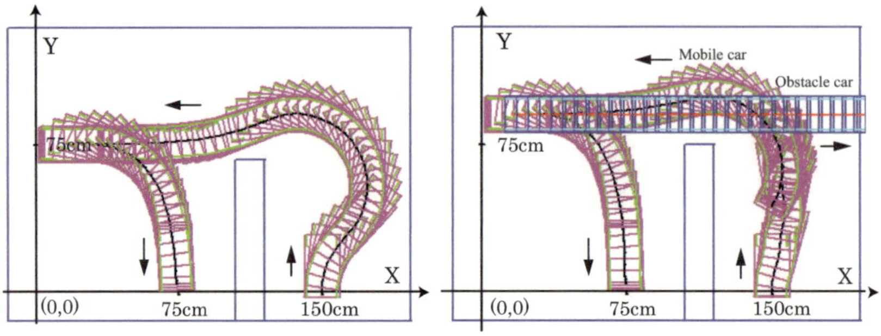


(a) With static obstacle (b) With static and dynamic obstacles

Fig. 19. Simulation trajectories with static and dynamic obstacles

respond flexibly to different placement of obstacles, arbitrary setting of initial position and final target.

Though we mainly constructed parking experiments, it can be expanded to autonomous navigation easily because parking problem is in nature a highly autonomous navigation to achieve a special target or task.

A result example with static and dynamic obstacles is shown as Fig. 19. The initial position and final target are set as $(150\ cm, 0\ cm, 0.5\pi)$ and $(75\ cm, 0\ cm, 0.5\pi)$ respectively. The static obstacle is set as placed at $105\ cm \leq x \leq 120\ cm$ and $0\ cm \leq y \leq 75\ cm$. The dynamic obstacle is assumed as a mobile car with the same size of the controlled vehicle moving at a speed of $10\ cm/s$ from left to right with steering angle 0. (a) is the moving trajectories with only static obstacle, the vehicle moves from initial position by choosing soft target elements $(165\ cm, 30\ cm, 0.25\pi)$, $(150\ cm, 75\ cm, 0.75\pi)$, $(30\ cm, 60\ cm, \pi)$, $(30\ cm, 75\ cm, \pi)$, $(60\ cm, 60\ cm, 0.75\pi)$, and $(75\ cm, 0\ cm, 0.5\pi)$ in turn, and achieved final target smoothly. But when there are static and dynamic obstacles as (b), the vehicle tries to move like (a) firstly by selecting sub-targets $(165\ cm, 30\ cm, 0.25\pi)$, $(150\ cm, 75\ cm, 0.75\pi)$, then it detected the moving obstacle car and had to reverse to guarantee safety by selecting optimun target element $(165\ cm, 45\ cm, 0.25\pi)$. After it detected that the near range is safe (obstacle car have passed over), it moves to the final target again and achieved it successfully.

### 3.6 Experiment

In order to confirm the validity of proposed method further, we constructed the experiment system as denoted in Fig. 20. It is mainly composed of a control PC, a CCD camera, and a reconstructed vehicle from a RC car. The image of ceiling camera (KOCOM CCD camera) is transferred to PC by a USB capture board to calculate the vehicle's posture (position and orientation) and acquire obstacles' information with detection and tracking algorithm. Infrared sensors signals, tachogenerator signals and steering servo position are transmitted to PC by Bluetooth module. After integrating all acquired information, PC explores all available soft target elements for current state, and makes fuzzy decision to decide the next optimal via-target, then calculate the steering and speed instructions, and then sends them to controlled vehicle by Bluetooth module.

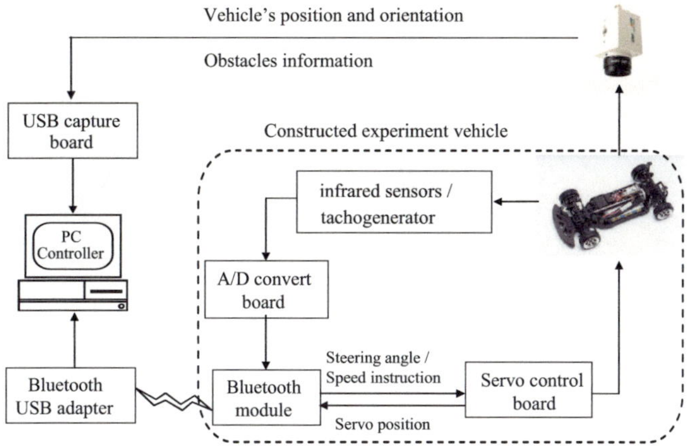


Fig. 20. Vehicle experiment system configuration

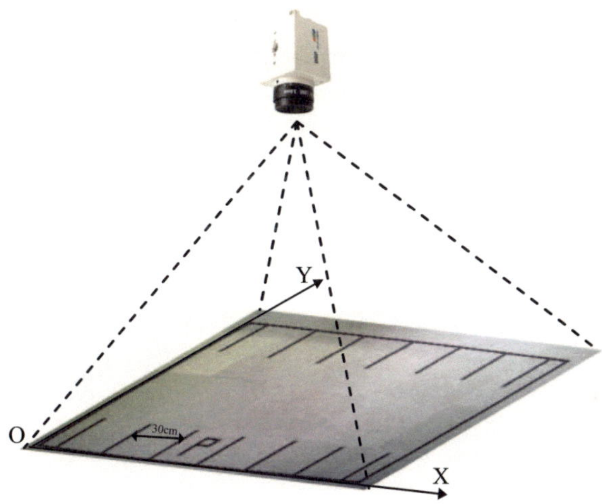


Fig. 21. Relationship of ceiling camera and world coordinate

### 3.6.1 Camera Sensing

In the experiment, a CCD camera hung on ceiling is used to capture the coordinate and orientation of controlled vehicle and all obstacles information in the space. As denoted in Fig. 21, the motion space is on the ground of ICS laboratory. It is marked by some black adhesive tapes with the same size 180 $cm \times 120$ $cm$ in simulation. Thus there are 12 units with the width 30 $cm$. The unit with a letter "P" means where final target lies in. And like the simulation, a 15 $cm$ margin distance is kept surrounding the edges of the space.

#### Camera Calibration

Camera calibration is a necessary step in 3D computer vision in order to extract metric information from 2D images. It is a process to determine the camera's parameters. Camera

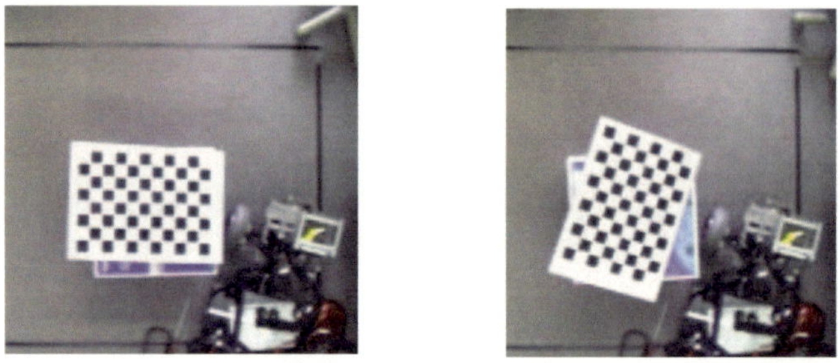

(a) Original sample images

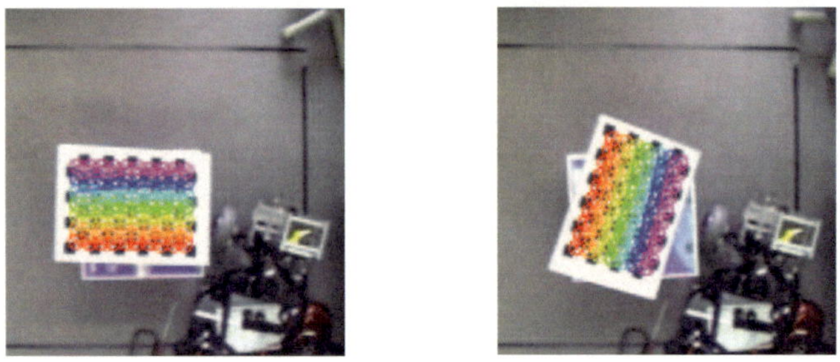

(b) Corners detection results

Fig. 22. Corners' detection results of some sample images

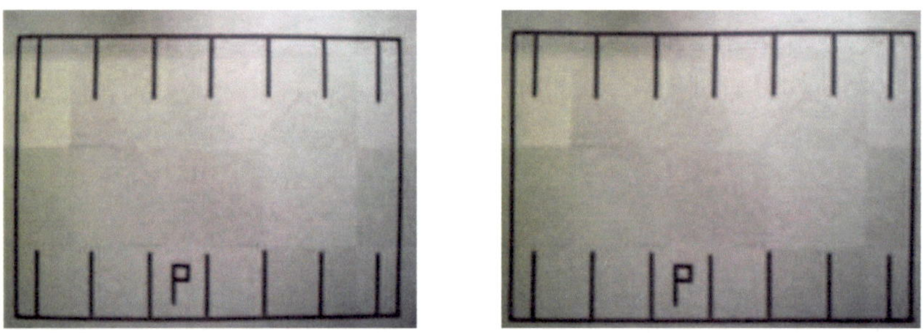


(a) Before calibration (b) After calibration

Fig. 23. Camera calibration

parameters are divided into intrinsic parameters and extrinsic parameters. Intrinsic parameters are used to specify the camera's characteristics. These parameters are: (1) focal length, i.e., the distance between the camera lens and the image plane; (2) the location of the image center in pixel coordinates; (3) the radial distortion coefficients of the lens. Extrinsic parameters are used to describe the relative spatial position and orientation between the camera and the world. These are the rotation matrix and translation vectors which determine the transformation between the camera and world reference frames (Wang & Sugisaka, 2003).
Here photogrammetric calibration method is used, which is performed by observing a calibration object whose geometry in 3-D space is known with very good precision. 25 pictures with different placements of a $7 \times 10$ black and white chess board are used as calibration sample images. Some examples of chess board's corners detection image are shown in Fig. 22. The

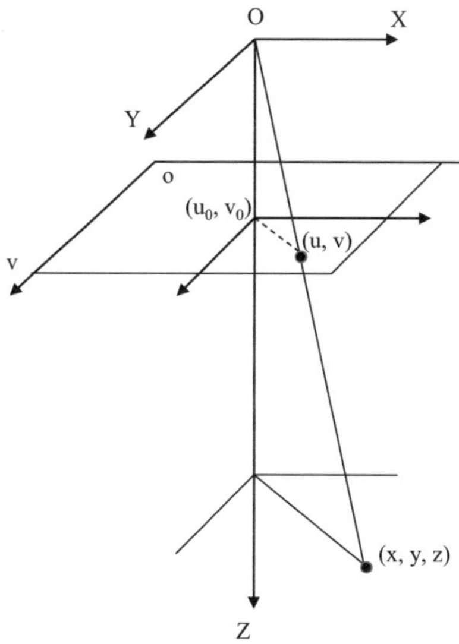


Fig. 24. Relationship of image coordinate and world coordinate

acquired intrinsic parameters matrix is

$$\begin{bmatrix} f_x & 0 & u_0 \\ 0 & f_y & v_0 \\ 0 & 0 & 1 \end{bmatrix} = \begin{bmatrix} 427.96 & 0 & 123.75 \\ 0 & 426.74 & 109.35 \\ 0 & 0 & 1 \end{bmatrix} \tag{11}$$

Rotation matrix $R = [2.61, -0.72, -0.14]$. Translation matrix $T = [710.91, 249.38, 2.19e+3]$. Lens distortion matrix is $[-0.23, 0.32, -2.94e-3, 5.86e-4]$. In the process of camera calibration, a distortion correction is also carried out. This is because straight lines in the world coordinate system are not straight in the image after perspective mapping. This would create errors when locating a point from the 2D image to 3D space. In the mapping process to correct any distortion, the distorted image taken directly with a CCD video camera is transformed into an undistorted image based on the radial distortion coefficient of the lens in lens distortion matrix. A comparison result of before calibration and after calibration with the acquired camera parameters are showed in Fig. 23. It is noted that the image distortions are greatly rectified after calibration, which guarantees the high precise of detection and tracking in following section.

Based on the acquired camera parameters, transform equation between world coordinate and image coordinate can be achieved from acquired translation rotation matrix, camera intrinsic matrix and lens distortion matrix. As shown in Fig. 24, the axis of the camera is thought of as being perpendicular, so the angle between the optical axes is not considered. Thus, the transform equation between world coordinate $(x, y, z)$ and image coordinate $(u, v)$ can be expressed as

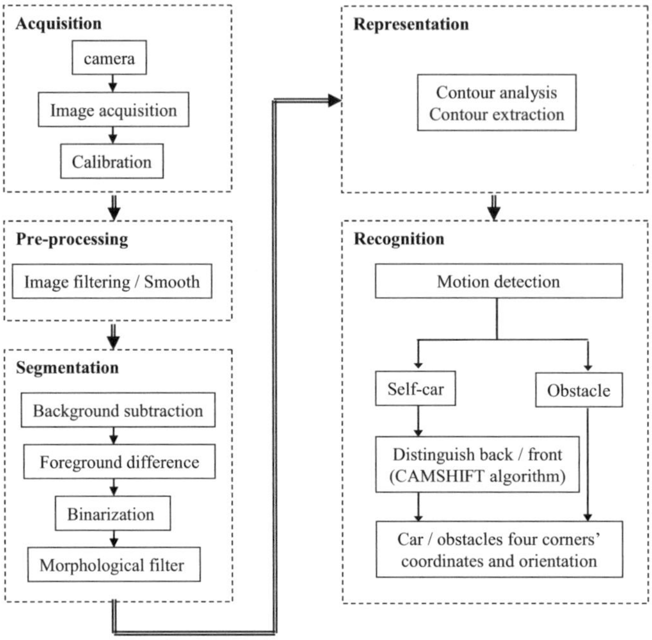


Fig. 25. Image processing flow chart

$$s\begin{bmatrix} u \\ v \\ 1 \end{bmatrix} = \begin{bmatrix} f_x & 0 & u_0 & 0 \\ 0 & f_y & v_0 & 0 \\ 0 & 0 & 1 & 0 \end{bmatrix} \begin{bmatrix} R & T \\ 0^T & 1 \end{bmatrix} \begin{bmatrix} x \\ y \\ z \\ 1 \end{bmatrix} \tag{12}$$

where, s is a scalar factor. $(f_x, f_y)$ is focal length of the camera. $R$ is rotation orthogonal matrix. $T$ is translation matrix. $(u_0, v_0)$ is image center coordinate.

### Detection and Tracking

To capture vehicle posture and all obstacles information, the camera is necessary to have the abilities to detect static or moving objects in the selected area, distinguish vehicle itself and static or dynamic obstacles, and distinguish back-front of the vehicle. The flowchart of detection and tracking algorithm is shown in Fig. 25. It is composed of five fundamental operations — acquisition, pre-processing, segmentation, representation, and recognition (Gini & Marchi, 2002). Acquisition includes capturing the original image (320 × 240 pixels) of problem domain and calibrating it to rectify distortions and acquire the intrinsic and extrinsic parameters matrix. Pre-processing is used for image filtering and smooth processing. Segmentation covers background subtraction, foreground difference, binarization, and morphological filtering. Representation is for contour extraction — identifying objects shaped contours from edge data. Recognition includes objects detection (either static or dynamic), distinguishing vehicle itself and static or dynamic obstacles, and distinguishing back-front of the vehicle. Opencv[1]

[1] Refer to http://opencv.jp/.

Fig. 26. Snapshot of detection and tracking

(a) Rebuilded vehicle internal structure (b) Vehicle's back-front differentiation mark

Fig. 27. Rebuilded bluetooth vehicle for experiment

(a computer vision library originally developed by Intel.) is used as a base to construct the camera detection and tracking algorithm. And a precision with coordinate error less than $1cm$ and orientation angle error less than $1^\circ$ was acquired. Fig. 26 displays a sample snapshot of detection and tracking result, which indicates both static and moving objects in the area are detected and tracked accurately.

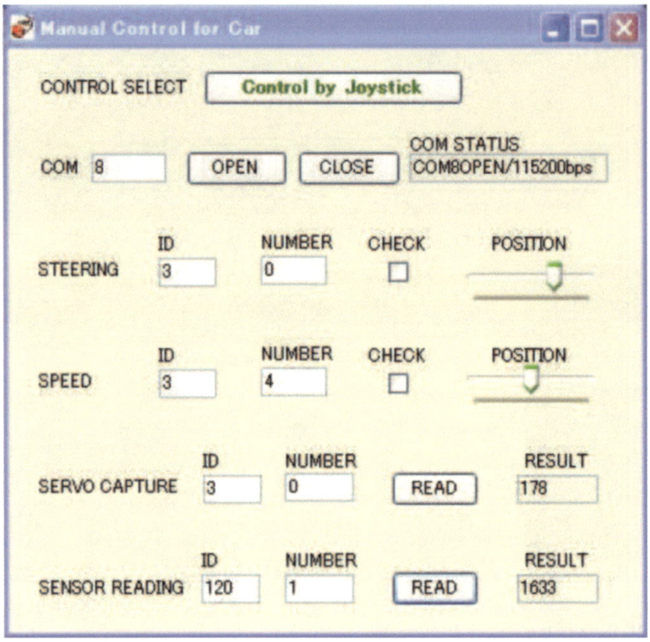


Fig. 28. Manual control interface

### 3.6.2 Constructed Experiment Vehicle

The experimental vehicle model (Fig. 27) is constructed by rebuilding a RC car (1/10 Calsonic Skyline GT-R Gr.A). It is mainly composed of a Bluetooth module, a servo control board and a A/D convert board. The original RC steering servo motor is replaced by KONDO's KRS-788HV servo motor with position capture function. The drive motor is replaced by a ESCAP motor-tachogenerator unit with a high gear ratio to guarantee the vehicle's speed unable to accelerate too high. Four infrared sensors (GP2D12) are equipped around vehicle's body to detect obstacles as a complementarity for camera sensing. A magenta mark is used to judge the detected object is the vehicle itself or obstacle, and distinguish back-front of the vehicle by CAMSHIFT (Continuously Adaptive Mean Shift) algorithm which is a simple color tracking algorithm widely used in motion tracking.

### 3.6.3 Manual Control Interface

To learn the vehicle control knowledge, including steering operation knowledge and speed control knowledge, a manual control interface (Fig. 28) is constructed. By this interface, the vehicle can be controlled by a joystick for steering, accelerating, braking, moving forward and backward, and the infrared sensors' value and steering servo motor position can be observed real-timely.

### 3.6.4 Experiment Results

As the same with simulations, we conducted experiments including same initial position and final target but different placement of obstacles, same initial position but different final target, and same final target but different initial position. A PC with the specifications 2.8 *GHz*

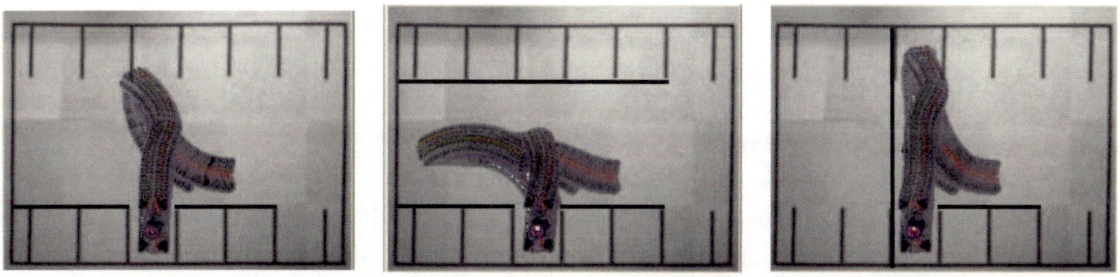

(a) Obstacles in parking side (b) Obstacles in parking and oppsite side (c) Obstacles in parking and left side

Fig. 29. Parking experiment trajectories with different placement of obstacles

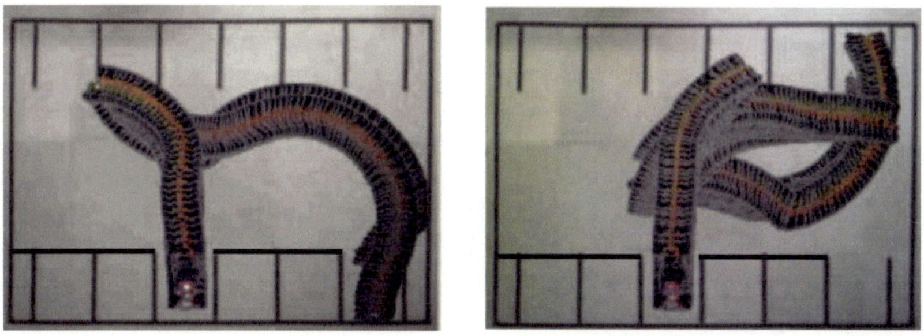

(a) Initial position (165 cm, 0 cm, $\pi/2$) (b) Initial position (165 cm, 90 cm, $\pi/2$)

Fig. 30. Experiment trajectories with different initial position

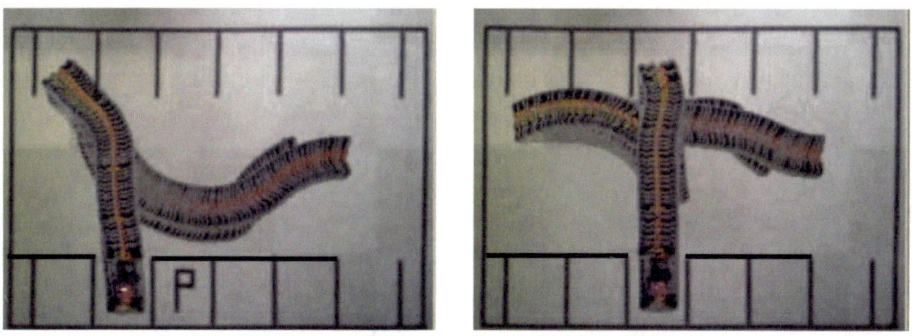


(a) Final target (45 cm, 0 cm, $\pi/2$) (b) Final target (75 cm, 90 cm, $\pi/2$)

Fig. 31. Experiment trajectories with different final target

Celeron(R) CPU, 512 *M* memory is used for simulation as well as experiment. The sampling time is set as 0.1 *s*. The first time possible targets exploring for the whole space consumes about 100 *ms*. And the possible targets exploring for the near domain of current position expends less than 20 *ms*. Experiment test indicates the PC specifications is enough for current calculation amount including image processing.

The experiment results are shown in Figs. 29, 30, and 31. They are about the same with simulation results. For same initial position and final target but different placement of obstacles,

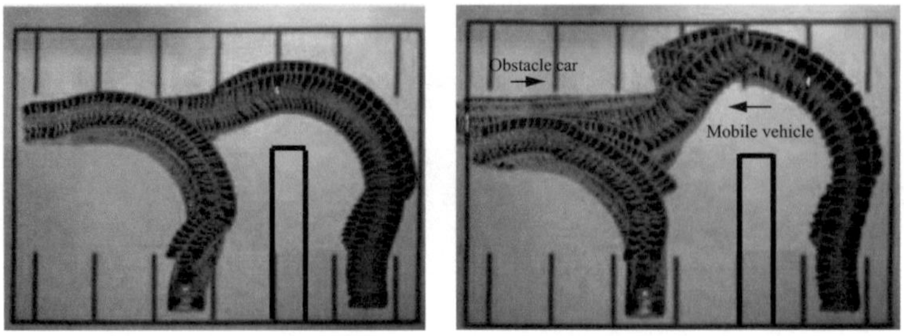


(a) With static obstacle (b) With static and dynamic obstacles

Fig. 32. Experiment trajectories with static and dynamic obstacles

the elapsed times until finally reaching are 11.468 $s$, 17.515 $s$, and 15.219 $s$ respectively. The results indicate the controlled vehicle achieved task without restriction of obstacles' placement. For same initial position but different final target, the times of reaching the final target are 23.909 $s$ and 33.502 $s$ respectively. The controlled vehicle with flexible ability to reach any final target is demonstrated. For same final target but different initial position, the elapsed times until reaching final target are 18.314 $s$ and 22.593 $s$ respectively. It is clear that the vehicle can reached final target successfully from arbitrary initial position.

And from the results, we will find, in Fig. 29 (b) and Fig. 31 (b), the vehicle achieved the final target by selecting one more sub-target $(75\ cm, 45\ cm, 0.5\pi)$ and $(75\ cm, 90\ cm, 0.5\pi)$ than the simulations respectively. And in Fig. 30, the vehicle even had to choose two more sub-targets to achieve the final target. It is considered that the camera sensing error and steering operation error which resulted in this tiny difference, but it didn't affect the achievement of the final target successfully. And on the other hand, it exactly demonstrated the flexibility of the proposed method more powerfully because of its ability to tolerate these errors and respond to uncertainties.

Fig. 32 denotes the running trajectories with static and dynamic obstacles. Here, the moving obstacle car is controlled by person with a remote controller. As it is powered by a line source rather than battery, the power line on the ground is easy to become a barrier for mobile vehicle. Thus the obstacle car is just controlled to move from left to center, and then reverse to return it start point. The mobile vehicle achieved the final target by few times reverses. The moving time until achieved final target is 27.249 $s$ for with static obstacles and 49.686 $s$ for with both static and moving obstacles. The results verify the vehicle can respond to dynamic environment flexibly.

These results verified the validity and flexibility of the proposed soft-target-based PFC fuzzy decision-making method for motion control of car-like mobile vehicle in changing situations more effectively.

### 3.6.5 Discussions

If there is no change in the environment, it is simple for vehicle making decision just to select the one with the highest membership value as control target until it reached the final target because the membership values will not change too. While in changing situation (with dynamic obstacle, change obstacle placement or add new static obstacle), the dynamic constraints are very possible to make it difficult to achieve the target (either sub-targets or final target) having the highest membership value because its evaluation value may decrease as the influence of constraints. It means the one with the highest membership value is unlikely the best one. Thus it is necessary to make decision to confirm which one is the best one in the alternative target set. We use PFC fuzzy decision-making method to simulate human's decision process to make dynamic fuzzy decision like human. Its advantages include:

- It is a multi-target-based rather than the conventional single-target-based method,
- Flexibility in responding to different optimization goals,
- Flexibility in responding to uncertainties,
- It simulated human's decision process completely and realized responding to changing situation flexibly,
- Vehicle is always supposed to choose the best one in the target set as control target, thus guaranteed the motion is collision-free and low cost,
- The soft-target-based PFC fuzzy decision-making control idea provides a new method for mobile body to realize dynamic environment or constraints self-adaptation.

### 3.7 Summary

Soft-target-based PFC fuzzy decision-making control for motion control of mobile vehicle (both two-wheeled differential drive vehicle and four-wheeled car-like Ackerman steering vehicle) with nonholonomic constraints in change situations is developed. From the simulation and experiment results (including same initial position and final target but different placement of obstacles, same initial position but different final target, same final target but different initial position, and with static and dynamic obstacles), it is learned that it is possible to achieve any final target and avoid obstacles (either static or dynamic) in the space flexibly without restriction of obstacle's placement and shape. A collision-free and low cost motion control of mobile vehicle with the ability of dynamic environment self-adaptation was achieved. A new method emulating the decision process of human for mobile body in dynamic environment was explored.

## 4. Conclusions

A soft-target-based fuzzy decision-making control for motion control of mobile vehicle with dynamic constraints was proposed. It emulates human's action decision process composed of prediction, evaluation and optimal decision-making based on multi-targets. The soft target defined as a fuzzy target set of all possible target candidates is learned based on final target, current state and constraints or environment information in real-time. For acquired soft target, fuzzy decision is made by predictive fuzzy control to select the best one as control target corresponding to the current state and constraints.

The proposed method was applied to motion control of a car-like mobile vehicle in a dynamic environment. The effectiveness and flexibility of this method was demonstrated by the simulation and experiment results. A collision-free and low cost motion control with the ability of dynamic environment self-adaptation was achieved. Parking in change situations (with different placement of static obstacles) and autonomous run with static and dynamic obstacle are realized. It is possible to achieve any final target from arbitrary initial position in a fixed narrow space without restriction of obstacle's placement and shape.

This soft-target-based fuzzy decision-making idea provides a new method for mobile body motion control in changing situation by simulating human's multi-target-based dynamic decision process.

## 5. References


Barraquand, J., Langlois, B. & Latombe, J. C. (1992). Numerical potential field techniques for robot path planning, *IEEE Trans. Syst. Man Cybernet* **Vol. 22**(No. 2): 224–241.

Bellman, R. E. & Zadeh, L. A. (1970). Decision-making in fuzzy environment, *Management Science* **Vol. 17**(No. 4): B–141–164.

Chen, Y. G. & Yasunobu, S. (2006). Soft target based intelligent controller for a system with dynamic restriction, *Joint 3rd International Conference on Soft Computing and Intelligent Systems and 7th International Symposium on advanced Intelligent Systems*, Tokyo, pp. 1949–1954.

Chen, Y. G. & Yasunobu, S. (2007a). Soft target based obstacle avoidance for car-like mobile robot in dynamic environment, *IEEE International Conference on Fuzzy Systems*, London, pp. 1351–1356.

Chen, Y. G. & Yasunobu, S. (2007b). Soft-target-based predictive fuzzy control for a cart-pendulum system with dynamic constraints, *Journal of Advanced Computational Intelligence and Intelligent Informatics* **Vol. 11**(No. 8): 931–936.

Chen, Y. G. & Yasunobu, S. (2008). Fuzzy target based soft decision for mobile vehicle in dynamic environment, *7th World Congress on Intelligent Control and Automation*, Chongqing, pp. 800–805.

Chen, Y. G. & Yasunobu, S. (2009). Soft decision with soft target for car-like mobile vehicle in dynamic environment, *IEEJ Transactions on Electrical and Electronic Engineering* **Vol. 4**(No. 4): 561 – 569.

Gini, G. & Marchi, A. (2002). Indoor robot navigation with single camera vision, *Proc. of Pattern Recognition in Information Systems*, PRIS, Alicante, Spain, pp. 67–76.

Huang, Y. & Yasunobu, S. (1999). Cascade fuzzy control using to large scale system, *Trans. IEE of Japan* **Vol. 119-C**(No. 12): 1548–1553.

Jing, X. J. (2005). Behavior dynamics based motion planning of mobile robots in uncertain dynamic environments, *Robotics and Autonomous Systems* **Vol. 53**(No. 23): 99–123.

Kim, K. H. & Cho, H. S. (2006). An obstacle avoidance method for mobile robots based on fuzzy decision-making, *Robotica* **Vol. 24**(No. 5): 567–578.

Ordonez, R. & Zumberge, J. (1997). Adapitive fuzzy control: Experiment and comparative analyses, *IEEE Transactions on Fuzzy Systems* **Vol. 5**(No. 2): 167–188.

Wang, J. & Sugisaka, M. (2003). Camera calibration for a mobile robot prototype, *Artificial Life and Robotics* **Vol. 7**: 91–94.

# 21

# Vision-Based Path Following Without Calibration[1]

Zhichao Chen and Stanley T. Birchfield,
*Clemson University*
*United States of America*

## 1. Introduction

Route-based knowledge, in which the spatial layout of an environment is recorded from the perspective of a ground-level observer, is an important component of human and animal navigation systems (Shelton & Gabrieli, 2002). In this representation, navigating from one location to another involves comparing current visual inputs with a sequence of views captured along the path in a previous instance. Applications that would benefit from such a path-following capability include courier and delivery robots (Burgard et al., 1999), robotic tour guides (Shen & Hu, 2006), or reconnaissance robots following a scout (Crawford et al., 2004). Furthermore, a solution to this problem would be useful for the general problem of navigating between two arbitrary locations in an environment by following a sequence of such paths.

One approach to path following is visual servoing, in which the robot is controlled to align the current image with a reference image, both taken by an onboard camera (Hutchinson et al., 1996). Such an approach generally employs a Jacobian to relate the coordinates of world points to their projected image coordinates (Burschka and Hager, 2001), a homography or fundamental matrix to relate the coordinates between images (Sagüés & Guerrero, 2005, Remazeilles & Chaumette, 2007, Liang & Pears, 2002, Šegvić et al., 2007), or bundle adjustment to minimize the reprojection error over multiple image frames (Royer et al., 2007). As a result, the camera usually must be calibrated (Burschka and G. Hager, 2001, Remazeilles & Chaumette, 2007, Royer et al., 2007, Šegvić et al., 2007), and even uncalibrated systems require lens distortion to be removed. Alternative vision-based algorithms make strong assumptions about the environment or the sensor, such as a flat ground plane (Burschka & Hager, 2001, Liang & Pears, 2002, Guerrero & Sagüés 2001, 2005), a man-made environment in which vertical straight lines are present (Kosaka & Kak, 1992, Sagüés & Guerrero, 2001, 2005, Tang & Yuta 2001), or an omnidirectional camera (Gaussier, 1997, Ulrich & Nourbakhsh, 2002, Tang & Yuta 2001, Kröse et al., 2001).

---

[1] Based on "Qualitative Vision-Based Path Following", by Z. Chen and S. Birchfield, which appeared in *IEEE Transactions on Robotics*, 25(3):749-754, June 2009. 

To overcome these limitations, we consider the problem from a novel viewpoint in which there is no equation relating image coordinates to world coordinates. Such a direct approach is motivated by the observation that the problem is vastly overdetermined, with tens of thousands of image pixels available to determine a single turning command output. We present a simple algorithm that uses a single, off-the-shelf camera attached to the front of the robot. The technique follows the teach-replay approach (Burschka & Hager, 2001) in which the robot is manually led through the path once during a teaching phase and then follows the path autonomously during the replay phase. Without any camera calibration (even calibration for lens distortion), the robot is able to follow the path by making only qualitative comparisons between the feature coordinates in the two phases. All that is needed is a single controller gain parameter to convert pixel coordinates to turning angles. We demonstrate the technique on several indoor and outdoor experiments, showing its robustness with respect to slanted surfaces, changing lighting conditions, and dynamic occluding objects. This paper extends the applicability and improves upon the robustness of our earlier work (Chen & Birchfield, 2006) by incorporating odometry information and correcting for camera roll. We also demonstrate the ability of the technique to work with wide-angle and omni-directional cameras, with only slight modification in the latter case to ignore the bottom half of the image which views the scene behind the robot.

The proposed approach falls within the category of mapless algorithms (DeSouza & Kak, 2002). As such, it is closely related to the view-sequenced route representation (VSRR) of Matsumoto et al. (Matsumoto et al., 1996, 2000, Jones et al., 1997) in which the turning angle is computed by cross-correlating images acquired during the replay phase with those captured during training. However, VSRR requires large amounts of memory to store the views and is sensitive to occlusions by dynamic objects. Along with a homography-based extension using vertical lines (Sagüés & Guerrero, 2005), it has only been demonstrated for short sequences on flat terrains. An alternate mapless approach is to learn the mapping from images to turning commands based on their classification (Weng & Chen, 1996, Ackerman & Itti, 2005). While this method can successfully follow a specific pattern such as a road or hallway, it will have difficulty generalizing to environments in which the images cannot be categorized into a small number of classes known at training time. Another approach that has received considerable attention (Gaussier, 1997, Wolf et al., 2002, Zhou et al., 2003, Sim & Dudek, 2001, Košeckā, 2003, Ulrich & Nourbakhsh, 2002, Horswill, 1993) is to store an example image with each specific location of interest. At run time, the image database is searched to find the image that most closely resembles the current one (or, alternatively, the current image is projected onto a manifold learned from the database (Nayar, 1994, Kröse et al., 2001) ). Such approaches require extensive training and have difficulty providing sufficient spatial resolution to determine actual turning commands in large environments. Similarly, sensory-motor learning has been used to map visual inputs to turning commands, but the resulting algorithms have been too computationally demanding for real-time performance (Giovannangeli et al., 2006). Other researchers have developed mapless algorithms for low-level functionality like corridor following or obstacle avoidance (Nelson & Aloimonos, 1988, Santos-Victor et al., 1995, Barrows et al., 2002, LeCun et al., 2005, Michels et al., 2005, Murali & Birchfield, 2008), but these techniques are not applicable to following a specific arbitrary path.

## 2. Qualitative mapping from feature coordinates to turning direction

Consider a mobile robot equipped with a camera whose optical axis is parallel to the heading direction of the robot. Suppose we wish to move the robot from location $C=(x_C, y_C, \theta_C)$ to a previously encountered location $D=(x_D, y_D, \theta_D)$, where $(x_i, y_i)$ and $\theta_i$ are the position and orientation, respectively, in the $xy$ plane, $i \in \{C, D\}$. The robot has access to a current image $I_C$, taken at *C*, and a destination image $I_D$, taken previously at the destination *D*.

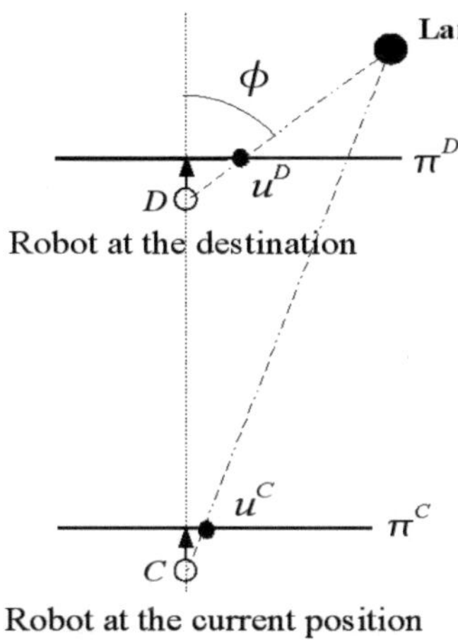


Fig. 1. The robot is at ***C*** moving toward the destination ***D*** with the same heading direction. The open circle coincides with both the camera focal point and the robot position, the arrow indicates the heading direction, $\pi$ is the image plane, and $\phi$ is the angle between the optical axis and the projection ray from the landmark.

We start with a simple observation. Suppose the robot views a fixed landmark in both images yielding image feature coordinates of $u^C$ and $u^D$, as shown in Fig. 1. The features are computed with respect to a coordinate system centered at the principal point (the intersection of the optical axis and the image plane), so that positive coordinates are on the right side of the image while negative coordinates are on the left side. If the robot moves toward the destination in a straight line with the same heading direction as that of the destination (i.e., $\theta_C = \theta_D$), then the point $u^C$ will move away from the principal point toward $u^D$, reaching $u^D$ when the robot reaches *D*. This observation is made more precise in the following theorem.

*Theorem 1:* Let a mobile robot move in a straight line toward location *D* on a flat surface. Let $u^j$ be the horizontal image coordinate, relative to the principal point, of a monotonic projection at location *j* of a fixed landmark. For any location *C* along the line such that $\theta_C = \theta_D$, $|u^C| < |u^D|$ and $\operatorname{sign}(u^C) = \operatorname{sign}(u^D)$.

The theorem can be easily proved by geometry. Note that the image projection function is only required to be monotonic (i.e., perspective projection is not necessary), so the result applies equally to a camera with radial lens distortion. The primary assumption is that the optical axis of the camera passes through the axis of rotation of the robot. Other assumptions include the alignment of the optical axis with the robot heading direction, zero roll and tilt angles of the camera with respect to the robot, and a flat ground plane. In practice, misalignment is not an issue because the camera alignment can be learned automatically by estimating the focus of expansion as the robot drives forward. Similarly, rough terrain is easily handled by measuring image rotation to compensate for a non-zero roll angle of the robot and by recognizing that a non-zero tilt angle has a negligible effect on the horizontal feature coordinates.

### 2.1 The funnel lane

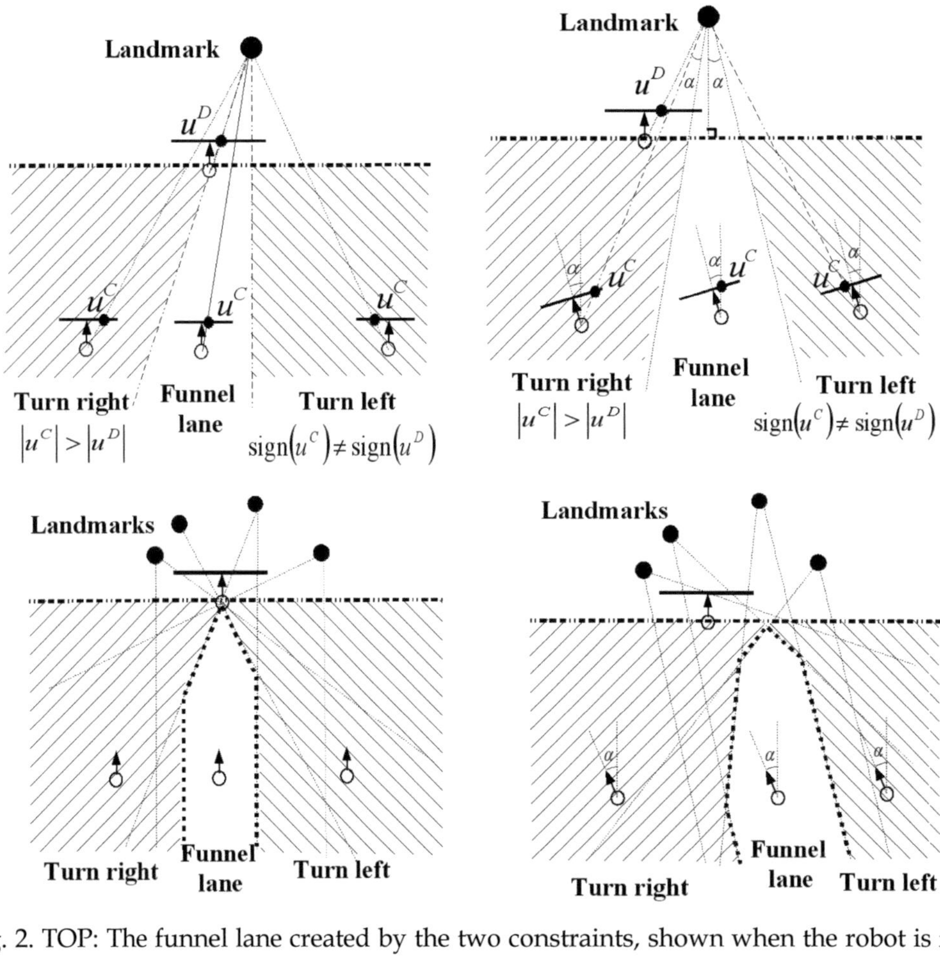


Fig. 2. TOP: The funnel lane created by the two constraints, shown when the robot is facing the correct direction (left) and when it has turned by an angle α (right). BOTTOM: The combined funnel lane created by multiple feature points, shown when the robot is facing the correct direction (left) and when it has turned by an angle α (right).

According to the preceding theorem, if the robot is on the path toward the destination with the same heading direction, then two constraints are satisfied. Conversely, as shown in Fig. 2, if the constraints are satisfied then the robot lies within a trapezoidal region (assuming perspective projection) for any given relative robot angle $\alpha = \theta_C - \theta_D$. For $\alpha = 0$, the sides of the trapezoid are defined by two lines passing through the landmark, one through $D$ and another that is parallel to the destination direction. These lines are rotated about the landmark by α if the relative angle is nonzero. We call the trapezoidal region the *funnel lane* associated with the landmark, destination, and relative angle. The terminology arises from the analogy of pouring liquid into a funnel: The liquid moves in a straight line until it hits the sides of the funnel, which cause it to bounce back and forth until it eventually reaches the spout. In a similar manner, the sides of the trapezoid act as bumpers, guiding the robot toward the goal. The notion of the funnel and the funnel lane are captured in the following definitions.

*Definition 1:* The *funnel* of a fixed landmark $\lambda$ and a robot location $D$ is the set of locations $\mathbb{F}_{\lambda,D}$ such that, for each $C \in \mathbb{F}_{\lambda,D}$, the two funnel constraints are satisfied:

$$|u^C| < |u^D| \qquad \text{(Constraint 1)}$$

$$\operatorname{sign}(u^C) = \operatorname{sign}(u^D) \qquad \text{(Constraint 2)}$$

where $u^C$ and $u^D$ are the coordinates of the image projection of $\lambda$ at the locations $C$ and $D$, respectively.

*Definition 2:* The *funnel lane* of a fixed landmark $\lambda$, a robot location $D$, and a relative angle α is the set of locations $\mathbb{F}_{\lambda,D,\alpha} \subset \mathbb{F}_{\lambda,D}$ such that $\theta_C - \theta_D = \alpha$ for each $C \in \mathbb{F}_{\lambda,D,\alpha}$.

Multiple features yield multiple funnel lanes, the intersection of which is the set of locations for which both constraints are satisfied for all the features. This intersection, which we call the *combined funnel lane,* is depicted in Fig. 2. Notice the importance of having features on both sides of the image in order to narrowly constrain the path of the robot, thus achieving more robust and accurate results. Features can be at any depth, and there need not be any relationship between the depths of the various features as long as they remain visible.

## 2.2 Qualitative control algorithm

The funnel constraints lead to a simple control algorithm, illustrated in Fig. 3. The robot continually moves forward, turning to the right whenever Constraint 1 is violated and to the left whenever Constraint 2 is violated, given a feature on the right side of the image ($u^D > 0$). If the feature is on the left side ($u^D < 0$), then the directions are reversed.

For each feature $i$, a desired heading is obtained by

$$\theta_d^{(i)} = \begin{cases} \gamma \min\left\{u^C, f\left(u^C, u^D\right)\right\} & \text{if } u^C > 0 \text{ and } u^C > u^D \\ \gamma \max\left\{u^C, f\left(u^C, u^D\right)\right\} & \text{if } u^C < 0 \text{ and } u^C < u^D \\ 0 & \text{otherwise} \end{cases}$$

where $f\left(u^C, u^D\right) = \frac{1}{\sqrt{2}}\left(u^C - u^D\right)$ is the signed distance to the line $u^C = u^D$. Here we approximate the conversion of pixels to radians with a constant gain $\gamma$.

At any given time, the desired heading of the robot is given by

$$\theta_d = \eta \frac{1}{N} \sum_{i=1}^{N} \theta_d^{(i)} + (1-\eta)\theta_0 \tag{1}$$

where $N$ is the total number of feature points, $\theta_0$ is the desired heading obtained by sampling a third-order polynomial that is fit to the initial and destination odometry measurements of the segment in the teaching phase, and the factor $0 \le \eta \le 1$ determines the relative importance of visual measurements versus odometry measurements. We set $\eta$ = 0.5 in our system.

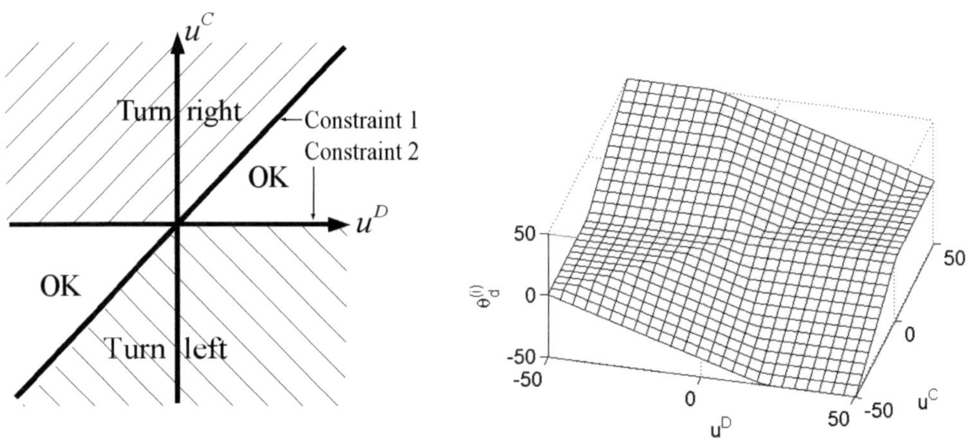


Fig. 3. Qualitative control decision space. The horizontal coordinates of the feature point in the current and destination images ($u^C$ and $u^D$, respectively) are compared to determine whether to turn the robot to the right, to the left, or not at all. LEFT: Top-down view of decision space. RIGHT: 3D view of decision space, showing the desired angle $\theta_d^{(i)}$ versus $u^C$ and $u^D$.

### 2.3 Analysis of qualitative control algorithm

Fig. 4 illustrates the qualitative control algorithm with an example. In its initial position the robot is outside the funnel lane, violating Constraint 1 (Fig. 4a). The robot turns to the right, causing the funnel lane to rotate as well, and the robot moves forward a small amount until the constraint is violated again (Fig. 4b). The robot turns a second time to the right, finds itself with a much clearer opening, and moves forward until the constraint is violated (Fig. 4c). Finally, the robot turns again and moves forward until it reaches a point close to the goal (Fig. 4d).

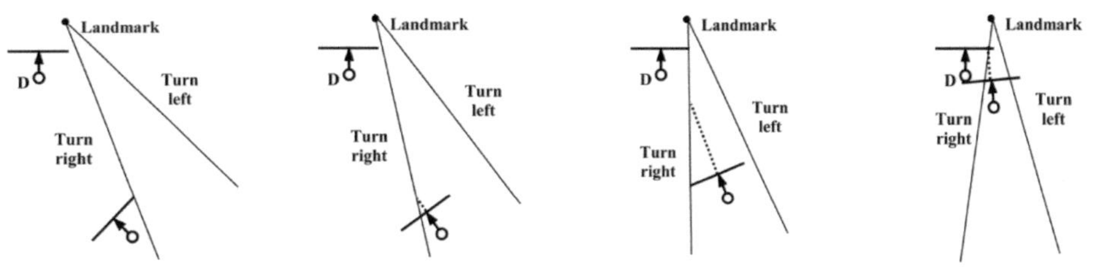


Fig. 4. Four snapshots of a robot making progress toward a destination D using the qualitative control algorithm. The two solid lines indicate the funnel lane, while the dashed line indicates the path of the robot.

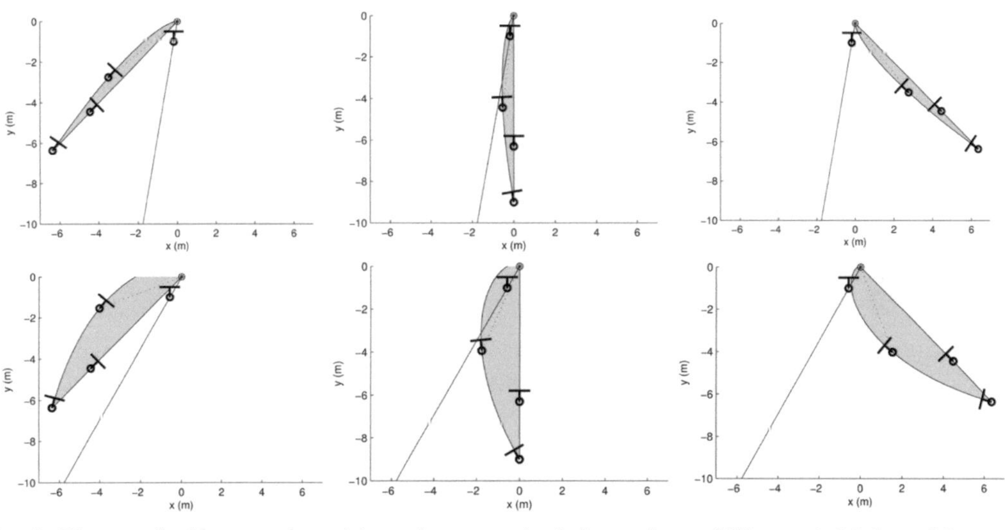


Fig. 5. The reachable set of positions (gray region) from three different initial positions (left, middle, and right) and two different values of $\phi$ (top: 10 degrees, bottom: 30 degrees). The landmark is at (0, 0), the initial position of the robot is at the bottom tip of the gray region, and the projection ray from the landmark to the destination is the angled line. Two possible robot locations along edges of the reachable set are shown, along with a possible location for the destination.

To better understand the behavior and accuracy of the algorithm, simulations were ran in Matlab, the results of which are shown in Fig. 5. A single landmark was placed at the origin, and the robot was placed at various initial positions for different values of $\phi$ (the angle of the landmark with respect to the optical axis). From any initial position the robot may turn and drive straight toward the landmark, in which case it will barely satisfy Constraint 2. Alternatively it may turn away from the landmark and drive along a curve so that Constraint 1 is always barely satisfied. In both cases the other constraint is automatically satisfied. This line and curve define a region of positions, shown as gray in the figure, that are reachable from the initial position by a non-holonomic vehicle without violating either constraint. Notice that the actual location of the destination along the projection ray is irrelevant for the plots, which depend only upon the starting location, the landmark location, and the angle $\phi$ that the light ray makes with respect to the destination optical axis. The reachable set is wider for increasing values of $\phi$.

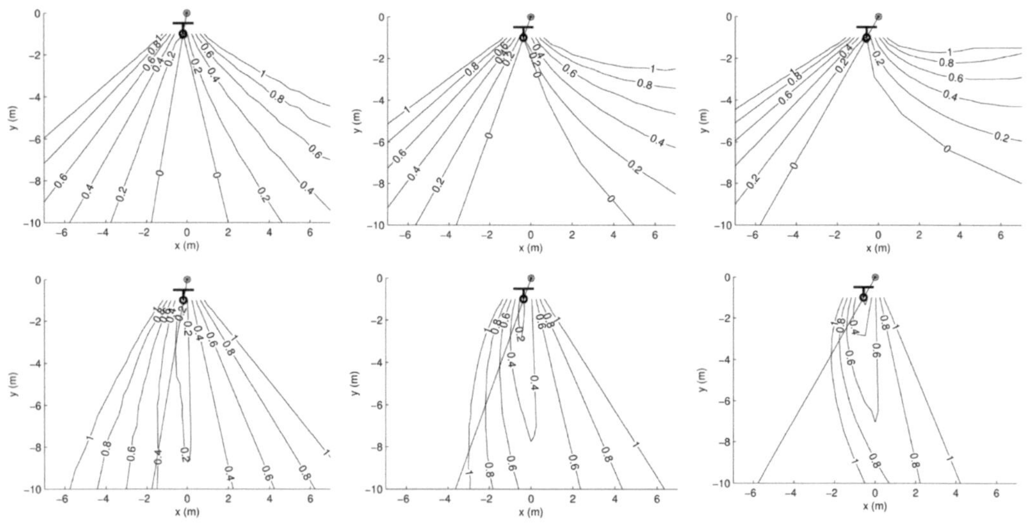

$\phi = 10$ $\phi = 20$ $\phi = 30$

Fig. 6. Contour plots of the minimum (top) and maximum (bottom) error in reaching a destination from any point in the plane. The destination is 1 m in front of the landmark, which is placed at the origin. To reduce clutter, contours with a value greater than 1.0 are not shown. The curves labeled 0 enclose the region of zero error.

Fig. 6 displays the data in a different format, showing the minimum and maximum error in reaching the goal from various initial positions. With the landmark still at the origin, we placed the destination at the intersection of the line y = −1 m with the projection ray from the landmark at a given $\phi$. The set of possible initial locations was densely sampled in order to generate contour plots of the error, as shown. The error was computed as the distance from the robot to the destination when the robot crossed the line y = −1 m. As can be seen, the probability of reaching the destination with zero error increases with larger values of $\phi$, but the probability of large errors increases as well. As long as the robot starts from a position nearly behind the destination at a reasonable distance, the minimum error is zero and the maximum error is approximately 20-50% of the distance from the destination to the landmark. Keep in mind that in a real system, the use of multiple landmarks dramatically reduces this error.

## 4. Tracking feature points

Feature points are automatically selected and tracked using the Kanade-Lucas-Tomasi (KLT) feature tracker (Birchfield, 1997), which computes the displacement $\mathbf{d} = \begin{bmatrix} d_x & d_y \end{bmatrix}^T$ that minimizes the sum of the squared differences between consecutive image frames $I$ and $J$:

$$\iint_W \left[ I\left(\mathbf{x} - \frac{\mathbf{d}}{2}\right) - \left(\alpha\, J\left(\mathbf{x} + \frac{\mathbf{d}}{2}\right) + \beta\right)\right]^2 d\mathbf{x}, \tag{2}$$

where $W$ is a window of pixels around the feature point and $\mathbf{x} = \begin{bmatrix} x & y \end{bmatrix}^T$ is a pixel in the image. This nonlinear error is minimized by repeatedly solving its linearized version by Taylor series expansion:

$$\begin{aligned} Z &= \sum_{\mathbf{x} \in W} g(\mathbf{x}) g^T(\mathbf{x}), \\ \mathbf{e} &= \sum_{\mathbf{x} \in W} g(\mathbf{x}) \left[ I(\mathbf{x}) - (\alpha\, J(\mathbf{x}) + \beta) \right], \end{aligned} \tag{3}$$

where $g(\mathbf{x}) = \frac{1}{2} \frac{\partial \left[ I(\mathbf{x}) + \alpha\, J(\mathbf{x}) \right]}{\partial \mathbf{x}}$ is the spatial gradient of the weighted average image. These equations are the standard Lucas-Kanade equations (Baker & Matthews, 2004, Shi & Tomasi, 1994, Tomasi & Kanade, 1991) with geometric symmetry between the two images and an affine model of brightness to model the dynamic lighting conditions encountered by the mobile robot, particularly when moving outdoors (Chen & Birchfield, 2006, Negahdaripour & Yu, 1993). A coarse-to-fine pyramidal strategy is used to allow large image motions. As in (Shi & Tomasi, 1994, Tomasi & Kanade, 1991), features are automatically selected as those points in the image for which both eigenvalues of Z are greater than a specified minimum threshold. This feature selection mechanism is a slight variation of the Harris corner detector which has been shown to be effective for both its repeatability rate, information content, and theoretical properties (Harris & Stephens, 1988).

## 5. Teach-and-replay navigation

The navigation system involves two phases. In the teaching phase, an operator manually moves the robot along a desired path to gather training data. The path is divided into a number of non-overlapping segments. Within each segment, feature points are automatically detected in the first image and tracked in subsequent images. When the percentage of features that have been successfully tracked falls below 50% of the original features in the segment, a new segment is declared. For each feature that is successfully tracked throughout a segment, its gray-level intensity pattern and $x$-coordinate in the first and last images of the segment are stored in a database for use in the replay phase. We also store the length of each segment and the change of heading direction of the robot in each segment by odometry, which are used in determining the desired heading and the segment transitions.

In the replay phase, the robot automatically proceeds sequentially through the segments starting from approximately the same initial location as that of the teaching phase. At the beginning of each segment, correspondence is established between feature points in the current image and those of the first teaching image of the segment. Then, as the feature points are tracked in the incoming images, their coordinates are compared with those of the *milestone image* (i.e., the last teaching image of the segment) in order to determine the turning direction for the robot. Prior to comparison, feature coordinates are warped to compensate for a non-zero roll angle about the optical axis by applying the RANSAC algorithm (Fischler & Bolles, 1981) to pairs of random features. This compensation removes the undesirable in-plane image rotation that occurs due to unpaved, rough terrain. Note that this is the only place in the algorithm where the $y$-coordinates of the features are used.

A crucial component of the technique is determining when to transition to a new segment. To solve this problem, we continually monitor the probability that the robot at time $t$ is at the end of the current segment:

$$\delta(t) = \underbrace{\exp\left\{-\frac{\varepsilon_f^2(t)}{2\sigma_f^2}\right\}}_{\text{feature}} \underbrace{\exp\left\{-\frac{\varepsilon_d^2(t)}{2\sigma_d^2}\right\}}_{\text{distance}} \underbrace{\exp\left\{-\frac{\varepsilon_h^2(t)}{2\sigma_h^2}\right\}}_{\text{heading}} \tag{4}$$

assuming that the feature, distance, and heading measurements are independent. In this equation $\varepsilon_f(t)$ is the mean squared error of the feature coordinates between the current and milestone images; $\varepsilon_d(t)$ is the difference between the distance traveled in the current segment and the corresponding segment in the teaching phase, calculated by odometry; and $\varepsilon_h(t)$ is the difference between the current heading and the heading at the end of the teaching segment. These errors are normalized by values computed automatically by the system: $\sigma_f$ is the mean squared error of the feature points at the beginning of the segment; $\sigma_d$ is the length of the segment calculated by odometry in the teaching phase; and $\sigma_h$ is the maximum variation in heading encountered during the teaching segment.

Two values are actually computed: $\delta(t)$ using the current milestone image and $\delta^-(t)$ using the previous milestone image. If $\delta(t-1)-\delta(t)>\tau$ and $\delta^-(t-1)-\delta^-(t)>\tau$, where $\tau$ = 0.05, then the system advances to using the next milestone image. The rationale is that both $\delta(t)$ and $\delta^-(t)$ increase as the robot approaches the end of the segment then decrease afterward. Therefore, when both values have decreased by a significant amount, the end has been reached. We have found that using both values yields improved results compared with using a single value. To reduce the effects of noise, both signals are first smoothed by a low-pass nonlinear filter.

## 6. Experimental results

The qualitative algorithm was implemented in Visual C++ on a Dell Inspiron 700m laptop (1.6 GHz) controlling an ActivMedia Pioneer P3-AT mobile robot with an inexpensive Logitech QuickCam Pro 4000 webcam mounted on the front. The 320×240 images were acquired at 30 Hz and processed by the KLT algorithm with the default 7 × 7 feature window size (Birchfield, 1997). In all experiments a maximum of 60 features were detected and tracked throughout each segment. On average 85% of the features survive the initial correspondence in the first image of the segment during replay.

The algorithm was tested in a number of indoor and outdoor environments. (Videos of the results can be found at http://www.ces.clemson.edu/~stb/research/mobile_robot .) Figs. 7 and 8 show two typical runs in which the robot successfully navigated between chairs and desks along a 10 m path in our laboratory, as well as along a 380 m loop trajectory in a parking lot of our university campus. The driving speed of the robot was 100 mm/s and the turning speed was 4 degrees per second during both the teaching and replay phases of the indoor experiments. Outdoors, the additional maneuvering room enabled the driving and turning speeds to be increased to 750 mm/s (the maximum driving speed of the robot) and 6 degrees per second, respectively. The error was less than 1 m for two-thirds of the sequence and remained below 3.5 m for the entire sequence.

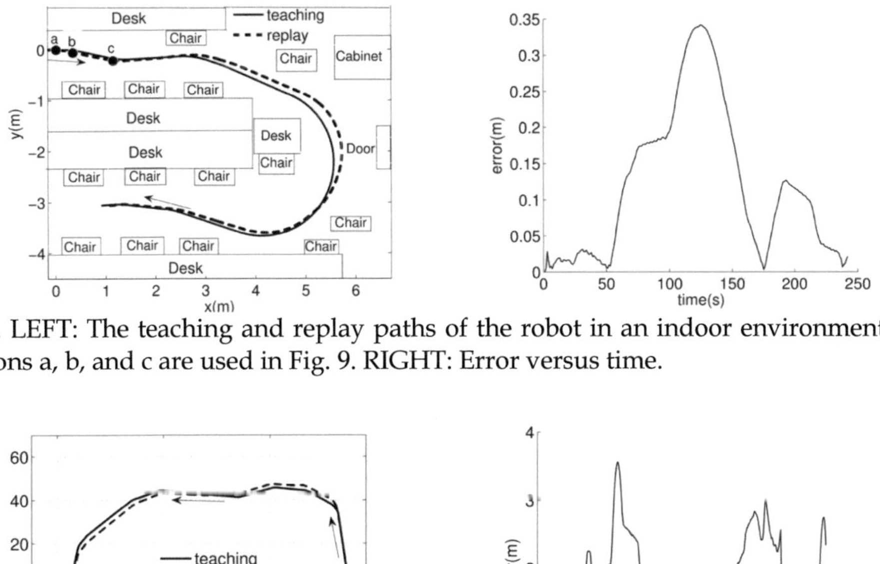


Fig. 7. LEFT: The teaching and replay paths of the robot in an indoor environment. The locations a, b, and c are used in Fig. 9. RIGHT: Error versus time.

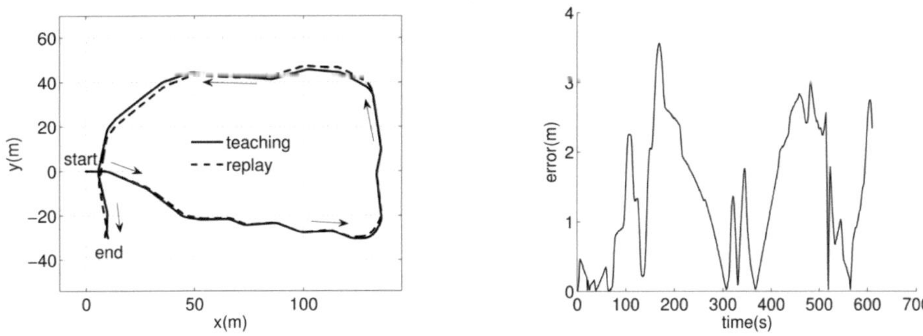


Fig. 8. LEFT: The teaching and replay paths of the robot in an outdoor environment (parking lot). RIGHT: Error versus time.

The decision process during three segments of the experiment is shown in Fig. 9. In the first segment all the features lay in the OK region, so the robot drove forward. In the remaining two segments a majority of the features told the robot to turn right and left, respectively, which caused the features to move toward the OK region. Notice the near unanimity in voting: Except for a lone feature in the second segment that votes incorrectly to turn left, all the features were in agreement in all segments. In this figure the display is simplified by assuming $\tau = 0$ and by normalizing the feature coordinates:

$$\xi = \begin{cases} u_i^C / u_i^D & \text{if } u_i^D \geq 0 \\ 1 - u_i^C / u_i^D & \text{otherwise} \end{cases} \tag{5}$$

so that the interval $0 \leq \xi \leq 1$ indicates "do not turn", larger values ($\xi > 1$) indicate "turn right", and smaller values ($\xi < 0$) indicate "turn left".

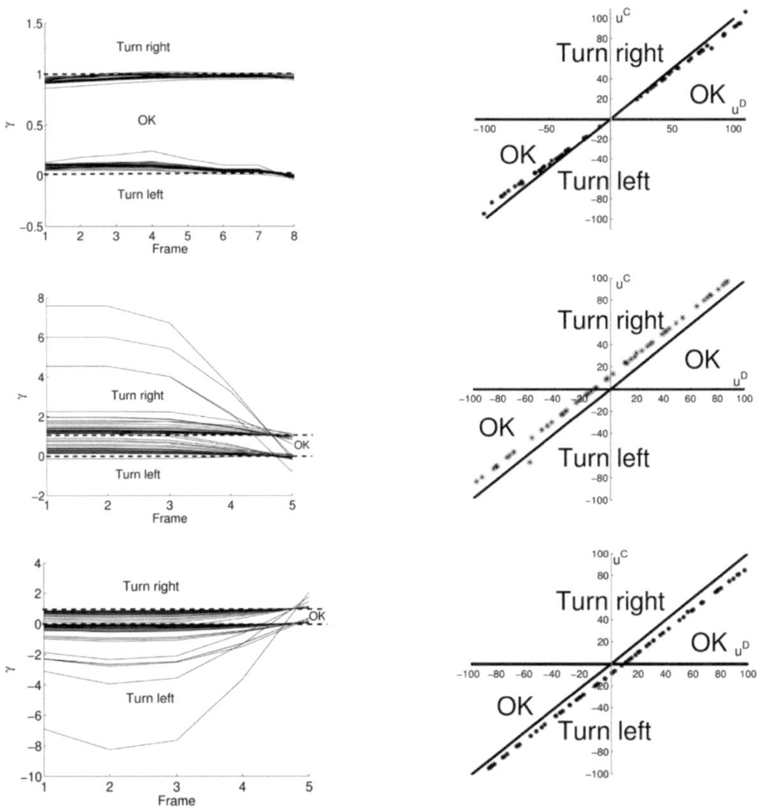


Fig. 9. LEFT: The normalized feature coordinates of all the features plotted versus the image frame number for three segments of the indoor experiment. Features below 0 vote for "turn left", while those above 1 vote for "turn right". RIGHT: A snapshot of the features from frame 2 of each segment plotted on the qualitative control decision space to show the instantaneous decision. The three segments correspond to the points a, b, and c from Fig. 7, in which the robot respectively did not turn, turned right, and turned left.

Fig. 10 shows sample images from two experiments demonstrating the robustness of the algorithm. In the first, the robot navigated a slanted ramp in a 40 m run, thus verifying that the algorithm does not require a flat ground plane. In the second, the robot navigated a narrow road for 80 m while a pedestrian walked by the robot and later a van drove by it. Because the milestone images change frequently, the algorithm quickly recovered from the loss of features due to the occlusion caused by the dynamic objects.

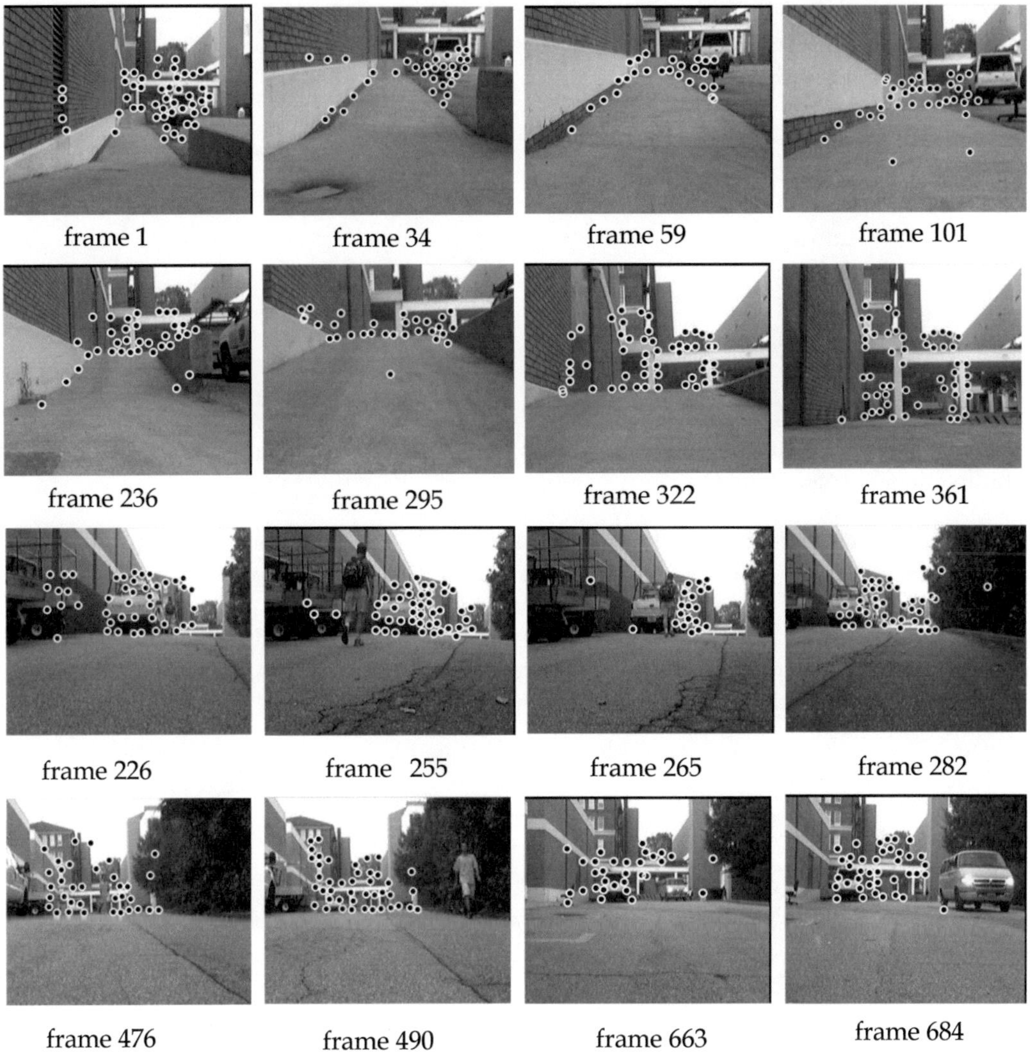


Fig. 10. Sample image frames from two different sequences, one in which the robot traveled down and up a ramp (top 2 rows), and the other containing dynamic objects (bottom 2 rows). The circles indicate the features.

Similarly, Fig. 11 shows the results of the approach using cameras with severe lens distortion. In one experiment we used a wide-angle camera with a 3.5 mm focal length and

110-degree field of view. The other experiment utilized an omnidirectional camera with a 360-degree field of view. For both experiments we used the same parameters as the previous experiments. The only change made to the code was to discard the bottom half of the omnidirectional donut image. This step was necessary because features behind the robot (whether viewed by an omnidirectional or standard camera) move in a way that violates the fundamental assumptions of our approach. In contrast, features in front of the camera obey the funnel constraints sufficiently to be of use in keeping the robot on the path, despite their moving in curved image paths due to the severe lens and catadioptric distortion. The average error of the two experiments was 0.04 m and 0.04 m, respectively, while the maximum error was 0.13 m and 0.09 m.

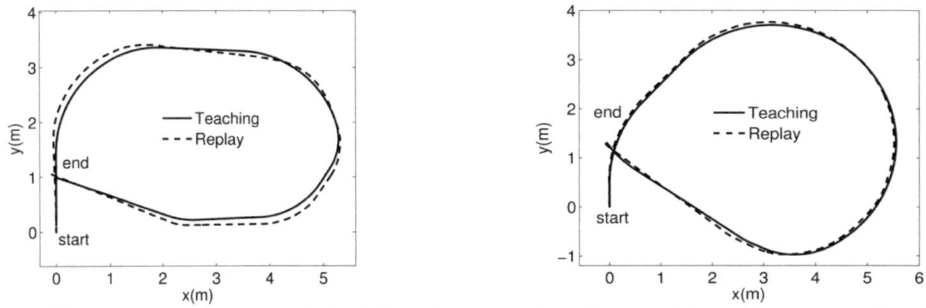


Fig. 11. The approach successfully following a path using a wide-angle camera (left) and an omnidirectional camera (right).

To further illustrate the lack of calibration, we conducted an outdoor experiment in which the robot navigated the same 50 m path twice. In the first run the robot used the Logitech Quickcam Pro 4000 camera, while in the second run it used an Imaging Source DFK21F04 Firewire camera with an 8.0 mm F1.2 lens. The same camera was used for both teaching and replay. As shown in Fig. 12, the algorithm was able to successfully follow the path using either camera, without changing any parameters between runs.

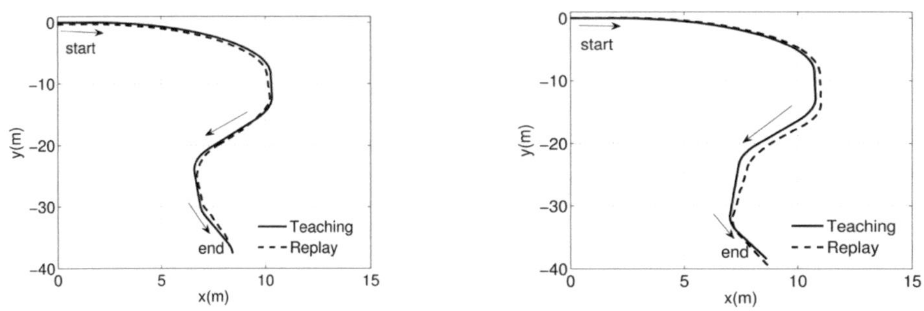


Fig. 12. Teaching and replay paths for the robot using two different uncalibrated cameras, with the same system parameters. LEFT: Logitech QuickCam Pro 4000 USB webcam, RIGHT: Imaging Source DFK 21F04 Firewire camera.

Three additional experiments are shown in Fig. 13. In the first, a scout robot was sent along an outdoor path. Another robot, which received the transmitted path information, was then able to follow the same path as the scout. This demonstrates the natural application of swarm robotics, where calibrating dozens or hundreds of cameras would be prohibitive, especially if recalibration is needed whenever the lenses are refocused or the cameras adjusted. The second experiment shows the robot following a path along rough terrain, in which roll and tilt angles up to 5 degrees were encountered. The roll angle compensation described earlier was sufficient to enable the robot to remain on the path. In the third, a path with several sharp turns is demonstrated. This ability is achieved by setting the replay driving speed to be that of the teaching driving speed, which is decreased during a turn.

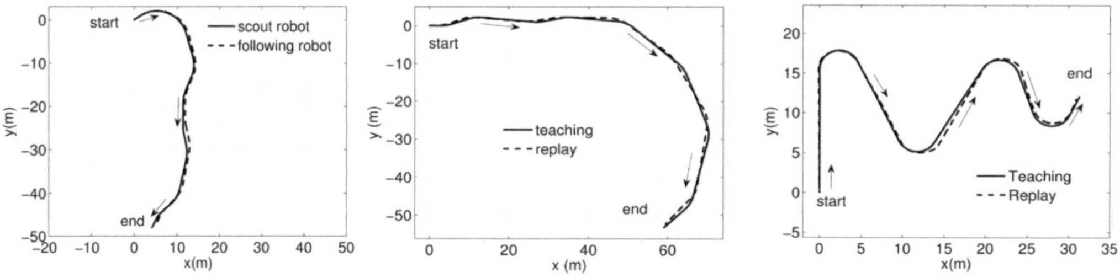


Fig. 13. LEFT: The robot followed a path taken earlier by a scout robot. MIDDLE: A path on rough terrain. RIGHT: A path with sharp turns.

Additionally, the algorithm was tested in various scenarios to quantitatively measure its accuracy and repeatability. Table I displays the results of the algorithm compared with those of the earlier version (Chen & Birchfield, 2006) which did not use odometry, relied upon a bang-bang control scheme, and did not compensate for the camera roll angle. The algorithms were tested in three environments: a 15 m path in an indoor laboratory environment with rich texture for feature tracking, a 60 m trajectory in an outdoor paved parking lot, and a 40 m path along unpaved terrain. In each case, we conducted ten trials and recorded the final 2D location of the robot for each trial: $\{\mathbf{x}_i\}_{i=1}^{n}$, where $\mathbf{x}_i \in \Re^2$ and $n$ = 10. Accuracy was measured as the RMS Euclidean distance to the final ground truth location: $\sqrt{\frac{1}{n}\sum_{i=1}^{n}\left\|\mathbf{x}_i - \mathbf{x}_{gt}\right\|^2}$. Repeatability was measured as the standard deviation of the final locations: $\sqrt{\frac{1}{n}\sum_{i=1}^{n}\left\|\mathbf{x}_i - \mu\right\|^2}$, where $\mu = \frac{1}{n}\sum_{i=1}^{n}\mathbf{x}_i$. While the earlier algorithm works well when the ground is paved and the scenery is rich in texture, the improved algorithm is more robust, achieving maximum errors of only 0.23 m, 1.20 m, and 1.76 m, respectively, compared with 0.45 m, 1.20 m, and 5.68 m for the earlier algorithm.

| Algorithm | indoor acc. / rep. (m) / (m) | outdoor paved ground acc. / rep. (m) / (m) | outdoor rough terrain acc. / rep. (m) / (m) |
|---|---|---|---|
| Vision only (Chen & Birchfield, 2006) | 0.30/0.18 | 0.77 / 0.74 | 3.87 / 1.85 |
| Combination (this work) | **0.14 / 0.08** | **0.60 / 0.55** | **1.47 / .0.66** |

Table 1. Comparison of the accuracy and repeatability of the algorithm with an earlier version, in three different scenarios. The lowest number in each case is in bold.

The algorithm assumes that the robot is placed in the same initial location in both the teaching and replay phases. To test the sensitivity to this assumption, we conducted an experiment with a fairly straight teaching path outdoors, with the background approximately 50 m from the initial location. The robot was then placed at different initial locations for the replay phase, deviating laterally from the initial teaching location by 0 m, 0.5 m, 1.0 m, 1.5 m, and 2.0 m. To ensure overlap between the teaching and replay features, the initial robot orientation was adjusted accordingly. The results, shown in Fig. 14a, show that the robot converged to the teaching path in all cases, reducing the error in half after approximately 20 m. With closer backgrounds, the convergence was faster. In a similar experiment, the robot was placed 0 m, 0.5 m, 1.0 m, 1.5 m, and 2.0 m ahead and behind the initial teaching location. Fig. 14b plots the deviation in the estimate of the position of the robot along the path versus the driving distance along the teaching path. Although these results exhibit more noise, the errors reduce over time, requiring about 6 m for convergence for locations in front, and approximately 20 m for locations behind.

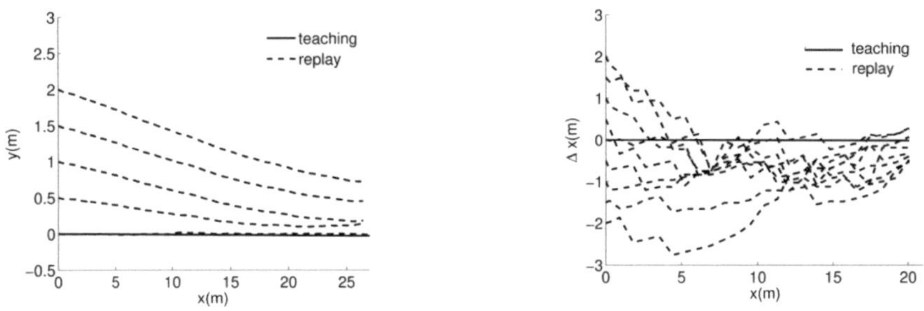


Fig. 14. Sensitivity to the initial location. (a) The replay path of the robot versus time, starting from different locations deviating laterally from the initial teaching location. (b) The deviation of the robot along the replay path versus time, starting from different deviations along the path from the initial teaching location.

Another way for the robot to deviate from the desired path is the presence of an obstacle. While avoiding an obstacle, the robot will rotate in such a way that there may be no overlap between the current image and the milestone image. To solve this problem, we use odometry to return the robot to the original path, after the obstacle has been avoided. (Obstacle detection is accomplished using a ring of sonars.) By comparing the current features with the milestone features, the robot is then able to locate the segment which has the minimum mean squared error between the current features and the milestone features. Once this segment has been determined, the path-following algorithm resumes. An experiment of this capability is shown in Fig. 15.

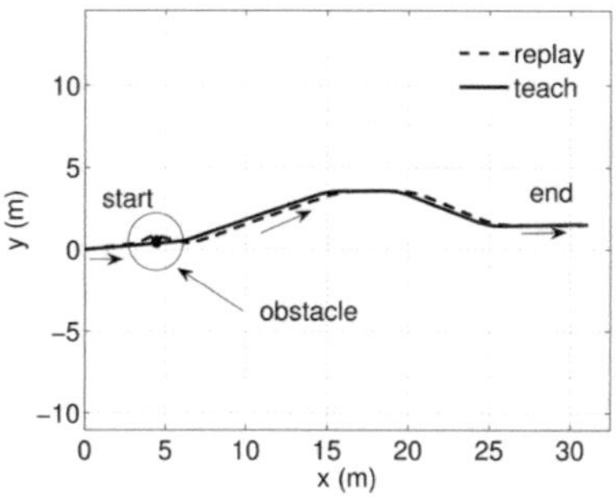


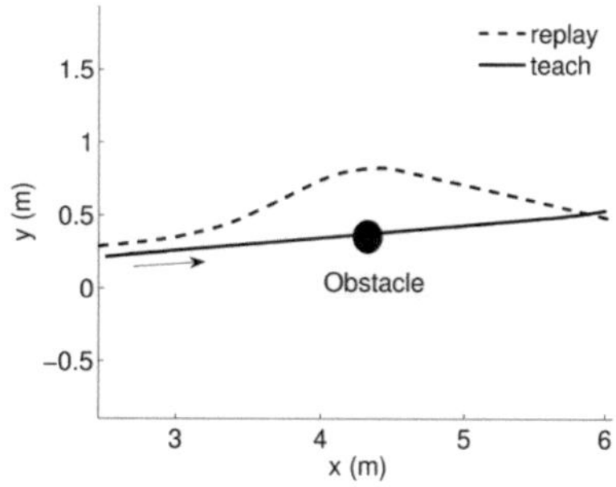


Fig. 15. LEFT: The teaching and replay paths of the robot in an outdoor environment (parking lot), with an obstacle blocking the path. RIGHT: Close-up of the section of the path containing the obstacle.

## 6. Discussion

Because our system does not explicitly model the geometric world, its geometric accuracy is limited. Therefore, when compared with map-based approaches using calibrated cameras (Royer et al., 2007), the errors exhibited by the simple control scheme of our algorithm are rather large. Nevertheless, the remarkable flexibility and versatility of the system offer some important advantages over more precise techniques. With our approach, one can literally take an off-the-shelf camera, attach it to the robot, align it approximately in the forward direction, and start the system. The algorithm is not perfect, and there are scenarios in which it will fail. For example, occasionally the algorithm does not properly transition to the next milestone image, in which case the overlap between the current and milestone image can decrease to the point that an insufficient number of features are matched. Also, untextured scenes containing distant trees, bushes, or undecorated indoor hallways sometimes prevent the KLT algorithm from successfully tracking enough features to accurately compute the heading direction. While only a handful of features are necessary for the algorithm to succeed, it is important that features exist on both sides of the image, and that some number of features remain visible throughout the milestone. Another source of error is due to distant features. Although features near the center of the image produce a narrow funnel lane even when they are far from the camera, distant features near the side of the image produce much larger funnel lanes which are less useful for navigation. Moreover, image parallax is inversely proportional to the distance to a feature. As a result, distant features are primarily useful for correcting the rotation of the robot and are quite incapable of informing the robot about minor translation errors. This problem is compounded by the inherent ambiguity between rotation and translation in the funnel lane itself. Even though this ambiguity has little effect when the robot is near the path, it hinders the ability of the visual information to correctly determine the correct amount of rotation when the robot has deviated significantly. Odometry helps to overcome this limitation, and we have conducted experiments in which the robot consistently returns to the path after deviating by several meters. However, much larger deviations either initially or during replay cannot be handled by our present system. At any rate, it should be noted that odometry drift is not an issue because we only store odometry values local to the segment, not in a global coordinate frame.

## 7. Conclusion

In this chapter we have presented a novel approach to the problem of vision-based mobile robot path following using a single off-the-shelf camera. The robot navigates by performing a qualitative comparison of feature coordinates across the teaching and replay phases, utilizing the novel concept of a *funnel lane*. Vision information is combined with odometry for increased robustness. The algorithm does not make use of the traditional concepts of Jacobians, homographies, fundamental matrices, or the focus of expansion, and it does not require any camera calibration, including lens calibration. It only requires implicit calibration in the form of a controller gain. Experimental results on both indoor and outdoor scenes demonstrate the effectiveness of the approach on trajectories of hundreds of meters, along with its robustness to effects such as dynamic objects, slanted surfaces, and rough terrain. The versatility of the algorithm in working with wide-angle and omnidirectional cameras with only minor modification has also been shown. Future work should be aimed at incorporating higher-level scene knowledge to enable obstacle avoidance and terrain characterization, as well as connecting multiple teaching paths in a graph-based framework to enable autonomous navigation between arbitrary points.

## Acknowledgements

This work was partially supported through a Ph.D. fellowship from the National Institute for Medical Informatics.

## 8. References

Ackerman, C. & Itti, L. (2005). Robot steering with spectral image information. *IEEE Transactions on Robotics*, Vol. 21, No.2, pp. 247–251.

Baker, S. & Matthews, I. (2004). Lucas-Kanade 20 years on: A unifying framework. In *International Journal of Computer Vision*, Vol. 56, No. 3, pp. 221–255.

Barrows, G. L.; Chahl, J. S. & Srinivasan, M. V. (2002). Biomimetic visual sensing and flight control. In *Bristol Conference on UAV Systems*.

Birchfield, S. (1997). KLT: An implementation of the Kanade-Lucas-Tomasi feature tracker, http://www.ces.clemson.edu/~stb/klt/.

Burgard, W.; Cremers, A. B.; Fox, D.; Hähnel, D.; Lakemeyer, G.; Schulz, D.; Steiner, W. & Thrun, S. (1999). Experiences with an interactive museum guide-robot. *Artificial Intelligence*, Vol. 114, No. 1-2, pp. 3–55.

Burschka, D. & Hager, G. (2001). Vision-based control of mobile robots. In *Proceedings of the International Conference on Robotics and Automation*, pp. 1707–1713.

Chen, Z. & Birchfield, S. (2006). Qualitative vision-based mobile robot navigation. In *Proceedings of the IEEE International Conference on Robotics and Automation (ICRA)*, pp. 2686–2692.

Crawford, S.; Cannon, M.; Letourneau, D.; Lepage, P. & Michaud, F. (2004). Performance evaluation of sensor combinations on mobile robots for automated platoon control. In *ION GNSS Conference*, pp. 706–717.

DeSouza, G. N. & Kak, A. C. (2002). Vision for mobile robot navigation: A survey. *IEEE Transactions on Pattern Analysis and Machine Intelligence*, Vol. 24, No. 2, pp. 237–267.

Fischler, M. A. & Bolles, R. C. (1981). Random sample consensus: A paradigm for model fitting with applications to image analysis and automated cartography. *Communications of the ACM*, Vol. 24, No. 6, pp. 381-395.

Gaussier, P.; Joulain, C.; Zrehen, S.; Banquet, J. P. & Revel, A. (1997). Visual navigation in an open environment without map. In *Proceedings of the IEEE/RSJ International Conference on Intelligent Robots and Systems. (IROS)*, pp. 545–550.

Giovannangeli, C.; Gaussier, P. & Désilles, G. (2006). Robust mapless outdoor vision-based navigation. In *Proceedings of the IEEE/RSJ International Conference on Intelligent Robots and Systems*, pp. 3293–3300.

Guerrero, J. J. & Sagüés, C. (2001). Uncalibrated vision based on lines for robot navigation. *Mechatronics*, Vol. 11, No. 6, pp. 759–777.

Harris, C. G. & Stephens, M. (1988). A combined corner and edge detector. In *Proceedings of the 4th Alvey Vision Conference*, pp. 147–151.

Horswill, I. D. (1993). Polly: A vision-based artificial agent. In *Proceedings of the National Conference on Artificial Intelligence*, pp. 824–829.

Hutchinson, S.; Hager, G. & Corke, P. (1996). A tutorial on visual servo control. *IEEE Transactions on Robotics and Automation*, Vol. 12, No. 5, pp. 651–670.

Jones, S. D.; Andersen, C. S. & Crowley, J. L.(,1997). Appearance based processes for visual navigation. In *Proceedings of the IEEE/RSJ International Conference on Intelligent Robots and Systems (IROS)*, pp. 551–557.

Kosaka, A. & Kak, A. C. (1992). Fast vision-guided mobile robot navigation using model-based reasoning and prediction of uncertainties. *Image Understanding*, Vol. 56, No. 3, pp. 271–329.

Košeckā, J.; Zhou, L.; Barber, P. & Duric, Z. (2003). Qualitative image based localization in indoors environments. In *Proceedings of the IEEE Conference on Computer Vision and Pattern Recognition (CVPR)*, pp. 3–8.

Kröse, B.; Vlassis, N.; Bunschoten, R. & Motomura, Y. (2001). A probabilistic model for appearance-based robot localization. *Image and Vision Computing*, Vol. 19, No.6, pp. 381–391.

LeCun, Y.; Muller, U.; Ben, J.; Cosatto, E. & Flepp, B. (2005). Off-road obstacle avoidance through end-to-end learning. In *Advances in Neural Information Processing Systems (NIPS)*, pp. 739–746.

Liang, B. & Pears, N. (2002) Visual navigation using planar homographies. In *Proceedings of the International Conference on Robotics and Automation*, Vol. 1, pp. 205–210.

Matsumoto, Y.; Inaba, M. & Inoue, H. (1996). Visual navigation using view-sequenced route representation. In Proceedings of the International Conference on Robotics and Automation, Vol. 1, pp. 83–88.

Matsumoto, Y.; Sakai, K.; Inaba, M. & Inoue, H. (2000). View-based approach to robot navigation. In Proceedings of the IEEE/ RSJ International Conference on Intelligent Robots and Systems (IROS), Vol. 3, pp. 545–550.

Michels, J.; Saxena, A. & Ng, A. Y. (2005). High-speed obstacle avoidance using monocular vision and reinforcement learning. In Proceedings of the Twenty-second International Conference on Machine Learning, pp. 593–600.

Murali, V. N. & Birchfield, S. T. (2008). Autonomous navigation and mapping using monocular low-resolution grayscale vision. In Workshop on Visual Localization for Mobile Platforms (in association with CVPR).

Nayar, S. K.; Murase, H. & Nene, S. A. (1994). Learning, positioning, and tracking visual appearance. In Proceedings of the International Conference on Robotics and Automation, pp. 3237–3244.

Negahdaripour, S. & Yu, C. H. (1993). A generalized brightness change model for computing optical flow. In *Proceedings of the International Conference on Computer Vision*, pp. 2–11.

Nelson, R. C. & Aloimonos, J. (1998). Using flow field divergence for obstacle avoidance towards qualitative vision. In Proceedings of the International Conference on Computer Vision, pp. 188–196.

Remazeilles, A. & Chaumette, F. (2007). Image-based robot navigation from an image memory. *Robotics and Autonomous Systems*, Vol. 55, No.4, pp. 345–356.

Royer, E.; Lhuillier, M.; Dhome, M. & Lavest, J.-M. (2007). Monocular vision for mobile robot localization and autonomous navigation. *International Journal of Computer Vision*, Vol. 74, No. 3, pp. 237–260.

Sagüés, C. & Guerrero, J. J. (2005). Visual correction for mobile robot homing. *Robotics and Autonomous Systems*, Vol. 50, No. 1, pp. 41–49.

Santos-Victor, J.; Sandini, G.; Curotto, F. & Garibaldi, S. (1995). Divergent stereo in autonomous navigation: From bees to robots. *International Journal of Computer Vision*, Vol. 14, No. 2, pp. 159–177.

Šegvić, S.; Remazeilles, A.; Diosi, A. & Chaumette, F. (2007). Large scale vision based navigation without an accurate global reconstruction. In Proceedings of the IEEE Conference on Computer Vision and Pattern Recognition (CVPR), pp. 1–8.

Shelton, A. L. & Gabrieli, J. D. E. (2002). Neural correlates of encoding space from route and survey perspectives. *The Journal of Neuroscience*, Vol. 22, No. 7, pp. 2711–2717.

Shen, J. & Hu, H. (2006). Visual navigation of a museum guide robot. In *Proceedings of the* 6th World Congress on Intelligent Control and Automation (WCICA), Vol. 2, pp. 9169–9173.

Shi, J. & Tomasi, C. (1994). Good features to track. In *Proceedings of the IEEE Conference on Computer Vision and Pattern Recognition*, pp. 593–600.

Sim, R. & Dudek, G. (2001). Learning environmental features for pose estimation. *Image and Vision Computing*, vol. 19, No. 11, pp. 733–739.

Tang, L. & Yuta, S. (2001). Vision based navigation for mobile robots in indoor environment by teaching and playing-back scheme. In Proceedings of the International Conference on Robotics and Automation, pp. 3072–3077.

Tomasi, C. & Kanade, T. (1991). Detection and tracking of point features. In *Technical Report CMU-CS-91-132, Carnegie Mellon University.*

Ulrich, I. & Nourbakhsh, I. (2002). Appearance-based place recognition for topological localization. In *Proceedings of the International Conference on Robotics and Automation*, pp. 1023–1029.

Weng, J. & Chen, S. (1996). Incremental learning for vision-based navigation. In *Proceedings of the IAPR International Conference on Pattern Recognition*, pp. 45–49.

Wolf, J.; Burgard, W. & Burkhardt, H. (2002). Using an image retrieval system for vision-based mobile robot localization. In *Proceedings of the International Conference on Image and Video Retrieval (CIVR)*, pp. 108–119.

Zhou, C.; Wei ,Y. & Tan, T. (2003). Mobile robot self-localization based on global visual appearance features. In Proceedings of the International Conference on Robotics and Automation, pp. 1271–1276.

# 22

# Motivation and Local Image Entropy Based Measures in Evolutionary Mobile Robot Navigation

Tomás Arredondo and Wolfgang Freund
*Universidad Técnica Federico Santa María*
*Chile*

## 1. Introduction

Robotic navigation in unknown and unstructured environments is a complex and difficult task fraught with potential problems and peril to the navigator. Extensive research has been made and is ongoing into this problem with the aid of various robotic architectures, sensors and processors. Behavior based robotics is an approach that in general does not promote the use of complex world models or symbolic knowledge. This design philosophy promotes the idea that robots should be low cost, built incrementally, capable of withstanding sensor and other noise, and without complex computers and communication systems. Behavior based learning systems typically include reinforcement learning, neural networks, genetic algorithms, fuzzy systems, case and memory based learning (1). These biologically based mechanisms are capable of novel complex behaviors which avoid local minima and have the ability to extrapolate from training information.

One area of research in behavior based robotics has focused on providing more natural and intuitive interfaces between robots and people (2; 3). One recent investigation (4), decouples specific robot behavior using an intuitive interface based on biological motivations (e.g. curiosity, hunger, etc) (5). Training the robot to behave according to said motivations requires optimization of the robot inference system (e.g. neural network) which in our approach is implemented using a genetic algorithm (GA).

It is a well known fact in machine learning (6), that having diversity during training can provide for the emergence of more robust systems which are capable of coping with a variety of environmental challenges (7; 8). Early studies have shown that information theory can be used as an aid in analyzing robotic learning performance in terms of the diversity of information received during training (4; 9). Performing a more extensive analysis of environment training diversity using such information theoretic based measures was something of interest to us. Toward this goal we investigate the capability of the entropy based environmental and motivation measures toward analyzing the outcome of several robotic navigation experiments.

In our work, robot navigation is performed in a simulator (10) by providing sensor values directly into a neural network inference engine that drives left and right motors. The robot uses infrared sensors which give limited information about the surroundings in which the robot is located. In order to reduce the complexity of the action space, action-based environmental modeling (AEM) (11) is implemented with a small action set of four basic actions (e.g. go straight, turn left, turn right, turn around) in order to encode a sequence of actions based on sensor values. The search space of behaviors is huge and designing suitable behaviors by hand is very difficult (11) therefore we use a genetic algorithm (GA) within the simulator in order to find appropriate behaviors. The GA selects from a population of robots (neural networks) using a fuzzy fitness function that considers various robotic motivations such as: the need for exploration (curiosity), the need to conserve its battery (energy), the desire to determine its location (orientation), and the capacity to return to its initial position (homing). This paper is organized as follows. Section 2 gives a description of the robotic system. Section 3 describes the entropy measures used for diversity evaluation. Section 4 introduces the experiments performed. In section 5 we describe and summarize our results. Finally, in section 6 some conclusions are drawn.

## 2. Robotics System Description

This section presents the robotic system used for these studies. The system has several different elements including: the robot simulator, action based environmental modeling, neural networks, GA, and fuzzy logic based fitness (Fig.1).

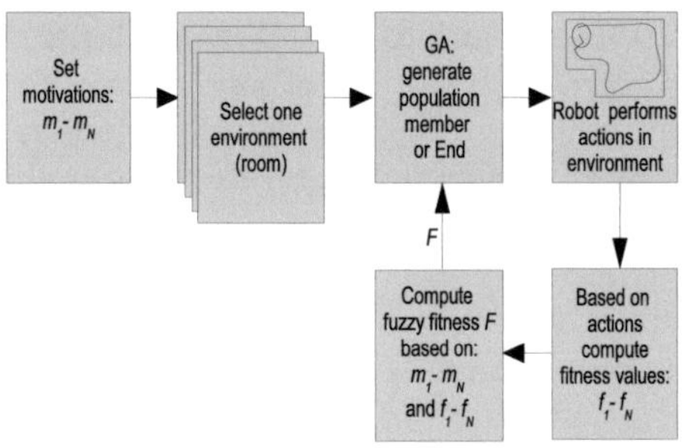


Fig. 1. System Overview

### 2.1 Robot Configuration

Our project uses a small mobile robot which has two DC motors and eight (six front and two back) infrared proximity sensors used to detect nearby obstacles. These sensors provide 10 bit output values (with 6% random noise). These readings allow the robot to know in approximate form the distance to local obstacles. The simulator (10) provides the readings for the robot sensors according to the robot position and the map (room) it is in. The simulator also has information for the different areas that the robot visits and the various obstacles detected in the room.

The robot internally generates a zone map in which the zones are marked with various values: obstacles are indicated with a value of -1, those not visited by the robot are marked with a 0 and the visited ones with a 1. During navigation, the robot executes 500 steps in each experiment, but not every step produces forward motion as some only rotate the robot. The robot is not constrained by its battery given that a 100% charge level allows more than 1000 steps.

### 2.2 Artificial Neural Network

As seen in (Fig.2), the robot uses a feed-forward neural network with eight input neurons (one per sensor), five neurons in the hidden layer and two output neurons directly connected to the motors that produce the robot movement. The transfer function used by the neurons is of the sigmoid type.

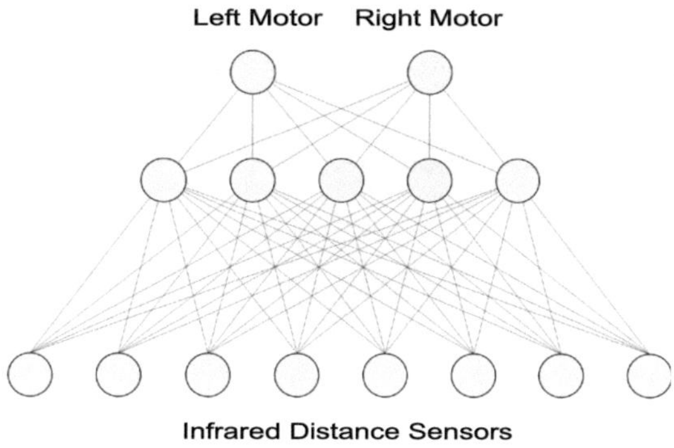


Fig. 2. Neural Network

### 2.3 Action-based Environmental Modeling

In order to reduce the vast behavior search space, we use AEM (11) based encoding to select the basic actions used by the robot to perform navigation:

- A1: Go 55 mm straight on.
- A2: Turn 30°left.
- A3: Turn 30°right.
- A4: Turn 180°right.

AEM uses a SOM (12) network so that the robot may determine the room it is navigating in (localization). Robot navigation produces actions which are saved as action sequences. These action sequences are converted using chain coding into an environment vector. These vectors are fed into the SOM network for unsupervised learning. After training the SOM network associates a vector with one of its output nodes (*r*-nodes)(11). We used inputs of 1000 steps for all rooms used in training, these were alternatingly presented to the SOM network for 10000 iterations, the network had a linear output layer of 128 *r*-nodes indicating the maximum possible number of rooms that could be identified.

### 2.4 Genetic Algorithm

A GA is used to find an optimal configuration of weights for the neural network. Each neural network is encoded as an individual in the GA by a concatenation of all the weights in the neural network. The GA evolves the neural network with the passing of different generations using the fuzzy fitness measure described below (Fig.3). The GA uses the following parameters:

- Population size: 200.
- Crossover operator: Random crossover.
- Selection method: Elite strategy selection.
- Mutation rate $P_{mut}$: 1%.
- Generations: 90.

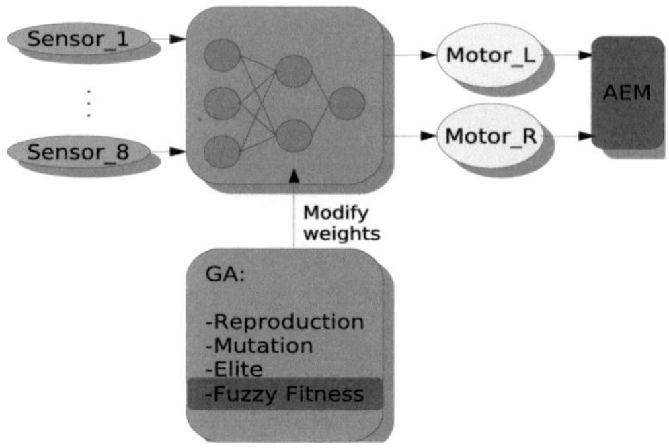


Fig. 3. GA Overview

### 2.5 Fuzzy Fitness Calculation

Fuzzy logic (Fig.4) is used toward implementing a motivation based interface and towards determining the fitness of robotic behavior. The motivation set ($M$) considered includes: homing ($m_1$), curiosity ($m_2$), energy ($m_3$), and orientation ($m_4$). These motivations are used as input settings (between 0 and 1) prior to running each experiment (4).
The set of fitness criteria and the fuzzy variables that correspond to them are: proper action termination and escape from original neighborhood area ($f_1$), amount of area explored ($f_2$), percent of battery usage ($f_3$) and environment recognition ($f_4$). The values for these criteria are normalized (range from 0 to 1) and are calculated after the robot completes each run:

- $f_1$: a normalized final distance to home
- $f_2$: percentage area explored relative to the optimum
- $f_3$: estimated percent total energy consumption considering all steps taken
- $f_4$: determined by having the robot establish which room he is in ($r$-node versus the correct one), using the previously trained SOM network.

Four fuzzy variables with five triangular membership functions each are used ($5^4 = 625$ different fuzzy rules) toward calculating robotic fitness. The membership functions used are given in Fig. 4. Sample fuzzy rules (numbers 9 and 10) are given as follows (here the K array is a simple increasing linear function):

if ($f_1$ == H) and ($f_2$ == L) and ($f_3$ == V.L.) and ($f_4$ == V.L.) then
f[9] = $m_1 f_1 K[4] + m_2 f_2 K[2] + m_3 f_3 K[1] + m_4 f_4 K[1]$
if ($f_1$ == V.H.) and ($f_2$ == L) and ($f_3$ == V.L.) and ($f_4$ == V.L.) then
f[10] = $m_1 f_1 K[5] + m_2 f_2 K[2] + m_3 f_3 K[1] + m_4 f_4 K[1]$

During training, a run environment (room) is selected and the *GA* initial robot population is randomly initialized. After this, each robot in the population performs its task (navigation and optionally environment recognition) and a set of fitness values corresponding to the performed task are obtained. Finally, robotic fitness is calculated using the fitness criteria information provided by the simulator and the different motivations at the time of exploration.

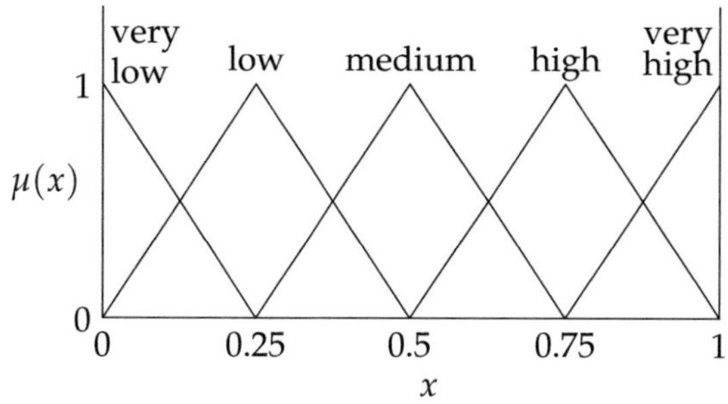


Fig. 4. Fuzzy membership functions

## 3. Entropy Measures

Two entropy based measures are used in this study to measure environmental and motivation diversity (4). These measures of diversity come from the well established definition of entropy as a measure of the uncertainty (which generates a diversity of outcomes) in a system (13; 14).

### 3.1 Environmental Entropy

Average local image entropy is used as a measure of environmental uncertainty (i.e. uncertainty being a measure of an environmental diversity) given the shape and distribution of obstacles in a room by using an extension of the local image entropy method (4). In the local image entropy method an estimation of the information content of a pixel $(x, y)$ based on a histogram of its neighborhood $(w, h)$ pixel values is calculated as:

$$E_{w,h}(x,y) = -\sum_k p_{w,h}(k) \log p_{w,h}(k).$$

A region of an image (e.g. neighborhood) is interpreted as a signal of $k$ different states with the local entropy $E_{w,h}(x,y)$ determining the observer's uncertainty about the signal (15).
To obtain a measure of an environment's diversity (using a neighborhood window as given by $w$ and $h$), we compute the average local image entropy for the room. Here the size of $w$ and $h$ are set using an uncertainty criteria of interest given the scale of the robot and obstacles in its path (e.g. a robots uncertainty of crossing a zone). In our experiments we study this diversity measure to find neighborhood width and height settings which are successful in helping us predict better training environment sets.
Clearly if a neighborhood is empty or full the robot will have either no difficulty or no chance of traversing the terrain and hence the uncertainty for that neighborhood will be zero but if the neighborhood has some obstacles then its maximum uncertainty should be when $E_{w,h}(x,y) = 1.0$.
Obstacles were chosen at least as large as the robot size so as to preclude small checkered patterns which could also result in a local entropy value near 1.0 but would not make any sense in terms of our definition of traversing uncertainty (the robot clearly could not cross such a neighborhood).

### 3.2 Motivation Entropy

In order to calculate motivation entropy we first define a motivation set $M$ as $\{m_1, m_2, \ldots, m_n\}$. Toward the calculation of motivation diversity $H(M)$, we consider the cor-

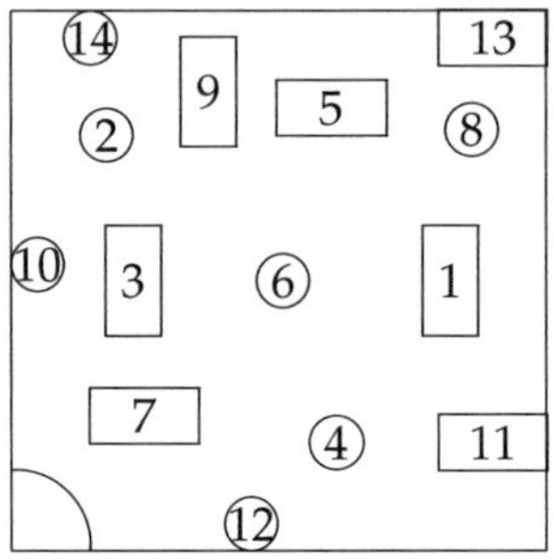


Fig. 5. Experiment 1 room layout ($r_1$-$r_{15}$).

responding probabilities for $\{m_1, m_2, \ldots, m_n\}$ as $\{p_1, p_2, \ldots, p_n\}$. We compute the entropy of the random variable $M$ using:

$$H(M) = -\sum_{i=1}^{n} p_i \log(p_i), \ where \ \sum_{i=1}^{n} p_i = 1.$$

Note that when $H(M) = 0$ the motivations are considered to have no diversity and for $n = 4$ the maximum diversity is when $M$=$\{0.25, 0.25, 0.25, 0.25\}$ and $H(M) = 2$.

## 4. Experimental Evaluation

Of interest to us was to study the impact of the entropy based diversity measures and their capability toward predicting a robot's ability to learn behaviors based on environmental and motivation diversity. Our results were verified based on an empirical model using multiple complete test runs given random motor and sensor noise.

All experiments had a training phase, in order to obtain the best weights for the NN, followed by a testing phase. The GA settings used in the training phase are given in section 2.4. The number of steps in each iteration was set to 1000 for the training phase, and 500 for the testing phase. The robot radius was 68.75 mm with a forward step size of 55 mm. The rooms are square (sides of 2750 mm) with various internal configurations of walls and obstacles. For mapping purposes rooms are divided into 2500 zones (each 55 mm by 55 mm).

Toward studying the diversity measures, we tested training in 15 different square rooms: $r_1$ - $r_{15}$. These rooms have an increasing number of obstacles in a fairly random distribution with the condition that any one obstacle should not preclude the robot from reaching any area of the room. Room $r_1$ has no obstacles, $r_2$ has only obstacle 1, $r_3$ has obstacle 1 (in the same place as in $r_2$) plus obstacle 2, and so on until $r_{15}$ which contains obstacles 1 - 14. A room layout with the position of the obstacles (identified by its number) is shown in Fig. 5.

### 4.1 First experiment: environmental diversity measure

In order to analyze the sensitivity of the environmental diversity metric we first computed the average local image entropy of each training room given different square window sizes (given as $w$ or $h$). These values are seen in Fig.6 which shows the average local entropy of each room given a range of window sizes from 0 to 1000.

To evaluate the impact of environmental training diversity over the robots navigation behavior, we trained the robot in each room for 90 generations. We set the fuzzy motivations

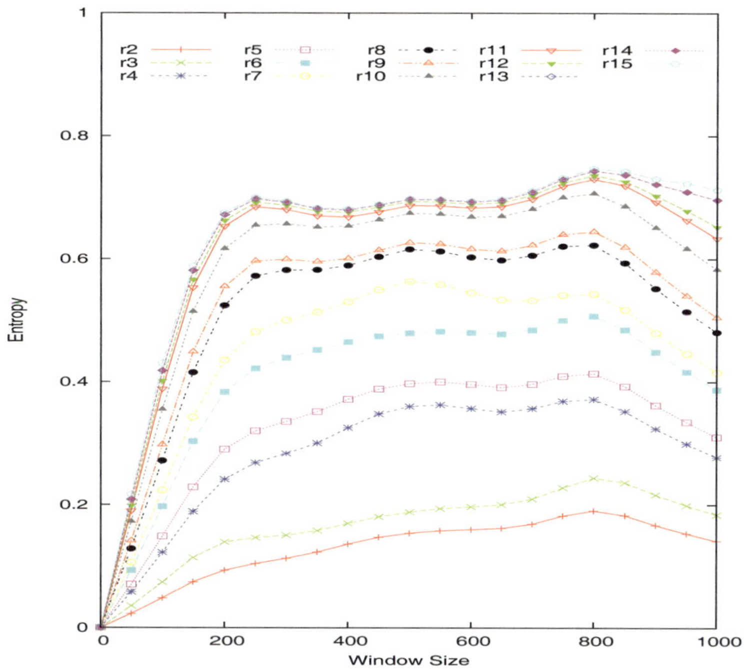


Fig. 6. Average local entropy for all rooms.

$(m_1, m_2, m_3, m_4)$ as $(0, 1, 0, 0)$. After the training phase, the best individual from the GA population was selected to run its respective testing phase in rooms $r_1 - r_{10}$. During testing, rooms $r_{11} - r_{15}$ produced low fitness in all experiments, because of this small contribution we discarded these results from our analysis. A window size of 100 was used given that it is the minimum window size which can still discern changes in the environment and provides the best small region granularity (after analyzing window size results in Fig.6). Even so other possible window sizes could be used for example a window size of 200 would give greater separation between rooms but lower granularity.

### 4.2 Second experiment: motivation diversity measure

For this experiment, in order to see the effect of motivation diversity, we used 66 sets of fuzzy motivation criteria. Motivations ($M$) ranged from highly focused to more diverse. Average fitness values are given for motivations $(m_1, m_2, m_3)$ using all combinations ranging from $(0, 0, 1)$ to $(0.4, 0.3, 0.3)$ with values changing in increments of 0.1. Minimum entropy for $M$ is 0 and maximum entropy is 1.571

The population was trained for 90 generations in each of the rooms. During testing the best individual of the 90 generations was tested in all rooms and average fuzzy fitness values were calculated using the various fitness values $f_1 - f_3$.

## 5. Experimental Results

Ten complete runs were performed of each experiment (each run consisting of one training and 10 test executions) and only average values are reported in our results. Only ten runs are used given that obtained tests results were similar between runs.

### 5.1 First experiment: environmental diversity

In Fig.7 we show the results of the testing phase after applying the respective training method specified for the environmental diversity experiment.

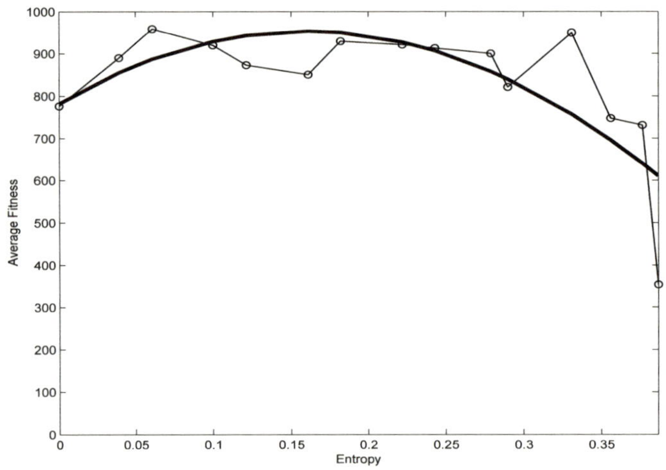


Fig. 7. Learning behavior given training rooms entropy.

### 5.2 Second experiment: motivation diversity

In Fig.8 we show the results of the testing phase after applying the respective training method specified for the motivation diversity experiment.

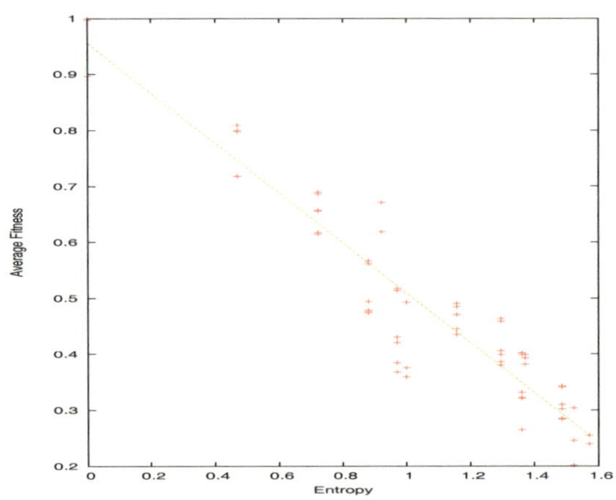


Fig. 8. Learning behavior given motivation diversity.

## 6. Discussion

Average local image entropy can be seen as an effective measure of a training environments potential toward producing highly fit robots. As seen in Fig.6 the measure is sensitive to different window values but in our application a wide range of values could have been used. Our selection of neighborhood size was a reasonable as only very low values (around 10) show poor discriminancy.
Fig. 7 shows how the average fitness obtained during testing is clearly dependent on the diversity of the training room used. A training environment with too much environmental diversity is as unsuitable as one with not enough diversity.
Higher motivation diversity ($H(M)$) caused lower average score values. This is possibly due to the different and conflicting (e.g. orthogonal) requirements of the different motivations upon the robotic neural brain. These scores are also negatively impacted by the fitness function used (4) given that it includes a motivation satisfaction product ($m_n \times f_n$) in each rule that further penalizes lower behavior fitnesses ($f_n$). Even though obtained fitness was generally lower with more diverse motivations, the obtained behaviors demonstrated very good capability (e.g. room exploration, battery usage, etc) and were in agreement with the specified motivations. Because of this, motivation diversity results are somewhat counter intuitive in that by diversifying motivation values one could naively expect higher overall fitness but this is clearly not the case due to system constraints.

## 7. Conclusions

Our entropy based measures are useful toward providing some understanding of complex systems such as robotic training environments. As seen by our test results, the average local image entropy measure is not very sensitive to differences in window size as long as they are above a minimum value. In general the window size used should be the minimum value possible that can still differentiate between environments. The average local entropy measure also shows promise as an image diversity metric with potential usage in signal processing and pattern recognition. Fuzzy motivations are confirmed as an intuitive and user friendly mechanism for obtaining a wide range of different robotic behaviors. Using these entropy based measures it was possible to evaluate the effect of diversity on robotic fitness.
Future work includes hardware implementation and testing of these methods, testing other fuzzy fitness functions, utilizing fuzzy motivations within hybrid architectures, including environment recognition in motivation studies, parametric studies (e.g. linkages between motivations, testing different groups of GA parameters), adding different actions to the robot and potential application in other areas of research.

## Acknowledgements

This research was partially funded by the DGIP of UTFSM (230808).

## 8. References

[1] Arkin, R.: Behavior-Based Robotics, MIT Press, Cambridge, (1998).
[2] Park, H., Kim, E., Kim, H.: Robot Competition Using Gesture Based Interface. LNAI,Vol. 3353. Springer-Verlag, Berlin (2005) 131-133

[3] Jensen, B., Tomatis, N., Mayor, L., Drygajlo, A., Siegwart, R.: Robots Meet Humans - Interacion in Public Spaces. IEEE Transactions on Industrial Electronics, Vol. 52, No. 6 (2006) 1530-1546
[4] Arredondo, T., Freund,W., Muñoz, C., Navarro, N., and Quirós, F.: Learning Performance in Evolutionary Behavior Based Mobile Robot Navigation. LNAI, Vol. 4872. Springer-Verlag, Berlin (2007) p. 811-820
[5] Huitt, W.: Motivation to learn: An overview. Educational Psychology Interactive. http://chiron.valdosta.edu/whuitt/col/motivation/motivate.html, (2001).
[6] Mitchell, T.: Machine Learning, McGraw Hill, Boston, (1997)
[7] Tan, K.C., Goh, C.K, Yang, Y.J., Lee, T.H.: Evolving better population distribution and exploration in evolutionary multi-objective optimization, European Journal of Operations Research 171 (2006) 463-495
[8] Chalmers, D.J.: The evolution of learning: An experiment in genetic connectionism. Proceedings of the 1990 Connectionist Models Summer School, p. 81-90. San Mateo, CA.: M. Kaufmann, 1990
[9] Arredondo, T., Freund,W., Muñoz, C.: Entropy Based Diversity Measures in Evolutionary Mobile Robot Navigation. LNCS, Vol. 5027. Springer-Verlag, Berlin (2008) p. 129-138
[10] YAKS simulator website: http://www.his.se/iki/yaks
[11] Yamada,S.: Recognizing environments from action sequences using self-organizing maps, Applied Soft Computing, 4 (2004) 35-47
[12] Teuvo, K.: The self-organizing map. Proceedings of the IEEE, 79(9), (1990) 1464-1480
[13] Cover, T., Thomas, J.: Elements of Information Theory, Wiley, New York, (1991)
[14] Neelakanta, P. S.: Information-Theoretic Aspects of Neural Networks, CRC Press, Boca Raton, (1999)
[15] Handmann, U., Kalinke, T., Tzomakas, C., Werner, M., Weelen, W.v.,: An image processing system for driver assistance. Image and Vision Computing, 18, (2000) 367-376

# 6-DoF Navigation Systems for Autonomous Underwater Vehicles

Andrew Lammas, Karl Sammut and Fangpo He
*Flinders University*
*Australia*

## 1. Introduction

Navigation is one of the most critical factors in determining the operational suitability of any unmanned vehicle for its designated environment. In fully autonomous vehicles, due to the lack of a human operator to perform the navigation task, there is a fundamental requirement to incorporate estimation techniques that can provide the desired information necessary for navigation. Such information includes position, attitude, and velocity of the vehicle. In most vehicles, estimating this information is a relatively straight forward process due to the information available from the measurement sensors, particularly position information from Global Positioning System (GPS) sensors.

This chapter will focus on the more difficult problem of navigation, particularly positioning, where GPS reception is limited or non-existent. One example where GPS is only partially available is the case of an Autonomous Underwater Vehicle (AUV) where the vehicle is forced to use dead-reckoning in between GPS sightings in order to navigate accurately. Though an underwater environment is used within the context of this example, the techniques examined can be equally applied anywhere where GPS reception is compromised by other electromagnetically opaque mediums, satellite availability or even deliberate denial.

When using dead reckoning, due to the integrative nature of the position estimate, the error of the position estimate will grow unbounded. The growth in this error is caused by errors in the velocity and attitude estimates which are in turn affected by the accuracy of the navigational filter and the accuracies of the measurements observing these states. The accuracy of the measurement sensors is typically related to cost, thus some of the effects can be negated through more precise (expensive) equipment. For a given cost the single biggest factor affecting the accuracy of the navigational estimate is the navigation filter itself.

There have been numerous papers on the accuracy of various algorithms employed in the generic problem of state estimation (Arulampalam et al., 2001). Research pertaining to navigation algorithms for underwater vehicles have almost exclusively concentrated on ways to improve positioning accuracy below the surface through the use of ranging information other than GPS (Caiti et al., 2004). The methodology most prevalent in the literature is Simultaneous Localization And Mapping (SLAM) (Choset & Nagatani, 2001; Dissanayake et al., 2001)where information used for mapping, such as sonar 'pings', is fed back into localization estimate as the map is being built. This process correlates the position of the vehicle with the map, thereby improving both positioning accuracy and map information.

When an AUV is in an environment that is devoid of any distinctive seabed features, SLAM may perform poor, as it is the motion of the AUV relative to these features that provides most of the information in the localization aspect of SLAM. In this context the navigation algorithm has to estimate the position of the vehicle solely using dead reckoning and thus the filter becomes the most significant factor in determining the accuracy of the position estimate. This brings out the critical question addressed in this chapter, which filters are best at minimizing the growth error in the position estimate using dead reckoning and which are best at reacquiring an accurate position estimate when GPS or other positioning information become available again after a period of dead reckoning.

The Bayesian filter algorithm, specifically the recursive Bayesian estimation, is chosen as the class of filters to be examined in this chapter for the following reasons. Bayesian filtering in general is advantageous in that the filter represents an estimate of a state's probability distribution function (pdf) rather than a particular estimate of a state (Ristic et al., 2004), and as such can inherently accommodate uncertain models and incorporate uncertainties of noisy measurements thus making the filter more robust. This robustness comes at the cost of increased computational demand due to the necessity of estimating the whole error distribution rather than a single solution. The recursive Bayesian filter algorithm is in comparison to its batch method equivalents more memory and computationally efficient since, due to its recursive nature, only information pertaining to the current estimate needs to be stored and processed. Recursive methods also have the capability of achieving higher accuracy since the estimate at a particular time is based on all the previous measurements as opposed to a fixed window of measurement, as used in batch methods. The computational efficiency, robustness to noise, and improved accuracy of the recursive Bayesian filter make it the ideal filter architecture for real-time state estimation for a navigation system.

The recursive Bayesian estimation algorithm in itself is, however, intractable due to the infinite number of solutions in a continuous solution space. In this chapter the filters that will be considered are all implementations that are based on the recursive Bayesian filter architecture, yet the algorithms are made tractable by applying simplifying assumptions or approximations to the pdf being estimated. These implementations consist of three main classes: Kalman Filters (KF), Particle Filters (PF), and Grid-based Filters (GF). The filters that are assessed in this chapter belong to the KF and PF classes since the GF class is more applicable for discrete or bounded solution spaces (Doucet et al., 2001).

This chapter aims to present a comprehensive analysis of a variety of KF and PF implementations by implementing these filters to estimate the navigational state of an AUV in a simulated environment. In the test-scenario, shown in Fig. 1, the simulated vehicle first manoeuvres on the surface before diving down to 50 meters to perform a raster scan covering an area of 500 m x 500 m and finally returning to the surface at the point where it had originally submerged. GPS position measurements are only available on the surface. This scenario has been chosen to compare the two most critical aspects of any navigational estimator, namely the error growth and integration of new observations after a period of absence. In the case of AUV navigation this increased estimation error can cause some filters to perform poorly or not at all when presented with new position information from the GPS that is inconsistent with their own estimate.

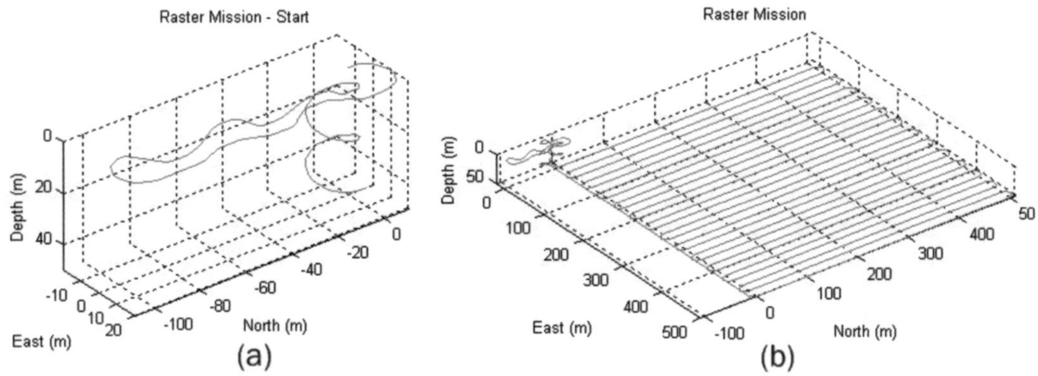


Fig. 1. Simulated raster-scan mission trajectory

This chapter will first cover the requirements of and limitations placed on a navigation system, and then review the concepts and algorithms used by the Bayesian filters in the particular context of the navigation estimation case. This will include:

- Requirements - what information is required of the navigation system by other systems (e.g., motion control systems) on the vehicle, and what limitations are placed on the system due to measurement and computational constraints.
- Dynamics - the kinematic and kinetic models used in defining the motion of a free moving rigid body vehicle, i.e., a vehicle with six degrees of freedom (6-DoF).
- Sensors - the instruments used to acquire the set of observation measurements from which the vehicle's state vector can be estimated, as well as the mathematical models that relate the sensor measurements to the states being estimated.
- Estimators - a representative set of KF and PF variants that have been used for vehicular navigation.

The chapter will then conclude with a comparison of the selected filters based on the aforementioned test scenario illustrated in Fig. 1, using the vehicle dynamics and sensor models provided.

## 2. Requirements

A navigation system is required to maintain a navigational solution at all times during a mission. A navigation solution, $x$, in (3) is an estimate of the pose (state) of the vehicle in (1). This estimate consists of translational, $p^n$, and rotational, $\Theta$, position (i.e., position and attitude) coordinates $\eta$, in (2), and translational, $v_o^b$, and rotational, $\omega_{nb}^b$, velocity coordinates, $\upsilon$, in (2).

$$v_o^b = \begin{bmatrix} u \\ v \\ w \end{bmatrix}, \quad \omega_{nb}^b = \begin{bmatrix} p \\ q \\ r \end{bmatrix}, \quad p^n = \begin{bmatrix} x_n \\ y_n \\ z_n \end{bmatrix}, \quad \Theta = \begin{bmatrix} \phi \\ \theta \\ \psi \end{bmatrix} \tag{1}$$

$$\upsilon = \begin{bmatrix} v_o^b \\ \omega_{nb}^b \end{bmatrix}, \quad \eta = \begin{bmatrix} p^n \\ \Theta \end{bmatrix} \tag{2}$$

$$x = \begin{bmatrix} \upsilon \\ \eta \end{bmatrix} \quad (3)$$

This estimate is built on the stream of measurements being acquired by sensors that can be related to the states of interest. These measurements may be consistent or inconsistent over the duration of a mission. If the measurements are inconsistent then the navigation system maintains estimates of the states using knowledge of the dynamics of the system and the remaining measurements. This most commonly occurs while estimating position when GPS is unavailable, but can also affect velocity estimates when velocity measurements are unavailable due to the limited range of the acoustic based velocity sensors.
On an AUV there are a variety of power and physical size constraints on the vehicle due to mission requirements, and these constraints present themselves as computational restrictions on any embedded algorithms being used onboard, including the navigation system. Most of the vehicle power budget is highly prioritised for propulsion and control actuation, making the budget for computation limited and tightly constrained. This situation has been mitigated somewhat since the introduction of Li-ion battery technology in that the amount of power available on a vehicle has now increased significantly. There have also been significant improvements in the performance and efficiency of the computing platforms used, through advancements in computer architectures and silicon technology. Over the past few years several alternatives that could be used as suitable computational platforms for AUVs have been developed. These platforms include:
CPU - Central Processing Unit - the ubiquitous x86 based compact PCI 104 architecture which is currently the most commonly used architecture for mobile robotics.
DSP - Digital Signal Processors - microprocessor optimised for digital signal processing which constitutes the majority of calculations in a navigation system.
FPGA - Field Programmable Gate Arrays - reconfigurable integrated circuits enabling parallel architectures and optimal custom-built hardware implementation.
GP-GPU - General Purpose - Graphic Processor Units - essentially highly parallel data processing units which are suitable for digital signal processing and control applications.

## 3. Dynamics

To estimate the dynamic state of an AUV, first the equations that describe its motion must be defined and as such the dynamic navigation equations of a 6-DoF vehicle are presented. These equations are used to define the vehicle's kinematic and kinetic behaviour. They cover the co-ordinate frames used to represent the state information and the transformations that convert information represented in one frame to another frame. Rigid body dynamics and equations that model the hydrodynamic forces and moments acting on the vehicle are also included.

### 3.1 Reference Frames

Before reviewing the principles of navigation in underwater environments, certain reference frames need to be defined. A reference frame is a co-ordinate frame in which information can be represented. The frames that are used in underwater navigation are:

- ECEF - The Earth Centered Earth Fixed reference frame is a three dimensional Euclidean space with an $x_e$, $y_e$, $z_e$ cartesian representation whose origin is at the centre of the earth and is stationary in reference to the earth's surface. As illustrated in Fig. 2a, the $x_e$ axis of this reference frame projects out of the earth's surface at $0^0$N $0^0$E, the $y_e$ axis out at $0^0$N $90^0$E, and the $z_e$ axis out at $90^0$N the geographic north pole. This frame is used in long distance navigation over significant portions of the earth surface, as any point in the vicinity of the earth can be defined in this frame.

- ECI - The Earth Centered Inertial reference frame represents the true inertial frame in the confines of the earth. It is much like ECEF but does not rotate with the earth, rather the $x_e$ and $y_e$ axes point to specific points in the celestial sphere. Although the ECI frame is used for inertial measurements, in the case of slow moving vehicles and for short missions ECEF can be considered as inertial.
- NED - The North East Down reference frame is defined as a 'flat earth' frame in that it makes use of the assumption that for relatively small distances the surface of the earth can be considered to be flat. The origin of the NED is placed at some arbitrary point on the earth reference ellipsoid. The axes of this frame, as the name suggests, points north, east and down for axes $x_n$, $y_n$, and $z_n$, respectively, as shown in Fig. 2a. This frame is used when the flat earth assumption holds as it simplifies the equations in comparison to ECEF.
- BODY - The body-fixed frame is a reference frame which is fixed to the vehicle. The origin of the frame is located at a convenient position in relation to the vehicle usually the centre of gravity (or buoyancy in the case of underwater vehicles). The axes of this frame project out of the front (bow), right (starboard) and bottom (keel) of the vehicle for the $x_b$, $y_b$, and $z_b$ axes, respectively, as presented in Fig. 2b. The body-fixed frame is used to define the dynamics of the vehicle as well as the measurements as many of the measurement sensors are attached to the vehicle.

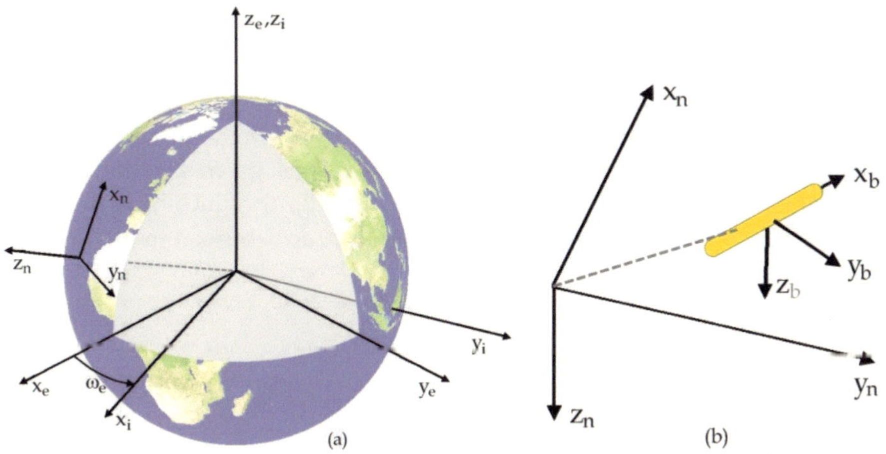

Fig. 2. Reference frames ECI, ECEF and NED (a), NED and BODY frame (b)

### 3.2 NED to Body-Fixed Frame Conversion

While it is convenient to represent the vehicle's velocities in the body-fixed frame, it is more convenient to represent the vehicle's position in the NED frame. The consequence of using two frames is the necessity of transforming information between the frames. In the case of 6 DoF dynamics of a free moving body, the velocity information represented in the body-fixed frame, υ, must be rotated such that the same velocity information is represented in the NED frame, $\dot{\eta}$. To rotate the velocity information, each of the vectors, $v_o^b$ and $\omega_{nb}^b$, are applied to two rotational matrices, $R_b^n(\Theta)$ and $T_\Theta(\Theta)$, respectively.

### 3.2.1 Linear Velocity Transformation

The linear velocity transformation matrix $R_b^n(\Theta) \in \mathbb{R}^{3\times3}$ transforms the translational velocities defined in the body-fixed frame, $v_o^b$, into the velocities in the NED frame, $\dot{p}^n$, based on the rotational differences between the two frames represented using Euler angle notation Θ (4). $R_b^n(\Theta)$ consists of three rotations, one for each Euler angle (5), and is defined in (6) using the common convention zyx (Titterton & Weston, 2004), where $v_o^b$ is rotated by ϕ, then by θ, and then by ψ. The order of rotations as shown in (6) is not arbitrary, due to the compounding effect of the rotational order.

$$\dot{p}^n = R_b^n(\Theta)v_o^b \tag{4}$$

$$R_{x,\phi} = \begin{bmatrix} 1 & 0 & 0 \\ 0 & \cos\phi & -\sin\phi \\ 0 & \sin\phi & \cos\phi \end{bmatrix}, R_{y,\theta} = \begin{bmatrix} \cos\theta & 0 & \sin\theta \\ 0 & 1 & 0 \\ -\sin\theta & 0 & \cos\theta \end{bmatrix}, R_{z,\psi} = \begin{bmatrix} \cos\psi & -\sin\psi & 0 \\ \sin\psi & \cos\psi & 0 \\ 0 & 0 & 1 \end{bmatrix} \tag{5}$$

$$R_b^n(\Theta) := R_{z,\psi}R_{y,\theta}R_{x,\phi} \tag{6}$$

$$R_b^n(\Theta) = \begin{bmatrix} \cos\psi\cos\theta & -\sin\psi\cos\phi + \cos\psi\sin\theta\sin\phi & \sin\psi\sin\phi + \cos\psi\sin\theta\cos\phi \\ \sin\psi\cos\theta & \cos\psi\cos\theta + \sin\phi\sin\theta\sin\psi & -\cos\psi\sin\phi + \sin\theta\sin\psi\cos\phi \\ -\sin\theta & \cos\theta\sin\phi & \cos\theta\cos\phi \end{bmatrix} \tag{7}$$

### 3.2.2 Angular Velocity Transformation

The angular velocity transformation matrix $T_\Theta(\Theta) \in \mathbb{R}^{3\times3}$ transforms the body-fixed rotational velocity information, $\omega_{nb}^b$, into the NED frame, $\dot{\Theta}$, based on the rotational difference between the frames (8). Similar to (4), the transformation conforms to the zyx rotational ordering as shown in (9) for the same reasons given for the linear velocity transformation. For clarity, consider $T_\Theta^{-1}(\Theta)$, given in (10), in which $\dot{\psi}$ is rotated by $R_{y,\theta}$, added to $\dot{\theta}$, then the sum rotated by $R_{x,\phi}$ and added to $\dot{\phi}$, as described in (9). This effect is in contrast to (4) where the whole vector is rotated.

$$\dot{\Theta} = T_\Theta(\Theta)\omega_{nb}^b \tag{8}$$

$$\omega_{nb}^b = \begin{bmatrix} \dot{\phi} \\ 0 \\ 0 \end{bmatrix} + R_{x,\phi}^T \begin{bmatrix} 0 \\ \dot{\theta} \\ 0 \end{bmatrix} + R_{x,\phi}^T R_{y,\theta}^T \begin{bmatrix} 0 \\ 0 \\ \dot{\psi} \end{bmatrix} := T_\Theta^{-1}(\Theta)\dot{\Theta} \tag{9}$$

$$T_\Theta^{-1}(\Theta) = \begin{bmatrix} 1 & 0 & -\sin\theta \\ 0 & \cos\phi & \cos\theta\sin\phi \\ 0 & -\sin\phi & \cos\theta\cos\phi \end{bmatrix} \Rightarrow T_\Theta(\Theta) = \begin{bmatrix} 1 & \sin\phi\tan\theta & \cos\phi\tan\theta \\ 0 & \cos\phi & -\sin\phi \\ 0 & \sin\phi/\cos\theta & \cos\phi/\cos\theta \end{bmatrix} \tag{10}$$

### 3.2.4 Unit Quaternion

Unit quaternion is a four element replacement of the three element Euler angle representation. The main motivation for quaternions is to avoid singularities, commonly called 'gimbal lock', that can occur in the three element representation when θ=±90°. When the pitch attains these values, the rotational matrix $T_\Theta(\Theta)$ become undefined due to the trigonometric functions used in the rotation. A quaternion is a four element array that is described as a four dimensional vector in a complex space (Fossen, 2002), or alternatively a complex number (11) with one real and three complex components. Replacing the Euler infomation in (3) with (11) yields an alternative state vector, $x$ (12). Using a complex space representation is advantageous due to the implied orthogonality of a complex number.

$$q = \eta_0 + \varepsilon_1 i + \varepsilon_2 j + \varepsilon_3 k = [\eta_0 \quad \varepsilon_1 \quad \varepsilon_2 \quad \varepsilon_3]^T \tag{11}$$

$$x = [v \quad p^n \quad \mathrm{q}]^T \tag{12}$$

A unit quaternion is essential for representing frame rotations as any non-unit magnitudes will scale the transformations rather than a pure rotational transformation. A unit quaternion must satisfy (13) and as such a normalization would consist of (14).

$$q^T q = 1 \tag{13}$$

$$\tilde{q} = \frac{q}{q^T q} \tag{14}$$

When applying the same principle to unit quaternions as to Euler angles in defining (4), (15) results. As evident in (15), an advantage of the unit quaternion representation is the absence of trigonometric relationships in the rotation matrix $R_b^n(\mathrm{q})$ thus simplifying the computation. When applying unit quaternion definition to (8), (16) is produced. Equation (16), similar to (15), produces a result that contains no trigonometeric functions and likewise simplifies the computation.

$$R_b^n(\mathrm{q}) = \begin{bmatrix} 1-2(\varepsilon_2^2+\varepsilon_3^2) & 2(\varepsilon_1\varepsilon_2-\varepsilon_3\eta_0) & 2(\varepsilon_1\varepsilon_3+\varepsilon_2\eta_0) \\ 2(\varepsilon_1\varepsilon_2+\varepsilon_3\eta_0) & 1-2(\varepsilon_1^2+\varepsilon_3^2) & 2(\varepsilon_2\varepsilon_3-\varepsilon_1\eta_0) \\ 2(\varepsilon_1\varepsilon_3-\varepsilon_2\eta_0) & 2(\varepsilon_2\varepsilon_3+\varepsilon_1\eta_0) & 1-2(\varepsilon_1^2+\varepsilon_2^2) \end{bmatrix} \tag{15}$$

$$T_\mathrm{q}(\mathrm{q}) = \begin{bmatrix} -\varepsilon_1 & -\varepsilon_2 & -\varepsilon_3 \\ \eta_0 & -\varepsilon_3 & \varepsilon_2 \\ \varepsilon_3 & \eta_0 & -\varepsilon_1 \\ -\varepsilon_2 & \varepsilon_1 & \eta_0 \end{bmatrix} \tag{16}$$

### 3.3 Kinematics

Kinematics are the equations of motion that relate the state of any body at one point in time to another point in time due the geometric nature of space and the motion of the object in that space. For a free body in three dimensional space, the 6-DoF equations of motion comprising the three translational and three rotational degrees of freedom (Titterton & Weston, 2004), are defined as (17).

$$\dot{\eta} = J(\eta)\upsilon \quad \Rightarrow \quad \begin{bmatrix} \dot{p}^n \\ \dot{\Theta} \end{bmatrix} = \begin{bmatrix} R_b^n(\Theta) & 0_{3\times 3} \\ 0_{3\times 3} & T_\Theta(\Theta) \end{bmatrix} \begin{bmatrix} v_o^b \\ \omega_{nb}^b \end{bmatrix} \tag{17}$$

Alternatively, applying (17) to a quaternion form (11) yields (18).

$$\dot{\eta} = J(\eta)\upsilon \quad \Rightarrow \quad \begin{bmatrix} \dot{p}^n \\ \dot{\mathrm{q}} \end{bmatrix} = \begin{bmatrix} R_b^n(\mathrm{q}) & 0_{3\times 3} \\ 0_{3\times 3} & T_\Theta(\mathrm{q}) \end{bmatrix} \begin{bmatrix} v_o^b \\ \omega_{nb}^b \end{bmatrix} \tag{18}$$

Equation (17) or (18) can be used to define the motion of any freely moving object in three dimensional space as it does not consider how the motion is created. This results in (18) being valid for any mobile platform regardless of context.

### 3.4 Kinetics

Kinetics, now commonly known as analytical dynamics or just dynamics, can be described as a branch of mathematics that deals with the motion of bodies and the influences that create the motion, namely forces and torques. Kinetics, in contrast to kinematics, incorporates the causes of the motion as well as the motion itself.

The basic premise of dynamics is expressed in Newton's Second Law $F = ma$. Expanding Newton's Second Law to three dimensional space, as found in a freely moving body with mass $M$ and force $F_i$ exerted upon it, and using the principle of superposition of forces result in the generalized kinetic equation (19).

$$M\dot{\upsilon} = \sum_{i=1}^{n} F_i \tag{19}$$

The components of (19), $M$ and $F_i$, are defined by two distinct effects, rigid body dynamics and hydrodynamics.

#### 3.4.1 Rigid Body Dynamics

Rigid body dynamics apply Newtonian mechanics using only forces involved with the 'rigid' body of the vehicle and do not include forces produced by motion of the vehicle through the medium. In (20), the effect from the intrisic interia of the vehicle, $M_{RB}$, and the forces apparent due to the motion of the reference frame, namely the 'coriolis force', $C_{RB}(\upsilon)\upsilon$, and the control forces, $\tau_{RB}$, are modelled.

$$M_{RB}\dot{\upsilon} + C_{RB}(\upsilon)\upsilon = \tau_{RB} \tag{20}$$

#### 3.4.2 Vehicle Hydrodynamics

When describing the motion of a vehicle through a fluid, in this case an AUV through water, there is a generalized equation given in (21) which expands from (20) and parameterizes this motion (Fossen, 2002).

$$M\dot{\upsilon} + C(\upsilon)\upsilon + D(\upsilon)\upsilon + g(\eta) = \tau + w \tag{21}$$

where $M$ and $C(\upsilon)\upsilon$ add the effect of 'added mass' (Fossen, 2002) to (20), i.e. $M = M_{RB} + M_A$ and $C(\upsilon) = C_{RB}(\upsilon) + C_A(\upsilon)$, while $D(\upsilon)\upsilon$ characterises hydrodynamic drag, $g(\eta)$ does the same for buoyancy and gravitational effects, and $w$ models external disturbances.

### 3.4.3 Navigation Equations

Collating (21) and (17) or (18), a concise representation of the 6-DoF dynamics of an underwater vehicle, accurate to the second order, can be represented in (22).

$$\dot{\upsilon} = M^{-1}\big(\tau - C(\upsilon)\upsilon - D(\upsilon)\upsilon - g(\eta)\big), \\ \dot{\eta} = J(\eta)\upsilon \tag{22}$$

## 4. Sensors

An AUV employs a variety of exteroreceptive and proprioceptive sensors to obtain information about the surrounding environment and the vehicle itself. This section will review some of the most common sensors that can provide information related to (3). The review will cover the information available from these sensors, limitations of these sensors and sources of noise. AUVs are particularly susceptible to limitations in sensors since the ranges and capabilities of the sensors are limited to what can be carried by the vehicle based on power, size, and cost constraints.

### 4.1 IMU

An Inertial Measurement Unit (IMU) is an orthogonal triad of accelerometers and gyroscopes that can detect the three translational accelerations and rotational velocities experienced by any free moving body in three dimensional space. Integrating this information yields position and attitude in a process called dead reckoning. These sensors are defined as inertial sensors as they use the inertia of an internal reference mass or structure to generate the measurements and, as a consequence, they are completely self-contained.

There are two main configurations, namely 'inertial-frame' IMUs and 'strapdown' IMUs. An inertial-frame IMU mechanically maintains its orientation using the gyroscopes, while the accelerometers stay in the same frame effectively mechanically integrating the rotational rates. Being mechanical, these systems are large and power hungry and thus unsuitable for small AUVs. In strapdown IMUs, all the sensors are attached to the vehicle and the information is computationally integrated. The advent of the microprocessor makes this integration a relatively trivial process.

#### 4.1.1 Accelerometers

An accelerometer measures translational acceleration in the inertial frame of the sensor (vehicle). In principle, accelerometers measure acceleration by measuring the deflection of a mass attached to a spring. The most suitable technology for use in an IMU is MEMS accelerometers. MEMS accelerometers use either a pendulous mass or a resonant beam structure. In the pendulous mass architecture, as illustrated in Fig. 3a, a capacitance is generated that is related to the displacement of the proof mass. In the resonant beam structure, variations in the resonant frequency of the beam due to acceleration induced loading are measured. These values are converted to voltages and subsequently converted to digital representations.

#### 4.1.2 Gyroscopes

Gyroscopes are devices that can measure rotational velocity in the inertial frame. For use in small underwater vehicles, gyroscopes can be classified into two classes, optical based and MEMS technology based. Optical based technologies, such as ring-laser gyroscopes and fibre-optic gyros, all use the same principal. A structured light beam is first split and then sent in opposing directions around a circular path, resulting in a phase difference between the two beams whenever a rotational change occurs – this is known as 'the Sagnac effect'. The intensity of the recombined beams is related to the rotational rate. MEMS gyroscopes use a resonant beam or ring structure. A resonant structure will tend to vibrate in the same direction and any vibration not parallel or orthogonal to this axis is related to rotational motion – this is known as 'the coriolis effect' and is shown in Fig. 3b. This motion can be detected as a capacitance changes.

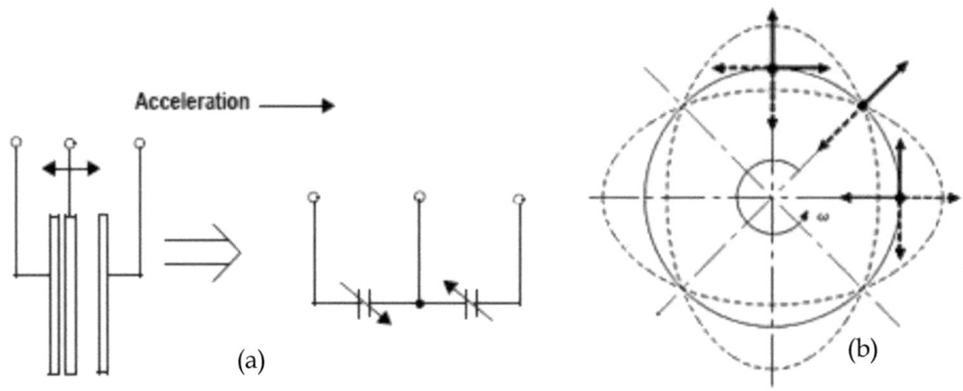


Fig. 3. MEMS Accelerometer (a), and Gyroscope (b) Structures

#### 4.1.3 MEMS Sensors

While MEMS sensors are smaller, cheaper, and use much less power than their counterparts, they have one downside, i.e., reduced accuracy. The errors generated from the non-idealities in MEMS sensors include:

- Fixed-Bias ($b_i$) - a bias in the measurement due to offset in the zero of the sensor and/or A/D biases (m/s$^2$).
- Scale Factor ($a_i$) - a scaling constant which defines the relationship between the measured voltage or binary value and the perceived acceleration (m/s$^2$/V or m/s$^2$/LSB). This can include second or higher order terms.
- Cross Coupling ($\epsilon_i$) - is a matrix that represents the sensitivity of a sensor to motion or fields orthogonal to its desired axis of sensitivity.
- Temperature based ($c$) - some of the errors mentioned previously, specifically bias and scale errors, can also be affected by the ambient temperature.
- Quiescent Noise ($w$) - is the noise that occurs when the instrument is sensing a zero state. This value is used to define the random error, *noise*, in the output signal.

Combining these effects to define a relationship between (3) and the measurement produces (23).

$$\beta_{INS} = \begin{bmatrix} a_1 & 0 & 0 \\ 0 & a_2 & 0 \\ 0 & 0 & a_3 \end{bmatrix} \begin{bmatrix} 1 & -\epsilon_\psi & \epsilon_\theta \\ \epsilon_\psi & 1 & -\epsilon_\phi \\ -\epsilon_\theta & \epsilon_\phi & 1 \end{bmatrix} \beta + \begin{bmatrix} b_1 \\ b_2 \\ b_3 \end{bmatrix} + \begin{bmatrix} c_1 \\ c_2 \\ c_3 \end{bmatrix} T + w \tag{23}$$

where

$$\beta_{INS} = \begin{bmatrix} \omega_x \\ \omega_y \\ \omega_z \end{bmatrix} \; and \; \beta = \begin{bmatrix} p \\ q \\ r \end{bmatrix} \text{ for gyroscopes, and}$$

$$\beta_{INS} = \begin{bmatrix} \alpha_x \\ \alpha_y \\ \alpha_z \end{bmatrix} \; and \; \beta = \begin{bmatrix} I_{3\times3} & \begin{matrix} 0 & -x_a^b & y_a^b \\ x_a^b & 0 & -z_a^b \\ -y_a^b & z_a^b & 0 \end{matrix} \end{bmatrix} \dot{v} \text{ for accelerometers}$$

Most modern MEMS sensors have factory calibration for many of these errors, however, some of the errors are not static but change very slowly over the lifetime of the device in a process known as 'ageing'. These residual errors can be accounted for using various offline or online parameter estimation techniques. Assuming the sensors are calibrated, (23) simplifies to (24).

$$\beta_{INS} = \beta + w \tag{24}$$

### 4.2 Compass

A compass is a sensor that measures the earth's magnetosphere. The core of most modern IC based magnetometers uses the 'Hall Effect' principle. The Hall Effect dictates that in presence of a magnetic field, the electron flow in a conductor will be distorted orthogonally to the current flow, and the magnetic field producing a voltage difference across the conductor that is proportional to the magnetic field.

MEMS based magnetometers have many of the same errors as MEMS IMU components such as scale, $C_{sf}$, and misalignment, $C_m$, but have two extra sources of error due to the interaction of the sensed magnetic field with nearby magnetic materials:

- Hard Iron Errors ($\delta B^b$) - The largest error induced in the magnetic field tends to be unwanted magnetic fields produced by surrounding electrical equipment or magnetic abnormalities in the environment. These hard iron errors produce a bias in the resulting measurement due to the summation of the desired and interfering magnetic fields.
- Soft Iron Errors ($C_{si}$) - When non-magnetized magnetic materials are exposed to an external magnetic field, they will distort the field in vicinity of the material.

Combining these errors to define a relationship between the measurement and (3) produces (25).

$$B^b = C_m C_{sf} C_{si}\left(R_b^{n^T}(\Theta)B^n + \delta B^b\right) + w \tag{25}$$

Most of the errors that occur in MEMS based magnetometers can be removed through calibration. The only effect that cannot be calibrated for are hard iron errors due to local magnetic abnormities which must be removed using online estimation (Gebre-Egziabher et al., 2001) if the error is significant enough to be of concern. Once calibration or estimation has removed the above errors (25) simplifies to (26).

$$B^b = R_b^{n^T}(\Theta)B^n + w \tag{26}$$

### 4.3 GPS

The Global Positioning System (GPS) is a global navigation satellite system (GNSS) which consists of an array of satellites that is used by a receiver to triangulate the receiver's position anywhere in the world depending on satellite availability (Tsui, 2005).

GPS operates on the principle that, given a common time reference, a range to a satellite can be calculated by the time difference between transmission from the satellite and reception by the receiver of a radio signal multiplied the speed of light. This measurement is called a pseudo-range (PR), as the measurement is a range calculated from a time difference rather than a direct range measurement. Using three satellite range measurements and positions, $p_{s_i}^e$, the three values that constitute a position in 3D space, $p^e$, can be solved using simultaneous equations (27).

A common time reference does not exist, however, as the receiver only has a quartz crystal oscillator whereas each of the satellites have atomic clocks that are accurate to a few nanoseconds being periodically corrected by the GPS control segment. This problem can be solved by including a receiver clock offset, $\delta t_r$, into the triangulation problem (27). Since four unknowns now exist, a minimum number of four satellites are required for a position fix (Tsui, 2005).

$$\begin{aligned} PR_1 &= \delta t_r + \left\| p^e - p_{s_1}^e \right\|, \\ PR_2 &= \delta t_r + \left\| p^e - p_{s_2}^e \right\|, \\ &\vdots \\ PR_n &= \delta t_r + \left\| p^e - p_{s_n}^e \right\|. \end{aligned} \tag{27}$$

The main sources of error in this system are:

- Clock Inaccuracies ($\delta t_r$) - as mentioned previously there exists a clock bias due to clock drift that is compensated for in the positioning algorithm. Remaining inaccuracies in the clock can create ranging errors of up to two meters.
- Satellite Geometry – the apparent angle between satellites as perceived by the receiver can have a significant effect on the triangulation problem. If the angles are small, any error in the range measurement from one of the satellites can significantly affect the resolved position with errors as large as 100 - 150 meters.
- Satellite Orbits – Even though GPS satellites are very accurately placed in orbit, errors in this orbit can still occur. 'Ephemeris data' transmitted within the GPS signal contains correction information, limiting the effect of orbit errors to less than two meters.
- Multipath Effect –The line of sight 'desired' signal between satellite and receiver can be corrupted by delayed multipath reflections of itself. Typically the error for multipath is no more than a few meters.
- Atmospheric Effects – The speed of the GPS signals are affected by their passage through the atmosphere, particularly the ionosphere which slows down electromagnetic waves proportional to $1/f^2$ (Tsui, 2005). This has motivated the use of dual frequency receivers

which receive a GPS signal on two frequencies. The amount that each signal is slowed by can be calculated using the time difference between the reception of the GPS signals on the two frequencies.

- Relativistic Effects – Due to relativistic effects, the atomic clock onboard the satellite runs slower than on earth and so is tuned before launch to account for these effects.
- Selective Availability (SA) – This refers to the inclusion of pseudo noise by the US government to restrict access to high quality position information. SA was removed in May 2000.

Although the addition of a second frequency can be used to compensate for ionspheric errors, it does nothing to compensate for some of the other effects or for ionospheric errors in single frequency receivers. In such cases, the use of systems, like WAAS or EGNOS (Tsui, 2005), can be used to provide coarse corrections for atmospheric, clock, and orbit related errors.

### 4.4 SONAR

SOund Navigation And Ranging (SONAR) is an acoustic method for determining the range and bearing of a target or of the sonar device itself, if multiple targets are at known locations.

There are two types of sonar: active and passive. In this chapter, the discussion will be restricted to active sonar systems as they are the most pertinent in relation to navigation.

Active sonar assumes that sound has a finite speed, $c$, in any medium, thus a range to a feature, $p_t^n$, can be determined if the time difference between transmission and reception, $\delta t$, can be measured. By using a transceiver to transmit a focused sound pulse (ping) and receive the return signal (echo), target information can be extracted from the amplitude, round trip time, and transceiver azimuth (28) of the echo.

$$\frac{1}{2}c\delta t = \|p^n - p_t^n\| \tag{28}$$

There are several varieties of active sonar systems, the most common are scanning sonar, side-scan sonar, and multibeam (bathymetric) sonar systems.

The errors in any sonar system are highly nonlinear, the most significant being:

- Refraction – the path of the sonar ping can become curved, changing the ensonified area, due to the non uniformity of the temperature and density of the water over the trajectory of the ping.
- Scattering – when the angle of incidence of the ping to the target is not ideal, the reflection is diffuse rather than specular, resulting in a poor return with the possibility of multipath reflections producing errant echoes.
- Speed of sound – any error in the value of the sound speed due to temperature, salinity, or density fluctuations, will produce a proportional error in the range calculation.

Many of the above errors cannot be calibrated for and must be accounted for using robust signal processing and estimation algorithms within the sonar.

### 4.5 DVL

A Doppler Velocity Log (DVL) is a form of downward looking sonar that rather than simply measuring the round-trip time of the sonar ping, it can also measure the Doppler shift, $f_0 - f_r$, in the return signal which is related to the speed of the vehicle relative to the seabed, $v_{s,r}$, and the speed of the wave in the medium, $c$, via (29).

$$v_{s,r} = \frac{(f_0 - f_r)c}{f_0} \tag{29}$$

DVLs typically use four transponders that all approximately point down but are each offset in a different direction by a given angle, usually 30°, as illustrated in Fig. 4. These four measurements provide enough information to determine the velocity of the transponder in three dimensions relative to some other reference (30). If the seafloor is in range of the DVL's beams, the velocity relative to the ground can be determined. If not, the DVL can determine its velocity to a given body of water below the vehicle.

$$v_{s,r} = \begin{bmatrix} \cos\psi_{t1}\sin\theta_{t1} & \sin\psi_{t1}\sin\theta_{t1} & \cos\theta_{t1} \\ \cos\psi_{t2}\sin\theta_{t2} & \sin\psi_{t2}\sin\theta_{t2} & \cos\theta_{t2} \\ \cos\psi_{t3}\sin\theta_{t3} & \sin\psi_{t3}\sin\theta_{t3} & \cos\theta_{t3} \\ \cos\psi_{t4}\sin\theta_{t4} & \sin\psi_{t4}\sin\theta_{t4} & \cos\theta_{t4} \end{bmatrix} v_n^b \tag{30}$$

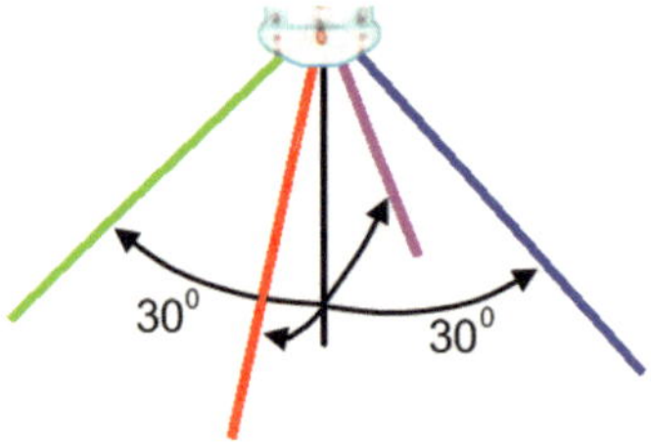


Fig. 4. DVL Transponders

DVLs suffer from the same error sources as sonar. As a consequence, many of these errors cannot be calibrated for and must be accounted for using robust signal processing and estimation algorithms.

### 4.6 CTD

A CTD is a sensor triad comprising a conductivity, a temperature, and a pressure sensor. The CTD can thus determine the salinity and density, as well as the temperature of the surrounding water. The first two parameters are crucial for correcting the sonar range information. AUV depth, $h$, below the surface of the water can be determined using (31) which is dependent on the pressure above atmospheric pressure, $P - P_0$, and density of the water, $\rho$. Depending on required accuracy and operating environment, the density of water can be assumed to be constant making (31) purely proportional to pressure. For more accurate results, the surrounding water density and salinity can be determined from CTD measurements using high-order polynomials fitted to experimental data as described in 'The International Equation of State of Seawater 1980' (IESS 1980).

$$h = \frac{P - P_0}{\rho g} \tag{31}$$

### 4.7 Acoustic Positioning

There have been several attempts to augment the onboard navigation sensors by the use of acoustic beacons. Typical schemes include long baseline (LBL), short baseline (SBL), and ultra-short baseline (USBL) positioning systems. In LBL systems, four GPS positioned beacons are located at the corners of an AUV's operating site. The AUV determines its position by transmitting a single pulse and then working its distance from each beacon from the time taken for the signal to be retransmitted by the beacons and received by the AUV. SBL systems use three or more separate beacons that are centrally connected to a surface vessel to provide synchronised timing. An initiator signal is sent by one of the beacons and a reply, generated by the AUV, is received by all beacons. The position of the AUV is then calculated and sent back to the AUV. USBL systems, conversely, use only a single beacon which is rigidly mounted on a precisely located object, and a receiver array mounted on the AUV. Distance from the beacon is determined from the signal return time, and direction by measuring the phase shift of the reply signal on the receive array.

Acoustic positioning systems have their drawbacks, the most significant of which is that such systems restrict the autonomous nature of the vehicle since the vehicle has to remain in an area that is covered by the beacons system. Other problems with acoustic positioning occur due to multipath reflections, occlusion and interference from other transponders in the vicinity.

## 5. Estimators

Navigation estimation systems are filtering mechanisms that process the available measurements to obtain an estimate of the required information with minimal noise corrupting the solution. This section will review the use of filters that employ a recursive Bayesian technique. This set of filters encompasses the Kalman filter family which includes the Extended Kalman Filter (EKF) and the Unscented Kalman Filter (UKF), as well as the particle filter family which includes the Sequential Importance Re-sampling (SIR) filter, the Regularized Particle filter, the Monte Carlo Localizer (MCL), and the Measurement-Assisted Partial Re-sampling (MAPR) Particle filter.

### 5.1 Bayesian Filtering

The general Bayesian filter approach defines how the pdf (probability distribution function) of a given state's estimate evolves over time based on the state dynamics (i.e., 'state model'), observations of the state (measurements), and the uncertainties of each 'observation model'(Ristic et al., 2004). The principle of Bayesian filtering states that:

If state and observation models are known as (32) and (33), respectively, then the state estimate, $p(x|Z)$, can be propagated from time k-1 to k via (34) and (35).

$$x_k = f_{k-1}(x_{k-1}, v_{k-1}) \tag{32}$$

$$z_k = h_k(x_k, w_k) \tag{33}$$

$$p(x_k|Z_{k-1}) = \int p(x_k|x_{k-1})p(x_{k-1}|Z_{k-1})dx_{k-1} \quad (34)$$

$$p(x_k|Z_k) = \frac{p(z_k|x_k)p(x_k|Z_{k-1})}{p(z_k|Z_{k-1})} \quad (35)$$

where

$$p(z_k|Z_{k-1}) = \int p(z_k|x_k)p(x_k|Z_{1:k-1})\, dx_k$$

These equations can conveniently be defined as a two-phase system in which a prediction of the new state value is made based on the previous value and the state dynamics (34) and the prediction is then corrected by any available observations (35).

### 5.2 Kalman Filter

The Kalman filter is an implementation of the general recursive Bayesian filtering approach. It is a recursive filter for linear systems and is generally considered to be a Linear Quadratic Estimator (LQE). The Kalman filter is a computable approximation of the general equation (32)-(35) and achieves this by placing linear and Gaussian assumptions on (32)-(33), thus resulting in (36)-(37). These assumptions ensure that for all times, (34)-(35) will be Gaussian and thus can be represented using the mean and covariance (Ristic et al., 2004).

$$x_k = F_{k-1}x_{k-1} + v_{k-1} \quad (36)$$

$$z_k = H_k x_k + w_k \quad (37)$$

The Kalman filter applies Bayesian filtering to (36)-(37), and defines (38)-(43).

$$\hat{x}_{k|k-1} = F_{k-1}\hat{x}_{k-1|k-1} \quad (38)$$

$$P_{k|k-1} = Q_{k-1} + F_{k-1}P_{k-1|k-1}F_{k-1}^T \quad (39)$$

$$\hat{x}_{k|k} = \hat{x}_{k|k-1} + K_k\left(z_k - H_k\hat{x}_{k|k-1}\right) \quad (40)$$

$$P_{k|k} = P_{k|k-1} - K_k S_k K_k^T \quad (41)$$

where

$$S_k = R_k + H_k P_{k|k-1} H_k^T \quad (42)$$

$$K_k = P_{k|k-1} H_k^T S_k^{-1} \quad (43)$$

#### 5.2.1 EKF

The Extended Kalman Filter (EKF) is a modification of the standard Kalman Filter to accommodate the use of nonlinear equations in defining (32)-(33). This is accommodated by the use of the Jacobian operator to provide a linear approximation, (52)-(53), of (44)-(45) at x. An important note in relation to accuracy is that, the Jacobians of (52)-(53) are accurate to a first-order Taylor series expansion of (44)-(45). Replacing (44)-(45) with (52)-(53), where required, produces (46)-(53).

$$x_k = f_{k-1}(x_{k-1}) \quad (44)$$

$$z_k = h_k(x_k) \quad (45)$$

$$\hat{x}_{k|k-1} = f_{k-1}(\hat{x}_{k-1|k-1}) \tag{46}$$

$$P_{k|k-1} = Q_{k-1} + \hat{F}_{k-1} P_{k-1|k-1} \hat{F}_{k-1}^T \tag{47}$$

$$\hat{x}_{k|k} = \hat{x}_{k|k-1} + K_k \left( z_k - h_k(\hat{x}_{k|k-1}) \right) \tag{48}$$

$$P_{k|k} = P_{k|k-1} - K_k S_k K_k^T \tag{49}$$

where

$$S_k = R_k + \hat{H}_k P_{k|k-1} \hat{H}_k^T \tag{50}$$

$$K_k = P_{k|k-1} \hat{H}_k^T S_k^{-1} \tag{51}$$

$$\hat{F}_{k-1} = \left[ \nabla_{x_{k-1}} f_{k-1}^T(x_{k-1}) \right]^T \Big|_{x_{k-1} = \hat{x}_{k-1|k-1}} \tag{52}$$

$$\hat{H}_k = \left[ \nabla_{x_k} h_k^T(x_k) \right]^T \Big|_{x_k = \hat{x}_{k|k-1}} \tag{53}$$

The EKF has been heavily used and has become the de facto standard for nonlinear filtering/estimation. Navigation and in particular vehicular navigation has been the most defining application for the EKF. The EKF despite its proliferation is not an optimal filter (Ristic et al., 2004) due to the accuracy of $(\bar{x}, P)$ being dependent on the error between (44)-(45) and (52)-(53). When this error is zero, the equations breakdown to (36)-(37).
The EKF has proliferated in navigation applications as it provides reasonable estimates despite its non-optimality, and it is relatively simple to implement as well as computationally cheap to support.

### 5.2.2 UKF

The Unscented Kalman Filter (UKF) is a modification of the standard Kalman Filter. Unlike the EKF, which makes a linear approximation of the nonlinear system (52)-(53), the UKF makes no approximation or assumption of the system or measurement model. Rather, the UKF approximates the Gaussian distribution via a set of appropriately placed sample points, sigma points. This method is motivated by the logic that it is possible to make a more accurate approximation to a Gaussian distribution using well placed sample points, creating a sample mean and covariance, rather than a linear approximation of a nonlinear function using a Jacobian operator (Ristic et al., 2004). These points are placed according to the unscented transform (Ristic et al., 2004), giving the Kalman Filter variant its name. Replacing $(\bar{x}, P)$ with a set $\chi$ which are related by (55)-(56), and using the unscented transform (54), produce (57)-(66). The unscented transform places sigma points, $\chi^i$, according to the scaled (by κ) matrix square root of $P_a$. $P_a$ is the covariance matrix $P$ augmented with the control or measurement covariance matrices $Q$ and $R$ respectively. P is augmented such that the sigma points are placed as to capture the most information about the effect of the control or measurements. $n_a$ is the size of the augmented state vector, $x_a$, which is the x augmented with zeros to accommodate the size $P_a$. This augmenting can be performed in two different ways, either $Q$ augments $P$ for (57)-(59) and $R$ augments $P$ for (60)-(66), or $Q$ and $R$ all augment $P$ for (57)-(66).

$$\chi^0 = x_a = \begin{bmatrix} \bar{x} \\ 0_{Q\times 1} \\ 0_{R\times 1} \end{bmatrix} \qquad W^0 = \frac{\kappa}{n_a + \kappa}$$
$$\chi^i = x_a \pm \left(\sqrt{(n_a+\kappa)P_a}\right)_{\mathrm{i}} \quad W^i = \frac{1}{2(n_a+\kappa)} \quad i = 1,..,2n_a \tag{54}$$

$$\bar{x} = \sum_{i=0}^{2n_a} W^i \chi^i \tag{55}$$

$$P = \sum_{i=0}^{2n_a} W^i [\chi^i - \bar{x}][\chi^i - \bar{x}]^T \tag{56}$$

$$\chi^i_{k|k-1} = f_{k-1}(\chi^i_{k|k-1}) \tag{57}$$

$$\hat{x}_{k|k-1} = \sum_{i=0}^{N-1} W^i_{k-1} \chi^i_{k|k-1} \tag{58}$$

$$P_{k|k-1} = Q_{k-1} + \sum_{i=0}^{2n_a} W^i_{k|k-1} [\chi^i_{k|k-1} - \hat{x}_{k|k-1}][\chi^i_{k|k-1} - \hat{x}_{k|k-1}]^T \tag{59}$$

$$\hat{z}_{k|k-1} = \sum_{i=0}^{2n_a} W^i_{k-1} h_k(\chi^i_{k|k-1}) \tag{60}$$

$$\hat{x}_{k|k} = \hat{x}_{k|k-1} + K_k(z_k - \hat{z}_{k|k-1}) \tag{61}$$

$$P_{k|k} = P_{k|k-1} - K_k S_k K_k^T \tag{62}$$

where

$$K_k = P_{xz} S_k^{-1} \tag{63}$$

$$S_k = R_k + P_{zz} \tag{64}$$

$$P_{xz} = \sum_{i=0}^{2n_a} W^i_{k-1} [\chi^i_{k|k-1} - \hat{x}_{k|k-1}][h_k(\chi^i_{k|k-1}) - \hat{z}_{k|k-1}]^T \tag{65}$$

$$P_{zz} = \sum_{i=0}^{2n_a} W^i_{k-1} [h_k(\chi^i_{k|k-1}) - \hat{z}_{k|k-1}][h_k(\chi^i_{k|k-1}) - \hat{z}_{k|k-1}]^T \tag{66}$$

The UKF has one drawback in that, in the navigational estimation context, it is not as computationally efficient as the EKF, due to the requirement for multiple evaluations of the state dynamic equation (57) and a large matrix squareroot operation (54) which, however, can be offset somewhat with Cholesky decomposition (Ristic et al., 2004). The UKF is not affected by the nonlinearities in the dynamics, and as such produces an estimate that captures the 'true' mean and covariance more accurately than that of the EKF. The unscented transform used for defining the relationship between sample points and $(\bar{x}, P)$ is accurate to the third order Taylor series expansion for Gaussian distributions, as opposed to only first-order for the Jacobians in the EKF.

### 5.3 Particle Filtering

The particle filter is an alternative means of implementing the Bayesian filtering algorithm. The hypothesis of the particle filter is that an arbitrary distribution can be approximated by a multinomial distribution with a sufficiently large number of sample points in the solution space. Alternatively, a particle filter represents the pdf of the state estimate as a weighted set of sample points, i.e., 'particles' in the solution space (67). Particle filtering is also referred to as the Sequential Monte Carlo method as it is a recursive sequential analogue to the Markov Chain Monte Carlo (MCMC) batch methods (Doucet et al., 2001). A particle filter uses Sequential Importance Sampling (SIS), which is a recursive version of importance sampling. Equations (32)-(35) are applied to the multinomial representation (67) to produce (68)-(69). In the particle filter, a sample, $x_k^i$, is drawn from the proposal density (68) whose weight is modified based on the support from the proposal, likelihood, and transitional densities (69).

$$p(x_k|Z_{1:k}) \approx \sum_{i=1}^{N} \omega_k^i \delta(x_k - x_k^i) \tag{67}$$

so

$$x_k^i \sim q(x_k|x_{k-1}^i, z_k) \tag{68}$$

$$\omega_k^i \propto \omega_{k-1}^i \frac{p(z_k|x_k^i)p(x_k^i|x_{k-1}^i)}{q(x_k^i|x_{k-1}^i, z_k)} \tag{69}$$

As a consequence of the finite number of particles being evaluated, the drawback of the basic algorithm becomes evident due to a shortcoming of the SIS algorithm. If allowed to run unattended, the weights will converge to a point where all but one of the weights will have a mass of zero with the remaining particle having all the mass, a weight of one. This condition is defined as particle degradation and has the consequence that only one particle is contributing to the solution voiding the principle of the particle filter hypothesis. All particle filters employ a re-sampling stage (Douc et al., 2005) to redistribute the particles so as to avoid degradation, and produce a proposal density $q(x_k|x_{k-1}, z_k)$ that ensures x is drawn from an appropriate subset of the solution space. Two methodologies are available for resampling in a particle filter. The first is to only resample when it is needed via some fitness measure of the particle weights (70), and this methodology produces an architecture that conforms to the SIS with Resampling (SISw/R) frame work. Alternatively resampling can occur at every epoch, regardless of particle fitness, fitting under the definition of Sequential Importance Resampling (SIR).

$$\widehat{N}_{eff} = \frac{1}{\sum_{i=1}^{N} (\omega_k^i)^2} \tag{70}$$

An appropriate choice of proposal density is crucial as it can, if optimal, minimize the variance of the sample weights introduced in the sampling stage. An example of this is the first practical particle filter implementation, the 'Bootstrap' particle filter (Ristic et al., 2004), whose SIR architecture is one of two essential factors for the filter's successful operation. The other factor is that the bootstrap filter makes convenient use of (71) for the proposal distribution, which can easily be sampled from and which simplifies (69) to produce (72) but is not optimal.

$$q\left(x_k \middle| x_{k-1}^i, z_k\right) \triangleq p\left(x_k \middle| x_{k-1}^i\right) \quad (71)$$

$$\omega_k^i \propto \omega_{k-1}^i p\left(z_k \middle| x_k^i\right) \quad (72)$$

Sequential Importance Resampling (SIR) is the principal architecture employed in effectively all particle filter implementations. Applications related to mobile robots include target identification and tracking both airborne and marine (Ristic et al., 2004), terrain navigation , airborne navigation , underwater navigation , and Simultaneous Localization And Mapping (SLAM) . The bootstrap architecture however has a drawback, that is when the system noise is small or the likelihoods are peaked, the particles within the bootstrap filter can still degenerate due to lack of diversity and poor proposal choice, respectively. There are numerous variations of the SIR architecture, as used in the original bootstrap particle filter, that are designed to address this problem. Most of the variants consist of modifying two stages of the particle filter: the proposal stage, thus changing the proposal density used, and the resampling stage, thus changing the resampling densities used.

### 5.3.1 Resampling

Resampling consists of creating a new unweighted set **X**, from a weighted set **X***, whose distribution reflects **X*** resulting in the weights being normalized and avoiding degradation. All current resampling algorithms use the same technique to sample from the posterior. They all create a new set **X** by applying uniformly generated random numbers to an inverse of the cumulative distribution function (cdf) of the state estimate (73) (Doucet et al., 2001). This is not a cdf in the conventional sense in that it does not accumulate over the solution space but over the particle indices indicated in Fig. 5.

A uniform number applied to (73) produces a value defining the index of a particle. The indicated particle's state then constitutes a sample in the new distribution. There are four common resampling algorithms: Multinomial, Stratified, Systematic, and Residual. These four algorithms are differentiated by how the uniformly distributed numbers are generated.

$$\text{for } i = 1,..,N, \text{set } I^i = D_\omega^{-1}\left(u^i\right) \text{ and } \tilde{\xi}^i = \xi^{I^i} \quad (73)$$

| | | | |
|---|---|---|---|
| Multinomial - | for i = 1,..,N, | $u_i \sim u((0,1])$ | |
| Stratified - | for i = 1,..,N, | $u_i \sim u((N^{-1}(i-1), N^{-1}i])$ | |
| Systematic - | for i = 1,..,N, | $u_i = N^{-1}(i-1) + u_1$ | where $u_1 \sim u(0,N^{-1}]$ |
| Residual - | for i = 1,..,m | $N^i = \lvert N\omega^i \rvert + N'^i$ | where $\lvert N\omega^i \rvert$ is the integer part and denotes the number of samples from $X^i$ |
| then | | | |
| for i = 1,..,(N-R), | $u_i \sim u((0,1])$ | | where $R = \text{nm} \lvert N\omega^i \rvert$ |

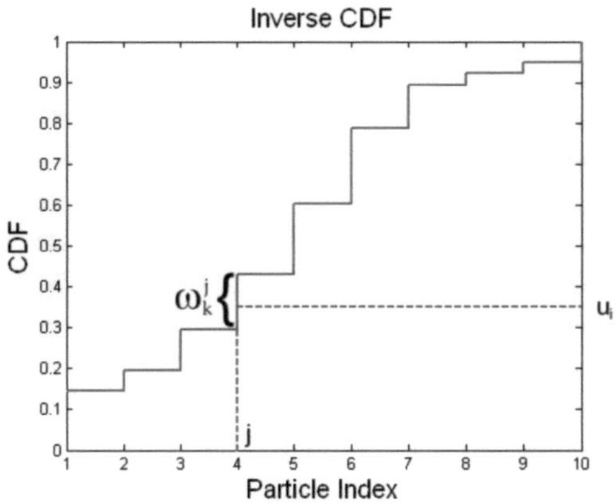


Fig. 5. CDF Example

#### 5.3.2 Regularized Particle Filter

The Regularized particle filter uses regularization of the empirical distribution by means of a kernel method (74) in the resampling stage. Regularization avoids particle degradation by generating diversity through exploiting the smoothness that is contained in the distribution in a continuous space. This diversity makes the filter more robust at the expense of the theoretical disadvantage that the samples are no longer asymptotically approximate to those of the posterior.

$$x_k^{i^*} = x_k^i + h_{opt} D_k \epsilon^i \tag{74}$$

#### 5.3.3 MCL Particle Filter

The Monte-Carlo Localizer was developed specifically for robots that operate in buildings (Doucet et al., 2001) to offset the drawbacks of the case dependent quality of the bootstrap filter's proposal choice. The MCL employs a modified resampling stage that uses the observations in the generation of the resampled distribution. The resampling stage of the MCL includes the observation by drawing samples from not only the prior, as in the bootstrap filter, but a fixed percentage from the likelihood as well (75). This indirectly improves the proposal distribution when the likelihoods are peaked, by placing particles at suitable points in the solution space according to the likelihood before applying to (71), thus indirectly producing (68).

$$\begin{array}{ll} \text{for } i = 1,..,N-R, & \text{set } I^i = D_\omega^{-1}(u^i) \text{ and } \tilde{\xi}^i = \xi^{I^i} \\ \text{for } i = 1,..,R, & \text{set } \tilde{\varsigma}^i = z_k^i \\ & \text{where} \quad z_k^i \sim p(z_k) \end{array} \tag{75}$$

#### 5.3.4 Auxiliary Particle Filter

The Auxiliary particle filter uses a simulation based method to define the proposal distribution. The proposal density is generated in a three stage process. First, a deterministic projection of the set $x_k$ from the set $x_{k-1}$ is generated, as shown in (76), based on some characteristic of x. Second, the weights are modified in (77) based on this prediction and current observations. Last, the resultant multinomial distribution is resampled according to (73). This resampling creates a set of indices that denote the collection of particles at k-1 which if propagated according to the bootstrap filter (71)-(72) create a distribution equivalent to $q\left(x_k^i \middle| x_{k-1}^i, z_k\right)$.

$$q(x_k, i|Z_k) \propto p\left(z_k \middle| \mu_k^i\right) p\left(x_k \middle| x_{k-1}^i\right) \omega_{k-1}^i \tag{76}$$

$$\omega_k^j \propto \omega_{k-1}^{i^j} \frac{p\left(z_k \middle| x_k^j\right) p\left(x_k^j \middle| x_{k-1}^{i^j}\right)}{q\left(x_k^j, i^j \middle| Z_k\right)} = \frac{p\left(z_k \middle| x_k^j\right)}{p\left(z_k \middle| \mu_k^{i^j}\right)} \tag{77}$$

### 5.3.5 Rao-Blackwellised Particle Filter (RBPF)

Rao-Blackwellisation is a method of marginalizing states with Gaussian error distributions resulting in two systems: a linear Gaussian system (78) which can be processed by Kalman filter algorithms (36)-(43), and a nonlinear non-Gaussian system (79) which can be estimated by particle filter algorithms (67)-(69). This split approach avoids the 'curse of dimensionality' by lowering the number of dimensions to be estimated by the particle filter. With less dimensions, the particle filter will require significantly less particles to obtain a given accuracy. However, this comes at the expense of significantly increased computational cost per particle, since each particle will need to propagate a Kalman representation of the marginalized states. The RBPF has been studied for navigational purposes but has received most attention in tracking applications (Ristic et al., 2004).

$$x_k^a = F_{k-1} x_{k-1}^a + v_{k-1} \tag{78}$$

$$B_k^m = B_k^m + G_{k-1} x_{k-1}^a + w_{k-1} \tag{79}$$

### 5.3.6 Hybrids

As can be seen, the above set of variants is all concerned with either indirectly or directly defining the proposal density for the given problem. This has led to the proliferation of hybrid filters in which a Kalman Filter variant is used to estimate (80) for a particle filter. Examples of this include the Extended Particle Filter and the Unscented Particle Filter.
In these filters, the EKF (44)-(53) or UKF (54)-(66) equations are used to generate a mean and covariance estimate, $\hat{x}_k^i\ P_k^i$, for each particle, $x_k^i$, such that they can be used as the proposal density (80).

$$q\left(x_k \middle| x_{k-1}^i, z_k\right) \triangleq N\left(\hat{x}_k^i, P_k^i\right) \tag{80}$$

As revealed in (80) N particles, i.e., N Kalman filter representations need to be calculated at each time step k and as such present a significant computational burden but with the benefit of significantly increased accuracy. This trade off is much like the RBPF in that although more computationally intensive per particle, it requires fewer particles to achieve a given accuracy.

### 5.3.7 MAPR Particle Filter

The Measurement Assisted Partial Resampling (MAPR) particle filter is a novel new particle filter algorithm which is based on the bootstrap algorithm (71)-(72) with a modified resampling stage (Lammas, 2008). In place of the resampling scheme described in (73), MAPR uses a heuristic resampling scheme which nominates a proportion of the particles for resampling. This heuristic resampling scheme draws on influences from the Regularised (74) and MCL (75) particle filters and is designed to minimise variance in the particle weights introduced due to the resampling process.

## 6. Implementation and Analysis

To compare the algorithms detailed in the previous section, the algorithms are implemented in a navigational system and tested in the scenario described in the introduction and illustrated in Fig. 1. To implement this navigational system, the state (25) and measurement (26) models must be defined. In this scenario, the vehicle dynamics (22) defines the state models (44), (57), and (69), and the measurement equations (23)-(31) define the observation models (45), (58), and (69). The filters are run synchronously at 100 Hz in order to mitigate some of the nonlinearity effects of the Kalman Filter variants. This sampling rate is chosen under the consideration that, within a reasonable range, the smaller the sample period, the better the linear approximation to a nonlinear function. The particle filters are run using a particle population of 1000, i.e., N=1000, which is generally considered as insufficient, for all but the MAPR filter, to estimate the given scenario which has a state vector with 10 states.
For the sake of clarity and conciseness, the results presented will be restricted to four of the aforementioned filters, namely the EKF, UKF, Bootstrap, and MAPR filters. The measurements available in this scenario are taken from the IMU, DVL, Compass, and GPS modules described in Table 1. In the scenario, it is assumed that the seafloor is visible to the DVL transponders from the surface, and as such provides a continuous stream of velocity measurements. This premise means that only the position states needed to be estimated via dead reckoning, i.e., for the period of time the vehicle is submerged.

| Part No. | Sensors | | | | | Biases | |
|---|---|---|---|---|---|---|---|
| | Vendor | Description | Data | Hz | RMS | Case 1 | Case 2 |
| MMA7260Q | Freescale | 3 axis accelerometer* | $a_x,a_y,a_z$ | 100 | 0.0063 | 0 | 0.01 |
| CRS03-04 | Silicon Sensing | 3x 1 axis Gyroscopes* | $\omega_x,\omega_y,\omega_z$ | 100 | 0.0045 | 0 | 0.01 |
| HMR3000 | Honeywell | Compass | $\phi,\theta,$ $T_x,T_y,T_z$ | 20 | 0.002 0.001 | N/A | |
| LEA-4T | uBlox | GPS | $PR_{1:N}$ $\delta_{1:N}$ | 10 | 20 0.03 | | |
| Explorer | Teledyne | DVL | $V_{1:4}$ | 7 | 0.0055 | | |
| * the 3 axis accelerometer and the 3 single axis gyroscopes aligned orthogonally form an Inertial Measurement Unit (IMU) | | | | | | | |

Table 1. Sensors and Intrinsic Errors

Two variants of this scenario are analysed based on Table 1. The first variant (Case 1) assumes a perfectly calibrated IMU, with only Gaussian noise left as the source of errors. The second variant (Case 2) assumes a more realistic IMU which, apart from Gaussian noise, has a bias of 0.01 rad/s for the gyroscopes and 0.01 m/s$^2$ for the accelerometers. This arrangement allows analysis of the robustness of the four filters to 'non-ideal' measurements in comparison to 'ideal' measurements.

A Monte Carlo trial of the scenario is performed for each of the four filters to compare the systematic characteristics. Each trial consists of 400 runs, following the same trajectory with independently identically distributed (i.i.d.) measurements. The results of these trials for unbiased (ideal) IMU and the biased (non-ideal) IMU are presented in Fig. 6 and Fig. 7, respectively. Figs. 8a and 8b show the cross-track errors of the estimates presented in Fig. 6a and Fig. 7a, respectively.

Figs. 6a and 6b show the top down view of the estimated raster trajectory and the estimate and the residual of the surge, $u$, respectively. As can be seen in Fig. 6b, the EKF and UKF produce a smaller error compared to the particle filter variants. This behaviour is predictable since the errors in this scenario assume only Gaussian noise and for the given sampling rate, the EKF and UKF with their associated limited nonlinearities in (22) and (23)-(32) should theoretically produce an optimal solution. It should also be noted that although the Kalman variants are optimal in this case, the MAPR's position estimate is still comparable to those of the Kalman Filters, as shown in Fig. 8a, and its surge estimate though degraded is still less than 5mm/s.

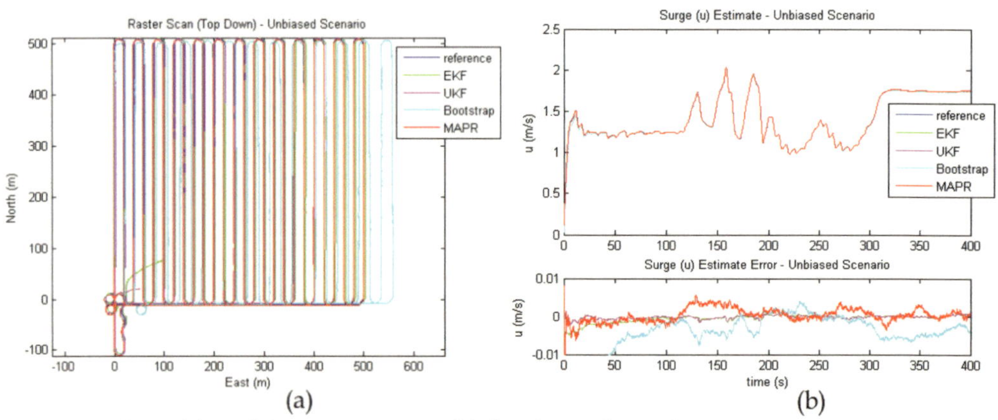


Fig. 6. Raster Scan (a) and Surge Estimates (b) for the Unbiased Case

Fig. 7 shows the result of the more realistic scenario which includes residual bias errors in the IMU. In Fig. 7a, it can be seen that the EKF and Bootstrap filters suffer significantly in their position estimates from these biases when compared with Fig. 6a. Similarly, significant distortions in the velocity estimate due to the biases are also evident in Fig. 7b. Though the UKF suffers in a similar way to the EKF and Bootstrap filters in estimating the velocity, the UKF can still estimate the position accurately. The reason for the deterioration of the EKF estimates is because, despite there being minimal nonlinearities in the measurements, GPS still contains non-negligible nonlinearities which is compounded by its limited availability and by the critical impact of poor initial estimates on the dead reckoning process. Likewise, for the Bootstrap filter, limited GPS availability and the impact of initial estimates are also compounding factors, however, in this case the initial error is due to inadequacies in the bootstrap filter for small particle population, i.e., N=1000. As evident in Fig. 7b, the best performance in term of estimating velocity in this case is the MAPR filter which has an error six times less than that of the next accurate estimate, UKF.

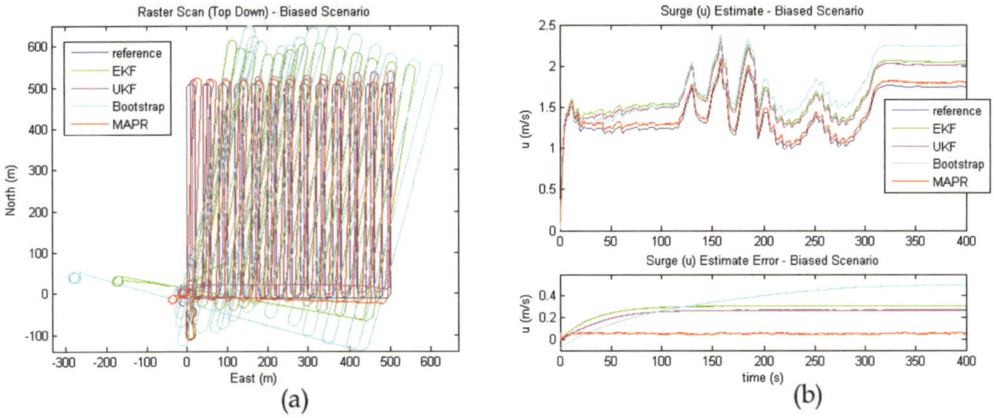


Fig. 7. Raster Scan (a) and Surge Estimates (b) for the Biased Case

In Fig. 7a, as in Fig. 6a, the position error of the UKF and MAPR filter appear to be comparable. However, in Fig. 8, the difference between the unbiased and biased cases becomes more evident. In Fig. 8a, with no biases and only Gaussian noise, it can be seen that, apart from the Bootstrap filter, all the filters have a comparable cross-track error, with the MAPR filter being marginally less accurate than the Kalman filters. In Fig. 8b, with the inclusion of biases, the MAPR clearly has a smaller cross-track error than that of the UKF in almost all parts of the trajectory except for a small section at the start which exhibits a larger error.

In conclusion, although some navigation filters may work more accurately than others with 'ideal' sensors, the performance degradation of these filters in the presence of increasing sensor bias is evident as demonstrated in the test scenario presented here. This has the consequence that filters that may appear not to be desirable in the ideal test may actually perform better than the perceived optimal choice under realistic conditions. As such, the use of high-fidelity simulation and field testing of proposed navigation algorithms is essential to identify the optimal selection. As shown in this section, although the merits of the MAPR filter may not seem apparent for ideal or very expensive sensor suites, the MAPR filter performs better than the UKF which is considered to be the best of the filters in the ideal situation. Moreover, the MAPR filter retains many of the desirable advantages of particle filters, such as multimodal distributions which are commonly encountered when using SLAM-based positioning. The multimodal distributions result from multiple-position hypothesises due to the similarity of different features or different angles of the same feature, as viewed by the sensors. This multimodal support is increasingly desirable as the use of SLAM for positioning becomes more prolific in mobile robotic applications.

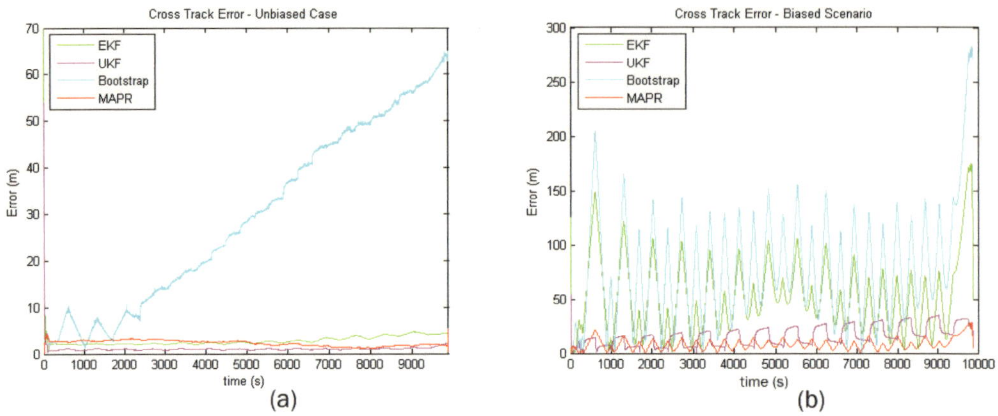


Fig. 8. Cross-Track Error for Unbiased Case (a) and Biased Case (b)

## 7. Conclusion

The chapter has provided a review of various sensing modalities and filtering techniques used in AUV navigation. Specifically, a range of suitable navigation sensors as carried by modern AUVs is discussed, as a basis for selecting a sensor platform that is appropriate for the intended navigation algorithm and application. A set of algorithms, based on the Bayesian filter family, is reviewed. These filters include various versions of Kalman and particle filters, as well as the new MAPR particle filter developed by the authors. Implementations and analyses of these filters, in the context of a navigation estimation system for an AUV, are provided. These implementations have been tested based on the growth in the filters' state estimate errors, particularly with respect to position estimates, during a raster scan mission, typical for survey class AUVs. The information provided will allow vehicle designers to consider the use of these concepts for the purpose of improving performance and efficiency, and reducing vehicle instrumentation costs.

## Acknowledgements

This work is supported by the CSIRO Wealth from Oceans Flagship program through the Subsea Pipelines Cluster project.

## 8. References

Arulampalam, S.; Maskell, S.; Gordon, N. & Clapp, T. (2001). A Tutorial on Particle Filters for Online Nonlinear/Non-Gaussian Bayesian Tracking. *IEEE Transactions on Signal Processing*, 50. 1. (February 2002), page numbers 15)

Caiti, A.; Garulli, A.; Livide, F. & Prattichizzo, D. (2004). Localization of Autonomous Underwater Vehicles by Floating Accoustic Bouys: A Set Membership Approach. *IEEE Journal of Oceanic Engineering*, 30. 1. (January 2005), page numbers 13)

Choset, H. & Nagatani, K. (2001). Topological simultaneous localization and mapping (SLAM): towardexact localization without explicit localization. *IEEE Transactions on Robotics and Automation*, 17. 2. numbers 125-137)

Dissanayake, M.; Newman, P.; Clark, S.; Durrant-Whyte, H. F. & Csorba, M. (2001). A solution to the simultaneous localization and map building (SLAM) problem. *IEEE Transactions on Robotics and Automation*, 17. 3. numbers 229-241)

Douc, R.; Cappe, O.; Polytech, E. & Palaiseau, F. (2005). Comparison of resampling schemes for particle filtering.

Doucet, A.; Freitas, N. d.; Gordon, N.; Smith, A.; Crisan, D.; Moral;, P. D. & Jacob, J. (2001). *Sequential Monte Carlo Methods in Practice*, Springer, 0-387-95146-6, New York

Fossen, T. I. (2002). *Marine Control Systems: Guidance, Navigation, and Control of Ships, Rigs and Underwater Vehicles*, Marine Cybernetics, 82-92356-00-2, Trondheim, Norway

Gebre-Egziabher, D.; Elkaim, G. H.; Powell, J. D. & Parkinson, B. W. (2001). A nonlinear, two-step estimation algorithm for calibrating solid-state strapdown magnetometers.

Lammas, A. (2008). Improving Navigational Accuracy for AUVs using the MAPR Particle Filter. Oceans'08. Quebec, Canada.

Ristic, B.; Arulampalam, S. & Gordon, N. (2004). *Beyond the Kalman Filter; Particle Filters for Tracking Applications*, Artech House, 1-58053-631-x, Boston

Titterton, D. H. & Weston, J. L. (2004). *Strapdown inertial navigation technology*, Peter Peregrinus Ltd,

Tsui, J. B. Y. (2005). *Fundamentals of global positioning system receivers: a software approach*, Wiley-Interscience

# Quantitative Performance Metrics for Mobile Robots Navigation

Nelson David Muñoz Ceballos*, Jaime Alejandro Valencia** and Nelson Londoño Ospina**
**Polytechnic Jaime Isaza Cadavid, **University of Antioquia Medellin, Colombia*

## 1. Introduction

Mobile robots navigation involves the design of navigation algorithms. There are several navigation methods for mobile robots and every time a new navigation algorithm is proposed, some comparisons with traditional algorithms are usually made, but there isn't set of benchmarks globally accepted to assess the performance of algorithms. Despite the wide variety of studies and research on robot navigation systems, performance measures are often neglected in this research area, which makes it difficult to make an objective performance comparison (Wong et al., 2002); in general, use of quantitative metrics is limited to measuring the length of the path or the time needed by the robot to complete the task. Additionally, the lack of consensus on how to define or measure these systems impedes rigor and prevents evaluation of progress in this field and compare its different capabilities (Evans & Messina, 2000). As the degree of complexity in robotics is increasing, becomes necessary to establish proper approaches and benchmarking procedures, mainly for two reasons: first, reliable benchmarking is necessary in order to allow the comparison of the many robotics research results seeking this way to enable their industrial application. Second, to position the robotics as a serious science is important to consider the replication of experiments, i.e., it is important to verify whether a new procedure or algorithm proposed is really a breakthrough that can be used in new applications. This would be achieved only if it is determined which are the appropriate benchmarking procedures that allow to compare the actual practical results with reference to standard accepted procedures (Eurongemsig, 2008). However, by applying navigation comparison metrics of a mobile robot, such as the length of trajectory (path), collision risk and smoothness of trajectory, using a protocol, that is in a systematic and ordered way, works on mobile robots navigation algorithms can be systematized, and this will help researchers to decide which algorithm should be implemented in the vehicle.

This chapter describes aspects related with a procedure used for the performance evaluation of mobile robots navigation algorithms. First, several performance metrics used in the navigation of mobile robots are described, then, the protocol to be followed in the performance evaluation is defined, finally, the results are presented. To illustrate the procedure, we used simulation software.

## 2. Performance Metrics for Robot Navigation

Robot navigation is based in sensing of world (scenario), to compute the next motion, and actuation. (Minguez, 2008) cites a concise summary of several navigation methods:
The potential field methods addressed the first sensor-based motions (Khatib, 1986; Krogh & Thorpe, 1986), the vector field histogram was the first alternative to do obstacle avoidance with uncertain sensor like ultrasounds (Borenstein & Koren, 1991), Elastic bands was the first technique combining planning and reaction schemas in a unified framework (Quinlan & Khatib, 1993), the dynamic window was the first technique to address kinematics and dynamics to carry out motion at high speeds (Fox et al., 1997), Curvature-Velocity method was a similar method developed alternatively (Simmons, 1996), Nearness diagram navigation was the first technique to address motion in troublesome scenarios (Minguez & Montano, 2004), etc
Additionally, extensions of the previous techniques were developed, for example:
Based in different potential functions (Tilove, R.B., 1990; Koren & Borenstein, 1991; Chenqing et al., 2000; Azarm & Schmith, 1994; Borenstein & Koren, 1989). Based in the vector field histogram (Ulrich & Borenstein, 1990; Ulrich & Borenstein, 2000; Borenstein & Raschke, 1992; Yang et al., 2000; Bell et al., 1994). Based in the elastic bands (Brock & Khatib, 2000; Khatib et al., 1997). Based in the dynamic window approach (Brock & Khatib, 1999; Stachniss & Burgard, 2002). Based in the nearness diagram (Minguez et al., 2004; Minguez, 2005; Marques, 2001), etc.
Also, there are hybrid methods that combine these techniques with tactical planners (Yang et al., 2000; Brock & Khatib, 1999; Minguez et al., 2001; Stachniss & Burgard, 2002; Philipsen & Siegwart, 2003).
The navigation system gives to robot the capability to move between given locations. There are several metrics that can be used to evaluate the performance of a navigation system, but none of them are able to indicate the quality of the whole system. Therefore it is necessary to use a combination of different indexes that quantify different aspects of the system. Having a good set of performance metrics is useful for: Optimizing algorithm parameters, testing navigation performance within a variety of work environments, making a quantitative comparison between algorithms, supporting algorithm development and helping with decisions about the adjustments required for a variety of aspects involved in system performance (Cielniak et al., 2005).
In autonomous navigation and obstacle avoidance, typical performance criteria are: (Minguez, 2008), (Álvarez, 1998)

1. Mission success: number of successful missions.
2. Path length: distance traveled to accomplish the task.
3. Time: time taken to accomplish the task.
4. Collisions: number of collisions per mission, per distance and per time.
5. Obstacle clearance: minimum and mean distance to the obstacles.
6. Robustness in narrow spaces: number of narrow passages successfully traversed.

7- Smoothness of the trajectory: relative to control effort.

This performance metrics can be classified in the following importance order:
A- Metrics that consider the security in the trajectory or proximity to obstacles
B- Metrics that consider the dimensions of the trajectory towards the goal
C- Metrics that evaluate the smoothness of the trajectory
In the following, a set of different performance metrics are described.

### 2.1 Security metrics (Álvarez, 1998)

These metrics express the relationship between the security with which the robot travels through a trajectory, taking into account the distance between the vehicle and the obstacles in its path.

Security Metric-1 (SM1): Mean distance between vehicle and the obstacles through the entire mission measured by all the sensors; the maximum value will be produced in an obstacle free environment. If the deviation of the index from its maximum value is low, it means that the chosen route passed through obstacles free area.

Security Metric-2 (SM2): Minimum mean distance to obstacles. This is taken from the average of the lowest value of the n sensors. This index gives an idea of the risk taken through the entire mission, in terms of the proximity to an obstacle. In an obstacles free environment SM1 = SM2 is satisfied.

Minimum Distance (Min): Minimum distance between any sensor and any obstacle through the entire trajectory. This index measures the maximum risk taken throughout the entire mission.

### 2.2 Dimension metrics

The trajectory towards the goal is considered in its time and space dimensions. In general, it is assumed that an optimal trajectory towards the goal is, whenever possible, a line with minimum length and zero curvature between the initial point ($x_i$,$y_i$) and the finishing point ($x_n$,$y_n$), covered in the minimum time.

Length of the Covered Trajectory ($P_L$): is the length of the entire covered trajectory by the vehicle from the initial point to the goal. For a trajectory in the x-y plane, composed of n points, and assuming the initial point as ($x_1$, f($x_1$)) and the goal as ($x_n$, f($x_n$)), $P_L$ can be calculated as:

$$P_L = \sum_{i=1}^{n-1} \sqrt{(x_{i+1} - x_i)^2 + (f(x_{i+1}) - f(x_i))^2} \tag{1}$$

Where ($x_i$, f($x_i$)), i = 1, 2, . . . , n are the n points of the trajectory in cartesian coordinates (Guo & Wang, 2003).

The length of a trajectory given by y = f(x), in the x-y plane between the points (a, f(a)) and (b, f(b)), can also be calculated as (Selekwa et al.,2004)

$$P_{Laprox} \cong \int_a^b \sqrt{1 + (f'(x_i))^2}\, dx \tag{2}$$

Mean distance to the goal (Mgd): This metric can be applied to robots capable of following reference trajectories. An important aspect when determining the quality of the robot navigation system is the ability to follow a trajectory that aims to reach a goal, so, to evaluate the quality of the execution of the trajectory, the mean distance between the vehicle and goal is analyzed. The difference is more significant if the covered distance is shorter (Rosenblatt, 1997). The mean distance to the goal is defined by the square of the proximity to the goal distance $l_n$, integrated across the length of the trajectory and normalized by the total number of points n:

$$l_n = \min\left(\forall n\left(\sqrt{(x_i - x_n)^2 + (f(x_i) - f(x_n))^2}\right)\right) \tag{3}$$

$$Mgd = \frac{\int_0^l l_n^2 ds}{n} \tag{4}$$

Control Periods (LeM): It is the amount of control periods. This metric relates to the number of decisions taken by the planner to reach the goal. If the robot moves with constant lineal speed (v), this gives an idea of the time needed to complete the mission (Álvarez, 1998).

### 2.3 Smoothness metrics

The smoothness of a trajectory shows the consistency between the decision-action relationship taken by the navigation system, and also, the ability to anticipate and to respond to events with sufficient speed (Rosenblatt, 1997). The smoothness in the way a trajectory is generated is a measure of the energy and time requirements for the movement; a smooth trajectory allows translates into energy and time savings (Dongqing, 2006). Additionally a smooth trajectory is also beneficial to the mechanical structure of the vehicle.
Bending Energy ($B_E$): This is a function of the curvature, k, used to evaluate the smoothness of the robot's movement. For curves in the x-y plane, the curvature, k, at any point $(x_i, f(x_i))$ across a trajectory is given by:

$$k(x_i, f(x_i)) = \frac{f''(x_i)}{(1 + (f'(x_i))^2)^{\frac{3}{2}}} \tag{5}$$

The bending energy can be understood as the energy needed to bend a rod to the desired shape (Aguirre & Gonzales, 2000). $B_E$ can be calculated as the sum of the squares of the curvature at each point of the line $k(x_i, f(x_i))$, along the length of the line L. So, the bending energy of a robot trajectory is given by:

$$B_E = \frac{1}{n}\sum_{i=1}^{n} k^2(x_i, f(x_i)) \tag{6}$$

Where $k(x_i, f(x_i))$ is the curvature at each point of the robot trajectory and $n$ is the number of points in the trajectory.
The value of $B_E$ is an average and does not show with clarity enough that some trajectories are longer than others. Therefore, $TB_E$ can be used instead; this metric takes into account the smoothness and length of the trajectory simultaneously.

$TB_E$ is defined by

$$TB_E = \int_a^b k^2(x)dx \tag{7}$$

And numerically,

$$TB_E = \sum_{i=1}^{n} k^2(x_i, f(x_i)) \tag{8}$$

In a straighter trajectory, the values $B_E$ and $TB_E$ will be lower, which is desirable since the energy requirement is increased according to the increase in the curvature of the trajectory.
Smoothness of Curvature (Smoo): is defined by the square of the change in the curvature k of the trajectory of a vehicle with respect to the time, integrating along the length of the trajectory and normalized by the total time t (Rosenblatt, 1997).

$$Smoo = \frac{\int_0^l \left(\frac{dk}{dt}\right)^2 ds}{t} \tag{9}$$

## 3. Simulation Framework

In this section it's describe a simulation framework used for evaluation of mobile robot navigation algorithms. Simulation is one of the most important tools in robotic development. It enables the evaluation of different alternatives during the design phase of robot systems and may therefore lead to more general solutions. Also, the simulation supports the process of software development by providing a replacement for robots that are currently not available (e.g. broken or used by another person) or not able to endure long running experiments (e.g. learning tasks). Finally, it is much easier to build scenarios using a simulator. The execution of robot programs inside a simulator offers the possibility to perform an easier and faster debugging phase before the first real experiment. (Calisi et al, 2008)
The simulation framework is compound for a Giraa_02 mobile robot model (see Fig 1a), and a 6m x 4m structured environment with static obstacles (see Fig. 2).
The Giraa_02 robot has a cylindrical structure of 30cm diameter and approximately 20cm height. It has 8 ultrasound and 8 infrared sensors distributed equally around the robot's circumference, for the simulation, only 8 infrared sensors were taken into account, these have a range of d= 26.5cm and a= 15 degree detection cone; figure 1b. The vehicle has an odometric system and a differential locomotion system (Muñoz et al., 2006).
Data acquisition in the mobile robot, which occurs during each control period, consists of the current position of the robot and its orientation ($x_i$,$y_i$,$\theta_i$). The eight (8) proximity sensors are also read, the maximum reading being 26.5 cm, so that, if the robot spends n control periods reaching the goal, there is an array of n x 11, and n sampling points per 11 pieces of data (3 coordinates and 8 sensors).

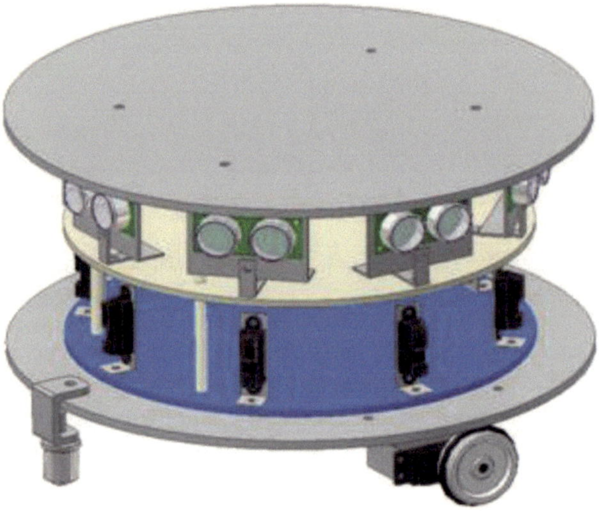

Fig. 1a. Mobile Robot Giraa_02

Fig. 1b. Infrared Sensor Configuration

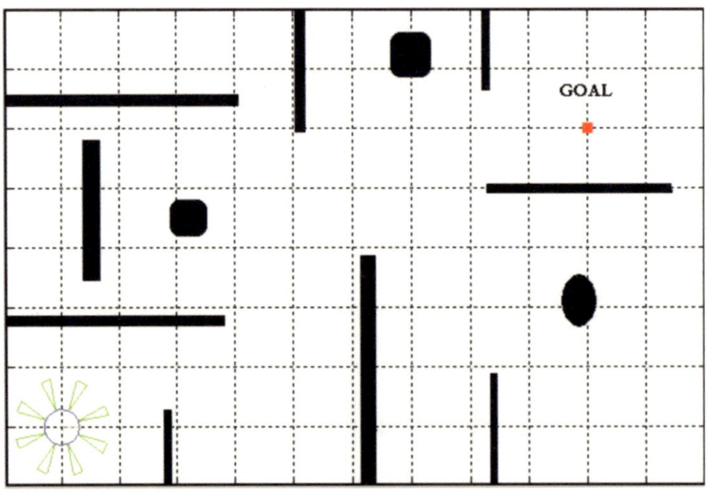


Fig. 2. Test Scenario for Mobile Robot Navigation

## 4. Definition of the evaluation protocol

In this section it describe a protocol used for evaluation of mobile robot navigation algorithms, and it's explains the proposed benchmark for navigation algorithms.
An evaluation protocol a systematic and ordered way to make the test. As a navigation benchmark, it's common to choose a "towards a goal" mission; the objective is to execute a navigation mission from a starting point to a final point (goal).
An order of importance for evaluating the navigation characteristics can be established as follows:

1. The mean distance between the vehicle and the obstacles during the trajectory
2. The distance covered by the vehicle between the starting point and the goal
3. The time needed to complete the mission
4. The smoothness of the trajectory

The first point considers the security of the trajectory and measures the risk taken by the robot in its movement towards the goal. The second and third points measure aspects related to the planning of the trajectory and the fourth point considers the quality of the trajectory according to the energy and time required for the movement. These characteristics can be analyzed using the following set of performance metrics:

1. SM1, SM2 and Min are proposed for evaluating security.
2. PL and LeM are proposed for evaluating the trajectory
3. $TB_E$ is proposed for evaluating the smoothness of the trajectory.

For general purposes, only one metric is required for each one of the 3 categories described in section 2, but the use of various metrics helps to improve the analysis. In this case, the indexes were selected according to the capabilities of the GIRAA_02 mobile robot, considering the information provided by its data acquisition system; the readings from all the sensors are available for each point of the path, allowing the calculation of SM1, SM2, and Min. The Mgd index does not apply in this navigation mission since it applies when a trajectory is followed; $TB_E$ is proposed because it analyses the smoothness and length of the path. Also, this metric is numerically simpler and more precise, making it easier to calculate than the other metrics.

## 5. Tests and Results (Muñoz et al, 2007)

The control algorithms provide basic capabilities for the mobile robot, such as the ability to evade obstacles and to generate a trajectory towards a goal. In this case, the framework is used for the evaluation of a navigation mission between two points (towards a goal). For this mission, it can use several navigation methods, like mentioned in section 2. As example, we go to compare the performance of two methods: potential field and AFREB.

### 5.1 Control Algorithm 1

This is a reactive algorithm based on a potential field method (Khatib, 1986; Krogh & Thorpe, 1986), which produces two different behaviors: first, goal attraction, and second, obstacles repulsion (keep away from objects). The planning of the movement consists in the proper combination of both behaviors in such a way that the robot reaches the goal without collisions (See fig 3.). This combination is achieved using a vector sum (Latombe, 1996).

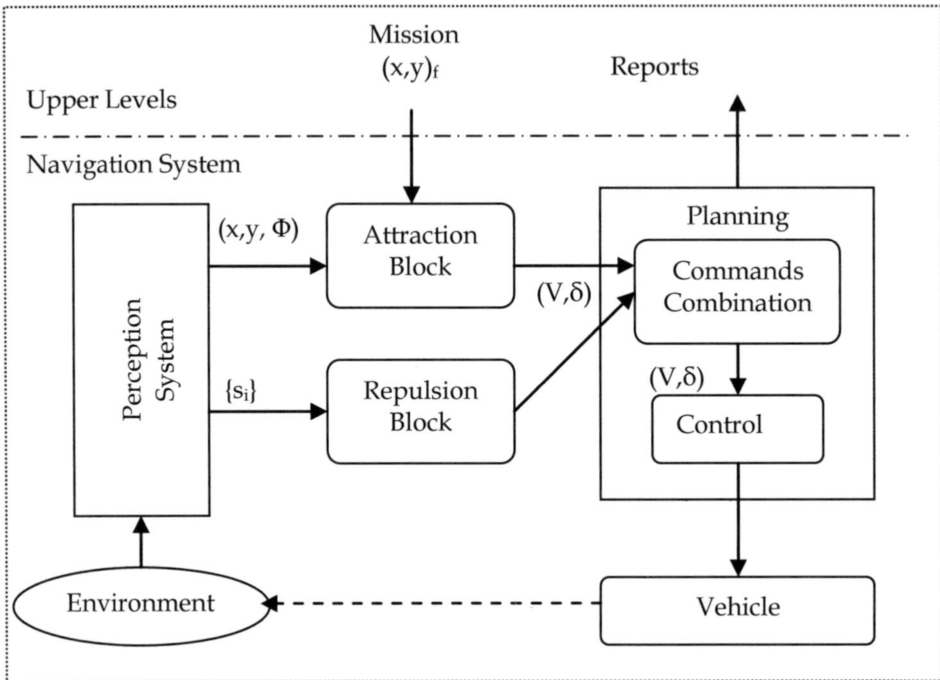


Fig. 3. Diagram of navigation based in potential field method

Where
$(x, y, \Phi)$ : current position and orientation
$(x,y)_f$ : final goal position
$S_i$ : sensors information (length measurement sensors)
$(V, \delta)$ : velocity and orientation angle commands
$(v, w)$ : linear and angular velocity

### 5.2 Control Algorithm 2

This control algorithm is based on reactive behaviors, denominated AFREB "adaptive fusion of reactive behaviors" (Zalzala & Morris, 1996), (Gachet et al, 1994). By using a neural net, an appropriate combination of the behaviors can be achieved, so that the system is able to perform tasks, such as navigation towards a goal, while evading obstacles in its path. The AFREB control architecture is depicted in Fig 4, basically consists of the following modules: behavioral fusion, fusion supervisor, behavior primitives (1, 2,…n), and executor.
A primitive behavior can be characterized by a temporal sequence of appropriate values for linear velocity v(i), and curvature k(i) which cause the robot to exhibit the prespecified response to sensorial information. Thus it is defined the output of a primitive behavior c(i) as a vector:

$$c(i) = (v(i), k(i))^T \tag{10}$$

Where

The variable i denotes the i-th cycle of the robot controller
$(x,y,\Phi)$ : current position and orientation
$(x,y)_f$ : final goal position
c1..cn: Behavior primitives output
ai: Behavior weighs (coefficients)
c: Emergent behavior (mission), linear and angular velocity vector (v,w)
The primitive behaviors implemented are:
c1: goal attraction
c2: perimeter following (contour left - CW)
c3: perimeter following (contour right - CCW)
c4: free space
c5: keep away (from objects)

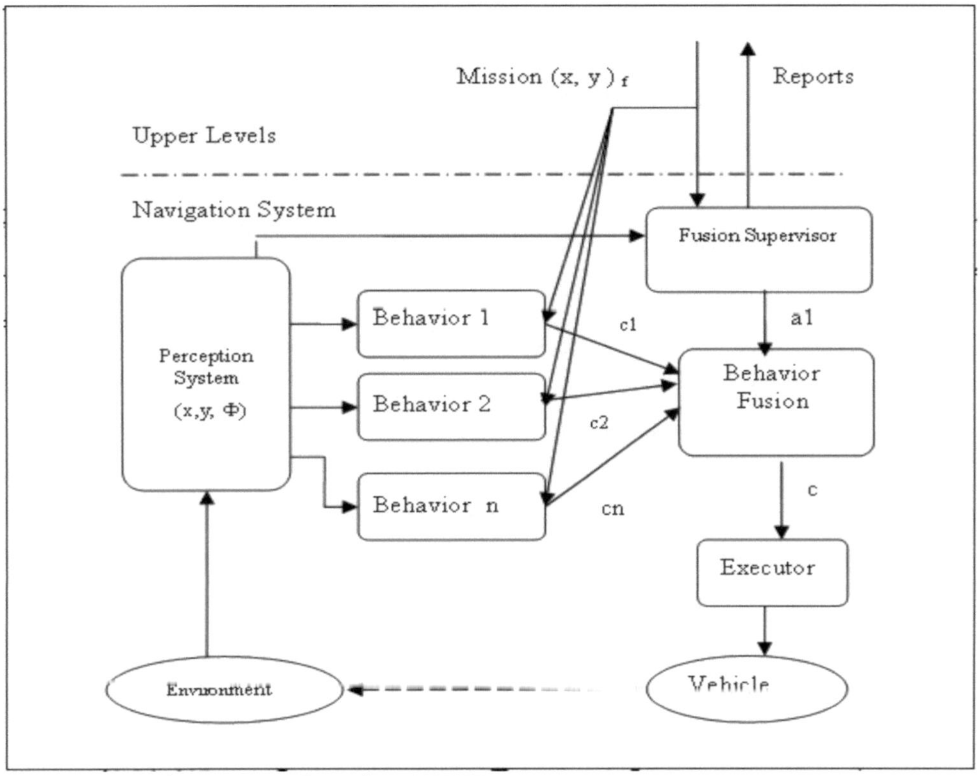


Fig. 4. Diagram of navigation based in AFREB

If c1..cn are the output of each primitive behavior, then the output of an emergent behavior (i.e., towards a goal, other mission, etc) is:

$$c = \sum_{i=1}^{N} a_i c_i \tag{11}$$

Where $a_i$ coefficients, with $0 \leq a_i \leq 1$, are found by an appropriate combination of measurement information provided by the perception system.

### 5.3 Results

The paths generated by the algorithms, in the scenario 1, are shown in figure 5. The table 1 summarizes the results obtained from the simulation using both control algorithms for all scenarios, according to the quality metrics proposed.

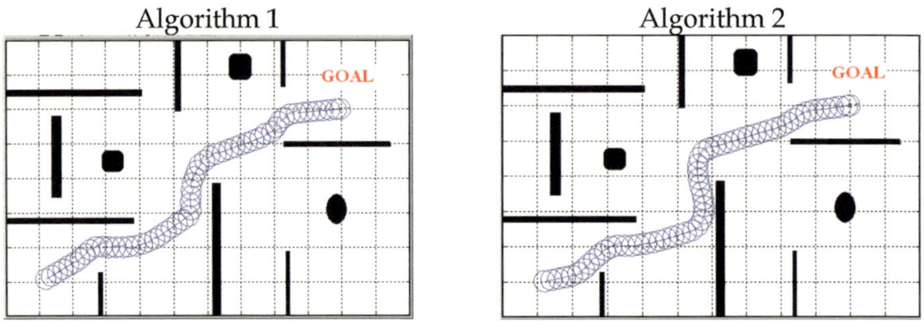


Scenario 1. Start point (50,50) Goal (500,300)

Fig. 5. Paths generated by the control algorithms

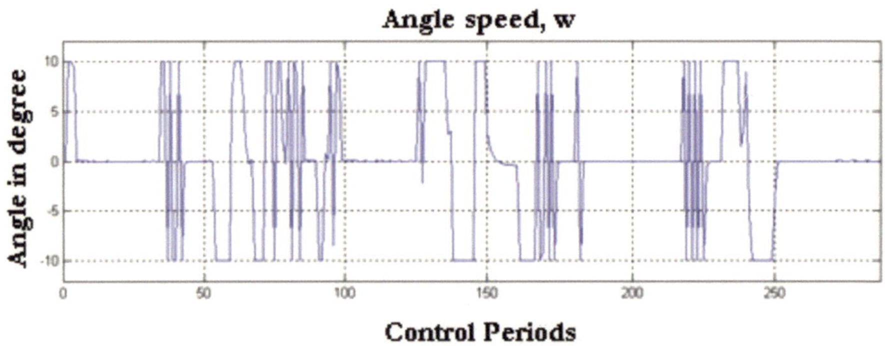


Fig. 6. Smoothness of the trajectory, change in the robot heading each control period, generated by algorithm 1.

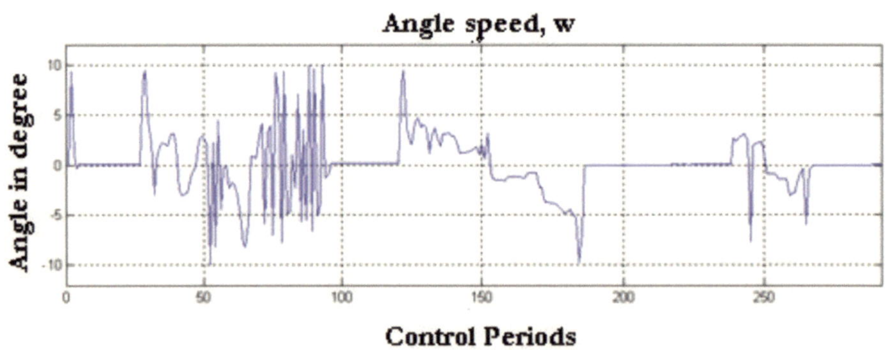


Fig. 7. Smoothness of the trajectory, change in the robot heading each control period, generated by algorithm 2

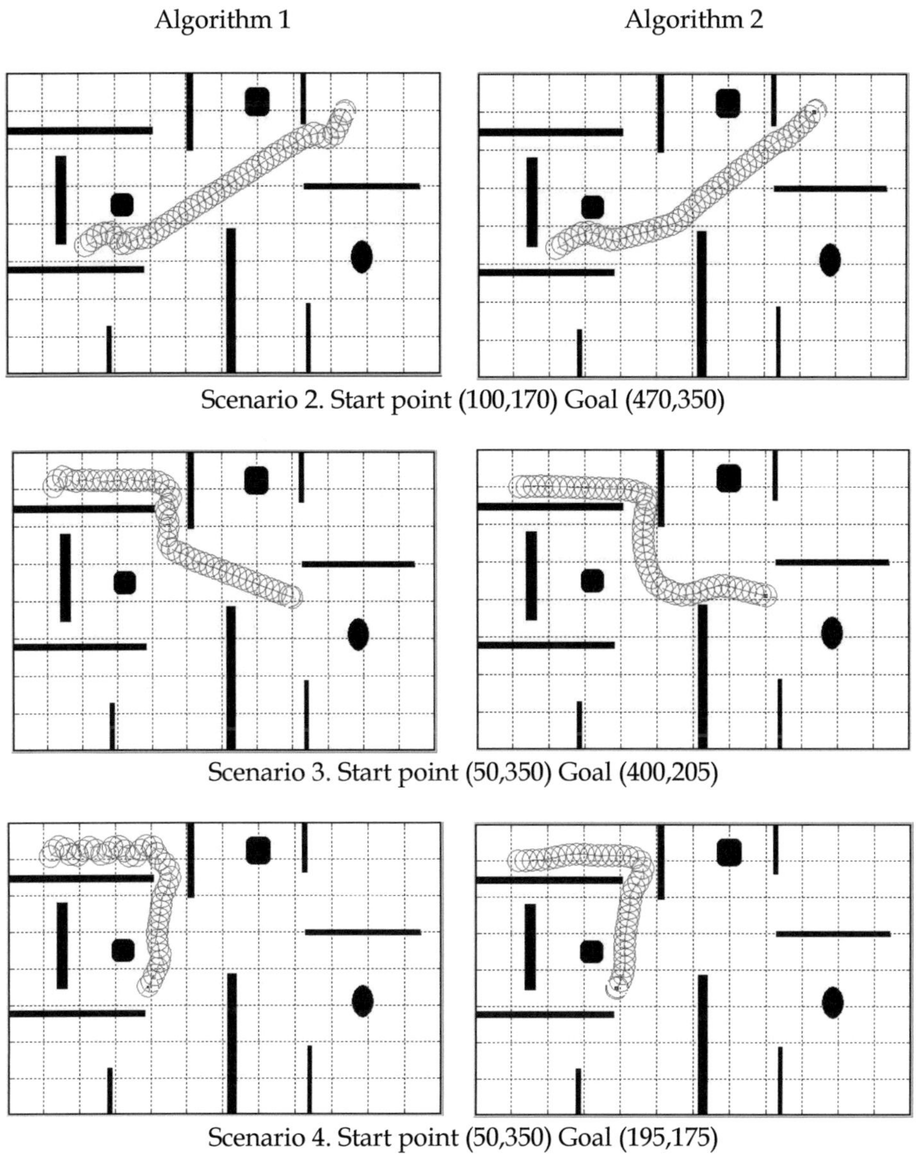


Fig. 8. Paths generated by the control algorithms in scenarios 2, 3, 4.

Algorithm 1 Algorithm 2

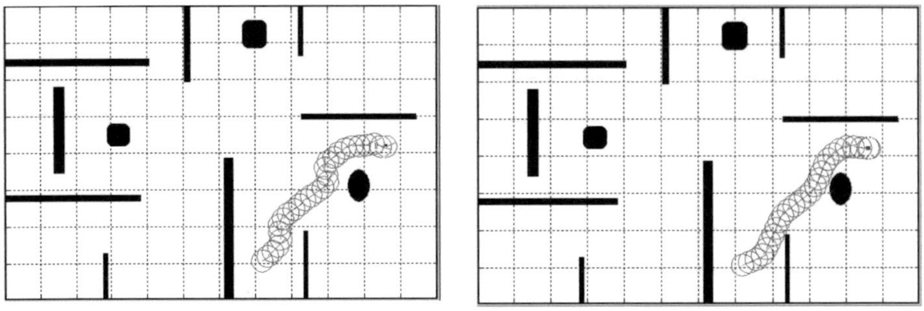

Scenario 5. Start point (350,50) Goal (530,210)

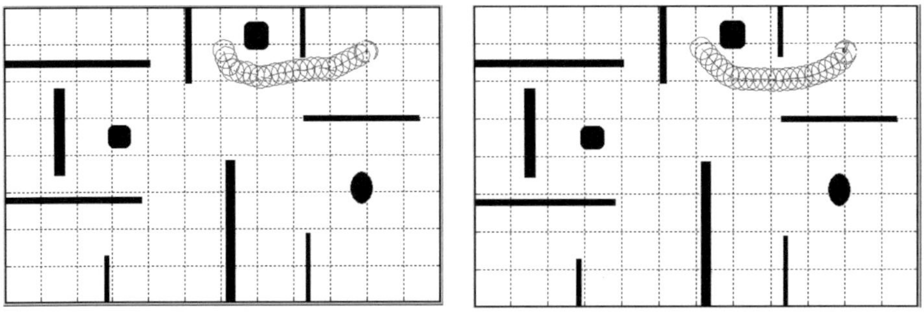

Scenario 6. Start point (300,350) Goal (500,340)

Fig. 9. Paths generated by the control algorithms in scenarios 5 and 6.

| Metric | SM1 [cm] | | SM2 [cm] | | Min [cm] | | PL [cm] | | LeM | | $TB_E$ | |
|---|---|---|---|---|---|---|---|---|---|---|---|---|
| Scenario | Alg. 1 | Alg. 2 | Alg. 1 | Alg. 2 | Alg. 1 | Alg. 2 | Alg. 1 | Alg. 2 | Alg. 1 | Alg. 2 | Alg. 1 | Alg. 2 |
| 1 | 26.1 | 25.6 | 18.3 | 17.3 | 11 | 7 | 562.7 | 581.9 | 283 | 292 | 0.2463 | 0.0846 |
| 2 | 25.9 | 25.7 | 13.0 | 14.0 | 3 | 7 | 441.8 | 429.9 | 222 | 216 | 0.2810 | 0.0718 |
| 3 | 25.4 | 23.9 | 10.0 | 8.9 | 3 | 3 | 456.7 | 462.9 | 234 | 235 | 0.5873 | 0.0120 |
| 4 | 25.0 | 24.4 | 13.0 | 12.4 | 7 | 3 | 395.7 | 359.9 | 199 | 181 | 0.4007 | 0.0140 |
| 5 | 25.9 | 24.9 | 19.4 | 16.4 | 15 | 3 | 275.8 | 259.9 | 139 | 131 | 0.1626 | 0.0394 |
| 6 | 26.0 | 25.9 | 19.7 | 22.6 | 7 | 11 | 229.9 | 229.9 | 116 | 116 | 0.1722 | 0.0469 |

Table 1. Robot performance

### 5.4 Analysis of results

In scenario 1, the algorithm 1 uses less control periods, and consequently takes less time to complete the mission, and covers a safer and shorter path, the figure 6 shows that algorithm 1 produces a great orientation change for each control period. Algorithm 2 covers a smoother path, the figure 7 shows a smaller change in the orientation during each control period, with consequent energy saving and less structural effort on the robot.

From table 1, it can be deduced that the difference between both algorithms in the trajectory and time taken is only 3.3% and 3.1% respectively. The robot programmed with algorithm 2 passed a minimum 7 cm from any obstacle, which is acceptable for a 30cm diameter robot; also, it showed approximately 65% less bending energy than algorithm 1. For these reasons, algorithm 2 can be considered the best choice.
The figures 8 and 9, shows other sceneries. The algorithm 1 has the tendency to generate safer trajectories, because the robot transits, on average, through zones that are farther from the obstacles, it is explained because the closer the robot is to the obstacles, the higher the repulsion potential, this way, the robot succeeds in keeping away from them. Despite Algorithm 2 is ruled by the same repulsion principle, the command that finally guides the robot, is a combination of 5 different behaviors, which reduces the role of the repulsion potential, but without having collisions. The main difference between all simulations is that Algorithm 2 generates smoother trajectories than Algorithm 1.
The wall following behavior CW and CCW present in Algorithm 2, makes the robot able to transit through narrow zones like corridors, keeping a safe distance from the obstacles and also generating smooth trajectories, which doesn't happen with Algorithm 1 as it is seen in sceneries 3 and 4. In general, Algorithm 2 exhibits a better performance, the bending energy index is always smaller than in the Algorithm 1, even in the sceneries 2, 4 and 5, generates shorter trajectories and uses less time to complete the mission, and in the scenario 6, generates a safe trajectory.

## 6. Conclusion

This chapter provides an analysis on several performance metrics to contrast mobile robots navigation algorithms including safety, dimension and smoothness of the trajectory. The suggested metrics are quite straight forward, however, it has been shown that they can be used together to systematize simulated or experimental studies on control algorithms for mobile robot navigation.
A very simple application example was presented. The obtained results demonstrate the need to establish a procedure that can be used to analyze and compare control algorithms using several performance metrics. This is an open topic of research. It is becomes necessary to establish proper approaches and benchmarking procedures, for example, using standard framework of benchmarks for navigation algorithm and performance evaluation.
This metrics can be applied in simulated environments, but the performance metrics evaluation is more important in real environments. Many of the challenges in robot navigation come from the challenges of real environments, such an uncertainty in the sensors and the errors in odometry, and this, in general, is not considered in simulation.

## 7. References


Aguirre, E. & Gonzales, A. (2000). Fuzzy Behaviors for mobile robot navigation, design, coordination and fusion. *International Journal of Approximate Reasoning,* vol. 25, 2000, pp. 255-289.

Álvarez, J. (1998). Planificación del movimiento de vehículos autónomos basada en sensores. *Tesis doctoral,* Universidad de Oviedo, Oviedo, España, pp. 178.

Azarm, K. & Schmith, G. (1994). Integrated mobile robot motion planning and execution in changing indoor environments. *Proceedings of IEEE/RSJ International Conference on Intelligent Robots and Systems,* Munchen, Germany, 1994, pp. 298–305.

Bell, D.; Levine, S.; Koren, Y.; Jaros, L. & Borenstein, J. (1994). Design criteria for obstacle avoidance in a shared-control system. *Proceedings of RESNA Conference,* Nashville, 1994.

Borenstein, J. & Koren, Y. (1989). Real-Time Obstacle Avoidance for Fast Mobile Robots. *Proceedings of IEEE Transactions on Systems , Man and Cybernetics,*Vol. 19, pp. 1179–1187.

Borenstein, J. & Koren, Y. (1991). The Vector Field Histogram–Fast Obstacle Avoidance for Mobile Robots. *Proceedings of IEEE Transactions on Robotics and Automation,* pp. 278–288.

Borenstein, J. & Raschke, U. (1992). Real-time obstacle avoidance for non-point mobile robots. *Proceedings of Fourth World Conference on Robotics Research,* Michigan, USA.

Brock, O. & Khatib, O. (1999). Real High-Speed Navigation Using the Global Dynamic Window Approach. *Proceedings of IEEE International Conference on Robotics and Automation*, Detroit, MI, pp. 341–346

Brock, O. & Khatib, O. (2000). Real-Time Replanning in High-Dimensional Configuration Spaces using Sets of Homotopic Paths. *Proceedings of IEEE International Conference on Robotics and Automation*, San Francisco, USA, pp. 550–555.

Calisi, D., Iocchi, L., & Nardi, D. (2008). A unified benchmark framework for autonomous mobile robots and vehicles motion algorithms (MoVeMA benchmarks). In RSS workshop on experimental methodology and benchmarking, Zurich, Switzerland.

Chenqing, L.; Ang, M.; Krishman, H. & Yong, L. (2000). Virtual Obstacle Concept for Local-minimum-recovery in Potential-field Based Navigation. *Proceedings of IEEE International Conference on Robotics and Automation*, San Francisco, USA, pp. 983–989.

Cielniak, G.; Treptow, A. & Duckett, T. (2005). Quantitative Performance Evaluation of a People Tracking System on a Mobile Robot. *Proceedings of the European Conference on Mobile Robots (ECMR)*, Ancona, Italy.

Dongqing, S. (2006). Aerial robot navigation in cluttered urban environment, *PhD Thesis,* The Florida State University, Florida, USA, pp. 87.

Eurongemsig, (2008). Special Interest Group on Good Experimental Methodology in Robotics European Robotics Research Network (EURON), [Online]. Available: www.heronrobots.com/EuronGEMSig/

Evans, J. & Messina, E. (2000). Performance Metrics for Intelligent Systems, *Proceeding of the Performance Metrics for intelligent Systems Workshop,* August 2000, MD, Gaithersburg.

Fox, D.; Burgard, W. & Thrun, S. (1997). The Dynamic Window Approach to Collision Avoidance. *Proceedings of IEEE Robotics and Automation Magazine.* Vol. 4, Num. 1, 1997.

Gachet D.; Salichs, M. A.; Moreno, L.; Pimentel, J. R. (1994). Learning Emergent Tasks for an Autonomous Mobile Robot. *Proceedings of the International Conference on Intelligent Robots and Systems (IROS '94).*

Guo, Y & Wang, J. (2003). A new performance based motion planner for nonholonomic mobile robots, *Proceedings of the 3rd performance metrics for the Intelligent Systems Workshop (PerMIS'03) NIST,* September 2003, Gaithersburg, MD.

Khatib, M.; Jaouni, H.; Chatila, R. & Laumond, J. (1997). Dynamic Path Modification for Car-Like Nonholonomic Mobile Robots. *Proceedings of IEEE International Conference on Robotics and Automation*, Alburquerque, Mexico, pp. 2920–2925.

Khatib, O. (1986). Real-Time Obstacle Avoidance for Manipulators and Mobile Robots. *Proceedings of International Journal of Robotics Research*, Vol. 5, pp. 90-98, Spring, 1986.

Koren, Y. & Borenstein, J. (1991). Potential Field Methods and Their Inherent Limitations for Mobile Robot Navigation. *Proceedings of IEEE International Conference on Robotics and Automation*, Vol. 2, Sacramento, CA, pp. 1398–1404.

Krogh, B.H. & Thorpe, C.E. (1986). Integrated Path Planning and Dynamic Steering control for Autonomous Vehicles. *Proceedings of IEEE International Confrence on Robotics and Automation*, San Francisco, USA, pp. 1664–1669.

Latombe, J.C. (1996). Robot Motion Planning, *Kluwer Academic Publishers*, 4th Edition, 1996, Boston.

Marques, C. (2001). Multi-sensor navigation for soccer robots. Master's thesis. In: *Instituto Superior Tecnico*, Portugal, 2001.

Minguez, J. (2005). The Obstacle Restriction Method (ORM): Obstacle Avoidance in Difficult Scenarios. *Proceedings of IEEE International Conference on Intelligent Robot and Systems*, Edmonton, Canada, 2005.

Minguez, J. (2008). Robot Obstacle Avoidance Papers Using Experiments, In: *Good experimental methodology guidelines*, Bonsignorio, F.; Hallam J., & del Pobil, A. P. (Ed), Special Interest Group on Good Experimental Methodology in Robotics European Robotics Research Network (EURON), Tech. Rep., 2008. [Online].Available: www.heronrobots.com/EuronGEMSig/Downloads/GemSigGuidelinesBeta.pdf

Minguez, J. & Montano, L. (2004). Nearness Diagram (ND) Navigation: Collision Avoidance in Troublesome Scenarios. *Proceedings of IEEE Transactions on Robotics and Automation*, Vol. 20, Num. 1, pp. 45–59.

Minguez, J. ; Montano, L.; Simeon, N. & Alami, R. (2001). Global Nearness Diagram Navigation (GND). *Proceedings of IEEE International Conference on Robotics and Automation*, pp. 33–39, Seoul, Korea .

Minguez, J.; Osuna, J. & Montano, L. (2004). A Divide and Conquer Strategy to Achieve Reactive Collision Avoidance in Troublesome Scenarios. *Proceedings of IEEE International Conference on Robotics and Automation*, Minessota, USA.

Muñoz, N.; Andrade, C. & Londoño, N. (2006). Diseño y construcción de un robot móvil orientado a la enseñanza e investigación, *Ingenieria & Desarrollo Ed. 9*, 2006.

Muñoz, N.; Valencia, J. & Londoño, N. (2007). Evaluation of Navigation of an Autonomous Mobile Robot, *Proceedings of the Performance Metrics for Intelligent Systems (PerMIS) Workshop*, Washington, DC, EEUU.

Philipsen, R. & Siegwart, R. (2003). Smooth and efficient obstacle avoidance for a tour guide robot. *Proceedings of IEEE International Conference on Robotics and Automation*, Taipei, Taiwan.

Quinlan, S. & Khatib, O. (1993). Elastic Bands: Connecting Path Planning and Control. *Proceedings of IEEE International Conference on Robotics and Automation*, Vol. 2, pp. 802–807, Atlanta, USA.

Rosenblatt, J. (1997). DAMN: Distributed Algorithm for Mobile Navigation. PhD. Thesis, *Carnegie Mellon University Robotics Institute*, Pittsburg, PA, 1997.

Simmons, R. (1996). The Curvature-Velocity Method for Local Obstacle Avoidance. *Proceedings of IEEE International Conference on Robotics and Automation*, pp. 3375–3382, Minneapolis, USA.

Selekwa, M.; Collins, E. & Combey, J. (2004). Multivalue Verus univalued Reactive Fuzzy Behavior Systems for Navigation Control of Autonomous Ground Vehicles, *Proceedings from the 17th annual Florida Conference on the Recent Advances in Robotics FCRAR2004*, May 2004.

Stachniss, C. & Burgard,W. (2002). An Integrated Approach to Goal-directed Obstacle Avoidance under Dynamic Constraints for Dynamic Environments. *Proceedings of IEEE-RSJ International Conference on Intelligent Robots and Systems*, pp. 508–513, Switzerland, 2002.

Tilove, R.B. (1990). Local Obstacle Avoidance for Mobile Robots Based on the Method of Artificial Potentials. *Proceedings of IEEE International Conference on Robotics and Automation*, Vol. 2, pp. 566–571, Cincinatti, OH, 1990.

Ulrich, I. & Borenstein, J. (1990). VFH+: Reliable Obstacle Avoidance for Fast Mobile Robots. *Proceedings of IEEE International Conference on Robotics and Automation*, pp. 1572–1577.

Ulrich, I. & Borenstein, J. (2000). VFH*: Local Obstacle Avoidance with Look-Ahead Verification. *Proceedings of IEEE International Conference on Robotics and Automation*, pp. 2505–2511, San Francisco, USA, 2000.

Wong, S; Middleton, L. & MacDonald, B. (2002). Performance metrics for robot coverage task, *Proceedings Australasian Conference on Robotics and Automation ACRA*, Auckland, New Zealand, 2002, pp. 7-12

Yang, H.; Borenstein, J. & Were, D. (2000). Double-vfh: Reliable obstalce avoidance for large, non-point, omni-directional mobile robots. *Proceedings of ANS Conference on Robotics and Remote Systems*, Pitsburg, PA, 1999.

Zalzala, A. & Morris, A. (1996). *Neural Networks for Robotic Control,* Ellis Horwood, 1996, pp.278.

# 25

# Testing performance of current video codecs in teleoperated mobile robot applications: a practical experience

Pablo Piñol, Otoniel López, Miguel Martínez-Rach,
M.P. Malumbres, José Oliver and Carlos Calafate
*Universidad Miguel Hernández de Elche*
*Spain*

## 1. Introduction

Mobile robots are used in a wide variety of applications and environments. We have a lot of examples like spatial missions, planet surveying, industrial cleaning, people transport, surveillance and security, material storage, underwater tasks, bomb deactivation, tunnel works, building industry, leisure toys, …

One of the problems a mobile robot faces to is "knowing what is around". To solve this problem the robot uses perception. There are many different types of "perceptions" of the surrounding that a robot can get using sonar, laser, infrared, pressure sensors, temperature measurements, sound analysis and of course real-time video. Real-time video can be used by the robot itself, if we have an autonomous mobile robot or by an operator if we have a teleoperated robot.

Real-time video produces a huge amount of data which requires a great deal of resources of storaging capacity (for saving video data) and network bandwidth (for transmission). But, even though storage capacity and network bandwidth keep growing in present-day devices, video compression is mandatory.

When autonomous robots have to communicate with each other or when a teleoperator needs feedback of the environment or wants to send navigation commands to the robot we use a wireless link. In the first stages of mobile robot communication, proprietary RF systems were used. Nowadays WLANs are a good choice for implementing communications in a mobile robot due to its good characteristics: standardization, low cost, flexibility, high bandwidth, a certain degree of security, …

IEEE's 802.11 (IEEE-WG-802.11, 1999) standard is being increasingly used throughout corporations worldwide due to its good balance of cost, range, bandwidth and flexibility and along with 802.11e (IEEE-WG-802.11, 2002), that supplies QoS at MAC layer, form a good base to support multimedia traffic.

Although the main application area of wireless networks is related with user access to existing wired network infrastructure, other applications, like telerobotics can benefit from WLANs technology. The development of teleoperated systems has gained considerable attention due to the new potential applications, such as remote production monitoring (Luo et al., 1999), remote exploration and manipulation in inhospitable environments (Hirzinger et al., 1993), tele-surgery (Kwon et al., 1999), and remote training. Important issues concerning communication channels, random propagation delays, bandwidth limitations, fault-tolerance, synchronization, tele-presence, and the stability of the robotic systems involving human operators have all been taken into account in different works across the literature (Brady & Tarn, 1998); (Elhajj et al., 2001). Most of them consider Internet as the interconnection network between telecontrolled systems and control stations (Goldberg & Siegwart, 2002).

The use of wireless channels for teleoperating mobile robots is not new. Most of the existing wireless-controlled robots use a specific (non-standard) radio-modem for communications between control station and robot. But this wireless channel is usually a point-to-point dedicated wireless link which works under good signal quality environments. Therefore, in that case, the wireless link does not represent a drawback in terms of performance of the overall system.

Other works describe implementations of teleoperation systems with standard wireless devices. In (Fong et al., 2001) authors use a PDA with an IEEE 802.11 wireless interface for telecontrolling a robot, and in (Sgouros & Gerogiannakis, 2003) the same was done through a WAP connection. However, they do not analyze the behavior and performance of wireless link for the correct operation of this kind of applications.

In IEEE 802.11 networks working in infrastructure mode, the access points (base stations) provide the network connectivity between mobile and wired hosts. WLAN nodes share the medium through a CSMA-CA protocol, so network latency may be significant and dependent of current traffic load. The time-varying conditions of the wireless channel, roaming processes and certain node mobility patterns, may disturb established communications, increasing packet loss ratio and the average delay and jitter, up to intolerant levels for certain kind of multimedia applications.

On the other hand, under this kind of error-prone environments, delivery of encoded video is a challenging task if a minimum QoS is demanded from applications. Video encoders should be prepared to recover damaged bitstreams employing one or several error resilience tools. Video codecs will play an important role in the overall application performance, being critical the ability of concealing network delivery errors to assure the minimum required video quality.

In this chapter we will show the results of studying three present-day codecs (H.264/AVC, MPEG-4 (DivX), VP7) and how they manage with packet losses (due to node mobility and time-varying signal quality) in a real wireless environment. A prototype has been built consisting of a mobile robot with a laptop on it. This laptop is equipped with a webcam and a wifi network card. Both timing measurements and video quality assesment have been performed in several scenarios. Our experiments have been done at different rates of background traffic and with the robot moving from good signal areas to places where wireless signal gets lost. We have used different packet sizes, different codec modes (*intra/inter*) and different video frame sizes (QCIF/CIF/D1). We have employed DirectShow technology and a client/server scheme. Some simulation has also been done to tune up our prototype.

The chapter is organized as follows: in section 2 we describe the teleoperated framework that will be used in this work. Section 3 shows a detailed study of end-to-end delays in both video and navigation data flows. In section 4 we analyze commercial video codecs and how they recover damaged bitstreams in scenarios where packets are lost. In section 5 we devise some real scenarios to measure the impact of different network conditions in our telecontrolled robot system. Finally, in section 6, some conclusions are drawn.

## 2. Mobile Robot System

In Fig. 1 we show the client/server framework implemented for this study. The server hardware consists of an ER1 mobile robot from Evolution Robotics (ER1, 2009) which holds a laptop on it (see Fig. 2(a)). The laptop has a webcam for capturing a video stream and a 802.11 WLAN network adapter which sends the video stream to the client (control station) and receives navigation orders to operate the robot. The robot hardware is controlled by the laptop through a USB port. The client hardware consists of another laptop connected to a Fast Ethernet network. We also have two workstations for injecting background traffic. The traffic sender is connected via a Fast Ethernet link and the traffic receiver is connected via a WLAN link.

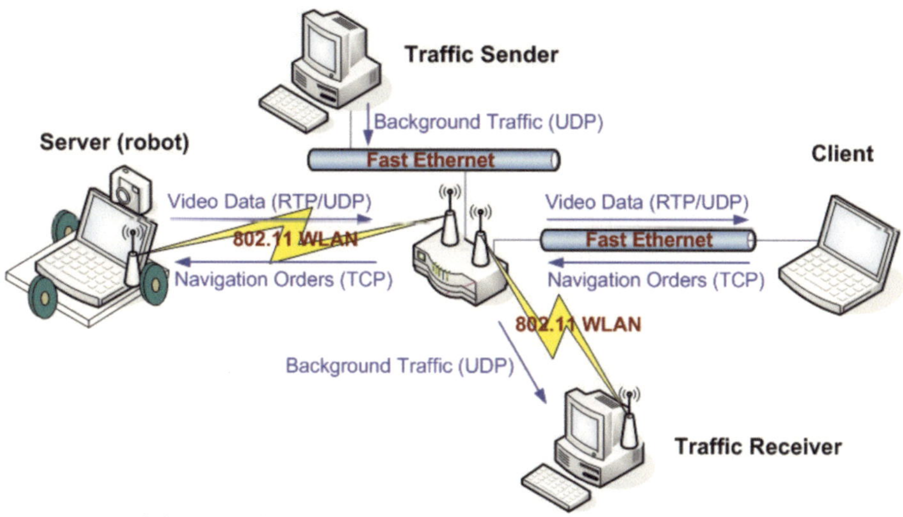


Fig. 1. Overview of the telerobotic experimental platform.

The software applications developed (see Fig. 2(b,c)) are based on DirectX/DirectShow architecture (DirectShow, 2009). This enables us to use any of the commercial codecs available for this technology. DirectShow applications handle data through different "boxes" called filters which process and transform information. So, for encoding and decoding a video frame we will use a codec filter. Let's talk about data flows in our framework. There are two different kind of data flows between client and server: video stream and navigation commands.

Video stream consists of RTP/UDP (Real-time Transport Protocol) (RFC-3550, 2003) video packets which are sent by the server and collected by the client. First, the laptop in the robot captures a frame using the webcam. This frame is compressed using one of the codecs installed in the laptop (we can also select "uncompressed" if we do not want to apply compression). After this is done, the RTP filter, which is based on LIVE555 Streaming Media Library (Live Networks, 2009) packetizes the compressed frame and then sends the packets to the client through the IEEE 802.11 link. The client receives the packets, arranges them and reassembles the compressed frame. This frame is decoded and finally rendered on the user interface window.

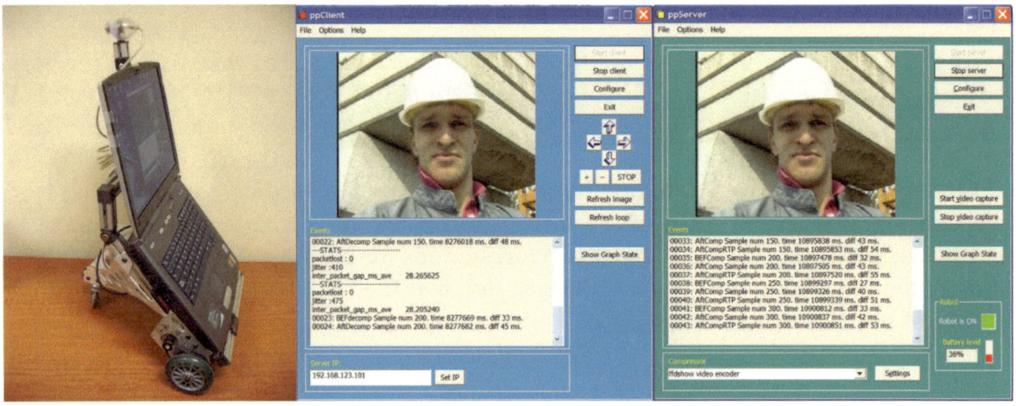


a b c

Fig. 2. (a) ER1 and laptop, (b) Client application, (c) Server application.

The other data flow, navigation commands, is sent by the user in order to control the robot. When the user issues a navigation command using the GUI client software, a short TCP packet is sent to the server. After receiving the navigation command, the server processes it in order to control the robot motion subsystem.

## 3. Delay Analysis

In this section we are going to analyze the control loop delay sources in both video and navigation flows. For that purpose we have selected three well known video test sequences: *Foreman* video sequence (400 frames) at QCIF format (176x144 pixels), *Coastguard* video sequence (300 frames) at CIF format (352x288 pixels) and *Mobile* video sequence (320 frames) at D1 format (720x576 pixels). For all video sequences we have fixed a common frame rate of 30 fps and a 4:1:1 YUV color space (12 bits/pixel).

We have established three different compression levels: (1) the lowest image quality (a low-but-acceptable quality, around 27 dB of PSNR - Peak Signal-to-Noise Ratio), (2) a medium video quality (around 33 dB), and (3) the highest image quality, offering around 39 dB (where further improvements on the quality are nearly imperceptible).

We have chosen three DirectX/DirectShow codecs under evaluation: MPEG-4 (MPEG-4, 2009), H.264/AVC (H.264, 2009) and VP7 (VP7, 2009). There are two available coding modes: *intra* and *inter*. In *intra* coding mode every video frame is individually encoded. However, when *inter* coding mode is configured, some video frames are individually encoded (key frames) and the rest are encoded taking as reference other frames. For *inter* mode we have configured them to produce a key frame every 15 frames (so we label it as *inter-15*).

### 3.1 Video stream delay

Video stream delay is the time elapsed since the robot's webcam captures a single frame until the client renders it to the user. The first source of delay is generated by the robot's webcam. By *vd1* we represent the time the webcam takes to capture a video frame. From the webcam specifications, it takes from 5 to 20 ms to capture an image. Our webcam has resolutions that range from 160x120 to 640x480 pixels. So, we have lineally determined the capturing delay for each video format.

When compressing video, we will encounter delay *vd2* which measures the time the video encoder takes to compress one video frame. This delay will depend on different variables such as the encoder used, the requested bitrate (opposite to the reconstructed video quality) and other compression parameters. If we do not compress video then *vd2* will have a value of 0. In Table 1 we show the average delay for encoding and decoding a single video frame using a medium level of compression. *Intra* mode is usually faster than *inter* mode but produces a greater number of packets for delivery.

By *vd3* we represent the time required to perform the packetization and delivery of one frame. We have used Wireshark (Lamping et al., 2006) network analyzer for measuring network delays. In Table 2 we show this delay using a packet size of 1272 bytes and a medium level of compression. We have measured it without background traffic (no traffic in the IEEE 802.11 and Fast Ethernet segments). We have also measured uncompressed video delivery at a resolution of 160x120 pixels. Here, the average delivery delay is 24.68 ms (with 23 packets per frame).

| | | QCIF - Foreman | | CIF - Coastguard | | D1 - Mobile | |
|---|---|---|---|---|---|---|---|
| | | Encode | Decode | Encode | Decode | Encode | Decode |
| H.264 | intra | 7.48 | 3.47 | 27.78 | 13.13 | 105.01 | 42.20 |
| | inter-15 | 7.60 | 1.80 | 35.36 | 8.20 | 113.31 | 17.82 |
| MPEG-4 | intra | 0.95 | 1.42 | 2.73 | 8.04 | 9.79 | 14.90 |
| | inter-15 | 2.68 | 0.96 | 10.43 | 2.35 | 36.50 | 7.81 |
| VP7 | intra | 14.61 | 2.43 | 26.07 | 8.66 | 74.53 | 58.50 |
| | inter-15 | 24.12 | 5.50 | 23.32 | 14.92 | 35.98 | 26.69 |

Table 1. Encoding and decoding delay (ms)

| | | QCIF - Foreman | | CIF - Coastguard | | D1 - Mobile | |
|---|---|---|---|---|---|---|---|
| | | pkt/fr | ms/fr | pkt/fr | ms/fr | pkt/fr | ms/fr |
| H.264 | intra | 2.25 | 0.88 | 8.94 | 8.63 | 28.08 | 29.70 |
| | inter-15 | 1.09 | 0.65 | 3.75 | 2.36 | 5.06 | 3.91 |
| MPEG-4 | intra | 2.27 | 0.80 | 8.02 | 7.71 | 26.91 | 27.11 |
| | inter-15 | 1.08 | 0.51 | 3.53 | 2.32 | 5.08 | 3.59 |
| VP7 | intra | 2.22 | 0.85 | 8.88 | 8.25 | 27.97 | 29.64 |
| | inter-15 | 1.07 | 0.56 | 3.45 | 2.30 | 4.96 | 3.29 |

Table 2. Delivery delay (ms) per frame and packets per frame

Once a packetized frame has been assembled at the client, it is delivered to the uncompressing module to get the reconstructed version of the original frame. The time required by the decoder to reconstruct the original frame is represented by *vd4*. If we do not compress video then *vd4* is 0. See Table 1 for average decoding delay.

The overall video flow delay will be the sum of all these delays (*vd1* + *vd2* + *vd3* + *vd4*). At a same level of compression (similar bitrate) all codecs produce a similar delay (*vd3*) when delivering a single frame through the network. So the difference between the codecs will be determined by the time needed to encode and decode one frame (*vd2* + *vd4*). MPEG-4 has always proved to be faster than the other two codecs (Fig. 3).

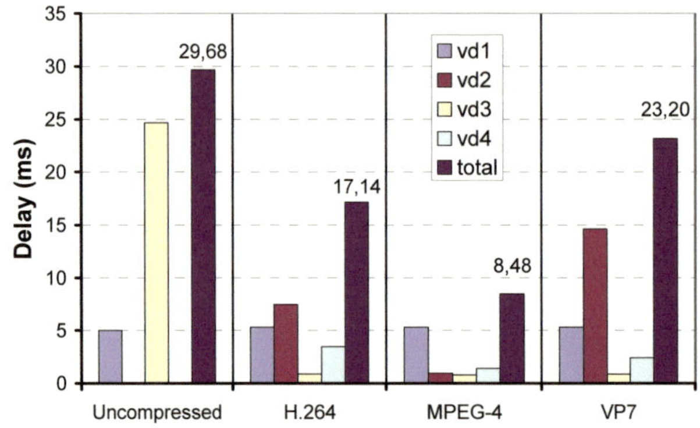


Fig. 3. Video stream delay for different codecs (QCIF resolution, *intra* mode).

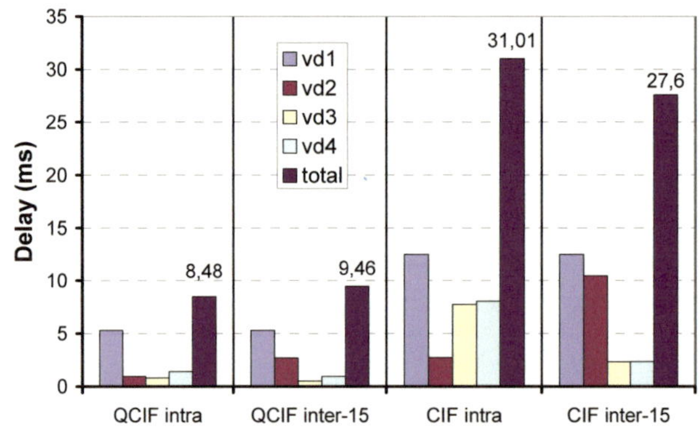


Fig. 4. Video stream delay for different resolutions and modes (MPEG-4 codec).

We have noticed that coding-decoding delay in *inter-15* mode is usually higher than in *intra* mode. At low resolution images (QCIF), the number of packets generated by both methods is very low. So, in the overall video flow delay, using *intra* mode is faster (Fig. 4). However, for medium and large resolution images (CIF, D1) the number of packets generated with *intra* mode is significantly greater than *inter-15*. So, in terms of overall video delay, *inter-15* mode is faster than *intra* mode.

### 3.2 Navigation commands delay

The other flow of data carries navigation commands from the client to the server in order to teleoperate the robot. The available navigation commands are the following: move forward/backward, turn left/right, increase/decrease speed, and stop. We send navigation

commands in very small TCP packets. We use TCP protocol instead of UDP protocol because it guarantees reliable and in-order delivery of navigation commands.
The first delay we find corresponds with the time elapsed since the user issues a navigation command until the server receives it. We label this delay as *nd1*. When the server receives a navigation command it immediately sends the order to the robot hardware. By *nd2* we indicate the time elapsed since the server sends the order until the robot begins to execute it (and it can proceed to wait for another command). We have measured the average time for executing one navigation command, resulting in the following delays: 22.3, 12.2, and 18.3 ms required by forward/backward, turn and stop commands, respectively.

### 3.3 Overall delay

The overall delay indicates the time the prototype takes for a complete teleoperation action (control loop). It will depend on several factors: frame size, codec used, compression level, navigation command, network status, etc. For example, if we send a forward moving command using an MPEG-4 codec in *intra* mode, a medium level of compression and QCIF video format, we will have an overall delay of 30.78 ms (video delay + navigation command delay). The same configuration but with CIF video format will give an overall delay of 53.31 and 49.9 ms for *intra* and *inter-15* coding modes, respectively.

## 4. Simulation Tests

Now, we will analyze the video codec behavior under a scenario with and without packet losses. Bit error patterns are not considered here since IEEE 802.11 network technology uses ACKs for every packet, so the only delivery error is packet loss. In the scenario represented in Fig. 1, the access point (US Robotics USR8054) uses an MTU of 1300 bytes (when working in turbo mode) so removing the IP and UDP headers, we will have a maximum RTP packet size of 1272 bytes. We have done our tests with a pair of packet sizes, 1272 and 512 bytes (512 bytes is the default payload size used by some traffic generators). RTP library can be configured to randomly drop packets with a specific PLR (Packet Loss Ratio), so we have used this feature to simulate an error prone environment.

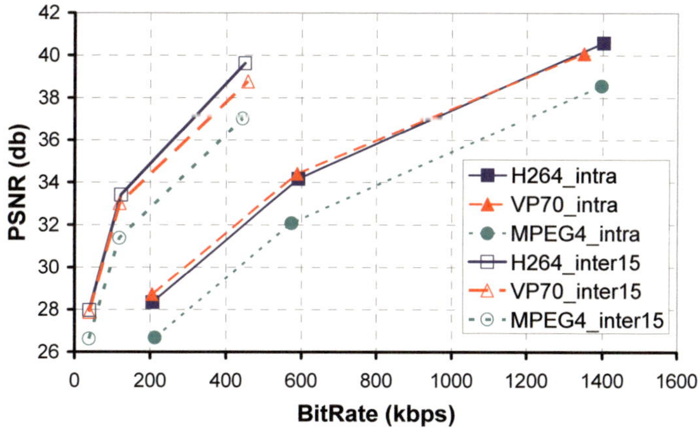


Fig. 5. R/D performance: objective PSNR quality comparison.

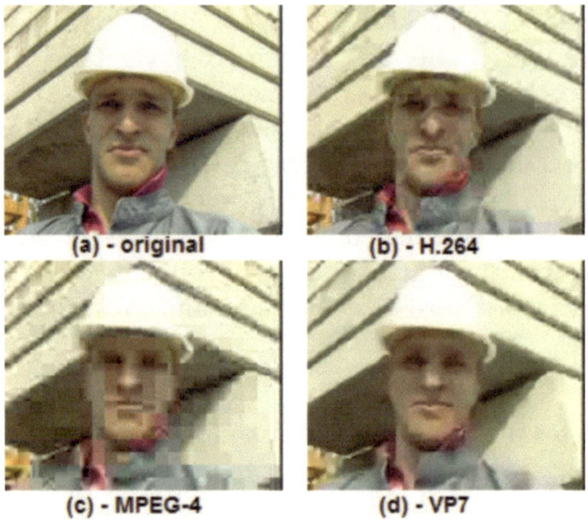


Fig. 6. R/D performance: subjective quality at lowest bitrate.

In Fig. 5 we show the R/D (Rate/Distortion) performance of the evaluated codecs with Foreman (QCIF) video sequence and no packet losses. This figure shows what we obtained in most of our loss-free tests: at similar bitrates H.264/AVC offers the best quality (nearly followed by VP7) and MPEG-4 offers the worst quality with difference. In Fig. 6 we can perceive the subjective quality difference between the codecs for the first frame of the video sequence, compressed at the lowest bitrate.

### 4.1 Simulated packet losses (evaluation)

In Fig. 7 we evaluate the behavior of the selected video codecs when the video delivery system begins to lose packets. We have used Foreman (QCIF) video sequence working in *intra* mode, a medium compression level, and a fixed packet size of 1272 bytes. We show the R/D results with different PLR values. It can be seen that in absence of packet losses H.264/AVC and VP7 get similar results, much better than the ones obtained by MPEG-4. It can also be seen that H.264/AVC does not fall as much as VP7 when PLR increases.

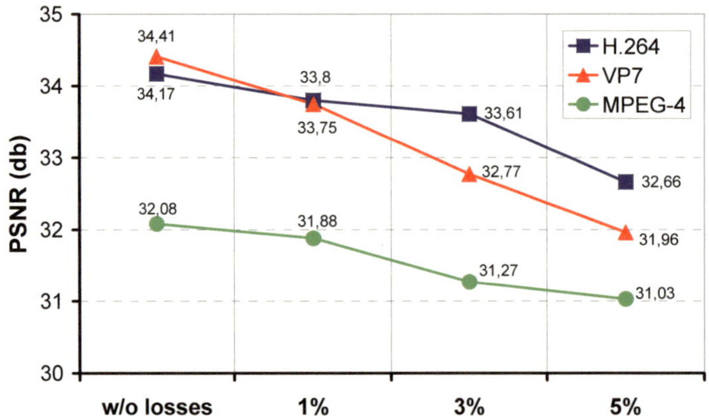


Fig. 7. R/D performance of codecs under packet losses.

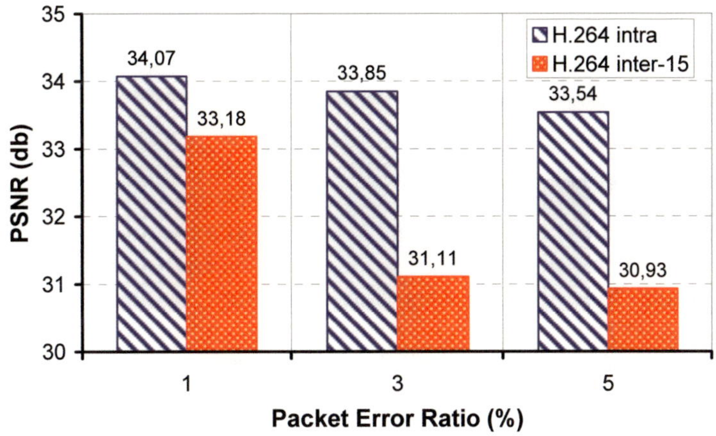


Fig. 8. Resilience of *intra* and *inter* compression modes.

Fig. 8 shows the video quality measures for Foreman (QCIF) video sequence using H.264/AVC codec with a medium compression level and a packet size of 512 bytes. Here, we show the difference between *intra* and *inter* modes in terms of error resilience. The main reason that explains the lower error resilience behavior of *inter* mode coding is due to the propagation of errors to several frames.

So, if we find an error when decoding one frame, this error will propagate to those frames that require the former one to be reconstructed. As a consequence, the reconstructed video quality will be significantly reduced.

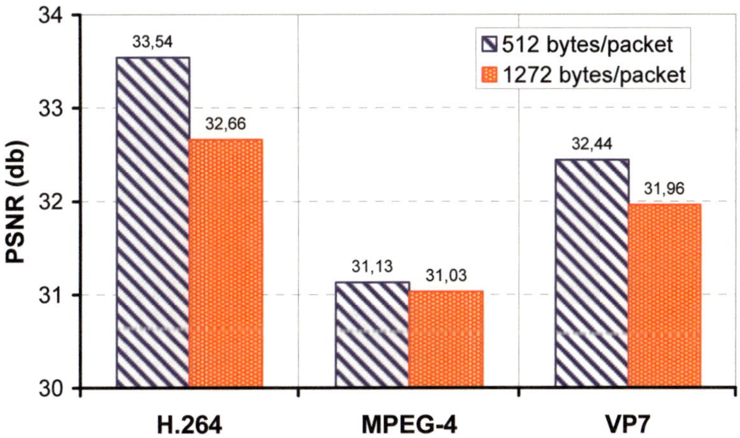


Fig. 9. R/D comparison for two packet sizes (*intra* mode, PLR=5%).

In our tests we realized that, at a fixed PLR, the use of big packet sizes introduces more distortion than using small packet sizes (Fig. 9). But, as we will see later, at real scenarios, the use of small packets produces more packets to send and therefore there are more probabilities for them to get lost, so small packets provide a higher PLR and, as a consequence, a worse video quality.

The most outstanding outcomes for the tests we have performed are that, in general, at the same bitrates H.264/AVC and VP7 offer better quality than MPEG-4 (remember that in section 3 we observed that MPEG-4 was the faster codec) and *intra* mode is more error resilient than *inter* mode.

## 5. Real Scenarios Tests

We have performed two test sets under real scenarios. The first one was in an environment with no background traffic, and the second one was in an environment with different levels of background traffic load. All the tests have been done indoor with the robot moving forward at a constant speed of 55 cm/s. It starts moving at the beginning of a corridor nearby the access point (where it has 100% signal strength) and goes away during 110 seconds up to the end of the corridor, where the signal gets lost (see Fig. 10). Traffic Sender and Traffic Receiver are both static and near the access point. We have used Coastguard (CIF) video sequence concatenated ten times (so it has a length of 3000 frames and is 100 seconds long) with a frame rate of 30 fps. We have setup the codecs to provide a bitrate of 2 Mbps in *intra* mode.

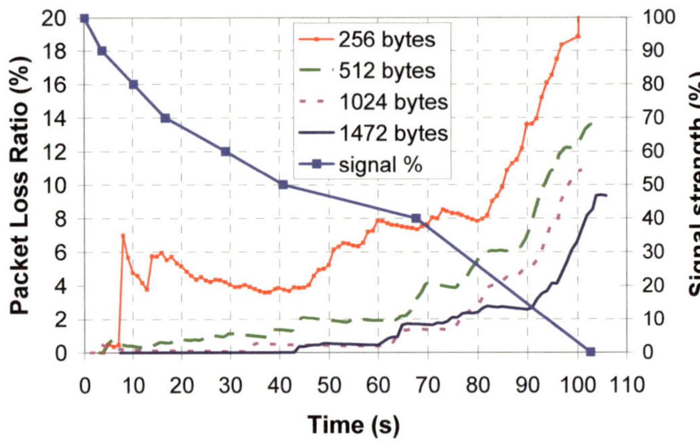


Fig. 10. Signal strength and PLR for different packet sizes.

First, we have done some tests with MPEG-4 codec to evaluate the system performance for different packet sizes without background traffic (unloaded network). In Fig. 10, the packet size that gets the lowest PLR is 1472 bytes (maximum size for an Ethernet MTU of 1500 bytes). So, from now on, we will fix the packet size to 1472 bytes.

The second set of tests was done introducing different background traffic loads in the system network. We have used D-ITG (D-ITG, 2009) traffic generator and D-ITG GUI (D-ITG-GUI, 2009) graphical interface in both sender and receiver to inject background traffic (CBR UDP stream) to our WLAN. For these tests we have used all codecs in *intra* mode with a packet size of 1472 bytes (as mentioned above). We have injected seven different background traffic loads: from 2048 to 3277 Kbps. Traffic loads under 2048 Kbps have a negligible incidence on the video flow delivery and those over 3277 Kbps drive the network to a saturation state.

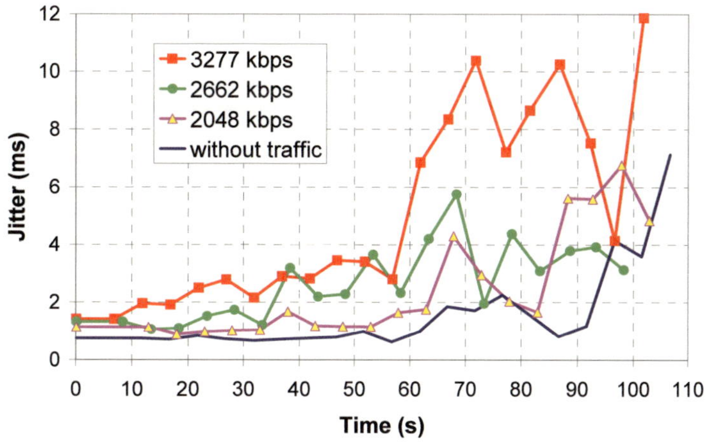


Fig. 11. Jitter comparison at different background traffic loads (MPEG-4 video codec).

In Fig. 11, a comparison of experienced jitter (average packet delay variance) during the session is shown under different background traffic loads. The observed jitter shows a different behavior with respect background traffic (network contention) and received signal strength.

When the signal strength is above 50%, the jitter fluctuations are mainly due to network contention (background traffic). However, when signal strength is below 50%, the jitter wildly changes, amplifying the effect that traffic background has over it.

In Fig. 12, we show the time-average PLR in a session where MPEG-4 was used with different background traffic loads (as reference we show the results without background traffic). As it can be seen, packet losses are directly influenced by signal strength and traffic load for all the region of coverage, in a similar fashion than in Fig. 10.

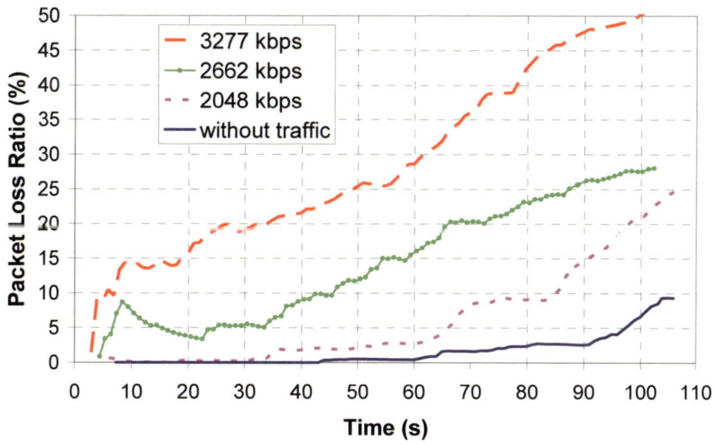


Fig. 12. PLR evolution using MPEG-4 codec with different background traffic loads.

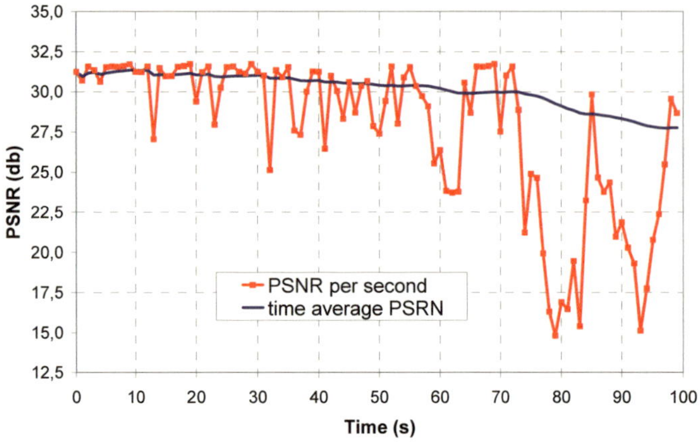


Fig. 13. Quality of MPEG-4 reconstructed video with a background traffic load of 2259 Kbps.

In Fig. 13 we show the reconstructed video quality of the received video sequence using the MPEG-4 codec and a background traffic load of 2259 Kbps. We show the evolution of PSNR (each point represents the average PSNR of 30 frames) and the time-average PSNR. It can be seen that when a packet or a burst of packets is lost the PSNR falls drastically (we have found a PSNR decay of up to 15 dB for a single frame). This situation appears in those areas where the received signal strength is under 50%. In those situations there are a lot of lost packets, typically in bursty fashion. So, in most cases, the frame can not be reconstructed (the frame is declared as lost) or, if the codec has enough information for decoding the frame, the quality of the frame recovered is too low.

In Fig. 14 we compare the selected codecs and the video quality they provide with two different traffic loads. As it can be seen, the H.264/AVC video codec always obtains better results than the other two, showing the same behavior than the one observed without packet losses. However, VP7 shows lower performance than MPEG-4, since this codec is not able to reconstruct a frame when the first part of it is missing. So, to avoid VP7 crashing, we drop the incomplete frame at RTP receiver process. This is the main reason of VP7 low performance.

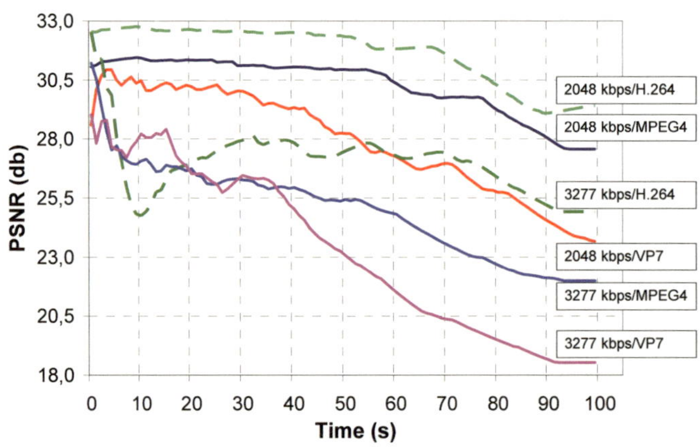


Fig. 14. Quality of reconstructed video with different codecs and traffic loads.

In Fig. 15 we can see the subjective quality of a reconstructed video. Fig. 15(a) shows frame number 2191 of the video sequence without any compression. This corresponds to second 73 of the sequence. In Fig. 15(b) we can see the same frame after it has been encoded and decoded using VP7 codec (without any loss). For this frame we get a value of 31.47 dB of PSNR. In Fig. 15(c) we can see the same frame after reconstruction in a test where packet loss has occurred. This frame offers a value of 14.50 dB of PSNR.

Fig. 15. Subjective quality: (a) Coastguard original frame number 2191, (b) VP7 encoded-decoded, (c) VP7 reconstruction with packet losses.

## 6. Conclusions

In this chapter we have analyzed the performance of current commercial video codecs for video streaming through 802.11 networks running in a remotely teleoperated mobile robots testbed. We have developed a client-server application based in DirectX/DirectShow architecture for testing video compression, data delivery and error resilience behavior in a real scenario.

A first analysis was performed to study the teleoperation control loop delays and the performance of video delivery process in order to get a first impression of overall system behavior. For this kind of applications, compression is mandatory for proper operation in order to cope with the minimum video quality and bounded delay demanded by them. The use of video compressors allows us to adjust the required quality without exhausting network resources like available bandwidth and they significantly reduce the control loop delay improving the functionality of this kind of applications.

With respect to packet losses we have observed that H.264/AVC codec is the one that best performance results achieves. Also, we have checked that *intra* coding mode gets better results than *inter* mode, since it avoids error propagation. In the packetization process, the RTP packet size determines the resulting application bandwidth (goodput). Results show that the longer the packet size is, the higher goodput the application gets.

Finally, we think that telerobotic applications running under standard wireless network technologies like IEEE 802.11 can benefit from the use of state-of-the-art video encoders, like H.264/AVC, to improve throughput, error resilience and real-time feedback.

## 7. References


Brady, K. & Tarn, T.J. (1998). Internet-Based Remote Teleoperation, *Proceedings of the IEEE Conference on Robotics and Automation*, pp. 65-70, 0-7803-4300-X, Leuven, Belgium, May-1998

DirectShow. (2009). Microsoft Developer Network, DirectShow Application Programming Interface , *http://msdn.microsoft.com/en-us/library/dd375454(VS.85).aspx*

D-ITG. (2009). Universita' degli Studi di Napoli "Federico II" (Italy), Distributed Internet Traffic Generator, *http://www.grid.unina.it/software/ITG*

D-ITG-GUI. (2009). Volker Semken, Graphical User Interface for D-ITG, *http://www.semken.com/projekte*

Elhajj, I.; Xi, N.; Fung, W.K.; Liu, Y. H.; Li, W.J.; Kaga T. & Fukuda T. (2001). Supermedia in Internet-Based Telerobotic Operations, *Proceedings of the 4th IFIP/IEEE International Conference on Management of Multimedia Networks and Services*, pp. 359-372, 3-540-42786-4, Chicago, IL, USA, November-2001

ER1. (2009) Evolution Robotics Personal Robot System, *http://www.evolution.com/er1*

Fong, T.W.; Conti, F.; Grange, S. & Baur, C. (2001). Novel interfaces for remote driving: gesture, haptic and PDA. *Proceedings of SPIE Mobile Robots XV and Telemanipulator and Telepresence Technologies VII*, pp. 300-311, 0-8194-3860-X, Boston, MA, USA, November-2001

Goldberg, K. & Siegwart, R. (2002). *Beyond webcams: An introduction to online robots*, MIT Press, 0-2620-7225-4, 2002

H.264. (2009). H.264/MPEG-4 AVC free software library, *http://x264.nl*

Hirzinger, G.; Brunner, B.; Dietrich, J. & Heindl, J. (1993). Sensor-Based Space Robotics-ROTEX and Its Telerobotic Features. *IEEE Transactions on Robotics and Automation*, Vol. 9, No. 5, October-1993, pp. 649-663, 1042-296X

IEEE-WG-802.11. (1999). IEEE Standard for Wireless LAN Medium Access Control and Physical Layer Specifications, *Technical Report*, IEEE WG 802.11, August-1999

IEEE-WG-802.11. (2002). Medium Access Control (MAC) Enhancements for Quality of Service (QoS), *Technical Report*, IEEE WG 802.11, November-2002

Kwon, D.; Woo, K.Y. & Cho, H.S. (1999). Haptic Control of the Master Hand Controller for a Microsurgical Telerobot System. *Proceedings of the IEEE International Conference on Robotics and Automation*, pp. 1722-1727, 0-7803-5180-0, Detroit, MI, USA, May-1999.

Lamping, U.; Sharpe, R. & Warnicke, E. (2006). Wireshark User's Guide. 2006

Live Networks. (2009). LIVE555 Library for Multimedia Streaming, *http://www.live555.com/liveMedia*

Luo, R.; Lee, W.Z.; Chou, J.H. & Leong, H.T. (1999). Tele-Control of Rapid Prototyping Machine Via Internet for Automated Tele-Manufacturing. *Proceedings of the IEEE International Conference on Robotics and Automation*, pp. 2203-2208, 0-7803-5180-0, Detroit, MI, USA, May-1999

MPEG-4. (2009). MPEG-4 library for coding/decoding, *http://sourceforge.net/projects/ffdshow*

RFC-3550. (2003). RTP: A Transport Protocol for Real-Time Applications. *IETF document*, July-2003

Sgouros, N.M. & Gerogiannakis, S. (2003). Robot Teleoperation Environments Featuring WAP-based Wireless Devices. *Journal of Network and Computer Applications*, Vol. 26, No. 3, August-2003, pp. 259-271, 1084-8045

VP7. (2009). VP7 proprietary video compression codec, *http://www.on2.com/technology/vp7*

# 26

# Virtual Simulator for Design of Mobile Robot Control and Navigation Systems

Leonimer F Melo , Jose F Mangili Jr and Fernando C Dias Neto
*State University of Londrina*
*Brazil*

Joao M Rosario
*State University of Campinas*
*Brazil*

## 1. Introduction

The development of control systems for independent mobile robots has appear as a great challenge for the researchers until the current days. Different platform for the project of control system for independent mobile robots come being used in diverse research areas. Per many years the researchers have constructed control systems that present an intelligent behavior in controlled environments, with ideal situations, but that normally does not keep the same performance in the real world. Innumerable systems of control exist to be used in the real world, but generally these systems are limited and they do not present an independent or intelligent behavior.

Diverse possible applications for the mobile robots already exist. In the transport, monitoring, inspection, cleanness of houses, space exploration, aid the physical deficient, among others. However, the independent mobile robots had not yet caused much impact in domestic or industrial applications, mainly had the lack of a system with robust, trustworthy and flexible control that it would allow these robots operated in dynamic environments, less structuralized, and inhabited by human beings. The development of a mobile robotic model system with open architecture and flexible control, with robust control system, that incorporates what exists of modern in terms of embedded hardware technology and that makes possible the operation of a mobile robotic systems in a real world environment is one of the motivations of this work.

The locomotion planning, under some types of restrictions, is a very vast field of research in the area of the mobile robotics (Graf, 2001). The basic planning of trajectory for the mobile robots imply the determination of a way in the space-C (*configuration space*) between an initial configuration of the robot and a final configuration, in such a way that the robot does not collide with no obstacle in the environment, and that the planned movement is consistent with the kinematic restrictions of the vehicle. In this context, one of the boarded points in this work was development of a trajectory calculator for mobile robots.

One of the main motivations of this work is to propitiate a virtual environment that facilitates the development of archetypes of embedded systems, emphasizing the implementation of tools that allow the simulation of the kinematic, dynamic and control conditions, with real time monitoring of all important points of the system. In this way, the proposal of a virtual simulator of mobile robotic systems is presented together with techniques of rapid prototyping.

## 2. The Mobile Robot Platform

Platforms for knowledge consolidation could be used in several educational and research areas, such as modeling, control, automation, power systems, sensors, transmission of data, embedded electronics and software engineering. In fact, the use of the mobile robots as base for knowledge consolidation, has been successful adopted in many educational and research institutions mainly because they appear to be a quite attractive low cost solution that allow the integration of several important areas of knowledge. Mobile robots also become a better solution for practical problem in modern society. These appointments shows a large applicability of mobile robots and a crescent request in the modern world. The proposal of this project is developing a generic open system for control a mobile robot and supply this need. The system emphasizes the control structure, the supervision and the transfer of information. In a context of educational and research aims the project aspects and integration solution compose the desired know-how acquired during the development of the system, which one certainly would not to be approached if a commercial mobile robot was acquired.

Fig. 1. Mobile robot platform prototype.

An embedded processor, with a dedicated control software, to be used on a platform mobile robotics, is considered. In addition to this platform another one, a commercial platform, coupled to a communication net, is analyzed. The set of platforms, whose objectives are making use of the existing communication interfaces and providing am embedded user interface alternative in the mobile robot, allows the creation of a powerful link with the external world. The objective of this platform is to make use of the existing communication interfaces, as well

as to provide an embedded user interface alternative in the mobile robot. Another aspect considered, is the flexibility of the hardware project, which allows the expansion of mobile robot facilities. New sensor combinations should be used. Different supervision and control models should equally be used to carry out the mobile robot tasks.
This work aim to presents the implementation of a virtual environment for project simulation and conception of supervision and control systems for mobile robots and focus on the study of the mobile robot platform, with differential driving wheels mounted on the same axis and a free castor front wheel, whose prototype used to validate the proposal system is depicted in Fig. 1 and Fig. 2 which illustrate the elements of the platform.

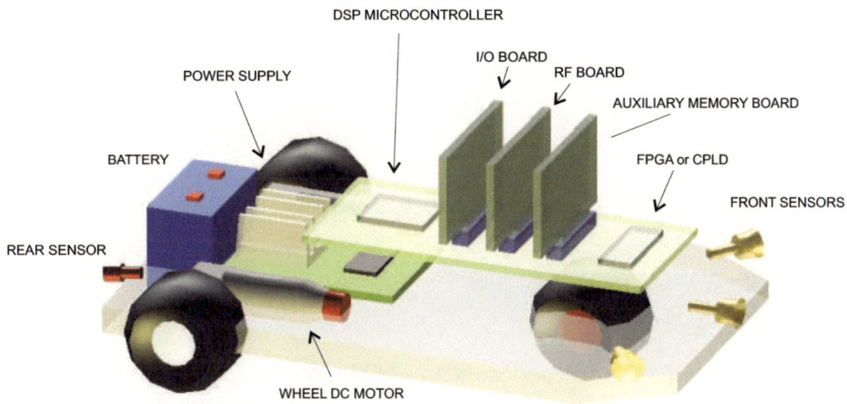


Fig. 2. Mobile robot platform and elements.

## 3. Mobile robot modeling

It's necessary to make the mobile robot mathematics modeling in order to extract the equation and algorithms that will be used in the simulator's blocks. This section due with that, observing the fact our physics model is a mobile robot with differential driving wheels with nonholonomic restrictions.
Suppose that a robot is in a position $(x, y)$ and "facing" along a line making an angle $\theta$ with the $x$ axis (Fig. 3). Through manipulation of the control parameters $v_e$ and $v_d$, the robot can moves to different positions. Determining the positions that is reachable at given control parameters is know as the forward kinematics problem for the robot (Siegwart & Nourbakhsh, 2004). Because of $v_e$, $v_d$ and hence $R$ and $\omega$ are functions of time, is straightforward to show (Fig. 3) that, if the robot has position $(x, y, \theta)$ at some time $t$, and if the left and right wheels have ground-contact velocities $v_e$ and $v_d$ during the period $t \rightarrow t + \delta t$, the ICC (Instantaneous Center of Curvature) is given by

$$\mathrm{ICC} = [x - R\sin(\theta),\ y + R\cos(\theta)]. \tag{1}$$

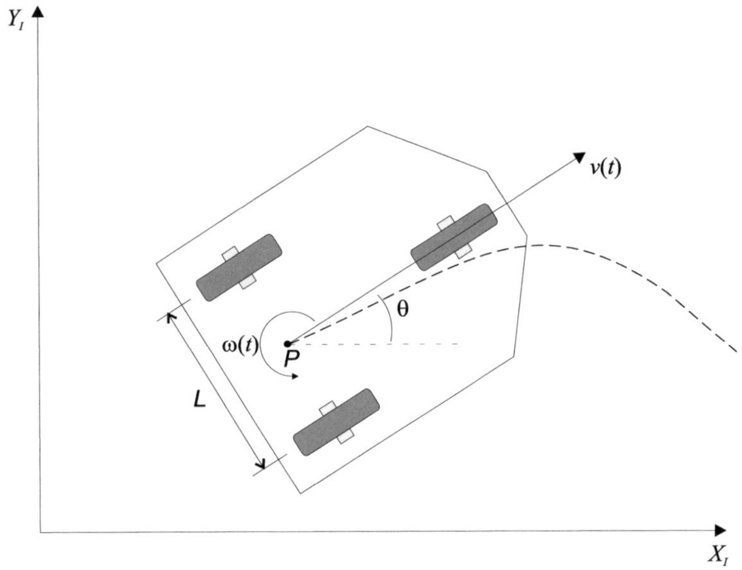


Fig. 3. Forward kinematics geometry.

For better comprehension of the equation, it's possible to simplifying $ICC = I$, $\cos(\omega\delta t) = C$ and $\sin(\omega\delta t) = S$, then, at time $t \rightarrow t + \delta t$, the position of the robot is given by

$$\begin{bmatrix} x' \\ y' \\ \theta' \end{bmatrix} = \begin{bmatrix} C & -S & 0 \\ S & C & 0 \\ 0 & 0 & 1 \end{bmatrix} \begin{bmatrix} x - \mathrm{I}_x \\ y - \mathrm{I}_y \\ \theta \end{bmatrix} + \begin{bmatrix} \mathrm{I}_x \\ \mathrm{I}_y \\ \omega\delta t \end{bmatrix}. \tag{2}$$

The Eq. (2) describes the motion of a robot rotating a distance $R$ about its ICC with an angular velocity given by $\omega$ (Dudek & Jekin, 2000). Different classes of robots will provide different expressions for $R$ and $\omega$ (Shim et al., 1995).

The forward kinematics problem is solved by integrating (2) from some initial condition $(x_0, y_0, \theta_0)$, it is possible to compute where the robot will be at any time $t$ based on the control parameters $v_e(t)$ and $v_d(t)$. For the special case of a differential drive vehicle, it is given by

$$\begin{aligned} x(t) &= \frac{1}{2}\int_0^t [v_d(t) + v_e(t)] \cos[\theta(t)] dt, \\ y(t) &= \frac{1}{2}\int_0^t [v_d(t) + v_e(t)] \sin[\theta(t)] dt, \\ \theta &= \frac{1}{L}\int_0^t [v_d(t) - v_e(t)] dt. \end{aligned} \tag{3}$$

A question more interesting, and at same time more difficult to answer, is: How can the control parameters could be selected in a way the robot obtain a specific global position or follow a specific trajectory? This is known as the task of determining the vehiclet's *inverse kinematics*: inverting the kinematic relationship between control inputs and behavior. It is also related to the problem of trajectory planning.

### 3.1 Inverse kinematics for differential drive robots

The Eq. (3) describe a constraint on the robot velocity that cannot be integrated into a positional constraint. This is known as a *nonholonomic constraint* and it is in general very difficult to solve, although solutions are straightforward for limited classes of the control functions $v_e(t)$ and $v_d(t)$ (Zhao & BeMent, 1992). For example, if it is assumed that $v_e(t) = v_e$, $v_d(t) = v_d$ and $v_e \neq v_d$, then (3) yields

$$\begin{aligned} x(t) &= \frac{L}{2}\frac{v_d + v_e}{v_d - v_e}\sin\left[\frac{t}{L}(v_d - v_e)\right], \\ y(t) &= -\frac{L}{2}\frac{v_d + v_e}{v_d - v_e}\cos\left[\frac{t}{L}(v_d - v_e)\right] + \frac{L}{2}\frac{v_d + v_e}{v_d - v_e}, \\ \theta(t) &= \frac{t}{L}(v_d - v_e), \end{aligned} \tag{4}$$

where $(x, y, \theta)_{t=0} = (0, 0, 0)$. Given a goal time $t$ and goal position $(x, y)$. The Eq. (4) solves for $v_d$ and $v_e$ but does not provide a solution for independent control of $\theta$. There are, in fact, infinity solutions for $v_d$ and $v_e$ from Eq. (4), but all correspond to the robot moving about the same circle that passes through $(0, 0)$ at $t = 0$ and $(x, y)$ at $t = t$; however, the robot goes around the circle different numbers of times and in different directions.

## 4. Control Architecture System

The control architecture system can be visualized at a logical level in the blocks diagram of Fig. 4.

The system was divided into three control levels, organized in different degrees of control strategies. The levels can be described as:

- **Supervisory control level**: This represents a high level control. In this level it was possible to carry out the supervision of one or more mobile robots, through the execution of global control strategies.
- **Local control level**: In this level the control was processed by the mobile robot embedded software implemented in a 8 bits microcontroller. The control strategies allowed decision making to be done at a local level, with occasional corrections from the supervisory control level. Without communication with the supervisory control level, the mobile robot just carried out actions based on obtained sensor data and on information previously stored in its memory.
- **Interface control level**: This control level is restricted to strategies of control associated with the interfaces of the sensor and actuators. The strategies in this level were implemented in hardware, through PLD (Programmable Logic Devices).

The hardware architecture, from the point of view of the mobile robot, was organized into several independent blocks, connected through the local bus which is composed by data, address and control bus (Fig. 5). A master manager block operates several slave blocks. Blocks associated with the interfaces of sensors and actuators, communication and auxiliary memories were subjected to direct control from the manager block. The advantage of using a common bus was the facility to expand the system. Despite of the limitations of resources, it was possible to add new blocks, allowing an adapted configuration of the robot for each task.

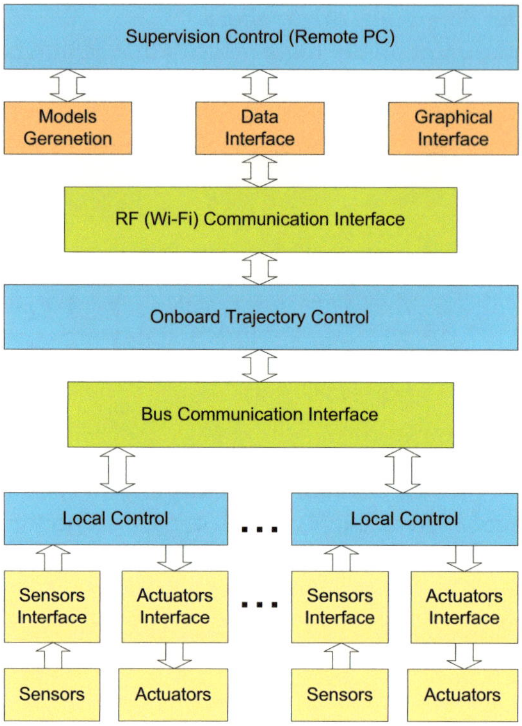


Fig. 4. Different control levels of the proposed system.

### 4.1 Description of Blocks

- **Supervisory control block**: Is the highest level of control. In this block, the supervision of one or more mobile robots is managed through the execution of global control strategies. It is implemented in an IBM PC platform and is connected with the local control level, in the mobile robot, through Ethernet wireless WI-FI link. This protocol uses IEEE 802.11a standard for wireless TCP/IP LAN communication. It guarantees up to 11 Mbps in the 2.4 GHz band and requires fewer access points for coverage of large areas. Offers high-speed access to data at up to 100 meters from base station. 14 channels available in the 2.4 GHz band guarantee the expansibility of the system with the implementation of control strategies of multiple robots.
- **Master manager block**: It is responsible for the treatment of all the information received from other blocks, for generation of the trajectory profile for the local control blocks and for communication with the external world. In communication with the master manager block, through a serial interface, a commercial platform was used, which implemented external communication using an Ethernet WI-FI wireless protocol. The robot was seen as a TCP/IP LAN point in a communication net, allowing remote supervision through supervisory level. It's implemented with Texas Instrument TMDSDSK6416 DSP board Kit that uses the TMS320C6416 DSP, a 1 GHz device delivering up to 8000 million instructions per second (MIPs) with highest performing.
- **Sensor interface block**: It is responsible for the sensor acquisition and for treatment of data in digital words, to be sent to the master manager block. The implementation of

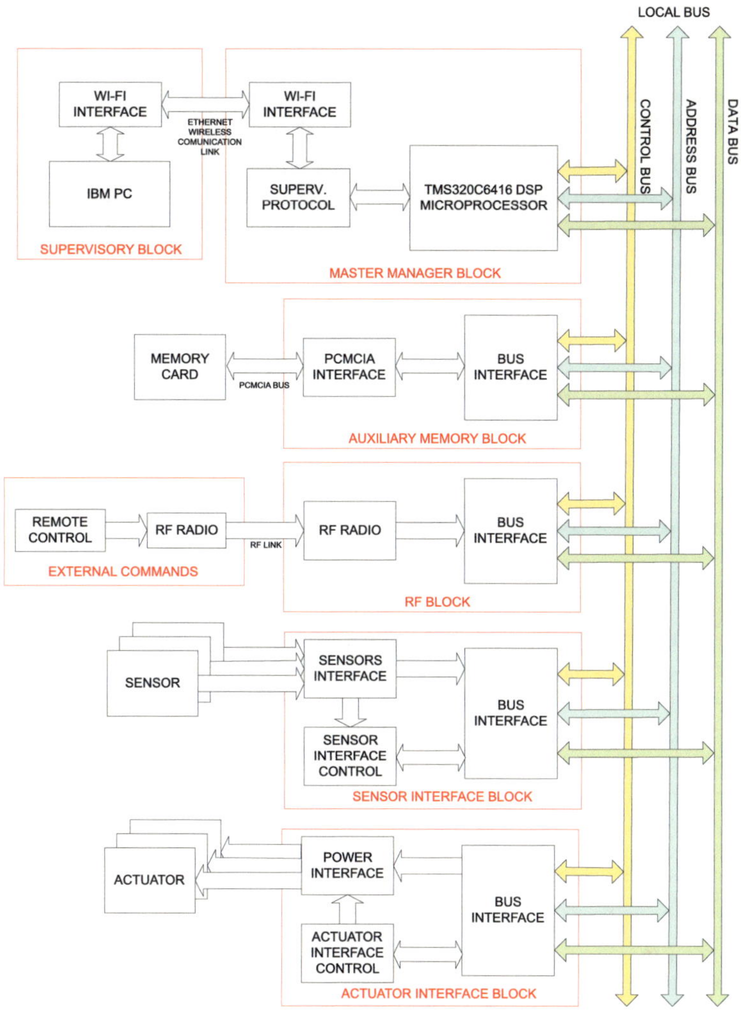


Fig. 5. Hardware architecture block diagram of the proposed system.

that interface through PLD allowed the integration of information from sensors (sensor fusion) locally, reducing manager block demand for processing. In same way, they allowed new programming of sensor hardware during robot operation, increasing sensor treatment flexibility.

- **Actuator interface block**: This block carried out speed control or position control of the motors responsible for the traction of the mobile robot. The reference signals were supplied through bus communication in the form of digital words. Derived information from the sensor was also used in the controller implemented in PLD. Due to integration capacity of enormous hardware volume, PLD was appropriate to implement state machines, reducing the need for block manager processing. Besides the advantage of the integration of the hardware resources, PLD facilitated the implementation and debugging. The possibility of modifying PLD programming allowed, for example, changes in control strategies of the actuators, adapting them to the required tasks.

- **Auxiliary memory block**: This stored the information of the sensor, and operated as a library for possible control strategies of sensors and actuators. Apart from this, it came with an option for operation registration, allowing a register of errors. The best option was an interface PCMCIA, because this interface is easily accessible on the market, and being well adapted for applications in mobile robots, due to low power consumption, low weight, small dimensions, high storage capacity and good immunity to mechanical vibrations.
- **RF communication Block**: It allows the establishment of a bi-directional radio link for data communication. It operated in parallel with the commercial platform WI-FI link. The objective of these communication links was to allow the use of remote control. The remote control has a high trajectory priority from other blocks, like supervisory control block, and can take the control of the mobile robot to execute, for example, emergency necessary movements or stop. To implement this block was used a low power UHF data transceiver module BiM-433-40.

## 5. Mobile Robot Simulator

The use of the system has beginning for the captation of main points for generation of the mobile robot trajectory. The idea is to use a system of photographic video camera that catches the image of the environment where the mobile robot navigates. This initial system must be able to identify the obstacles of the environment and to generate a matrix with some strategic points that will be good for input data for the system of trajectory generation. Figure 6 presents a general vision of the considered simulator system.

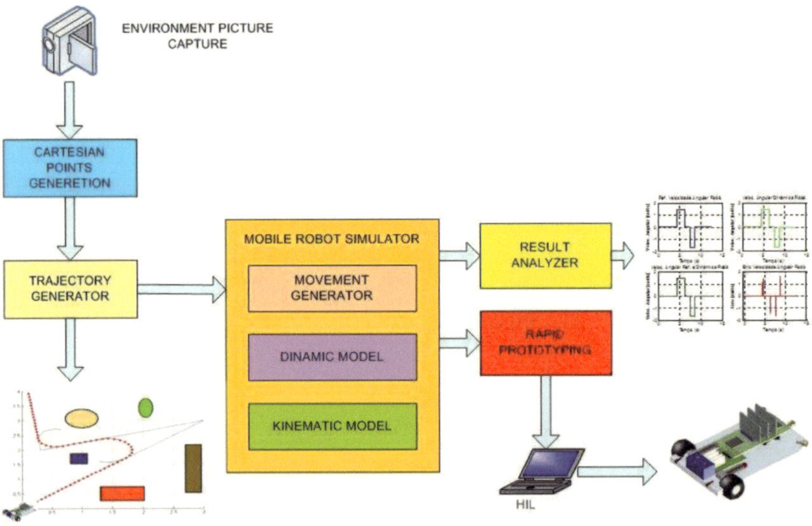


Fig. 6. General vision of the simulator system.

Fig. 7 illustrates an example of an environment with some obstacles where the robot must navigate. In this environment, the robot is located initially in the P1 point and the objective is to reach the P4 point. The generating system of initial cartesian points, must supply to the module of trajectory generation, with the cartesian points P1, P2, P3 and P4, that are the main points of the traced route.

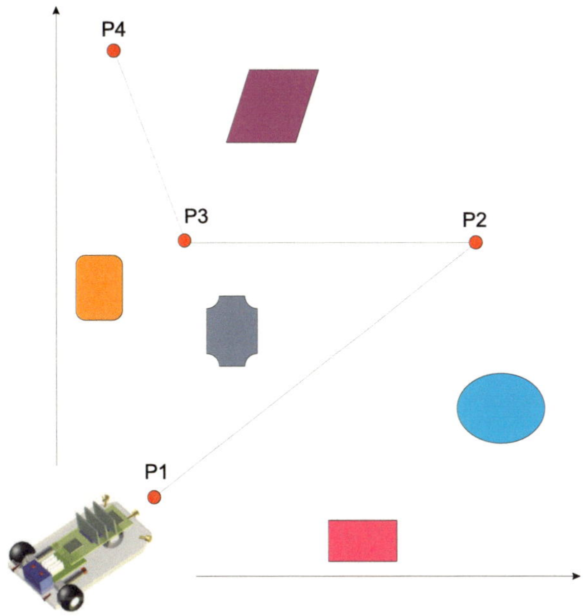


Fig. 7. Example of an environment with some obstacles where the robot must navigate.

This system is particularly interesting and can be used, for example, in robotic soccer games, where the navigation strategies are made from images of the environment (soccer field) and the obstacles are the other robots players. As it's described follow, with this system, the best trajectory can be defined and traced, respecting always the kinematics holonimics or nonholonomics constraints of the robotic systems in question, and to make all the simulation of the system foreseeing imperfections and analyzing results before the final implementation of the control system in the mobile robot (Melo & Rosario, 2006).

### 5.1 Mobile Robot Control Structure

The tasks carried through for the mobile robots are based on the independent movement of each degree of freedom, coordinated from a trajectories plan based in its kinematic model. In the most of the cases, the tasks programming is planned with anticipation and a map of the environment is loaded in the robot memory board. The mobile robot accomplish the trajectory with sequence of independent movements of each axle, until reaching the desired final position. From the knowledge of these articulated positions, easily is implemented a generator of references (profile of speeds) based in the kinematic characteristics of each joint.

For accomplishment of tasks in level of cartesian coordinates system and for generation of the reference signals for the position controller of each robotics joint of the mechatronics system in study, the establishment of mathematical model based in the kinematics of the system becomes necessary, what is described in section 3. Therefore, the control of a robot needs procedures to transform the data of positioning reference, such as the linear speed and the bending radius, in cartesian coordinates, when it is desired to realize the control through a cartesian referential. The Fig. 8 illustrates the mobile robot structure of control with the representative blocks of the trajectory generation, dynamic and kinematic model of the system.

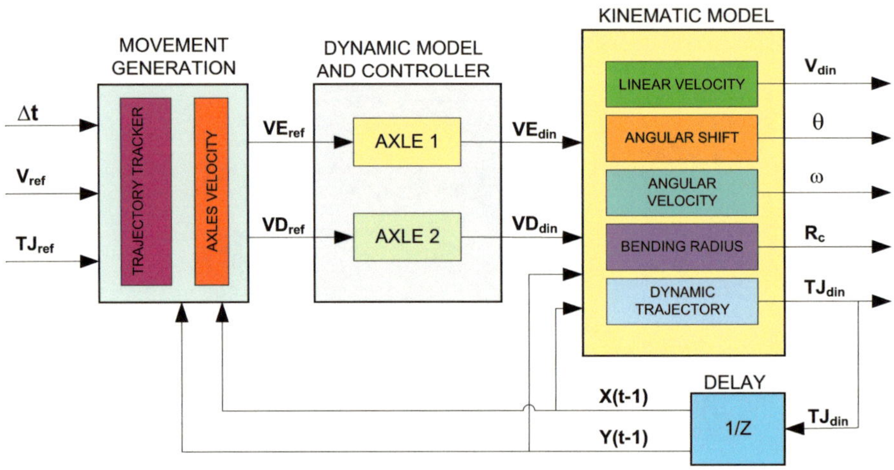


Fig. 8. The mobile robotic control structure.

The trajectory generator receives the references data, such as the positioning vector $\mathbf{X}_{ref} = [\mathbf{x}_{ref}, \mathbf{y}_{ref}, \theta_{ref}]$, the robot reference linear speed $V_{ref}$ and the robot instantaneous trajectory radius $R_{curv}$, that are converted into $VE_{ref}$ (linear speed of the left wheel) and $VD_{ref}$ (linear speed of the right wheel). These differentiated speeds are received by the controller, and in the dynamic model of the system, they are sent to the respective wheels of the robot, through its actuators. Then are generated by the controller the vectors $VE_{din}$ (dynamic linear speed of the left wheel) and $VD_{din}$ (dynamic linear speed of the right wheel). Into the block of the kinematic model, these data are converted into the vector final positioning of the robot $\mathbf{X} = [\mathbf{x}, \mathbf{y}, \theta]$.

### 5.2 Trajectory Generation Block

The Trajectory Generation Block of trajectory, receives some important points from the camera system so that the trajectory can be traced to be realized by the mobile robot. These points form a cartesian matrix containing more or less points, depending on the complexity of the environment. For a reason or purpose tests and for validation of the system, the number of points to be fed the system was fixed in four. Nevertheless, the number of points can be increased depending on the complexity of the environment where the mobile robot will navigate. Another data important to be used by the system have relation with the holonomics constraints of the modeled mobile robot. The bending radius must be informed to the system to be performed in the trajectory. A time that, for a reason or purpose tests, was fixed in four the number of cartesian input points, must also be supplied the radius of the two curves to be

executed for the robot. The information of distinct radius makes the system more flexible, being able the trajectory to be traced with different bendings depending on the angle of direction displacement and on the restrictions of the robot.
The graphic of the Fig. 9(a) illustrates the initial points for the trajectory generator. In this example, it was captured from the camera system and transmitted for the simulator the vector $\mathbf{x} = [.1 \quad 3 \quad .3 \quad .1]$ and the vector $\mathbf{y} = [.1 \quad 3 \quad 2 \quad 4]$, with the bending radius for first and the second semicircle given by the vector $\mathbf{r} = [.4 \quad .3]$. All the measures are in meters.

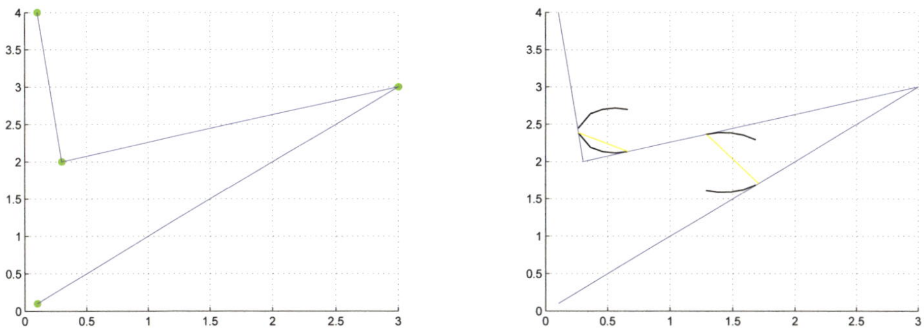


(a) Initial points given to trajectory generation. (b) Assay to the delineated trajectory.

Fig. 9. Initial points given by camera system and Assay to the delineated to trajectory.

In the Fig. 9(b) it's possible to see the tracing of the straight lines, the semicircles and an intermediate segment of straight line indicating the start and the end of the tracing of each circular movement to be executed by the mobile robot.
The final tracing of the mobile robot trajectory can be observed in the Fig. 10, represented by red spots.

### 5.3 The Virtual Simulator Implementation

Now are presented the main characteristics of a simulator of mobile robotic systems. It was implemented from the kinematic and dynamic model of the mechanical drive systems of the robotic axles, for the simulation of different control techniques in the field of the mobile robotics, allowing to deepen the concepts of navigation systems, trajectories planning and embedded control systems. This simulator, designed in modular and opened architecture, as presented in section 4, allows the direct application of some concepts into of the mobile robotics area, being used for its validation, and as main objective of this study, the model of an prototype of mobile robot with nonholonomic kinematic constraint and differential drive with two degrees of freedom (movement of linear displacement and rotation).
For the simulator development, the constructive aspects of the mobile robot prototype had been considered, including the kinematics and dynamics modeling of drive and control systems. The simulator presents the trajectories generating module that is the first block of the system and was implemented with the functionality to generate a trajectory for the mobile robot from a matrix of points supplied initially. Another presented block is the controller, implemented in the PID traditional form.
Fig. 11 shows the initial page of the Virtual Mobile Robot Simulator. The user can choose one of the captured trajectory for analyzing, take a look in the graphical results of the simulation

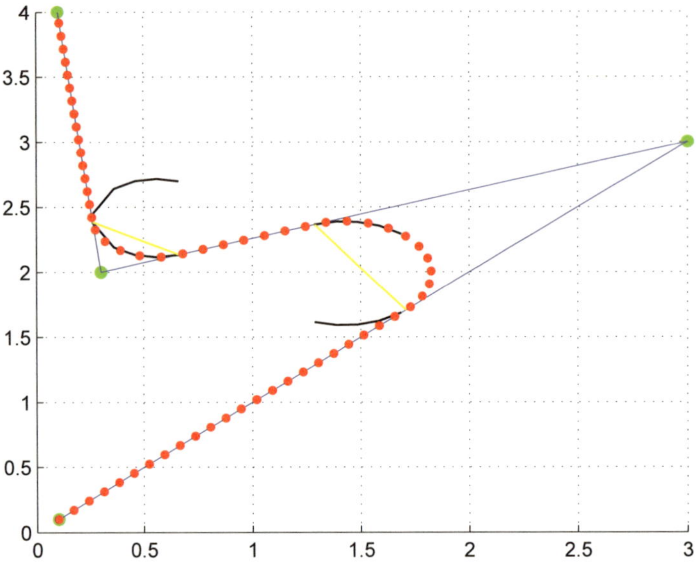


Fig. 10. The final tracing of the mobile robot trajectory.

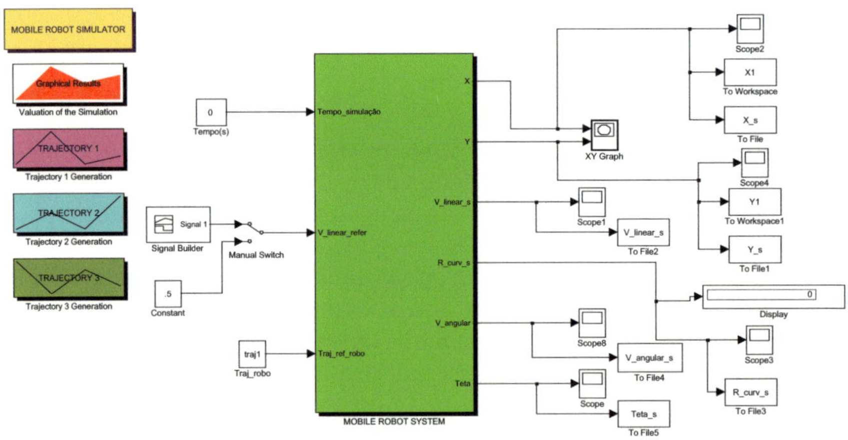


Fig. 11. The Virtual Simulator first page.

or implement changes in the robot model, by clicking on the mobile robot system (green main block).

The virtual simulator system of the mobile robot is composed of three main blocks. The first one is called movements generation block. The second one is the block of the controller and dynamic model of the mobile robot and the third one is the block of the kinematic model. The Fig. 12 illustrates the mobile robot simulator implemented into Matlab Simulink® blocks.

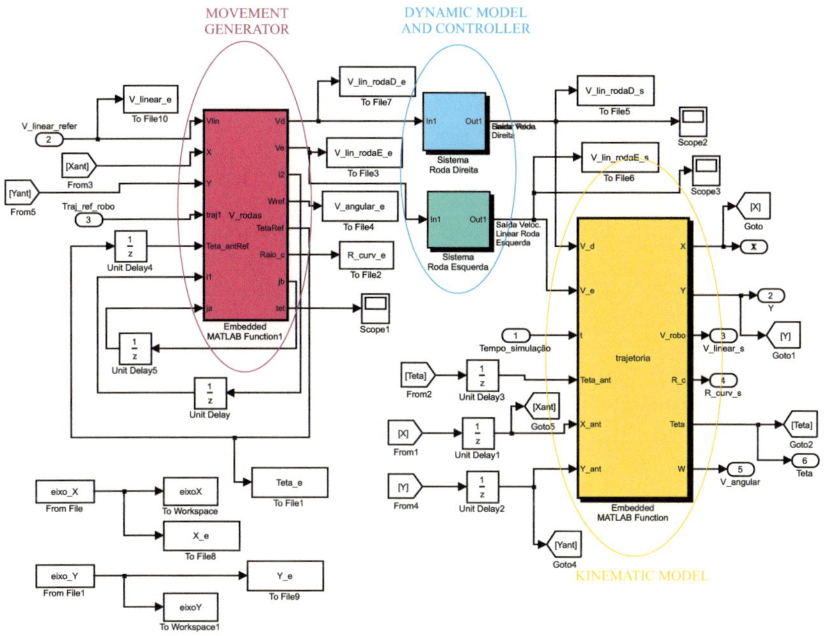


Fig. 12. Mobile robot simulator implemented into Simulink®.

### 5.4 Results Graphical Analyzer

The simulator implemented in Simulink® environment allows the visualization of the inputs and outputs of the system in study.

For a better understand and analysis of the behavior of the system the implementation of a results graphical analyzer becomes essential. In this way, after realizing the simulations in the domain of the time, timings data archives are got corresponding to the study variables (angular and cartesian position, linear and angular speed and control signals), that after convenient treatment, becomes possible to verify important results for better analysis of the system behavior. The Fig. 13 illustrates a menu of the graphical analyzer of the mobile robotic system in study with an example of generated graphic.

One kind of analysis that is made is in relation to the linear displacement of the robot in axis X and Y. In Fig. 14, the dynamic behavior of the robot with regard to these parameters, as well as the presented errors is illustrated.

Another important graphic generated for the system, in the *cartesian trajectory* sub-menu, that is the graphic of the cartesian trajectory kinematics and dynamics of the mobile robotic system

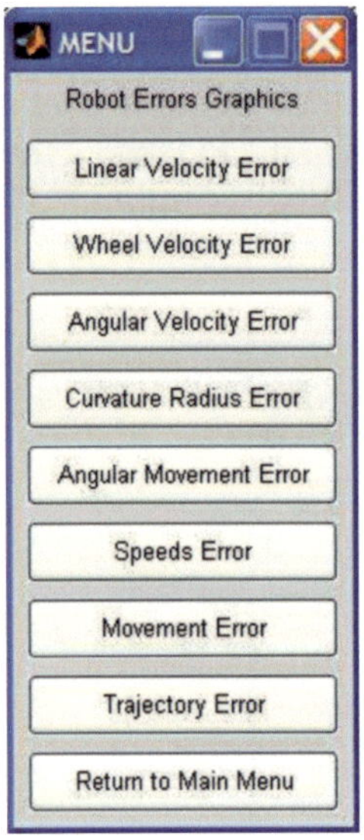


(a) Robot errors analyzer submenu.

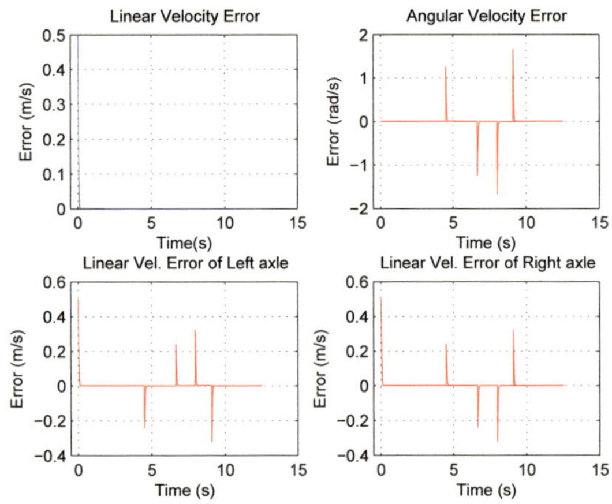


(b) A robot speeds errors graphics.

Fig. 13. Submenu of the mobile robot graphical analyzer with an example of generated graphic.

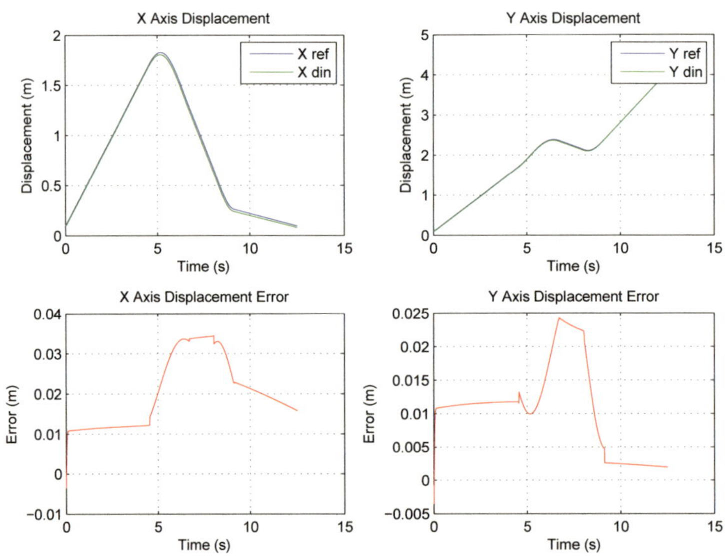


Fig. 14. Dynamic behavior graphics of the robot in the X and Y axis with their errors.

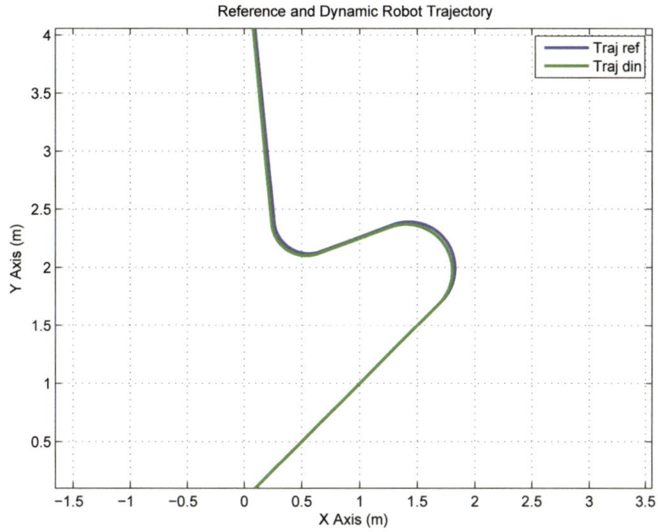


Fig. 15. Cartesian trajectory kinematics and dynamics of the mobile robotic.

in plan XY. The Fig. 15, shows the dynamic tracing of reference and of the trajectory of the mobile robot.
The Fig. 16 illustrates the graphic of the trajectory error.

## 6. Mobile Robot Rapid Prototyping

The use of the rapid prototyping technique in mobile robotic systems differs from the traditional target used in mechanics engineering and enters in new field of research and development for projects of mobile robots mechatronics systems. In this way, the rapid prototyping of these systems is associated not only with the project of the physical system, but mainly with the experimental implementations in the fields of *hardware* and *software* of the robotic system. It is fundamental that the architecture of *hardware* of the considered system be opened and flexible in the way of effecting the necessary modifications for system optimization. A proposal of open architecture system was presented in (Melo *et al.*, 2003). The *software* of the embedded control system of the mobile robot, in the context of the rapid prototyping, can be elaborated in simulators and tested all the parameters for adjustments that makes necessary in accordance with the physical system to be implemented, the hardware architecture, the actuators and the sensors. In this way, in the context of this work, the rapid prototyping is then the methodology that allows the creation of a virtual environment of simulation for the project of a controller for mobile robots. After being tested and validated in the simulator, the control system is programmed in the control board memory of the mobile robot. In this way, a economy of time and material are obtained, sooner validating all the model virtually and later operating the physical implementation of the system.

### 6.1 HIL (Hardware-in-the-loop) Simulation

The HIL technique of simulation is used in development and tests for real time embedded systems. HIL simulations provide a platform accomplish of development for adding the com-

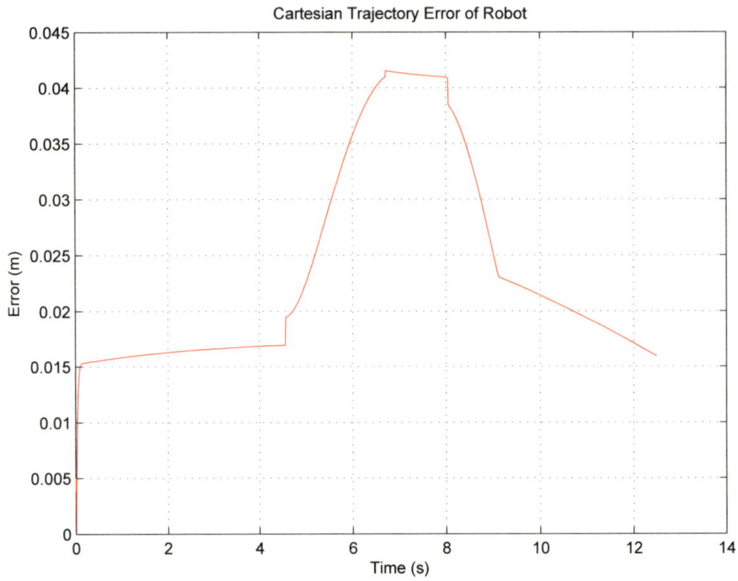


Fig. 16. Error of the kinematics and dynamics trajectory of the mobile robot.

plexity of the plant under control to the tests platform. The control system is enclosed in the tests and developments through its mathematical models representations and all the respective dynamic model (Melo & Rosario, 2006).
The Fig. 17 illustrates the use of the HIL simulation technique for real time simulation of the considered mobile robotic system.

## 7. Experimental Validation

The Aedromo, a didactic experimental environment for mobile robots, is an ambient used to test and validate the trajectory virtual simulator, the onboard robot control and the rapid prototyping system, all of them describe above. This environment is utilized for many others applications, like robotic soccer game, two or more robots interaction, etc. The supervision and the robots movements coordination are made through a close loop architecture based in a satellite camera over the arena. The information supplied for a video camera is sent for one or two computers for processing. The data obtained from this process are used to generate a sequence of instructions that are sent for the robots. The robots receive the instructions and carry through the actions obeying the predetermined tasks. The instructions are results of developed programmers strategies to execute the tasks and to realize the robot navigation into the environment. The Fig. 18 depicts this environment.
The trajectory to be executed by the robot, from trajectory generation software, is illustrated on Fig. 19. This trajectory allows to get parameters of comparison between the real system, represented by de robot prototype, and the virtual system. On the onboard control of the mobile robot was implemented a dynamic and kinematic strategy, in order to find best trajectory, avoiding obstacles on the way, like can be seen in Fig. 7.

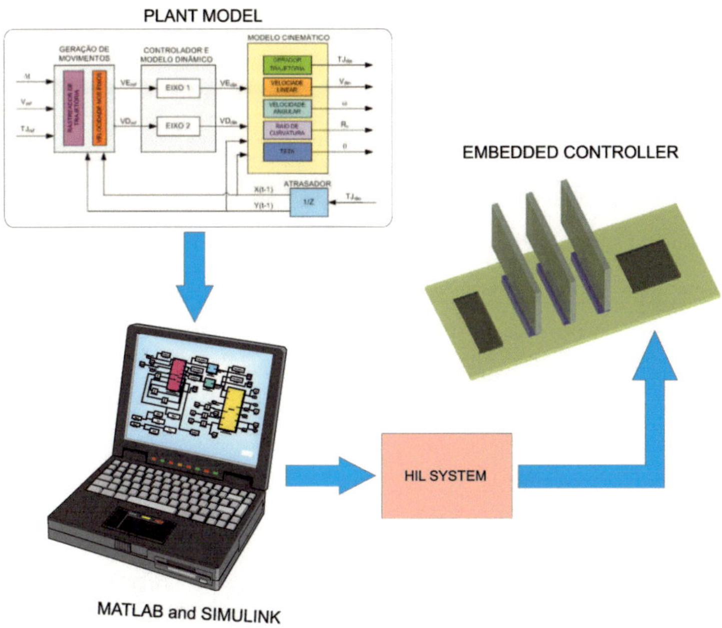


Fig. 17. HIL simulation for mobile robot system.

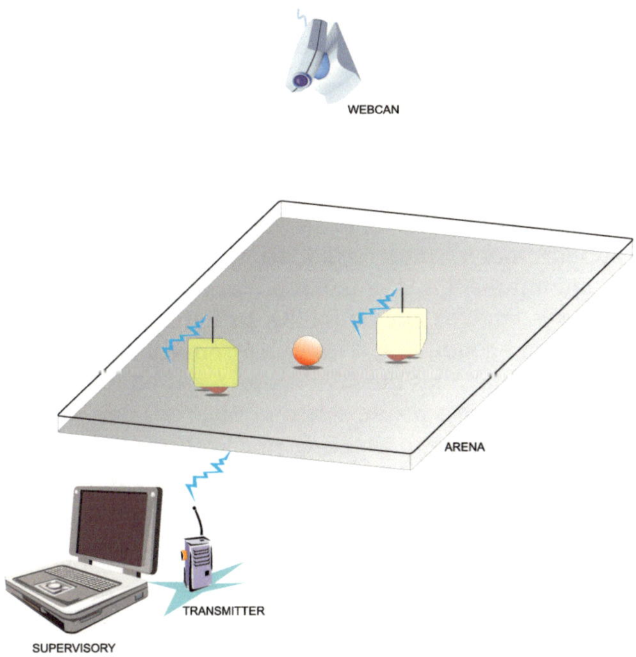


Fig. 18. The Aedromo environment.

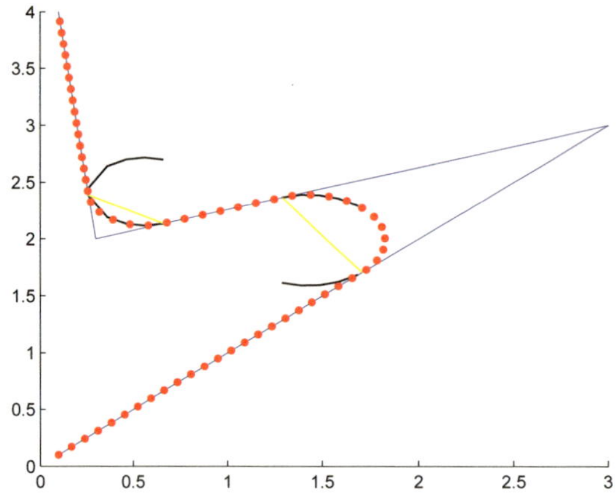


Fig. 19. Best trajectory to be executed by the robot.

The graphic presented on Fig. 20 illustrates the difference of the reference trajectory, showing by the slim line with square blocs (violet) and the dynamic trajectory executed by the robot, showing by the thick line (blue).
Even so the robot executes the proposed trajectory with success, we can see errors between the two trajectory, mainly on curves. It was foreseen on simulator of the cartesian trajectory kinematics and dynamics of the mobile, depicted on Fig. 15 with the trajectory error on Fig. 16. These errors are predictable for the simulator because of the intrinsic dynamic characteristic of nonholonomic mobile robotics systems with differential drive wheels, which the prototype are made, and for the PID close-loop controller of wheels velocity axles. It's demonstrated, in this example, the necessity and efficacy of the virtual system, proposed by this work.

## 8. Conclusion

The main objective of this work was to propose a generic platform for a robotic mobile system, seeking to obtain a support tool for under-graduation and graduation activities. In this way, it presents the virtual environment implementation for simulation and design conception of supervision and control systems for mobile robots, that are capable to operate and adapting in different environments and conditions. This came from encountering the growing need to propose to the research that integrates the knowledge acquired in several domains that stimulates teamwork in order to reach a result. Another objective was to to improve knowledge in the mobile robotic area, aiming at presenting practical solutions for industrial problems, such as maintenance, supervision and transport of materials. Some promising aspects of this platform and simulator system are:

- Flexibility: there are a great variety of possible configurations in the implementation of solutions for several problems associated with mobile robots.
- Great capacity of memory storage allowing implementation of sailing strategies for maps.
- The use of the rapid prototyping technique in mobile robotic systems.

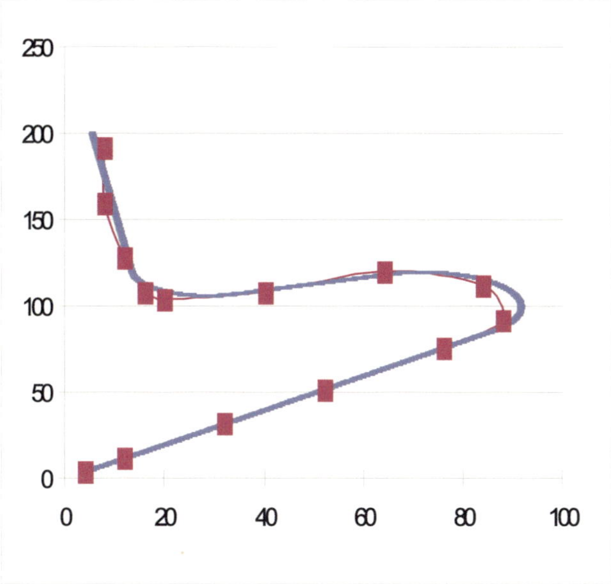


Fig. 20. Reference and dynamic trajectory executed by the robot.

- Possibility of modification of control strategies during the operation of the mobile robot in special mechatronics applications.

## 9. References


Dudek, G. & Jenkin, M. (2000). *Computational Principles of Mobile Robotics*, Cambridge University Press, UK.

Graf,B; Wandosell, J. M. H. & Schaeffer, C. (2001). Flexible Path Planning for Nonholonomic Mobile Robots. *Proceedings of Angenommen zur Eurobot'01*. pp 456–565, 2001, Stuttgart, Germany.

Melo, L. F.; Lima, C. R. E. & Rosario, J. M. (2005). A Reconfigurable Control Architecture for Mobile Robots. *Proceedings of 2nd. Internacional Symposium on Mutibody and Mechatronics - MuSMe 2005*, v. 1. pp. 1–8, september 2005.

Melo, L. F. & Rosario, J. M. (2006). A Proposal for a hybrid opened archtecture with hardware reconfigurable control applied in mobile robots. *Proceedings of IEEE International Conference on Robotic and Bionemetics - ROBIO 2006*,v. 1. pp. 1101-1106, october 2006.

Shim, H.-S.; Kim, J.-H. & Koh, K. (1995). Variable structure control of nonholonomic wheeled mobile robot. *Proceedings of IEEE International Conference on Robotics and Automation*. vol. 2, pp. 1694–1699, April 1995.

Siegwart, R.; & Nourbakhsh, I. R. (2004) *Introduction to Autonomous Mobile Robots*. The MIT Press, Cambridge, Massachusetts.

Zhao, Y. & BeMent, S. L. (1992). Kinematics, dynamics and control of wheeled mobile robots. *Proceedings of IEEE International Conference on Robotics and Automation*, vol. 1, pp. 91–96, June 1992.

# 27

# Robot Topological Mapping and Goal-Oriented Navigation Based on Rat Spatial Cognition

Alejandra Barrera
*Mexico's Autonomous Technological Institute (ITAM)*
*Mexico*

## 1. Introduction

Spatial mapping has been subject of extensive research in the robotics field. Approaches to map building have been proposed, such as topological (Franz et al., 1998), metric (Movarec & Elfes, 1985), and hybrid maps combining these two approaches (Guivant et al., 2004; Kuipers et al., 2004; Bosse et al., 2004; Zivkovic et al., 2005). Diverse issues have arisen as critical to implement practical and robust solutions to the SLAM (Simultaneous Localization And Mapping) problem in real, large and static environments, such as data association, which relates to whether or not two features observed at different points in time correspond to one and the same object or place in the physical world (Hähnel et al., 2003), and perceptual ambiguity that occurs when trying to distinguish between places in the environment that may provide equivalent visual patterns (Frese, 2006).

In general terms, relevant aspects of concern are related to the evaluation of SLAM algorithms according to their computational complexity, their approach to the data association problem, and the sort of environment representation built. However, few efforts have been documented relative to the use of spatial representations to navigate directly to designated goal locations (Stentz & Hebert, 1995; Thrun et al., 2000). Even more, we have not found reports of attempts dealing with the "unlearning" process of previously learnt target locations. Therefore, one of the purposes of our research consists on addressing the imbalance between spatial mapping and map exploitation detected in the SLAM literature, as well as the lack of target unlearning research, by taking inspiration from these abilities in animals. Indeed, research cycles involving animal studies, computational modeling, and robotic experimentation, have inspired for many years the understanding of the underlying neurophysiology and neuromechanics of biological systems while also inspiring new robotic architectures and applications. Specifically, the study of behavioral and neurophysiological mechanisms involved in rat spatial cognition provides a basis for the development of computational models of robot mapping and robot experimentation during goal-oriented navigation tasks. These models and robotic architectures offer neurobiologists and neuroethologists alternative platforms to study, analyze and predict spatial cognition based behaviors.

This chapter presents a computational system-level model of rat spatial cognition addressing aspects relative to cognitive map generation, adaptation, and exploitation during navigation. This model relates rat learning and memory processes by (i) interaction of different brain structures to demonstrate skills associated with global and relative positioning in space, (ii) integration of traveled path, (iii) use of kinesthetic and visual cues during orientation, (iv) generation of topological-metric spatial representation of the unknown environment, (v) pattern association using Hebbian learning, (vi) representation of the rat internal motivational state based on the hunger drive, (vii) management of rewards implemented by reinforcement learning using an Actor-Critic architecture allowing both learning and unlearning of goal locations, (viii) propagation of reward information through the cognitive map enabling both learning and unlearning of routes to a goal in the unknown environment, (ix) exploitation of maximum reward expectations after environment exploration to enable navigation towards the goal from any given departure location, and (x) on-line adaptation of the cognitive map to changes in the physical configuration of the environment.

The chapter starts by introducing relevant research background and related work. After providing a detailed description of the spatial cognition model, the manuscript presents results from its evaluation through mobile robot experimentation during tasks such as (i) learning and unlearning of reversal behaviors in a T-maze and in an 8-arm radial maze, (ii) place recognition and goal-oriented navigation in multiple T-mazes having external landmarks where the robot is trained to learn the correct route to a goal from a fixed location in the maze, and tested to reach the goal from different starting positions, and (iii) on-line adaptation of the cognitive map in those multiple T-mazes where landmark configuration is modified after having trained the robot to find the goal. Finally, the chapter concludes by addressing contributions of the presented work from biological and robotics perspectives, and discussing next directions to extend computational modeling of spatial cognition in rodents and corresponding experimentation in robots.

## 2. Research Background

The position in space is part of the fundamental information that animals and people learn. In addition to the memory of the position of objects or places, the memory to move through space is equally important to reach a point B from a point A at which the individual is located. We use the term spatial cognition to refer to the process of coding, storing and exploiting spatial information for successful orientation and movement toward specific goals in space (Roberts, 1998).

### 2.1 Cognitive Mapping

In understanding the brain mechanisms involved in the processing of spatial information, O'Keefe and Nadel (1978) discovered critical participation of the hippocampus in (i) the development of high-level internal representations of allocentric spatial relations, i.e. representations of the full context and not just of the current animal position, and (ii) spatial learning allowing the animal to solve navigation problems that require memory of such representations. The neural substrate of such internal spatial representations was the prior discovery made by O'Keefe and Dostrovsky (1971) from individual records of place cells found in hippocampal substructures CA3 and CA1. Each of these cells exhibits high rate activation when the animal is in a continuous and compact area defined as the place field of the cell. The activity derived from the place cell population codifies the current location of

the animal within a familiar environment, and it is stored in the internal spatial representation referred to as cognitive map by Tolman (1948).
Place cell activity appears to be dependent on the location of visual cues in the environment, since rotation of such cues causes a corresponding rotation in place fields (McNaughton et al., 1994a). However, place cells maintain their fields when some of the visual cues are removed from the environment (O'Keefe & Conway, 1978), and even continue to respond in the dark (Quirk et al., 1990). Hence, it is assumed that the response of place cells is derived from combining kinesthetic and visual cues information (Jeffery & O'Keefe, 1999).
According to Poucet (1993), the cognitive map is generated by the acquisition of topological and metrical information from space. Topological information is related to the knowledge of the spatial relationships between places or objects, whereas metrical information is related to quantitative information about specific angular directions and distances between locations or objects. Poucet suggests that initial cognitive maps are basic topological maps that are refined as the animal explores the environment with the acquisition of metrical information provided by cells in the entorhinal cortex (Moser et al., 2008).

### 2.2 Motivation and Learning

The motivated behavior is usually oriented towards a goal that in animals may be associated with a drive such as hunger, and can also be stimulated by external incentives such as food smell. The hypothalamus is regarded as the main brain area where information about the internal state of the rat is combined with incentives (Risold et al., 1997). Specifically, it is assumed that food pursuit and intake are activities controlled by the lateral hypothalamus (Kelley, 2004), which determines the primary reward these activities produce in animals.
In addition to the hypothalamus, the striatum in the basal ganglia is also involved in the extraction of information related to rewards from environmental stimuli, and in the use of such information in the generation of goal-oriented behaviors (Schultz et al., 1998).
Reward information is processed in the basal ganglia by dopaminergic neurons responding to primary and secondary rewards. These responses can reflect "errors" in the prediction of rewards, thus constituting teaching signals for reinforcement learning. On the other hand, neurons in the ventral striatum (nucleus accumbens) are activated when animals predict rewards and adapt expectation activity to new reward situations (Schultz et. al, 1998). Houk et al. (1995) suggested the striatum implements an Actor-Critic architecture (Barto, 1995), where an Adaptive Critic predicts expectations of reward values and produces an error signal to adapt reward expectations associated to different rat actions represented in this learning architecture by multiple Actor units.
In goal-oriented behaviors, it is assumed that rats are able to learn spatial tasks by associating rewards with locations in the environment (spatial learning), and rewards with procedures (procedural learning) (O'Keefe & Nadel, 1978). Neurophysiological and neuroanatomical studies have shown that procedural learning relies on the striatal system, whereas spatial learning employs the hippocampal system (Hartley & Burgess, 2005).

### 2.3 Experimental Basis for Spatial Cognition Processes in Rats

Interested in demonstrating the use of both learning strategies, spatial and procedural, in normal rats and rats with hippocampal lesions, O'Keefe (1983) implemented two discrimination tasks.

The first task took place in an 8-arm radial maze without external signals (see Fig. 1(a)), where five arms were removed to form a T-shaped maze (see Fig. 1(b)) and a reward (food) was placed at the end of the left arm of the T. Every five trials during training, the maze was modified so that other three arms constitute the T, preventing the animal to base its choice of the correct arm on signals outside the maze, but on the spatial or procedural strategies. As a result of training, rats with injured hippocampus learned the task faster than normal rats.
To determine the type of learning strategy used by animals, O'Keefe continued the experiment after reaching the learning criterion now alternating between testing trials in the radial maze and in the T maze. During tests, all rats chose arms in the radial maze located on the same side of the reward arm in the T, thus showing the use of a procedural strategy since rats learned to rotate left.

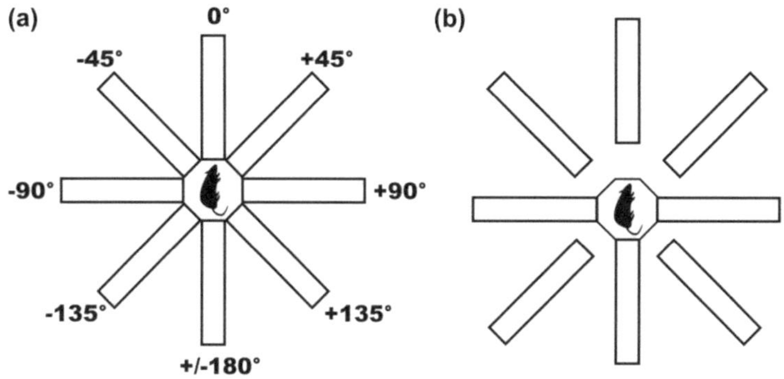


Fig. 1. Diagrams of mazes employed by O'Keefe during discrimination tasks. (a) The 8-arm radial maze showing the orientation of arms according to an egocentric (local) reference frame, which specifies the rat behavior in terms of rotations relative to its body midline. (b) A T-maze resultant from separating five arms of the original 8-arm radial maze.

The second task consisted on a reversal discrimination training all animals to turn left in the T-maze, and testing their orientation decisions when changing the reward location at the end of the right arm of the T. During tests, O'Keefe alternated between two trials in the T and one trial in the radial maze. Results obtained from injured animals showed an abrupt change in the rotation performed at the T-junction from the left to the right arm, but in the radial maze, the rat orientation change was continuous and incremental from the training quadrant (-90° -45°), through straight ahead (0°) until the testing quadrant (+45°, +90°) (see Fig. 1(a)). On the other hand, the performance of healthy rats in the T-maze proceeded in the same manner as that of injured animals, but their orientation in the radial maze did not change on a continuous and incremental manner, but randomly.
The outcome of the reversal task allowed O'Keefe to ratify the following:

- Injured rats used a procedural learning strategy during training, associating the reward with the response of turning left. During tests, extinction of the reward on the left T-maze arm gradually modified the animals' orientation until turning right.
- Healthy rats employed a spatial learning strategy during training before reaching the learning criterion. However, after this event, they based their learning on a procedural strategy that abruptly ended with the reward extinction due to the intervention of the hippocampus to shoot again the environment exploration and the corresponding use of the spatial learning strategy.

According to O'Keefe and Nadel (1978), the cognitive map is responsible for estimating the rat's position within the environment and its relative direction to the goal location, enabling the animal to reach this location from any other one. One classical spatial task showing this goal-directed navigation behavior is the water maze introduced by Morris (1981).
Under the original Morris experiment (Morris, 1981), the maze consisted of a circular tank filled with an opaque mixture of milk and water including a platform in a fixed location. In each training trial, normal rats and rats with hippocampal lesions were independently placed at one of various departure locations near the tank's circumference, and required to swim until they located the platform, upon which they could stand and escape from the cold water. After training, rats were tested in two situations: (i) with the platform visible and (ii) with the platform submerged inside the tank and visual cues placed around the arena. In the first case, all rats were able to find the platform immediately, whereas in the second, only normal rats found it from any known or novel starting location at the periphery of the tank.
The experiment also tested situations where the platform was removed from the tank or placed in a random location from trial to trial. In the first case, healthy rats looked for the platform persistently in its previous location, whereas in the second one, healthy rats could not learn to navigate directly to the platform.
Successful navigation in the water maze task by hippocampus healthy animals involved the use of a spatial learning strategy that allowed them to relate the goal location with external landmarks, store this relation in a cognitive map, and use the map to find a correct route.

### 2.4 Related Work

Taking inspiration from spatial cognition in rats, several computer-simulated or robot-tested navigation models have been proposed. From this point, we use the term "animat" to refer the modeled rat.
The models by Burgess et al. (1994) and Brown and Sharp (1995) include location and head direction signals, as well as reward-related changes in the efficacy of synapses onto motor cells that control the animat behavior. In these studies, different from our model, hippocampal place cells provide a code for space that is not stored explicitly in a cognitive map. Burgess et al. (1994) use metric information, such as distances to identified visual cues, as the exclusive and direct input to their system. In contrast, place units in our model codify the integration of visual and kinesthetic information, and we interpret metric properties of visual cues by means of neurons sensitive to specific landmarks information patterns. Our model builds and maintains a topological map storing place cell population activity and enabling navigation between locations in the environment. Brown and Sharp (1995) associate place cells with motor responses, and perform a trace-like learning rule where a given animat behavior immediately followed by reward has a higher probability of occurrence the next time the animat experiments the same situation. In contrast, our model implements reinforcement learning through an Actor-critic architecture enabling the animat to learn as well as unlearn reward locations.
Actually, prior use of this reinforcement learning method to solve goal-search tasks has been documented (Guazzelli et al., 1998; Foster et al., 2000). The work by Foster et al. (2000), for instance, implements an Actor-Critic architecture to enable a simulated rat to solve the reference memory task in a Morris water maze providing the animat with a reward at any given time of the experiment. If the animat moves towards the hidden platform (goal) in the water maze, the reward is set to 1; otherwise, it is set to 0. The animat is able to learn correct

actions precisely because such a reward signal is given as input at each time step. On the contrary, the animat in our model does not receive any information related to its physical proximity to the hidden goal during the exploration of the environment, but it is the animat who traces the correctness of its followed route once it finds the goal location. This route learning is achieved by propagating reward information through nodes in the topological map after every trial in the given experiment.

Our work is partially inspired on the computer-simulated model by Guazelli et al. (1998). One of our main extensions to this system includes a map exploitation process to enable goal-directed navigation in a mobile robot. Their original model endowed the simulated rat with the ability to learn goal locations from a fixed departure position within mazes that included just one decision point from where the goal was visible, and the animat was unable to find the target in more complex mazes having two or more decision points, and also to reach it from arbitrary starting positions. Specifically, in the model by Guazzelli et al., the action selection process considers the addition of reward expectations derived from affordances with reward expectations from their world graph model. As reward expectations from affordances were computed over the assumption of having different rotation possibilities at every decision point in the environment, the simulated rat fails to find the goal when executing the same reinforced rotation at two or more decision points offering the same rotation possibilities. We also extend the original model by providing a map adaptation process that permits on-line representations of changes in the physical configuration of the environment perceived by the robot (Barrera & Weitzenfeld, 2007b).

Our proposal differs from the model of Gaussier et al. (2002) in that they employ only visual information as input, hippocampal cells do not encode places but transitions between states, and the place recognition process is carried out by the entorhinal cortex rather than by the hippocampus. As in our model, they build a topological space representation; however, nodes in this map do not correspond to places, but to transitions between states. They implement a sort of map exploitation to determine sequences of transitions between states that lead to a goal location. Nonetheless, they do not model the animal's motivation and the prediction of reward expectations.

The main components of the neural architecture proposed by Arleo et al. (2004) are similar to those found in our model: integration of allothetic (visual) information and idiothetic (kinesthetic) signals at the level of hippocampal representation, use of Hebbian learning to correlate these inputs, mapping of the place cell population activity into spatial locations, and application of reinforcement learning to support goal-oriented navigation. We add to this model the use of affordances information instead of population vector coding to map the ensemble dynamics of place cells into spatial locations, an explicit construction of a topological map of places and their metric relations, and the implementation of an Actor-Critic reinforcement architecture that predicts, adapts and memorizes reward expectations during exploration to be exploited during goal-oriented navigation.

The focus of our approach differs from the one followed by Milford et al. (2007). Whereas they are concerned with the effectiveness of the hippocampus models in mobile robot applications exploring large environments with natural cues, our interest consists on endowing mobile robots with spatial cognition abilities similar to those found in rodents in order to produce comparable behavioral results and eventually provide experimental neuroscience with valuable feedback. Nevertheless, our model coincides with Milford et al.'s in some aspects related to mapping, map exploration and map adaptation, and contrasts with it in map exploitation. Specifically, in the model by Milford et al., a topological map of experiences is built with each experience representing at a given time a snapshot of the activity within pose cells, which codify physical localization and orientation, and local view cells that encodes for visual information. In this map, transitions between experiences are associated with locomotion information. Nodes in our topological map also represent associations between visual-kinesthetic information patterns and the place cell population activity. Transitions between nodes are associated with metric information derived from the animat locomotion. During the exploration of the environment, previously unused movement behaviors are implemented by the animat in trying to experience new routes. Similarly, motion in our model is partially influenced by the curiosity to execute rotations not yet explored by the animat. Additionally, the map of experiences can be adapted to physical changes in the environment, which involves the elimination/creation of experiences and the update of transitions between experiences. We have demonstrated that the map built by our system is adapted on-line to represent changes in the physical configuration of landmarks (Barrera & Weitzenfeld, 2007b). The map exploitation process involves the storage of temporal information in the experiences map to find the fastest route to the goal, and the storage of behavioral information in transitions between experiences to navigate to the goal. In contrast, our model provides map exploitation based on reward expectations of locations predicted and adapted during environment exploration and used to guide the animat towards the goal from any given departure point.

## 3. A Bio-inspired Computational Model of Spatial Cognition

The spatial cognition model comprises distinct functional modules, shown in Fig. 2 and described below, that capture some properties of rat brain structures involved in learning and memory. A detailed mathematical depiction of each module is presented in (Barrera & Weitzenfeld, 2008).

### 3.1 Motivation

Motivation module relates to the rat lateral hypothalamus (LH) computing the value of hunger drive and producing the primary reward (*r*) the animat gets by the presence of food (goal). The reward *r* depends on the animat internal drive experimented at any given time.

### 3.2 Kinesthetic Processing

Rats carry out path integration processes using kinesthetic information derived from two systems: (i) vestibular organs in the semicircular canals of the inner ear, and (ii) feedback from muscles controlling movement. During path integration, rats update the position of their point of departure each time they move in relation to their current position (Mittelstaedt & Mittelstaedt, 1982; Etienne, 2004). The posterior parietal cortex (PPC), a sensory structure receiving multimodal information such as kinesthetic, visual, and relative to affordances, has been suggested to mediate path integration (Parron & Save, 2004) involving also the retrosplenial cortex (RC) (Cho & Sharp, 2001).

In our model, we attribute to PPC the representation of the updated animat position in relation to its point of departure by integrating past rotations and translations through a dynamic remapping perceptual schema (*DR*), and to RC, the generation of groups of neurons in the path integration feature detector layer (PIFDL) of the model that respond to specific kinesthetic information patterns (*PI*) due to the use of Hebbian learning (Hebb, 1949). *PI* is defined as a matrix of activation values registered by all neurons in PIFDL at any given time.

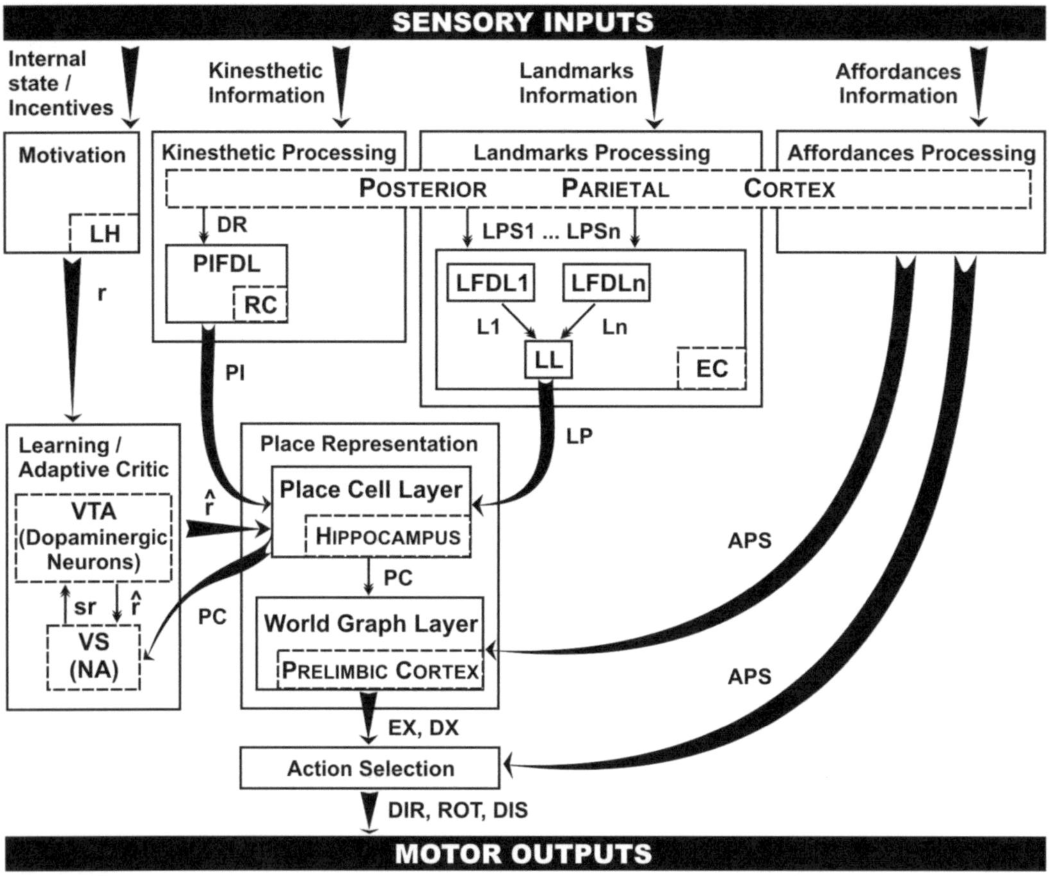


Fig. 2. The modules of the spatial cognition model and their interaction. Glossary: LH - Lateral Hypothalamus; RC - Retrosplenial Cortex; EC - Entorhinal Cortex; VTA - Ventral Tegmental Area; VS - Ventral Striatum; NA - Nucleus Accumbens; PIFDL - Path Integration Feature Detector Layer; LFDL - Landmark Feature Detector Layer; LL - Landmarks Layer. Inputs/Outputs: r= primary reinforcement; sr= secondary reinforcement; $\hat{r}$ = effective reinforcement; DR= dynamic remapping perceptual schema; LPS= landmark perceptual schema; APS= affordances perceptual schema; PI= kinesthetic information pattern; L1, Ln= information pattern for landmark 1, landmark n; LP= landmarks information pattern; PC= place information pattern; EX= maximum reward expectations and their corresponding directions (DX); DIR= next animat direction; ROT= animat rotation; DIS= next animat moving displacement.

### 3.3 Landmarks Processing

Landmark processing module computes landmark-related spatial positioning, i.e. distance and relative orientation of each landmark to the animat. According to Redish and Touretzky (1997), the entorhinal cortex (EC) is involved in landmarks processing by receiving spatial information about landmarks from PPC. In our model, spatial information about each landmark is encoded in a landmark perceptual schema (*LPS1, …, LPSn*) serving as input to a specific landmark feature detector layer (LFDL1, …, LFDLn) that produces a landmark information pattern (*L1, …, Ln*). Then, different LFDLs are combined into a single landmarks layer (LL). Hebbian learning updates connection weights between layers LFDLs and LL producing groups of neurons in LL that respond to specific landmark information patterns (*LP*) derived from the integration of all landmarks presented in the environment. In this way, *LP* is defined as a matrix of activation values of all neurons in LL representing the egocentric view from the animat at any given time.

### 3.4 Affordances Processing

McNaughton et al. (1994b) suggested that preceding the rat motion, nearly half of the cells in PPC exhibit movement-related activity discriminating among basic modes of locomotion: left turns, right turns, and forward motion. The affordances processing module represents PPC cell discrimination among different possible orientations for navigation by use of an affordances perceptual schema (*APS*) encoding possible egocentric (local) turns from -180° to +180° in 45° intervals at any given time from any given animat location (see Fig. 1(a)).

### 3.5 Place Representation

As shown in Fig. 3(a), place representation module receives input from kinesthetic (*PI*), landmark (*LP*) and affordances (*APS*) modules in addition to its interaction with the learning module ($\hat{r}$, later defined in Section 3.6). Place representation module comprises a place cell layer (PCL) and a world graph layer (WGL).

Place cell layer (PCL) corresponds to the rat hippocampus encompassing regions CA3, CA1 and dentate gyrus (DG). Overlapping place fields in the collection of neurons in PCL are associated with a physical area in the environment that is identified directionally by the ensemble place cell activity pattern (*PC*), and whose extension is determined by affordances changes sensed by the animat during exploration. Specifically, neurons in the path integration feature detector layer (PIFDL) and in the landmarks layer (LL) of the model are connected to neurons in PCL. Synaptic efficacy between layers is maintained by Hebbian learning producing groups of neurons in PCL that respond to specific place cell information patterns (*PC*) derived from kinesthetic and egocentric visual information sensed by the animat while being at certain location and orientation. In this way, *PC* is defined as a matrix of $1 \times n$ activation values registered by the collection of $n$ neurons in PCL at any given time $t$:

$$PC(t) = PI(t)\ W_{PI}(t) + LP(t)\ W_{LP}(t), \tag{1}$$

where *PI* and *LP* are matrices of $1 \times n$ input signals from layers PIFDL and LL to PCL, $W_{PI}$ and $W_{LP}$ are matrices of $n \times n$ connection weights between neurons in PIFDL, LL and PCL.

Associations between overlapping place fields and physical areas are represented by world graph layer (WGL) through a topological map enabling navigation between locations in the

environment. Specifically, nodes in this map represent associations between kinesthetic and visual information patterns and the place cell population activity, whereas transitions between nodes are associated with metric information derived from animat locomotion such as its moving direction and displacement. Besides this mapping process, WGL also performs place recognition, and we assume that its functionality could be corresponded to the prelimbic cortex (Granon & Poucet, 2000).
Actor units in WGL store place cell activation patterns (*PC*) generated by PCL when the animat is oriented to diverse directions. These directions vary from 0° to 315° in 45° intervals, according to an allocentric (global) reference frame that is relative to the animat departure location in the exploration process as illustrated by Fig. 3(b). Hence, every node in the map (a place) connects to a maximum of eight Actor units (eight possible orientations at each place). Every Actor connection is associated with a weight (representing the expectation of reward when orienting to a particular direction from the current location), and an eligibility trace (marking the connection eligible to be reinforced later in time). In this way, Actor units compete to select the next moving direction from the current location or node that allows the animat to get the greatest reward, thus WGL analyzes Actor weights to obtain the biggest ones (*EX*) and their corresponding directions (*DX*).

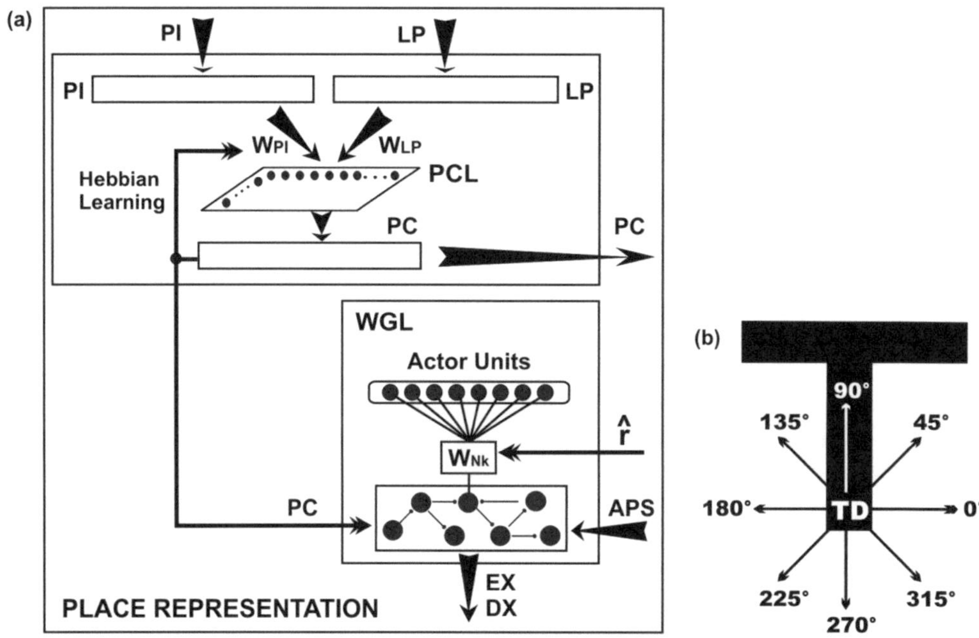


Fig. 3. (a) The place representation module of the spatial cognition model. Glossary: PCL= place cell layer; WGL= world graph layer; PI= kinesthetic information pattern; LP= landmarks information pattern; $W_{PI}$= connection weights between PIFDL and PCL; $W_{LP}$= connection weights between LL and PCL; $W_{Nk}$= connection weights between any given map node *k* and its corresponding Actor units; PC= place information pattern; APS= affordances perceptual schema; $\hat{r}$ = effective reinforcement signal; EX maximum reward expectations and their directions (DX). (b) Allocentric (global) reference frame representing possible directions from 0° to 315° in 45° intervals to be adopted by the animat. This reference frame is relative to the animat departure location (TD) in the exploration process of a given maze.

**Map Creation**

To determine whether or not the animat recognizes a place, WGL searches the current activity pattern *PC* produced by PCL within all Actor units in the map. This search involves the computation of the similarity degree *SD* between *PC* and every stored place cell activity pattern *pat* as follows:

$$SD = \sum_{i=1}^{n} \min(pat_{1,i}, PC_{1,i}) \Big/ \sum_{i=1}^{n} PC_{1,i} \quad , \tag{2}$$

where *pat* and *PC* are matrices of $1 \times n$ activation values registered by the collection of *n* neurons in PCL, *i* is the matrix column index, and *min* is a function that computes the minimum value between its two arguments.

The model distinguishes among two cases:

(i) If at least one *SD* exceeds certain threshold close to 1, the Actor unit storing the activation pattern with the biggest *SD* is considered the winner.

(ii) If there is no winner, WGL creates a new Actor unit storing pattern *PC* associated to the current animat orientation.

Then, WGL activates or creates a node in the map depending on the following considerations:

(i) If affordances encoded by *APS* at time *t* are different from those at time *t-1* and a new Actor unit was created, then a new node is created in WGL, connected with that Actor unit, and set as the new active node in the map (see Fig. 4(a)).

(ii) If affordances did not change and a new Actor unit was created, then WGL averages the activation pattern stored in the new Actor unit and the pattern stored in the Actor unit of the active map node that is associated to the current animat orientation (see Fig. 4(b)).

(iii) If there was an Actor unit winner, an arc from the active node to the node connected to that Actor unit is created if necessary, and this node becomes the new active one (see Fig. 4(c-d)).

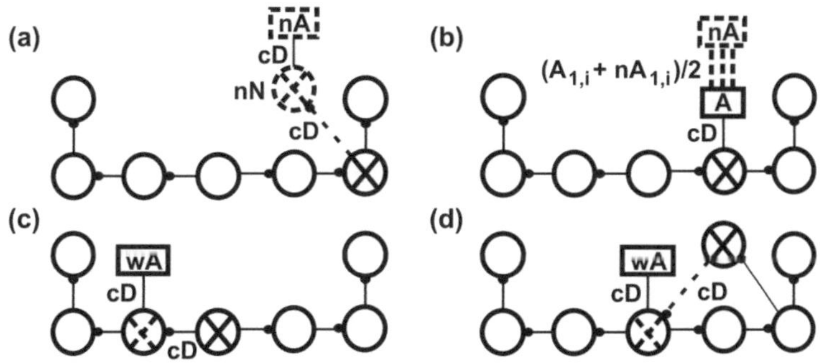


Fig. 4. Creation/activation of nodes in the map. Dotted lines illustrate new components, and the crossed node is the active one. Glossary: nA= new Actor unit; nN= new node; cD= rat's current direction; A= existing Actor unit; wA= winner Actor unit. (a) The creation of a new node. (b) The average between the activation patterns of two Actor units. (c) The activation of an existing node. (d) The connection between two existing nodes.

**Map Adaptation**

During exploration, an existing topological map may be adapted according to kinesthetic and visual information perceived by the animat in its current orientation due to two reasons:

(i) Late recognition of already visited locations. When the animat is exploring a given maze location represented by a certain map node *nX* in a particular direction *d1*, then the animat executes a rotation orienting itself to direction *d2* and an existing node *nY* is activated at that time, an integration process occurs between both active nodes *nX* and *nY* since they are representing the same maze location in two different directions. This integration process may or may not involve the creation of a new node in the map:

  a) If node *nX* was created at time *t-1* and the activation of node *nY* occurs at time *t*, the arc pointing to node *nX* in direction *d1* is now updated to point to node *nY*, Actor unit associated to direction *d1* of node *nX* is assigned to node *nY*, node *nX* is eliminated, and node *nY* prevails as the only one active (see Fig. 5(a)).

  b) If node *nX* was not created but just activated in direction *d1* at time *t-1* and the activation of node *nY* occurs at time *t* in direction *d2*, a new node *nZ* is created at this time integrating all input/output arcs to/from nodes *nX* and *nY*, as well as all Actor units of both original nodes. Merged nodes *nX* and *nY* are removed from the map, and node *nZ* becomes the active one (see Fig. 5(b)).

(ii) Dynamic changes in the physical configuration of the environment. When the environment is modified during the exploration process, the animat stops recognizing locations from where a different view is now obtained. In this case, the animat adds new map nodes establishing appropriate connections, and removes all those nodes that are no longer linked to the map (see Section 4.3 for detailed examples).

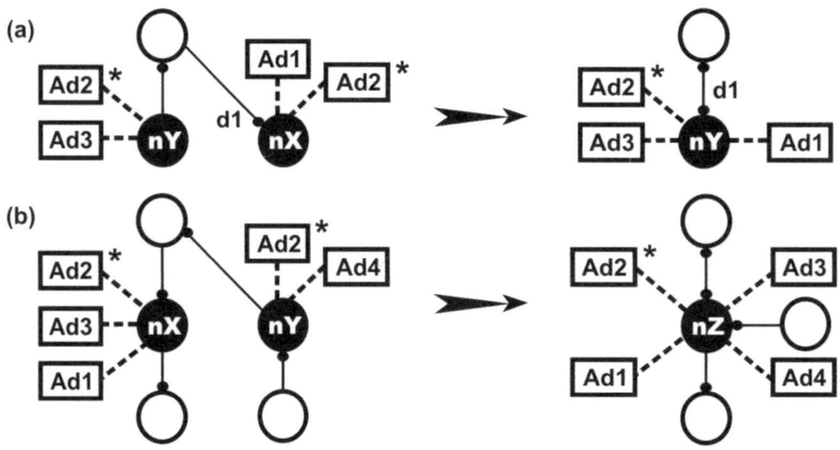


Fig. 5. Illustration of map adaptation resultant from late recognition of already visited locations. Glossary: Ad1 – Ad4= existing Actor units of different nodes; *= active node in direction *d2*. (a) The incorporation of node *nX* created at time *t-1* into the existing node *nY* activated at time *t* in animat direction *d2*. (b) The integration of existing nodes *nX* and *nY* into the new node *nZ* with both, *nX* and *nY*, active at time *t* in animat direction *d2*.

### 3.6 Learning

Learning module is related to dopaminergic neurons in the ventral tegmental area and to ventral striatum, processing reward information by using an Actor-Critic architecture. As illustrated in Fig 6, the Adaptive Critic (AC) includes a Prediction Unit (PU) that estimates the future reward value of any particular location at a given time. To do this, every neuron in PCL is connected to PU, and these connections are associated with weights $W$ and eligibility traces $E$. At each time step $t$ in a trial of an experiment, PU computes the future value $P$ of the activity pattern $PC$ generated by PCL according to (3):

$$P(t) = PC(t) \cdot W_{PCL}(t), \tag{3}$$

where $PC$ is a matrix of $1 \times n$ input signals from PCL to PU, and $W_{PCL}$ is a matrix of $1 \times n$ connection weights between neurons in PCL and PU.

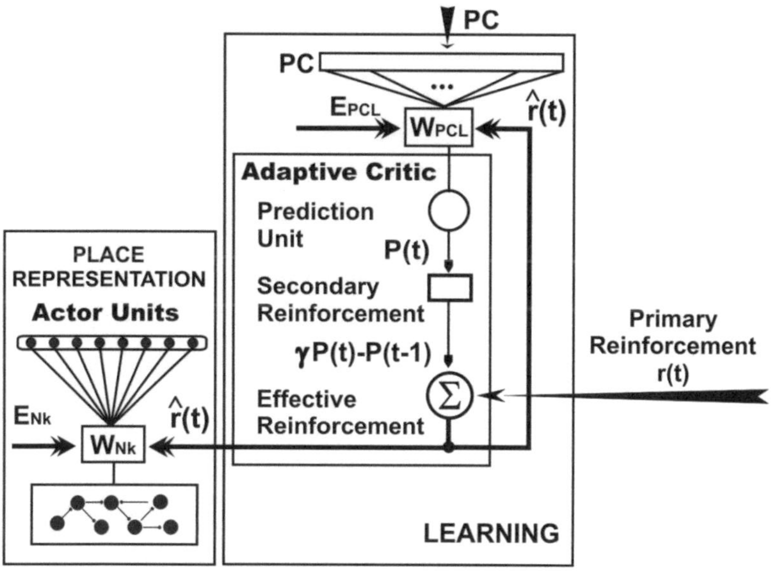


Fig. 6. The learning module of the spatial cognition model. Glossary: PC= current activation pattern in the PCL layer; $W_{PCL}$= connection weights between PCL and Prediction Unit; $E_{PCL}$= connection eligibility traces between PCL and Prediction Unit; $P(t)$ and $P(t-1)$ correspond to predictions of the future value of $PC$ at time $t$ and $t-1$ respectively; $W_{Nk}$= connection weights between any given map node $k$ and its corresponding Actor units; $E_{Nk}$= connection eligibility traces between any given map node $k$ and its corresponding Actor units.

AC uses predictions computed at times $t$ and $t-1$ to determine the secondary reinforcement, discounting the current prediction at a rate $\gamma$ to get its present value. The addition of the secondary reinforcement with the primary reinforcement $r$ computed by the motivation module of the model constitutes the effective reinforcement $\hat{r}$ as described by (4):

$$\hat{r}(t) = r(t) + \gamma P(t) - P(t-1). \tag{4}$$

The effective reinforcement is used to update the connection weights between PCL and PU in AC (i.e., reward expectation associated to a place), and also between Actor units and map nodes (i.e., reward expectations associated to different orientations). In the first case we use

$$W_{PCL}(t+1) = W_{PCL}(t) + \beta \hat{r}(t) E_{PCL}(t), \tag{5}$$

where $\beta$ is the learning rate, and $E_{PCL}$ is the matrix of $1 \times n$ eligibility traces corresponding to connections between PCL and PU in AC. In the second case we use

$$W_{Nk}(t+1) = W_{Nk}(t) + \beta \hat{r}(t) E_{Nk}(t) \quad \forall\, map\ node\ k, \tag{6}$$

where $W_{Nk}$ is the vector of connection weights between map node $k$ and a maximum of eight Actor units, and $E_{Nk}$ is the vector of eligibility traces corresponding to a maximum of eight Actor units. As shown in (5) and (6), both learning rules depend on the eligibility of the connections. At the beginning of every trial in a given experiment, eligibility traces in AC and in Actor units are initialized to 0. At each time step $t$ in a trial, eligibility traces in AC are increased in the connections between PU and the most active neurons within PCL only when the action executed by the animat at time $t-1$ allowed it to perceive the goal:

$$E_{PCL}(t) = E_{PCL}(t-1) + \chi PC(t), \tag{7}$$

where $\chi$ is the increment parameter, and $PC$ stores the activity pattern registered by the collection of neurons in PCL. Also at time step $t$, the eligibility trace $e$ of the connection between the active map node $na$ and the Actor unit corresponding to the current animat orientation $dir$ is increased by $\tau$ as described by (8):

$$e_{na}^{dir}(t) = e_{na}^{dir}(t-1) + \tau. \tag{8}$$

Finally, after updating connection weights between PCL and AC, and between Actor units and map nodes at any time step $t$ in the trial, all eligibilities decay at certain rates $\lambda$ and $\sigma$ respectively, as shown in (9):

$$\begin{aligned} E_{PCL}(t) &= \lambda E_{PCL}(t-1) \\ E_{Nk}(t) &= \sigma E_{Nk}(t-1) \quad \forall\, map\ node\ k. \end{aligned} \tag{9}$$

The use of the Actor-Critic architecture enables the estimation of reward expectation values of different locations in the environment, where maximum expectations correspond to locations from where the goal is perceptible and to orientations needed to be performed at those locations to reach the goal. Since the model allows the animat to recognize the goal just one step away from it in order to prevent a taxon or guidance navigation strategy, and the animat does not receive any information related to its physical proximity to the hidden goal during the exploration of the environment, it is the animat who traces the correctness of its followed route once it finds the goal location. This route learning is achieved by implementing backward reward propagation through nodes in the topological map after every trial in the given experiment.

The backward reward propagation involves updating eligibility traces of Actor units in the direction of the arcs connecting the map nodes that represent the complete route followed by the animat. The strategy involved in this process is based on the existence of a factor referred to as goal gradient by Hull (1932), according to which the reinforcement effect is the most at the goal location and diminishes progressively as the animal moves backward through the environment. Specifically, in case the animat finds the goal at the end of the path, each eligibility trace is updated in a given positive amount of reinforcement divided by the amount of steps the animat performed to move from one node to the next one. Likewise, in case the animat does not find the goal at the end of the path, the amount of reinforcement used is negative. The reinforcement is initialized to a certain amount at the beginning of every training trial in the experiment, and this amount decreases as the distance from a node to the goal increases.

The update on eligibility traces that occurs when the given trial is concluded at time $t$ renders a corresponding update on the connection weights between map nodes (representing the complete route) and Actor units (associated to animat orientations) at time $t+1$ according to (6), enabling the animat to learn as well as unlearn performed routes.

The positive update registered when the target is found resembles the activation pattern of rat striatal cells. Mulder et al. (2004) showed that after having learnt to locate the reward in a maze, striatal cells of the rat respond continuously for the complete path followed by the animal, implying that striatum cells are able to modify their synaptic weights to the inputs received from the place cells responding to locations in the traveled route.

### 3.7 Action Selection

Action selection module computes the motor outputs of the model consisting on the next animat direction (*DIR*), the required rotation to point to that direction (*ROT*), and the moving displacement (*DIS*).

Motion is determined by considering:

(i) all possible affordances to execute from current location and orientation (*APS*),
(ii) one random rotation between possible affordances (*RPS,* internally computed),
(iii) curiosity to execute rotations not yet explored (*CPS,* internally computed), and
(iv) maximum reward expectation (*EMR,* internally computed by using *EX* and *DX*).

These four "signals" are computed by using one or more Gaussian functions whose values are stored in vectors. Specific positions distributed throughout a vector correspond to particular relative rotations between -180° and +180° in 45° intervals, and the biggest value of a Gaussian function (the height of the curve's peak) is stored in one of these positions.

The influence of each signal in the final action selection depends on the biggest possible value of the Gaussian functions representing it. Varying the parameter that regulates the height of the curve's peak, the model assigns the following priority to signals: (i) *EMR,* (ii) *APS,* (iii) *CPS,* and (iv) *RPS.*

Vectors derived from the representation of those signals are added, and the specific relative rotation associated with the position of the resultant vector storing the biggest value is used to determine the next animat direction from 0° to 315° in 45° intervals.

In the course of an experiment, while the *EMR* signal is weak, the animat executes a rotation not yet experimented at its current location or a random rotation in case it had tried all possible rotations earlier. Regarding this last situation, it is feasible for the animat to show a "hesitation" behavior, in the sense of performing two or more body rotations at its current location before restarting navigation.

## 4. Robot Experimentation

The rat cognitive model was designed and implemented using the NSL system (Weitzenfeld et al., 2002). The computational model interacts with a real robotic environment through an external visual processing module getting as input three non-overlapping snapshots (0°, +90°, -90) taken by the robot at each step using its local camera, and a motor control module that executes rotations and translations on the robot. Refer to (Barrera & Weitzenfeld, 2008) for further detail on the robotic implementation of the model.

This section presents results from the evaluation of the model through robot experimentation in a single T-maze, an 8-arm radial maze, and multiple T-mazes, performing spatial reversal behaviors, place recognition and goal-oriented navigation, and map creation and adaptation.

### 4.1 Reversal Behavior in a T-maze and in an 8-arm Radial Maze

The experiment carried out with the robot in the T-maze and in the 8-arm radial maze is inspired on the reversal task implemented by O'Keefe (1983) and described in Section 2.3, and was performed in both mazes separately.

In the T-maze and in the 8-arm radial maze shown in Fig. 7, the robot departs from location TD and navigates to any arm extreme. This process is repeated in every experiment's trial, at the end of which the robot is manually placed again at the departure location TD.

During the training phase of the experiment, the goal is placed at the end of the arm oriented to 180° (see Fig. 7). As abovementioned in Section 3.6, the system allows the robot to recognize the goal just one step away from it in order to prevent a taxon or guidance navigation strategy. At the beginning, the decisions of the robot at the choice location are determined by the curiosity for unexecuted rotations. After having visited each arm once, the curiosity level decreases notably, thus the robot executes random rotations visiting different maze arms. Eventually, the robot meets the learning criterion once the expectation of finding reward in orienting to 180° at the choice point becomes bigger than the random factor. When this event occurs, the training phase ends.

The average duration of the training phase obtained from experimenting separately with six robots in both mazes was 12 trials performed in less than 20 minutes.

When the testing phase begins, the goal is moved to the arm oriented to 0° (see Fig. 7). This situation constitutes, in the words by O'Keefe, "a reversal discrimination problem that involves an unlearning process giving up a previously correct hypothesis and switching to a new one" (O'Keefe and Nadel, 1978).

During reversal, the expectation of future reward for the arm oriented to 180° decreases continuously each time the robot reaches the end of that arm not finding reward, since this event is coded as frustrating by the learning module of the model. When the expectation of reward becomes smaller than the random factor, the robot starts visiting other arms. Each time it visits the 0° arm that provides reward, the expectation for this arm increases. Eventually, the robot meets the learning criterion when the expectation of reward for turning right at the choice point is large enough to avoid exploring any other maze arm.

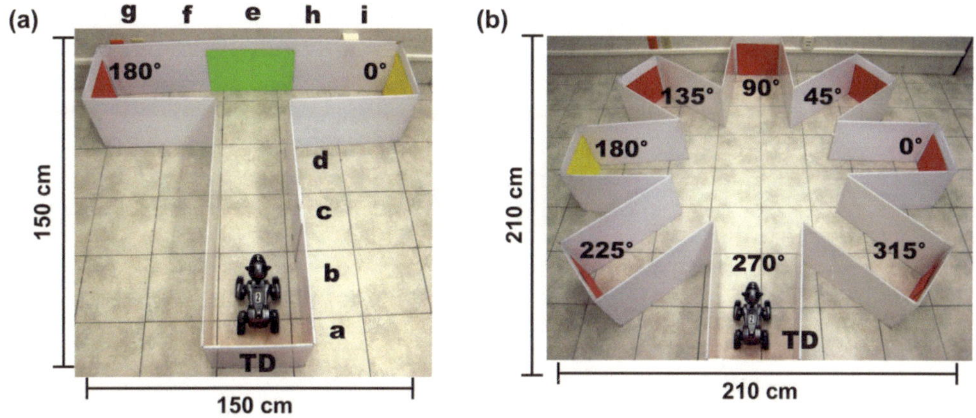


Fig. 7. (a) The physical T-maze used in the reversal task. Different locations are labeled with letters. (b) The physical 8-arm radial maze used in the reversal task. The picture shows the allocentric (global) directions of the arms in the maze. In (a) and (b), the AIBO robot is located at the starting position TD.

The average performance of six robots during reversal in the T-maze is shown in Fig. 8(a), where the robot's reversal behavior is expressed in terms of percentage of correct choices at the T-junction. Likewise, the average performance of those six robots during reversal in the 8-arm maze is shown in Fig. 8(b), which presents the robot's orientation decisions at the choice location of the maze in egocentric (local) terms (see Fig. 1(a)).

As a result from training in the T-maze, 100% of the robot's choices were correct (tag "control" in Fig. 8(a)). The graph in Fig. 8(a) shows 32 testing trials. As can be seen, the robot takes 12 trials to unlearn (i.e., decrease reward expectations) the previously correct hypothesis (i.e., arm oriented to 180°) (tag "criterion" in Fig. 8(a)). From trial 13, the robot has learnt (i.e., has increased reward expectations) the new one (i.e., arm oriented to 0°). In this way, the percentage of correct choices shifts from 36% in trial 12 to 95% in trial 16 and 100% from there. On the other hand, as a result from training in the radial maze, the robot chose consistently to turn left at the center location of the maze (tag "control" in Fig. 8(b)). During reversal, the robot's orientation did not reveal any systematic shift. As in the T-maze, the criterion occurred around trial 12, when the average orientation crosses the midline of the graph in Fig. 8(b).

Comparing our results with those reported by O'Keefe with normal rats in (O'Keefe, 1983), we can appreciate a behavioral similitude with robots in the T-maze. O'Keefe presented the average results obtained from four rats. His graph also shows 32 testing trials and a control measure of 100% of correct choices at the T-junction after training. In this case, rats reached the criterion in trial 20, where the percentage of correct choices was between 20% and 40%. By trial 24, rats chose the new reward arm in more than 90% of the times until performing 100% of correct decisions. As described by O'Keefe (1983), we could confirm that there was an abrupt change from the incorrect to the correct arm.

We also appreciate in our results a similitude with those reported by O'Keefe with four normal rats in the radial maze (O'Keefe, 1983). He also presented a graph showing 32 testing trials, where rats reached the criterion around trial 20. He explained that the abrupt shift from turning left to turning right in the T-maze was revealing the moment when the average orientation crosses the midline in the radial maze. According to our tests, the same fact applies to the robot's behavior.

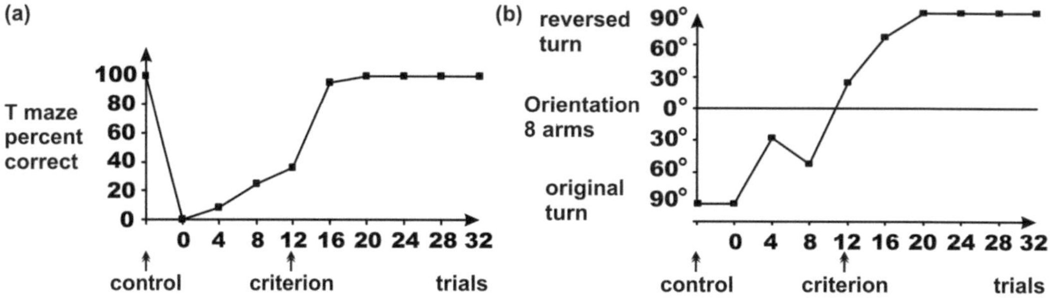


Fig. 8. The performance of six robots during the reversal task. Each graph was obtained by averaging the graphs of the individual robots. The graph in (a) shows the percentage of correct choices in the T-maze averaged over periods of four trials. The graph in (b) presents robot decisions in the radial maze also averaged over periods of four trials. To compare these results with those obtained by O'Keefe with four normal rats, refer to (O'Keefe, 1983). In our graphs, as in O'Keefe's, the abrupt shift from turning left to turning right in the T-maze reveals the moment (criterion) when the average orientation crosses the midline in the radial maze.

During the experiment, the robot builds and maintains the topological spatial representations shown in Fig. 9 for the T-maze and the 8-arm radial maze. The ensemble activity of the neurons found in the place cell layer of the model was determined only by the use of kinesthetic information, since no landmarks were available. The activity pattern of the collection of place cells registered when the robot is at any given maze location is stored in a particular Actor unit associated with the current robot's orientation and linked to the active map node.

As described in Section 3.5., the relevance of locations in a maze relies on the presence of rewards or on affordances changes sensed by the robot during exploration. In Fig. 7(a), for instance, the robot considers locations "a", "b", "e", "f", "g", "h" and "i" as relevant, and that is why the map in Fig. 9(a) includes seven nodes. When the robot reaches location "b" in direction 90°, the current ensemble activity of the place cell layer is stored in node 2, and although the activity pattern could slightly vary at locations "c" and "d", the affordances sensed by the robot did not change from "b" to "c" or "d", thus activity patterns registered in these locations are averaged and stored in the same node 2, defining in this way its physical extension. It should be pointed out that, different from rats, the robot was programmed to avoid 180° rotations when there exist other possible rotations (0°, +90°, -90°), thus optimizing the exploration process to find the goal. For this reason, arc directions between nodes in the map are one way.

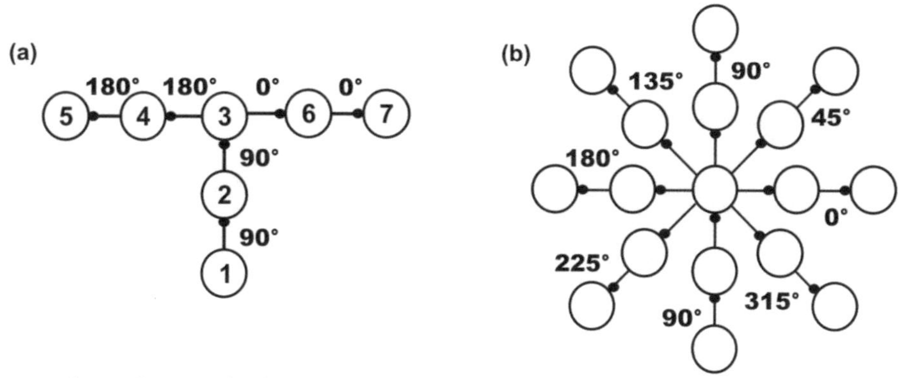


Fig. 9. Topological maps built and maintained during the reversal task in (a) the T-maze, and (b) the 8-arm radial maze. In (a) nodes are numbered in order of creation. In (a) and (b) arcs are associated to the robot direction when it moved from one node to the next one.

### 4.2 Place Recognition and Goal-oriented Navigation in Multiple T-Mazes

In (Barrera & Weitzenfeld, 2007a) we describe a variant of the classical Morris water maze procedure (Morris, 1981). On the basis of the results presented by Hollup et al. (2001), we supported our decision of using a land-based maze, and not implementing open-field navigation to exploit the affordances module of the spatial cognition model. Therefore, we designed and employed in this task a maze formed by multiple Ts surrounded by three colored cylinders representing landmarks as shown in Fig. 10(a).

In the reversal experiment just described in Section 4.1, the primary objective was to test the ability of the robot to learn the correct route to the goal from a fixed location by using just kinesthetic information, and to unlearn that route while learning the new one that leads to the target. In the task presented in this section, the objective is two-fold: (i) to test the place recognition process carried out by the robot employing not only kinesthetic but also visual information while exploring the maze, and (ii) the goal-directed navigation to find the goal from different starting locations.

The training phase proceeds as in the previous experiment in the T-maze and in the radial maze; i.e., in any given trial, the robot explores the maze until it finds the goal or the end of a corridor. Different from the usual Morris procedure, where animals start from a different location each training trial, the robot in our task, instead, starts from the same fixed location each time (TD in Fig. 10(a)). This variation has also been implemented in other existing studies such as that by Eichenbaum et al. (1990) with successful results.

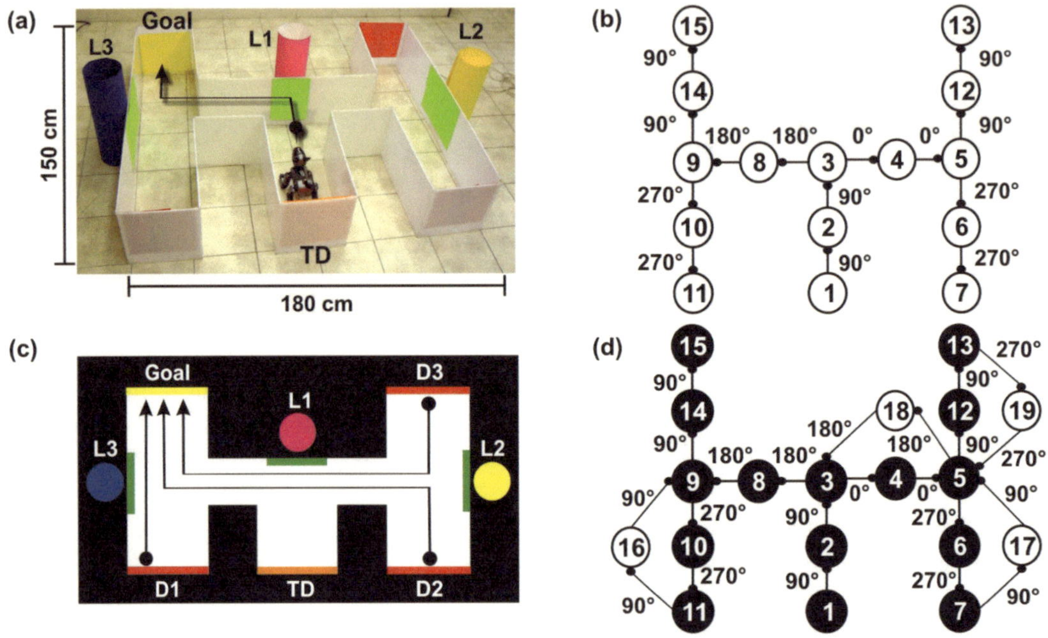


Fig. 10. (a) The physical maze formed by multiple Ts surrounded by landmarks L1, L2 and L3. The AIBO robot is located at the training starting position (TD), and the path to the goal is marked with an arrow. (b) The map built by the robot during training. (c) Routes followed by the robot in three testing trials departing from locations D1 to 90°, D2 to 90° and D3 to 270°. (d) The map updated by the robot during tests, showing existing nodes in black and new nodes in white. In (b) and (d) nodes are numbered in order of creation and arcs are associated to the robot direction when it moved from one node to the next one.

During maze exploration, the robot built a map similar to the one shown in Fig. 10(b), and took in average five training trials reaching the goal to learn the route from location TD. For six robots, the average duration of the training phase was 9 trials, i.e. 23 minutes approximately.

During testing trials, we placed the robot at different departure locations and orientations: D1, D2 and D3 in orientations 90°, 90° and 270°, respectively (see Fig. 10(c)). All six robots found the goal successfully from all starting positions tested, following (in most cases) direct routes as shown in Fig. 10(c), and updating the map of the environment as illustrated by Fig. 10(d). The average duration of trials starting at D1, D2 and D3 were 50 seconds, 120 seconds and 140 seconds, respectively.

To reach the target from any given location, the robot performs place recognition and map exploitation processes. Let's consider the case starting at location D2 (see Fig. 10(c)). The robot begins the trial being oriented to 90°. Since no landmarks were visible from there, neurons in the place cell layer of the model respond according to kinesthetic information only. The current ensemble activity of the layer is searched within the nodes in the map and found in Actor unit 270° of node 7. Although the robot's current direction is not 270°, both activity patterns are similar since none of them encodes for visual information. Thus, node 7 is activated in the map, indicating in this way that the robot recognizes location D2 as previously visited. Then, the robot moves forward and perceives some part of landmark 2

(L2). This time, the current activity pattern of the collection of place cells is not similar to any of those found in the map. This is because the robot had previously visited that location being oriented only to 270° and not to 90°, thus visual information sensed in both directions is different. As a result, the current activity pattern is stored in a new Actor unit 90°, which is associated with the new number 17 node. When the robot reaches the choice point of the corridor, the current ensemble activity of place cells is found within Actor unit 90° of node 5, thus this node is activated. At this location, the robot chooses to turn left by curiosity since reward expectation values are not significant. The current activity pattern of the collection of place cells is stored in a new Actor unit 180°, which is associated with node 5. Being oriented to 180°, the robot then moves forward not recognizing the places represented by node 4 in the map since they were previously visited only in direction 0°, thus node 18 is created. When reaching the choice location at the center of the maze, the current ensemble activity of place cells is found within Actor unit 180° of node 3, thus this node is activated and the link from node 18 to node 3 is added. At this location, the expectation of future reward in direction 180° was increased during training in an important manner; therefore, the robot exploits this information when decides to go ahead instead of turning left. From this location, the robot moves forward recognizing the places represented by nodes 8 and 9 in the map. At this point, the expectation of future reward in direction 90° was increased during training in an important manner; hence, the robot exploits this information when decides to turn right instead of turning left. From this location, the robot moves forward recognizing the places represented by node 14 in the map until it reaches the target, whose location is represented by node15.

As Morris (1981) and Eichenbaum et al. (1990) did with normal rats, we could confirm that to find a "hidden" goal in the maze independently of the starting location, the robot requires to exploit its cognitive map by recognizing reward expectations that make it navigate towards the target location successfully.

### 4.3 On-line Map Adaptation in Multiple T-mazes

In order to evaluate the on-line adaptation of the cognitive map carried out by the model when the environment is altered, we tested two different scenarios of the experiment performed in the maze formed by multiple Ts (see Fig. 10(a)) when landmark configuration is modified after having trained the robot to find the goal. Specifically, in (Barrera & Weitzenfeld, 2007b), we describe results from removing and interchanging landmarks in the maze.

Removing one landmark

The training phase of the task proceeds as in the original experiment just described in Section 4.2. After the robot learns to find the goal departing from the fixed location TD, there are 12 testing trials where one particular landmark was removed (L1, L2 or L3) and the robot departs from locations D1, D2, D3 and TD.

Considering tests from locations D2 and TD when landmark L1 was removed, Fig. 11(a) shows trajectories followed by one of the robots, whereas Fig. 11(b) depicts the map built during training and adapted during both tests.

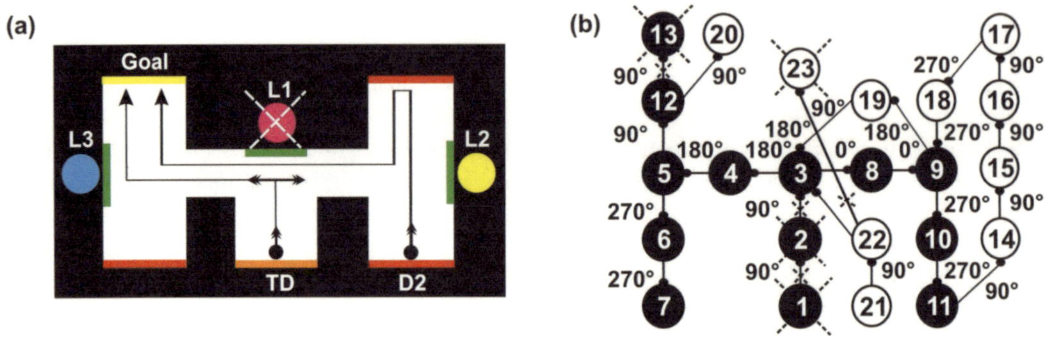


Fig. 11. (a) Robot trajectories traveled during tests departing from locations D2 and TD in direction 90° when landmark L1 was removed. (b) The map built and adapted by the robot during the experiment. Black nodes were created during training, whereas white nodes were added during both tests. Nodes are numbered in order of creation. Nodes and arcs marked as "X" were eliminated.

When starting at location D2, node 10 and 9 are not recognized in direction 90° since corresponding locations in that corridor were previously explored in direction 270°, hence the robot adds new nodes 14 and 15. Being at location represented by new node 15 oriented to 90°, the robot is curious for executing any possible rotation, thus it chooses to go ahead until the end of the corridor while creating nodes 16 and 17. While returning in direction 270°, the robot does not recognize node 16 thus adding node 18. Then, the robot recognizes node 9 and defines link 18-19 in direction 270°. At this time, the robot decides to turn right orienting itself to 180° and moves forward in that direction creating node 19 since node 8 was previously explored in direction 0°. When reaching the choice location at the center of the horizontal corridor in direction 180°, the robot still recognizes node 3 in spite of the absence of L1 because the visual pattern stored in its Actor unit 180° includes mainly information about L3 and the similarity degree between this visual pattern and the current one results close to 1. Exploiting the reward expectation value associated to Actor unit 180° of node 3, the robot goes ahead recognizing nodes 4 and 5, and exploits the reward expectation value associated to direction 90° while deciding to turn right at node 5. When reaching the goal location, the current visual information pattern is different from the one stored in Actor unit 90° of node 13 due to the absence of L1, thus the robot creates the new number 20 node, defines the new link between nodes 12 and 20, and removes connection 12-13 and node 13 since it is not feasible to reach two different places in the same direction 90° from the location represented by node 12 in the map.

On the other hand, during the test starting at location TD in direction 90°, the robot does not recognize map nodes 1, 2 and 3 due to the absence of L1, thus new nodes 21, 22 and 23 are added and connected in direction 90°. Being at the central choice location of the horizontal corridor now represented by new node 23, the curiosity for exploring the place in any orientation makes the robot turn right orienting itself to 0°. At this point, the robot is able to recognize that the current visual information pattern is quite similar to the one stored in Actor unit 0° of node 3 since the position of L2 has remained constant. The activation of node 3 while node 23 is also active involves the redundant representation of the same physical location, and hence the integration of both nodes. As node 23 has just been created, there is no need to add a new node in the map. Actor unit 90° of node 23 is assigned to node

3, arc 90° pointing to node 23 is updated to point to node 3, and node 23 is eliminated as well as connection 2-3, node 2, connection 1-2 and node 1. After the integration process of redundant nodes, the robot orients itself to 180° at the location represented by node 3 due to the large reward expectation value associated to its Actor unit 180°. From node 3, the robot follows directly the correct route to the goal exploiting reward expectation values of recognized map nodes 4, 5 and 12.

<u>Interchanging two landmarks</u>

After performing the training phase of the task as in the original experiment, the robot was tested during 12 trials departing from locations D1, D2, D3 and TD where landmarks L2 and L3 were interchanged. Fig. 12(a) shows trajectories followed by one of the robots during the test starting at D2, whereas Fig. 12(b) depicts the map built during training and adapted during the test.

Due to the new landmarks configuration, the robot does not recognize the previously explored nodes 5 (in direction 90°), 3 (in directions 180° and 90°), 8 (in direction 180°), and 9 (in direction 180°), thus it creates new nodes 17, 19, 22 and 23. In particular, relevant decision locations now corresponding to nodes 19 and 23 are not associated with reward expectations; therefore, robot motor actions are determined by the curiosity and randomness factors during the complete testing trial. Finally, the robot recognizes the goal location represented by node 11 from where L1 is perceptible and has remained in the same location during the entire experiment.

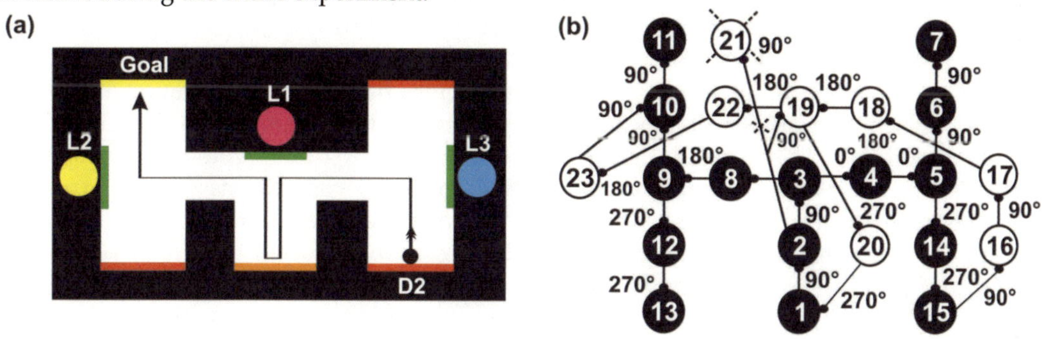


Fig. 12. (a) Robot trajectories traveled during test departing from location D2 in direction 90° when landmarks L2 and L3 were interchanged. (b) The map built and adapted by the robot during the experiment. Black nodes were created during training, whereas white nodes were added during the test. Nodes are numbered in order of creation. Nodes and arcs marked as "X" were eliminated.

## 5. Conclusions and Discussion

This chapter has presented a computational system-level model of rat spatial cognition whose performance has been demonstrated through mobile robot experimentation during diverse learning and memory tasks within different a priori unknown environments.

Discrimination experiments carried out in the T-maze and in the 8-arm radial maze showed that robots, after appropriate training, are not only able to learn the target location and the route to reach it directly from a fixed starting point, but also to "unlearn" both, the previously correct target location and the corresponding route towards it, when the target is moved to another location in the maze. During this "unlearning" behavior, the robot adapts the cognitive map built during training by decreasing the expectation of future reward for the previously correct hypothesis while increasing the reward expectation for the new hypothesis until meeting the criterion of only traveling the appropriate route towards the goal.

We reported the behavioral similitude between results from robots and those obtained by O'Keefe (1983) with normal rats performing the classical reversal task. As stated within the manuscript, both subjects showed an abrupt shift in their response from turning left to turning right in the T-maze revealing the moment when their average orientation crosses the midline in the radial maze from the original quadrant to the reversed one.

In the spatial task performed in the maze composed of multiple Ts, allocentric information derived from landmarks was used by robots firstly to build the cognitive map and then to recognize locations previously explored. During training in the task, robots learned the route leading to the target location by reinforcing motor actions executed at every place belonging to that route. In locating the goal from any given departure location, robots navigated until recognizing a place belonging to the previously learnt route, and exploited properly the information stored in Actor units to reach the goal location by executing rotations associated with maximum reward expectations.

Even though all tested robots found the goal successfully from all starting positions used, we should point out that two of them took longer paths from locations D2 and D3 since they did not explore the environment exhaustively during training. Specifically, one of these robots did not explore the route from location TD to D2, thus not recognizing the decision location at the center of the corridor when departing from D3 oriented to 270°. The opposite occurs in the case of the other robot. It did not explore the route from location TD to D3, thus not recognizing the decision location at the center of the corridor when departing from D2 oriented to 90°. Therefore, an efficient performance during tests would be obtained by extending the number of training trials enabling robots to visit all possible direct routes from the fixed departure location to the goal or the end of each corridor.

An appropriate comparison between our results and those reported by Morris (1981) and Eichenbaum et al. (1990) with normal rats turns out difficult since we experimented with a land-based maze having corridors instead of an open field arena. Nevertheless, as both authors did with rats, we could confirm that to find a hidden goal in the maze independently of the starting location, robots require to exploit their cognitive map by recognizing reward expectations that make them navigate towards the target location successfully.

By means of testing scenarios where landmarks are removed or interchanged in the maze formed by multiple Ts after having trained the robot to find the goal, it was feasible to evaluate the robot's ability to adapt, in real-time, the map previously built during training by adding and/or eliminating nodes in order to reveal changes in the physical configuration of the maze.

As described within the manuscript, variations implemented in the landmarks configuration of the maze may alter robot place recognition and successful navigation to the goal, rendering the generation of a partially new cognitive map. Repeating the training process in such modified environment would enable the robot to relearn appropriate reward expectations and store them in the new map.

From a robotics perspective, this research constitutes one attempt to develop an integrated computational system addressing, from a biological approach, not only the spatial mapping problem, but also the problems of map adaptation and exploitation during successful navigation.

From a biological perspective, our aim is to provide the computational model of spatial cognition in rats to neurobiologists/neuroethologists as a technological platform to test with robots biological experiments whose results can predict rodents' spatial behavior. In this way, experimentalists could test neuroscientific hypotheses in the robotic model and obtain results in a few hours instead of spending weeks experimenting directly with animals. Nevertheless, to achieve this objective, our model needs to be sophisticated by incorporating aspects such as the function of head-direction cells (Ranck, 1984) providing information about the rats' rotation magnitude and movement direction, and the functional differences between hippocampal substructures CA1, CA3 and DG relative to their capabilities of pattern completion and pattern separation (Leutgeb & Leutgeb, 2007). Extending the model in these directions will enable robots to navigate in different spatial contexts (i.e., mazes with corridors as well as open field environments), and improve its adaptive capability in conditions where environmental changes promote new navigational behaviors.

## 6. References

Arleo, A., Smeraldi, F., and Gerstner, W. (2004). Cognitive navigation based on nonuniform Gabor space sampling, unsupervised growing networks, and reinforcement learning. *IEEE Transactions on Neural Networks* 15 (3): 639-652.

Barrera, A., Weitzenfeld, A. (2007a). Bio-inspired Model of Robot Spatial Cognition: Topological Place Recognition and Target Learning, *Proceedings of the 7th IEEE International Symposium on Computational Intelligence in Robotics and Automation – CIRA*, Jacksonville, Florida, USA.

Barrera, A., Weitzenfeld, A. (2007b). Rat-inspired Model of Robot Target Learning and Place Recognition, *Proceedings of the 15th Mediterranean Conference on Control and Automation – MED*, Athens, Greece.

Barrera, A., Weitzenfeld, A. (2008). Biologically-inspired Robot Spatial Cognition based on Rat Neurophysiological Studies. *Autonomous Robots*, Vol 25, No. 1-2, pp. 147-169.

Barto, A. G. (1995). Adaptive critics and the basal ganglia, In: *Models of information processing in the basal ganglia*, Houk, J. C., Davis, J. L., Beiser, D. G. (Eds.), pp. 215-232, MIT Press, Cambridge, MA.

Bosse, M., Newman, P., Leonard, J., Teller, S. (2004). SLAM in large-scale cyclic environments using the Atlas Framework. *International Journal on Robotics Research* – 23(12):1113–1139.

Brown, M. A., Sharp, P. E. (1995). Simulation of Spatial Learning in the Morris Water Maze by a Neural Network Model of the Hippocampal Formation and Nucleus Accumbens. *Hippocampus* 5: 171-188.

Burgess, N., Recce, M., and O'Keefe, J. (1994). A model of hippocampal function. *Neural Networks* 7 (6-7): 1065-1081.

Cho, J., and Sharp, P. (2001). Head direction, place, and movement correlates for cells in the rat retrosplenial cortex. *Behavioral Neuroscience* 115 (1): 3-25.

Eichenbaum, H., Stewart, C., Morris, R. G. M. (1990). Hippocampal representation in place learning. The Journal of Neuroscience 10(11): 3531-3542.

Etienne, A., Jeffery, K. (2004). Path integration in mammals. *Hippocampus* 14(2): 180-192.

Foster, D. J., Morris, R. G. M., Dayan, P. (2000). A Model of Hippocampally Dependent Navigation, Using the Temporal Difference Learning Rule. *Hippocampus* 10: 1-16.

Franz, M. O., Schölkopf, B., Mallot, H. A., Bülthoff, H. (1998). Learning view graphs for robot navigation. *Autonomous Robots* - 5: 111-125.

Frese, U. (2006). A discussion of Simultaneous Localization and Mapping. *Autonomous Robots* - 20: 25-42.

Gaussier, P., Revel, A., Banquet, J. P., Babeau, V. (2002). From view cells and place cells to cognitive map learning: processing stages of the hippocampal system. *Biological Cybernetics* 86, pp. 15-28.

Granon, S., and Poucet, B. (2000). Involvement of the rat prefrontal cortex in cognitive functions: A central role for the prelimbic area. *Psychobiology* 28 (2): 229-237.

Guazzelli, A., Corbacho, F. J., Bota, M. and Arbib, M. A. (1998). Affordances, motivation, and the world graph theory. *Adaptive Behavior* 6 (3-4): 435-471.

Guivant, J., Nebot, E., Nieto, J., Masson, F. (2004). Navigation and mapping in large unstructured environments. *International Journal of Robotics Research* - 23(4): 449-472.

Hähnel, D., Burgard, W., Wegbreit, B. and Thrun, S. (2003). Towards lazy data association in SLAM, *Proceedings of the 11th International Symposium of Robotics Research (ISRR)*, Sienna, Italy.

Hartley, T., Burgess, N. (2005). Complementary memory systems: competition, cooperation and compensation. *Trends in Neuroscience* 28(4):169-170.

Hebb, D. O. (1949). *The organization of behavior: a neuropsychological theory*, Wiley-Interscience, New York.

Hollup, S. A., Molden, S., Donnett, J. G., Moser, M. and Moser, E. I. (2001). Place fields of rat hippocampal pyramidal cells and spatial learning in the watermaze. *European Journal of Neuroscience* 13: 1197-1208.

Houk, J. C., Adams, J. L., and Barto, A. G. (1995). A model of how the basal ganglia generate and use neural signals that predict reinforcement, In: *Models of information processing in the basal ganglia*, Houk, J. C., Davis, J. L., Beiser, D. G. (Eds.), pp. 249-270, MIT Press, Cambridge, MA.

Hull, C. L. (1932). The goal gradient hypothesis and maze learning. *Psychological Review* 39: 25-43.

Jeffery, K. J., and O'Keefe, J. M. (1999). Learned interaction of visual and idiothetic cues in the control of place field orientation. *Experimental Brain Research* 127: 151-161.

Kelley, A. (2004). Ventral striatal control of appetitive motivation: role in ingestive behavior and reward-related learning. *Neuroscience and Biobehavioral Reviews* 27 (8): 765-776.

Kuipers, B., Modayil, J., Beeson, P., MacMahon, M., Savelli, F. (2004). Local metrical and global topological maps in the Hybrid Spatial Semantic Hierarchy, *Proceedings of IEEE International Conference on Robotics and Automation (ICRA)*, New Orleans, USA.

Leutgeb, S., Leutgeb, J.K. (2007). Pattern separation, pattern completion, and new neuronal codes within a continuous CA3 map. *Learning and Memory* 14(11):745-57.

McNaughton B. L., Knierim J. J., and Wilson, M. A. (1994a). Vector encoding and the vestibular foundations of spatial cognition, In: *The cognitive neurosciences*, Gazzaniga, M. (Ed.), pp. 585-595, MIT Press, Boston.

McNaughton, B., Mizumori, S., Barnes, C., Leonard, B., Marquis, M., and Green, E. (1994b). Cortical representation of motion during unrestrained spatial navigation in the rat. *Cerebral Cortex* 4: 27-39.

Milford, M., Wyeth, G. (2007). Spatial mapping and map exploitation: a bio-inspired engineering perspective, In: *Spatial Information Theory*, Winter, S., Duckham, M., Kulik, L. and Kuipers, B. (Eds.), pp. 203-221, Springer-Verlag, Germany.

Mittelstaedt, M., Mittelstaedt, H. (1982). Homing by path integration in a mammal, In: *Avian Navigation*, Papi, F. and Wallraff, H. G. (Eds.), pp. 290-297, Springer Verlag, Berlin.

Morris, R. G. M. (1981). Spatial localization does not require the presence of local cues. *Learning and Motivation* 12: 239 - 260.

Moser, E. I., Kropff, E., Moser, M-B. (2008). Place cells, grid cells, and the brain's spatial representation system. *Annu. Rev. Neurosci.* 31: 69-89.

Movarec, H. P., Elfes, A. (1985). High resolution maps from wide angle sonar, *Proceedings of IEEE International Conference on Robotics and Automation (ICRA)*, pp. 116-121.

Mulder, A. B., Tabuchi, E., Wiener, S. I. (2004). Neurons in hippocampal afferent zones of rat striatum parse routes into multi-pace segments during maze navigation. *European Journal of Neuroscience*, Vol. 19, pp. 1923-1932.

O'Keefe, J. (1983). Spatial memory within and without the hippocampal system, In: *Neurobiology of the Hippocampus*, Seifert, W. (Ed.), pp. 375-403, Academic Press, NY.

O'Keefe, J., Conway, D. H. (1978). Hippocampal place units in the freely moving rat: why they fire where they fire. *Experimental Brain Research* 31: 573–590.

O'Keefe, J., Dostrovsky, J. (1971). The hippocampus as a spatial map: preliminary evidence from unit activity in the freely moving rat. *Brain Research* 34(1): 171 - 175.

O'Keefe, J., Nadel, L. (1978). *The hippocampus as a cognitive map*, Oxford University Press.

Parron, C., Save, E. (2004). Evidence for entorhinal and parietal cortices involvement in path integration in the rat. *Experimental Brain Research* 159 (3): 349-359.

Poucet, B. (1993). Spatial cognitive maps in animals: new hypotheses on their structure and neural mechanisms. *Psychological Review* 100 (2): 163–182.

Quirk, G. J., Muller, R. U., Kubie, J. L. (1990). The firing of hippocampal place cells in the dark depends on the rat's recent experience. *Journal of Neuroscience* 10(6): 2008-2017.

Ranck, J. B., Jr. (1984). Head-direction cells in the deep layers of dorsal presubiculum in freely moving rats. *Soc. Neurosci. Abstr.* 10: 599.

Redish, A., Touretzky, D. (1997). Cognitive maps beyond the hippocampus. *Hippocampus* 7(1): 15-35.

Risold, P., Thompson, R., and Swanson, L. (1997). The structural organization of connections between hypothalamus and cerebral cortex. *Brain Research Reviews* 24 (2-3): 197-254.

Roberts, W. A. (1998). *Principles of animal cognition*, McGraw Hill, USA, pp. 201 - 230.

Schultz, W., Tremblay, L., and Hollerman, J. (1998). Reward prediction in primate basal ganglia and frontal cortex. *Neuropharmacology* 37 (4-5): 421-429.

Stentz, A., Hebert, M. (1995). A complete navigation system for goal acquisition in unknown environments. *Autonomous Robots* 2(2): 127-145.

Thrun, S., et al. (2000). Probabilistic algorithms and the interactive museum tour-guide robot Minerva. *The International Journal of Robotics Research* 19(11): 972-999.

Tolman, E. (1948). Cognitive maps in rats and men. *Psychological Review* 55: 189-208.

Weitzenfeld, A., Arbib, M., Alexander, A. (2002). Neural Simulation Language, MIT Press.

Zivkovic, Z., Bakker, B., Kröse, B. (2005). Hierarchical map building using visual landmarks and geometric constraints, *Proceedings of IEEE/RSJ International Conference on Intelligent Robots and Systems (IROS)*, pp. 7-12, Edmonton, Canada.

# Topological Mapping and Navigation using a Developmental Learning Approach based on Imitation through Sensory-motor Maps

Raquel Frizera Vassallo[1], Hans Jörg Andreas Schneebeli[1]
and José Santos-Victor[2]
[1]*Universidade Federal do Espírito Santo and* [2]*Instituto Superior Técnico*
*Brazil and Portugal*

## 1. Introduction

Over the past decades, we have witnessed a massive utilization of robots in industrial settings. The use of robotic manipulators with high precision and repeatability were fundamental for the increase of efficiency of many manufacturing processes and mass production. Due to the complete predictability of the factory environment, each robot is programmed a priori, and repeats the same set of movements for weeks or even months. In spite of the limited sensory information available, these robots can complete their tasks in an extremely efficient manner.
Recent progress in computer vision, robot navigation, estimation theory, reasoning and computing have afforded engineers with the tools to build more flexible robots able to "live" in more unpredictable environments and interact with (non-technical) humans in a rich and sophisticated manner. In the long-run, this research effort will contribute to the ubiquitous presence of robots in our homes and offices to assist, entertain and work together with humans.
For this to become true, there are two very challenging questions that need to be addressed. Firstly, there is the issue of complexity, as such systems are likely to be much more complex than industrial robots with regard to the number of motor and perceptual degrees of freedom and will have to operate in a highly unpredictable environment. Secondly, there is the question of programming or commanding these robots, as they will interact over extended periods of time with non-technical users, who are not expected to write sophisticated computer programs each time the robot is required to perform a new task. As some of these challenges are common to many biological social species, we will borrow inspiration from biology, developmental psychology and neuroscience in our approach.
Many animals can be seen as complex and flexible systems, able to learn and adapt through a simultaneous motor, sensorial and cognitive development. Studies in developmental psychology clearly show how human infants progressively learn how to acquire more sophisticated skills over time, through interaction with their own body, the environment and caretakers. At birth, many motor and perceptual capabilities are not yet in place and the acquisition of some skills is a pre-requisite to acquire more complex ones. It would otherwise be very difficult to handle the overwhelming flow of sensory-motor information.
In this work, we adopt the approach of developmental robotics Fitzpatrick et al. (2003); Lungarella & Pfeifer (2001); Metta, Sandini, Natale, Manzotti & Panerai (2001) to build systems

able to acquire complex sensory, motor and cognitive skills, in a progressive manner, guided through different levels of interaction with the environment.

With regard to the second challenge of "programming" such complex robotic systems without the need to explicitly writing computer programs, we can again see how knowledge and skills are transferred amongst individuals of some animal species. Imitation seems to be a powerful social learning and adaptation modality in social species, Dautenhahn (1995); Dautenhahn & Nehaniv (2002). Imitation avoids the need to undergo through extensive trial and error, since the imitator can learn directly from the teacher's experience.

Similarly, endowing a robot with the ability to imitate tasks performed by a demonstrator could become an intuitive and yet extremely powerful way of "programming" such robots by non-technical users. Learning by imitation has been addressed by several researchers with interesting results for task learning, skill acquisition and communication Billard (2002); Billard & Hayes (1997); Matáric (2002). Other works in robotics inspired by imitation in animals can be seen in Dautenhahn & Nehaniv (2002).

It is clear how imitation can be a powerful learning mechanism and how it can avoid having to pre-program the robot for every new task. Then, the developmental approach would define the set of skills the robot needs to acquire, in an incremental manner, modulated by the interaction with the environment, objects and people.

Recent findings in neuroscience have shed new insight as to the brain mechanisms possibly implied in imitative behavior. We are referring to the discovery of the *mirror neurons* in area F5 of monkey's brain Fadiga et al. (2000); Gallese et al. (1996), that fire during the execution of a goal directed (grasp) action by the animal as well as with the observation of similar actions performed by others Rizzolatti & Arbib (1998); Rizzolatti et al. (2002).

These findings suggest that the recognition of actions may be facilitated by the ability to execute that same action, the same brain circuitry being used for both purposes. In addition, these results show the importance of motor information for perception and action understanding tasks. It indicates that recognition is done with motor data as opposed to visual information.

In our work, we explore the inspiration from mirror neurons and rely on motor representations for our robot to execute or recognize "actions". We will show how this becomes a very natural way of representing, learning and executing robotic tasks. The fact that mirror neurons are located in the motor area of the brain, suggests that observed actions are first "converted" to motor information, before recognition may take place.

The existence of a visuo-motor mapping that converts visual information into motor measurements is also supported by biology. When newborns look to their hands and own movements, they are probably learning the relationship between motor action and visual stimuli (sensory-motor coordination). Through this visuo-motor mapping, children become able to recognize and repeat movements. Some works have explored these visuo-motor mappings for learning how to grasp objects with manipulators, Fitzpatrick et al. (2003); Lopes & Santos-Victor (2003a;b); Metta, Sandini & Natale (2001).

In this work we propose an approach for building robots able to learn and adapt in an open-ended manner, while interacting with humans. The approach is based on three main ideas: (i) use of artificial ontogenetic development as a means for the robot to acquire complex skills under controlled complexity, (ii) use of imitation learning as the main form of "programming" such robots and (iii) exploiting the use of motor representations for action recognition, learning and executions. The proposed developmental architecture is illustrated in Figure 1.

The first stage in this developmental pathway is that of sensory-motor coordination or learning sensory-motor maps. Sensory-motor maps relate motor actions to sensor (visual, propri-

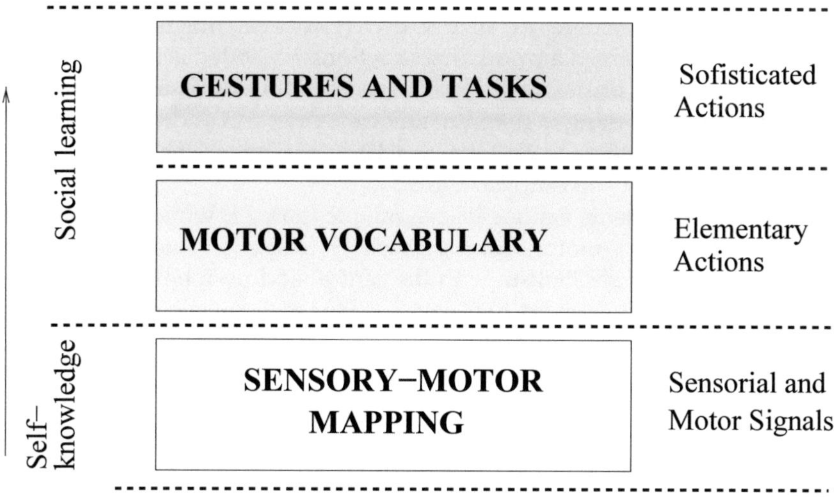


Fig. 1. Robot learning and adaptation approach.

oceptive or auditive) percepts and vice-versa. The forward map allows to predict the new sensory observations resulting from a certain motor action (e.g. the arm configuration in the retina arising from the activation of a set of muscles). Conversely, the inverse map is useful to determine which motor actions need to be taken to produce given sensory percept (e.g. determine the muscle activations to reach a certain point in space) or to recognize movements from sensory information. Figure 2 illustrates the use of the forward and inverse versions of sensory-motor maps.

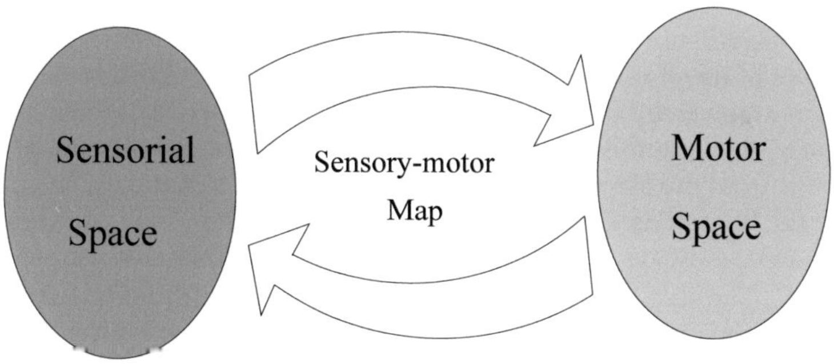


Fig. 2. Using the sensory-motor map in a dual way.

Sensory-motor maps can either be computed analytically (whenever calibration information is available) or learned directly from data through the observation of the robot's own movements.

Once the sensory-motor maps are in place, the system can move to the next developmental stage and learn a repertoire (or a vocabulary) of elementary motor actions that require the coordination of various degrees of freedom. It is important to notice that this repertoire results from two types of constraints. It is tightly related to the robot's own kinematic constraints but, more importantly, it is modulated by the interaction with the environment (co-development) and other agents (either humans or robots).

Hence, interaction allows the robot to organize its own (low-level) movements and learn task-specific actions that are not defined a priori. These actions are coded according to the robot's motor capabilities (using the learned sensory-motor maps) and this approach can thus be applied to different robots, working in different environments and addressing different tasks. Whenever the application changes, a new set of actions can be taught. There is no need to program every action needed to accomplish a task.
At this level, the robot represents actions in a symbolic manner, while the fine motor control is handled by the sensory-motor maps previously learned. This hierarchy of movements/actions follows a close parallelism with the central and peripheral nervous system of many animals, where the brain triggers high-level actions and the spinal cord is responsible for the individual muscle activation (e.g. in locomotion).
In the next stage of development, the system will be able to learn how to use this action vocabulary to solve specific tasks. The idea is that the system can observe actions performed by others and map the observed actions to those it knows how to perform (the action vocabulary). Usually, this requires combining these elementary actions in a task-specific manner, to solve different day to day tasks.
This strategy for learning and adaptation is a very flexible way of mapping the desired task (and the way it should be performed), to the robot's motor capabilities. Different robots can learn the same action vocabulary and different vocabularies can be taught if very different environments or task domains are required.

### 1.1 Application to mapping and navigation

In order to illustrate the approach applicability, we chose the task of topological mapping and navigation for a mobile robot, where vision is the primary sensor modality.
The level of sensory-motor coordination corresponds to the system's ability to relate visual stimuli (image motion) to motor/proprioceptive data (egomotion). The process to estimate the motor information from image flow is based on the particular geometry of an omnidirectional camera and benefits from the use of enlarged fields of view.
In the second level of development, the system relies on vision to follow a person. Through this stage, an egomotion estimation process is utilized to convert visual motion to motor data. In the motor space, the system will learn how to categorize the most frequent actions. Those actions result both from the system's kinematics as well as the shape of the environment and the guidance of the human assistant. For example, if the system would be taught in a world composed of circular hallways, then rotational motion would be very dominant in the action vocabulary.
In the final level of development, the system engages in social learning through imitation. The idea is that one person can now act as a guide and show the workspace to the robot. By following the person, the robot will then be able to create a topological map of the environment, expressed in terms of the motor vocabulary. Then, for a navigation task like "going to a place B", the system will just need to localize itself using the visual information and invoke the motor programs (actions) to move to the next sub-goal (e.g. turn right, go ahead, etc). The implementation steps are illustrated in Figure 3.
By using the proposed learning and adaptation approach, the robot performs mapping and navigation without the need for programming every action required for the task. The basic actions are learned while interacting with an user. Later on, these actions are used for mapping and navigating. Besides illustrating the proposed methodology, the task of topological mapping has not been addressed in many previous works done on imitation.

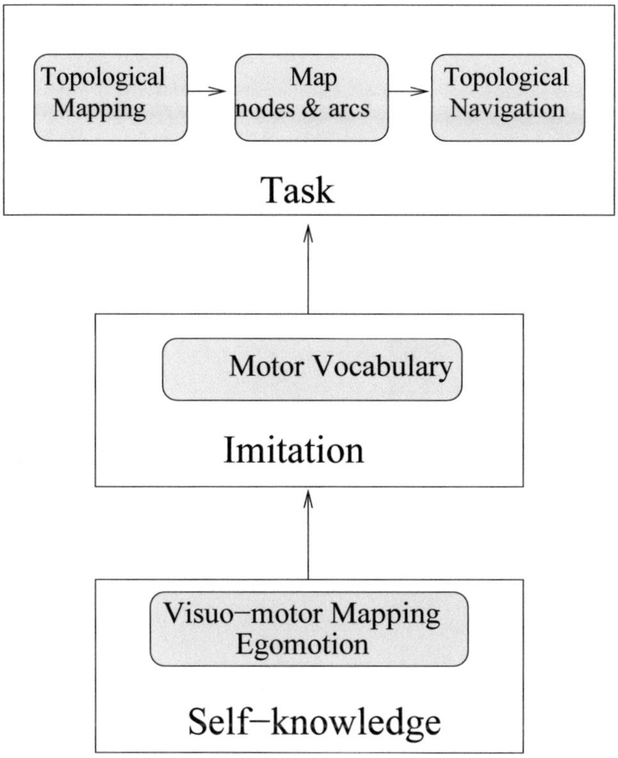


Fig. 3. Topological mapping and navigation application.

### 1.2 Structure of this chapter

In Section 2 we present the motor and sensory spaces defined for the used robot and the proposed egomotion estimation process responsible for the visuo-motor mapping. Sections 3 and 4 describe how to learn the motor vocabulary and the mobile robot application. Experiments and results are listed in Section 5 while our conclusions and future work are discussed in Section 6.

## 2. Visuo-motor coordination: egomotion estimation

As we discussed previously, the goal of the visuo motor coordination is to build maps that relate visual stimuli to motor data. In the task of mapping and navigation for a mobile robot, this corresponds to estimating the camera egomotion from optical flow measurements obtained from a sequence of images.

The robot used in this work is a differential plataform, capable of moving in the ground plane and rotating about the $Z$-axis by receiving motor commands for the linear velocity $T_y$ and angular velocity $\omega_z$. Thus its motor space can be defined by its linear velocity on the $XY$ plane and the angular velocity about the $Z$-axis.

The robot is equipped with an omnidirectional camera aligned with the platform's rotation axis. It is a catadioptric system formed by a B&W camera and a spherical mirror Baker & Nayar (1998). The robot also has a color (perspective) camera pointing to the forward direction, that is used for the person-following behavior. The robot and both vision systems can be seen in Figure 4.

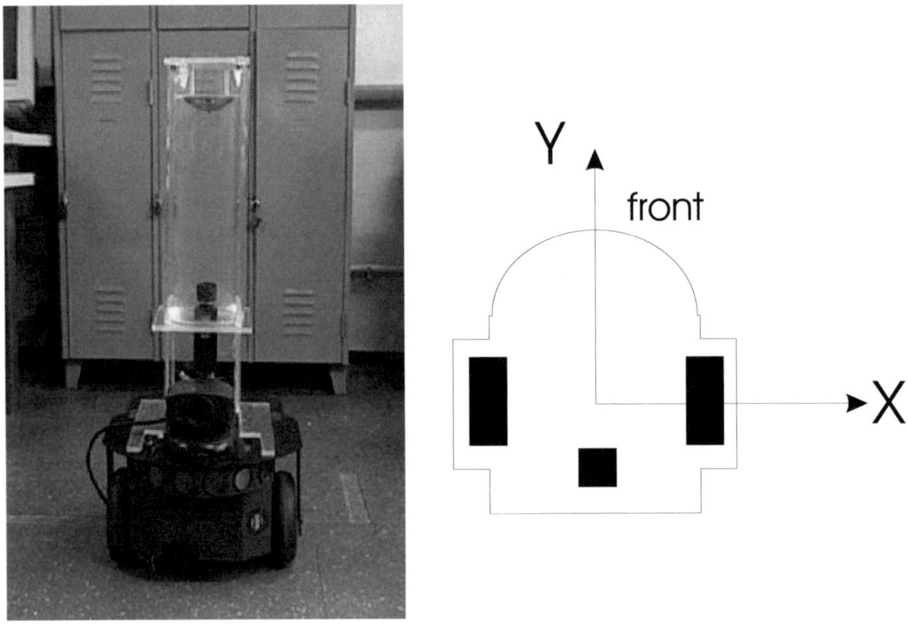


Fig. 4. The robot, the vision systems and the adopted robot reference frame.

The sensory space is defined by spherical optical flow measurements, calculated from omnidirecional images and used for egomotion estimation. In what follows, we will show how to obtain a spherical optical flow from a sequence of omnidirectional images, and how the visuo-motor mapping converts visual information into motor data.

The reason for using omnidirectional images for egomotion estimation is that the problem becomes easier if a spherical motion field is used instead of a planar field obtained with perspective cameras Nelson & Aloimonos (1988). We start by calculating the optical flow field from a sequence of omnidirectional images. Then, the image flow vectors are remapped to the surface of the unit sphere, through an image Jacobian matrix. On such hemispherical motion field, we know that either the focus of expansion (FOE) or the focus of contraction (FOC) must be visible. Finally, motor information is estimated from the motion field adapting an egomotion algorithm designed for planar projection to spherical projection Vassallo et al. (2002a).

In previous works, the Jacobian matrix needed to remap image flow vectors was defined according to the system projection model Gluckman & Nayar (1998). Instead, we use a general projection model defined by Geyer & Daniilidis (2000a;b) to define a general Jacobian. This projection model can represent different omnidirectional systems (with a single projection center) by combining a mapping of a 3D point $P$ to a sphere, followed by a projection to the image plane. The center of the sphere $C(0,0,0)$ lies on the optical axis of the projection to the plane and represents the origin of the reference frame. The general projection model is illustrated in Figure 5.

The parameters $l$ and $m$ can be adjusted to model different camera types (mirror shapes) and correspond to the distances from the sphere center $C$ to the projection center $O$ and to the

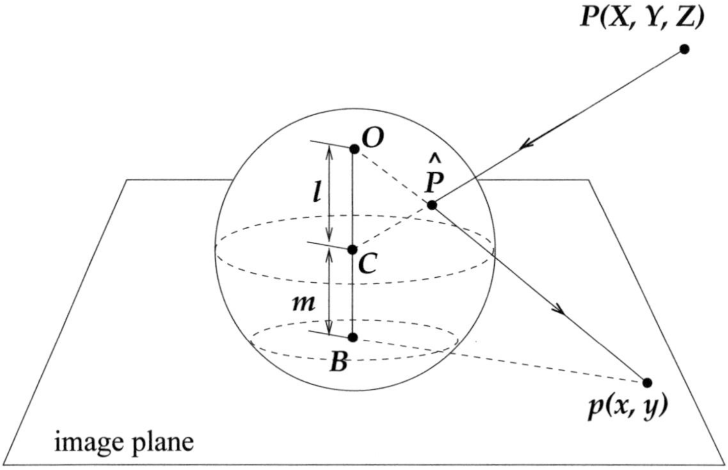


Fig. 5. The general projection model for single projection center cameras.

projection plane. According to the model, the image projection $p(x, y)$ of a 3D point $P(X, Y, Z)$, can be defined by:

$$\begin{bmatrix} x \\ y \end{bmatrix} = \frac{l+m}{lR-Z} \begin{bmatrix} X \\ Y \end{bmatrix} \quad \text{with} \quad R = \sqrt{X^2+Y^2+Z^2} \tag{1}$$

Back-projecting an image point $(x, y)$, we can obtain a point on the unit sphere $\hat{P}(\hat{X}, \hat{Y}, \hat{Z})$ corresponding to the direction of the incoming ray from the original 3D point $P(X, Y, Z)$ Gaspar et al. (2001). The back-projection is described by Equation 2.

$$\begin{aligned} \begin{bmatrix} \hat{X} \\ \hat{Y} \end{bmatrix} &= \frac{lA + sign(A)\sqrt{(x^2+y^2)(1-l^2)+(A)^2}}{x^2+y^2+(A)^2} \begin{bmatrix} x \\ y \end{bmatrix} \\ A &= l+m \quad \hat{Z} = \pm\sqrt{1-\hat{X}^2-\hat{Y}^2} \end{aligned} \tag{2}$$

where $\hat{Z}$ becomes negative if $|l+m|/l > \sqrt{x^2+y^2}$ and positive otherwise. Note that the camera intrinsic parameters, image center and focal length are not considered in the above expression.

To reproject flow vectors from the image plane to the sphere surface, a general *Jacobian* matrix $J$ is defined by differentiating the spherical coordinates $(\hat{X}, \hat{Y}, \hat{Z})$ on the back-projection equation with respect to the image coordinates $(x, y)$ Vassallo et al. (2002a). It maps image velocity vectors to the unit sphere surface, transforming a planar flow field to a hemispherical motion field that will help estimate egomotion (see Equation 3 and Figure 6).

$$\begin{bmatrix} \dot{\hat{X}} & \dot{\hat{Y}} & \dot{\hat{Z}} \end{bmatrix}^T = J \begin{bmatrix} \dot{x} & \dot{y} \end{bmatrix}^T \quad \text{with} \quad J = \begin{bmatrix} \frac{\partial \hat{X}}{\partial x} & \frac{\partial \hat{Y}}{\partial x} & \frac{\partial \hat{Z}}{\partial x} \\ \frac{\partial \hat{X}}{\partial y} & \frac{\partial \hat{Y}}{\partial y} & \frac{\partial \hat{Z}}{\partial y} \end{bmatrix}^T \tag{3}$$

Egomotion is estimated by adapting the Bruss & Horn (1983) algorithm designed for planar perspective projection to spherical projection. The motion field $U$ at a point $\hat{P}$ on the unit

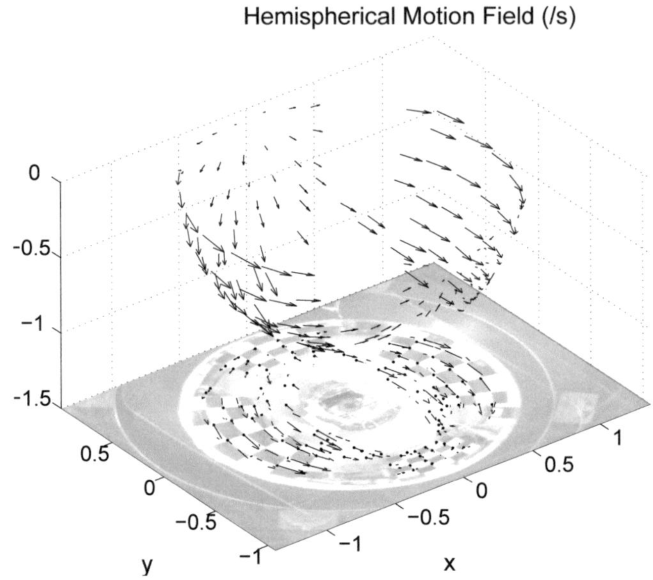


Fig. 6. Image velocities remapped to the unit sphere surface.

sphere is a function of the camera rotation $\Omega$, translation $T$ and the corresponding 3D point depth $R = \sqrt{X^2 + Y^2 + Z^2}$ (Eq. 4).

$$U(\hat{P}) = \frac{1}{R}((T \cdot \hat{P})\hat{P} - T) - \Omega \times \hat{P} \tag{4}$$

Depth dependency is removed by taking the cross product with $\hat{P}$ and the dot product with $T$, resulting in Equation 5. Estimation was done through an iterative process using non-linear minimization considering $|T| = 1$, since the linear velocity can only be determined up to a scale factor.

$$T \cdot (\hat{P} \times (U + (\Omega \times \hat{P}))) = 0 \tag{5}$$

The process for egomotion estimation that we have just described can be interpreted as a visuo-motor map that converts visual information to motor measurements, $T(T_x, T_y, T_z)$ and $\Omega(\omega_x, \omega_y, \omega_z)$. It is important to stress that this process is intimately related to the robot's motor and visual capabilities.

Some egomotion results obtained with the mobile robot used for the experiments are shown in Figure 7. Image flow vectors were calculated using Lucas & Kanade (1981) method and then remapped to the unit sphere by the general Jacobian (Equation 3). The first example corresponds to a translation, the second is a rotation and the third a combined motion, translation plus rotation. The robot can translate in the $XY$ plane (the ground plane) and rotate around the $Z$-axis. The vectors $(T_x, T_y)$ and $\omega_z$ obtained by egomotion estimation can also be seen in Figure 7.

Table 1 presents estimated values and errors calculated by comparing the results with the nominal values. For the translation vectors, just the error on direction was computed, once it is not possible to recover the absolute velocity values from egomotion estimation. For the angular velocity, the errors were mesured along the Z-axis in $^o/s$.

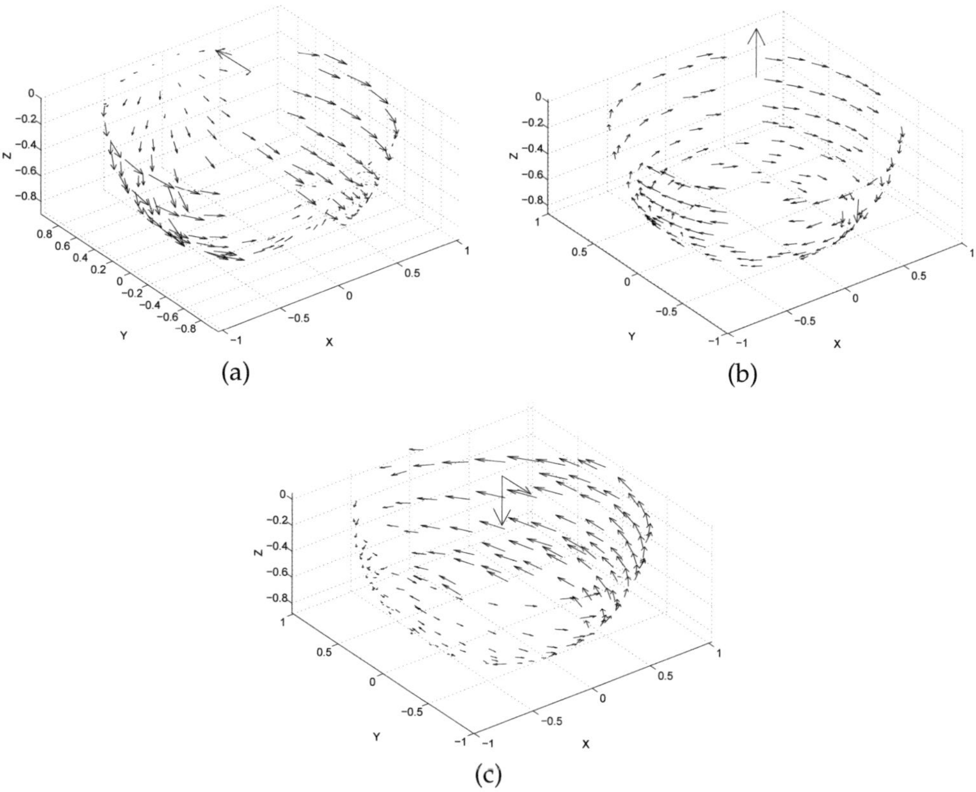


Fig. 7. Examples of (a) Translation, (b) Rotation and (c) Combined move. Egomotion vectors $(T_x, T_y)$ and $\omega_z$ indicated in the middle of the hemispherical flow.

| | Translation | Rotation |
|---|---|---|
| (a) | $T\ (T_x\ \ T_y)$ | $\Omega\ (\omega_z)\ (^o/s)$ |
| nominal | $T = [0\ \ 1]$ | $\Omega = [0]$ |
| estimation | $\hat{T} = [0.0128\ \ 0.9999]$ | $\hat{\Omega} = [-0.0132]$ |
| error | $e_t = 0.734^o$ | $e_\Omega = [-0.0132]$ |
| (b) | $T\ (T_x\ \ T_y)$ | $\Omega\ (\omega_z)\ (^o/s)$ |
| nominal | $T = [0\ \ 0]$ | $\Omega = [-3.1255]$ |
| estimation | $\hat{T} = [0\ \ 0]$ | $\hat{\Omega} = [-3.4965]$ |
| error | $e_t = 0^o$ | $e_\Omega = [-0.371]$ |
| (c) | $T\ (T_x\ \ T_y)$ | $\Omega\ (\omega_z)\ (^o/s)$ |
| nominal | $T = [0\ \ -1]$ | $\Omega = [3.8776]$ |
| estimation | $\hat{T} = [0.0252\ \ -0.9997]$ | $\hat{\Omega} = [4.3284]$ |
| error | $e_t = 1.45^o$ | $e_\Omega = [0.4508]$ |

Table 1. Egomotion estimations/errors.

A set of 608 different movements were executed for testing the egomotion method used as visuo-motor mapping. All the results were compared to the nominal values and the errors analysed. The mean error found for the direction of translation was $1.44^o$ and for the angular velocity was $0.56^o/s$. The errors for translation and rotation are shown in Figures 8 and 9.

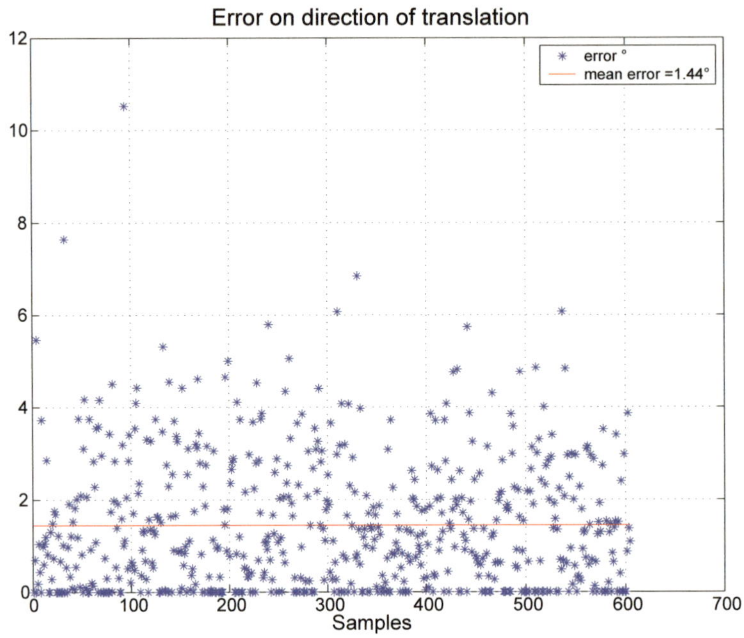


Fig. 8. Errors on direction of translation.

Although the translation and rotation estimation errors were in the order of $5^o$ and $2^o/s$ for some experiments, the results were satisfactory for the envisaged application. The mean errors do not have a large impact on the direction of translation and on the angular velocity estimation, when the robot travels small distances (instantaneous measurements).
At this point, we have defined the nature of the visuo-motor map that will be used throughout the remainder of the chapter. It allows the system to convert purely visual information to its motor variables, the egomotion.

## 3. Learning a purposive motor vocabulary through imitation

The first stage of development endows the system with the ability to map the motion of the visual world onto the robot's motor variables. In this second level of development, this motor information will be organized onto more complex actions. From that point onwards, such actions can be used to solve more sophisticated tasks, without the need to consider the details of motion execution.
The way motor signals are organized to build a vocabulary of actions depends on the environment where the robot usually operates and will be modulated by the typical tasks it will execute. Hence, we rely on imitation, in the form of a person-following mechanism, to drive this process.
The idea is that the robot is guided through the workspace by a human operator. The correspondence problem for imitation was solved by a person-following behavior. While following the guide, the robot continuously maps the observed visual motion to its own motor variables

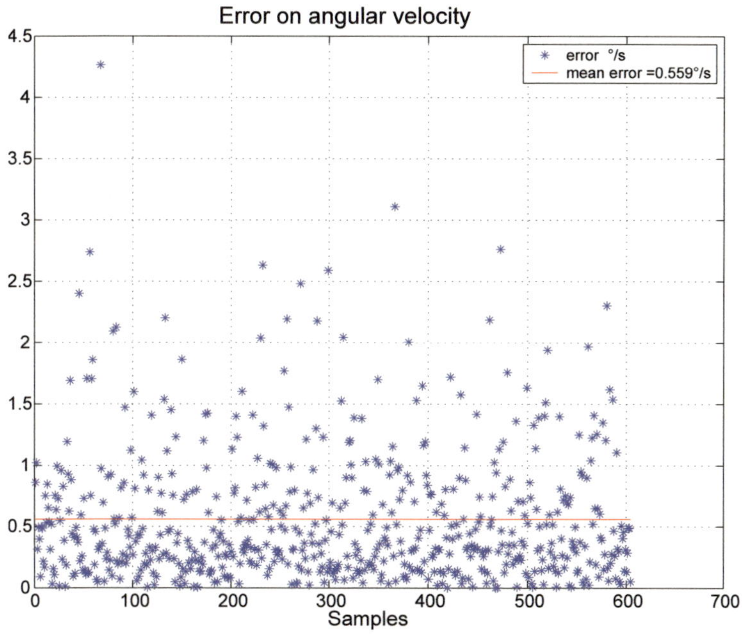


Fig. 9. Errors on angular velocity.

using the visuo-motor map (egomotion estimation). In this motor space, one can identify what are the most typical executed movements to create a basis of elementary actions.

For simplicity, we assume that the person-to-follow carries a distinctive green-colored rectangle. The target is first detected using the hue channel of frontal images captured by a color camera. Noise in the resulting binary image is filtered through morphological operators and the largest remaining blob is selected. The contour is detected and the rectangle lines are estimated by a robust fitting procedure Fischler & Bolles (1981). Finally, the corners coordinates are determined from the lines intersection, as shown in Figure 10. A visual servoing strategy was used so that the robot could follow the green rectangle at a predefined distance (1m) and oriented toward its center Vassallo et al. (2002b).

While the robot follows a person, it uses its visuo-motor map (egomotion) to perceive its own movements. This motor information is classified into clusters by an unsupervised learning method based on K-means. In our implementation, the number of desired clusters is defined by the user but it could be learned in an automatic manner. The different clusters (and associated centroids) represent the learned characteristic movements and will constitute the motor vocabulary. If needed, labels can also be associated to each movement as they were motor words.

During the tests, a set of person-following experiments were done to learn this motor vocabulary. Egomotion values were normalized and classified into clusters defining the motor words. As we discussed before, the motor variables for a robot with differential kinematics, are the angular velocity $\omega_z$ and the robot forward velocyty, $T_y$. We estimate both the values for $T_x$ and $T_y$. The values for $T_x$ are usually non-zero (albeit small) due to sliding of the robot wheels or noise in the estimation process. Once these values are small, they will not be considered for defining the motor vocabulary, but just $T_y$ and $\omega_z$.

Fig. 10. The green rectangle's corners detection: the rgb-image, the hue channel, the image segmentation and processing until the corners estimation.

The clusters found in the $(T_y, \omega_z)$ space are shown in Figure 11. The mean values are represented by asterisks and Voronoi lines separating the various clusters are drawn. The centroids represent the purposive motor vocabulary. Names were given to each cluster defining motor words which are detailed in Table 2.

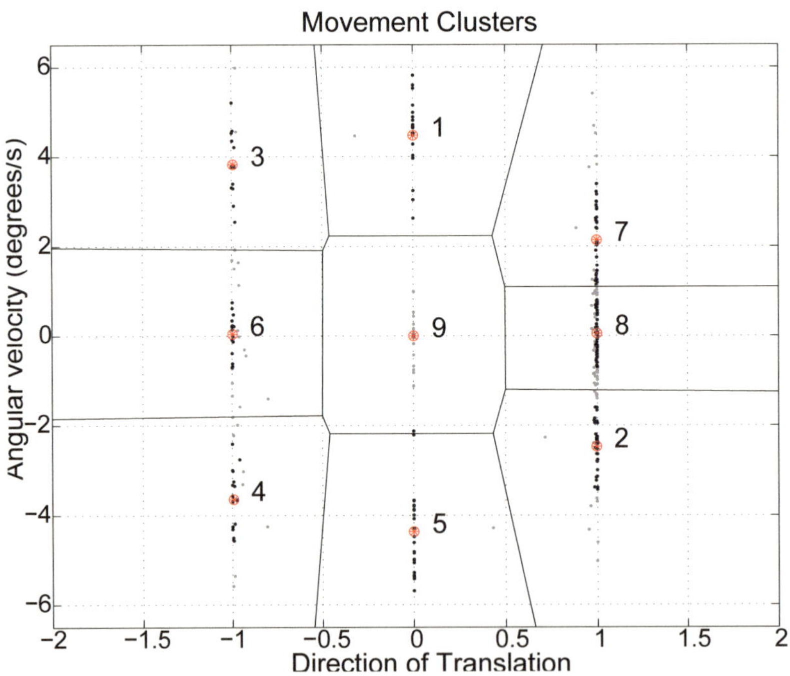


Fig. 11. Clusters of the created vocabulary. Black/gray points indicate inliers/outliers.

| N | Motor Word | $T_y$ | $\omega_z$ (degress/s) |
|---|---|---|---|
| W 1 | left turn | 0 | 4.4722 |
| W 2 | front right | 0.9948 | -2.4749 |
| W 3 | back left | -0.9927 | 3.8223 |
| W 4 | back right | -0.9927 | -3.6440 |
| W 5 | right turn | 0 | -4.3646 |
| W 6 | back | -0.9972 | 0.0386 |
| W 7 | front left | 0.9973 | 2.1350 |
| W 8 | front | 0.9974 | 0.0488 |
| W 9 | stopped | 0 | 0 |

Table 2. Motor words for each cluster.

It is important to stress that this level of development allows the system to organize movements onto more complex motor actions. Motor actions can be seen as the result of several motor programs and treated in a more symbolic level when compared to individual movements.
The motor vocabulary is determined by two main factors: the environment and the user/demonstrator. If the environment were exclusively constituted of curved sections, rectilinear movements would not be part of the vocabulary. Similarly, if the demonstrator always uses right-turns, then left-turns would not be represented in the vocabulary. This is advantageous because the vocabulary adapts to the structure of the environment as well as to the "behavior" of the demonstrator.
Now that the motor action vocabulary is created, the robot can define and recognize movements of interest and use those to perform a desired task. Movement recognition is done in motor space, using the Euclidean distance as the discriminant function, instead in visual space, which depends on frame orientation and position, and environment illumination.

## 4. Solving a task: Topological Mapping and Navigation

This section shows how we used the visuo-motor map and the learned motor vocabulary for a mobile robot application: topological mapping and navigation. This task appears at the end of a sequence of (developmental) steps where the robot acquires progressively more sophisticated skills:

1. Definition of a visuo-motor map for the mobile robot, in the form of an egomotion estimation process.
2. A person-following (imitation) behavior was used to allow the robot to learn a purposive motor action vocabulary. These actions depend both on the environment structure and the demonstrator pattern of actions. These actions are "discrete" and are built upon the elementary motor signals acquired though the visuo-motor mapping process.
3. Finally, the learned vocabulary is used for building topological maps and for navigation.

During the map building phase, we assume that the robot is guided through the environment by a user or a demonstrator. While moving, the robot creates a topological map, in the form of a graph. Nodes of this graph correspond to omnidirectional images captured during the

robot motion, while the links are associated to motor actions, resulting from the egomotion estimates projected onto the motor vocabulary.
The decision of whether or not inserting a new node in the map is taken based on two criteria: (i) a comparison with the previously stored reference image or (ii) abrupt changes in motion. A new node is added to the map whenever the *sum of squared differences* (SSD) between successive images exceeds a threshold or the robot motion changes suddenly. Once a new node is stored, the most frequently recognized motor word (to introduce robutsness) is attributed to the link between the current and previous nodes.
Navigation with the created map comprises several steps. First, the robot determines its initial position. Then, the robot searches a path in the graph leading from the initial position to the goal destination. During navigation, progress along the route is monitored by comparing the captured images against the current image node and the subsequent one in the path. Image correlation is based on the same SSD metric that is used for map building. To determine whenever the robot position should be updated, we adopt a hill-climbing strategy in such a way that a new node is accepted whenever a persistent increase in the SSD values is detected after a clear-cut minimum ("valley").
Every time a new map position (node) is reached, the motor word stored in the subsequent link determines the next motor command. This behavior goes on until the final location is reached. Navigating through the map can be seen as executing a concatenation of the motor actions sampled from the motor vocabulary. Such commands are directly retrieved from the map, as that information was stored during the mapping phase. In other words, it can be considered a "motor program" formed by "motor words". Some experiments and results are shown in the next section, a natural way of expressing and representing navigation tasks.

## 5. Experiments and Results

The robot used in the experiments described in this paper is a Pioneer DX2 equipped with an onboard computer (Pentium II MMX - 266 MHz - 128 MRAM). An omnidirectional camera is mounted on the top of the robot with its axis aligned with the platform's rotation axis. It is a catadioptric system formed by a B&W camera and a spherical mirror Baker & Nayar (1998). In addition, the robot is equipped with a color (perspective) camera pointing to the forward direction. The robot and both vision systems were already shown in Figure 4.
The omnidirectional system is used for defining the visuo-motor mapping, based on the ego-motion estimation process, and for performing the topological mapping task. The color camera is used for implementing the imitation/following behavior.
As we have seen before the mobile robot imitation (person following) behavior was used in two distinc ways. First, it is used to build a visuo-motor mapping and for the robot to learn elementary motor actions, coded according to the robot's motion repertoire. Secondly, the following behavior is used to combine such basic actions for topological mapping and navigation. These maps become motor representations of the environment, which main advantage is to be adapted to the used robot and environmental structure.
Several mapping and navigation experiments were conducted for testing the proposed method. For that purpose, we have used an experimental arena adapted to the robot size and camera height. One of the mapping tests is shown in Figure 12. The executed path for map building is plotted and the places where images were selected as topological map nodes are indicated by asterisks
In this experiment, a total of 16 omidirectional images were selected for map nodes. These images were subsampled twice down to 166 x 166 pixels. As described, motor words extracted

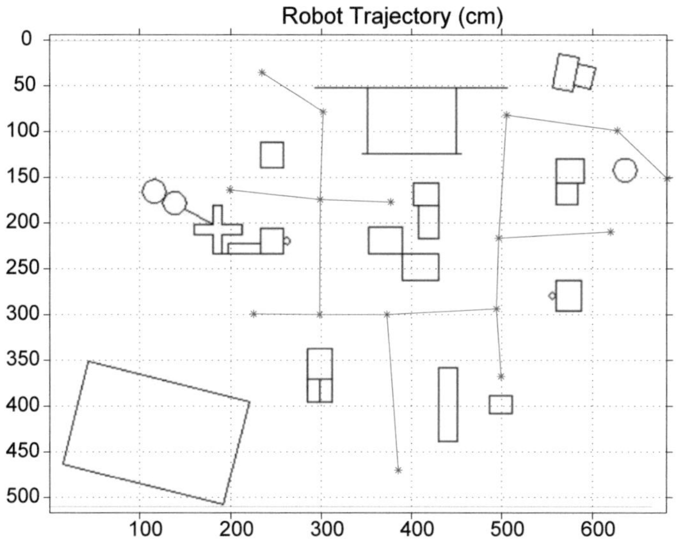


Fig. 12. Robot trajectory (odometry) during map building. Asterisks indicate map nodes.

from the learned vocabulary were associated to the links between images, according to robot motion. Whenever possible, the dual (symmetric) motion of each motor word was also associated to the map link. This strategy allows the robot to navigate in each link in opposite directions. Figure 13 shows the omnidirectional images overlaid the map nodes.
Some navigation experiments for the created map are shown in Figure 14. The robot was asked to navigate between different points accross the map. It involved traversing some of the map links in the direction opposite to that learned during mapping.
The places where the robot updated its (qualitative) position are indicated by asterisks. Although some points correspond to the same node in the map, the asterisks did not happen exactly in the same coordinates but in the same region, associated to a common "qualitative" location. The robot solved the navigation tasks succesfully in about 90% of the trials. It only fails in cases where it encountered obstacles during the mission, since we did not include any obstacle avoidance scheme. Instead, our efforts were directed to testing the task learning method and the use of motor information for topological map building and navigation. We therefore found the results very encouraging.

## 6. Conclusions and Future Work

In this work, we have proposed an approach that allows a robot to use motor representations for learning a complex task through imitation. Our approach is inspired after developmental psychology and some findings in neuroscience. The framework relies on development as the process allowing the robot to acquire sophisticated capabilities over time, as a sequence of simpler (learning) steps. At the first level, the robot learns about sensory-motor coordination. Then, motor actions are identified based on lower level, raw signals. Finally, these motor actions are stored in a topological map and retrieved in an efficient way during navigation. The entire methodology is grounded on hierarchical motor representations as suggested by experiments in Mirror Neurons.

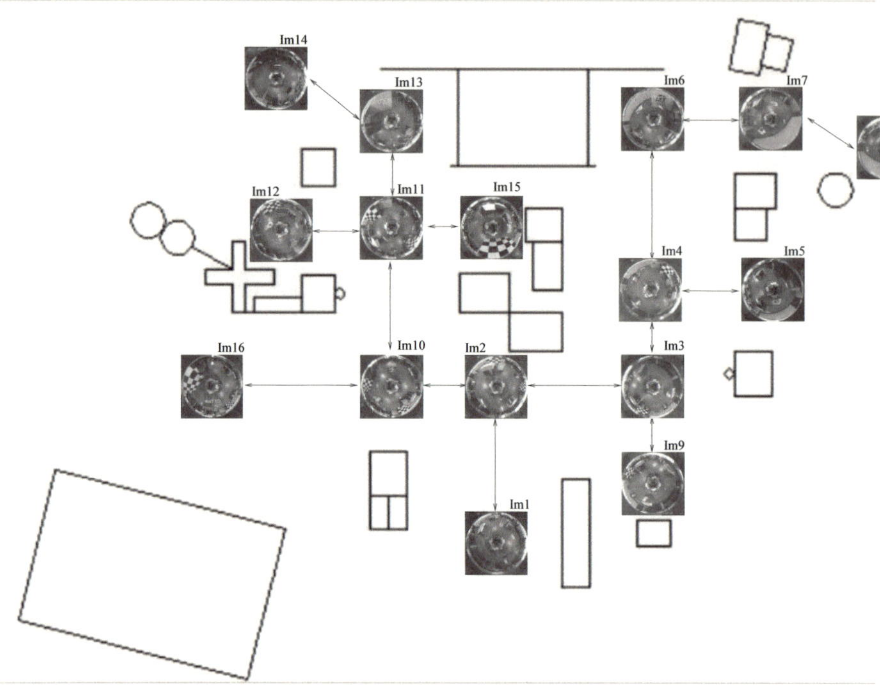

Fig. 13. The 16 omnidirectional images of each map node.

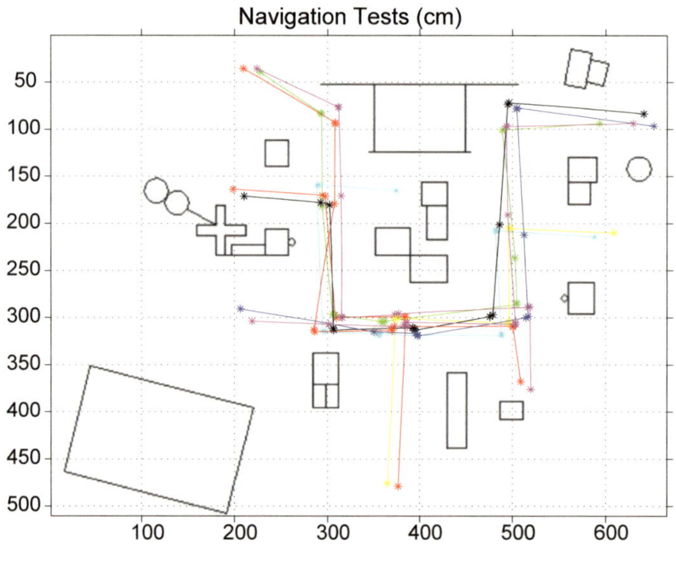

Fig. 14. Robot trajectories (odometry) when map navigating. Asterisks are position updates.

We illustrated the approach with an application with a mobile robot. An egomotion estimation process was used to perform visuo-motor mapping. It transforms optical flow measurements on omnidirectional images to motor information, matching the motor capabilities (kinematics) of the differential robot chosen for the task. Extensive tests were performed to test and analyze the egomotion implementation validating the visuo-motor map.

In a subsequent stage of development, a person following behavior was implemented. The robot was able to learn a set of elementary actions, based on the visuo-motor mapping, that form a motor vocabulary.

Finally, the robot builds a topological map of the workspace, while it is guided through the environment by a human operator. During this process, the robot selects omnidirectional images to store as map nodes and uses the motor vocabulary to associate basic actions to links. After that, navigating through the map was considered a higher level action, composed by basic motor movements.

It is important to stress that the learning process occurs throughout the entire action hierarchy: during the self-knowledge phase when the visuo-motor map was defined; learning through imitation when basic actions are encoded in motor space and, finally, executing a desired task by composing more complex actions based on the elementary ones. This approach thus avoids the need to program every action or situation that robot must execute or encounter during its lifetime.

We consider that the main contributions of this work are multi-fold: (i) to make a robot able to rely on interaction with a user to learn sets of movements that were not programmed or defined a priori; (ii) to provide a natural manner of adapting the desired application and the way it is executed to the robot's motion repertoire; (iii) a methodology for open-ended robot learning and adaptation that can be applied to different robots, applications and environments. The sensory-motor maps and the created motor vocabulary are defined according to robot's body and sensory system. Whenever the application or environment changes, one can either teach the robot a new set of movements or modify the current motor vocabulary.

More specifically, we show that the use of omnidirectional images is beneficial both for egomotion estimation as well as for the mapping and navigation processes. Large fields of view help finding and tracking objects and allows more of the environment to be caught in just one image. Also, using motor representations to define the links in a topological map allows representing the environment in a motor way, adapted to the robot used for the task.

The results obtained are encouraging. Future work will focus on further extending these ideas and conducting tests with other robots and applications. We are also working on a way of learning the visuo-motor mapping instead of computing it explicitly through egomotion equations. We have already got some results using neural networks based on perceptrons Vassallo et al. (2004).

## Acknowledgements

This work is partly supported by CAPES - Coordenação de Aperfeiçoamento de Pessoal de Nível Superior via a cooperation between Instituto Superior Técnico, Lisbon - Portugal, and Universidade Federal do Espírito Santo, Vitória - Brazil; and also by CNPq - Conselho Nacional de Desenvolvimento Científico e Tecnológico, Brazil, through the project 472727/2003-6.

## 7. References

Baker, S. & Nayar, S. K. (1998). A theory of catadioptric image formation, *Proc. of 6th International Conference on Computer Vision - ICCV98*.

Billard, A. (2002). *Imitation in Animals and Artifacts*, The MIT Press, chapter 11 - Imitation: A Means to Enhance Learning of a Synthetic Protolanguage in Autonomous Robots, pp. 281–310.

Billard, A. & Hayes, G. (1997). Learning to communicate through imitation in autonomous robots, In Proc. of ICANN97 - 7th International Conference on Artificial Neural Networks.

Bruss, A. & Horn, B. K. P. (1983). Passive navigation, *Compuyer Vision, Graphics and Image Processing* **21**: 276–283.

Dautenhahn, K. (1995). Getting to know each other - artificial social intelligence for autonomous robots., *Robotics and Autonomous Systems* **16**: 333–356.

Dautenhahn, K. & Nehaniv, C. L. (eds) (2002). *Imitation in Animals and Artifacts*, The MIT Press.

Fadiga, L., Fogassi, L., Gallase, V. & Rizzolatti, G. (2000). Visuo-motor neurons: ambiguity of the discharge or 'motor' perception?, *Int. J. Psychophysiology* **35**: 165–177.

Fischler, M. A. & Bolles, R. C. (1981). Random sample consensus: A paradigm for model fitting with applications to image analysis and automated cartography., *Communications of ACM* **24**(6): 381–395.

Fitzpatrick, P., Metta, G., Natale, L., Rao, S. & Sandini, G. (2003). Learning about objects through action - initial steps towards artificial cognition, *IEEE International Conference on Robotics and Automation - ICRA 2003.*

Gallese, V., Fadiga, L., Fogassi, L. & Rizzolatti, G. (1996). Action recognition in the premotor cortex, *Brain* **119**: 593–609.

Gaspar, J., Grossmann, E. & Santos-Victor, J. (2001). Interactive reconstruction from an omnidirectional image, In Proc. of the 9th International Symposium on Intelligent Robotic Systems - SIRS2001.

Geyer, C. & Daniilidis, K. (2000a). Equivalence of catadioptric projections and mappings of the sphere, Proc. of IEEE Workshops on Omnidirectional Vision.

Geyer, C. & Daniilidis, K. (2000b). A unifying theory for central panoramic systems and practical implications, *ECCV 2000*, Vol. 2, pp. 445–461.

Gluckman, J. & Nayar, S. K. (1998). Ego-motion and omnidirectional cameras, *Proc. of 6th International Conference on Computer Vision - ICCV98*, pp. 999–1005.

Lopes, M. C. & Santos-Victor, J. (2003a). Motor representations for hand gesture recognition and imitation, *IROS Workshop on Robot Programming by Demonstration.*

Lopes, M. C. & Santos-Victor, J. (2003b). Visual transformations in gesture imitation: what you see is what you do, *IEEE Intl. Conference on Robotics and Automation - ICRA2003.*

Lucas, B. & Kanade, T. (1981). An iterative image resgistration technique with an application to stereo vision., IJCAI'81.

Lungarella, M. & Pfeifer, R. (2001). Robot as cognitive tools: Information theoretic analysis of sensory(-motor) data, *EDEC 2001 - Symposium at the International Conference on Cognitive Science.*

Matáric, M. J. (2002). *Imitation in Animals and Artifacts*, The MIT Press, chapter Sensory-Motor Primitives as a Basis for Imitation: Linking Perception to Action and Biology to Robotics, pp. 391–422.

Metta, G., Sandini, G. & Natale, L. (2001). Sensorimotor interaction in a developing robot, *1st Intl. Workshop on Epigenetic Robotics: Modeling Cognitive Development in Robotic Systems*, pp. 18–19.

Metta, G., Sandini, G., Natale, L., Manzotti, R. & Panerai, F. (2001). Development in artificial systems, *EDEC 2001 - Symposium at the International Conference on Cognitive Science.*

Nelson, R. & Aloimonos, J. (1988). Finding motion parameters from spherical motion fields (or the advanntages of having eyes in the back of your head), *Biological Cybernetics* **58**: 261–273.

Rizzolatti, G. & Arbib, M. A. (1998). Language within our grasp, *Trends in Neurosciences* **21**(5): 188–194.

Rizzolatti, G., Fadiga, L., Fogassi, L. & Gallese, V. (2002). *The Imitative Mind: development, evolution, and brain bases*, Cambridge University Press, chapter 14 - From mirror neurons to imitation: facts and speculations, pp. 247–266.

Vassallo, R. F., Santos-Victor, J. & Schneebeli, H. (2002a). A general approach for egomotion estimation with omnidirectional images, *IEEE Workshop on Omnidirectional Vision 2002 (with ECCV)*, pp. 97–103.

Vassallo, R. F., Santos-Victor, J. & Schneebeli, H. J. (2004). Task learning and adaptation based on visuo-motor mapping, *CBA 2004 - XV Congresso da Sociedade Brasileira de Automática*.

Vassallo, R., Santos-Victor, J. & Schneebeli, H. (2002b). Using motor representations for topological mapping and navigation, *Intl. Conference on Intelligent Robots and Systems - IROS2002*, pp. 478–483.

# A mechatronic description of an autonomous mobile robot for agricultural tasks in greenhouses

Julián Sánchez-Hermosilla[1], Francisco Rodríguez[2], Ramón González[2], José Luís Guzmán[2] and Manuel Berenguel[2]
[1]*Departamento de Ingeniería Rural,* [2]*Departamento de Lenguajes y Computación, University of Almería, Spain*

## 1. Introduction

Today, greenhouses are one of the main productive sectors of many areas of the world (Spain, the Netherlands, Australia, Morocco, etc.). One extraordinary area is the province of Almeria (SE Spain), where production exceeds 2.7 million tn on a surface area of 48000 ha of greenhouses (Cámara Oficial de Comercio, 2007). This constitutes not only economic wealth but also a source of research and innovation.

In recent years, greenhouse techniques have undergone many improvements in irrigation processes, phytosanitary treatments, and climatic-control systems (van Henten, 2006). Nevertheless, many agricultural tasks are conducted manually, such as harvesting, spraying and farming. The environmental conditions inside greenhouses, characterized by high temperatures and humidity, make work harsh and sometimes hazardous for people, especially in applying toxins (pesticides) with little air renewal. This has recently given rise to the development of different automatic machinery to perform greenhouse work.

Some projects and references related to the application of robots in greenhouses have appeared in the literature. On one hand, manipulator robots, usually being controlled by vision systems, have been successfully tested (Sandini et al., 1990) and (Dario et al., 1994) presented the *Agrobot Project*, a robotic system for greenhouse cultivation of tomatoes. This involved a mobile robot with a colour stereoscopic vision system plus an anthropomorphic arm with a gripper/hand and six degrees of freedom. (Acaccia et al., 2003) described a robotic system (robotic arm and mobile platform) which was able to analyse the plant to evaluate its state of health. (Kitamura and Oka, 2005) presented a robotic system based on vision for navigation control, composed of a cutting system to harvest sweet peppers in greenhouses. (Belforte et al., 2006) presented a fixed-position robot. This was interfaced to a standard belt-conveyor displacement system that provided the robot with pallets containing the crops. The main drawback of this solution was that since the robot remained in a fixed position within a restricted workspace and thus had limited applications.

Another solution is to use automated guided vehicles (AGVs). These vehicles follow a trail fixed on the ground of the greenhouse. (Sammons et al., 2005) described an autonomous spraying robot with navigation control based on inductive sensors which detect metal pipes buried in the soil. (Van Henten et al., 2002) presented an autonomous robot for harvesting cucumbers in greenhouses; it was guided using heating steel pipes. The disadvantage of these types of vehicles is that they require an extensive and costly modification of the greenhouse. Few papers have addressed the autonomous navigation problem of a mobile robot in greenhouses. (Mandow et al., 1996) described an autonomous vehicle (*Aurora*) for spraying tasks. The navigation control of this robot depends on a previous sequence of behaviour established by an operator. (Subramanian et al., 2005) and (Singh et al., 2005) also described a mini-robot to perform spraying activities, for which navigation is controlled by algorithms based on fuzzy logic. The sensorial system uses vision and ladar (laser + radar) sensors. The main drawback of these two autonomous robots is that they have reduced dimensions and a limited power system. For these reasons, they can operate only with a small payload and over small distances. For a complete review of robotic systems in agriculture, see (Kondo and Ting, 1998) and the references therein.

This article presents a project developed at the University of Almeria (Spain) aimed at designing, implementation, and testing a multi-use autonomous vehicle with safe, efficient, and economic operation which moves through the crop lines of a greenhouse and which performs tasks that are tedious and/or hazardous for people, it is called *Fitorobot.* First, it has been equipped for spraying activities, but other configurations have also been designed, such as: a lifting platform to reach high zones to perform tasks (staking, cleaning leaves, harvesting, manual pollination, etc.), and a forklift to transport and raise heavy materials (Sánchez-Gimeno et al., 2006).

The mobile robot developed to operate in greenhouses involves a synergetic integration of mechanical engineering with electronics and automatic control. From the first step of construction to the final tests in greenhouses, all the processes were supervised by a mechanical, electronic, and information technology combination (Isermann, 2003), (Bishop, 2006).

Regarding a mobile platform used to carry on a spraying system, there are some circumstances where it is impossible to maintain a constant velocity due to the irregularities of the soil, different slopes of the ground, and the turning movements between the crop lines. Thus, for work at a variable velocity (Guzmán et al., 2008), it is necessary to spray using a variable-pressure system based on the vehicle velocity, which is the proposal adopted and implemented in this work. This system presents some advantages, such as the higher quality of the process, because the product sprayed over each plant is optimal. Furthermore, this system saves chemical products because an optimal quantity is sprayed, reducing the environmental impact and pollution as the volume sprayed to the air is minimized.

This chapter is organized as follows. An overall mechatronic description is outlined in Section 2, presenting the mechanical, electronic, sensor, and hardware systems. Section 3 examines the control architecture, the navigation control, and the low-level controllers or servocontrollers. Experimental results of sensors, servocontrollers and navigation are reported in Section 4. Conclusions and future works are discussed in Section 5.

## 2. Mechatronic Description

The mobile robot presented in this work has been designed and built following the paradigm of Mechatronics. According to (Bolton, 2003) a mechatronic system is not just a combination of electrical and mechanical systems and is more than just a control system; it is a complete integration of all of them. For these reasons this project has been supported by engineers of different areas of specialization (Mechanics, Robotics, Automatic Control, Agronomy, Computer Science, and Electronics).

Fig. 1 shows the Mechatronic decomposition of the steps followed:

- *Mechanical systems*: CAD/CAE tools have been employed to design the prototype. The design took into account several requirements, for example the environment conditions, the appropriate position of sensors on the platform, the position of the control system, the maximum desired velocity, the range of pressure of the spraying system, etc. Then the prototype was built and assembled using the CAD planes designed in the first stage.
- *Electronic systems:* Some sensor systems were evaluated, and the most appropriate were acquired. Furthermore, one computer was selected to run the control programs that autonomously govern the vehicle. Coupled with the computer, appropriate input/output cards receive/send commands from/to sensors/actuators.
- *Information technology*: Simultaneously to the previous phases, autonomous navigation strategies and spraying controllers were analysed and studied. These control structures were tested and calibrated when the real vehicle was built.

As detailed in Fig. 1 these three areas are linked. Information technology is related to Mechanics because the design of the mechanical structure of the vehicle has been realized using CAD/CAE tools. This design has taken into account the features and requirements of the physical elements. On the other hand, Information technology is also related to Electronics, because sensors and electronic actuators are required by the controllers implemented in the computer, in order to feedback to the control algorithms and to send the appropriate control signals to the actuators. Finally, Electronics and Mechanics are linked through electromechanical elements such as electronic valves in the track motors, electronic systems to control the hydraulic pressure system, etc. Furthermore, sensors are used to measure these electromechanical systems.

### 2.1 Mechanical Systems

In the present work, the vehicle works primarily with a spray system but includes such configurations as: a lifting platform (Fig. 2a) to reach high zones to perform special tasks (staking, cleaning leaves, harvesting, manual pollination, etc.), and a forklift (Fig. 2b) to transport and raise heavy materials (Sánchez-Hermosilla et al., 2003), (Sánchez-Gimeno et al., 2006).

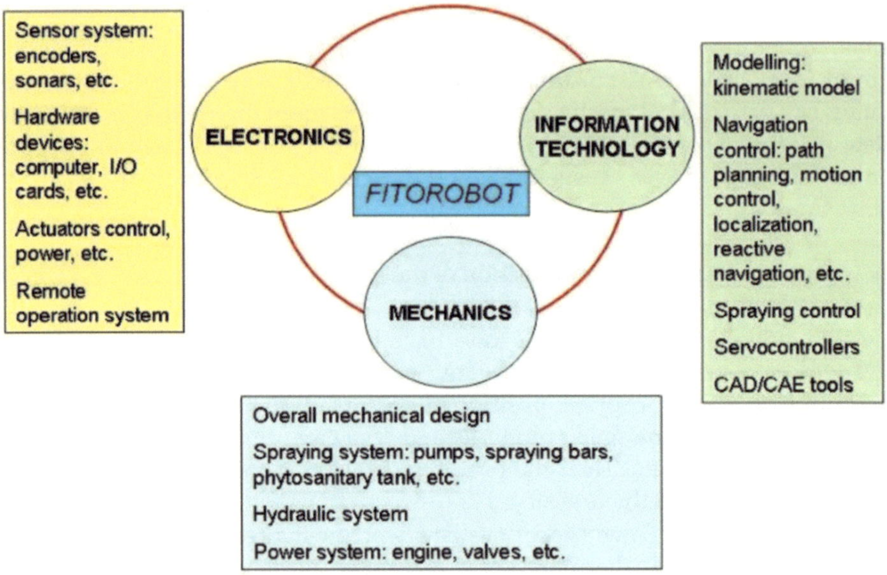


Fig. 1. Mechatronics paradigm followed to develop the *Fitorobot*

The configuration and the position of the greenhouses impose limits to the mechanical structure of the platform:

- It is necessary for the vehicle to be compact (small dimensions) with high maneuverability in an environment having numerous obstacles and tight spaces.
- The vehicle should have a roller system to guarantee mobility on loose soils and to reduce soil compaction that might harm crop development. Compaction is one of the most important problems to solve in modern agriculture, many studies have determined the negative effect of soil compaction on crop performance (e.g. Kirkegaard et al., 1992, Radford et al., 2001, Hamza and Anderson, 2003, Hamza and Anderson, 2005, Sadras et al., 2005, Chan et al., 2006).
- The vehicle should be autonomous and have sufficient charge capacity for optimal work performance.
- The vehicle should have good flexibility to adapt the work velocity to the requirements of each task.

Satisfying these characteristics, a rubber-track vehicle with differential steering was developed for agricultural tasks in greenhouses. The rubber tracks exert low pressures on the soil (Brown et al., 1992), providing strong traction on uneven ground (Bashford et al., 1999).

CAD/CAE technology was used to design the mechanics of the vehicle, adapting previously established morphology and minimizing volumes and weights (taking into account the manufacturing materials), and optimising the disposition of the elements in its interior. For these tasks the CAD/CAE tool Solidworks (Solidworks Corp., Massachusetts, USA) was used.

(a) Lifting platform.

(b) Forklift.

Fig. 2. Other configurations of the *Fitorobot*

The dimensions of the vehicle were fixed at 0.7 m wide and 1.7 m long. This ensured movement through the crop rows, which in full development leave a free path of approximately 0.8 cm, and guaranteed the possibility of turning in the greenhouse lanes, which are normally 2 m wide. A load capacity of 400 kg was established to ensure autonomy and proper work performance.

The rubber tracks were powered with hydraulic fixed-displacement gear motors (TF0204, Parker Hannifin Corp., Cleveland, USA) that were placed in the wheel sprockets. The hydraulic motors are driven by a double variable-displacement axial piston pump (M4PV21, Bondioli & Pavesi, Bologna, Italy) with electronic proportional control. Pump displacement is proportional to the electric current feeding one of the two proportional control electrovalves.

All the energy needed for the vehicle to function comes from a petrol-combustion motor. A 20 HP air-cooled 4-stroke petrol engine (Honda GX 620 K1, Honda Motor Co., Ltd., Tokyo, Japan) was selected, taking into account the vehicle weight (800 kg loaded), a maximum velocity of 2 m/s, under the assumption that the vehicle will be travelling on a sandy surface at a slope of 5°. It was also assumed that the vehicle would have a fixed-volume pump to provide energy for the drive of the attached implement (e.g., sprayer pump of the spraying system).
A maximum velocity of 2 m/s has been considered because this is the appropriate velocity to spray a typical greenhouse with the current tank available in the mobile robot (Guzmán et al., 2004) and this relatively slow velocity permits partially to neglect lateral-slip phenomena (González et al., 2009b).

Steering is accomplished by changing the velocities of the hydraulic gear motors, achieving turning radii of nearly zero. For straight trajectories, the two motors turn at the same velocity. The turn is achieved by reducing the velocity of one of the motors with respect to

the other. Under situations of limited space, turning radii close to zero can be achieved by making the two motors turn in opposite directions.
A chassis was designed to joint the different elements that form part of the vehicle and that should support the different actions that are generated during the functioning. It is constituted by different pieces in 8-mm steel sheeting (Fig. 3a). On the lower part it fits into the rubber tread which makes up the roller system. The upper part is equipped with a system to transport and connecting accessory equipment, characterized by rectangular guides and self-locking pins. Furthermore, in order to study the robustness and stability of the chassis, a finite element analysis (FEA) of the chassis was made (Fig. 3b).

(a) Front view of the chassis

(b) FEA analysis of the chassis

Fig. 3. Chassis of the vehicle

The rest of the main elements making up the vehicle are indicated in Fig. 4. This figure also includes the spraying system designed for the vehicle, composed of:

- A 300-litre fibreglass-reinforced polyester resin tank (92x61x60 mm)
- A 26.75 l/min spray tank (Comet MP30, COMET S.p.A. Reggio Emilia, Italy), 15 bar maximum pressure.
- A vertical boom sprayer with ten nozzles (Teejet DG 9502 EVS, Spraying Systems, Co., Wheaton, Ill.).
- A frame made of hollow rectangular steel tubes of 30x30x1 mm

The arrangement of the different elements gives the vehicle a wide zone to transport equipment; the connection and disconnection of the equipment proves easy, as the guides of the transport system are situated near the soil level (approx. 50cm).

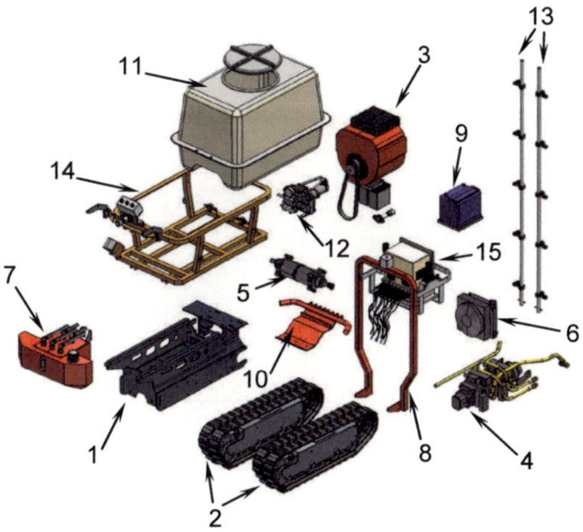


1. Chassis
2. Rubber tracks
3. Petrol engine
4. Double variable-displacement axial piston pump
5. Hydraulic gear motors
6. Hydraulic heat exchangers
7. Oil tank
8. Security frame
9. Battery
10. Back protective frame
11. Phytosanitary tank
12. Spraying Puma
13. Spraying bars
14. Frame-system spraying
15. Control system

Fig. 4. Exploded drawing of the prototype

## 2.2 Electronic Systems

### 2.2.1. Sensors

Autonomous control and operation of the mobile robot relies heavily on external sensor information. Therefore, the performance of the navigation and spraying controller will depend heavily on the installed sensors on the platform. For this, several sensors are installed on the platform; some sensors are redundant for the purpose of testing different configurations, and thus, in a future commercial version of the mobile robot, only the most useful and appropriate sensors will be installed. The position of each sensor has also been studied, in order to determine the best location of the sensors depending on the mechanical structure and the environment.

One middle-range sonar and four short-range sonars have been located in front and in each side of the platform, respectively. These sensors enable to the robot to sense the environment and the greenhouse corridors. Odometry establishes the position and velocity of the robot, using two incremental optical encoders attached to the axle of rotation of the track motors. One radar and one magnetic compass measure the linear velocity and orientation of the vehicle, respectively. It is protected from unexpected obstacles in the environment by a security sensor composed of four tactile bars around the vehicle. Finally, a pressure sensor has been installed in the spraying hydraulic system to regulate the spraying controllers. The main features of these sensors are summarized in Table 1 and their positions on the platform are shown in Fig. 5.

| Sensor | Mark, Model | Range | Precision / Resolution | Signal Output |
|---|---|---|---|---|
| Pressure | WIKA, ECO-1 | 0-250 bar | 0.50% | 0-10 VDC |
| Sonar (middle dist.) | Siemens, Bero III | 40-300 cm | 2.5 % | 0-10 VDC |
| Sonar (short dist.) | Siemens, Bero M18 | 15-100 cm | 1.5 % | 0-10 VDC |
| Magnetic Compass | KVH, C100 | 0-360° | 0.1° | 0.1-1.9 VDC |
| Radar | LH Agro, Compact II | 0.08-17.3 m/s | 128.4 pulse/m | Pulse |
| Encoders | Sick, DRS61 | 0-360° | 1024 pulse /rev. | Pulse |
| Security sensor | SafeWork, SKL25-40 | On-off | - | Digital (0-1) |

Table 1. Features of the installed sensors.

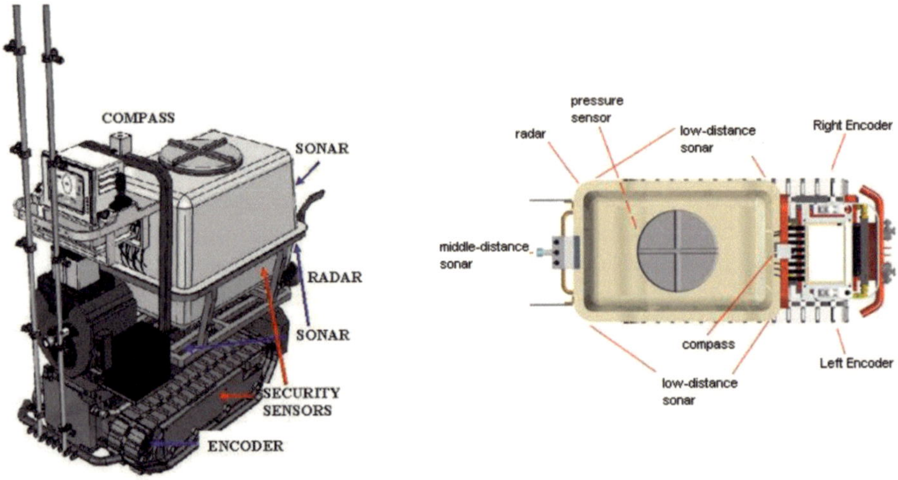


(a) Lateral view of the sensorial system (b) Top view of the sensorial system

Fig. 5. Sensors on the experimental platform

### 2.2.2 Hardware devices

The mobile robot is governed by an embedded system based on the PC-104 standard (PC-104 Embedded Consortium, 2008) (PCM-9371, Advantech, Irvine, USA). This system is based on the ISA specifications to work in embedded devices. The main reasons of why PC-104 system has been selected are: It has reduced dimensions; low power supply and, ease to integrate new input/output cards. The main characteristics of our PC-104 are summarized in Table 2.

| Feature | Description |
|---|---|
| CPU | Intel Pentium III - 933 MHz |
| Memory | Compact Flash of 1 GB |
| Cache | 512 KB |
| Chipset | VIA PN133T (133 MHz FSB) |
| Interfaces | USB, Ethernet, Audio, Serial, PS2 |
| Graphic Card | VIA VT8606 |
| Power | 12 VDC |
| Temperature | 0 - 60° |

Table 2. Main characteristics of PC-104

Some input/output digital/analogue boards have been connected to the PC-104 bus, so that the signals of the different sensors onboard the vehicle can be read and commands can be sent to the actuators of the vehicle. An analog input board (PCM-3718H, Advantech, Irvine, USA) with digital outputs is used for sonar, compass, and pressure sensing; furthermore, the digital output data determine the rotation of the tracks (forward or backward); a counter board (PC104-3126, Nagasaki IPC, Taiwan) with digital inputs governs the security sensors, encoders and radar; two analog outputs boards (PCM-3712, Advantech, Irvine, USA) are used for the track engines and spaying valves. The composition of the system is shown in Fig. 6.

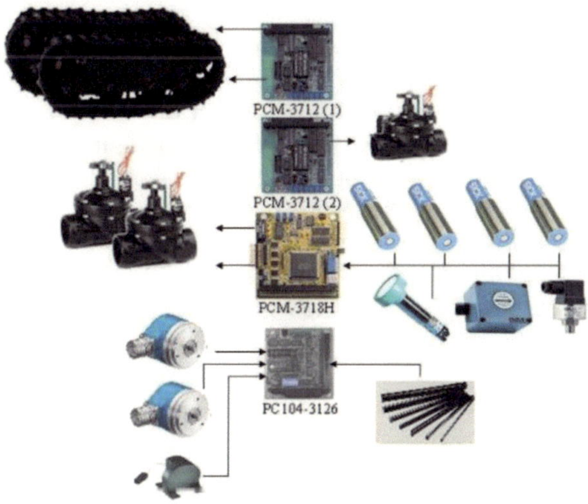


Fig. 6. Input/Output boards

The system is completed by a tactile screen (7" LCD Screen, Nagasaki IPC, Taiwan) connected to the PC-104. This monitor permits to the user to define the parameters and options of the programs to make the desired tests. Fig. 7 shows a real image of the hardware system.

The software used to develop the control and supervision programs is LabVIEW® of National Instruments® and Matlab® of MathWorks®. These programs were selected for their simplicity and familiarity with program control algorithms, as well as their appropriate connection with real systems: input/output boards, engines, etc. Furthermore, they have appropriate graphic interfaces, facilitating the use of their programs.

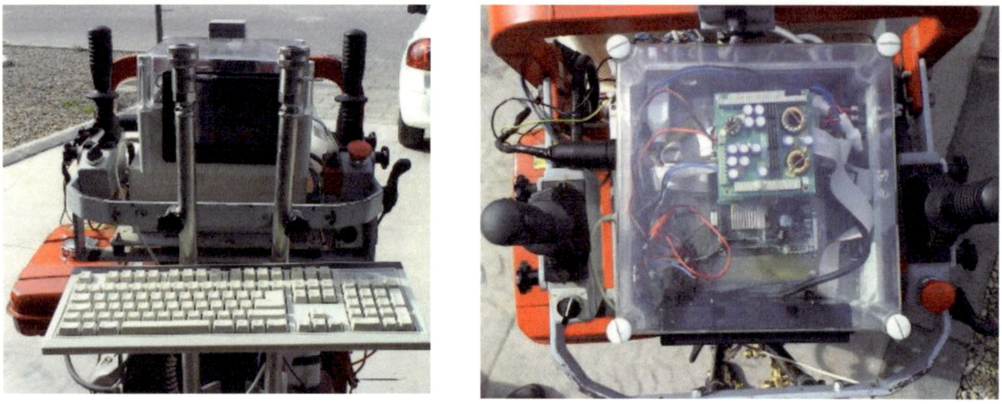

(a) Front view (b) Top view

Fig. 7. PC-104 on *Fitorobot*

#### 2.2.3 Remote operation system

The mobile robot *Fitorobot* can also be controlled remotely by a teleoperation system in complex situations. The remote operation system is composed of two segments (Fig. 8a). One segment is the user keypad or user interface that provides to the operator the remote control of the vehicle's position and velocity (Fig. 8b). The operator can also regulate the pressure of the spraying system and open/close the spraying bars. The keypad has a selection switch that establishes the velocity program (maximum velocity, joysticks functioning, etc.) and the working pressure. Data communications are implemented via an RS-232 serial link, half-duplex radio modem operating at the free band of 868 MHz. The second part is defined by a program running in the computer installed on the mobile robot. This program interprets the messages sent by the user keypad and sends the appropriate signals to the actuators.

### 2.3 Spraying Systems

As commented above, the mobile robot is equipped with some spraying devices for autonomous spraying tasks to be controlled by programs running in the main computer. As shown in Fig. 9, the spraying system is composed of a tank containing the chemical treatment, vertical bars with several nozzles, two on/off electrovalves to activate each spraying bar, a proportional electrovalve to regulate the output pressure, and a pressure sensor to close the control loop. This arrangement of the bars and the nozzles was established in previous studies (Guzmán et al., 2004; Guzmán et al., 2008).

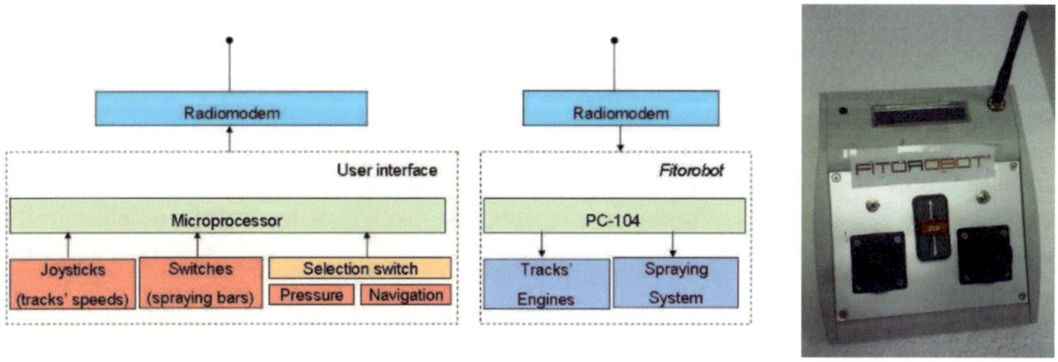


(a) Scheme of the remote operation control (b) User keypad

Fig. 8. Remote operation system

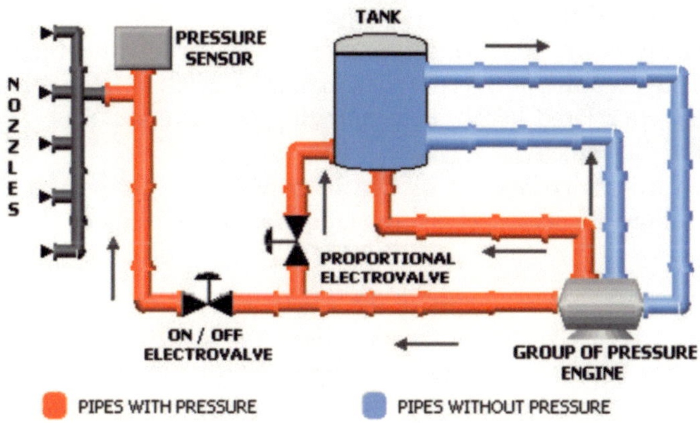


Fig. 9. Spraying system used to control the output pressure

## 3. Control System

### 3.1 Modes of operation

The vehicle can be operated on three different modes: autonomous control, remote operation, and manual control. As commented above, remote human operation is performed by radio communication. The platform can be manually controlled using two joysticks directly connected to the two valves that move each track. In the case of autonomous control, the motion of the vehicle is governed by controllers running on a computer.

The control architecture of the mobile robot is studied in (González et al., 2009a) and it is detailed in (Fig. 10). Greenhouses are built using digital maps, which can be used by robotic navigation algorithms to control and steer the vehicle. Deliberative approaches are based on these maps to construct an obstacle-free path which will act as the reference path that the robot will follow in real time. On the other hand, reactive navigation strategies focus on controlling the robot's trajectory using information provided by its sensors during robot motion. One navigation algorithm of each technique has been implemented in the mobile robot *Fitorobot.* The main results are discussed in (González et al., 2009a).

The architecture follows a hierarchical decomposition on five layers (Fig. 10). The function of each layer is the following:

- The *user layer* is composed of the interface by which the user can calibrate the program; for example, selecting between reactive or deliberative navigation algorithms.
- The *path-planning level* generates path following from the map information (this map will be built using the reactive algorithm) and gives instructions to the motion-control layer and spraying controller.
- The *motion control level* gives the setpoints to the servo layer depending on the current position of the vehicle. Furthermore, it sends the setpoints to the spraying controller, depending of the vehicle's velocity. Additionally, in this layer a new filter has been added for the sonar readings.
- The *spraying controller level* consists on an appropriate robust controller which controls the pressure of the bars (Guzmán et al, 2008). Motion control and spraying controller are related because the spraying pressure should be appropriate to the velocity of the vehicle.
- The *servo level* is composed of two PI controllers that ensure that the actuators (tracks) follow the setpoints given by the motion-control level.

### 3.2 Path-planning level

Greenhouses are structured environments where the distribution of plants is at least partially known. The plants are usually arranged in parallel straight rows with narrow corridors for the operation of humans and machines. The navigation challenge for a robot operating in a greenhouse involves planning a reference trajectory and reacting to unforeseen events (workers, boxes, tools, etc).

In order to solve this problem, a hybrid solution has been developed, using the most dominant paradigms for robot control (Fig. 10) (González et al., 2009a):

- *Deliberative strategies.* This technique uses a map to calculate an obstacle-free path of the greenhouse before the robot moves along it. In this work, the selected path-planning algorithm is a modified Voronoi Diagram, obtaining a graph with all the possible paths in the greenhouse without obstacles. Based on the task that the robot must do (spraying, transport, etc.), one of this paths is selected. Once the path is chosen, a navigation control strategy is used to follow it. The navigation control strategy used in this work will be described in the next section.
- *Reactive strategies.* This technique focus on control of the robot's trajectory using information provided by its sensors during robot motion. As previously mentioned, a greenhouse is a structured environment and therefore a priori knowledge can be added to the reactive navigation strategy. For example, when the mobile robot is at the end of a corridor, it could turn to the left or to the right side. This decision will be made relying on the sensorial system and on the previous turn. The key idea is that the second turn will be to the same side as the

previous one. When it has turned two consecutive times, the next two turns will be to the opposite side. This can be seen as a zigzag movement, but taking into account that the robot can react to unexpected events and to different configurations of greenhouses where zigzag is not appropriate. Really, this technique should be considered pseudo-reactive, because the navigation algorithm has been implemented knowing that there are parallel corridors arranged in rows in the greenhouses.

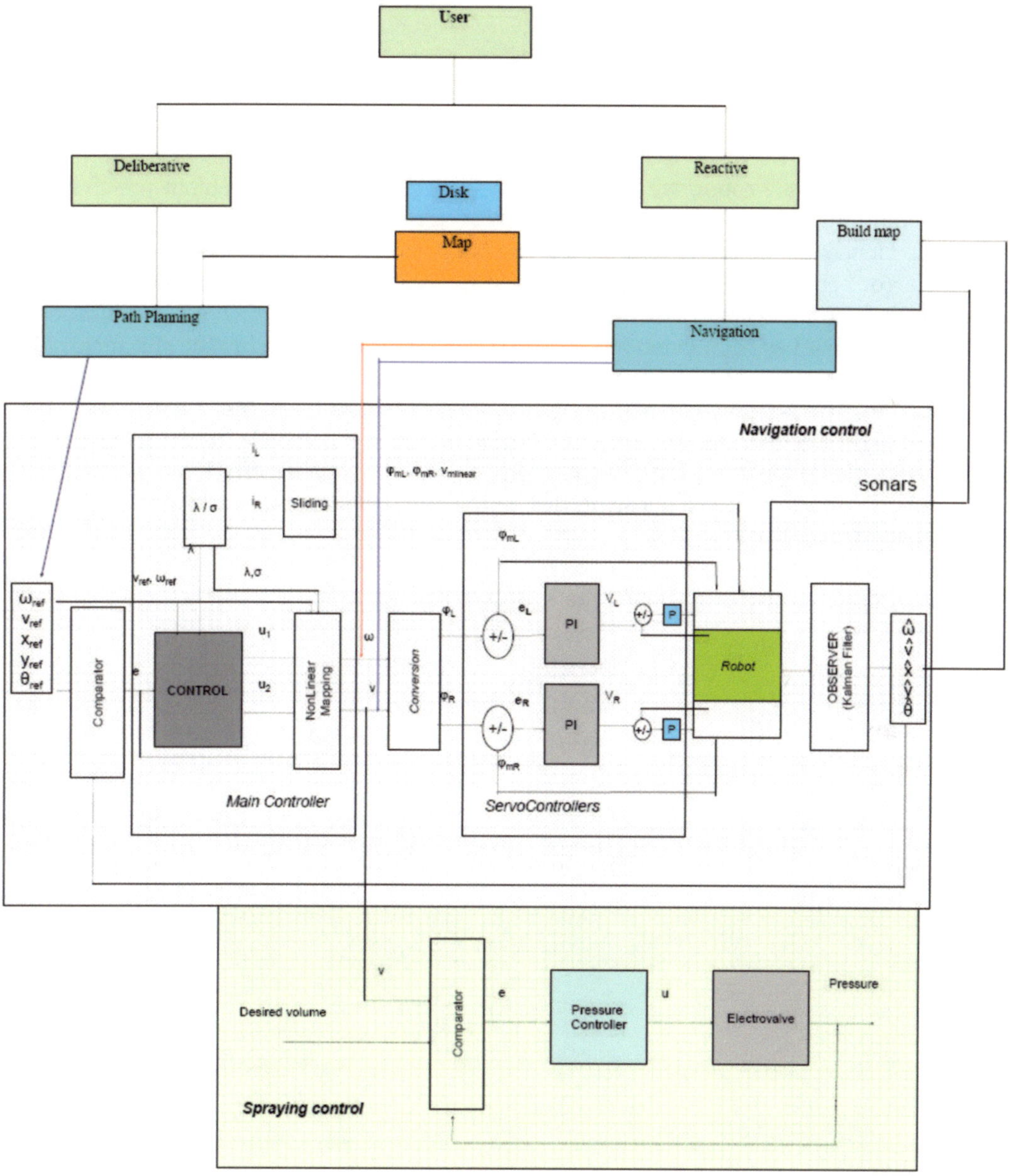


Fig. 10. Motion-control architecture

Therefore, the first time that the robot navigates the greenhouse if a map exists, it is employed by a deliberative method. On the other hand, when there is no map, a pseudo-reactive method is used. Moreover, a sensorial map is built along the path to be employed by the deliberative module in later runs.

The two previous approaches utilize a security layer to avoid collisions. The main obstacle to the movement of mobile robotics in greenhouses is related to the fact that navigation algorithms should take into account unexpected events (humans working in the greenhouse). This layer uses on/off sensors.

### 3.3 Navigation control

As discussed above, when a deliberative strategy is used and a fixed path is selected, a control strategy must be used to follow this path. In order to address this navigation control problem of the mobile robot, we have used the trajectory tracking problem which consists that a robot must follow a *virtual mobile robot* representing the desired positions and velocities. Hence, the objective is to find a feedback control law (Canudas et al., 1997), such that, the error between the desired location and the real location of the mobile robot is close to zero (regulation problem). For that purpose, the kinematic model (KM) of a differential-drive mobile robot has been extended with a parameter which weights the slip factor of the terrain (González et al., 2009b). In our case, we suppose that the mobile robot will operate at low velocities, and we only consider longitudinal slip. As stated in (Le, 1999), lateral slip is zero for straight line motions and it can be neglected when the vehicle turns "on the spot" or at low velocities. However, longitudinal slip is an unavoidable effect of pneumatic tire compression / reaction to soil shearing due to the own characteristics of wheeled / tracked locomotion (Wong, 2001).

As detailed in (González et al, 2009b) the KM under longitudinal slip can be expressed as

$$\begin{bmatrix} \dot{x} \\ \dot{y} \\ \dot{\theta} \end{bmatrix} = \begin{bmatrix} \cos\theta & 0 \\ \sin\theta & 0 \\ 0 & 1 \end{bmatrix} \begin{bmatrix} v_m \\ \omega_m \end{bmatrix} - \begin{bmatrix} \cos\theta & 0 \\ \sin\theta & 0 \\ 0 & 1 \end{bmatrix} \begin{bmatrix} \sigma \\ \lambda \end{bmatrix}, \qquad (1)$$

where (x, y) is the position of the robot, $\theta$ is the orientation, $v_m$ is the linear or translational velocity, $\omega_m$ is the angular or rotational velocity of the vehicle, and $\sigma$ and $\lambda$ are terms depending of slip,

$$\sigma = \frac{v_r i_r + v_l i_l}{2}, \qquad (2)$$

$$\lambda = \frac{v_r i_r - v_l i_l}{b}, \qquad (3)$$

where $v_r$ and $v_l$ are the linear velocities of the right and left track, respectively; $i_r$ and $i_l$ are the longitudinal slip of the right and left track, respectively, and b is the distance between the track centers.
This formulation of the kinematic model is used into the trajectory tracking problem (González et al., 2009b) to defined a new control law which is based on the work of (Canudas et al., 1997) but now including the additional terms compensating the slip effects (which affect to the tracked vehicles in sandy soils),

$$\begin{aligned} u_1 &= -k_1 e_x, \\ u_2 &= -k_2 sign(v_{ref}) e_y - k_3 e_\theta. \end{aligned} \tag{4}$$

This control law tries to reduce the error in the driving or forward direction $(e_x)$ using a gain $(k_1)$, the orientation error $(e_\theta)$ can be efficiently manipulated using gain $k_3$, and finally, the error orthogonal to the driving direction can be reduced using the gain $k_2$. The controller gains $(k_1, k_2, k_3)$ are determined using the procedure described in (Canudas et al., 1997) and adding new parameters to compensate for the slip effects, as is discussed in (González et al. , 2009b), it holds

$$\begin{aligned} k_1 &= 2\delta_c(\omega_{ref}^2 + \beta_c v_{ref}^2)^{\frac{1}{2}}, \\ k_2 &= \beta_c \,|\, v_{ref} \,|+ \omega_{ref}\lambda, \\ k_3 &= 2\delta_c(\omega_{ref}^2 + \beta_c v_{ref}^2)^{\frac{1}{2}}, \end{aligned} \tag{5}$$

where $\beta_c > 0$ and $\delta_c > 0$ are constants that are used to determine the desired closed-loop behaviour of the system; $v_{ref}$ and $\omega_{ref}$ are the linear and angular reference velocities, respectively.

### 3.4 Servocontrollers

PI controllers are used at the lowest level; the objective is that the main controller gives the setpoints to the low-level controllers. The function of these PI controllers is to control the valves which regulate the track motors. Due to saturation of the actuators, an anti-windup strategy has been implemented (Åström and Murray, 2008). As a means of determining the parameters of the controllers, a step-response test was made to characterize the valves as well as to obtain the empirical model of the valves. Then, the method of cancellation of poles (Åström and Murray, 2008) was used for tuning the PI parameters.

### 3.5 Spraying Controllers

A robust control system aimed at regulating the output pressure of the spraying system has been implemented (Guzmán et al., 2004; Guzmán et al., 2008). The pressure setpoints are calculated based on the actual velocity of the vehicle at each sampling time and on a predefined volume of pesticides to apply based on the crop-growth state, and are then used in a feedback loop for pressure control, where the controller accounts for uncertainty in the system. Quantitative Feedback Theory (QFT) is used as a robust control technique to cope with system uncertainties. This is a methodology to design robust controllers based on frequency domain (Horowitz, 2001), enabling the design of robust controllers which fulfil some minimum quantitative specifications considering the presence of uncertainties in the plant model and the existence of perturbations. With this theory, the final aim of any control design must be to establish an open-loop transfer function with the suitable bandwidth in order to sensitize the plant and reduce the perturbations (Guzmán et al., 2008). As shown in Fig. 11, the spraying-control system is divided into two steps. In the first one, the pressure reference is calculated based on the vehicle velocity and the desired spray volume (algebraic relationship). The second phase consists of performing the necessary actions to control the pressure in order to reach the desired set point in a robust way. The typical profile for the pressure references calculated in the first stage is given by combinations of ramps. When the robot starts to move or to break, the pressure must go up or down, respectively, and while the robot velocity remains constant the pressure set point also does (Guzmán et al., 2008).

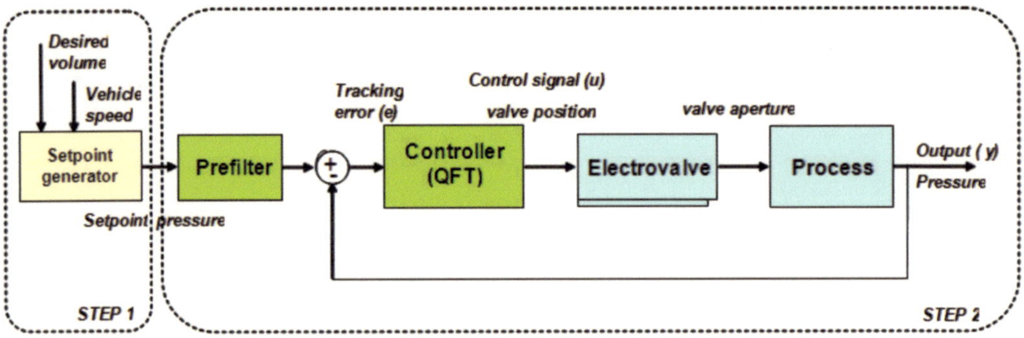


Fig. 11. Full spraying-control system taking into account the velocity and volume

## 4. Results and Discussion

Before the navigation techniques were attempted, several experiments were made to calibrate and test the sensor system and the low-level or servocontrollers. Later, two navigation strategies were tested in a real greenhouse (deliberative and reactive navigation strategies). Afterwards, the motion controller with slip compensation was also tested through simulations. A comparative with the original controller is discussed.

### 4.1 Sensor system validation

Several experiments were carried out to test the sensor system, and sonars were checked to obtain a right filter. These tests allowed us to design and calibrate an appropriate filter. As shown in Fig. 12a the sonar readings are homogeneous on greenhouse wall (plastic film).
On the other hand, when the sonar works on a greenhouse corridor their measurements are very heterogeneous due to the irregularity in the shape of plants. Fig. 12b shows the good performance of the implemented filter. It is advised that raw data presents multiple peaks due to the previously discussed holes in the plants. The implemented filter cleans these undesirable peaks and attenuates the rest of the rough data.

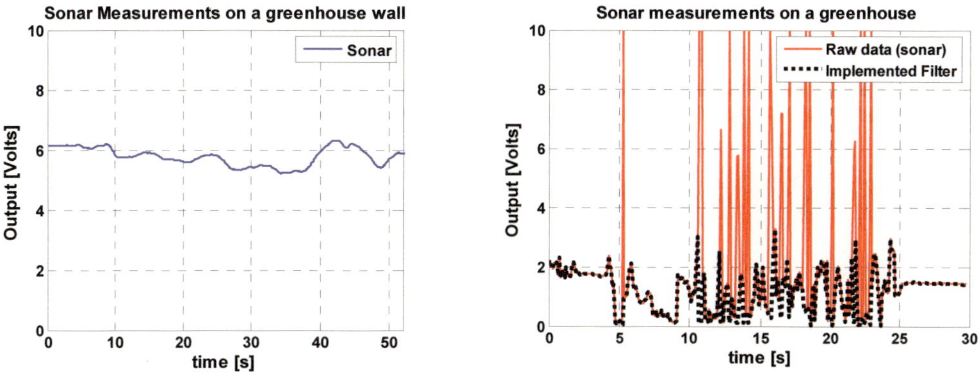


(a) Measurements of an ultrasound sensor on a greenhouse wall

(b) Measurements and filter of an ultrasound sensor on a greenhouse corridor

Fig. 12. Tests for sonars to check the response of the implemented filter

Additionally, some compass experiments were made. As example, Fig. 13a shows a test in which two consecutive 90° turns were made. It is possible to see the low noise of this sensor. In the same experiment the measurements of encoders and radar were recorded (Fig. 13b). This figure shows the effect of slip, because the velocity measured with radar is less than that of the encoders, and the difference determines the slip online (González et al., 2009b).

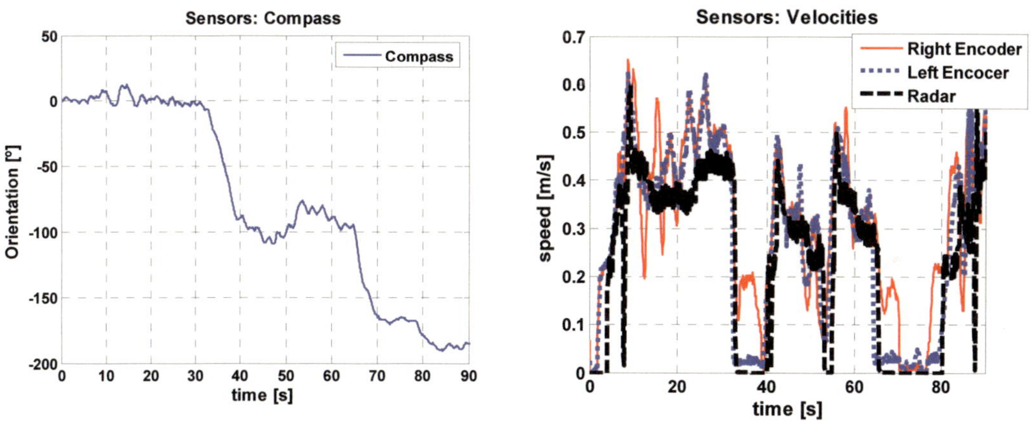


(a) Measurements of the compass

(b) Measurements of velocity sensors

Fig. 13. Real tests of the sensors on the mobile platform

### 4.2 Experimental tests of the Servocontrollers

The low-level controllers or servocontrollers have also been tested in several experiments. In this case, we discussed two experiments to check the single PI and the PI with the anti-windup improvement. Experiments were sampled at 50 ms, and the parameters of the PI controllers were: gain ($K_p$) = 2 m/s/V, constant time ($\tau_i$) = 0.58 s, and anti-windup constant time = 0.76 [s]. These values were determined using the procedure previously described in Section 3.4.

Fig. 14 shows the velocity and voltage of a track controlled with the PIs. It is possible to check the appropriate behaviour of the controllers. The peak at instant time 12 s is due to a fault in the encoder readings, because the mobile robots traversed a bump on the ground.

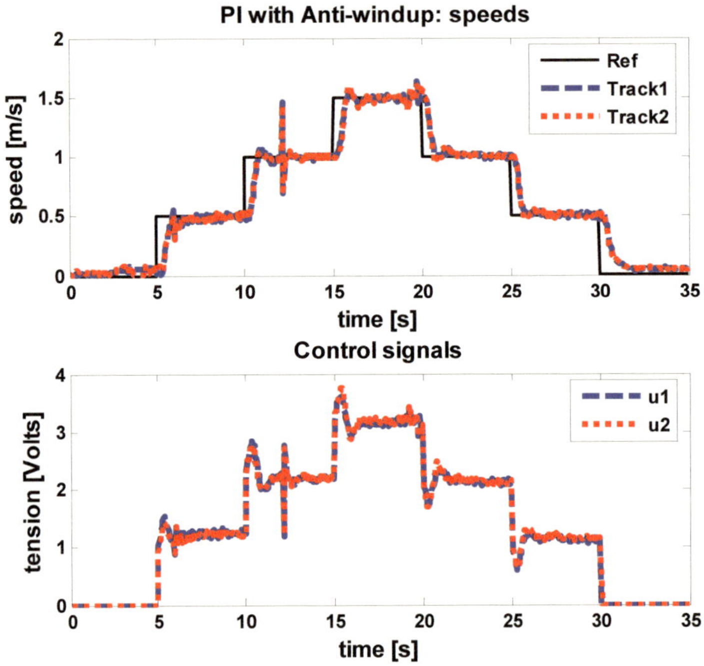


Fig. 14. Servocontroller test with Anti-windup

### 4.3 Navigation tests

Several real tests were performed in a greenhouse of the Experimental Centre "*Cajamar -Las Palmerillas*" in Almeria (Spain). Fig. 15 shows the mobile robot in the greenhouse. The greenhouse is only for experimentation its dimensions are middle-size, having an area of 450 $m^2$. Future new experiments will be made in larger greenhouses. The programs were executed at a sampling time of 100 ms. Velocities in the experiments were between 0.3 and 1.7 m/s. The position of the robot was determined using a relative localization technique (odometry). The initial position of the mobile robot was always the point (0, 0).

First, we tested the reactive navigation strategy. In this case, we show a test on a greenhouse corridor of 20 m length.

Fig. 16 display the sensorial map built (blue asterisks) and the real trajectory followed by the mobile robot (solid red line). In this figure, it is possible to discern the heterogeneous distribution of plants and the slight curvature of the corridor. Furthermore, the trajectory followed is quite appropriate because the mobile robot moves approximately in the centre of the corridor. For that reason, the plants are sprayed uniformly at each side.

(a) Entering in a corridor (b) Between lines of crop

Fig. 15. Mobile robot *Fitorobot* during tests

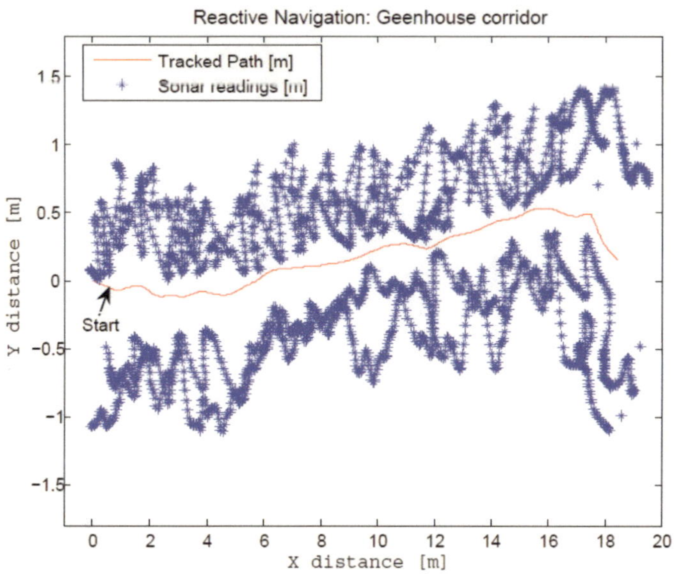


Fig. 16. Reactive navigation test in a greenhouse corridor and sensorial map built

After several experiments with the reactive navigation strategy, we also checked the deliberative approach. As commented in previous works (González et al., 2009a), we used for the motion control a simple controller, *Pure Pursuit*. For that reason, in Fig. 17a the trajectory followed by the mobile robot is slightly different from the reference, above all, in the turns. It can be checked in the errors plots (Fig. 17b) where the maximum error is close to

0.7 m. Furthermore, in Fig. 17c and Fig. 17d is possible to observe the appropriate behaviour of the low-level PI controllers, which assure that the tracks follow the set-points.

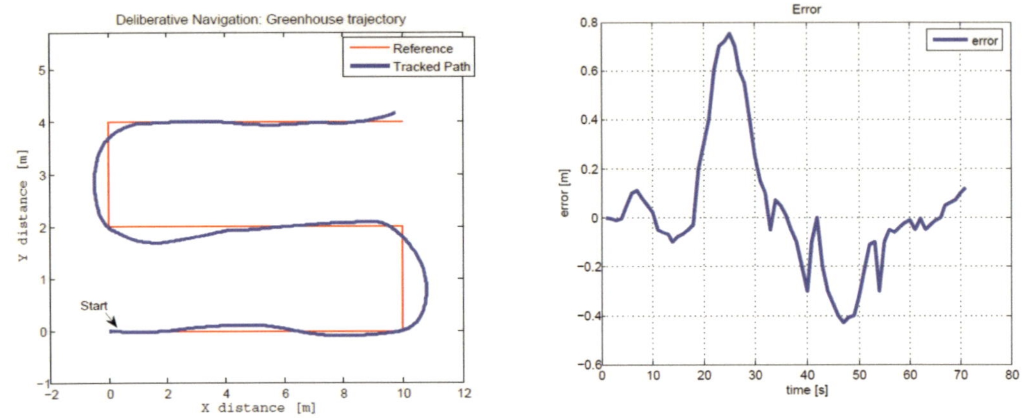


(a) Trajectory (b) Lateral error

Fig. 17. Deliberative navigation test in a greenhouse

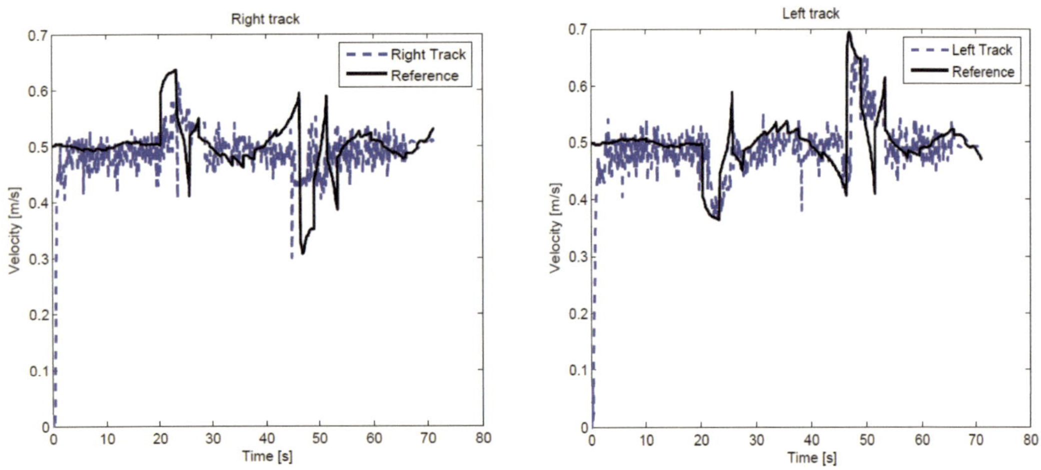


(a) Right track (b) Left track

Fig. 18. Reference velocities and measured velocities from the navigation test

Finally, the control approach with slip compensation was simulated and compared with the original controller. In this case, we have simulated a typical greenhouse trajectory (Fig. 19a). The controller parameters were set $\beta_c = 5$ and $\delta_c = 0.6$ to reach a soft overdamped closed-loop behavior (Canudas et al., 1997). For this test, the slip varies between 10-30 %. Furthermore, in order to do more realistic the simulations a small random noise has been added to the position estimation of the robot. Fig. 19b, c, d plot the errors. As expected, the controller with slip compensation achieves smaller errors than those obtained with the original controllers. Longitudinal, lateral and orientation errors are close to zero.

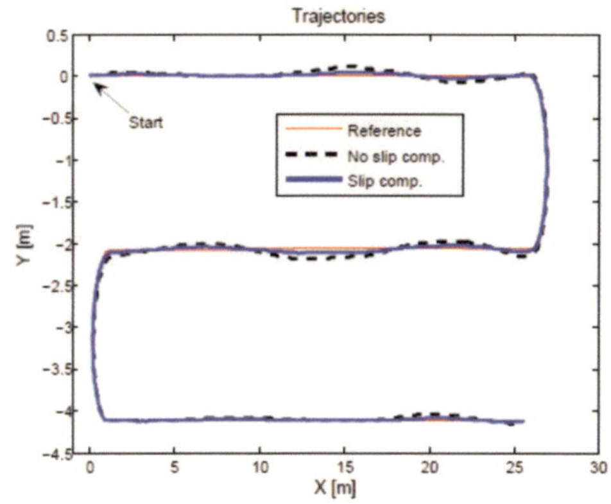


(a) Simulation of a greenhouse trajectory

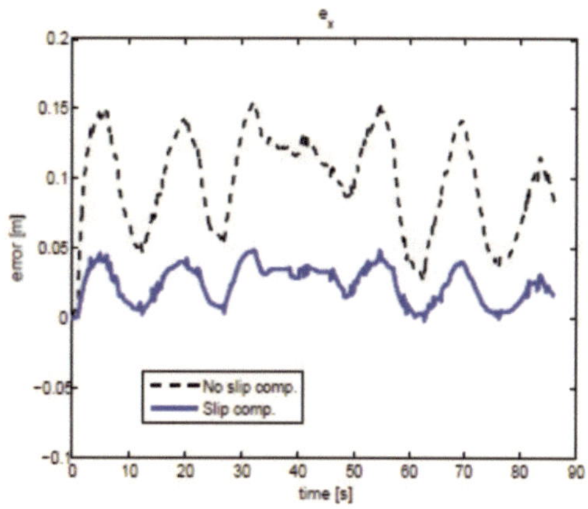


(b) Longitudinal errors

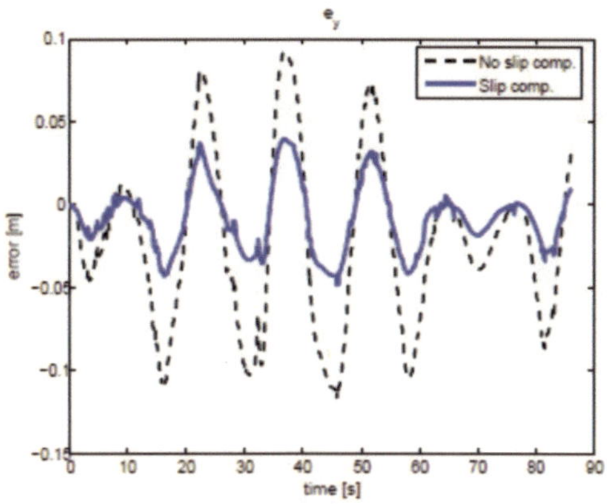


(c) Lateral errors

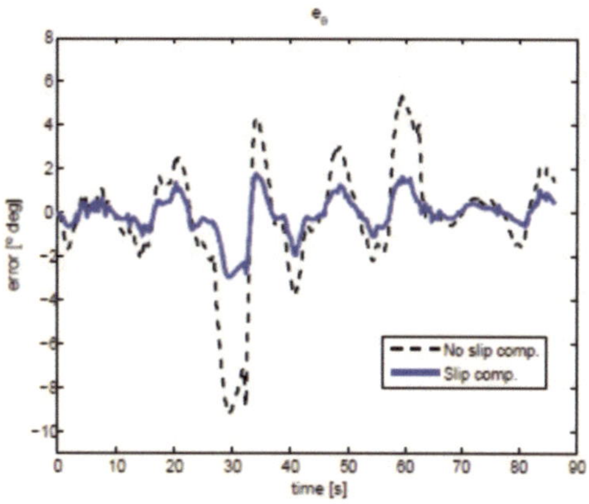


(d) Orientation errors

Fig. 19. Simulation in a typical greenhouse trajectory of the controller with slip compensation and the original controller

### 4.4 Spraying tests

The spraying system was tested under a combination of ramps and steps along different operating points. In this way, the system was validated in order to handle all kinds of possible input data. Fig. 20 (Guzmán et al., 2008) shows that the system responds correctly at the different operating points faithfully following the proposed *setpoint* profile. The robust-control system developed has presented a good disturbance rejection in these cases, modifying the control signal in order to help the system track the desired reference.

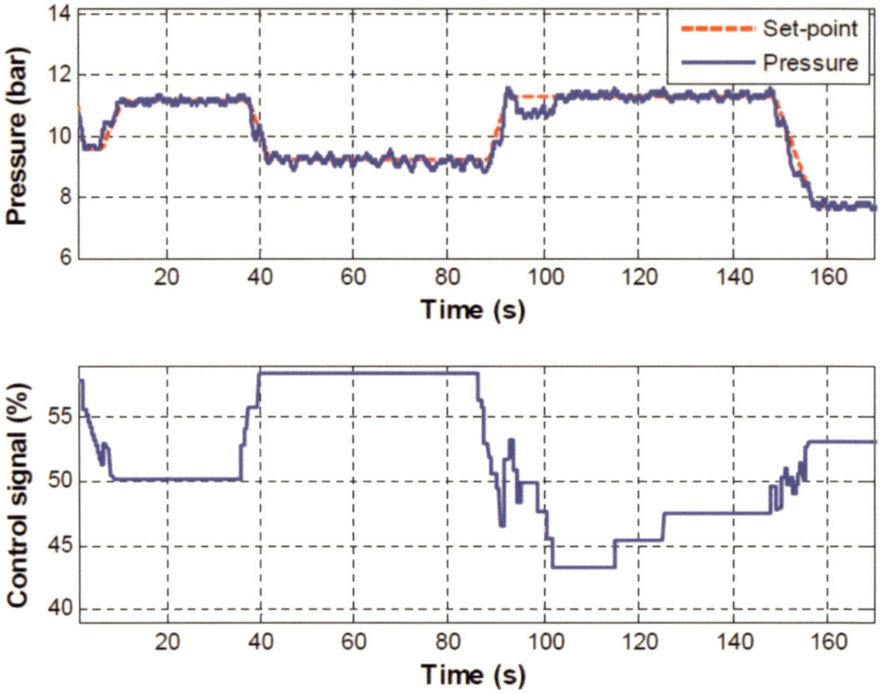


Fig. 20. Spraying control of the test (Guzmán et al., 2008)

## 5. Conclusions and Future research

This chapter presents a mechatronic description of an autonomous mobile robot for agricultural tasks in greenhouses. The mobile robot developed to operate in greenhouses has been supported by a synergetic integration of mechanical engineering with electronics and automatic control.

As commented above, few vehicles exist which navigate autonomously in greenhouses, many of these being static manipulators or semiautonomous vehicles. Because the importance of the greenhouse sector a whole mobile robot has been developed which has been designed for realize some of the most important tasks in greenhouses.

The mechanical design of the mobile robot has been carried out using CAD/CAE technologies in which the main features of greenhouses, the electronic components have been considered. Later some types of sensors were installed on the platform. These sensors are used by the spraying and navigation controls. Finally, navigation and spraying control strategies were implemented in order to govern the vehicle. Successful results have been obtained using this mobile robot, along four years of the project. In this chapter, some real tests have shown the appropriate behaviour of the vehicle working in greenhouses.

Future works involve a dynamic model which will be tested in order to model and control the vehicle. This model is intended to provide better performance of the navigation control.

Finally, we will try advanced navigation control techniques as adaptive and predictive control policies.

## Acknowledgments

This work has been supported by the Spanish CICYT under grants DPI 2007-66718-C04-04 and AGR2005-00848. The authors are also grateful to the research centre "Las Palmerillas, Cajamar" for allowing us access to their installations. The authors further wish to thank the professor J.G. Donaire, and the PhD student A. Pawloski at the University of Almeria for their collaboration.

## 6. References

Acaccia, G.M.; Michelini R.C.; Molfino, R.M. & Razzoli, R.P. (2003). Mobile robots in greenhouse cultivation: inspection and treatment of plants. *International Workshop on Advances in Service Robotics*, pp. 1-6, ISBN 3-8167-6268-9, March 2003, Bardolino, Italy

Åström, K.J. & Murray, R.M. (2008). *Feedback systems: An introduction for scientists and engineers*, Princeton University Press, ISBN 978-0-691-13576-2, Princeton, USA

Bashford, L.L. & Kocher, M.F. (1999). Belts vs. tires, belts vs. belts, tires vs. tires. *Applied Engineering in Agriculture*, Vol. 15, No. 3, 175-181, ISSN 0883-8542

Belforte, G.; Deboli, R.; Gay, P.; Piccarolo, P. & Ricauda., D. (2006). Robot Design and Testing for Greenhouse Applications. *Biosystems Engineering*, Vol. 95, No. 3, 309-321, ISSN 1537-5110

Bishop, R.H. (2006). *Mechatronics: An Introduction*, CRC Press, ISBN 978-0849363580, USA

Bolton, W. (2003). *Mechatronics: Electrical Control Systems in Mechanical and Electrical Engineering*, Third Ed., Prentice Hall, ISBN 978-0131216334, USA

Brown, H.J.; Cruse, R.M.; Erbach, D.C. & Melvin, S.W. (1992). Tractive device effects on soil physical properties. *Soil and Tillage Research*, Vol. 22, No. 1, 41–53, ISSN 0167-1987

Cámara Oficial de Comercio de Almería, (2007). *Informe Económico Provincial, Almería en Cifras 2007* [in Spanish]

Canudas, C.; Siciliano, B. & Bastin, B. (1997). *Theory of Robot Control*, Springer, ISBN 3540760547, The Netherlands

Chan, K.Y.; Oates, A.; Swan, A.D.; Hayes, R.C.; Dear, B.S. & Peoples, M.B. (2006). Agronomic consequences of tractor wheel compaction on a clay soil. *Soil and Tillage Research*, Vol. 89, No. 1, 13-21, ISSN 0167-1987

Dario, P.; Sandini, G.; Allotta, B.; Bucci, A.; Buemi, F.; Massa, M.; Ferrari, F.; Magrassi, M.; Bosio, L.; Valleggi, R.; Gallo, E.; Bologna, A.; Cantatore, F.; Torrielli, G. & Mannucci., A. (1994). The Agrobot Project for Greenhouse Automation. *Acta Hort. (ISHS)*, Vol. 361, 85-92

González, R.; Rodríguez, F.; Sánchez-Hermosilla, J. & Donaire, J.G. (2009a). Navigation techniques for mobile robots in greenhouses. *Applied Engineering in Agriculture*, Vol. 25, No. 2, 153-165

González, R.; Rodríguez, F.; Guzmán, J.L. & Berenguel, M. (2009b). Localization and Control of Tracked Mobile Robots under Slip Conditions, *IEEE International Conference on Mechatronics*, pp. 1-6, ISBN 978-1-4244-4195-2, April 2009, IEEE, Málaga, Spain

Guzmán, J.L.; Medina, R.; Rodríguez, F.; Sánchez-Hermosilla, J. & Berenguel, M. (2004). Pressure control of a mobile spraying system. *Spanish Journal of Agricultural Research,* Vol. 2, No. 2, 181-190, ISSN 1695-971X

Guzmán, J.L.; Rodríguez, F.; Sánchez-Hermosilla, J. & Berenguel, M. (2008). Robust Pressure Control in a Mobile Robot for Spraying Tasks. *Transactions of the ASABE,* Vol. 51, No. 2, 715-727

Hamza M.A. & Anderson, W.K. (2003). Responses of soil properties and grain yields to deep ripping and gypsum application in a compacted loamy sand soil contrasted with a sandy clay loam soil in Western Australia. *Aust. J. Agric. Res,* Vol. 54, No. 3, 273–282, ISSN 0004-9409

Hamza M.A. & Anderson., W.K. (2005). Soil compaction in cropping systems. A review of the nature, causes and possible solutions. *Soil & Tillage Research,* Vol. 82, No. 2, 121–145, ISSN 0167-1987

Horowitz, I. (2001). Quantitative feedback design theory (QFT). *International Journal of Robust and Non-linear Control,* Vol. 11, No. 10, 887-921

Isermann, R. (2005). *Mechatronic Systems. Fundamentals,* Springer, ISBN 978-1852339302, The Netherlands

Kirkegaard J.A.; So, H.B.; Troedson, R.J. & Wallis, E.S. (1992). The effect of compaction on the growth of pigeonpea on clay soils. I. Mechanisms of crop response and seasonal effects on a Vertisol in a sub-humid environment. *Soil Tillage Research,* Vol. 24, No. 2, 107–127

Kitamura, S. & Oka, K. (2005). Recognition and cutting system of sweet pepper for picking robot in greenhouse horticulture, *IEEE International Conference on Mechatronics and Automation,* pp. 1807-1812 , ISBN 0-7803-9044-X, July 2005, IEEE, Niagara Falls, Canada

Kondo, N. & Ting, K.C. (1998). *Robotics for Bioproduction Systems,* ASAE, ISBN 0-929355-94-6, USA

Le, A. (1999). *Modelling and control of tracked vehicles,* Ph.D. Thesis, University of Sydney, Sydney, Australia

Mandow, A.; Gómez de Gabriel, J.M.; Martínez, J.L.; Muñoz, V.F.; Ollero, A. & García, A. (1996). The Autonomous Mobile Robot Aurora for Greenhouse Operation. *IEEE Robotics and Automation Magazine,* Vol. 3, No. 4, 18-28, ISSN 1070-9932

PC104 Embedded Consortium, (2008). Available: http://www.pc104.org

Radford B.J.; Bridge, B.J.; Davis, R.J.; McGarry, D.; Pillai, U.P.; Rickman, J.F.; Walsh, P.A. & Yule, D.F. (2000). Changes in the properties of a Vertisol and responses of wheat after compaction with harvester traffic. *Soil Tillage Research,* Vol. 54, No. 3, 155–170, ISSN 0167-1987

Radford B.J.; Yule, D.F.; McGarry, D. & Playford, C. (2001). Crop response to applied soil compaction and to compaction repair treatments. *Soil Tillage Research,* Vol. 61, No. 3, 157–166, ISSN 0167-1987

Sadras V.O.; O'Leary, G.J. & Roget, D.K. (2005). Crop responses to compacted soil: capture and efficiency in the use of water and radiation. *Field Crops Research,* Vol. 91, No. 2, 131–148, ISSN 0378-4290

Sammons, P.J.; Tomonari, F. & Bulgin, A. (2005). Autonomous Pesticide spraying robot for use in a greenhouse. *Australian Conference on Robotics and Automation,* pp. 1-9, ISBN 0-9587583-7-9, December 2005, Sydney, Australia

Sánchez-Gimeno, A.; Sánchez-Hermosilla, J.; Rodríguez, F.; Berenguel, M. & Guzmán, J.L. (2006). Self-propelled vehicle for agricultural tasks in greenhouses. *World Congress - Agricultural Engineering for a better world*, September 2006, Bonn, Germany

Sánchez-Hermosilla J.; Medina R. & Gázquez J.C. (2003). Improvements in pesticide application in greenhouses. *Workshop in Spray Application Technique in Fruit Growing*, pp. 54-54, Cuneo, Italy

Sandini, G.; Buemi, F. ; Massa, M. & Zucchini, M. (1990). Visually Guided Operations in Greenhouses. *IEEE International Workshop on Intelligent Robots and Systems*, pp. 279-285, July 1990, IEEE, Ibaraki, Japan

Singh, S.; Lee, W.S. & Burks, T.F. (2005). Autonomous Robotic Vehicle Development for Greenhouse Spraying. *Transactions of the ASAE*, Vol. 48, No. 6, 2355-2361, ISSN 0001-2351

Subramanian, V.; Burks, T.F. & Singh, S. (2005). Autonomous Greenhouse Sprayer Vehicle Using Machine Vision and Ladar for Steering Control. *Applied Engineering in Agriculture*, Vol. 21, No. 5, 935-943, ISSN 0883-8542

Van Henten, E.J. (2006). Greenhouse Mechanization: State of the Art and Future Perspective. *Acta Horticulturae*, Vol. 710, 55-70

Van Henten, E.J.; Hemming, J.; Van Juijl, B.A.J.; Kornet, J.G.; Meuleman, J.; Bontsema, J. & Van Os, E.A. (2002). An Autonomous robot for harvesting cucumbers in greenhouses. *Autonomous Robots*, Vol. 13, No. 3, 241 –258, ISSN 0929-5593

Wong, JY. (2001). *Theory of Ground Vehicles*, John Wiley and Sons, ISBN 978-0-471-35461-1, USA

30

# A Mechatronics Vision for Smart Wheelchairs

Yoshiyuki Noda*, Akira Kawaguchi** and Kazuhiko Terashima*
**Toyohashi University of Technology*
*Japan*
***City College of New York*
*U.S.A.*

## 1. Introduction

People are living longer and surviving better. The size of the population of potentially limited by age or disability is increasing at a dramatic rate, and it is no longer an insignificant or silent part of the society. According to a report (United Nation, 2007) published by the Department of Economic and Social Affairs, Population Division, of the United Nations, by the year 2025 aged people, or those of 60 years old or over, will comprise 15% of the world's population, and by the year 2050 this figure will reach 22% compared with 11% in 2007, or nearly a third of the population in developed countries is projected to be in that age group. Moreover, a fact sheet published by the International Day of Disabled Persons (ILO, 2007) reports that around 10% of the world's population today is comprised of disabled people. This means that by the year 2025 around 1/5 of the world's total population will need economic and social benefits as well as some kind of artificial perambulatory assistance in their daily lives.

The U.S. and Japan are facing aging problems. Projections based on the U.S. Census Bureau estimates indicate that the number of persons ages 65 and over will grow to almost 40 million by the year 2010 (Jones and Sanford, 1996). More than 4 million people in the United States are over the age of 85 and about 60,000 topped age 100. By 2020, the Census Bureau further estimates that 7 million to 8 million people will be over age 85 and 214,000 will be over age 100. Japan has a very serious challenge of rapid pace of aging. Most experts predict more than 27 percent of the population to be age 65 or more by 2017, and this increase will continue to reach 35.7% in 2050 (Oe, 2006). It took only 24 years in Japan for the elderly population to grow from 7 percent to 14 percent. The speed of aging is the world's fastest and Japan's aging level will be the highest in 2010. Japanese society is in the midst of this rapid change, and community accessibility and social welfare have become major policy issues.

As life expectancy rises and modern medical treatment improves survival, there is a growing interest in building more advanced support structures for society. In particular, a desire for the enhanced quality of life with wheelchair is a steadily growing phenomenon (First Research Inc., 2007), and is generating a strong demand for a lesser restrictive environment of barrier-free accesses, safer and better route selections, more transportation alternatives, and so on. The motivation for this chapter is to pose a fundamental question on the handling of these challenging problems. We shall look into a missing part of today's transportation systems in the context of wheelchair's mobility, with our aims to explore the establishment of support systems to respond these foreseeable changes in demographics. The public acknowledgment of people with disabilities and progress toward enhanced cares have developed in the last few decades along three parallel tracks of activities (Longmore and Umansky, 2001) – these are legislation spurred by the disability rights movement, barrier-free design to universal design movement, and advances in rehabilitation and assistive technology. Nevertheless, there still remains an area that ought to be examined more seriously from an economic standpoint. We call this area a *mechatronics support* service, which lies as a support technology centered on mechanics, electronics, and computing.

The mechatronics support in our vision is meant to emphasize a synergistic integration of the latest techniques in cross-disciplinary fields. It is an approach for a pragmatic establishment of the society-driven transportation structure for the vulnerable people (Kawaguchi, et al., 2008). The so-called assistive technology (Cook, 2002) in computerized aid, care, and support for the people in mobile wheelchairs is evolving rapidly. It is in a technical branch that indeed spans autonomous process control, intelligent engineering systems, and information management. While transformative as well as establishing stronger through world-wide collaboration, there exists no widely accepted practice or reference model that leads to the integration of these assistive technical elements into mass transportation systems today. This chapter studies several key issues in bringing into an infrastructure and surrounding information system that hosts micro and macro control for the smooth and safe navigation of mobile wheelchairs in the public transportation environment. More specifically, we propose a high-level architecture that facilitates terrain surveillance and intelligence gathering through laser sensor implanted in the wheelchair, for the purpose to identify the obstacles and unusual patters of movement to predict accidents. This can be achieved by a number of sensors to map the environment into an appropriate perception-action model, which is then employed for planning and controlling the locomotion of the wheelchair. Information technology plays key roles for assisting the understanding of travel route and physical situations at the movement of wheelchair. All the technical elements highlighted in this paper are our on-going collaborative efforts between the U.S. and Japan.

The contributions of this chapter are threefold: first, a short survey on the accessibility issues related to today's mobile wheelchairs and various effort for the enhancement of public accessibility in transportation perspective is given to set the orientation of this research; second, foundations for and high-level specifications for micro and macro-control architecture developed based on the proposed mechatronics approach are explained; third, a product of a navigation guidance system using a haptic feedback joystick for an omni-directional wheelchair is introduced as a micro-control support, and an example of a macro-level information support system using New York City bus transit service is discussed. The widespread use of autonomous mobile wheelchairs has a socioeconomic impact. The significance of this research for the longer-term goals lies in its implications for adaptation of our intelligent support model into the future welfare activities.

## 2. Today's Mobile Wheelchairs in the U.S. and Japan

The demand for wheelchairs and other mobility devices in the U.S. is projected to increase 5.0 percent per year through 2010 to over $3 billion (Supplier Relations, 2007). This is partly due to the mobility impairing conditions accompanied by the aging of the baby boomer population in the U.S. (First Research Inc., 2007). The growth estimated will be similar for other devices designed primarily for disabled persons such as powered scooters and specialty elevators (Market Research, 2007). Advances in electronic controls, battery technologies, lightweight materials of construction and other areas are stimulating the growth in both the wheelchair- and non-wheelchair-related segments of the personal mobility devices industry.

### 2.1 Popular mobile wheelchairs in the U.S. and Japan

Electric wheelchairs, generally designed for both indoor and outdoor use, are prescribed for individuals who have difficulty using a manual chair due to arm, hand, shoulder or more general disabling conditions such as a lack of leg strength to propel a manual chair with their feet. Most electric wheelchairs consist of a seat and back, two small front wheels and two large rear wheels, a foot rest, and a joystick that allows directional movement such as driving forwards, backwards, and sideways, and turning round on the spot while moving (Yanco, 1998). Various parts often customized and added for the user's needs, such as the seat size, seat-to-floor height, footrests or leg rests, front wheel outriggers, adjustable backrests, and so on. Optional accessories are also available, such as anti-tip bars or wheels, tilt and/or recline features, adjustable backrests, safety belts, extra support for limbs or neck, mounts or carrying devices for crutches, walkers, drink holders, and clothing protectors.

Some of the leading manufacturers of mobile wheelchairs are AdaptaChair (U.K.), Balder Power Wheelchairs (U.S), Bromac Assistive Technology (U.S.), ConvaQuip (U.S.), Everest & Jennings (U.S.), Euroflex (Sweden), Falcon Rehabilitation Products (U.S.), Giraldin (Italy), Invacare Products (U.S.), IBOT (U.S.), Levo (Switzerland). Magic Mobility Extreme (Australia), Meyra (Germany), OmegaTrac (U.S.), Otto Bock, Permobil (U.S.), Quickie (U.S.), Rascal (U.S.), Redman Power Chair (U.S.), RGK Wheelchairs (U.S.), VERTRAN (U.S.), Kawamura (Japan), Yamaha (Japan), Suzuki (Japan), Matsunaga (Japan), etc.

These manufactures classify electric wheelchairs into transportable, powerbase, and heavy-duty types. Transportable type weights less than 100 lbs and has a compact size suitable for confined spaces or narrow doorways. The lightweight frames easily fold or disassemble into manageable pieces making them the lightest and easiest to carry. Powerbase type provides a range of options for suspension, driving, seating and handling, and has a low center of gravity and a tight turning radius with large battery range, thus, making the ride smooth and stable for outdoor and indoor use, but does not fold or disassemble. Heavy-duty type is for outdoor use, equipped with rear, front and mid wheel drive options, and larger wheels. Larger motors and longer frames allow the increase of torque over other types, thus providing a smooth ride over rough terrain and up some inclines.

### 2.2 Controlling electric wheelchairs

The mobility of these chairs relies on electric motors, usually powered by 4 or 5 amp deep-cycle rechargeable batteries available in wet (or liquid electrolyte), gel (or pasty low-moisture electrolyte), or AGM (Absorbed Glass Mat) types. Batteries are usually on-board rechargeable through a standard wall outlet. The adjustment of the speed and direction of wheelchairs can be mostly made by operating a joystick on a controller (see Fig.1). More complex devices, such as chin controls and puff/suck scanners, can be customized for those with severe and/or multiple disabilities, e.g., spinal cord lesions and lack of hands or fingers. There are three standard ways of propelling electric wheelchairs (Yanco, 1998; Market Research, 2007): rear wheel drive is most common for electric wheelchair, which makes the wheelchair fast but has a comparatively poor turning capability, mid wheel drive has best turning capability but is not suitable for uneven surfaces, and front wheel drive gives a lower top speed than rear wheel drive chairs, but offers a good turning capacity. Recent expensive model has gyroscopic circuitry that enables the chair to rise on two wheels. Various powered functions such as tilt, recline, leg elevation, and seat elevation are also available in most equipments.

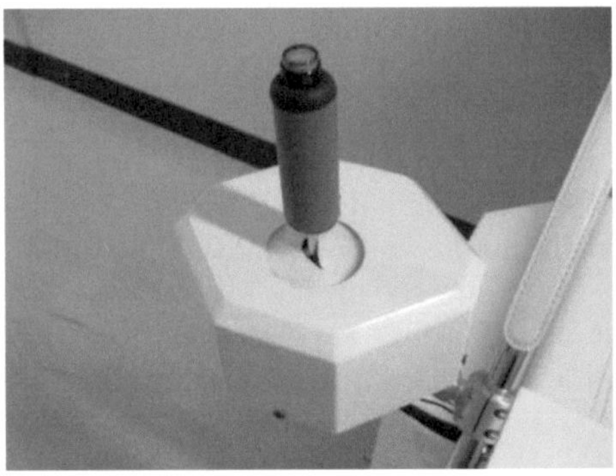

Fig. 1. Joystick

### 2.3 Support and regulation structures in the U.S. and Japan

Today's built environment shows a significant transformation that makes it more accessible to wheelchair users. This kind of transformation has been brought by the law and longstanding movement with aims of supporting people with disabilities to live as more

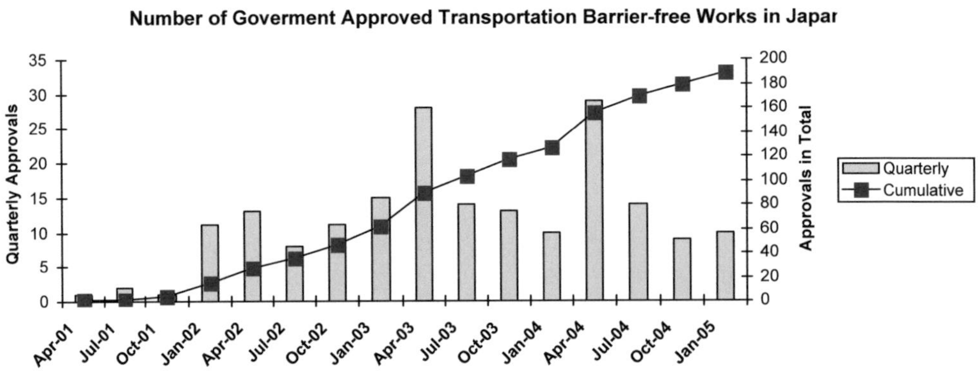


Fig. 2. Recently approved accessibility improvement works in Japan (Source from Ministry of Land, Infrastructure and Transport, 2006)

active participants in the society. Noticeable changes are visible in many cities and parts of the world, such as the installation of elevators, transit lifts, and wheelchair ramps in public areas. Especially in the advanced countries, such improvement is increasingly enforced for the access to public areas such as city streets and public buildings and restrooms, thus allowing people in wheelchairs and with other mobility impairments to use public sidewalks and public transit more easily and more safely (ILO, 2007). The demographic, legislative, economic, and social changes that brought us to this point are increasing the momentum that will propel us into a more modern transportation structure that enables maximum inclusion of all people.

In the U.S., the public law called the Americans with Disabilities Act of 1990 (ADA. (1990)) prohibits discrimination based on disability, in which disability is broadly defined as a physical or mental impairment that substantially limits a major life activity. In effect, all people regardless of disability are entitled to equal access to all parts of the society such as public transportation and buildings. The disability rights movement in the U.S. prevailed the nation during 1970s and grew into a significant force as a precursor of the ADA regulation to establish the civil-rights status for disable people. The ADA has uniform nationwide rules that ensure the accessibility regardless of local attitudes and governances. The Architectural and Transportation Barriers Compliance Board (Access Board) issued Accessibility Guidelines for accessible design in 1991, and these guidelines became the enforceable ADA Standards for Accessible Design. Today's most new construction for public use must be built to ADA standards of accessibility. Wheelchair users are now becoming free to access the environment without stairs, especially for more law-enforced places like education and employment.

Japan has been taking a number of legislative steps to address the challenges, chiefly led by the government, socially concerned architects, engineers, and advocates. In 1994, Japan implemented The Law for Promoting Easily Accessible and Useable Building for the Aged and the Disabled (so called Heart Building Law) to provide guidelines for accessibility to public facilities such as hospitals, theaters, and municipal buildings. Revised in 2002, the scope and size of buildings are expanded, and local governments such as cities, towns and villages are given full authorization to enforce stricter requirements and decide focus areas for improvement (see Fig. 2). The law is now mandatory if building is larger than 2000 square meters. Furthermore, The Law for Promoting Easily Accessible Public Transportation Infrastructure for the Aged and the Disabled (so called Barrier-Free Transportation Law) established in 2000 requires public transport to adhere to a series of accessibility regulations. These two laws are integrated into so-called New Barrier-Free Law in 2006. Japanese citizens strongly support the current systems.

### 2.4 Universal design in public transportation

Universal design is a design paradigm that aims to reduce the physical and attitudinal barriers between people with and without disabilities. This concept emerged from barrier-free design principles and assistive technology — the former provides a level of accessibility for people with disabilities, and the latter enhances the physical, sensory, and cognitive abilities of those people (Orpwood, 1990). The concept of universal design gives a broad-spectrum of solutions that help everyone, not just people with disabilities, and it is now becoming one of the most important design elements that range in scale from product

design to architecture and urban design, and from simple systems to complex information technologies (NCODH, 2005).
The forthcoming transportation facilities are therefore likely in compliance with the universal design principles. Especially in public transportation, universal design policy aims to create an environment that enables everyone to move freely and safely. Diverse parties responsible for public conveyance such as bus, rail, aircraft, and intercity or commuter rail transportation, must accomplish smooth and easier transfer at their connection points and flexible routing selection over those services. This must be done based on their cooperative efforts instead of competitive relationships (MLIT, 2006). As an example, the ADA requires new transit buses be accessible to people using wheelchairs (by lift or ramp) and have at least two spaces to secure wheelchairs in each bus. The construction of low floor trams and non-step buses is a trend and today's street furniture design incorporates better accessibility even for people with disabilities.

### 2.5 Paratransit services

Paratransit is another mode of passenger transportation that has been increasingly served for handicapped commuters and travelers, which in general does not follow fixed routes or schedules, but uses typically vans or mini-buses for the higher flexibility for picking up or discharging passengers on request. Most vehicles are specially equipped with wheelchair lifts or ramps to facilitate access, thus allowing people with disabilities to have jobs by providing transportation to and from their workplaces (Simon, 1997).
A sub-sector and private business are on the rise to meet the growing interest in the paratransit strategy, but the cost for serving low-density areas and dispersed trip patters does not balance with today's declining financial conditions, which may create an urgent need for public transit operators instead to maximize all available transportation resources. Veolia Transport and Laidlaw International, Inc. are two of the largest providers of contract paratransit services in North America (Veolia Transport is also successful in Europe and Australia). Paratransit in Japan is part of the health and welfare systems. Thus the service is small-in-scale from the transportation standpoint, while in Hong Kong, a territory-wide Rehabus transport network service is provided by the Hong Kong Society for Rehabilitation.

## 3. Mechatronics Approach for Enhanced Mobility

Imagine a great influx of people in wheelchairs having to navigate through crowded aisles and streets. Sidewalks can be harmful or even impassible by those wheelchairs. Many factors even including laws may need to be reconsidered for roadway traffic, parking space, and accident handling. As an example, since people with disabilities who use wheelchairs are considered pedestrians under many authorities in the U.S. and Japan, at serious traveling speeds the chairs can endanger other sidewalk users, raising a similar debate on whether bicyclists should be allowed to ride bicycles on sidewalk. Building more comprehensive support structures as well as better protections for wheelchair users becomes mandatory in the near future. This section looks into the so-called mechatronics approach (Mahalik, 2003) for coupling a wide range of hardware components and software functionalities using standard interfaces, which enables wheelchairs to possess a degree of intelligence, ultimately to predict and avoid risky situations and navigate safely in congested areas and confined spaces. Capabilities of sensor-based feature detection and

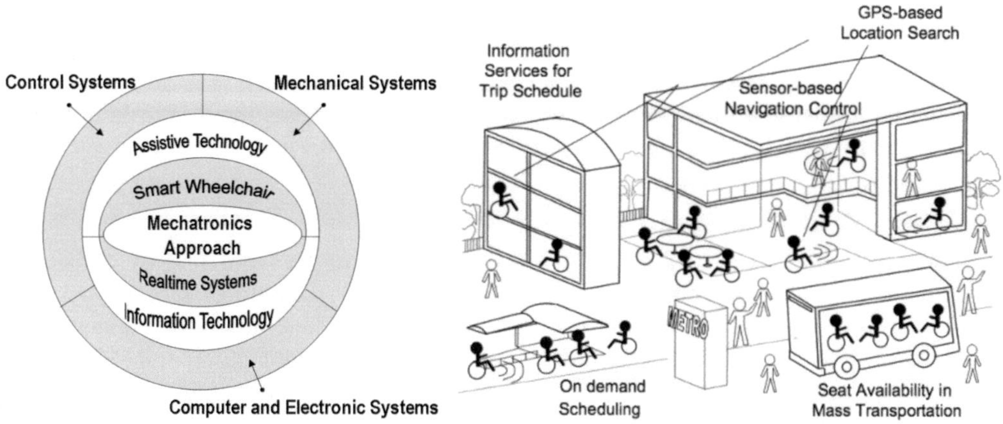


Fig. 3. Mechatronics approach for mobile wheelchair support in transportation

computerized navigation assistance available today make this kind of electromechanical cares possible.

### 3.1 Mechatronics approach

Mechatronics shares the common goal of systems thinking towards a synergistic combination of precision mechanical engineering, electronic control, and information processing. This is to achieve an optimal balance between basic mechanical structure and its overall control. In the scope of wheelchair's mobility enhancement, machatronics approach aims to an effective integration of assistive technology and information technology (Kawaguchi et al., 2008), by way of embedding real-time microprocessor control (micro-level control) for understanding the surrounding condition through laser sensor and applying on-demand communications to information points (macro-level control) for wheelchair to follow directions by extrapolating and comparing to the route pre-selected by the user (see Fig. 3).

The outcome of this effort can be broadly framed as an electrically powered wheelchair that is equipped with an intelligent assistance system being capable of (1) recognizing anticipatorily the user's intent according to a given situation and (2) guiding a trip to destination safely and reliably to a high degree based on given instructions such as travel routes and transportation methods. Notice that no single form of implementation exists due to the need of adaptations to various human factors such as severely handicapped or elderly people with heavily reduced physical and/or mental abilities, weakened perception, etc. For these cases, special control devices (e.g. sip-puff-device) with restricted command sets need to be developed.

Notice also the importance of community support in the development and deployment of the product. The mechatronics approach is a practice including a wide range of activities spanning design, instrumentation, computer integration, process and device control, and manufacturing. The successful implementation mandates not only interdisciplinary effort across the industrial and academic research spectrum, but also participation and cooperation of wheelchair users, transportation companies, disability advocates, and voluntary experts in public. Overall, it is a continuous process in which users in all parties participate actively from the preliminary study stage to the post evaluation stage, so that the knowledge and experience obtained through their participation process can be applied to the next level of innovation. In this regard, the concept we propose shares the philosophy and benefits of universal design – mechatronics approach serves the dual role of bringing greater recognition to this important area of engineering.

### 3.2 Assistive technology as micro control

Assistive technology (Cook, 2002) applies to devices for personal use to enhance the physical, sensory, and cognitive abilities of people with disabilities and to help them function more independently in environments oblivious to their needs. Its element closely related to realize mechatronics support is a smart wheelchair (Simpson, 2005) or an augmentative mobility aid that accepts a variety of different controls tailored to the rider's needs, and complements the rider's efforts by expanding and interpreting their limited control commands to provide safe transit (CALL Centre, 1994).

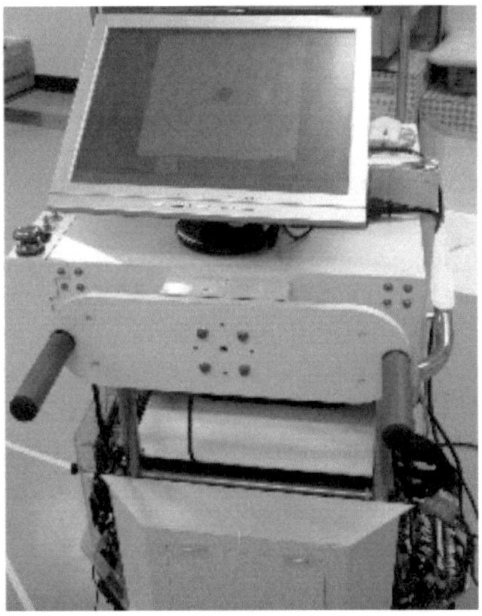

Fig. 4. Touch Panel

A smart wheelchair has a collection of sensors to work with several cognitive techniques similar to those developed in mobile robotics, but it is not necessarily acting autonomously because the aim is to complement and extend the user's abilities, not replace them. The typical interface consists of a conventional wheelchair joystick (or more complex input device) and a touch-sensitive display connected to a computer (see Fig. 4). Sensor techniques are chosen such as sonar, infrared radiation, and laser rangefinder, to understand the surrounding physical layout and to detect obstacles for navigation assistance and collision avoidance. The response is produced such that the system alerts and/or modifies user's choice of joystick drive command. Our recent attempt in this direction is discussed shortly. One drawback observed is that the majority of the smart wheelchairs that have been developed to date have been tightly integrated with the underlying power wheelchair, thereby requiring significant modifications to function properly (Simpson et al., 2004). The forthcoming research on the extension of smart wheelchairs, enhanced with path-planning, behavioral learning, and cognitive capabilities will have a significant impact on the outcome of the mechatronics implementation.

### 3.3 Information technology as macro control

Information technology is a key for providing flexible transportation services and increased choice for the users. In addition, useful information supplies psychological reassurance to vulnerable people to make them feel more safe and secure. A federal civil rights procurement law in the U.S. requires electronic and information technology to be accessible to people with disabilities. Flexible services are brought by a wide variety of innovative information services now in use increasingly in many countries. For instance, the presence of ubiquitous network technology that consists of IC tags, wireless tags, RFID tags, and other communication equipment such as portable information terminals makes it possible for elderly people and handicapped people to move freely and independently. The use of more economical PHS communication network generally available in Asian countries can afford precise tracking of the people in wheelchairs who move around dense urban areas like hospitals, libraries, and museums. GIS technology that allows conversion of geographical information into electronic form also facilitates the wide-scale integration of navigation assistance and tracking capability into various telecare information systems.
Up-to-date information of services such as availability, route guidance, cost, efficiency, and safety must be always available to improve the overall level of public transportation including railroad, bus, airline, and shipping. Traffic-aware routing based on portable GIS device is a new area in vehicle industries. At the same time, traditional internet-based software that can handle scheduling, dispatching and reservation, such as Ecolane DRT, PASS and NaviTrans, plays an important role to implement demand-responsive transportation schemes or flexible door-to-door paratransit service for wheelchair users.
Residents in metropolitan areas increasingly view the convenience of public transportation. They are becoming used to supply and retrieve information gathered for regional community. The accessibility improvement needs the exact same type of the cooperation of transportation companies and regional residents: transportation companies, private industries, public places, and any points of interest for residents such as restaurants, shopping centers, movie theaters, etc., must supply the accessibility information into a public database system, which in turn provides immediate retrieval for transportation and route selection, thereby giving great aids to realize macro control in wheelchair movement.

## 4. Available Technical Elements

A variety of electric wheelchairs with different options and special add-on features have been proposed to meet a wide range of needs in the drivability enhancement (Prassler, 2001; IEEE RAM, 2001). Amongst these capabilities, a driving mechanism called an *omni-directional movement* has been realized and demonstrated at the Toyohashi University of Technology to satisfy the demand for better mobility in narrow space and confined area. In this section, we describe several enhancements made to this prototype toward the fulfillment of the proposed mechatronics-based integration.

### 4.1 Omni-directional mobile wheelchairs

A wheelchair enhanced with omni-directional movement can have combination of forward, sideways, and rotational movement. Consequently, it yields better mobility and safety for the navigation of areas in the presence of many obstacles. The one designed and manufactured at the Toyohashi University of Technology (see Fig. 5) is based on the work of Wada and Asada (1999), in which each of the four balls to drive wheelchair is operated by an individual electric motor. The product shows the full capacity to move to any direction as well as rotate around while having abilities to change the angle between the beams holding these balls.

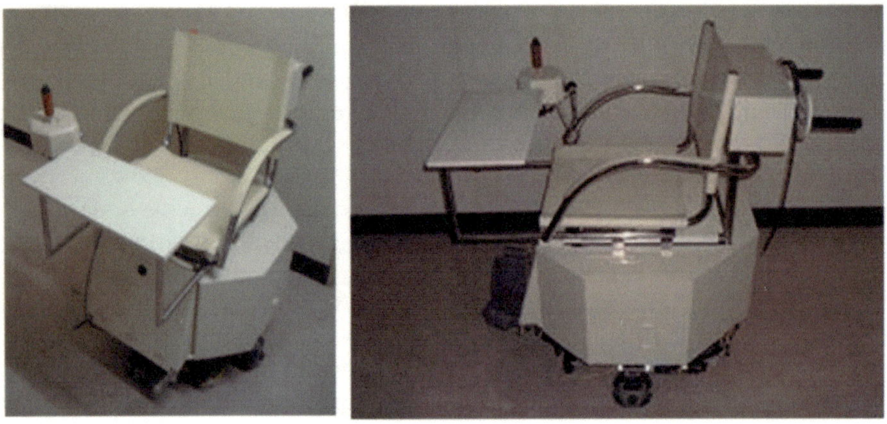

Fig. 5. Omni-directional wheelchair

Fig. 6(a) shows manufactured parts of omni-wheels. Omni-wheels are central to realize the arbitrary movement of the omni-directional mobile wheelchair (Urabano et al., 2005). As the assembly pictured in Fig. 6(b), the driving system consists of four omni-directional wheels, each of which connects to a motor. Notice that each wheel has passively driven free rollers at its circumference. The wheel that rolls perpendicularly to the direction of movement does not stop its movement because of the passively driven free rollers. These wheels thus facilitate the movement that is holonomic and omni-directional.

(a) Omni-wheel (b) Arrangement of omni-wheels and their motors

Fig. 6. Components of omni-directional drive system

### 4.2 Position sensitive device and laser sensing technology

An extension is made to find obstacles in the surrounding environment as well as to measure the distances to those objects. This can be done by the installation of range sensors, such as ultrasonic sensor, infrared sensor, and laser range scanner. In our first attempt, the ultrasonic sensors and the infrared sensors are attached to the omni-directional mobile wheelchair, as pictured in Fig. 7(a). This configuration, however, requires eight sensors in total to detect and measure the distance to the nearby obstacle as illustrated Fig. 7(b). In our second attempt, a laser range scanner is applied to reduce the number of sensors. The selected model, URG-X002 from Hokuyo Automatic Co., Ltd. (see Fig. 8(a)), is built as a high accuracy laser range scanner with a very compact design. It is not only the world's smallest range scanner of its class invented with advanced technology, but is highly reliable for the forefront use such as obstacle detection, environment recognition and path planning in robotics. The laser range sensor scans the surrounding environment by rotating the reflector (see Fig. 8(b)). The sensor has scanning speed of 100[ms/scan], angle resolution of 0.36[deg], measurement range of 270[deg], and measurement distance of 0.02-4[m]. This means that only two laser range sensors are required to measure the distance to the obstacles on the periphery of the omni-directional mobile wheelchair.

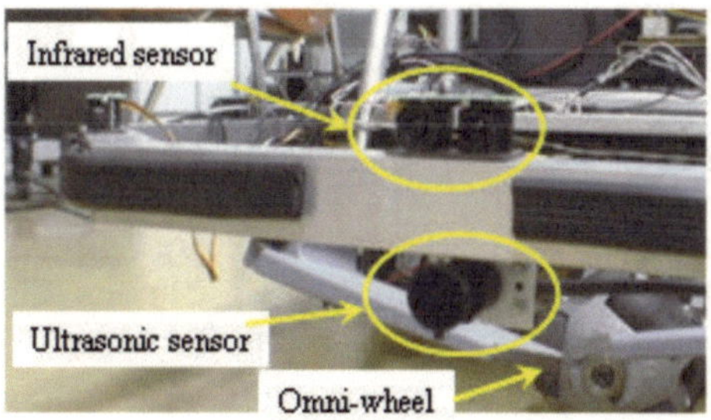


(a) Setting of ultrasonic sensors

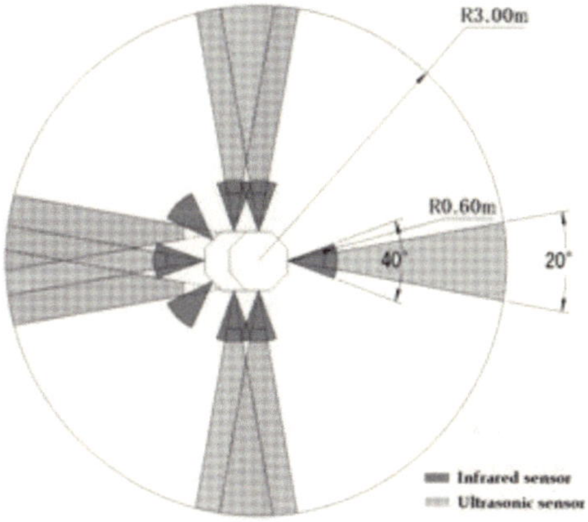


(b) Layout of sensors and measurement ranges

Fig. 7. Installation of ultrasonic sensors

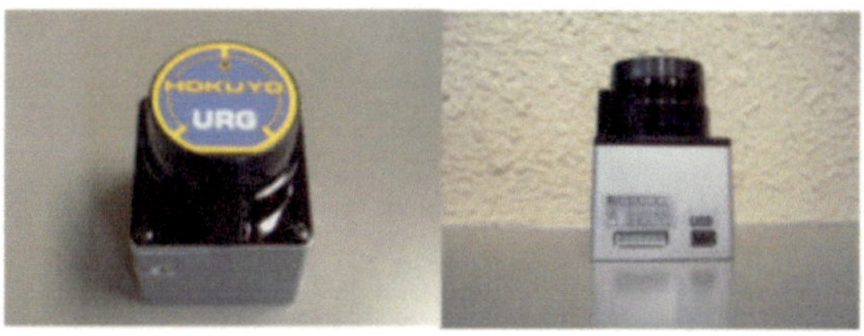


(a) External appearance

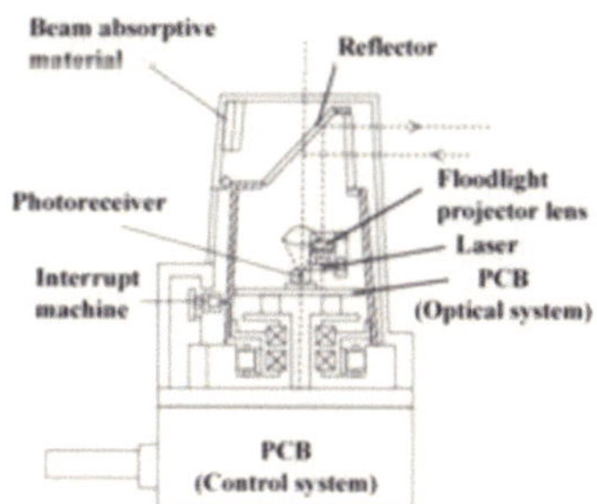


(b) Internal structure

Fig. 8. Laser range scanner, URG-X002

Fig. 9 shows the experimental performance of scanning environment by the use of the laser range scanners installed into the omni-directional mobile wheelchair. In this experiment, the omni-directional mobile wheelchair moves from point (a) to (f) as illustrated in the leftmost layout of Fig. 9. The developed system can successfully recognize obstacles and walls as the wheelchair proceeds to each point, and the map obtained at the point rotates at the same time the wheelchair changes its orientation.

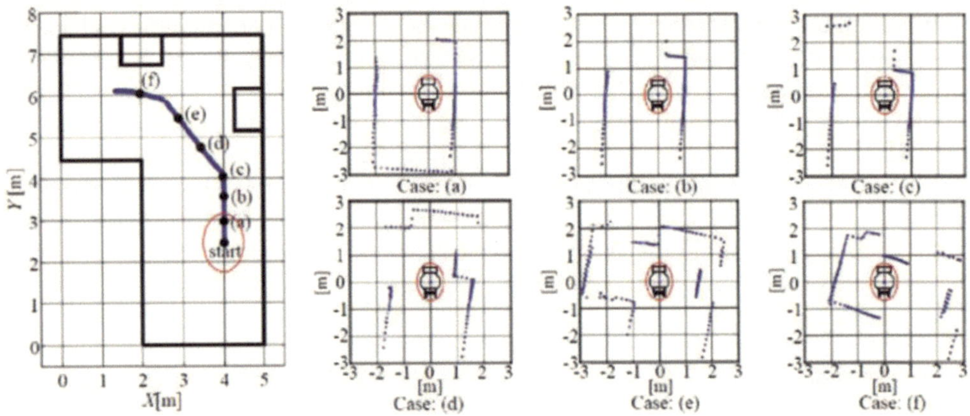


Fig. 9. Experimental results of recognizing environment

### 4.3 Spatial mapping and automated tours

Further extension is made to realize a semi-automated navigation system being combined with the obstacle detection capability described above. This is based on the real-time path planning of the tour to reach a goal point. The developed system is fully capable of avoiding the objects found in the path. Our approach uses potential field method (Arkin, 1987) constructed from a diffusion equation, in that the density of the target point is set 1 and the density of the region occupied by the obstacle is set 0 to establish boundary conditions. Given boundary conditions, the potential filed is generated from the iterative calculation of the discrete-diffusion equation. Accordingly, the wheelchair follows the path having an increasing order of densities to reach the target point. This is illustrated in Fig. 10 such that the target point holds density 1 and the area of obstacle holds density 0 as well as the potential field and trajectory of wheelchair's navigation path.

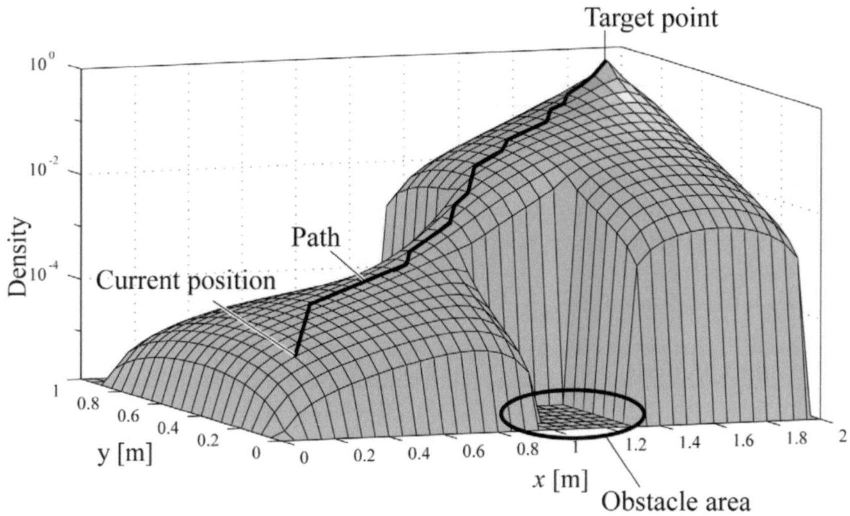


Fig. 10. Feature of potential field

### 4.4 Haptic feedback joystick system

The most recent extension made to omni-directional mobile wheelchair is a *haptic feedback joystick system* (Kondo et al., 2008). This is a control feedback system for which an alert is issued to the user by sending out a reaction force to the joystick when finding out obstacles in front. The system allows the user to navigate safely in narrow aisles, room entrances, and elevator rides. The signal to adjust the stiffness of joystick is determined based on the distance to the obstacle and the velocity of the wheelchair so that the maneuverability of joystick is to correspond to the drivability of wheelchair.

Another type of power-assisting system is built to help health-care attendants easily push and move heavyweight wheelchairs. Feedback control is similar to the haptic control mechanism. As shown in Fig. 11(a), a 6-axis force sensor is assembled into the supporting beam of the wheelchair's handles. The built system gives a boosting power to drive wheelchair, in which the magnitude of the power is computed by the reference velocity measured at the sensor from the attendant's force applied to the handles. The developed power-assisting system has shown effective support for wheelchair-cares.

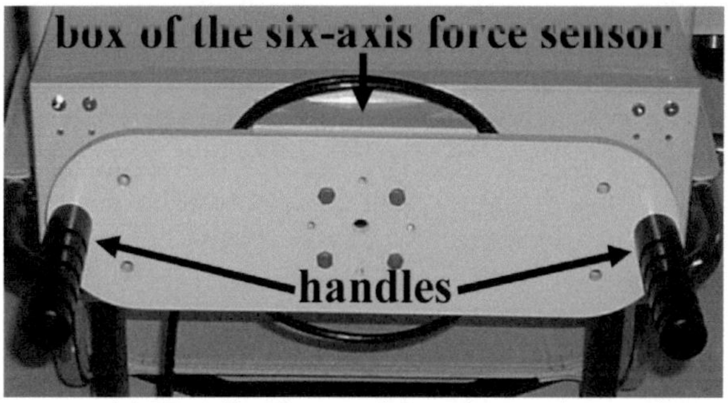


(a) Handles for applying force of attendant

(b) 6-axis force sensor

Fig. 11. Handles with power assisting system for attendant

### 4.5 Bus Transit Services of Wheelchair Riders

The New York City Transit in the Metropolitan Transportation Authority network operates the world's largest fleet of buses (4,373 public buses), serving over 666 million people per year for New York City to sustain its economy and attain projected job growth(MTA). These buses are equipped with wheelchair lifts either the front or back entrance of the vehicle, and have a "kneeling" feature that lowers the front entrance to within inches from the ground for easy access by any customer with mobility impairments or difficulty using the front steps. The Tokyo Metropolitan Bus Systems in Japan, by comparison, maintain the second largest scale of fleet, but do not run this level of services with lift-equipped vehicles (see Table 1 for the projected improvement from Japanese government standpoint).

| Transportation Media | 2003 | 2010 |
|---|---|---|
| Railway cars | 24% | 30% |
| Non-step buses | 9% | 20-25% |
| Passenger ships | 4% | 50% |
| Airplanes | 32% | 40% |

Table 1. Preset goals for the barrier-free implementations of Japanese public transportation (Ministry of Land. (2006))

The bus system is serving routes not served by the city subway system and outlying areas, and stopping every 2 blocks in almost 24 hours, The bus system is becoming a primary choice of transportation for wheelchair users living in the city. The city convention and visitors bureaus are offering guides that list wheelchair-accessible facilities. However, these brochures are rarely detailed enough to rely on and implement a full-scale mechatronics support for barrierfree accesses.

An on-line database system to facilitate the exchange of useful information among disabled bus riders and accessibility supporters in New York City is being built by our efforts. The system has a Web interface to find out the bus routing information on a trip from one point to another point in Manhattan. The capability beyond the "Trip Planner" web system (MTA Trip Planner System, 2009) implemented by the Metropolitan Transportation Authority is to respond using a map with appropriate paths (sequences of bus rides) to be taken to reach the destination along with the roadside information of toilet options, coffee shops and restaurants accommodating wheelchairs, quick repair services for motor trouble and battery replacement, and purchases of wheelchair equipments, etc., on the selected route. The system works with the GoogleMap viewing capability, thereby giving a quick way to narrow down the user choice of accessible sporting and cultural events.

The prototype system (see Fig. 12) currently stores all the traffic points (i.e., major stopping points and transfer points) of the MTA bus lines whose services are bound to Manhattan (thus the search capability is limited within Manhattan). A traffic point is a geographical location consisting of latitude and longitude, and all the points are structured as a directed graph in the database. A query transaction quickly computes acyclic paths between start location and destination. There can be multiple ways to travel by transferring to different bus lines. Thus, heuristics to avoid the explosive growth of traversal are devised for computing a maximal reachable set from a given point in the graph.

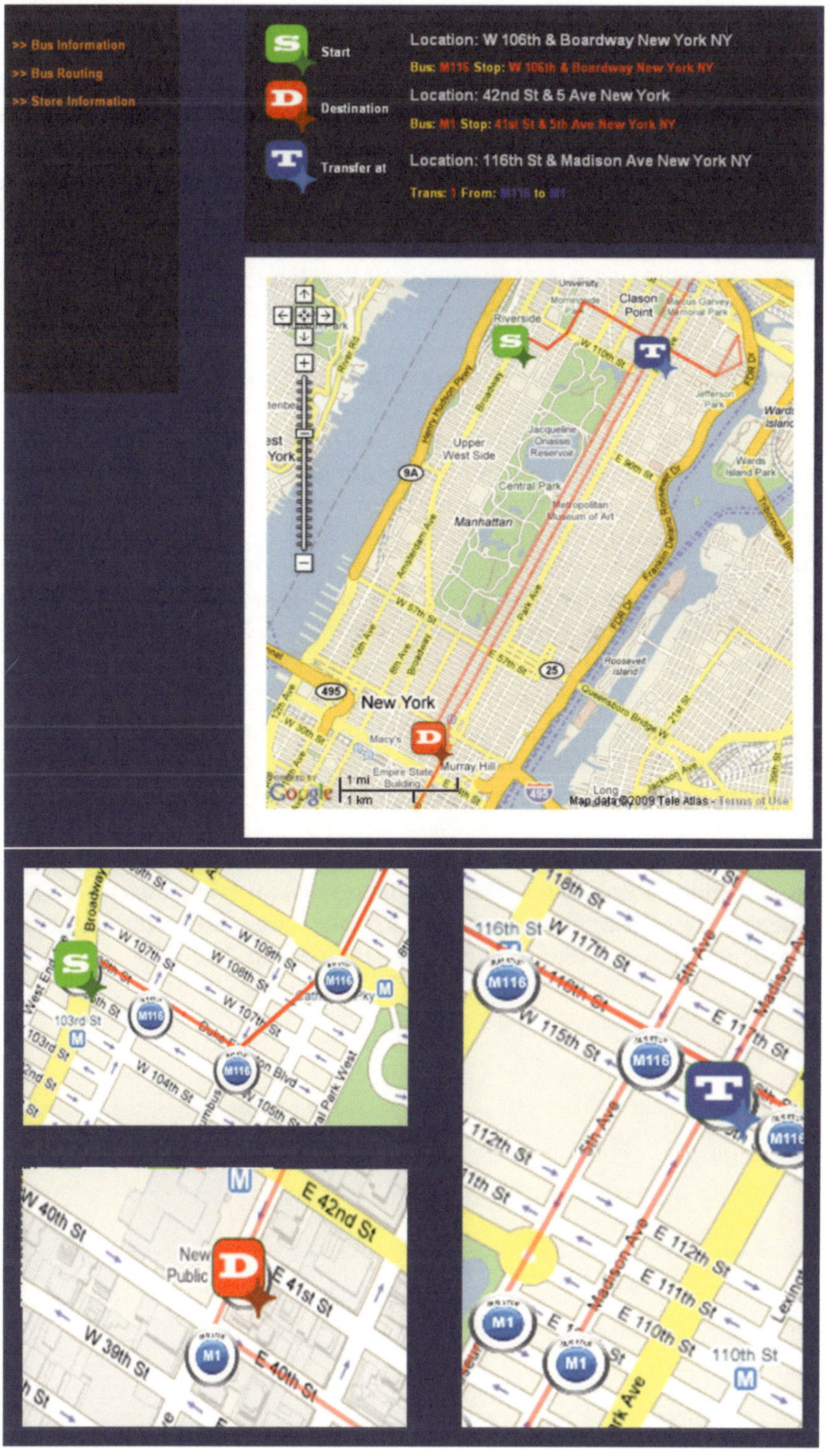


Fig. 12. Manhattan bus routing system

### 4.6 Community Support for Vulnerable People

For those aging or disabled people whose condition makes them unable to walk, advanced mobile wheelchairs provide many benefits, such as maintaining mobility, continuing or broadening community and social activities, conserving strength and energy, and enhancing their general quality of life. Especially residents in metropolitan areas increasingly view the convenience of public transportation. They are becoming used to supply and retrieve information gathered for regional community. The accessibility improvement needs the exact same type of the cooperation of transportation companies and regional residents: transportation companies, private industries, public places, and any points of interest for residents such as restaurants, shopping centers, movie theaters, etc., must supply the accessibility information into a public database system, which in turn provides immediate retrieval for transportation and route selection (Kawaguchi and Chan, 2009).

The Web interface of our system has additional capability to accumulate community information in the form of a point of interest and/or assistive ability. The people in the metropolitan areas can register and augment categorized information in the database. Specifically, a person (e.g., restaurant owner) can contribute to the system by registering his/her item (e.g., restaurant) into a set of suitable categories (e.g., bathroom services, eatery, etc.). The registration is done by filling out the owner's identity (full name, contact address, phone, and email) as well as detailing the item to register and the category to classify. The information on the item includes name, address, phone, web URL, geo-position, capacity, up to 3 pictures, textual comment, and a link to the owner. The category can be chosen from those already in the database or newly created at the registration. There may be multiple categories to which the item needs to belong. For instance, restaurant can be in the categories of bathroom services and eatery.

The registration of the item is not immediately reflected to the database, but becomes pending at first. An administrator of the system needs to inspect if the requested registration is valid or not. Email is generated to the owner right after the administrator's decision. Category classification is not static. The administrator can reorganize it by merging and splitting branches. The registered items whose locations are close to the suggested course of travel are selectively shown to the GoogleMap result (see Fig. 13).

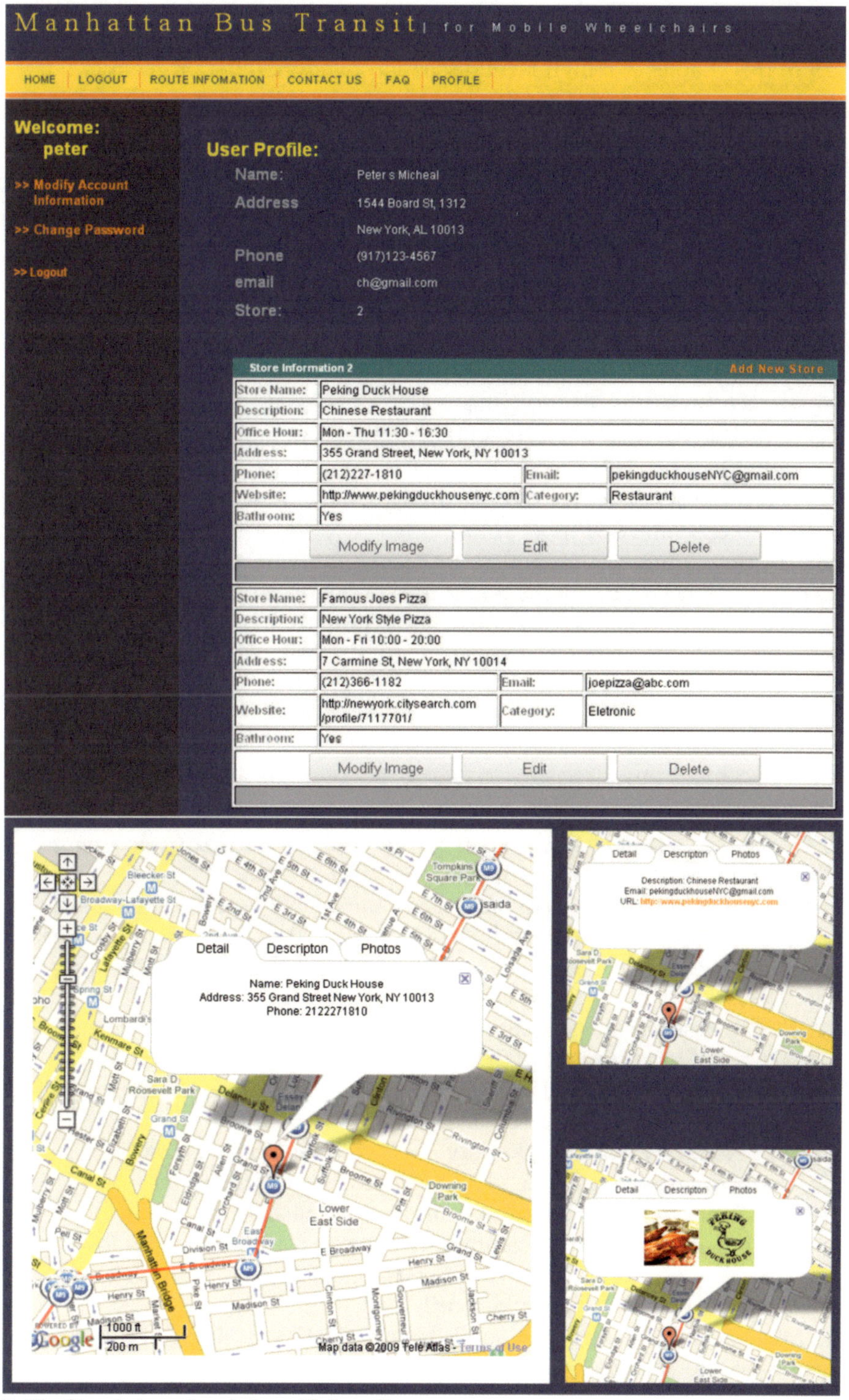


Fig. 13. Finding interest points with bus lines

## 5. Conclusions

The subject of this chapter is a high-level discussion to address the problem of our society in the coming age. The mechatronics approach outlined here is the framework and methodology to cope with the influx of vulnerable people in mobile wheelchairs. The concept of the micro and macro-control architecture based on the proposed mechatronics approach is hoped to give the opportunity to understand and promote resulting benefits in broader developmental contexts. In the transportation aspect, mobility issue has powerful effects on the living conditions of those with physical disadvantage. The ultimate goal of our on-going, collaborative work is to make electric powered wheelchairs predict and avoid risky situations and navigate safely through the congested areas and confined spaces. Presently, we are gathering our studies into the envisioned mechatronics support. Additional mechanisms are also being explored, so that the onmi-directional movement as well as rider's posture-sensitive movement can be further improved to achieve more comfortable travel to arbitrary places. For those aging or disabled people whose condition makes them unable to walk, advanced mobile wheelchairs provide many benefits, such as maintaining mobility, continuing or broadening community and social activities, conserving strength and energy, and enhancing their general quality of life. To accomplish a more future oriented view of this class of services, today's information systems must be expanded to incorporate the community support and to objectively evaluate whether or not the public transportation services are in the perspectives of the user's needs. The mechatronics approach we proposed encompasses these views and shares the same sprit of the universal design.

## 6. References

Arkin, R. (1987). Motor schema based navigation for a mobile robot, *The International Journal of Robotics Research*, Vol.8, No.4, pp.92-112.

ADA. (1990). *Americans with Disabilities Act of 1990* - ADA - 42 U.S. Code Chapter 126, U.S. Public Law 101-336, 104 Stat. 327, July 26, 1990.

CALL Centre. (1994). *Learning through smart wheelchairs.* Final report, Univ. of Edinburgh.

Chun-Ta Chen, Chieh-Chuan Feng and Yu-An Hsieh, (2004). Design and realization of a mobile wheelchair robot for all terrains. *Advanced Robotics,* Vol.17, No 8, December, pp. 739-760.

Cook, M.A. (2002). *Assistive technologies: principles and practice.* Mosby, 2nd Edition.

International Labour Organization. (2007). *Facts on disability in the world of work.*

Jones, M. and Sanford, J. (1996). People with mobility impairments in the United States today and in 2010. *Assistive Technology, 8.1, 43-53.*

Kawaguchi, A., Noda, Y., Sato, Y., Kondo, Y., and Terashima, K. (2008). A mechatronics vision for smart wheelchairs, *The 4th International Conference on Assistive Technologies (AT2008),* pp. 145-150, Baltimore, Maryland, April 2008.

Kawaguchi, A and Chan, C. (2009). Community Support for Disabled Bus Riders: What Can We Do? *International Conference of Computing in Engineering, Science and Informatics (ICC2009),* Fullerton, California, April, 2009.

Kondo, Y., Miyoshi, T., Terashima, K., and Kitagawa, H. (2008). Navigation guidance control using haptic feedback for obstacle avoidance of omni-directional wheelchair. *Proceedings of the 16th symposium on haptic interfaces for virtual environment and teleoperator systems,* pp. 437-444, March 2008.

Longmore, P.K. and Umansky, L. (2001). Editors, *The New Disability History: American Perspectives*, New York University Press.

*Medical Equipment and Supplies Manufacturing Industry in the U.S. and its Foreign Trade (1997-2009)*, Industry report by Supplier Relations US, LLC, August 28, 2007.

*Medical Supplies and Devices*, Industry report by First Research, Inc., August 27, 2007.

*Mechatronics principles, concepts and applications.* Mahalik, Nitaigour Premchand, Tata McGrow-Hill, 2003.

Ministry of Land, Infrastructure and Transport. (2006). *General Principles of Universal Design Policy*.

*MTA Guide to Accessible Transit*, Brochure of New York City Metropolitan Transportation Authority, 2006.

*MTA Trip Planner System.* New York City Metropolitan Transportation Authority, accessible at http://tripplanner.mta.info/ 2009.

Oe, Moriyuki. (2006). Problems and Implications of Japan's Aging Society for Future Urban Developments. *Policy and Governance Working Paper Series No.89*, March.

Orpwood, R. (1990) Design methodology for aids for the disabled. *Journal of Medical Engineering & Technology*. Vol.14, No.1, pp. 2-10.

Pin, F. and Killough, S. (1994). A new family of omni-directional and holonomic wheeled platforms for mobile robots, *IEEE Transactions on Robotics and Automation*, Vol.10, No.4, pp.480-489.

Prassler, E., Scholz, J., and Fiorini, P. (2001). A robotics wheelchair for crowded public environment. *IEEE Robotics & Automation Magazine*, Vol.8, No.1, pp. 38 - 45.

*Power wheelchair market opportunities, strategies, and forecasts, 2007 to 2013*, Market Research, April 2007.

*Reinventing the wheelchair - autonomous robotic wheelchair projects in Europe improve mobility and safety.* Issue of IEEE Robotics and Automation Magazine, 2001.

*Removing Barriers: Planning meetings that are accessible to all participants*, North Carolina Office on Disability and Health, 2005.

Richard C. Simpson, (2005) Smart Wheelchairs: A Literature Review *J. Rehabilitation Res. & Dev.* Vol.42, No.4, pp. 423-438.

Simon, R. M. (1997). Integrating Americans with disabilities act paratransit services and health and human services transportation. *Transportation Research Board*, April Number 10.

Simpson, R., LoPresti, E., Hayashi, S., Nourbakhsh, I., Miller, D. (2004) . *The smart wheelchair component system. J. Rehabilitation Res. & Dev.* Vol.41 No.3B, pp. 429—442.

Simpson, C. R. (2005) Smart Wheelchairs: A Literature Review, *J. Rehabilitation Res. & Dev.* Vol.42, No.4, pp. 423-438.

Urbano, J., Yang, Y., Terashima, K., Miyoshi, T., and Kitagawa, H. (2005). Navigation with comfort of omni-directional wheelchair driven by joystick. *Proc. of the IFAC World Congress*, Tu-M04-TP/14.

Wada, M. and Asada, M. (1999). Design and Control of a Variable Footprint Mechanism for Holonomic Omnidirectional Vehicles and its Application to Wheelchairs, *IEEE Transactions on Robotics and Automation*, Vol.15, No.6, pp.978-989.

*World population ageing*, United Nations, 2007.

Yanco, A. H. (1998). Wheelesley: A Robotic Wheelchair System: Indoor Navigation and User Interface. *Lecture notes in computer science*, Vol.1458, Springer Berlin.

# Communication and Artificial Intelligence systems used for the CAESAR robot

Riaan Stopforth (ZS5RSA) and Glen Bright
*Mechatronics and Robotics Research Group (MR²G), University of KwaZulu-Natal*
*South Africa*

R. Harley
*The school of Electrical and Computer Engineering*
*Georgia Institute of Technology*
*USA*

## 1. Introduction

Robots are necessary for search and rescue purposes, to access concealed places and environments that fire fighters and rescue personnel cannot gain entry to. Three hundred and forty-three firefighters died at the World Trade Center during the September 11 attacks in 2001 (Wiens, 2006). Often these rescuers unnecessarily entered an environment that had unstable structures as there were no live victims to rescue. Sixty-five of these rescuers died due to searching confined spaces that had flooded (Kleiner, 2006). Rescue workers have about 48 hours to retrieve victims (Gloster, 2007). Several hours are lost when rescuers are unsure of buildings stability. After a disaster the structures are often unstable and rescuers need to evacuate until the rubble has stabilized. Frequently the rescuers have to evacuate even though a body part of a possible survivor is seen, due to unstable surroundings (Roos, 2005). Robots can stay in the unstable area and continue searching for survivors. In the future, robots could possibly also be used to access mines after an accident prior to rescuers workers (Trivedi, 2001).

Urban Search And Rescue (USAR) Robots were first extensively tested at the collapsing of the World Trade Center site in 2001 (Greer et al., 2002). The University of South Florida were involved in these rescue attempts. The robots that they used are shown in figure 1. The advantage of these robots above rescue members is that the disaster areas can be entered immediately after a disaster.

Problems identified at the World Trade Center as well as at the testing grounds of the National Institution of Standards and Technology (NIST) are that the robot's traction system malfunctioned. (Greer et al., 2002). More research is needed for the robots to withstand the harsh conditions of a fire (Wiens, 2006). Other problems observed were unstable control system, chassis designed for narrow range of environmental conditions and limited wireless communication range in urban environment as well as unreliable wireless video feedback

(Calson et al., 2004). Some robots were either too large or not easily maneuverable (Gloster, 2007).

Further problems experienced were that the setup time of the robots was too extensive and the human to robot ratio for transport and controlling were not ideally 1:1 (Greer et al., 2002). Problems were identified regarding the communication with the robots (Remley et al., 2007).

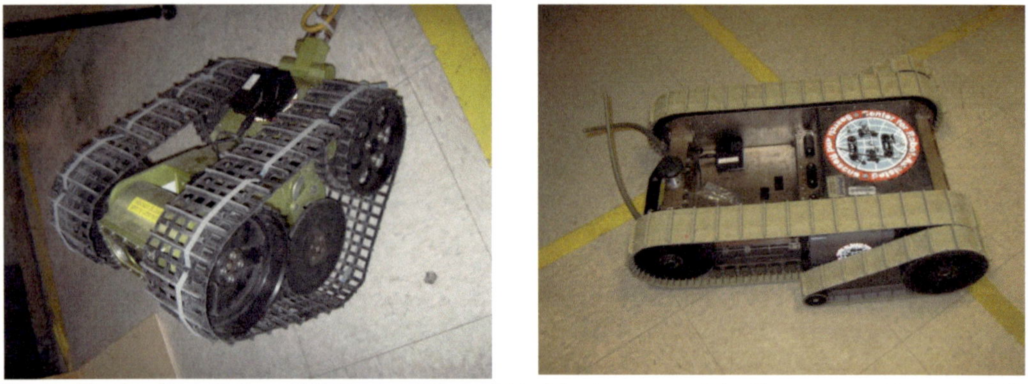

Fig. 1. The Inuktun MicroVGTV and I-Robot Packbot was used in the rescue attempts at the World Trade Center in September 2001

Communication is critical as the rescuers need to send instructions to the robots, but at the same time receive vital information about the environment. This could save lives as it could indicate poor structural areas, dangerous gases and extreme temperatures. Research has been performed to determine improvements and possible solutions to these problems experienced. These solutions include a combination of communication reliability in these environments, and a sensory system to allow the robots to maneuver across the terrain successfully.

The communication and sensory system is discussed as it was implemented on the CAESAR (Contractible Arms Elevating Search And Rescue) robot. These developments include communication protocols, hardware interface and artificial intelligence to indicate the safety and danger levels for both humans and the robot.

## 2. COMMUNICATION

The interferences that were experienced before are mainly due to the robots using Industrial, Scientific and Medical (ISM) bands. Many electronic communication units use the ISM bands which are unlicensed frequencies that have certain constraints. As USAR robots are used to save lives, it is suggested that licensed frequencies are utilized. This will significantly prevent interferences. The output power between the control unit and the robot can be constrained to prevent a signal from one unit overwhelming the signals from other units.

Another reason for failed robot communication is the loss of signals between the robot and its control unit. This is mainly caused by the frequency used. As wavelength is inversely proportional to the frequency and the antenna size is proportional to the wavelength therefore the higher the frequency, the smaller the antenna will be. Transmission efficiency decreases as higher frequencies are used. The signal penetration into buildings is also effected by the frequency used. Higher frequencies are capable of penetrating more dense materials that lower frequencies. The disadvantage of higher frequencies is that small items, such as dust particles, resonate at the high frequency therefore causing it to absorb the power of the signal. Therefore it is best to use a frequency in the center of the two extremes that will allow optimization for radio communication. The comparison of the different factors that are considered are shown in figure 2. Subsequently the decision is, to use UHF frequencies as these are able to penetrate with a relatively low power output and have a relatively good signal penetration property.

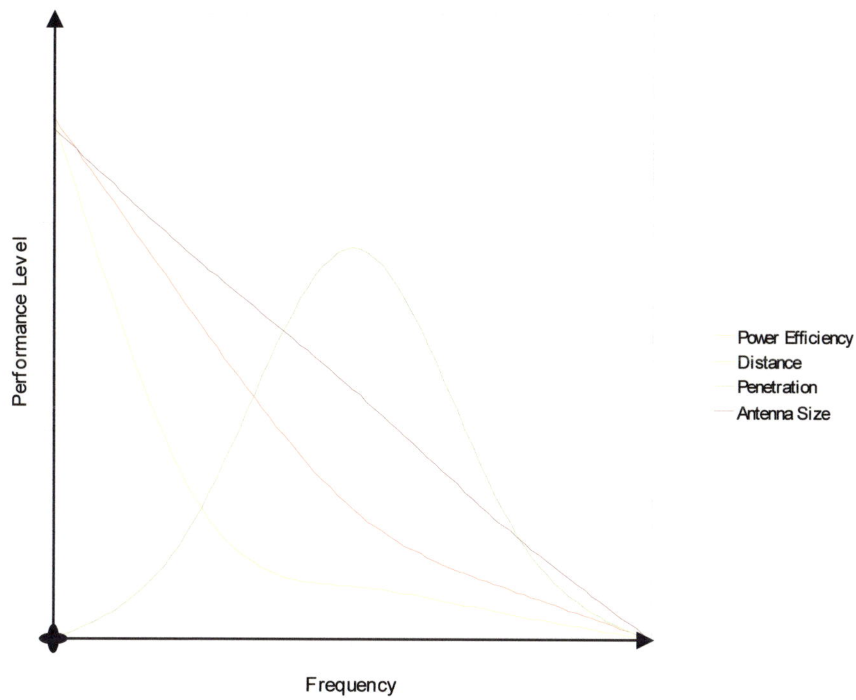


Fig. 2. Comparison of factors considered as frequency increases

### 2.1 Data Communication

Different modules and units are needed for the successful communication of video, audio and data. The modules that were used for the CAESAR robot, are discussed and explained.

#### 2.1.1 Radio Modules

The Radiometrix narrow band FM multi-channel UHF TR2M-433-5 transceiver modules is used for the data communication. A photograph of one of these modules is shown in figure 3 (Radiometrix, 2004).

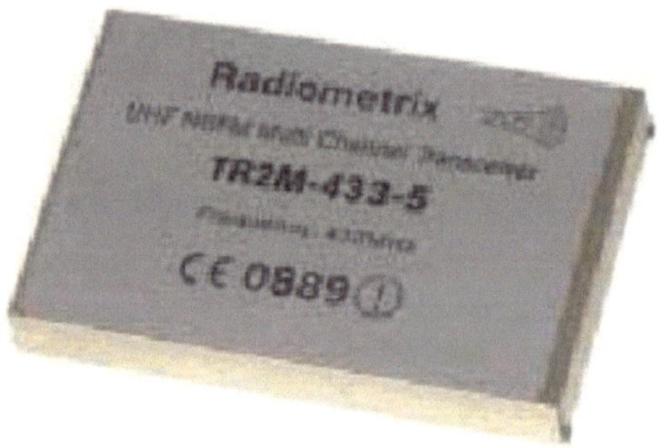


Fig. 3. Radiometrix TR2M radio module

The features of the TR2M modules are:

- Can be programmed to operate on any 5 MHz band from 420 MHz to 480 MHz.
- Fully screened
- 1200 baud dumb modem

Pertaining to the above features, these data modules will be valuable for the USAR robot. It enables the programming of the modules to operate on the frequencies supplied by the fire department. The power consumption is low which is vital for power saving. With this large range of operating temperatures the heat from the outside could be insulated and limited to the module.

The only problem that occurs regarding these modules is their inability to transmit more than 10 mW. An output power of 5 W is required for efficient communication with the restrictions of buildings and other power absorbers. A RF amplifier is needed to solve this problem.

RF amplifiers that amplify 10 mw to at least 5 W are either not readily available or they are expensive. In order to solve this problem, the final stages of Motorola MCX100 radios were used. The need arose for two of the three RF amplification stages as the amount of power that these final stages produce is sufficient, whereas the three final stages produce more than 5 W output power. Refer to figure 4 for the interconnection of these stages. The disassembly and reconstruction of these stages require the addition of discrete components. Not all the modules in the radio were used. These impedances of the missing modules are to be replaced. The circuit of the RF amplifier is traced with a probe to determine the amplification of each stage. There are two positive power supply points. Tracing the power point that was not powering the circuit of the first stage of the RF amplifier, it was found that there was a discontinuation for a closed loop circuit. This closed loop circuit was terminated to another module not used. By modifying the impedance on this point, a different output power was produced from the RF amplifier. It was discovered that a resistance of 300Ω made the RF amplifier produce 5W output.

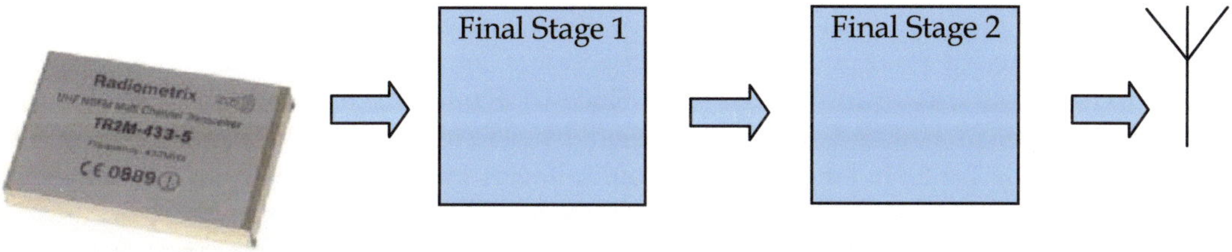


Fig. 4. Transmission process block diagram

A problem occurred in the reception, as the signal was not able to reach the TR2M module from the antenna due to the RF amplifier not being bi-directional. This could possibly be solved by connecting the antenna directly to the TR2M module and then reception would be possible, but the high output power from the RF amplifiers would terminate the operation of the TR2M module, as there is high power penetrating the sensitive module.

This problem was solved by the implementation of a switching circuit on the output to the antenna. Figure 5 illustrates the concept of this circuitry. While the two relays are in position 1, the TR2M module can receive data. Should the TR2M module need to transmit, then the relays are switched over to position 2, which will connect the TR2M module to the RF amplifier and in turn with the antenna. This prevents the need for two antennas and allows for only one radio module for data communication at each station.

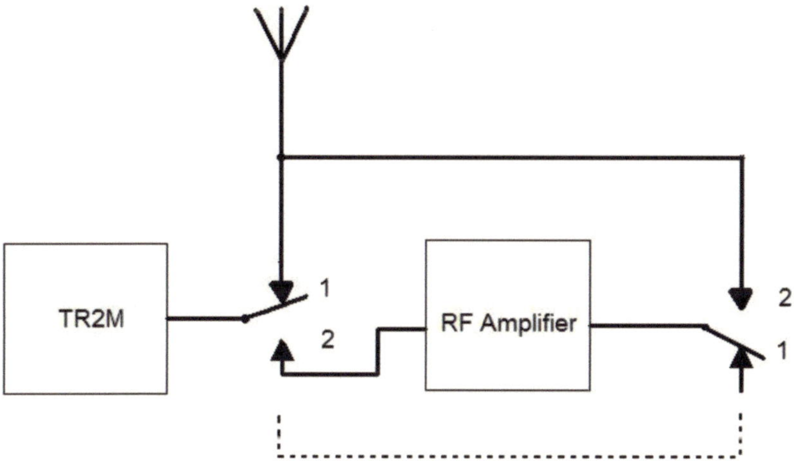


Fig. 5. TR2M and RF Amplifier with the appropriate switching

#### 2.1.2 Protocols

The use of protocols is important for data to be successfully transmitted. Using available protocols is an option, but the performance and efficiency must be considered. Most existing protocols have been developed over many years and by various people. These protocols are optimized for best performance for a specific task.

The IEEE 802.11 protocol could be used for communication between the robots, but there is not always an Access Point available for the wireless communication. The communication between the robots will be an Ad-Hoc style. Since UHF frequencies are being used, the data rate will be less in comparison to that used by wireless communication, as they use frequencies in the 2.4 GHz band and the quality factor bandwidth decreases as frequency decreases. Due to the bandwidth being decreased, additional collisions might occur and therefore smaller packet sizes are needed. More data transmission from other stations is able to occur when the packet sizes are smaller.

### 2.1.3 Robot Communication Protocol

The Robot Communication Protocol (RCP) uses different aspects from the wired and wireless LAN protocols. The problem when using wireless communication technology is that it uses the 2.4 GHz band which causes the small particles of buildings to resonate at this frequency and to absorb energy which can prevent penetration through buildings. A further problem with the use of the IEEE 802.11 protocol is that its packets contain header details that will not be utilized for the USAR robots. This is therefore unnecessary data that will be transmitted and will occupy the use of the medium. In view of the fact that the baud rate of the data communication modules can be low, unnecessary data must be prevented as this can saturate the medium.

Another problem pertaining to the existing protocols is that they may possibly contain non-printable characters that cannot be processed by certain computers and microcontrollers. The printable characters are those that have an ASCII value between 31 and 127.

A new wireless communication protocol is required for USAR robots to utilize. A decision was made to use callsigns to identify the robots and control units to prevent communication interference. A six character callsign that consist of letters of the alphabet and numbers is assigned to each robot and control unit. This gives a combination of $36^6 = 2.17 \times 10^9$ different callsigns available.

There are two types of protocols that need to be transmitted namely: a "one way packet" that is sent from one station to the other and that needs no confirmation (referred to from now on as a Robotic One-way (RO) packet) and a packet which is sent from one station to the other and which replies with an acknowledgment of reception packet (referred to from now on as a Robotic Confirmation (RC) packet).

There are four packets for the robotic network namely, Request-To-Send (RTS), Clear-To-Send (CTS), Acknowledgment (ACK) and Data packet. The different packets with their fields are explained below.

***RTS / CTS / ACK Packet***

The packet format for the RTS, CTS and ACK packets are shown in figure 6.

| **Size** | 1 byte | 1 byte | 2 bytes | 6 bytes | 6 bytes | 1 byte | 1 byte |
|---|---|---|---|---|---|---|---|
| **Field** | Start | Type | Duration | RA | TA | Checksum | End |

Fig. 6. RTS / CTS / ACK Packet

**Start:** The start character is for stations to identify the commencement of the packet. This is indicated with the hash (#) character. Should a station only start receiving in the middle of a transmission it will then recognize this and discard the packet. The purpose for the necessity of a start byte is that the transmission is asynchronous on a single channel.

**Type:** This field indicates the type of packet that is being sent. The indication for the RTS, CTS and ACK packets are the characters 0, 1 and 2 respectively.

**Duration:** The duration of the transmission is specified in this field. This provides the other stations with the time period to delay before attempting to transmit. The duration is specified by the number of characters. Time periods are calculated from the sum of the two bytes multiplied with x, where x is the time period for each character to transmit.

$$x = \frac{8bits}{baudrate} \tag{1}$$

Should these values be a "#" or "!", then the most significant byte must be incremented and the least significant byte must be decremented.

**RA:** This is the address of the receiving station. This field presents the opportunity for other stations to identify whether that the packet is for them or not. Should the packet not be intended for the station, the rest of the incoming packet can be disregarded and the station can start processing other incoming packets after the delay duration.

**TA:** This is the address of the transmitting station and is used by the receiving station to identify if the packet is from its approved station.

**Checksum:** This verifies the integrity of the packet. The field value consist of the sum of all ASCII values of all characters in packet modular 94 and the addition of 32. Should the receiving station receive a packet that is not approved then it is subsequently dropped. If the value of this field should be equal to "#" or "!" then the duration field is incremented and the checksum is recalculated. This field must be a printable character and not a control character (I.e. the character must have an ASCII value between 31 and 127)

**End:** This indicates the end of the packet with an exclamation mark (!) character.

***Data Packet***

The format of the Data packet is shown in figure 7.

| **Size** | 1 byte | 1 byte | 2 bytes | 6 bytes | 6 bytes | 0–255 bytes | 1 byte | 1 byte |
|---|---|---|---|---|---|---|---|---|
| **Field** | Start | Type | Duration | RA | TA | Data | Checksum | End |

Fig. 7. Data Packet

**Start:** The start character is for stations to identify the beginning of the packet. This is indicated with the hash (#) character. In the event that a station only starts receiving in the middle of a transmission, this will be identified and the packet will be discarded. The motivation for a start byte is that the transmission is asynchronous on a single channel.

**Type:** This field indicates the type of packet that is being sent. The identification of a RO Data packet is the character 3 while for a RC Data packet it is the character 4. The other possible values (except for the character values for # and !) for this field are reserved for future use.

**Duration:** The duration of the transmission is given here. This provides the other stations with the time period that they have to delay with before attempting to transmit. The duration is given by the number of characters. Time periods are calculated from the sum of the two bytes multiplied with x, where x is the time period for each character to transmit.

$$x = \frac{8\,bits}{baud\,rate} \tag{2}$$

Should these values exist of a "#" or "!", then the most significant byte must be incremented and the least significant byte must be decremented.

**RA:** This is the address of the receiving station. This gives the opportunity for other stations to identify whether the packet is meant for it or not. In the event that it is not, the station can ignore the rest of the incoming packet and start processing other incoming packets after the delay duration.

**TA:** This is the address of the transmitting station. This is used by the receiving station to identify that the packet is from its relative approved station.

**Data:** The data for specific instruction or information between the stations is stored in this field. The only characters that are not allowed in this field are the hash (#) and the exclamation mark (!) seeing that these are the start and end characters respectively. Control characters are also not allowed in this field.

**Checksum:** This verifies the integrity of the packet. The field value consist of the sum of all ASCII values of all characters in packet modular 94 and the addition of 32. Should the receiving station receive a packet that is not approved it is subsequently dropped. If the value of this field is equal to "#" or "!", the duration field is then incremented and the checksum is recalculated. Furthermore this field must be a printable character and not a control character (I.e. the character must have an ASCII value between 31 and 127)

**End:** This indicates the end of the packet with an exclamation mark (!) character.

#### 2.1.3.1 Communication Procedure

The description of the communication procedure is described by means of two stations; station A and station B. Should station A want to transmit, it would observe whether no transmissions are occurring. If none are detected, then station A starts transmitting a RTS packet. All the stations in the vicinity of station A will delay transmission for the period of the duration field in the RTS packet. The delay duration period consists of the sum of the following:

- the time period needed to transmit the RTS packet
- the time period needed to transmit a CTS packet
- the time period for the Data packet

- the time period to transmit an ACK packet (if this is needed)
- the sum of the processing time at each station

Station B receives the RTS packet and replies with a CTS packet which contains a delay duration period which is:

- the sum of the time period for the CTS packet
- the time period to transmit the Data packet
- the time period to transmit an ACK packet (if this is needed)
- the sum of the processing time at each station.

Station A responds with the Data packet that contains a delay duration period which is the sum of the time period for:

- the time period to transmit the Data packet,
- the time period to transmit an ACK packet if this is needed
- the sum of the processing time at each station.

Station B will reply with an ACK packet should the last received packet have a type value of 100. This packet will contain a delay duration period which is the sum of the time period to transmit the ACK packet as well as the processing time at each station.

Given that there is no Access point that is stationary, there is no station that controls communication within the network. In figure 8 four stations are shown with their respective radio coverage. C1 and R1 are control unit 1 and robot 1 respectively and C2 and R2 are control unit 2 and robot 2 respectively.

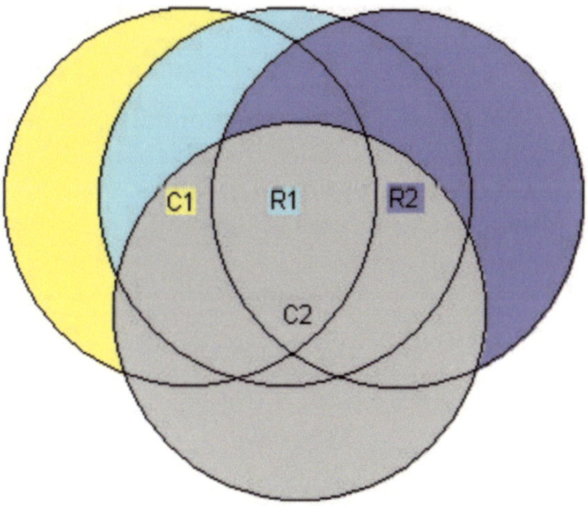


Fig. 8. Radio Coverage of two control units and two robots

As noted in figure 8, C1 is in radio coverage with R1 and C2; R1 is in radio coverage with R2 and C2; R2 is in radio coverage with C2. Since C1 and R2 are not in radio coverage packets to request transmission will not be received between these two stations. This is not a great disadvantage as the different stations operate in an ad hoc system. Of importance is the aspect that each robot is able to communicate with its own control unit.

Should a RTS packet be transmitted by C1 then R1 will subsequently receive the request and reply with a CTS packet. This CTS packet, which will contain a duration field, will be received by R2 as well. In view of the fact that R2 has received this packet, it will delay with any transmission for this time period before trying to transmit again.

In the instance that both C1 and C2 transmit a RTS at the same time, R1 will then receive data that will be a combination of data from the two control units. R1 will reject this data, as it will not recognize it or because it will not exist of an acceptable packet. After a time-out period C1 will realize that R1 has not responded and will transmit the RTS again if required.

As the RTS packets are relatively small, the overhead of retransmission would be small if two stations should transmit the same time. The sum of data being sent in the Data packet is limited to 128 characters and it need not be necessarily sent in a specific format, providing the format is understandable between the respective control units and the robots.

The advantage of the RCP is that a computer system could be connected to a modem that uses the same protocol and this modem could then transmit and receive instructions and data to a large network of robots. In this situation the computer will be the control unit and will not be dedicated to only a single robot. This network of robots could then be controlled to perform a task that could have a greater efficiency than a single robot.

The RCP packets that are used to control a robot have smaller packets sizes of at least 38 % compared to those used by hard-wired computer network protocols and 33 % compared to that used by IEEE 802.11 protocol. Communication between the robots and their control units are more reliable when used in a network scenario. The use of a computer network protocol could be valuable when the robots have to transmit data and information that involves more than just the basic instructions.

### 2.1.4 Modular Approach for Layered Model

A layered model similar to the OSI model is needed for data communication. Each layer has its unique task to optimize the communication. The advantage of having a layered model is that each layer can be modified and optimized without affecting the other layers. The layered model can be represented as indicated in figure 9.

| |
|---|
| Application |
| Data link / Transport / Session / Presentation |
| Physical |

Fig. 9. A three layered model

This model has been divided into three layers as each layer will be controlled by a separate module or microcontroller. The Physical layer consists of the hardware that will be used. In the case of the USAR robot, this will be the radio modules that will act as the transceivers.

The layer that is a combination of the Data link, Transport, Session and Presentation will be controlled by a single microcontroller. The Data link layer is in control of the packets that are being sent, while the Transport and Session layer is responsible for the packet's control and transmission permission respectively. All the received data must be presented in a format for the computer to understand. This is achieved by the Presentation layer.

The Application Layer is involved in the displaying of the information and with the interaction with the user. This layer is also involved in the output, being the movement of the motors and any other attachments of the robot. This layer will be controlled by a microcontroller which could be attached to other microcontrollers or modules, depending on the complex of the attached module.

### 2.2 Voice Communication

Voice communication between the robot and the rescuers is essential for the rescuers to get information from survivors. Rescuers can also calm the victims when the robot approaches and notify victims that help is on the way and of possible ways to save themselves. The voice communication between the robot and the rescuers will be achieved through the video communication but communication between the rescuers and the robot is still required.

Two radios are needed for this communication to occur. Since one of the assigned frequencies is used for the data communication, the other assigned frequency is to be used for the voice communication. It was thus decided to use Amateur radios for this communication as it was possible to purchase them due to a license obtained.

The decision was to use the Yaesu VX-7R and VX-3E transceivers. These radios can be modified to operate on these emergency bands and have different useful features. Diagrams of the Yaesu VX-7R (Vertex, 2002) and VX-3E (Vertex, 2007) are shown in figure 10.

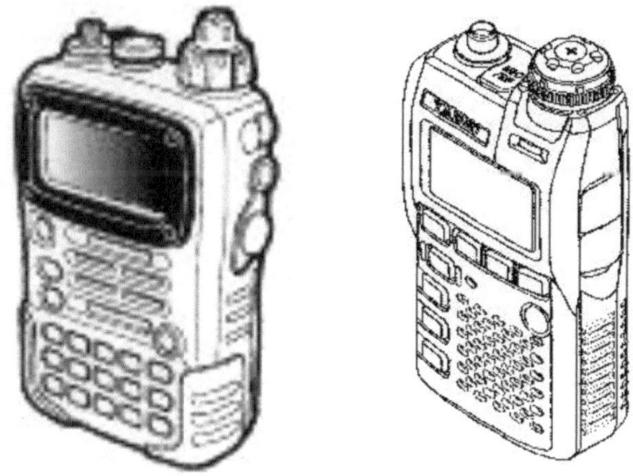

Fig. 10. Diagram of the Yaesu VX-7R and VX-3E radios

The Yaesu VX-7R have the feature to operate on UHF bands. The Yaesu VX-7R is used in the control unit. It has the useful characteristic of a keypad, allowing the rescuers to tune into frequencies other than those used for the robot, if so required. With this radio it is possible for the rescuers to tune into the audio frequency of the video transmission from the robot, should the sound from the television be unclear.

The Yaesu VX-3E has feature that it can receive between 420 and 470 MHz. The Yaesu VX-3E is used mainly for reception of audio in the robot. The useful characteristics of the Yaesu VX-3E is that it is small in size, light weighted and can operate at temperatures that could possibly occur in the robot.

### 2.2.1 Microphone and Speaker adapter

The audio input to the video transmitter, (discussed in Chapter 2.3 – Video Communication) needs to have an impedance of 600 Ω and a maximum voltage of $1V_{P\text{-}P}$ or 0.775 $V_{RMS}$. A 600 Ω dynamic microphone was initially connected to the input of the audio as there was no verification as to whether the transmitter had a build in preamplifier. This did not seem to work, so a mono microphone preamplifier is used to amplify the signal from the dynamic microphone. While the preamplifier is connected to the transmitter, the preamplifier output is tested on an oscilloscope and the gain is altered to get a maximum output of $1V_{P\text{-}P}$. The schematic of the microphone preamplifier is shown in figure 11 (Excellence, 1998).

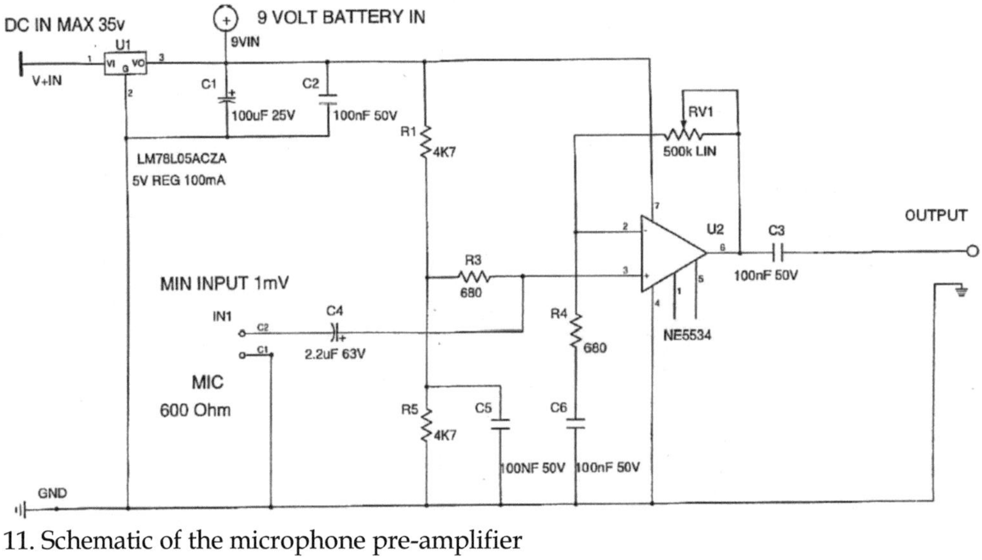


Fig. 11. Schematic of the microphone pre-amplifier

The dynamic microphone used is manufactured from plastic which will result in a problem at high temperatures. Research has been performed to determine the availability of high temperature microphones but the research proved unsuccessful. It is therefore decided to continue using the plastic microphone to enable the testing of the principles being discussed.

An ear piece with microphone is used in the Yaesu VX-7R radio, to allow the controller to communicate with any victims. The VOX-activation function could be set to allow transmission of spoken voice.

A 1.5mm earphone plug is used for the Yaesu VX-3E radio and connected to an 8Ω speaker. Research was performed to determine whether speakers were available that would be able to resist the high temperatures, but none were found. An ordinary speaker is used to prove the principle.

### 2.3 Video Communication

The video is from the FLIR PathFindIR thermal camera shown in figure 12 (FLIR, 2006). It has the following specifications:

- Size: (58mm x 57mm x 72mm)
- Input Voltage range: 6V - 16V
- Power dissipation: Less than 2W
- Weight: less than 0.4 kg

The PathFindIR is ideal for this project, as it is small, does not weigh much, and is affordable compared to other available thermal cameras. It has a low power dissipation and can operate from -40 °C to 80 °C. Should the temperature decrease below -40 °C, the heating element is switched on, therefore allowing images to be transmitted in cold environments.

The video from the PathFindIR needs to be transmitted. ICASA (Independent Communication Association of South Africa) and Sentech have given permission to use channel 54 (735 MHz) for video transmission, on the condition that the output power is less than 1W, and the transmitter is calibrated by one of their approved dealers.

Fig. 12. FLIR PathFindIR thermal camera

The modulator and IF converter is used to generate the video on the required frequency. This signal is then amplified to 1W. This amplifier is shown in figure 13 (Jackel, 2008).

Fig. 13. 1W UHF amplifier

These modules can operate between 470 – 862 MHz. It has been confirmed that all output power for communication must be at least 5W(Reynolds, 2008) for search and rescue reasons. As there is a restriction for the video output power, 1W is used to prove the concept for this robot. It is suggested that a video frequency is assigned for search and rescue purposes so that the output power can be increased to 5W.

A block diagram of the interconnection between the PathFindIR, converter/modulator, microphone, audio preamplifier, video amplifier and antenna is shown in figure 14.

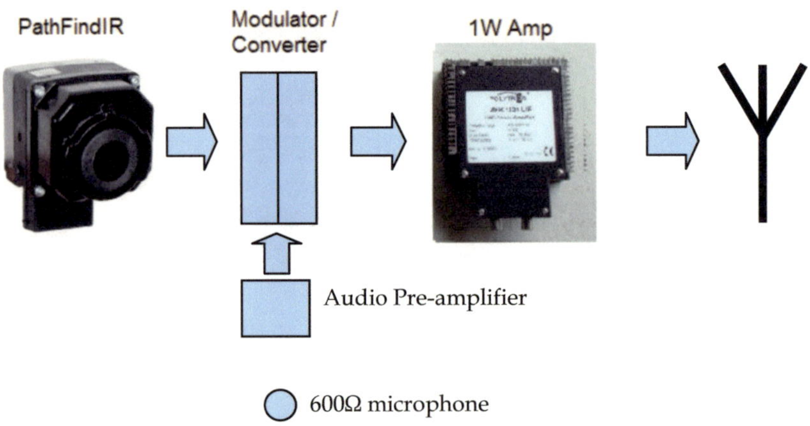


Fig. 14. PathFindIR connected to the modulator/converter, 1W UHF amplifier, audio preamplifier and antenna

### 2.4 Antennas

Antennas are the source of transmission into the medium of air and the absorber of signals from the medium of air. Different antennas have different properties of radiation patterns and polarization. This is a topic in communication that is often neglected, but the antenna used has an effect on the performance of transmission and reception of signals. The antenna of a radio can influence many factors that can be the cause of many problems. Calibrating and selecting an antenna influences the efficiency of output power and signal strength that will be radiated from a radio.

The antennas used were investigated. The orientation of the antenna effects the polarization of the transmitted waves. It would be ideal to have vertical and horizontal polarization. The best antenna for this purpose is the egg-beater type. It gives vertical and horizontal polarization, but it has the disadvantage of being relatively large, which is not ideal, as one of the objectives of a USAR robot is to design it as small as possible.

Vertical antennas were investigated and a problem encountered is that the base plane shields the signals from being transmitted through it. Different fractions of the wavelength antennas have got different properties. A ½ wavelength antenna has radiation lobes that are perpendicular to the antenna, while the ¼ wavelength antenna has radiation lobes that are at an angle of about 45 degrees. The use of the property of the ¼ wavelength antenna will work well as it was found that it has a degree of output power directed towards the end point of the antenna. The only disadvantage of this type of antenna is that there is no radiation past the base plane.

This problem is solved by removing the base plane and replacing it with a piece of coaxial cable that is longer than the ¼ wavelength. The reason for the need of the base plane or coaxial cable is that it produces the negative part of the modulated sine wave. With the removal of the base plane, the radiation from the antenna is relatively isotropic, with low radiation towards the end points of the antenna. The antenna then is seen as a ½ wavelength dipole antenna. This isotropic radiation pattern is caused by the minor lobes that are allowed to be radiated next to the main lobe. When the robot is a number of wavelengths above the ground, the radiation pattern will become more isotropic because of more lobes, and will lower the elevation angle of the lowest angle lobe (Roos, 2005). This antenna has the disadvantage in that it is not being vertically and horizontally polarized. This is solved by using an egg-beater type antenna that is scaled in size at the receiving unit. It will then be able to receive any polarized signal (discussed in section *2.4.2 Eggbeater Antenna Design*).

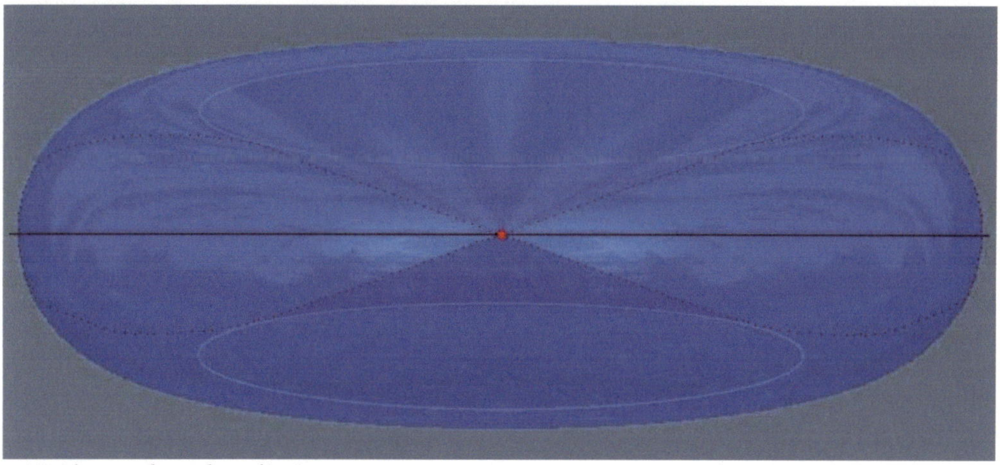

Fig. 15. ½ wavelength radiation pattern

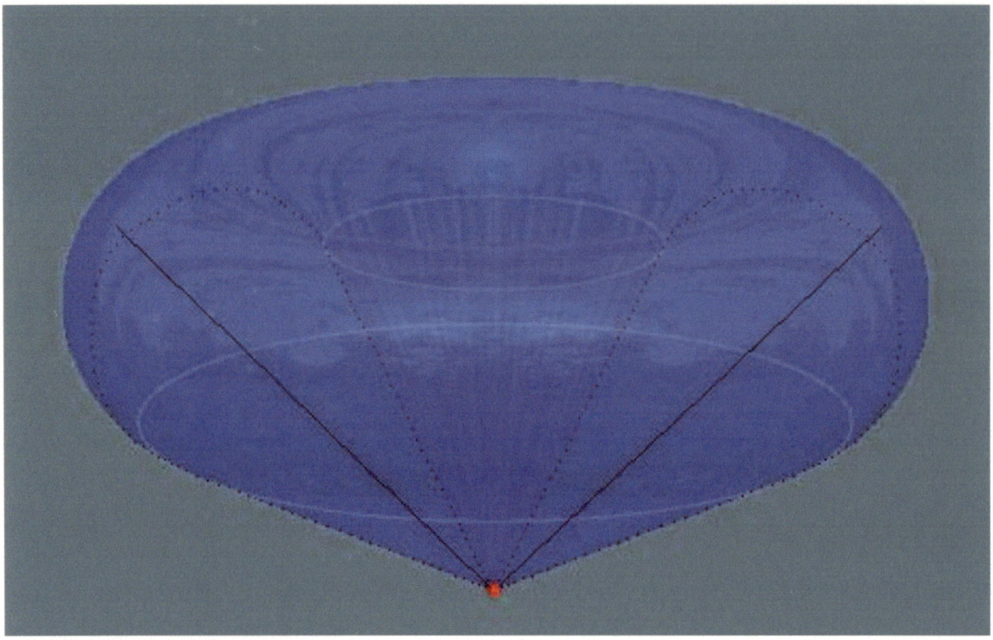

Fig. 16. ¼ wave radiation pattern

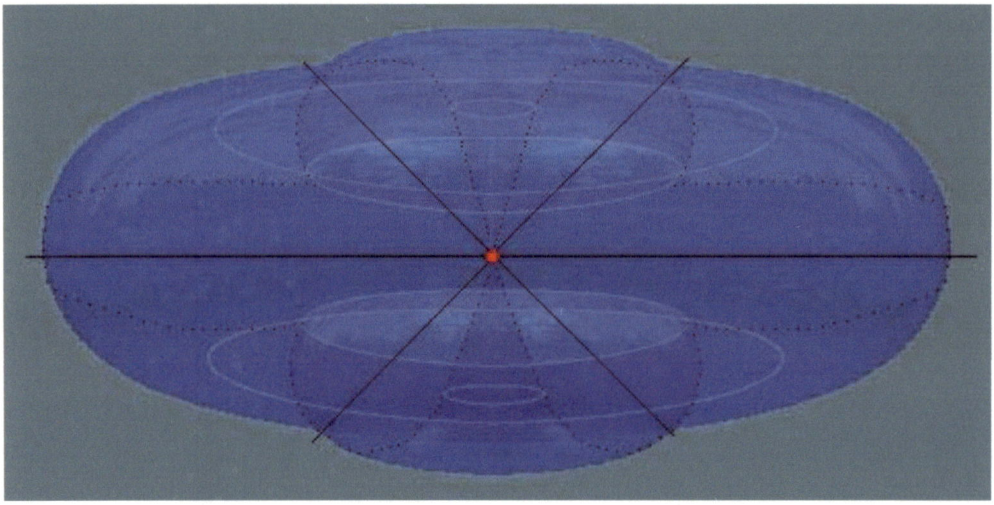

Fig. 17. Radiation pattern of antenna that is used.

Communication is improved with the use of UHF frequencies because, the penetration of the signal is increased, the antenna is relatively small and the transmission efficiency is still acceptable. With the use of a dipole antenna that has coaxial cable for the ground plane, the radiation pattern is increase by 100 % in terms of direction compared to an antenna that has a base plane. The radiation distance decreases as the output energy remains the same and is spread over a larger angle. The polarization of the radiated waves are in the same orientation as the antenna's orientation and can be received with an egg-beater antenna that is capable of receiving any polarized signal.

### 2.4.1 Quarter-wave Antenna Design

The length of the full wave antenna in free space is calculated from equation 3. This equation is valid for transmission in free space.

$$\lambda = \frac{300}{f} \tag{3}$$

where: $\lambda$ = wavelength in meters

f = frequency in MHz

The surrounding air has an effect on the antenna, so a factor η has to be multiplied with equation 4. The value of η is variable and depend on the antenna's surroundings. As the robot will be operated in conditions of smoke, heat and with various objects surrounding it, the value of η would vary.

Equation 4 is used to calculate the wavelength of the antenna. This length of antenna wire is then cut and connected to the radio with a Standing Wave Ratio (SWR) meter, which is connected in series with the feed line. Millimeters of the antenna is trimmed away until the SWR value is very close to a SWR ratio of 1:1.

SWR is the ratio of the forward and reflective power. Power is reflected back into the transmitter when the load does not have a matching impedance to that of the characteristic impedance. The SWR of a specific load can be calculated with equation 4 (Frenzel, 2001).

$$SWR = \frac{1+\sqrt{P_R / P_F}}{1-\sqrt{P_R / P_F}} \tag{4}$$

From equation 4, it is seen that as $P_R$ decreases, the SWR will tend to 1. To determine the SWR of a specific antenna, the meter is calibrated so that there is maximum deflection for the forward transmission of a signal, and then the reflective signal back into the system is read. This reading is performed every time an alteration of the antenna is made, until the SWR is close to 1:1. The ideal situation is to have a SWR of 1:1, but there are many factors that can influence this reading, such as surrounding objects.

An antenna tuned for a frequency in the UHF band is compatible for most frequencies in the UHF band. This characteristic is used to tune the antenna for a frequency of 450 MHz. With the use of equation 3, the antenna wavelength is calculated as:

$$\lambda = \frac{300}{450} = 667mm \tag{5}$$

The full wavelength is 667 mm, but since a quarter wavelength antenna is to be used, the antenna length required will be 166.75 mm. From this length, the antenna is lengthened or trimmed until a SWR of 1:1 is obtained. This is needed as the antenna is operated in an environment that is not free space. The final antenna length is 170 mm.

### 2.4.2 Eggbeater Antenna Design

Different forms of the eggbeater antenna design were considered. The testing of the antennas were performed with the use of a RF generator and a SWR meter. A receiver with a horizontal antenna was set up. The strength of the signal received by the receiver is displayed on a signal analyzer.

A folded dipole antenna was initially considered. This is a dipole antenna that is bent into a loop, bringing the ground and live point to each other, but not touching each other. The same configuration is used for another folded dipole antenna that is placed 90 degrees to the first one. The two loop antennas are separated with a quarter wave stub, so that the transmission between the two loops are 90 degrees out of phase and therefore prevent cancellation. The quarter wavelength coaxial cable stub must be shortened depending on the velocity factor of the transmission line. This value typically varies between 0.6 and 0.7. The velocity factor of a RG-174 coaxial cable is 0.66. This quarter wave coaxial cable length can be calculated by equation 6 (Frenzel, 2001).

$$\frac{1}{4}\lambda = F\frac{75}{f} \tag{6}$$

where F is the velocity factor of the coaxial cable.

It is very difficult to determine the exact length of the coaxial cable stub, as the theoretical value does not correlate to that of the practical assessment. Therefore a Dip Meter is used to cut the exact length of the coaxial stub. The Kenwood DM-81 Dip Meter was used for this. A photo of this Dip Meter is illustrated in figure 18.

The Dip Meter has a connection for a coil for the required frequency. A coil for a harmonic of 450 MHz was used. As the dial of the Dip Meter is not very accurate, the frequency counter that is on the Yaesu VX-7R was used to get the resonating frequency close to 450 MHz.

The Dip Meter is calibrated on the resonated frequency and a portion of coaxial cable is then placed next to the coil. A single loop then is made from a piece of wire and is soldered between the center conductor and the outside braid. Initially this loop is placed around the coil to get a broader band reading. The dial is cautiously turned until the Dip Meter is at full deflection. With this configuration, 2 mm pieces are cut from the coaxial cable, until it is detected that the Dip Meter is deflected towards zero. This is an indication that the coaxial cable being tested is absorbing most of the power at that frequency and that the coaxial cable is exactly a quarter wavelength which also incorporates the velocity factor.

The tests proved that the antenna is relatively omni-directional, but there is a couple of cancellations of signals where two parallel or perpendicular antennas cancel each other.

The eggbeater antenna was then considered. The eggbeater antenna consists of two loops that are perpendicular to each other. A quarter wavelength stub is placed between the two loops to cause the transmission between the two antennas to be 90 degrees out of phase. The tests confirmed that this type of antenna is more omni-directional and has relatively rare cases of signal cancellation. There is more dips in the signal strength than complete cancellation.

The problem with this type of antenna is that the loops must have a full wavelength circumference, making the diameter of the loop relatively large. This causes a space problem in the robot casing for this type of antenna (at 450 Mhz). The loop can be made smaller, but then higher frequencies must be used. As we want to use UHF frequencies, it would not be ideal to use smaller loops.

Resulting from this, the decision is to use the eggbeater antennas in the control unit where there is not as much constraints in size. Should the robot contain an antenna that is polarized in a single direction, then the eggbeater antenna (that has horizontal and vertical polarization) will be able to receive the transmitted signal. The tested eggbeater antenna is resonating between 440 MHz and 490 MHz, which is ideal for the available frequencies.

Fig. 18. Kenwood DM-81 Dip Meter

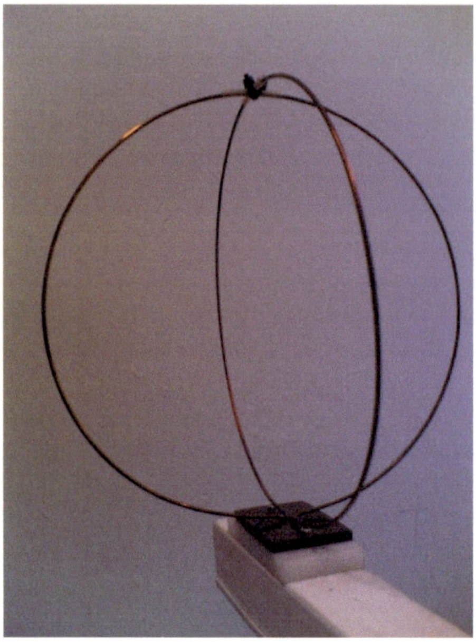

Fig. 19. Eggbeater Antenna

## 3. Gas Concentration Decisions

The gases that are of main importance in a search and rescue event is carbon dioxide, carbon monoxide, hydrogen sulphide, methane and oxygen (Gloster, 2007). Most sensors give an output of gas concentration, which is measured in parts per million (ppm). This data may be meaningless to the controller as it might hold no threat. An example will be the detection of methane gas. Should the robot detect 1ppm of methane, this could possibility not be dangerous, as it could be either a natural gas in the environment or of such a small quantity that it won't cause an explosion.

### 3.1 Gas Analysis

Further analysis of the gases and their respective properties need to be investigated. Different aspects of the gases were analyzed, to determine the concentrations that would be considered dangerous.

The Immediately Dangerous to Life or Health (IDLH) levels are used for non-emergency and emergency scenarios (Aerotech, 2009). These IDLH concentrations are determined with the following considerations (NIOSH, 2004):

- People must be able to escape the danger without the loss of life or irreversible health effects that could happen after exposure to that environment for a time period of 30 minutes.
- Prevention of severe eye or respiratory irritation, which will prevent a person from escaping the dangerous environment.

The compiled properties of the gases are represented in table 1.

| Substance | IDLH (ppm) | TLV* (ppm) | Smell | Flammability percentage | NFPA - Health / Flammability / Reactivity |
|---|---|---|---|---|---|
| $CO_2$ | 40 000 | 5 000 | Odorless | Non flammability | 3 / 0 / 0 |
| CO | 1 000 | 25 | Odorless | 12% - 75% | 3 / 4 / 0 |
| $H_2S$ | 100 | 10 | Rotten Egg | 4.3% - 46% | 4 / 4 / 0 |
| Methane | 5 000 *** | 1 000 | Odorless | 5% - 15% | 1 /4 / 0 |
| Oxygen | ** | ** | Odorless | Non flammability ** | N/A ** |

* Threshold Limit Value
** Oxygen is not flammable, but assist with combustion Oxygen level that are required to function mentally is 19.5% (NIOSH, 2004). Higher concentrations of Oxygen does not have serious effect on a person, but could cause sever explosions.
*** As methane is an asphyxiant, there is no IDLH data available (Stanford, 2008). A value for IDLH of five times that of the TLV is used.
Table 1. Properties of gases of interest.

Oxygen reacts with carbon to form carbon dioxide, therefore, as the carbon dioxide levels increase by x%, the oxygen levels will decrease by x% (U.S. OSHA et al., 2001). In the event that the oxygen levels decreases by x%, and the normal levels of oxygen in our atmosphere is considered to be 20.9%, then the oxygen level will decrease by (x / 20.9)%.

Concentrations of the gases up to the level of the Threshold Level Value (TLV) are considered to be safe. Any gas concentrations between the TLV and the IDLH are considered unsafe, while any concentration above the IDLH are dangerous. Table 2 shows a combination of all these properties in percentages.

| Substance | $Unsafe_{human}$(%) | $Dangerous_{human}$(%) | Flammable (%) |
|---|---|---|---|
| $CO_2$ | 0.5 | 4 | Non-flammable |
| CO | 0.0025 | 0.1 | 12% - 75% |
| $H_2S$ | 0.001 | 0.01 | 4.3% - 46% |
| $CH_4$ | 0.1 | 0.5 | 5% - 15% |
| $O_2$ | < 19.5 | < 16.9 | Non-flammable |

Table 2. Gas properties in percentages

Using the above data, it is possible to alert the rescuers in time of different possible conditions that could occur in the environment. These conditions could be either considered dangerous for humans or for the robot. With the information the rescuers could determine whether to risk their lives or the robot to enter a room with this environmental conditions.

### 3.1.1 CAESAR PC AI for Gases

Fuzzy logic is the way to determine logical expressions that are neither true or false. This type of reasoning is used to determine the unsafe, danger and flammable possibilities. The standard rules for fuzzy truth (T) are the following (Russell & Norvig, 2003):

$$T(A \wedge B) = \min(T(A), T(B)) \tag{7}$$

$$T(A \vee B) = \max(T(A), T(B)) \tag{8}$$

$$T(\neg A) = 1 - T(A) \tag{9}$$

For the above rules to be applied to the gas concentrations, it needs to be associated with a relationship referenced to the percentages in table 2, which are the boundaries. The value $g_n$ which is the specific gas concentration read from the sensors, is a value per million. This value has to be normalized with respect to 1 million to get a ratio. The unsafe value for humans ($u_h$), dangerous value for humans ($d_h$) and flammable value (f) are also normalized with respect to 1 million to get a ratio for the boundary values. Equations (10), (11) and (12) are used to determine A, B and C respectively.

$$A = g_n - u_h \tag{10}$$

$$B = g_n - d_h \tag{11}$$

$$C = g_n - f_{min} \tag{12}$$

In the event that the values of A,B and C are negative, the environment is safe for humans. Should any of the values become positive, it indicates that the gas concentration is either unsafe, dangerous or flammable.

Using equation (8) and equation (9), and only the positive numbers of A, B and C, (denoted by pos()), then

$$T(pos(\neg A) \vee pos(\neg B) \vee pos(\neg C) = \max(T(pos(\neg A)), T(pos(\neg B)), T(pos(\neg C))) \tag{13}$$

Combining equations (10), (11) and (12) with equation (13), results,

$$\begin{aligned} &T(pos(\neg(g_n - u_h)) \vee pos(\neg(g_n - d_h)) \vee pos(\neg(g_n - f_{min})) \\ &= \max(T(pos(\neg(g_n - u_h))), T(pos(\neg(g_n - d_h))), T(pos(\neg(g_n - f_{min})))) \end{aligned} \tag{14}$$

Let equation {14} relate to a function (K) that returns a solution to the logical expression. The relationship to determine if the gas concentration are unsafe, dangerous or flammable, the boundaries $u_h$, $d_h$ and $f_{min}$ are compared to ¬K, which concludes the possible decision (D). With this model, it specifies P(Safety of the environment | specific gas concentration).

Different solutions are expected from each gas analysis. All the solutions from the different gases are required to determine the safety of the environment. As the order of safety (being unsafe, dangerous and flammable) decreases and the gas concentration increases, the final decision is considered in respect to the worst solution from the different gases. This is expressed as the worst safety which is max ($D_n$). Should one gas concentration be flammable but another only unsafe for humans, then the flammability of the gas takes priority.

For the above equations to be valid, the gas concentration in table 2 needs to be converted to a ratio with respect to 1 million. As it becomes more dangerous for humans when the oxygen decreases, the values required are subtracted from 1 million. This allows for monitoring values that will be increasing throughout the table. The measurements for the oxygen concentration will also need to be deducted from 1 million to get an accurate decision. This is shown in table 3.

| Substance | $Unsafe_{human}$(ppm) | $Dangerous_{human}$(ppm) | $Flammable_{min}$ (ppm) |
|---|---|---|---|
| $CO_2$ | 500 | 4000 | Non-flammable |
| CO | 25 | 1000 | 120 000 |
| $H_2S$ | 10 | 100 | 43 000 |
| $CH_4$ | 1000 | 5000 | 50 000 |
| $O_2$ | 805 000 | 831 000 | Non-flammable |

Table 3. Gas properties in ratio with respect to 1 million

With the above information certain decisions can be made. Should the gas concentration be between $Unsafe_{human}$ and $Dangerous_{human}$ then the robot could continue to search for victims. In the event that the gas concentrations is higher than the $Dangerous_{human}$ level, the possibility for humans to survive in these conditions are decreasing and the rescuers must decide about entering the environment depending on other safety issues. These safety issues could be falling debris or unstable surfaces. As the gas concentration for the $flammable_{min}$ condition is much higher than that of the $Dangerous_{human}$ levels, it could imply that humans would not survive in these environments. The robot could make the decision to evacuate the environment and possibly save itself from an explosion, or searching for survivors in other areas of the disaster. These logical decisions will be determined from a weighting table shown in table 4.

There are two types of warnings that have to be considered. The unsafe / danger factor for humans and the danger factor for the robot. A model is required to determine the danger for humans. This is achieved with equation (15).

| Substance | $Unsafe_{human}$ | $Dangerous_{human}$ | $Dangerous_{robot}$ |
|---|---|---|---|
| $CO_2$ | 1 | 2 | 0 |
| CO | 1 | 2 | 1 |
| $H_2S$ | 1 | 2 | 1 |
| $CH_4$ | 1 | 2 | 1 |
| $O_2$ | 1 | 2 | 0 |

Table 4. Gas weighting factors

$$Danger = (100/2n)(w_u p + w_d q) \tag{15}$$

where n = number of gases being considered

p = number of gases that give an unsafe warning

q = number of gases that give a danger warning

$w_u$ = unsafe human weighting factor

$w_d$ = Dangerous human weighting factor

The above model will give a percentage of danger for humans. As the number of unsafe and dangerous factors for humans increases, the model increases the percentage value.

A model is also required for the danger of the robot. As seen in table 4, carbon dioxide and oxygen does not have a weighting factor, as these gases are not flammable. This danger for the robot is expressed by the model shown in equation (16).

$$Danger_{robot} = Gas_{HighestConcentrationPercentage} + \left( \left( \frac{100}{m} \sum_{n=1}^{n=m} \frac{x_n}{D_n} \right) \left( \frac{100 - Gas_{HighestConcentrationPercentage}}{100} \right) \right) \tag{16}$$

where m = number of gases not giving Danger warnings and that $w_n \neq 0$

$D_n$ = danger that gas n has on robot ($flammable_{min}$)

Should any one gas have a concentration that is higher than $flammable_{min}$, the environment is considered to be 100% dangerous for the robot. The danger for the robot could increase as other gas concentrations increase, but it will never decrease below the highest danger percentage.

Equation {28} could be used to determine the danger or unsafe value for humans, but $D_n$ will be the danger or unsafe value that gas n has on humans. This is a more accurate result compared to equation {27}, which only monitors the limits of the gas concentrations.

The models shown give a probability that humans would be able to survive in the surrounding environments and the probability that the robot is in a dangerous environment. All the decisions described above are performed by the control station, as it randomly requests for environmental status. The CAESAR robot responds to the request and awaits for it's next instruction. The control station considers the procedures that the rescuers and the robot must follow from the information received.

## 4. Conclusion

The Radiometrix TR2M modules with its related features, as well as the programming of the modules are discussed. Protocols and the basic procedure of the IEEE 802.11 protocol are explained, and a new robotic communication protocol, with its procedures of operation, are explained. The Robotics Communication Protocol (RCP) has a decreased size of 33 % to 38 % compared to IEEE 802.11 and hard-wired computer protocols respectively.

Specific features of voice communication with the Yaesu VX-7R and VX-3E radios are explained. The modification of the Yaesu VX-7R are also discussed, enabling communication over the allocated frequencies.

Further research should be performed in the development of microphones and speakers that are able to withstand high temperatures. An explanation is given for the video communication between the thermal camera and a television receiver for a successful observation of the robot's surrounding environment.

Radiation properties of feasible antennas are discussed, and the advantages and disadvantages of the vertical and eggbeater antenna are clarified. The testing procedures and verification of the different antennas and the radiation performance of the chosen antennas are also discussed.

Decisions are made from the analysis of gases and their concentrations in the environment. Models are developed that enable this analysis, determining if the environment is dangerous to humans or for the robot. This will assist rescuers to determine whether it is safe or worthwhile to risk their lives to enter the disaster.

With the improvements made on the communication and AI for gas detection it thereby allows a reliable control and communication interaction between the control station and the USAR robot. This also supplies the rescuers with critical information about the environment, before they enter and risk their lives in the unstable conditions.

## 5. References

Aerotech Environmental Laboratories. (2009). http://www.aeroenvirolabs.com/Resources/niosh.asp, 24 February 2009

Calson, J.; Murphy, R.; Valavanis, K. & Rundus, D. (2004). Analysis of How Mobile Robots Fail in the Field

Excellence in Audio. (1998). ProQuip Sound, Line & Mic Pre-Amplifiers datasheet

FLIR (2006). PathFindIR Brochure

Frenzel, L. (2001).Communication Electronics, Principles and Applications, Third Edition, McGraw-Hill Companies, Inc.

Gloster, A. (2007). Durban Metro Fire Department Training division Consultant

Greer, D.; McKerrow, P. & Abantes, J. (2002).Robots in Urban Search and Rescue Operations

Jackel, RF. (2008). South Africa

Kleiner, K. (2006) Better robots could help save disaster victims

NIOSH, National Institute for Occupational Safety and Health. (2004). Publication number: 2005-100: NIOSH Respirator Selection Logic 2004

Radiometrix TR2M Narrow Band FM Multi-channel UHF Transceiver data sheet. (2004)

Remley, K.; Koepke, G.; Messina, E. & Jacoff, A. (2007). Standards Development for Wireless Communications for Urban Search and Rescue Robots

Reynolds, D. Lt. (2008). City of Orlando Fire Department, Arson/Bomb Squad Bomb Squad Commander

Roos, A. (2005). South African Radio League Radio Amateur Examination Manual

Russell, S. & Norvig, P. (2003). Artificial Intelligence, A Modern Approach, second edition, Prentice Hall, 2003

Stanford TGO Data Tables. (2008). http://stanford.edu/dept/EHS/prod/researchlab/lab/tgo/tgodata.html, 25 February 2009

Trivedi, B. (2001).Search-and-Rescue Robots Tested at New York Disaster Site

U.S. OSHA, CMA, ANSI and Canadian WHMIS Standards (2001), Document number: 001033, Airgas, Material Safety Datasheet, Prepared by U.S. OSHA3

Vertex Standards Co., Ltd. (2002). Yaesu VX-7R operating manual, Japan

Vertex Standards Co., Ltd. (2007). Yaesu VX-3E operating manual, Japan

Wiens, K. (2006). Bread Crumbs and Robots May Save Your Life, SciTini

32

# Intelligent Control and Sensor Fusion of a Mobile Robot Based Monitoring System

Benítez-Read, Jorge S.[1,2] and Rojas-Ramírez, Erick[1,2]
[1]*Instituto Nacional de Investigaciones Nucleares and* [2]*Instituto Tecnológico de Toluca, State of Mexico, Mexico*

## 1. Introduction

The monitoring of some physical variables, such as radiation, corrosion, high humidity or temperatures above 70°C, are potentially hazardous for humans. In order to reduce these risks, it is desirable to have automatic or semi-automatic monitoring systems whose hardware platforms can be designed to reach areas of difficult physical access or to monitor variables potentially dangerous to the health of the personnel who manually realize the measurements. Monitoring systems based on mobile robots in which sensors, associated electronics, and computer equipment have been integrated, offer the versatility of navigation in environments of interest with certain degree of autonomy.

This mobility advantage has led some researches, in the last two decades, to focus on the development of control techniques and strategies to solve the mobile robot navigation problem in real non-structured or partially structured environments (Floreano et al., 2000), (Braünl, 2003), (Siegwart & Nourbakhsh, 2004). When using traditional approaches based on sequential tasks, some problems have come up in certain applications to achieve adequate real time responses. However, in recent years, many techniques are based on the control of navigation using genetic algorithms (Sung & Cho, 2004), fuzzy logic (Braunstingl & Sans, 1995), (Saffiotti et al., 1999), (Tunstel et al., 1997) and (Cuesta & Ollero, 2004) or the use of artificial neural networks (Sugihara et al., 2001) and (Kei et al., 2001).

Fuzzy logic based navigation control schemes are particularly interesting due to their inference and approximate reasoning capabilities. However, a major disadvantage arises when these schemes require the processing of a large number of inputs and outputs, and consequently, a large number of rules, since the design becomes very complex. Nonetheless, several multistage or hierarchical control schemes have been proposed to reduce the number of fuzzy rules and the complexity of the controllers (Wang, 1996).

In this work, the design, construction, and testing of a monitoring system based on a Khepera mobile robot are presented. The functions performed by the system are: (a) line following, (b) obstacle avoidance, (c) identification of test points along the path, and (d) recognition of the mark (bar code) located at each test point. For the task of navigation, a cascade fuzzy scheme is used, in which a first stage fusions the information collected by the optical sensors (fuzzy perception of the environment). This fusion process reduces the number of variables utilized by the second stage that controls the motion of the robot wheels. The navigation of the robot considers the presence of obstacles along the robot path and incorporates an algorithm based on fuzzy logic to avoid them and to return to the path. Along the trajectory, several light sources indicate the points where measurement of certain variables must be taken. The identification of these test points is carried out by means of a Kohonen neural network (Kohonen, 1996). The robot departs from the line and approaches the light source to a specified distance. Each test point has a bar code, which is identified by the processing of a one-dimensional image taken by a linear vision module. Once the measuring task at each point is completed, the data can be transferred to a PC using a radiofrequency module mounted on the robot.

## 2. The Khepera Robot

It is a mobile robot of 5.5 cm of diameter (Fig. 1). The manufacturer (K-team: www.k-team.com) developed this robot for research and didactical purposes. Algorithms for obstacle avoidance, object recognition using artificial vision, and robot arm manipulation can be designed and implemented on the robot microprocessor platform. More versatility is added to the robot with the following modules:

1. Linear vision, used for mark recognition and light source identification.
2. Analog and digital inputs and outputs which are used to add sensors and actuators to the robot system.
3. Radio communication between the robot and a PC.
4. One degree of freedom robot arm.

a) b)

Fig. 1. Khepera Robot, a) Front side and b) Back side

The robot control, data processing, and communication system is based on a 32-bit, 16 MHz Motorola MC68331 microcontroller (μC). In addition, the robot includes 8 infrared (IR) sensors, heterogeneously deployed around the robot. These sensors measure the distance between the robot and the objects around it, thus providing a proximity measure.

Furthermore, additional sensors and their associated electronics have been mounted on the robot (Fig. 1.b) for measuring the temperature at prespecified points and sensing a black line on the floor which represents the main robot trajectory.

## 3. Robot Navigation

The multi-stage fuzzy navigation scheme of the robot is shown in Figure 2. In this diagram, several fuzzy systems are used to describe the surroundings around the robot, which in turn leads to the definition of the motion of each robot's motor.

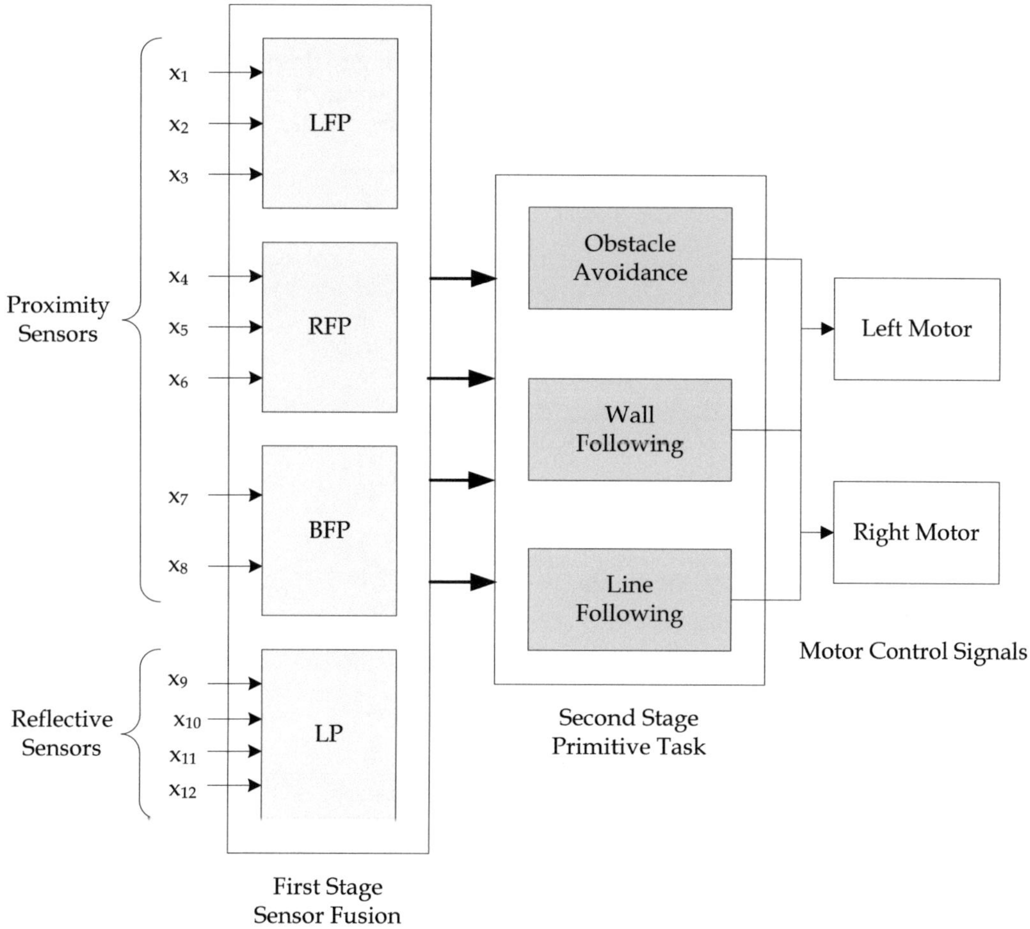


Fig. 2. Multi-stage fuzzy navigation scheme

The first stage shown in Fig. 2 performs a sensory fusion of eight signals of proximity and two signals from the sensors that detect a black line drawn on the floor surface. These signals are the inputs to four fuzzy systems (LFP, RFP, BFP, and LP) that characterize the perception of the environment around the robot. The first fuzzy system, LFP (Left Fuzzy Perception), receives the signals from the proximity sensors $x_1$, $x_2$ and $x_3$. The second system,

RFP (Right Fuzzy Perception), uses the signals coming from the proximity sensors $x_4$, $x_5$ and $x_6$. The third fuzzy system, BFP (Back Fuzzy Perception), processes the signals generated by the proximity sensors $x_7$ and $x_8$. Finally, system LP (Line Perception) receives the signals from the infrared reflective sensors $x_9$, $x_{10}$, $x_{11}$ and $x_{12}$. The first three perception systems describe approximately the surroundings of the robot (fuzzy description). The last system gives information of the line along which the robot moves. The four outputs, one per system, are used as the inputs to the following stage.

### 3.1. Second stage (Speed and Direction Control)

The second stage shown in Fig. 2 receives the four signals generated by the sensor fusion layer, processes the information and generates the signals that control the speed and direction of the robot's motors. The algorithm corresponding to this second stage and shown in Fig. 3 is to select one of four possible actions: Following of the line, wall following, obstacle avoidance or go to xy.

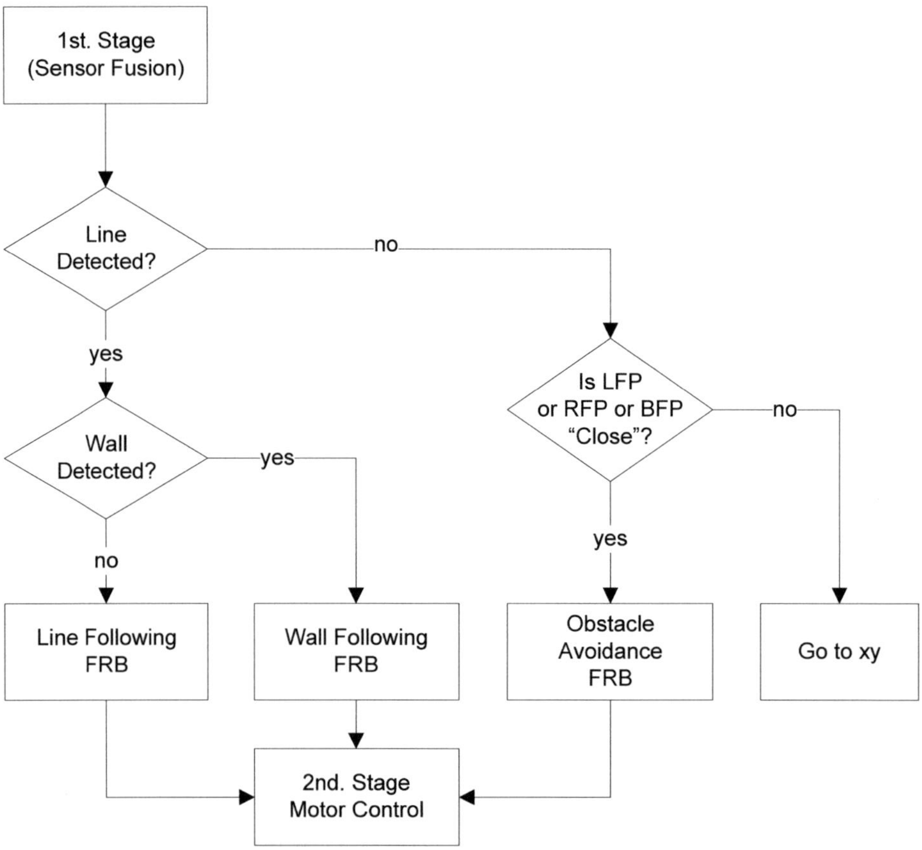


Fig. 3. Algorithm associated to the second stage whose tasks are: (a) Line following, (b) Wall following, (c) Obstacle avoidance and d) Go to xy

If the sensor fusion layer (see tables 1-3), generates values indicating that the robot is close to an obstacle, then only the fuzzy rule base corresponding to the obstacle avoidance algorithm will be executed (table 4).

| #Fuzzy Rule | Proximity Sensors | | | LFP: Perception Level |
|---|---|---|---|---|
| | $x_1$ | $x_2$ | $x_3$ | |
| 1 | Low | Low | Low | Far |
| 2 | High | High | Low | Close |
| 3 | | | High | Close |
| 4 | High | Low | | Medium |
| 5 | High | Low | Low | far |

Table 1. Fuzzy rule base of the LFP fuzzy system

| #Fuzzy Rule | Proximity Sensors | | | RFP: Perception Level |
|---|---|---|---|---|
| | $x_4$ | $x_5$ | $x_6$ | |
| 6 | Low | Low | Low | Far |
| 7 | Low | High | High | Close |
| 8 | High | | | Close |
| 9 | | Low | High | Medium |
| 10 | Low | Low | High | Far |

Table 2. Fuzzy rule base of the RFP fuzzy system

| #Fuzzy Rule | Proximity Sensors | | BFP: Perception Level |
|---|---|---|---|
| | $x_7$ | $x_8$ | |
| 11 | Low | Low | Far |
| 12 | Low | High | Close |
| 13 | High | Low | Close |
| 14 | High | High | Close |

Table 3. Fuzzy rule base of the BFP fuzzy system

| #Fuzzy Rule | Fuzzy Perception | | | Speed | |
|---|---|---|---|---|---|
| | LFP | RFP | BFP | Left Motor | Right Motor |
| 15 | Far | Close | Far | Backward | Forward |
| 16 | Close | Far | Far | Forward | Backward |
| 17 | Close | Close | Close | Stop | Stop |
| 18 | Close | Close | Far | Backward | Backward |

Table 4. Fuzzy rule base to perform the obstacle avoidance task

If the sensor fusion layer indicates that (a) no obstacle is close to the robot and (b) the black line below the robot has been detected, then the fuzzy rule base corresponding to the line following task will be executed (table 5). When the sensor fusion indicates that (a) an obstacle is around the robot and (b) the robot is on its main trajectory (black line detected), then the wall following fuzzy rule base (table 6) is executed. Finally, if the sensor fusion indicates that no obstacle is present around the robot and that the black line has not been detected, then the robot moves forward slowly until an obstacle or the black line is detected. The results of the robot's navigation are described in more detail in section 6.

| #Fuzzy Rule | Reflexive Sensors | | | | Speed | |
|---|---|---|---|---|---|---|
| | $x_9$ | $x_{10}$ | $x_{11}$ | $x_{12}$ | Left Motor | Right Motor |
| 19 | White | Black | Black | White | Forward | Forward |
| 20 | Black | Black | White | White | Backward | Forward |
| 21 | White | White | Black | Black | Forward | Backward |
| 22 | Black | White | White | White | Stop | Forward |
| 23 | White | White | White | Back | Forward | Stop |

Table 5. Fuzzy rule base for line following task

| #Fuzzy Rule | Fuzzy Perception | | | Speed | |
|---|---|---|---|---|---|
| | LFP | RFP | BFP | Left Motor | Right Motor |
| 15 | Far | Close | Far | Forward | Forward |
| 16 | Far | Far | Close | Forward | Backward |
| 17 | Close | Close | Far | Backward | Forward |
| 18 | Close | Close | Close | Stop | Stop |

Table 6. Fuzzy rule base corresponding to the wall following task

## 4. Test Point Detection

This stage, known as the detection of test points, is in charge of searching each test point along the robot main trajectory. One infrared light source is located at each test point. IR sensors around the robot measure the amount of IR light. A Kohonen type neural network (Haykin, 1999) is used to orientate and move the robot towards the light source. Using the array of six IR proximity sensors located on the front part of the robot, a computational algorithm that represents a Kohonen neural network was implanted on the robot processor. The function of this network is to detect the angular position between an infrared light source and the relative axis of the Khepera robot (Malmostrom & Munday, 1994).

### 4.1. Kohonen's Neural Network Characteristics

The neural network consists of an architecture of two layers. The inputs to the first layer (input layer) are the signals coming from the six photo-transistors of the proximity sensors located around the front of the robot. The second layer (competition layer) is composed of 90 neurons. Fig. 4 shows the angle detected by means of the artificial neural network algorithm implanted on the Khepera robot platform. The robot is supposed to approach the light source (test point) where the measurement of temperature has to be taken.

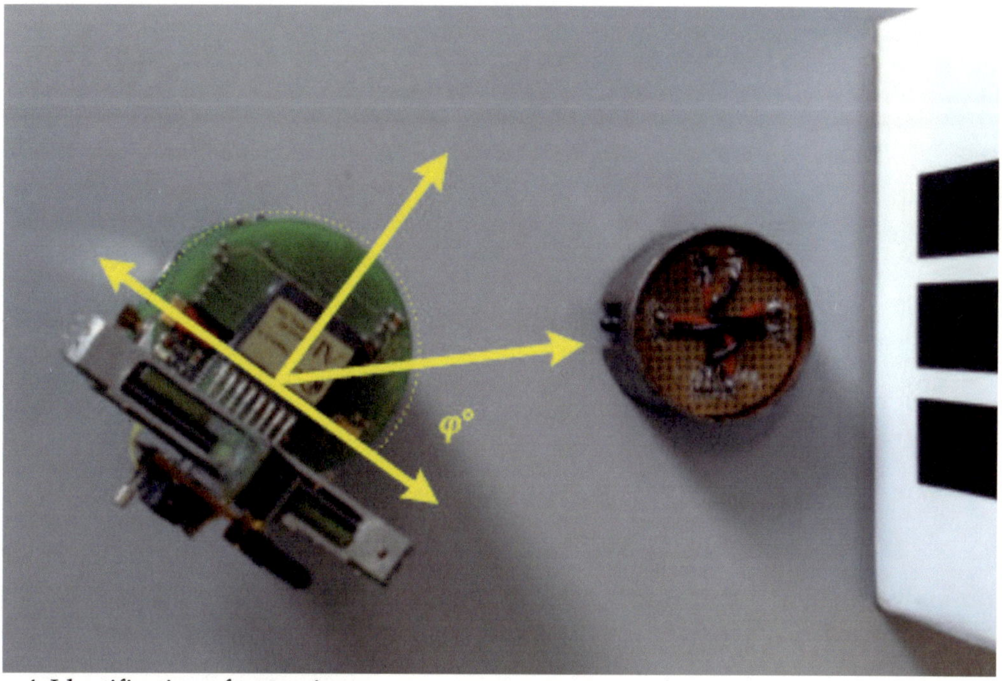


Fig. 4. Identification of test points

**4.2. Training Data and Training Phase**

For the training phase, a data set is obtained by placing the robot at a distance of 10 cm from a light source and then rotating the robot 180 degrees while recording both sensor vectors and associate angles. In the training phase, the network has been tuned using 1000 cycles. From one cycle to the next, the effective width of the neighborhood function is reduced from 45 to 1 neighboring neurons and the learning rate is reduced from 0.9 to 0.03.

Because the output of the neural network only indicates which neuron in the competition layer has won, it is necessary to interpret appropriately this output. The function that computes the angle between the light source and transversal axis of the robot is given by:

$$\varphi^\circ(x) = 2P_{i(x)} - 90^\circ \tag{1}$$

where $P_{i(x)}$ is the position of the i(x) winning neuron in the competition layer. The average error is 2 degrees (90 neurons are distributed over 180 degrees). The maximum error of 4 degrees occurs at the end points of the 180 degree range. The desired and the network responses are plotted in Fig. 5.

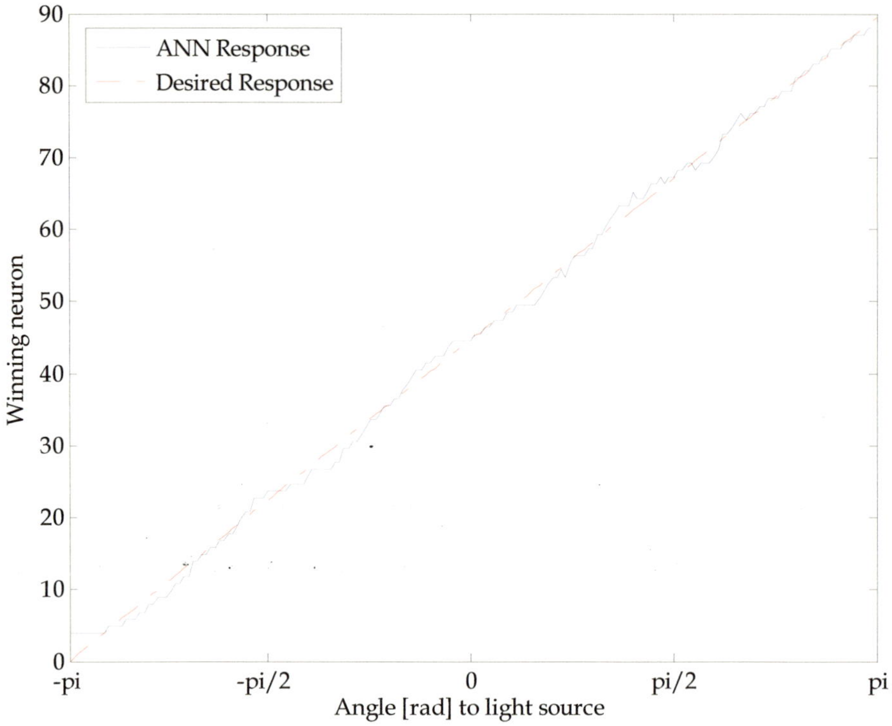


Fig. 5. Desired and ANN response

## 5. Test Point Recognition

At a predetermined distance of the source, a linear vision camera and the corresponding processing of the linear image are used to identify the type of test point. To perform this task, the linear vision module K213 of the Khepera robot is used. Each test point has a bar code that identifies the point. The scanning of a bar code using the vision module produces a matrix of data (M) with 1x64 pixels:

$$M = [p_1, p_2, p_3, \ldots, p_{64}] \tag{2}$$

This matrix is then processed pixel to pixel by the equation:

$$G_i = p_i - p_{i+1} \tag{3}$$

where $i$ is the i-th value associated to each pixel, and $G_i$ is the resultant gradient for the i-th pixel. Then, a threshold is used for each gradient value as follows:

$$G_i = \begin{cases} -1 & if \quad G_i < -40 \\ 0 & if \quad -40 < G_i < 40 \\ +1 & if \quad G_i > 40 \end{cases} \quad (4)$$

The purpose of this threshold is to reduce the noise inherently attached to the signals. The result of this processing is shown in Figure 6. The continuous black-line represents the scanning of the bar code. The values of the $G_i$ provide information of the changes between whites and blacks (bar codes) detected by the camera. This information is used to identify the type of test point.

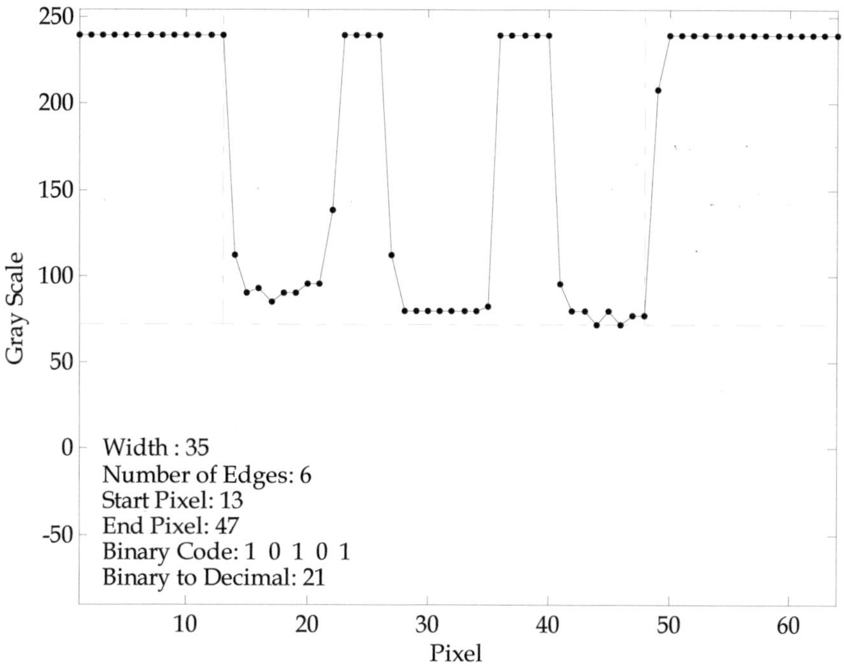


Fig. 6. Result of the processing of a scanned image

## 6. Experimental Results

The experimental tests have been carried out using a personal computer which has the following characteristics: Pentium 4 at 3.0 GHz processor, a RAM of 512 Mbytes, and Windows as the operating system. The communication between the Khepera robot and the PC is established through a Graphical User Interface. The speed of communication through a standard RS-232 allows the monitoring of the sensors of the robot that sense temperature, proximity, intensity of ambient light, line on the floor, voltage applied to every motor and the scanning of the resultant image of a bar code. Likewise, it is possible to visualize, using a compass, the orientation of a light source with respect to the robot, computed by the Kohonen neural network.

All the functions related to the robot navigation, IR light source identification, test point recognition (bar code reading), temperature measuring, and the graphical user interface are written in Matlab. The experimental setup is shown in Fig. 7. The mobile robot system has to navigate within its environment, following a main trajectory identified by a line painted on the floor, avoiding obstacles, searching for the test points, approaching each one of them to a certain distance, identifying each test point and measuring the temperature at each test point location. As can be seen in Fig. 7, there are three test points, M1, M2, and M3, in which IR light sources are located. Likewise, the bar codes behind the light sources can be appreciated on the setup. Relevant information associated to each bar code is stored on the PC. Finally, several obstacles were put along the main robot trajectory.

The robot departs from the initial position (Home), sensing and following the black line (main trajectory) on the floor. The control of the velocity and direction of the robot's left and right motors is carried out using the fuzzy rule base associated to the fuzzy system "line following" (see table 5). As a result, the robot follows the line, correcting the velocity of the motors at each control cycle, depending on the degree of "black" and "white" detected by the line IR reflective sensors. The green arrows shown in Fig. 7 represent the robot behavior under the "line following" commands.

This task, however, gets more complicated when an obstacle is encountered along the path (see point x2 in Fig. 7). At this point (obstacle present and line detected) the program jumps to the "wall following" block and executes the corresponding rule base (table 6) until the obstacle is left behind and the robot once again senses the line (see point x3). The robot behavior corresponding to the "wall following" is shown in Fig. 7 with blue arrows.

While approaching test point M1, the robot detects, through its front phototransistors, the IR light source located at M1. The robot is commanded to a halt, and the system proceeds, using the ANN, to compute the orientation of the light source with respect to the transversal axis of the robot. Next, the robot approaches (see "Go to xy" block in Fig. 3) this source without the supervision of the motor velocity fuzzy controllers since these are disabled once the ANN initiates its operation. This approach is shown in Fig. 7 with yellow arrows.

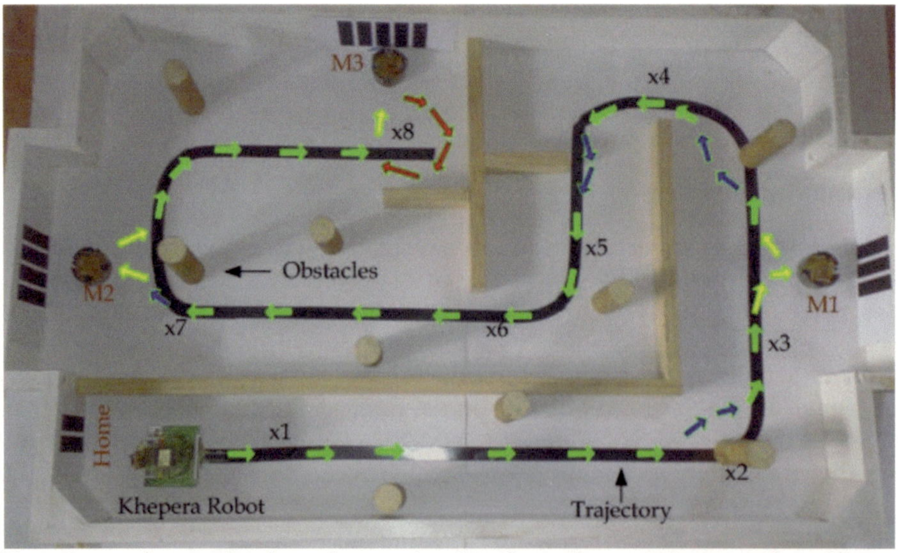


Fig. 7. Experimental setup for the robot navigation through an unstructured environment

Once the angle between the robot and the light source is within ±5°, the Khepera linear vision module starts the scanning of the bar code located behind the light source. The dimensions of the width, spacing and number of bars are such that an adequate vision scope of the linear vision module is obtained at a distance of 10 cm.
The linear image processing is performed as described in section 5. Once the bar code is identified to correspond to a valid test point, the control program sends the command to take the reading of temperature at M1.
Subsequently, the robot returns to the main trajectory (black line on the floor) executing the commands of the functional block "Go to xy" (see Fig. 3). The robot reassumes the following of the main path searching for more test points. The above described sequence is repeated for the other test points (M2 and M3).
If at some point the robot does not sense back the main path, as at point x8 in Fig. 7, the program instructs the robot to execute a line searching routine. During this search, the obstacle avoidance block prevents any collision, thus preserving the physical integrity of the robot. This behavior is shown in Fig. 7 with the red arrows.
The number of test points is a known parameter. Thus, once the robot has reached the last one, the robot takes the road back until it reads the bar code corresponding to the initial position (Home), time at which the system stops completely.

## 7. Conclusions

The use of a sensor fusion layer, which processes the data in a fuzzy manner, reduces the amount of information to be processed by the stage that controls the speed and direction of the robot. Matlab has been used to implement the control scheme on the PC. The signals (commands and measurements) to and from the robot are transmitted through radiofrequency modules mounted on both the PC and the Khepera robot. The processing is fast enough, preventing delays that could damage the robot. In addition, the fusion sensor layer reduces the disturbances caused by the noise, mainly originated by the light sources, thus giving certain degree of robustness to the system: the robot moves smoothly. On the other hand, the response of the neural network is very close to the ideal one. In fact, part of the error comes from the non homogeneous disposition of the sensors. Nevertheless, there offers a resolution of ±2 degrees, with an execution time of 10 ms, which is ideal for applications that require real time processing. At this moment, only temperature is being measured by the prototype. The temperature measured value is sent to the PC using a radiofrequency module.

## 8. References

Braünl, T. (2003). Embedded Robotics, Mobile Robots Design and Applications with Embedded Systems, Springer-Verlag, Berlin Heidelberg, 2003

Braunstingl, R. & Sans, P. (1995). Fuzzy Logic Wall Following of a Mobile Robot Based on the Concept of General Perception; Proc. 7th Intl. Conference on Advanced Robotics, Vol. 1, pp. 367-376; Spain

Cuesta, F. & Ollero, A. (2005). *Intelligent Mobile Robot Navigation*, Springer Tracks in Advanced Robotics, Springer-Verlag, ISBN 3-540-23956-1, Netherlands.

Floreano, D.; Godjevac, J.; Martinoli, A.; Mondada, F.& Nicoud, D.J. (2000). Design, Control, and Application of Autonomous Mobile Robots; Swiss Federal Institute of Technology in Lausanne

Haykin, S. (1999). *Neural Networks: A Comprehensive Foundations*, MacMillan, ISBN 0-02-352781-7, 1999

Kei, S., Masayoshi, T., Tatsuro, S. & Tetsuro, K.(2001). Control System of the Khepera robot by a Neural Network with Competition and Cooperation, Artificial Life and Robotics, Springer, Vol. 5, Num. 6, pp. 26-28, Japan, 2001

Kohonen, T. (1996). New Developments and Applications of Self Organising Maps, Int. Workshop on Neural Networks for Identification, Control, Robotics and Signal/image Processing, NICROSP-96, pp. 164-170,1996.

Lee, S. & Cho, S. (2001). Emergent Behaviors of a Fuzzy Sensory-Motor Controller Evolved by Genetic Algorithm; IEEE Transactions on Systems, Man, and Cybernetics, Vol. 31, No. 6, pp. 919-929

Malmostrom, K. & Munday L. (1994). A Simple Robust Robotic Vision System Using Kohonen Feature Mapping; Proccedings of the 2nd IEEE Australia and New Zealand Conference on Intelligent Information Systems, pp. 135-139

Saffiotti, A.; Ruspini, E., & Konolige, K. (1999). Using Fuzzy Logic for Mobile Robot Control, International Handbook for Fuzzy Sets, Kluwer Academic Publisher, 1999

Siegwart, R. & Nourbakhsh, I (2004); Introduction to Autonomous Mobile Robots, The MIT Press, London, England 2004

Sugihara, K.; Tabuse, M.; Shinchi, T. & Kitazoe, T. (2001). Control System for the Khepera Robot by a Neural Network with Competition and Cooperation, Artificial Life and Robotics, Springer Japan, Vol. 5, Num. 6. pp. 26-28

Tunstel, E.; Lippincott, T. & Jamshidi, M. (1997). Behavior Hierarchy for Autonomous Mobile Robots: Fuzzy-Behavior Modulation and Evolution, *Int. J. Of Intelligent Automation and Soft Computing*, Vol. 3, No. 1, pp. 37-50

Wang, L.X. (1996). *A course in Fuzzy Systems and Control*, Prentice Hall, ISBN 0-13-540882-2, New Jersey